Lecture Notes in Computer Science 1065

Edited by G. Goos, J. Hartmanis and J. van Leeuwen

Springer
Berlin
Heidelberg
New York
Barcelona
Budapest
Hong Kong
London
Milan
Paris
Santa Clara
Singapore
Tokyo

Bernard Buxton Roberto Cipolla (Eds.)

Computer Vision – ECCV '96

4th European Conference
on Computer Vision
Cambridge, UK, April 15-18, 1996
Proceedings, Volume II

Springer

Volume Editors

Bernard Buxton
University College London, Department of Computer Science
Gower Street, London WC1E 6BT, United Kingdom

Roberto Cipolla
University of Cambridge, Department of Engineering
Cambridge CB2 1PZ, United Kingdom

Cataloging-in-Publication data applied for

Die Deutsche Bibliothek - CIP-Einheitsaufnahme

Computer vision : proceedings / ECCV '96, Fourth European Conference on Computer Vision, Cambridge, UK, April 1996. Bernard Buxton ; Roberto Cipolla (ed.). - Berlin ; Heidelberg ; New York ; Barcelona ; Budapest ; Hong Kong ; London ; Milan ; Paris ; Santa Clara ; Singapore ; Tokyo : Springer.

NE: Buxton, Bernard [Hrsg.]; ECCV <4, 1996, Cambridge>

Vol. 2. - 1996
(Lecture notes in computer science ; Vol. 1065)
ISBN 3-54061123-1

NE: GT

CR Subject Classification (1991): I.3.5, I.5, I.2.9-10, I.4

ISBN 3-540-61123-1 Springer-Verlag Berlin Heidelberg New York

Typesetting: Camera-ready by author
SPIN 10512766 06/3142 – 5 4 3 2 1 0 Printed on acid-free paper

Preface

Following the highly successful conferences held in Antibes (ECCV'90), Santa Margherita Ligure (ECCV'92) and Stockholm (ECCV'94), the European Conference on Computer Vision has established itself as one of the major international events in the exciting and active research discipline of computer vision. It has been an honour and pleasure to organise the Fourth European Conference on Computer Vision, held in the University of Cambridge, 15–18 April 1996.

These proceedings collect the papers accepted for presentation at the conference. They were selected from 328 contributions describing original and previously unpublished research. As with the previous conferences each paper was reviewed double blind by three reviewers selected from the Conference Board and Programme Committee. Final decisions were reached at a committee meeting in London where 43 papers were selected for podium presentation and 80 for presentation in poster sessions. The decision to keep ECCV'96 a single track conference led to a competitive selection process and it is entirely likely that good submissions were not included in the final programme.

We wish to thank all the members of the Programme Committee and the additional reviewers who each did a tremendous job in reviewing over 25 papers papers in less than six weeks. We are also extremely grateful to the chairmen of previous conferences, Olivier Faugeras, Giulio Sandini and Jan-Olof Eklundh for their help in the preparations of the conference and to Roberto's colleagues and research students in the Department of Engineering for their patience and support. The conference was sponsored by the European Vision Society (EVS) and the British Machine Vision Association (BMVA). We are grateful to the Chairman of the BMVA and the executive committee for advice and financial support throughout the the organisation of the conference.

Cambridge, January 1996 Bernard Buxton and Roberto Cipolla

Conference Chairs

Bernard Buxton	University College London
Roberto Cipolla	University of Cambridge

Conference Board and Programme Committee

N. Ayache	INRIA, Sophia Antipolis
R. Bajcsy	University of Pennsylvania
A. Blake	University of Oxford
P. Bouthemy	IRISA/INRIA, Rennes
M. Brady	University of Oxford
H. Burkhardt	Technical University of Hamburg-Harburg
H. Buxton	University of Sussex
S. Carlsson	KTH, Stockholm
H. Christensen	Aalborg University
J. Crowley	INPG,Grenoble
R. Deriche	INRIA, Sophia Antipolis
M. Dhome	Blaise Pascal University of Clermont-Fd
E. Dickmanns	Universität der Bundeswehr, Munich
J.-O. Eklundh	KTH, Stockholm
W. Enkelmann	IITB, Karlsruhe
O. Faugeras	INRIA, Sophia Antipolis
G. Granlund	Linköping University, Sweden
W. Förstner	Bonn University
L. Van Gool	Katholieke Universiteit Leuven
D. Hogg	University of Leeds
R. Horaud	INPG,Grenoble
J.J. Koenderink	Universiteit Utrecht
H. Knutsson	Linköping University, Sweden
O. Kübler	ETH, Zurich
J. Malik	University of California at Berkeley
J. Mayhew	Sheffield University
R. Mohr	INPG, Grenoble
H. Nagel	IITB, Karlsruhe
B. Neumann	Hamburg University
S. Peleg	The Hebrew University of Jerusalem
B. M. ter Haar Romeny	Universiteit Utrecht
G. Sandini	University of Genova
W. von Seelen	Ruhr University, Bochum
F. Solina	Ljubljana University, Slovenia
G. Sparr	Lund University, Sweden
G. Sullivan	University of Reading
M. Tistarelli	University of Genova
V. Torre	University of Genova
S. Tsuji	Wakayama University, Japan
D. Vernon	Maynooth College, Ireland
A. Verri	University of Genova
J.J. Villanueva	Autonomous University of Barcelona
D. Weinshall	The Hebrew University of Jerusalem
A. Zisserman	University of Oxford
S. Zucker	McGill University, Canada

Additional Reviewers

Frank Ade
P. Anandan
Mats Andersson
Tal Arbel
Martin Armstrong
Minoru Asada
Kalle Astrom
Jonas August
Dominique Barba
Eric Bardinet
Benedicte Bascle
Adam Baumberg
Paul Beardsley
Marie-Odile Berger
Rikard Berthilsson
Jørgen Bjørnstrup
Jerome Blanc
Philippe Bobet
Luca Bogoni
Magnus Borga
Kadi Bouatouch
Pierre Breton
Joachim Buhmann
Andrew Bulpitt
Hans Burkhardt
Franco Callari
Nikolaos Canterakis
Carla Capurro
Geert de Ceulaer
Francois Chaumette
Jerome Declerck
Herve Delingette
Bernard Delyon
Jean-Marc Dinten
Leonie Dreschler-Fischer
Christian Drewniok
Nick Efford
Bernard Espiau
Hanny Farid
Jacques Feldmar
Peter Fiddelaers
David Forsyth
Volker Gengenbach
Lewis Griffin
Eric Grimson
Patrick Gros
Etienne Grossberg
Enrico Grosso
Keith Hanna
Mike Hanson
Friedrich Heitger
Fabrice Heitz
Marcel Hendrickx
Olof Henricsson
Anders Heyden
Richard Howarth
Steve Hsu
S. Iouleff
Michal Irani
Hiroshi Ishiguro
Peter Jürgensen
Frédéric Jurie
Ioannis Kakadiaris
Jørgen Karlholm
Hans-Joerg Klock
Lars Knudsen
Jana Kosecka
Steen Kristensen
Wolfgang Krüger
Rakesh Kumar
Claude Labit
Bart Lamiroy
Tomas Landelius
Jean-Thierry Lapreste
Ole Larsen
Fabio Lavagetto
Jean-Marc Lavest
Jean-Pierre Leduc
Chil-Woo Lee
Aleš Leonardis
Jean-Michel Létang
Tony Lindeberg
Oliver Ludwig
An Luo
Robert Maas
Brian Madden
Claus Madsen
Twan Maintz
Gregoire Malandain
Stephen Maybank
Phil McLauchlan
Etienne Memin
Baerbel Mertsching
Dimitri Metaxas
Max Mintz
Theo Moons
Luce Morin
Jann Nielsen
Mads Nielsen
Wiro Niessen
Alison Noble
Klas Nordberg
Eric Pauwels
Marcello Pelillo
Xavier Pennec
Patrick Perez
Bernard Peuchot
Nic Pillow
Paolo Pirjanian
Rich Pito
Marc Pollefeys
Marc Proesmans
Long Quan
Paolo Questa
Veronique Rebuffel
Ian Reid
Gérard Rives
Luc Robert
Karl Rohr
Bart ter Haar Romeny
Charlie Rothwell
Paul Sajda
Alfons Salden
Josè Santos-Victor
Gilbert Saucy
Harpreet Sawhney
Ralph Schiller
Cordelia Schmid
Christoph Schnoerr
Carsten Schroeder
Ulf Cah von Seelen
Eero Simmoncelli
Sven Spanne
Rainer Sprengel
H. Siegfried Stiehl
Kent Strahlén
Peter Sturm
Gerard Subsol
Gabor Szekely
Tieniu Tan
Hiromi Tanaka
Jean-Philippe Thirion
Phil Torr
Bill Triggs
Morgan Ulvklo
Dorin Ungureanu
Peter Vanroose
Jean-Marc Vezien
Uwe Weidner
Carl-Fredrik Westin
Johan Wiklund
Lambert Wixon
Gang Xu
Zhengyou Zhang

Contents of Volume II

Colour Vision and Shading

Image Features

Motion

Medical Applications

Tracking (2)

Applications and Recognition

Calibration/Focus/Optics

Tracking (3)

Applications

Structure from Motion (3)

Contents of Volume I

Structure From Motion (1)

Recognition (1)

Geometry and Stereo

Texture and Features

Tracking (1)

Grouping and Segmentation

Stereo

Recognition/Matching/Segmentation

Structure from Motion (2)

A Vision of the Future...
(or: Computer Vision à la Karel Čapek)

Colour Vision and Shading

Colour Constancy for Scenes with Varying Illumination

Kobus Barnard[1], Graham Finlayson[2], and Brian Funt[1]

[1]School of Computing Science, Simon Fraser University, Burnaby BC, Canada, V5A 1S6
kobus@cs.sfu.ca, funt@cs.sfu.ca

[2] Department of Computer Science, University of York, Heslington, York, YO1 5DD, United Kingdom
graham@minster.cs.york.ac.uk

Abstract. We present an algorithm which uses information from both surface reflectance and illumination variation to solve for colour constancy. Most colour constancy algorithms assume that the illumination across a scene is constant, but this is very often not valid for real images. The method presented in this work identifies and removes the illumination variation, and in addition uses the variation to constrain the solution. The constraint is applied conjunctively to constraints found from surface reflectances. Thus the algorithm can provide good colour constancy when there is sufficient variation in surface reflectances, or sufficient illumination variation, or a combination of both. We present the results of running the algorithm on several real scenes, and the results are very encouraging.

1 Introduction

Many colour constancy algorithms have been developed, but all are subject to quite restrictive assumptions and few have been tested on real images. Of the existing algorithms we believe that the one by Finlayson [8] is currently the most general and most thoroughly tested. Nonetheless, it is restricted to scenes in which the illumination is constant or at least locally constant. This assumption is more often violated than one might at first suspect given that the incident illumination from any fixed direction does generally vary slowly as function of position. The problem is that the surface orientation even of smooth surfaces can vary quite rapidly with position so that light at nearby surface locations may be received from very different regions of the illumination field. Since these different regions of the illumination field often posses substantially different spectral power distributions—such as is the case in a room in which where there is light from a light bulb mixed with daylight from a window—nearby points on the surface in fact can receive very different incident illumination.

This paper addresses the problem of colour constancy in scenes where the spectral power distribution of the incident illumination is allowed to vary with scene location. Finlayson et. al. [7], D'Zmura et. al. [18], and Tsukada et. al. [16] have shown that a difference in illumination, *once it has been identified*, provides additional constraints that can be exploited to obtain colour constancy, but they do not provide an automatic method of determining when such a difference exists. We present a new algorithm that first uncovers the illumination variation in an image and then uses the additional constraint it provides to obtain better colour constancy. The algorithm presupposes a diagonal model for illumination changes, and that the illumination varies spatially slowly. However it is quite robust to moderate violations of these assumptions.

The colour constancy problem is the retrieval of an illumination-independent description of a scene's surface colours. This is essentially equivalent to modeling the illumination incident on the scene, since if the illumination is known the surface colours can be calculated. Following Forsyth [9] we interpret colour constancy as taking images of scenes under unknown illumination and determining the camera response to the same scene under a known, canonical light. In a general context this problem has proven difficult to solve, so to make progress, restrictive assumptions are made. In particular, it is common to assume that the scene is flat [9, 13, 14], that the illumination is constant throughout [2, 3, 9, 10, 15], and that all reflectances are matte. Finlayson [8] has shown that if we focus on solving only for surface chromaticity and forego estimating surface lightness then the restriction to flat matte surfaces can be relaxed. However, the assumption that the chromaticity of the illumination does not change is more troublesome.

The Retinex algorithm [1, 11, 13, 14] partially addresses the issue of varying illumination. At least in principle—it does not in fact work in practice—Retinex eliminates the variation in illumination and computes surface lightnesses for each of the three colour channels independently. Since eliminating the illumination and recovering the illumination are equivalent problems [4], if Retinex worked it could be used to recover the incident illumination. Retinex operates on the principle that within a single colour channel small changes in image intensity arise from changes in illumination while large changes indicate changes in surface colour. The small changes are thresholded away and the big changes are preserved so that the surface lightness can be reconstructed, essentially by integration. Unfortunately any error in classifying the intensity changes can lead to serious errors in the recovered result. In essence the Retinex algorithm uses a primitive, gradient-based-edge-detection strategy to identify the reflectance edges, so given the long history of edge-detection research, it should not be surprising that it does not perform well.

To overcome the weaknesses of Retinex's edge detection method, we incorporate knowledge about the set of plausible illuminants and from this set derive information about the kinds of chromaticity change that a change in illumination can induce within a region of uniform reflectance. This constraint is more global than local edge detection and using both types of constraint together yields good results. Once the illumination variation is uncovered it is combined with the other constraints arising from the set of colours found in the image as will be discussed below.

2 The Colour Constancy Algorithm

Our colour constancy algorithm has two main components: one to extract the illumination field and another to combine the constraints provided by the a priori knowledge of the surface and illumination gamuts with those obtained from the observed surfaces and the extracted illumination field. The constraint part of the algorithm will be described first.

2.1 Surface and Illumination Constraints

In order to represent the constraints efficiently, we make the approximation that the effect of the illumination can be modeled by a diagonal matrix [5, 6]. Specifically, if [r, g, b] is the camera response of a surface under one illumination, then $[r, g, b]\mathbf{D} = [r\mathbf{D}_{11}, g\mathbf{D}_{22}, b\mathbf{D}_{33}]$, where $\mathbf{D}$ is a diagonal matrix, is the camera response to the same surface under a second illumination. In other words, each camera channel is scaled independently. The accuracy of the diagonal approximation depends on the camera sensors, which for the camera used in the experiments is within 10% (magnitude of [r, g, b] difference) of the general linear model. For sensors for which the diagonal model is too inaccurate, it is usually possible to improve it by spectrally sharpening the sensors [5].

Following Finlayson [8], we work in the chromaticity space [r/b, g/b]. This space preserves the diagonal model in the sense that if illumination was exactly modeled by a diagonal transform applied to [r, g, b], then it will also be exactly modeled by a diagonal transform (now 2D) applied to [r/b, g/b]. If either the illumination is spatially constant or pre-processing has removed all illumination variation, then transforming the input image to what it would have looked like under the canonical illuminant requires simply transforming it by a single diagonal matrix. The goal of the colour constancy algorithm is to calculate this matrix.

The algorithm's basic approach is to constrain the set of possible diagonal maps by adding more and more information so that only a small set of possible maps remains. The first constraints are the those due to Forsyth [9]. He observed that the set of camera responses that could be obtained from all combinations of a large set of surfaces viewed under a fixed illuminant is a convex set which does not fill all of the [r, g, b] colour space. This set is referred to as that illuminant's gamut, and in the case of the canonical illuminant is called the canonical gamut. For a typical scene under unknown illumination, the camera responses will lie in a subset of the unknown illuminant's full gamut. Since all possible surfaces are assumed to be represented within the canonical gamut, whatever the unknown illuminant is, it is constrained by the fact that it is a diagonal map projecting the scene's observed response set into the canonical gamut. There will be many possible diagonal maps satisfying this constraint because the scene's set is a subset of the full gamut and so it can 'fit' inside the larger gamut many different ways. Forsyth shows that the resulting constraint set of diagonal maps is convex. As shown in [8], all the required relationships hold in the [r/b, g/b] chromaticity space.

The second source of constraint arises from considering the set of common illuminants as has been formulated by Finlayson [8]. After applying Forsyth's surface constraints, the resulting set of diagonal maps typically includes many that correspond to quite atypical illuminants. The illumination constraint excludes all the illuminants that are not contained in the set of typical illuminants. Finlayson restricted the illumination to the convex hull of the chromaticities of the 6 daylight phases provided by Judd et al [12], the CIE standard illuminants A, B, C [17], a 2000K Planckian radiator, and uniform white. We have improved upon this sampling of illuminants by using 100 measurements of illumination around the university campus, including both indoor and outdoor illumination. Some inter-reflected light was included such as that from concrete buildings and light filtering through trees, but illumination that was obviously unusual was excluded. The set of chromaticities of the measured illuminants is larger in area than the original set, but it does not contain it entirely, as the 2000K Planckian radiator is more red than what is common.

It should be noted that the set of typical illuminants provides a constraint on mappings from the canonical to the unknown, which is the reverse of that for surfaces discussed above in which the restriction was on mappings from the unknown illuminant to the canonical illuminant. To make use of the constraint it must be inverted which means that the restriction on the set of illuminants becomes a non-convex set in the mapping space used for surface constraints. This potentially presents a problem since the sets must be intersected in order to combine constraints and in three-dimensions it is much faster to compute intersections of convex sets than non-convex ones. While in the two-dimensional case the set intersections can be directly computed, in practice the inverse of the measured illumination non-convex gamut was found to be close enough to its convex hull that for convenience the hull could be used anyway.

Varying illumination provides the third source of constraint. Our use of it here generalizes the algorithm presented in [7]. In that work the map taking the chromaticity of a single surface colour under an unknown illuminant to its chromaticity under the canonical illuminant is constrained to lie on a line. Effectively this amounts to assuming all the candidate illuminants are approximately Plankian radiators and their chromaticities lie roughly on the Plankian locus. The chromaticity of the same surface viewed under a second illuminant defines a second line. If the difference in the illuminations' chromaticities is non-trivial, the two lines will intersect, thereby constraining the surface's chromaticity to a unique value.

We extend the idea of using the variation in illumination in two ways. First we use the entire illumination gamut instead of simply the Plankian radiators. Second we exploit the illumination variation across the entire image, as opposed to just that between two points on one surface patch. Thus the illumination over the entire image is both used, and solved for. The details follow.

For the moment assume that we already have the relative illumination field for the image. The relative illumination field for each pixel P is defined by the diagonal transform required to map the illumination at some chosen base pixel B to the illumination at P. The relative illumination field describes all the pixels only with respect to one another, so given it, the remaining problem is to solve for the illumination at B and hence establish the illumination everywhere in absolute terms.

The approach is motivated by the following argument. Suppose that the left side of the image is illuminated by a blue light. This means that the relative illumination field at a pixel on the left side transforms illuminants so that they are more blue. However, the illumination at the center of the image cannot be so blue that making it even more blue produces an illumination that falls outside the set of possible illuminants. Thus the illumination at the center is constrained by the jump towards blue. All entries in the field contribute this sort of constraint. This will now be made more formal.

First we verify the intuitive claim that the constraint provided by one of the values **D** in the relative illumination field is the set of possible illuminants scaled by $\mathbf{D}^{-1}$. Consider the illumination gamut, I which is a convex set:

$$I = \left\{ \mathbf{X} \,\middle|\, \mathbf{X} = \sum_i \lambda_i \mathbf{X}_i \text{ where } \sum_i \lambda_i = 1 \right\} \text{ for hull points } \{\mathbf{X}_i\} \qquad (1)$$

We have the constraint that we can map the illumination by the diagonal map **D** and still be in this set:

$$\mathbf{XD} \in I \tag{2}$$

This means that:

$$\mathbf{XD} = \sum_i \lambda_i \mathbf{X}_i \quad \text{for some } \lambda_i \text{ with } \sum_i \lambda_i = 1,\ \lambda_i \geq 0 \tag{3}$$

And

$$\mathbf{X} = \sum_i \lambda_i \left(\mathbf{X}_i\ \mathbf{D}^{-1}\right) \text{ for some } \lambda_i \text{ with } \sum_i \lambda_i = 1,\ \lambda_i \geq 0 \tag{4}$$

So we define a new constraint set V as:

$$V = \left\{ \mathbf{X} \,\middle|\, \mathbf{X} = \sum_i \lambda_i \left(\mathbf{X}_i\ \mathbf{D}^{-1}\right) \text{ where } \sum_i \lambda_i = 1,\ \lambda_i \geq 0 \right\} \tag{5}$$

It is clear that for all $\mathbf{X} \in V$, $\mathbf{XD} \in I$. Furthermore, the argument is reversible. That is, if $\mathbf{Y} = \mathbf{XD} \in I$, $\mathbf{X} \in V$. It should be noted that the above also shows that we can identify the convex constraint set with the mapped hull points $\mathbf{X}_i\ \mathbf{D}^{-1}$.

Next we note that the convex hull of these constraints is just as powerful as the entire set. The motivation for using the hull is that it saves a significant amount of processing time. We are free to use the hull regardless, but it is comforting to know that doing so does not weaken the algorithm.

Despite the details, the additional constraint is very simple in that it says that we have to be able to scale the illuminant by a certain amount and *still* satisfy the illumination constraint. This constraint is realized by simply scaling the set of illuminants by the inverse. As a simple example, consider the one-dimensional line segment [0,1]. If we have a condition on these points that when they are scaled by a factor of two the result must still be in that segment, then the set of points in our constrained set must be $[0, \frac{1}{2}]$. In other words, the set was scaled by the inverse of the scale factor.

2.2 Combining the Constraints

Given the above formulation of the various constraints they can be easily combined into a forceful colour constancy algorithm. First the relative illumination field is used to remove the illumination variation from the image leaving an image which is of the scene with chromaticities as they would have appear if it had been illuminated throughout by the illumination at the base point. Starting from this intermediate result a constraint on the possible illumination maps is derived for each of the surface chromaticities. The illumination constraint provided by the set of plausible illuminants is fixed by the initial measurement of the various illuminants around the campus. Each hull point of the set of the relative illumination field furnishes yet another constraint; namely, the illumination constraint multiplied by the appropriate diagonal transform. The illumination constraint and the transforms due to the relative illumination field are intersected, and the result is inverted. As mentioned above, this inverted set was approximated well by its convex hull. The inverted set is then intersected with the intersection of all the surface constraints.

The final step of the algorithm is to chose a solution from the set of possible solutions. In [8, 9] the solution chosen maximizes the volume of the mapped set, which is equivalent to maximizing the product of the components of the mapping. In this work, however, we use the centroid of the solution set, which is more natural. This choice can be shown to minimize the expected error if all solutions are equally likely and error is measured by the distance from the choice to the actual solution. Furthermore, in both synthesized and real images, the centroid was found to give better results.

The colour constancy algorithm that incorporates all the different constraints was tested first on generated data. One thousand sets containing 1, 2, 4, 8, and 16 surfaces were randomly generated and used in conjunction with 1 of 5 illuminants, with 0 through 4 of the remaining lights playing the role of additional illuminants arising as a result of varying illumination. Table 1 gives the results which are exactly as wished. As either the number of surfaces or the number of extra lights increases, the answer consistently improves. Thus it was verified that varying illumination is a powerful constraint, and furthermore, it can be effectively integrated with the other constraints.

3 Finding the Relative Illumination Field

We now detail an algorithm for finding the relative illumination field describing the variation in the incident illumination. This algorithm can be divided into two parts. The first is a new technique for image segmentation appropriate for scenes with varying illumination. The second part uses the segmentation to determine the illumination map robustly.

Unless the illumination is known to be constant, it is essential that a segmentation method be able to accommodate varying illumination. In general, the segmentation problem is quite difficult, especially with varying illumination, as in this case an area of all one reflectance can exhibit a wide range of colour. Fortunately for our purposes, it is not critical if an occasional region is mistakenly divided into two pieces, nor if two regions which have almost the same colour are incorrectly merged. This is because the goal at this point is simply to identify the illumination, not the surfaces. Nonetheless, the better the segmentation, the more reliable and accurate the possible colour constancy.

One approach to segmentation is that used by Retinex theory [13, 14]. In Retinex small changes in pixel values at neighboring locations are assumed to be due to changes in the illumination and large changes to changes in surface reflectance. This idea can be used to segment an image into regions of constant surface reflectance properties by growing regions by including pixels only if they are less than some small threshold different from their neighbours. The threshold must be large enough to allow for both noise and the illumination changes and yet not admit small changes in surface reflectance—a balance which is of course impossible to establish.

We use this method as part of our algorithm, but alone, it is not sufficient. The problem is that two dissimilar regions will eventually mistakenly merge if there exists a sequence of small jumps connecting them. This can occur if the edge is gradual or because of noise. In essence, a threshold large enough to allow for noise (and varying illumination) allows for enough drift in the pixel values to include an entirely dissimilar region. Local information alone is insufficient, so we resolve the problem by adding a more global condition involving illumination constraints.

Number of Surfaces

Solution Method	1	2	4	8	16
BF	0.073	0.073	0.073	0.073	0.073
BDT	0.116	0.116	0.116	0.116	0.116
GW	1.62	1.01	0.69	0.513	0.428
RET	1.62	1.10	0.72	0.478	0.354
S	12.4	4.4	1.65	0.585	0.285
SI	2.275	1.68	0.99	0.480	0.271
SIV1	1.65	1.26	0.79	0.420	0.256
SIV2	1.154	0.896	0.620	0.351	0.242
SIV3	0.800	0.656	0.488	0.311	0.231
SIV4	0.384	0.36	0.317	0.274	0.228

Solution Method Key

BF	Error of best possible solution using full linear map
BDT	Error of best possible solution using a diagonal map
GW	Naive Grey World Algorithm (scale each channel by average)
RET	Naive Retinex Algorithm (scale each channel by maximum)
S	Surface constraints alone
SI	Surface and illumination constraints
SIV	Surface and illumination constraints with view under one extra illuminant
SIV2	Surface and illumination constraints with view under 2 extra illuminants
SIV3	Surface and illumination constraints with view under 3 extra illuminants
SIV4	Surface and illumination constraints with view under 4 extra illuminants

Table 1. Results of color constancy experiments for 1000 sets of 1, 2, 4, 8, and 16 surfaces under all combinations of test lights and extra lights for varying illumination. The canonical illuminant was a Philips CW fluorescent light. The values shown are the average magnitude of the chromaticity vector difference between the estimate and the desired answer, averaged over all results.

The new global segmentation condition is based on the idea that in order for two pixels—no matter how far apart they are spatially—to be considered part of the same region, a plausible illumination change between them must exist. The set of plausible illuminant changes can be derived in advance from the initial set plausible illuminants. This condition binding pixels within a single region based on plausible illumination changes is called patch coherence. The patch coherence condition differs from the Retinex condition in two important ways. First, the cumulative drift in pixel values along a path is limited, as opposed to growing linearly with the pixel distance. Second, the allowable drift is constrained more in certain directions due to the nature of the set of common illuminants. For example, green illuminants are rare, which means that the set of common illuminants is narrow in the green direction, and thus overall, the drift towards or away from green is more restricted than that towards or away from blue.

It was found to be useful to retain the Retinex condition as well as the patch coherence method described above for two reasons. First, the Retinex condition is faster to compute, and thus can be use to reject pixels that do not need to be tested further for inclusion. Second, if a comprehensive set of possible illuminants is used, then an occasional surface boundary change will also be a possible illumination change. Since the Retinex method by itself works much of the time, these exceptional cases in which a surface change mimics an illumination change generally will be covered by the Retinex condition.

In detail our segmentation algorithm begins with an arbitrary initial starting point in a region and assumes that the illumination at that point is constrained to be in the set of plausible illuminants. It is important to update the constraints on the illumination at the starting point each time a new point is added to the region. Each newly included point further constrains the possible illuminations at the starting point. Updating the constraints is similar to using the relative illumination field to solve for colour constancy as described above. The element-wise ratio of the chromaticities of the new point to that of the initial point induces a constraint set V defined by (5). Specifically, the illumination gamut is transformed by the inverse of the ratio interpreted as a diagonal transform. This convex set is intersected with the current constraint set. If the intersection is null, then the new point is excluded and the constraint set is left unchanged. If it is not null, then the intersection becomes the updated constraint set and the new point is added to the region.

Similar to the situation when solving for colour constancy, it is sufficient to perform the intersection only when the new transform to be applied to the illumination gamut is outside the convex hull of the preceding transforms. Although calculations are relative to the initial point, this procedure ensures that all points in the region can be assigned illuminants from the set of plausible illuminants which are consistent with the illumination jumps between them. Furthermore, the inclusion of any of the rejected points would violate this condition.

Given our segmentation we reduce the problem of finding the relative illumination field to that of finding the illumination at the center of each region relative to that at the center of the base region. Since the center of a region, as defined by the center of mass, need not be inside the region, the implementation uses a point in the region close to the center of mass, preferably a few pixels from the boundary. The illumination at a point relative to that of the region center is simply the ratio of its response to the response of the center point. This follows directly from the assumption that the pixels are from the same surface, given that we accept a diagonal model for illumination change. Thus the map at an arbitrary point is simply the map at the center, adjusted by this relative jump.

To determine the maps at the center points we make the assumption that illumination does not change significantly at the region boundaries. Thus every jump across a boundary gives a condition on the relative maps of the centers of the two adjacent regions. More specifically, consider two regions A and B, with centers C_A and C_B, and boundary points B_A and B_B close to each other. Denote responses by R subscripted by the point label and denote the diagonal map relative to the grand central point as D, also subscripted by the point label. Each channel or chromaticity component is dealt with independently, so the quantities in the equations are scalars. The assumption that the illumination does not change significantly at the boundary is simply:

$$D_{B_A} = D_{B_B} \tag{6}$$

Since we are assuming a diagonal model of illumination change, and C_A is on the same surface as B_A, and similarly for the center and boundary of surface B, we have:

$$D_{B_A} = D_{C_A}\left({R_{B_A}}/{R_{C_A}} \right) \quad \text{and} \quad D_{B_B} = D_{C_B}\left({R_{B_B}}/{R_{C_B}} \right) \tag{7}$$

Combining (10) and (11) yields:

$$D_{C_A}\left({R_{B_A}}/{R_{C_A}} \right) = D_{C_B}\left({R_{B_B}}/{R_{C_B}} \right) \tag{8}$$

Taking logarithms of both sides of (12), and rearranging terms gives:

$$\ln(D_{C_A}) - \ln(D_{C_B}) = \ln(R_{B_B}) - \ln(R_{B_A}) + \ln(R_{C_A}) - \ln(R_{C_B}) \tag{9}$$

This final equation is at the heart of the method. Here we have a condition on the component of the map for two of the regions. Other boundary point pairs produce additional equations. In order to have a robust method, one would like long boundaries to have more weight in the process than short ones, since the latter may due to a small region consisting entirely of noise. But this is exactly what we will get if we enter one equation for each boundary pair and solve the resulting system of equations in the least squares sense. Furthermore, some boundary pairs can be identified as being more reliable and these are weighted even more by scaling the equation by a number greater than one (typically five). In addition, some boundary pairs should contribute less, and their equations are scaled by a number less than unity.

In order to have a solution to the set of equations, it must be insured that all segments connect to each other through the boundary pairs. This might be accomplished simply by assigning a region to every point, and using each break in either the horizontal or vertical directions to produce a boundary pair. This is usually not an option because often some parts of the image should not be used; for example, when an area is too dark. Therefore the likelihood of connectedness between regions was increased in the following manner. Boundary pairs were assigned at each horizontal and vertical change of region. If one of the regions was to be ignored, a good region was sought in the same direction, taking as many pixels as required. The resulting equation was weighted inversely to the distance taken to find a good region. Thus such a boundary would contribute little to the solution, but connectivity was not a problem for reasonable images (it is still possible to construct an image which will lack connectivity).

Several additional steps were taken to improve robustness. First very small regions were excluded from the computations. Second, it was found to be better to use pixels one unit towards the insides of the respective regions, if these were available. This way the pixels would tend to have contributions that were solely due to a single surface, as opposed to the possibility that they straddled more than one surface. These boundary pairs were weighted by a factor of five compared to ones where it was necessary to use pixels exactly on the boundary.

The final step in determining the relative illumination field is to interpolate over any excluded areas.

4 Results

The algorithm has been tested on a set of images of real scenes. In all cases the 'unknown' illumination consists of light from an incandescent bulb coming from one direction mixed with light from a Philip's CW fluorescent tube covered by a pale blue filter coming from another. The latter is similar in colour temperature to sky light. Thus the scenes mimic a common real world situation—an office with a window.

Unfortunately, qualitative evaluation of the results requires access to colour reproductions which are not available here. However, the grey scale counterparts reproduced in Figure 1 should give the reader some idea of the nature of the input used. The first image shown is a three-dimensional "Mondrian" made by affixing coloured construction paper to a conical waste paper bin. The bin is lying on its side with the incandescent light shining from above and the blue light from below. The top has blue and green papers on it, in the middle is a grey patch, and near the bottom are red and yellow papers. This illumination causes a distinct reddish tinge at the top and a clear bluish tinge at the bottom.

The second image is a simple two-dimensional "Mondrian" made by attaching eight sheets of coloured construction paper to the lab wall such that substantial parts of the wall remained visible. The third is a multi-coloured cloth ball. The cloth ball is interesting because the cloth has more texture than the construction paper. The fourth image shown in Figure 1 is of a person in front of a grey background with the sky-like light on the left and the incandescent light on the right. Under this illumination the left side of the grey background appears quite blue, and the flesh tones on the left are noticeably incorrect. It is not possible to obtain a canonical image for comparison because people move too much in the time it takes to set up the new illumination conditions. However, the qualitative results are quite promising. An additional image used for quantitative results is of a single piece of green poster board (not shown).

The numerical results in Table 2 reflect the RMS difference (over the entire image) between the [r/b, g/b] chromaticities at corresponding pixels in the recovered and canonical images. There are few colour constancy algorithms designed to deal with scenes of the generality addressed here so it is difficult to make comparisons with existing algorithms without violating their assumptions. As a first measure, we compare the solution obtained by a straight least squares fit of the input image to the canonical using a full linear model and then with a diagonal model. Without accounting for the illumination variation no algorithm working on the 3D Mondrian image can do better than the 0.78 error of the full linear case. In contrast, by

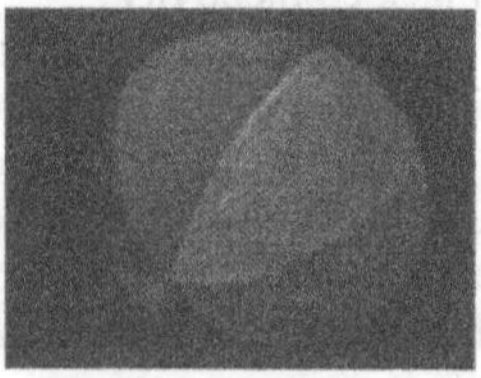

Figure 1. Some of the input images used to test the colour constancy algorithm. The bottom image was not used for quantitative results because people move too much in the time required to change illumination. The qualitative results for this image however were good.

Object

Algorithm		3D Mondrian	Ball	Green Card	Paper on Wall
	BT	0.78	0.536	0.195	0.265
	BDT	0.80	0.603	0.240	0.273
	N	0.96	0.93	0.279	0.405
	GW	0.88	1.37	0.312	0.349
	RET	0.87	0.75	0.246	0.290
	S	0.86	0.92	1.030	0.476
	SI	0.87	0.73	0.540	0.472
	VIR-BT	0.148	0.224	0.042	0.093
	VIR-BDT	0.176	0.227	0.044	0.133
	VIR-N	0.44	0.595	0.262	0.530
	VIR-GW	0.68	1.36	0.275	0.306
	VIR-RET	0.61	0.95	0.275	0.228
	VIR-S	0.33	1.11	1.929	0.211
	VIR-SI	0.39	0.38	0.269	0.163
	VIR-SIV	0.26	0.45	0.073	0.151

Error between the result and the canonical using the various solution methods

BT	Best possible linear map solution
BDT	Best possible diagonal map
N	No processing. Simply use the input image as the result.
GW	Naive Grey World Algorithm (scale each channel by average)
RET	Naive Retinex Algorithm (scale each channel by maximum)
S	Using Surface Constraints
SI	Using Surface and illumination constraints
VIR-BT	Best linear map applied after varying illumination is removed
VIR-BDT	Best diagonal map applied after varying illumination is removed
VIR-N	No processing applied after varying illumination is removed
VIR-GW	Naive Grey World Algorithm with varying illumination removed.
VIR-RET	Naive Retinex Algorithm with varying illumination removed.
VIR-S	Using surface constraints with varying illumination removed.
VIR-SI	Using surface and illumination constraints with varying illumination removed.
VIR-SIV	Complete new algorithm using surface, illumination, and varying illumination constraints with varying illumination removed.

Table 2. Results of color constancy algorithms applied to four images. The canonical illuminant was a Philips CW fluorescent light. The values shown are the RMS (over all pixels) magnitude of the chromaticity vector difference between the estimate and the desired answer, which is a view of the scene under the canonical light.

accounting for and utilizing the illumination variation our new algorithm reduces the error to 0.26.

Table 2 also shows the chromaticity error for the case of doing nothing at all. The grey world algorithm, which uses the average image chromaticity as an illumination estimate, and the Retinex normalization strategy of taking the maximum response from each colour band as an illumination estimate are tried even though the comparison is somewhat unfair because they both assume the illumination to be constant. Similar tests are run using surface constraints alone and surface constraints with the additional constraints on the set of plausible illuminants.

To make the comparison fairer, we also include similar tests with these algorithms but applied after the illumination variation has been discounted. In other words, we combined the first part of our algorithm (removal of the illumination variation) with each of the other algorithms. In this case, the other algorithms are applied to data which does not violate the constant illumination assumption, but they still do not exploit the information contained in the illumination variation. The Retinex normalization applied in this way gives an algorithm which is close, in theory, to the original Retinex idea.

The results show first that if the varying illumination is not accounted for, then all the colour constancy algorithms perform poorly. In all cases, the complete new VIR-SIV algorithm did better than any algorithm which assumed the chromaticity of the illumination to be constant. In fact, the performance is better than that of the best diagonal and best linear fits. The complete algorithm also performed better than the others applied to the data with the varying illumination removed, except when compared to applying the combination of surface and illumination constraints to the ball image. Most importantly, the algorithm performs better than the Retinex scaling applied to the data with the variation removed. As mentioned above, this procedure is close to the spirit of the original Retinex algorithm, which is unique as an alternative to our algorithm, even though its testing has been limited to scenes with more controlled illumination.

The results for the green card are included to illustrate that the varying illumination constraint can be very useful in the case when there is a paucity of other information. Most colour constancy algorithms require a good selection of surfaces for reliable performance. The ability of this algorithm to go beyond that in the case of varying illumination is encouraging.

5 Conclusion

We have presented a new algorithm for colour constancy which builds upon the recent gamut-based algorithms of Forsyth [9] and Finlayson [7, 8]. The new algorithm models the illumination via a diagonal transform or equivalently a coefficient rule model. Within the diagonal model framework, the algorithm combines the constraints provided by the observed gamut of image colours, by a priori knowledge of the set of likely illuminants, and by the variation in illumination across a scene. Existing algorithms that use the information inherent in illumination variation assume that some unspecified pre-processing stage has already identified the variation, and thus are not fully automated.

Identifying illumination variation is in itself a difficult problem. The Retinex algorithm is the only alternative colour constancy algorithm designed for scene conditions similar to those investigated in this paper. Nonetheless, it was restricted to flat Mondrian scenes and is known not to work very well for a variety of reasons.

The new algorithm is more powerful than Retinex both because it incorporates a new technique for identifying the relative illumination field and because the it actually uses the illumination variation when solving for colour constancy. While many improvements are still possible, tests using both synthetic and real image data for three-dimensional scenes verify that the algorithm works well.

6 References

1. A. Blake, "Boundary conditions for lightness computation in Mondrian world", Computer Vision, Graphics, and Image Processing, **32**, pp. 314-327, (1985)
2. G. Buchsbaum, "A spatial processor model for object colour perception", Journal of the Franklin Institute, **310**, pp. 1-26, (1980)
3. W. Freeman and David Brainard, "Bayesian Decision Theory, the Maximum Local Mass Estimate, and Color Constancy", in *Proceedings: Fifth International Conference on Computer Vision*, pp 210-217, (IEEE Computer Society Press, 1995)
4. B. V. Funt, M. S. Drew, M. Brockington, "Recovering Shading from Color Images", in *Proceedings: Second European Conference on Computer Vision*, G. Sandini, ed., 1992.
5. G.D. Finlayson and M.S. Drew and B.V. Funt, "Spectral Sharpening: Sensor Transformations for Improved Color Constancy", *J. Opt. Soc. Am. A*, **11**, 5, pp. 1553-1563, (1994)
6. G.D. Finlayson and M.S. Drew and B.V. Funt, "Color Constancy: Generalized Diagonal Transforms Suffice", *J. Opt. Soc. Am. A*, **11**, 11, pp. 3011-3020, (1994)
7. G. D. Finlayson, B. V. Funt, and K. Barnard, "Color Constancy Under Varying Illumination", in *Proceedings: Fifth International Conference on Computer Vision*, pp 720-725, 1995.
8. G. D. Finlayson, "Color Constancy in Diagonal Chromaticity Space", in *Proceedings: Fifth International Conference on Computer Vision*, pp 218-223, (IEEE Computer Society Press, 1995).
9. D. Forsyth, "A novel algorithm for color constancy", *Int. J. Computer. Vision*, **5**, pp. 5-36, (1990)
10. R. Gershon and A.D. Jepson and J.K. Tsotsos, "Ambient illumination and the determination of material changes", *J. Opt. Soc. Am. A*, **3**, 10, pp. 1700-1707, (1986)
11. B.K.P. Horn, "Determining lightness from an image", *Computer Vision, Graphics, and Image Processing*, **3**, pp. 277-299, (1974)
12. D.B. Judd and D.L. MacAdam and G. Wyszecki, "Spectral Distribution of Typical Daylight as a Function of Correlated Color Temperature", *J. Opt. Soc. Am.* , **54**, 8, pp. 1031-1040, (August 1964)
13. E.H. Land, "The Retinex theory of Color Vision", *Scientific American*, 108-129, (1977)
14. John J. McCann, Suzanne P. McKee, and Thomas H. Taylor, "Quantitative Studies in Retinex Theory", *Vision Research*, **16**, pp. 445–458, (1976)
15. L.T. Maloney and B.A. Wandell, "Color constancy: a method for recovering surface spectral reflectance", *J. Opt. Soc. Am. A*, **3**, 1, pp. 29-33, (1986)
16. M. Tsukada and Y. Ohta, "An Approach to Color Constancy Using Multiple Images", in *Proceedings Third International Conference on Computer Vision*, (IEEE Computer Society, 1990)
17. G. Wyszecki and W.S. Stiles, *Color Science: Concepts and Methods, Quantitative Data and Formulas*, 2nd edition, (Wiley, New York, 1982)
18. M. D'Zmura and G. Iverson, "Color constancy. I. Basic theory of two-stage linear recovery of spectral descriptions for lights and surfaces", *J. Opt. Soc. Am. A*, **10**, 10, pp. 2148-2165, (1993)

Color Angular Indexing

Graham D. Finlayson[1], Subho S. Chatterjee[2] and Brian V. Funt[2]

[1] Department of Computer Science, University of York, York YO1 5DD, UK
[2] School of Computing Science, Simon Fraser University, Vancouver, Canada

Abstract. A fast color-based algorithm for recognizing colorful objects and colored textures is presented. Objects and textures are represented by **just six numbers**. Let r, g and b denote the 3 color bands of the image of an object (stretched out as vectors) then the color angular index comprises the 3 inter-band angles (one per pair of image vectors). The color edge angular index is calculated from the image's color edge map (the Laplacian of the color bands) in a similar way. These angles capture important low-order statistical information about the color and edge distributions and invariant to the spectral power distribution of the scene illuminant. The 6 illumination-invariant angles provide the basis for angular indexing into a database of objects or textures and has been tested on both Swain's database of color objects which were all taken under the same illuminant and Healey and Wang's database of color textures which were taken under several different illuminants. Color angular indexing yields excellent recognition rates for both data sets.

1 Introduction

Various authors[13, 12, 10] (beginning with Swain and Ballard[13]) have found that the color distributions of multi-colored objects provide a good basis for object recognition. A color distribution, in the discrete case, is simply a three-dimensional histogram of the pixel values using one dimension for each color channel. For recognition, the color histogram of a test image is matched in some way (the matching strategy distinguishes the methods) against model histograms stored in the database. The closest model defines the identity of the image. Swain has shown that excellent recognition is possible so long as objects are always presented in the same pose and the illumination color is held fixed. Varying either the pose or illumination color can cause the color distribution to shift so as to prevent recognition.

Our goal is to develop a color-distribution-based descriptor that is concise, expressive and illuminant invariant. Swain and Ballard's histogram descriptors are expressive—they provide good recognition rates so long as the illuminant color is held fixed. However, they are not concise since each histogram is represented by the counts in 4096 bins. A descriptor requiring only a few parameters should speed up indexing performance and be useful in other more computationally intensive recognition algorithms (e.g. Murase and Nayar's [9] manifold method). Of course, we would prefer to decrease the match time without decreasing the match success.

Healey and Slater[4] have proposed a representation based on a small set of moments of color histograms[14]. They show that when illumination change is well described by a linear model (i.e., image colors change by a linear transform when the illumination changes) certain high-order distribution moments are illuminant invariant. Unfortunately, in the presence of noise (even in small amounts) this high-order information is not very stable and as such their representation may not be very expressive.

Requiring a full 3-by-3 linear model of illuminant change limits the kind of illuminant-invariant information that can be extracted from a color distribution. However, the linear model is in fact over general since only a small subset of the possible linear transforms correspond to physically plausible illuminant changes[2]. In particular if, as is usually the case, the sensitivities of our camera are relatively narrow-band then the images of the same scene viewed under two different illuminants are related (without error) by 3 simple scale factors. Each color pixel value (r_i, g_i, b_i) in the first color image becomes $(\alpha r_i, \beta g_i, \gamma b_i)$ in the second (where r_i, g_i and b_i denote the ith pixel in the red, green and blue image bands respectively and α, β and γ are scalars). This means that the images differing only in terms of the scene illuminant are always related by a simple 3-parameter diagonal matrix.

Under a diagonal model of illuminant change, the 3 angles between the different bands of a color image provide a simple illuminant independent invariant. To see this think of each image band as a vector in a high-dimensional space. When the illumination changes each vector becomes longer or shorter but its orientation remains unchanged. As well as being invariant to illumination change we show that the color angles encode important low-order statistical information.

Of course the camera sensors are not necessarily narrow-band and as such color angles might not be stable across a change in the illuminant. Nonetheless, good invariance can be attained if the angles are calculated with respect to a special *sharpened* color image. The sharpened color image, which exists for all sensor sets[2, 1], is created by taking linear combinations of the original color bands.

Angular indexing using just the 3 color angles suffice if the object database is small but performance breaks down for larger databases. To bolster recognition we develop a second angle invariant called the color-edge angle. Each band of the input color image is filtered by a Laplacian of Gaussian mask to generate a color edge map. The inter-band angles calculated with respect to this edge map are once again illuminant invariant and encode important information about the color edge distribution which when combined with the original 3 color angles leads to excellent recognition rates. Swain and Ballard demonstrated the richness of the color histogram representation for objects; and indexing on the 4096 histogram bins, they achieved almost flawless recognition for a database of 66 objects. Using color angular indexing, we attain very similar recognition rates indexing on **only 6 angles**.

As a second comparison, we evaluated recognition using Healey and Wang's color texture data set. It consists of 10 colored textures viewed under 4 illuminants, but we added images of the same textures viewed under 5 more orientations giving a grand total of 240 images. Indexing with the 6 combined angles delivers almost perfect recognition for the texture dataset.

2 Background

2.1 Color Image Formation

The light reflected from a surface depends on the spectral properties of the surface reflectance and of the illumination incident on the surface. In the case of Lambertian surfaces, this light is simply the product of the spectral power distribution of the light source with the percent spectral reflectance of the surface. For the theoretical development, we henceforth assume that most surfaces are Lambertian. Illumination, surface reflection and sensor function, combine together in forming a sensor response:

$$\rho_k^x = \int_w S^{x'}(\lambda) E^{x'}(\lambda) R_k(\lambda) d\lambda \tag{1}$$

where λ is the wavelength, R_k is the response function of the kth sensor class (r, g or b), $E^{x'}$ is the incident illumination and $S^{x'}$ is the surface reflectance function at location x' on the surface which is projected onto location x on the sensor array. We assume the illumination does not vary spectrally over the scene, so the index x' from $E(\lambda)$ can be dropped.

In the sections which follow we denote an image by I. The content of the jth color pixel in I is denoted (r_j^I, g_j^I, b_j^I). we will assume that there are M pixels in an image.

2.2 Color Histograms

Let $H(I_1)$ and $H(I_2)$ be color histograms of the images I_1 and I_2. $H_{i,j,k}(I_1)$ is an integer recording the number of colors in I_1 which fall in the ijkth bin. The mapping of color to bin is usually not 1-to-1 but rather color space is split into discrete regions. For example, Swain and Ballard[13] split each color channel into 16 intervals giving $16 \times 16 \times 16 = 4096$ bins in each histogram.

To compare histograms $H(I_1)$ and $H(I_2)$ a similarity measure is computed. Swain calculates similarity of a pair of histograms as their common intersection:

$$\sum_{i=1}^{N} \sum_{j=1}^{N} \sum_{k=1}^{N} \min(H_{i,j,k}(I_1), H_{i,j,k}(I_2)) \tag{2}$$

where N is the number of bins in each color dimension. While other methods of comparing histograms have been suggested, e.g. [10], they are all similar in the sense that they consist of many bin-wise operations. There are several problems with similarity calculated in this way. First, each distribution is represented by a feature vector of N^3 dimensions (the number of bins in a histogram), and this is quite large. The larger the feature space the slower the match. Secondly, it is unlikely that all the information in a distribution will be useful in calculating a match. Lastly the color histogram depends on the color of the light. Moving from red to blue illumination causes the color distributions to shift and results in poor match success[3].

2.3 Statistical Moments

If color distributions are described well by a small number of statistical features then comparing these features should suffice for determining distribution similarity. Suppose we must characterize the color distribution in an image I by one single color. A good candidate (and the obvious one) is the mean image color,

$$\underline{\mu}(I) = \frac{1}{M}\sum_{j=1}^{M}(r_j^I, g_j^I, b_j^I)^T \tag{3}$$

where M is the number of pixels in the image. The variance or spread of colors about the mean also captures a lot of information about the color distribution. The variance in the red channel $\sigma_r^2(I)$ is defined as

$$\sigma_r^2(I) = \frac{1}{M}\sum_{i=1}^{M}(r_i^I - \mu_r(I))^2 \tag{4}$$

The covariance $\sigma_r\sigma_g(I)$ between the red and green color channels is defined as

$$\sigma_r\sigma_g(I) = \frac{1}{M}\sum_{i=1}^{M}(r_i^I - \mu_r(I))(g_i^I - \mu_g(I)) \tag{5}$$

Similarly, $\sigma_g^2(I)$, $\sigma_b^2(I)$, $\sigma_r\sigma_b(I)$ and $\sigma_g\sigma_b(I)$ can be defined. These variances and covariances are usually grouped together into a covariance matrix $\Sigma(I)$:

$$\Sigma(I) = \begin{bmatrix} \sigma_r^2(I) & \sigma_r\sigma_g(I) & \sigma_r\sigma_b(I) \\ \sigma_r\sigma_g(I) & \sigma_g^2(I) & \sigma_g\sigma_b(I) \\ \sigma_r\sigma_b(I) & \sigma_g\sigma_b(I) & \sigma_b^2(I) \end{bmatrix} \tag{6a}$$

Suppose we represent an image I as an $M \times 3$ matrix, where the ith row contains the ith rgb triplet. If the mean image color is the zero-vector, $\underline{\mu}(I) = (0,0,0)^T$, then Equation (6a) can be rewritten in matrix notation:

$$\Sigma(I) = \frac{I^T I}{M} \tag{6b}$$

where ^{T} denotes matrix transpose. The covariance relationship given in Equation (6b) will prove very useful in subsequent discussion.

The mean color $\underline{\mu}(I)$ is called the first-order moment of the distribution of image colors and the covariance matrix $\Sigma(I)$ is composed of the second order moments. Third and higher order moments can be calculated in an analogous manner. For example the third order moment of the r color channel is

$s_r(I) = \frac{1}{M}\sum_{i=1}^{M}(r_i^I - \mu_r(I))^3$. This moment captures the *skew* of the red response distribution and is measure of the degree of symmetry about the mean red response. The green and blue skews $s_g(I)$ and $s_b(I)$ are similarly defined. In general the nth order moments of a color distribution is defined as:

$$\frac{1}{M}\sum_{i=1}^{M}(r_i^I - \mu_r(I))^{\alpha}(g_i^I - \mu_g(I))^{\beta}(b_i^I - \mu_b(I))^{\gamma} \tag{7}$$

where $\alpha, \beta, \gamma > 0$ and $\alpha + \beta + \gamma = n$.

Roughly speaking low-order moments give a coarse description of the distribution. More and more distribution details are unveiled as one progresses through the higher moments[11]. Two observations stem from this. First, for color-based object recognition low-order moments capture the most useful information. For example, low-order moments are less effected by confounding processes such as image noise or highlights.

Stricker and Orengo[12], have presented experimental evidence that color distributions can be represented by low-order moments. They show that the features $\underline{\mu}(I)$, $\sigma_r^2(I)$, $\sigma_g^2(I)$, $\sigma_b^2(I)$, $s_r(I)$, $s_g(I)$ and $s_b(I)$ provide an effective index into a large image database. Unfortunately these low-order moments of color distributions are suitable for recognition only if objects are always viewed under the same colored light. A white piece of paper viewed under reddish and bluish lights have predominantly red and blue color distributions respectively. What we really need is descriptors of the low-order information that do not depend on the illuminant in this way.

2.4 Finite Dimensional Models

Both illuminant spectral power distribution functions and surface spectral reflectance functions are described well by finite-dimensional models of low dimension. A surface reflectance vector $S(\lambda)$ can be approximated as:

$$S(\lambda) \approx \sum_{i=1}^{d_S} S_i(\lambda)\sigma_i \tag{8}$$

where $S_i(\lambda)$ is a basis function and $\underline{\sigma}$ is a d_S-component column vector of weights. Similarly each illuminant can be written as:

$$E(\lambda) \approx \sum_{j=1}^{d_E} E_j(\lambda)\epsilon_j \tag{9}$$

where $E_j(\lambda)$ is a basis function and $\underline{\epsilon}$ is a d_E dimensional vector of weights.

Given a finite-dimensional approximations to surface reflectance and illumination, the color response eqn. (1) can be rewritten as a matrix equation. A *Lighting Matrix* $\Lambda(\underline{\epsilon})$[7] maps reflectances, defined by the $\underline{\sigma}$ vector, onto a corresponding response vector:

$$\underline{p} = \Lambda(\underline{\epsilon})\underline{\sigma} \tag{10}$$

where $\Lambda(\underline{\epsilon})_{ij} = \int_\omega R_i(\lambda)E(\lambda)S_j(\lambda)d\lambda$. The lighting matrix depends on the illuminant weighting vector $\underline{\epsilon}$ to specify $E(\lambda)$ via eqn. (9). If surface reflectances are 3-dimensional then every $\Lambda(\underline{\epsilon})$ is a 3×3 matrix. It follows that response triples obtained under one light can be mapped to those of another by a 3×3 matrix.

$$\underline{p}^1 = \Lambda(\underline{\epsilon}^1)\underline{\sigma}\,,\ \underline{p}^2 = \Lambda(\underline{\epsilon}^2)\underline{\sigma} \ \Rightarrow\ \underline{p}^2 = \Lambda(\underline{\epsilon}^2)[\Lambda(\underline{\epsilon}^1)]^{-1}\underline{p}^1 \qquad (11)$$

Studies[6, 8] have shown that a 3-dimensional model is quite reasonable. Thus it follows that the color distributions of the same surfaces viewed under two illuminants are linearly related to a good approximation.

2.5 Illuminant Invariant Moments

Taubin and Cooper [14] have recently developed efficient algorithms for the computation of vectors of affine moment (or algebraic) invariants of functions. These vectors are invariant to affine transformations of the function which, as Healey and Slater observed, may make them a suitable illuminant-invariant representation for color distributions.

There are two steps in calculating Taubin and Cooper's invariants. First the distribution is manipulated such that its statistics are standardized in some sense. Second, features which are independent of the position of the standardized distribution are extracted.

Standardizing the distribution's statistics is best understood by example. Let I_1 and I_2 be $M \times 3$ matrices denoting the color images of some scene viewed under a pair of illuminants (the M *rgb* triplets in each image are placed in the rows of the matrices) where first the mean image color has been subtracted in both cases. Thus, $\underline{\mu}(I_1) = (0,0,0)^T$ and $\underline{\mu}(I_2) = (0,0,0)^T$. So long as reflectances are approximately 3-dimensional the two images are related by a 3×3 matrix $\mathcal{M}$

$$I_2 \approx I_1\mathcal{M} \qquad (12)$$

I_1 and I_2 are standardized by transforming them by the matrices $\mathcal{O}_1$ and $\mathcal{O}_2$ such that their column spaces are orthonormal,

$$\mathcal{O}_1^T I_1^T I_1 \mathcal{O}_1 = \mathcal{O}_2^T I_2^T I_2 \mathcal{O}_2 = \mathcal{I} \qquad (13)$$

Since $I_1\mathcal{O}_1$ and $I_2\mathcal{O}_2$ are orthonormal they differ only by a rotation and represent the same color distributions with respect to different coordinate axes. Thus the basic *shape* of the distributions is the same.

The second step in Taubin and Cooper's method is to extract features from the standardized distributions $I_1\mathcal{O}_1$ and $I_2\mathcal{O}_2$ which are independent of the coordinate frame. The precise details of their method do not concern us—it suffices that the invariants exist. However, we must ask whether these invariants are expressive; that is, do they convey useful information?

To explore this question let us examine the matrices in equation (13) more closely. From Equation (6b) it follows that:

$$\Sigma(I_1\mathcal{O}_1) = \frac{\mathcal{O}_1^T I_1^T I_1 \mathcal{O}_1}{M} = \frac{\mathcal{I}}{M} \quad (14)$$

where $\mathcal{I}$ denotes the identity matrix. That is the covariance matrix of all standardized images is equal to the scaled identity matrix. This is always the case regardless of the starting image statistics. Thus all the low-order statistics—those which convey the most useful information about the distribution—have been lost through the need to discount the effect of the illuminant. It follows that only high order moments can be extracted from $I_1\mathcal{O}_1$ and $I_2\mathcal{O}_2$. As discussed above, we expect that these will not suffice for reliable recognition and this prediction is borne out by experiment in section 4.

3 Distributions angles

While finite-dimensional models are a useful tool for investigating colors under a changing illuminant they do not tell the whole story. Indeed it turns out that a restricted subset of the possible linear transforms correspond to plausible illuminant changes. This observation allows us to extract useful illuminant-invariant statistics from color distributions.

Suppose that the sensor sensitivities of the color camera are delta functions, $R_k(\lambda) = \delta(\lambda - \lambda_k)$. In this case, the camera responses p_k and q_k generated by an arbitrary surface $S_j(\lambda)$ viewed under illuminants $E_1(\lambda)$ and $E_2(\lambda)$ are:

$$p_k = S_j(\lambda_k)E_1(\lambda_k) \ , \ q_k = S_j(\lambda_k)E_2(\lambda_k) \quad (15)$$

It is immediate that

$$q_k = \frac{E_2(\lambda_k)}{E_1(\lambda_k)} p_k \quad (16)$$

Since (16) no longer involves the reflectance function $S_j(\lambda)$ the camera responses induced by any surface are related by the same scaling factor $\frac{E_2(\lambda_k)}{E_1(\lambda_k)}$. Combining the scalings for each sensor class into a diagonal matrix, (16) can be expressed as:

$$\underline{q} = \mathcal{D}\underline{p} \quad (\mathcal{D}_{kk} = \frac{E_2(\lambda_k)}{E_1(\lambda_k)} \quad k = 1,2,3) \quad (17)$$

Thus for narrow-band sensors illuminant change is exactly modelled by a diagonal matrix and the full generality of a 3×3 linear model is not required.

Let us consider the problem of extracting invariant features from a color distribution under a diagonal model of illuminant change. We follow the basic approach of Taubin and Cooper in that we first standardize the statistics of the color distribution and then extract the statistical features. Under the diagonal model Equation (17), the relationship between a pair of images can be rewritten as:

$$I_2 = I_1 \mathcal{D} \tag{18}$$

where $\mathcal{D}$ is a diagonal matrix. By (18) the corresponding columns of I_1 and I_2 are vectors in the same direction but of different length so the distributions can be standardized by normalizing the lengths of the columns of I_1 and I_2. We define a function $N()$ for carrying out the column normalization:

$$N(I_1) = I_1 \mathcal{D}_N \tag{19}$$

where the ith diagonal entry of the diagonal matrix $\mathcal{D}_N$ is equal to the reciporical of the length of the ith column of I_1.

$$[\mathcal{D}_N]_{ii} = \frac{1}{|[I_1]_i|} \tag{20}$$

$[]_i$ and $[]_{ij}$ denote the ith column and ijth element of a matrix respectively. The normalized distributions of I_1 and I_2 are equal:

$$N(I_1) = N(I_2) \tag{21}$$

The covariance matrix of $N(I_1)$ equals:

$$\Sigma(N(I_1)) = \frac{[N(I_1)]^T N(I_1)}{M} = \frac{1}{M}\begin{bmatrix} 1 & M\sigma_r\sigma_g & M\sigma_r\sigma_b \\ M\sigma_r\sigma_g & 1 & M\sigma_g\sigma_b \\ M\sigma_r\sigma_b & M\sigma_g\sigma_b & 1 \end{bmatrix} \tag{22}$$

Note the off-diagonal terms are non-zero, so under a diagonal model of illuminant change the color distributions can be standardized while preserving 3 of the 6 second-order moments, namely the covariances $\sigma_r\sigma_g$, $\sigma_r\sigma_b$ and $\sigma_g\sigma_b$. This contrasts favourably with standardization under a linear model of illuminant where all second-order moments are lost. Note the covariances will not be the same as those defined for the pre-standardized distribution i.e. the covariance terms in (22) are not equal to those in (6a).

Consider the geometric meaning of the covariance terms. The ijth entry in $M\Sigma(N(I_1))$ equals the dot-product of the i and jth th columns of $N(I_1)$. Because each column of $N(I_1)$ is unit length it follows that each dot-product equals the cosine of the angle between the i and jth columns. The cosine function is non-linear which is inappropriate for indexing. Thus we calculate the inverse cosine of the covariance terms in (24) effectively linearizing the feature giving us the ***angles of a color distribution.***

$$\phi_{ij}(N(I_1)) = \cos^{-1}([M\Sigma(N(I_1)]_{ij}) \quad (i \neq j) \tag{23}$$

A distribution is represented by its three angles $\phi_{12}(N(I_1))$, $\phi_{13}(N(I_1))$ and $\phi_{23}(N(I_1))$. The 3-tuple of the 3 distribution angles for a color distribution I is denoted $\underline{\phi}(I)$. The distance between distributions I and J is calculated by:

$$||\underline{\phi}(I) - \underline{\phi}(J)||_F \tag{24}$$

where $_F$ denotes the Frobenius norm; that is, the distance between distributions is calculated as the root-mean square error between the respective vectors of color angles.

3.1 Relaxing the Narrow-band assumption

If the camera sensors are not narrow-band then the analysis (15) through (17) does not hold and a single diagonal matrix will not relate sensor responses across a change in illumination. However, Finlayson et al.[2, 1] have shown that in this case a generalized diagonal matrix can be used instead. A generalized diagonal matrix is defined as $\mathcal{T}\mathcal{D}\mathcal{T}^{-1}$, where $\mathcal{T}$ is fixed and $\mathcal{D}$ varies with illumination. Under the generalized scheme, images under different illuminants are related by

$$I_1 \approx I_2\mathcal{T}\mathcal{D}\mathcal{T}^{-1} \tag{25}$$

The relationship in (25) holds exactly if illumination and reflectances are well described by 2- and 3-dimensional linear models[2]. Because *2–3* conditions roughly hold in practice the generalized diagonal relationship describes illuminant change for all sensor sets. Equation (25) can be rewritten making the role of the diagonal matrix explicit:

$$I_1\mathcal{T} \approx I_2\mathcal{T}\mathcal{D} \tag{26}$$

It follows that the angles $\underline{\phi}(N(I_1\mathcal{T}))$ are approximately illuminant-invariant features of color distributions. Since the cameras used in the experiments reported later do in fact have quite narrow-band sensor sensor sensitivities, $\mathcal{T}$ is set to the identity matrix.

3.2 Color-edge distribution angles

Let us define a color edge map as an image convolved with a Laplacian of Gaussian filter in which the usual two-dimensional filter is replicated for each of the three image bands. Denoting the convolution filter $\nabla^2 G$ the edge map of the image I is written as $\nabla^2 G \star I$ where $\star$ represents convolution. Where, as before $\nabla^2 G \star I$ can be thought of as an $M \times 3$ matrix. Because convolution is a linear operator the edge maps of the same scene viewed under two illuminants are related by a diagonal matrix:

$$\nabla^2 G \star I_2 \;=\; \nabla^2 G \star I_1\mathcal{D} \tag{27}$$

It follows that the angles $\underline{\phi}(N(\nabla^2 G{\star}I))$ encode second-order moment information about the color-edge distribution and are illuminant invariant. If color and color-edge distribution angles encode distinct information then we can expect that used together they will out perform recognition using either alone.

3.3 Properties of distribution angles

Distribution angles (either of colors or color edges) do not depend on the spatial characteristics of an image. In particular they do not depend on the order of the rows in I_1 or $\nabla^2 G \star I_1$. That this is so is clear from the definition of a moment in Equation (7) since a moment is a sum of terms, with each term calculated on a per-pixel basis. Distribution angles are also independent of scale since $N(I_1) = N(I_1 k)$ for any non-zero k. Because distribution angles are independent of image spatial characteristics and image scale, we can expect angular indexing to recognize an object in different contexts such as when it is placed at different viewing distances or is rotated about the optical axis.

4 Results

The invariants described in section 3 were used as cues for object recognition. They were tested on two published sets of color images[5, 13]. Results are presented for color angle invariants, color-edge angle invariants and their combination. Three existing distribution-based techniques—color indexing, color constant color indexing and Healey and Slater's moment approach—are applied to the same data sets for comparison.

4.1 Swain's Database

Swain's model database consists of 66 images of objects. However, because ratio invariants are ill-defined for images containing saturated pixels, eleven of the images with saturated pixels were pruned from the data set leaving 55 images. The same whitish illumination was used for all objects. A set of an additional 24 images of the same objects but viewed in different poses and with small amounts of deformation (e.g. a rumpled T-shirt) is then used to test the recognition algorithm. The test images are shown (in black and white) in Figure 1. The recognition rankings for color indexing, color constant color indexing (denoted CCCI in Tables 1 and 2), and Healey and Slater's moment-based method are tabulated along with that of color angular indexing in Table 1. Rank is defined to be the position of the correct match in the sorted list of match values. Thus, a match rank of 1 indicates correct recognition, a rank of 2 means that the correct answer was the second best and so on.

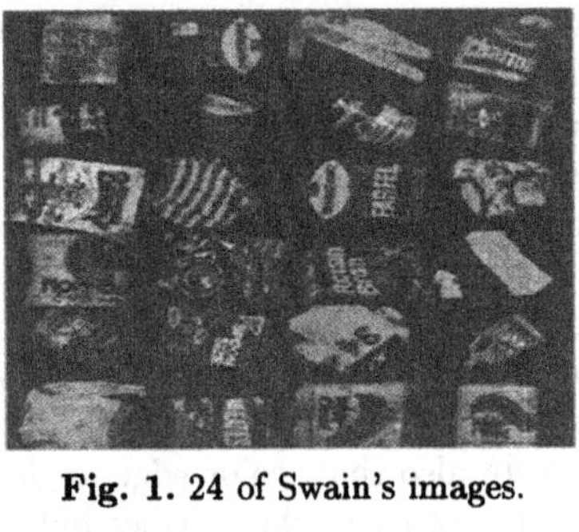

Fig. 1. 24 of Swain's images.

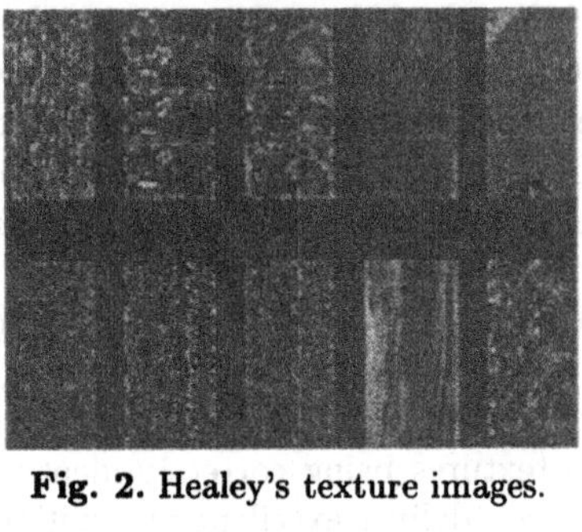

Fig. 2. Healey's texture images.

Algorithm	*Rankings*			
	1	*2*	*3*	*> 3*
Color angles	16	5	2	1
Edge angles	17	3	3	1
Color and edge angles	21	2	1	0
Color Indexing	23	1	0	0
CCCI	22	2	0	0
Healey's moments	7	7	3	7

Table 1. Object database performance.

Algorithm	*Rankings*			
	1	*2*	*3*	*> 3*
Color angles	124	45	29	32
Edge angles	222	8	0	0
Color and edge angles	224	6	0	0
Color indexing	74	21	27	108
CCCI	120	37	21	52
Healey's moments	121	40	20	49

Table 2. Texture database performance.

It is evident that Healey and Slater's higher-order moments based approach delivers poor performance. Note that 7 objects are matched with a greater than 3 rank. The color distribution angles and edge angles, used independently, give reasonable performance with 16 and 17 objects recognized correctly in each case. The combination of both, however, performs very well, comparable with the almost flawless recognition provided by color indexing and color constant color indexing. However, the latter two methods represent objects using a 4096 element feature vector (histogram bin counts).

4.2 Healey and Wang's texture database

Will color angular indexing successfully recognize the colored textures in Healey and Wang's texture data set[5]? This model data set contains ten images of natural textures viewed under white light; they are shown in (black and white) Figure 2. In addition to the model base set of 10 images, 30 other images were taken of the same textures but through 3 separate colored filters placed in front of the camera. This is equivalent to placing the filters in front of the illuminant so it models illumination change. The filters used had narrow pass bands in the blue, green and red regions of the spectrum. Such filters represent quite extreme illuminants and provide a stringent test for the illuminant invariance of the angular index.

Each of the 40 images (10 model and 30 test) was then rotated by 30°, 45°, 60°, 90° and 110° resulting in 240 images in total. Note the angle invariants of rotated textures are not trivially invariant because they are calculated with respect to a square image window so there is a windowing effect. The total test database consists of 230 images: the 30 test images in all 6 orientations and the model base in 5 orientations (all orientations except 0°). Results for the various algorithms are shown in Table 2.

Once again, recognition rates for color angle distributions alone are poor with almost half the textures not being recognized. Color angular indexing with the color and edge angle distributions yields the best results, with all but six of the textures being correctly identified. Note also that, color edge angles by themselves deliver excellent recognition. All the other methods, color indexing, color constant color indexing and Healey and Slater's moment based method, perform very poorly.

5 Conclusion

We have described a new color-based approach to object recognition called *color angular indexing* in which objects are represented by just 6 features: three color distribution angles and three color-edge distribution angles. Our experiments with real images on data bases of several hundred images show that colour angular indexing provides excellent recognition rates for a wide variety of objects and textures even under modest change in orientation and substantial change in illumination conditions.

References

1. G.D. Finlayson, M.S. Drew, and B.V. Funt. Color constancy: Generalized diagonal transforms suffice. *J. Opt. Soc. Am. A*, 11:3011–3020, 1994.
2. G.D. Finlayson, M.S. Drew, and B.V. Funt. Spectral sharpening: Sensor transformations for improved color constancy. *J. Opt. Soc. Am. A*, 11(5):1553–1563, May 1994.
3. B.V. Funt and G.D. Finlayson. Color constant color indexing. *IEEE transactions on Pattern analysis and Machine Intelligence*, 1995.
4. G. Healey and D. Slater. "Global color constancy: recognition of objects by use of illumination invariant properties of color distributions". *Journal of the Optical Society of America, A*, 11(11):3003–3010, November 1994.
5. G. Healey and L. Wang. The illumination-invariant recognition of texture in color images. *Journal of the optical society of America A*, 12(9):1877–1883, 1995.
6. L.T. Maloney. Evaluation of linear models of surface spectral reflectance with small numbers of parameters. *J. Opt. Soc. Am. A*, 3:1673–1683, 1986.
7. L.T. Maloney and B.A. Wandell. Color constancy: a method for recovering surface spectral reflectance. *J. Opt. Soc. Am. A*, 3:29–33, 1986.
8. D.H. Marimont and B.A. Wandell. Linear models of surface and illuminant spectra. *J. Opt. Soc. Am. A*, 9(11):1905–1913, 92.
9. H. Murase and S.K. Nayar. Visual learning and recognition of 3D objects from appearance. *International Journal of Computer Vision*, 14(1):5–24, 1995.
10. W. Niblack and R. Barber. The QIBC project: Querying images by content using color, texture and shape. In *Storeage and Retrieval for Image and Video Databases I, volume 1908 of SPIE Proceedings Series.* 1993.
11. R.J. Prokop and A.P. Reeves. A survey of moment-based techniques for unoccluded object representation and recognition. *CVGIP: Graphical Models and Image Processing*, 54(5):438–460, 1992.
12. M. A. Stricker and M. Orengo. Similarity of color images. In *Storage and Retrieval for Image and Video Databases III*, volume 2420 of *SPIE Proceedings Series*, pages 381–392. Feb. 1995.
13. M.J. Swain and D.H.. Ballard. Color indexing. *International Journal of Computer Vision*, 7(11):11–32, 1991.
14. G. Taubin and D. Cooper. "Object recognition based on moment (or algebraic) invariants". In J. Mundy and A. Zisserman, editors, *Geometric Invariance in Computer Vision*, pages 375–397. MIT Press, Cambridge, Mass., 1992.

Bidirectional Reflection Distribution Function expressed in terms of surface scattering modes

Jan J. Koenderink, Andrea J.van Doorn, and Marigo Stavridi

Helmholtz Instituut, Universiteit Utrecht, Princetonplein 5, PO Box 80 000, 3508 TA Utrecht, The Netherlands

Abstract. In many applications one needs a concise description of the Bidirectional Reflection Distribution Function (BRDF) of real materials. Because the BRDF depends on two independent directions (thus has four degrees of freedom) one typically has only a relatively sparse set of observations. In order to be able to interpolate these sparse data in a convenient and principled manner a series development in terms of an orthonormal basis is required. The elements of the basis should be ordered with respect to angular resolution. Moreover, the basis should automatically respect the inherent symmetries of the physics, *i.e.*, Helmholtz's reciprocity and (most often) surface isotropy. We indicate how to construct a set of orthonormal polynomials on the Cartesian product of the hemisphere with itself with the required symmetry and invariance properties. These "surface scattering modes" form a convenient basis for the description of BRDF's.

1 Introduction

Surface reflection (or rather: scattering) by natural materials[8] is conveniently described[13, 6, 7, 11] by the Bidirectional Reflection Distribution Function (or BRDF for short). The BRDF is the ratio of the radiance in the direction of the exit beam to the irradiance caused by the entrance beam. The BRDF depends on two directions, that is on four independent angles, two of them in the interval $(0°, 90°)$ and the other pair in the periodic range $(0°, 360°)$. The former two describe the deviation from the surface normal direction, the latter two the azimuth with respect to a fiducial direction on the surface. For an angular resolution of $\delta << 1$ one thus has to specify $4\pi^2/\delta^4$ independent samples; for an angular resolution of 10° that already amounts to more than fortythousand samples.

In view of these numbers it is perhaps not surprising that in practice few materials have been fully characterized in this way. For many important applications much lower resolutions suffice and then the BRDF description is a practical one. An example is graphics rendering, although hardly any attempt seems to exist where one uses empirical data. In practice one usually substitutes model approximations. In case empirical data are used the problem of interpolation or representation of the data comes up. One would prefer methods of representation that guarantee physically consistent results, *i.e.*, the BRDF should satisfy certain symmetries that reflect elementary physical constraints such as invariance under permutation of entrance and exit beams (so called Helmholtz's reciprocity[5]). Thus the problem arises of how to represent empirical data in a numerically advantageous and physically acceptable, principled manner.

Some type of series development in terms of an orhonormal set of basis functions that are ordered with respect to angular resolution appears the obvious choice. Then the desired angular resolution can be simply set by *truncating* the series, whereas the structure guarantees that the approximation is always optimal in the least squares sense. Thus one has to construct the desired orthonormal basis. We proceed to show how to do this.

2 A set of orthonormal polynomials on the hemisphere

The hemisphere (here denoted as $\mathbf{H}^2$) has the topology of the unit disk $\mathbf{D}^2$. Thus it makes sense to try to adapt known systems on the unit disk to our present problem. The unique set of polynomials that are complete and orthogonal on the unit disk and have the desired invariance properties with respect to rotations about the symmetry center of the unit disk are the well known Zernike polynomials[2]. This basis was introduced by Zernike in order to construct a principled method of describing wavefront aberrations for circular pupils (so called Zernike–Nijboer theory of aberrations).

We consider the upper hemisphere $\mathbf{H}^2$ of the unit sphere $\mathbf{S}^2$ with the usual coordinate systems. In Cartesian coordinates $\{x, y, z\}$ of $\mathbf{R}^3$ the upper hemisphere is given by $x^2 + y^2 + z^2 = 1$, $z \geq 0$. In polar coordinates $\{\vartheta, \varphi\}$ of $\mathbf{R}^2$ (ϑ the polar distance, φ the azimuth), it is $\vartheta \leq \frac{\pi}{2}$. The polar coordinates $\{\varrho, \varphi\}$ of the unit disk $\mathbf{D}^2$ will also be used.

The invariance we require is "invariant form" with respect to rotations about the origin of $\mathbf{R}^2$, or the z–axis of $\mathbf{R}^3$, that is, changes of azimuth. Thus if

$$\begin{aligned} x' &= x \cos\varphi + y \sin\varphi \\ y' &= -x \sin\varphi + y \cos\varphi, \end{aligned}$$

then the polynomial $V(x, y)$ should be taken to $V(x', y')$ such that

$$V(x, y) = G(\varphi) V(x', y'),$$

where $G(\varphi)$ is a continuous function with period 2π such that $G(0) = 1$.

We require that the transformation reflects the properties of the rotation group faithfully, thus $G(\varphi_1 + \varphi_2) = G(\varphi_1) G(\varphi_2)$. This determines the function G fully, we have

$$G(\varphi) = e^{il\varphi},$$

where l is any integer. Thus we have $V(\varrho\cos\varphi, \varrho\sin\varphi) = R(\varrho) e^{il\varphi}$. By hypothesis $V(x, y)$ is a polynomial in x, y of degree n (say). It follows that $R(\varrho)$ is a polynomial of degree n which contains no power of ϱ of degree lower than $|l|$. $R(\varrho)$ is even or odd as $|l|$ is even or odd. The Zernike polynomials are the unique choice that contains a member for each (n, l) and thus constitutes a complete basis.

The Zernike polynomials are denoted

$$V_n^l(\varrho\cos\varphi, \varrho\sin\varphi) = R_n^l(\varrho)e^{il\varphi}.$$

It has been shown that this is a complete basis for functions on the interior of the unit disk. This set contains $\frac{1}{2}(n+1)(n+2)$ linearly independent polynomials of degree $\leq n$. In the conventional normalization we have

$$\int_{\mathbf{D}^2} V_n^{l*} V_{n'}^{l'} \, dA = \frac{\pi}{n+1}\delta_{ll'}\delta_{nn'},$$

where δ_{pq} is the Kronecker symbol (*i.e.*, $\delta_{pp} = 1$ and $\delta_{pq} = 0$, $p \neq q$) and $dA = dx\,dy$.

The radial functions $R_n^l(\varrho)$ are closely related to Jacobi's polynomials, which are terminating hypergeometric series. A closed form formula is[2]:

$$R_n^{\pm m}(\varrho) = \sum_{s=0}^{\frac{n-m}{2}} (-1)^s \frac{(n-s)!}{s!(\frac{n+m}{2}-s)!(\frac{n-m}{2}-s)!}\varrho^{n-2s}.$$

The radial functions take the value unity on the boundary of $\mathbf{D}^2$ (due to the conventional normalization).

We show how to map $\mathbf{D}^2$ on $\mathbf{H}^2$ in such a way that Zernike's system can be mapped into a complete, orthonormal basis of functions on $\mathbf{H}^2$. The required system clearly has the general form

$$W(\vartheta, \varphi) = \Theta(\vartheta)G(\varphi),$$

for the very same reasons as discussed above. From elementary differential calculus we have $2\sin\frac{\vartheta}{2}\,d(2\sin\frac{\vartheta}{2}) = \sin\vartheta\,d\vartheta$. Thus we also have

$$\int_{\mathbf{D}^2} V(x,y)dA = \int_{\mathbf{H}^2} W(\vartheta,\varphi)d\Omega = \int_{\mathbf{H}^2} \frac{1}{2}R(\sqrt{2}\sin\frac{\vartheta}{2})G(\varphi)d\Omega,$$

where $d\Omega = \sin\vartheta d\vartheta d\varphi$. We use essentially the area true mapping of $\mathbf{S}^2$ on $\mathbf{R}^2$ due to Lambert.

When we define

$$K_n^l(\vartheta,\varphi) = \sqrt{\frac{2n+1}{2}}R_n^l(\sqrt{2}\sin\frac{\vartheta}{2})e^{il\varphi},$$

then we have

$$\int_{\mathbf{H}^2} K_n^{l*} K_{n'}^{l'} d\Omega = \delta_{nn'}\delta_{ll'},$$

i.e., the $K_n^l(\vartheta,\varphi)$ are a complete, orthonormal system on $\mathbf{H}^2$ with the desired invariance properties. Please notice that this system is different from that of the spherical harmonics $Y_l^m(\vartheta, \varphi)$ which are an orthonormal basis for functions on *the whole* of $\mathbf{S}^2$.

This system should have many uses in radiometry and photometry (and transport theory in general) since it allows you to expand arbitrary functions of direction at one side of a planar interface. It might seem unlikely that such a system has not been proposed earlier, yet we have not been able to find an instance in the literature on radiometry and/or photometry.

3 An orthonormal basis for the description of BRDF's

The bidirectional reflection properties of a surface are clearly specified by the Bidirectional Reflection Distribution Function (BRDF), originally due to Edwards[4] and effectively introduced by Nicodemus *et al*[13]. One defines

$$f(\vartheta_i, \varphi_i, \vartheta_r, \varphi_r) = \frac{dN_r(\vartheta_r, \varphi_r)}{dH_i(\vartheta_i, \varphi_i)},$$

(the subscript i denotes the incident beam, the subscript r the reflected beam), thus the BRDF is the ratio of the scattered radiance to the incident irradiance.

The BRDF may become singular, especially for the case of grazing incidence. This happens, for instance, for perfect specular reflection. In such cases it is advantageous to deal with the function

$$g(\vartheta_i, \varphi_i, \vartheta_r, \varphi_r) = f(\vartheta_i, \varphi_i, \vartheta_r, \varphi_r) \cos \vartheta_i,$$

which is the scattered radiance for irradiance by a collimated source of constant intensity, instead of the BRDF. We then develop g rather than f in terms of an orthonormal basis and thus avoid singular behavior. For the case of natural materials this will be seldom necessary though.

The most general form of the BRDF in terms of the aforementioned orthonormal basis functions is:

$$f(\vartheta_i, \varphi_i, \vartheta_r, \varphi_r) = \sum_{klk'l'} a_{klk'l'} K_n^l(\vartheta_i, \varphi_i) K_{n'}^{l'}(\vartheta_r, \varphi_r).$$

However, various symmetries severely constrain this general form. We consider these constraints below. Because of the orthonomality we find the coefficients $a_{klk'l'}$ by integration:

$$a_{klk'l'} = \int_{\mathbf{H}^2 \times \mathbf{H}^2} f(\vartheta_i, \varphi_i, \vartheta_r, \varphi_r) K_n^l(\vartheta_i, \varphi_i) K_{n'}^{l'}(\vartheta_r, \varphi_r) \, d\Omega \, d\Omega'.$$

3.1 Helmholtz's reciprocity

"Helmholtz's reciprocity" simply expresses the fact that in the approximation of geometrical optics

$$f(\vartheta_i, \varphi_i, \vartheta_r, \varphi_r) = f(\vartheta_r, \varphi_r, \vartheta_i, \varphi_i),$$

the idea being simply that one counts rays irrespective of their direction[10]. The extreme generality of the idea ensures that it applies under the most various conditions.

Helmholtz's reciprocity enables us to write

$$f(\vartheta_i, \varphi_i, \vartheta_r, \varphi_r) = \sum_{klk'l'} a_{klk'l'}(K_n^l(\vartheta_i, \varphi_i)K_{n'}^{l'}(\vartheta_r, \varphi_r) + K_{n'}^{l'}(\vartheta_i, \varphi_i)K_n^l(\vartheta_r, \varphi_r)).$$

The symmetrical functions

$$H_{nn'}^{ll'}(\vartheta_i, \varphi_i, \vartheta_r, \varphi_r) = K_n^l(\vartheta_i, \varphi_i)K_{n'}^{l'}(\vartheta_r, \varphi_r) + K_{n'}^{l'}(\vartheta_i, \varphi_i)K_n^l(\vartheta_r, \varphi_r),$$

may be called "Helmholtz surface scattering modes". Their azimuthal dependence is

$$e^{i(l+l')\varphi},$$

which for photometric purposes may be written in terms of (real rather than complex) trigonometric functions. In the case of isotropic surfaces (see below) we need only keep the cosine (or even) components, the sine (or odd) components describe the surface anisotropy.

3.2 Surface isotropy

A more special type of symmetry is surface isotropy. Although not completely general, this condition applies often to a good approximation. It yields a very strong constraint on the general form of the BRDF. Indeed, the BRDF may depend only on $|\varphi_i - \varphi_j|$. This implies that the azimuthal dependence is in terms of $\cos l(\varphi_i - \varphi_j)$, $l = 0, 1, \ldots$ We have

$$f(\vartheta_i, \varphi_i, \vartheta_r, \varphi_r) = \sum_{nml} a_{nml}(\Theta_n^l(\vartheta_i)\Theta_m^l(\vartheta_r) + \Theta_m^l(\vartheta_i)\Theta_n^l(\vartheta_r)) \cos l(\varphi_i - \varphi_r),$$

where

$$\begin{gathered} n \geq 0 \\ 0 \leq m \leq n \\ 0 \leq l \leq m \\ (n-l),\ (m-l) \text{ even.} \end{gathered}$$

This brings down the number of basis functions that have to be taken into account enormously.

We calculate the coefficients a_{nml} simply as

$$a_{nml} = \int_{\mathrm{H}^2 \times \mathrm{H}^2} f(\vartheta_i, \varphi_i, \vartheta_r, \varphi_r)(\Theta_n^l(\vartheta_i)\Theta_m^l(\vartheta_r) + \Theta_m^l(\vartheta_i)\Theta_n^l(\vartheta_r)) \cos l(\varphi_i - \varphi_r)\, d\Omega\, d\Omega'.$$

We will denote this system as

$$S_{nm}^l(\vartheta_i, \vartheta_r, \Delta\varphi_{ir}) = (\Theta_n^l(\vartheta_i)\Theta_m^l(\vartheta_r) + \Theta_m^l(\vartheta_i)\Theta_n^l(\vartheta_r)) \cos l\Delta\varphi_{ir}$$

where we have set $\Delta\varphi_{ir} = |\varphi_i - \varphi_r|$.

3.3 Resulting basis polynomials

Due to the symmetries the number of components in the orthonormal basis of functions on $\mathbf{H}^2 \times \mathbf{H}^2$ is much reduced. There are only five for order two, fourteen for order four, fiftyfive for order eight. Yet the number increases fast when one raises the maximum order: One has already to take 285 basis functions into account for order sixteen, even in the case of the isotropic surface. Order eight might be a limit for most practical work. The angular resolution will then be about $360°/2*8 = 22\frac{1}{4}°$ which is amply sufficient for purposes of graphics rendering of diffusely reflecting materials. For work in which sharply articulated functions (such as true specular components) have to be represented accurately one needs to draw much higher orders into account of course.

All surface scattering modes up to order 4 have been depicted in figure 1.

Explicit expressions for the basis functions up to order two are

$$\begin{aligned}
S_{00}^{0}(\vartheta_1,\vartheta_2,\Delta\varphi_{12}) &= \tfrac{1}{\pi}\\
S_{11}^{1}(\vartheta_1,\vartheta_2,\Delta\varphi_{12}) &= \tfrac{4}{\pi}\sin\tfrac{\vartheta_1}{2}\sin\tfrac{\vartheta_2}{2}\cos\Delta\varphi_{12}\\
S_{20}^{0}(\vartheta_1,\vartheta_2,\Delta\varphi_{12}) &= \tfrac{\sqrt{3}}{\pi}(\cos\vartheta_1+\cos\vartheta_1-1)\\
S_{22}^{0}(\vartheta_1,\vartheta_2,\Delta\varphi_{12}) &= \tfrac{3}{\pi}(1+\cos\vartheta_1)(1+\cos\vartheta_2)\\
S_{22}^{2}(\vartheta_1,\vartheta_2,\Delta\varphi_{12}) &= \tfrac{3}{\pi}(1-\cos\vartheta_1)(1-\cos\vartheta_2)\cos 2\Delta\varphi_{12}.
\end{aligned}$$

Although the basis functions become complicated for the higher orders it is easy enough to construct them automatically and the system is convenient enough for routine use.

4 Lambertian and specular components

The BRDF of a perfect Lambertian[9] surface has a BRDF that is constant, namely $f(\vartheta_1,\vartheta_2,\Delta\varphi_{12}) = 1/\pi$. The Lambertian BRDF is just the initial term of the series development, that is to say $S_{00}^{0}(\vartheta_1,\vartheta_2,\Delta\varphi_{12})$.

For a perfect mirror we have

$$g(\vartheta_i,\varphi_i,\vartheta_r,\varphi_r) = \delta(\vartheta_r-\vartheta_i)\delta(\varphi_i-\varphi_r+\pi)/\sin\vartheta_r.$$

The scattered radiance for a constant collimated source is easily expanded in terms of the basis, the coefficients are simply proportional with $(-1)^l\delta_{nm}$ for we have:

$$a_{nml} = \int_{\mathbf{H}^2\times\mathbf{H}^2} g(\vartheta_i,\varphi_i,\vartheta_r,\varphi_r)S_{nm}^{l}(\vartheta_i,\vartheta_r,\Delta\varphi_{ir})d\Omega_i d\Omega_r,$$

where we can immediately carry out one integration (because of the Dirac delta functions) and are left with the integral

$$\begin{aligned}
a_{nml} &= \int_{\mathbf{H}^2} S_{nm}^{l}(\vartheta,\vartheta,\pi)d\Omega =\\
a_{nml} &= (-1)^l 4\pi\delta_{nm}\int_{\mathbf{H}^2}\Theta_n^l(\vartheta)^2\sin\vartheta\, d\vartheta,
\end{aligned}$$

which again is immediate because of the properties of the $\Theta_n^l(\vartheta)$. Thus we obtain a constant angular spectrum, much like the Fourier development of a Dirac impulse.

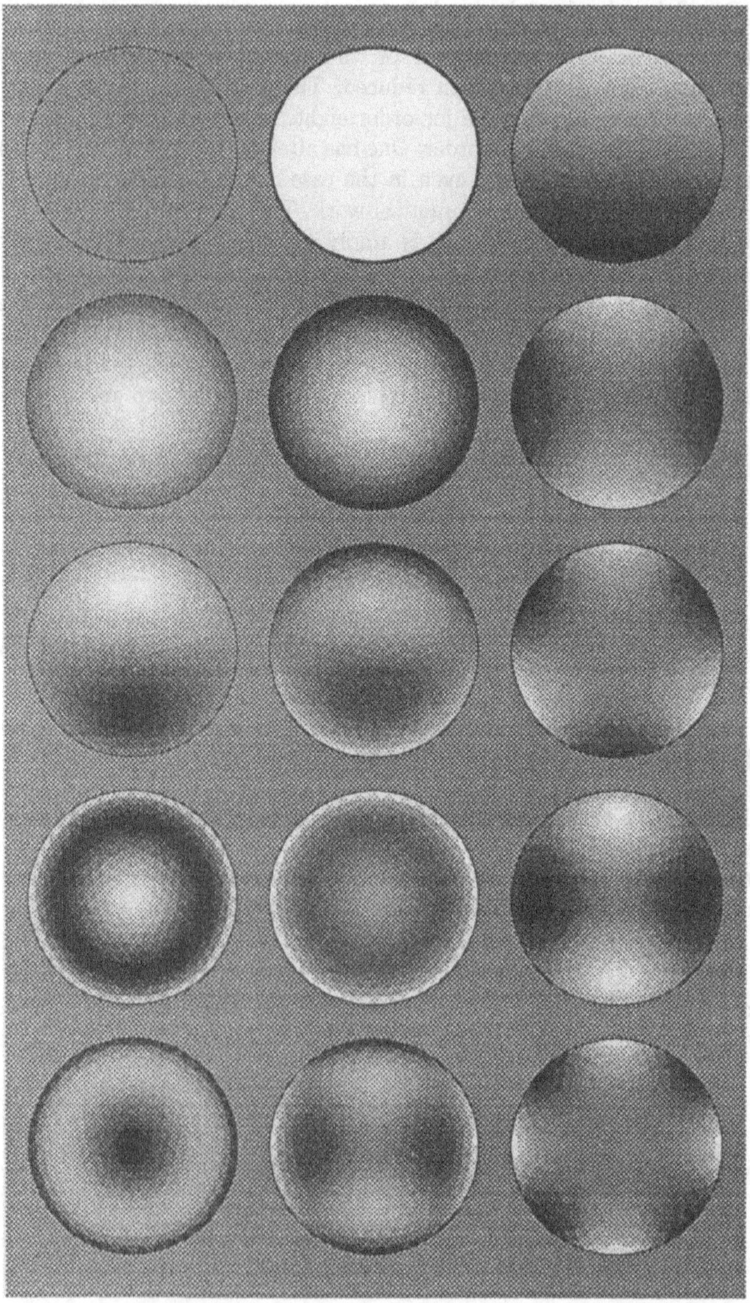

Fig. 1. *Plots of all surface scattering modes* $S^l_{nm}(\pi/4, \vartheta_r, \Delta\varphi_{ir})$ *up to order 4.*

By combination of the Lambertian and the specular expressions we can construct the BRDF for an ideal glossy paint layer. This is the classical description of glossy surfaces in terms of a purely diffuse and a purely specular component[3]. We assume that the pigment particles yield a Lambertian component, whereas the specular component is due to Fresnel reflection at the interface with air.

For the illustrations (figure 2) we took the specular reflection coefficient equal to 2% (thus we didn't take Fresnel's formulas into account). We show the scattering indicatrix for a collimated beam incident at $\vartheta_i = 45°$ in various approximations, orders 2 to 8. (Notice that order two preserves only the Lambertian component.) Even the low order approximations preserve the qualitative nature of the specularity quite well, the main degeneration being a loss of angular resolution. Adding higher order terms concentrates the forward scattering lobe more and more inside a small solid angle centered on the direction of the mirror reflection. For order 8 we have 55 degrees of freedom, thus the solid angle of a "pixel" is roughly $2\pi/55$, and we can estimate the diameter of the forward scattering lobe as $(180/\pi)\sqrt{(4/\pi)(2\pi/55)} \approx 22°$, that is the resolution of the order 8 approximation, *etc.*

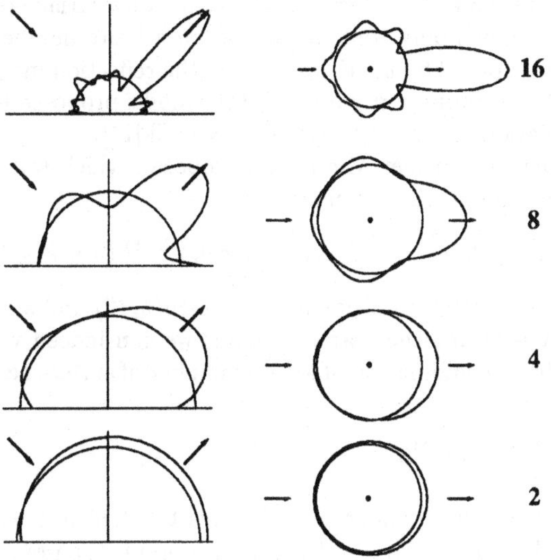

Fig. 2. *Various approximations (truncated series of Helmholtz surface scattering modes) of a model containing pure Lambertian and a pure mirror reflection term. The entrance beam enters at* $\vartheta_i = 45°$. *In the lefthand column we have meridional cross sections and in the righthand column azimuthal cross sections.*

5 Development of conventional expressions in terms of surface scattering modes

Many expressions have been proposed as models of the BRDF of generic materials[1, 12, 15, 16, 17]. Some of these fail to comply with Helmholtz's reciprocity, practically all are descriptions of isotropic surfaces. Such expressions can be roughly divided into two categories: First we have expressions derived on (phenomenological) physical principles for certain model surfaces such as randomly distributed micromirrors, *etc.* Secondly we have the category of formulas based on ease of numerical evaluation for datastructures readily available in graphics rendering pipelines. Although such expressions are designed not to be totally irrealistic, modelling any reasonable model surface on physical principles is a secondary objective.

One of the earliest and certainly simplest models is certainly the Lambertian diffusely scattering surface. We have seen that it can be described perfectly with only the initial term of a development into surface scattering modes. One may well ask how such a development for some of the other expressions turns out.

Many of the models proposed in the literature in fact are represented *exactly* (that is: the series expansion in terms of the Helmholtz surface reflection modes *terminates* and contains only contributions from a finite number of modes) by the development proposed by us. Examples are the reflection model proposed by Blinn[1] (popular in computer graphics) and the model proposed by Minnaert[10] (itself a generalization of a model proposed by Öpik[14]).

As an example we consider Minnaert's proposal, which was especially constructed to respect Helmholtz's reciprocity:

$$f(\vartheta_i, \varphi_i, \vartheta_r, \varphi_r) = \frac{k+1}{2\pi}(\cos\vartheta_i \cos\vartheta_r)^{(k-1)} \; (0 \leq k \leq 1).$$

Clearly this can be written as a polynomial in $\sin(\vartheta_i/2)$ and $\sin(\vartheta_r/2)$ of maximum degree $2(k-1)$ in either variable. Thus we can indeed write Minnaert's expression as a linear combination of a finite number of surface scattering modes.

6 Empirical BRDF's

In expressing the BRDF of materials that have to be measured in the laboratory we meet with at least two problems. First there is the practical problem of measuring the BRDF. Apart from the standard photometric problems we run here in the problem that the number of degrees of freedom is very large indeed. In principle we ought to sample a four–parameter space. Although Helmholtz's reciprocity (which may be expected to hold and can even be used as a check on the data) and surface isotropy (which need not pertain and has to be checked empirically) can be used to ameliorate this problem one still needs a great number of independent measurements. Then we have the problem of how to compute the development in terms of the basis.

The development in terms of the basis is primarily a problem of numerical integration. The coefficients of the development are defined as integrals over

$H^2 \times H^2$. We approximate the integrals by Riemann sums and therefore have to define a tesselation of the hemisphere, preferably one in which the tesseræ subtend identical solid angles. Since the hemisphere has to be evenly covered, a good method is to start with the tesselation defined by the faces of the regular icosahedron. We then can produce refinements by barycentric subdivision. Of course we only keep faces on the northern hemisphere. If we perform a single barycentric subdivision we obtain a triangulation of the hemisphere into 40 faces of identical area and of roughly equilateral shape.

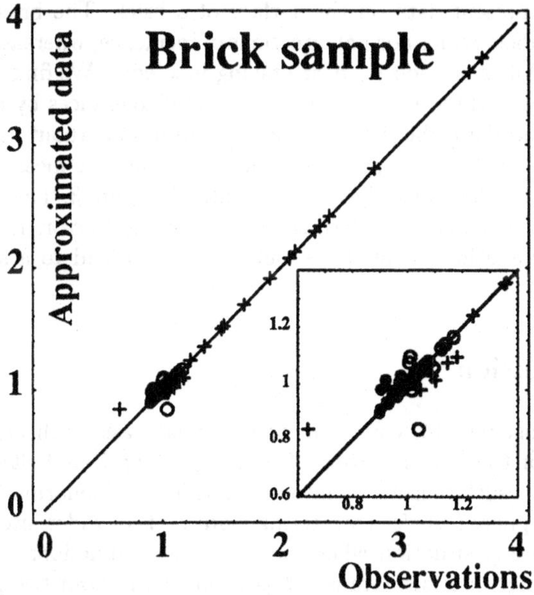

Fig. 3. *Scatterplot of the results from the* 8th *order approximation of the brick data against the observations.*

The full specification of a BRDF measurement on this tesselation would be a 40×40 data matrix, involving 1600 independent measurements. Thus the effort involved in actually measuring a BRDF is quite appreciable. This is perhaps the reason why full BRDF measurements are rarely reported in the literature. Most measurements are confined to the plane of incidence. Helmholtz's reciprocity brings down the number of degrees of freedom to 820, surface isotropy to 66, which is a number that might be considered practical. With 66 independent measurements we are able to construct an approximation of the 8th order (there are 55 independent functions in the 8th order basis).

Since the data slightly overspecify the 8th order approximation we use a pseudoinverse method to find the best fitting coefficients of the truncated series.

This representation of the BRDF by a 40×40 matrix (or larger by progressive barycentric subdivision) is of interest by itself and may prove useful in many computer vision contexts. Very high angular resolutions will in fact seldom be required in applications and the matrix can be used as a lookup table.

In collecting the data one should take care to use entrance and exit beams that are roughly centered on the barycentra of the faces of the triangulation and have an aperture of the order of the solid angle of the faces (about 20° in diameter). The latter condition is necessary to prevent aliassing problems. Likewise, the data should be fitted by truncated series where the highest frequency terms can still be sampled by the tesselation.

Here we present data on the surface of a brick. The brick surface scatters roughly Lambertian for normal directions of incidence, whereas both forward and backscatter lobes develop for near grazing incidence. We find that the measurements are represented within the experimental tolerances by a series expansion truncated at order eight (in fact order four would do about equally well).

That this analytical expression really represents the measurements very well is borne out by the scatterdiagram presented in figure 3. The correlation is very high, the residuals can be ascribed to experimental error. In this case a lower order approximation might do as well and would lead to a somewhat simpler expression.

7 Conclusion

We have constructed a complete, orthonormal basis with desirable invariance properties for the representation of BRDF's taking Helmholtz's reciprocity and (if applicable) surface isotropy into account. The orthonormal basis of functions on the hemisphere ("surface scattering modes") is of wider interest though, since it applies to any situation where one encounters functions of direction at one side of a planar interface (transport phenomena, radiometry, photometry). The orthonormal basis for the BRDF's is defined on the direct product of the hemisphere with itself.

We have shown how the basis can be applied to express empirical BRDF data. This has a number of advantages. First of all, the use of a truncated expansion in orthonormal functions ensures that we find the best approximation in the least squares sense among all linear combinations of elements in this basis. When the high order elements are practically of less importance than the low order ones (as is the case here), we obtain automatically desirable approximations. Secondly, the approximations are guaranteed to satisfy Helmholtz's reciprocity and thus are physically realistic. This is an advantage over *ad hoc* approximations which often turn out to violate Helmholtz's reciprocity and are thus not even physically possible. Thirdly, the approximations are guaranteed to express surface isotropy when so desired.

The method of subdivision of the hemisphere in equal solid area facets used in the data sampling, which leads to a matrix representation of surface scattering, might be of considerable utility by itself in computer graphics as it can conveniently be implemented as a lookup table.

We have outlined how the method can be applied to empirical data and have presented results for the BRDF of a real material sample (a piece of brick).

Acknowledgement: This work was done in the ESPRIT program REALISE of the European Commission.

References

1. Blinn, J.F., *Models of light reflection for computer synthesized pictures*, ACM Computer Graphics (SIGGRAPH 77), 19(10), pp. 542–547, 1977.
2. Born, M. and E.Wolf, *Principles of optics*, Pergamon Press, London, 1959.
3. da Vinci, L., *Treatise on painting*, A.P.McMahon, Trnsl., Princeton University Press, Princeton, 1959.
4. Edwards, D.K., Proc.Inst.Environmental Sci., pp. 417–424 (1963)
5. Helmholtz, H.L.F. von, Treatise on physiological optics, Vol.I, p.231, Dover, New York, 1962.
6. Horn, B.K.P. and R.W.Sjoberg, *Calculating the reflectance map*, Applied Optics **18**, pp. 1770–1779, 1979.
7. Horn, B.K.P. and M.J.Brooks, *Shape from shading*, MIT Press, Cambridge MA, 1989.
8. Kortüm, G., *Reflectance spectroscopy*, J.E.Lohr, transl., Springer, Berlin, 1969.
9. Lambert, J.H., *Photometria sive de mensura de gradibus luminis, colorum et umbræ*, Augsburg, Eberhard Klett, 1760.
10. Minnaert, M., *The reciprocity principle in lunar photometry*, Astrophysical Journal **93**, pp. 403–410, 1941.
11. Nayar, S.K, K.Ikeuchi and T.Kanade, *Surface reflection: Physical and geometrical perspectives*, IEEE PAMI–**13**, pp. 611–634, 1991.
12. Nayar, S.K. and M.Oren, *Visual appearance of matte surfaces*, Science **267**, pp. 1153–1156, 1995.
13. Nicodemus, F.E., J.C.Richmond, J.J.Hsia, I.W.Ginsberg and T.Limperis, *Geometrical considerations and nomenclature for reflectance*, NBS Monograph 160 (National Bureau of Standards, Washington, D.C., October 1977).
14. Öpik, E., *Photometric measures of the moon and the earth–shine*, Publications de L'Observatorie Astronomical de L'Université de Tartu, **26**, pp. 1–68, 1924.
15. Oren, M. and S.K.Nayar, *Generalization of the Lambertian model and implications for machine vision*, Int.J.Comp.Vision **14**, pp. 227–251, 1995.
16. Oren, M. and S.K.Nayar, *Generalization of Lambert's reflectance model*, ACM Computer Graphics (SIGGRAPH 94), pp. 239–246
17. Tagare, H.D. and R.J.P.deFegueiredo, *A theory of photometric stereo for a class of diffuse non-Lambertian surfaces*, IEEE PAMI–13, pp. 133–152, 1991.

Generalizing Lambert's Law For Smooth Surfaces

Lawrence B. Wolff

Computer Vision Laboratory, Department of Computer Science, The Johns Hopkins University,Baltimore, Maryland 21218

Abstract. **One of the most common assumptions for recovering object features in computer vision and rendering objects in computer graphics is that diffuse reflection from materials is Lambertian. This paper shows that there is significant deviation from Lambertian behavior in diffuse reflection from *smooth* surfaces not predicted by existing reflectance models, having an important bearing on any computer vision technique that may utilize reflectance models including shape-from-shading and binocular stereo. Contrary to prediction by Lambert's Law, diffuse reflection from smooth surfaces is significantly viewpoint dependent, and there are prominent diffuse reflection maxima effects occurring on objects when incident point source illumination is greater than 50° relative to viewing including the range from 90° to 180° where the light source is behind the object with respect to viewing. Presented here is a diffuse reflectance model, derived from first physical principles, utilizing results of radiative transfer theory for subsurface multiple scattering together with Fresnel attenuation and Snell refraction at a smooth air-dielectric surface boundary. A number of experimental results are presented demonstrating striking deviation from Lambertian behavior predicted by the proposed diffuse reflectance model.**

1 Introduction

A prevalent class of materials encountered both in common experience and in computer vision/robotics environments are inhomogeneous dielectrics which include plastics, ceramics, and, rubber. In computer vision a widely used assumption about diffuse reflection from materials is Lambert's law [13], namely the expression:

$$\frac{1}{\pi} L\rho \cos\psi d\omega$$

where light is incident with radiance L, at incidence angle ψ, and reflected through a small solid angle $d\omega$, and ρ is termed the *diffuse albedo* in the range $[0, 1.0]$. This reflectance model is typically instantiated into the implementation of a large number of algorithms such as shape-from-shading [9] and photometric-based binocular stereo [6], [19]. It is therefore important for researchers in the computer vision community who utilize assumptions about diffuse reflection to be aware of the conditions under which there is significant deviation from Lambert's law.

Almost all diffuse reflection from inhomogeneous dielectrics physically arises from subsurface multiple scattering of light caused by subsurface inhomogeneities in index of refraction. In this paper we model inhomogeneous dielectric material as a collection of scatterers contained in a uniform dielectric medium with index of refraction different from that of air. An expression is derived for diffuse reflected radiance resulting from the process of incident light refracting into the dielectric medium across a smooth surface boundary, producing a subsurface diffuse intensity distribution from multiple internal scattering, and then refraction of this subsurface diffuse intensity distribution back out into air. See Figure 1. Also accounted for is the infinite progression of internal specular reflection at the air-dielectric boundary and sub-surface scattering. A common property of diffuse reflection from smooth inhomogeneous dielectric surfaces is that such reflection is azimuth independent with respect to viewing about the surface normal, regardless of the fixed direction of incident light. We formally derive and empirically verify that for smooth inhomogeneous dielectric surfaces that exhibit such azimuth symmetric diffuse reflection, that

$$\varrho L \times (1 - F(\psi, n)) \times \cos\psi \times (1 - F(\sin^{-1}(\frac{\sin\phi}{n}), 1/n))\, d\omega \qquad (1)$$

describes the reflected radiance into viewing angle, ϕ, (i.e., angle between viewing and the surface normal, also known as *emittance angle*). The terms $F(,)$ refer to the Fresnel reflection coefficients [18], n, is the index of refraction of the dielectric medium, and, ϱ, is the *total diffuse albedo*. We show that the total diffuse albedo, ϱ, is directly related to both the *single scattering albedo* describing the proportion of energy reradiated upon each subsurface single scattering, and, the index of refraction n. An initial first order derivation of expression 1 was presented in [23], however without the accounting for higher order effects which is presented below.

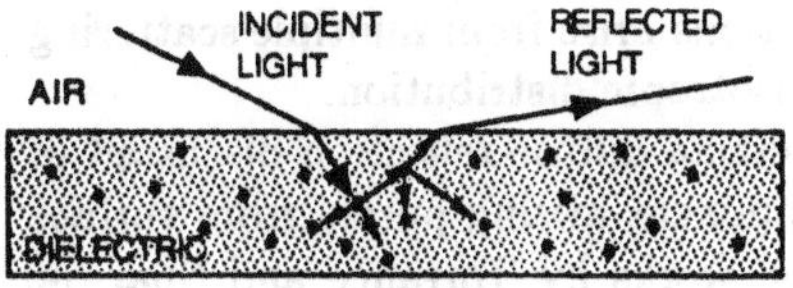

Figure 1

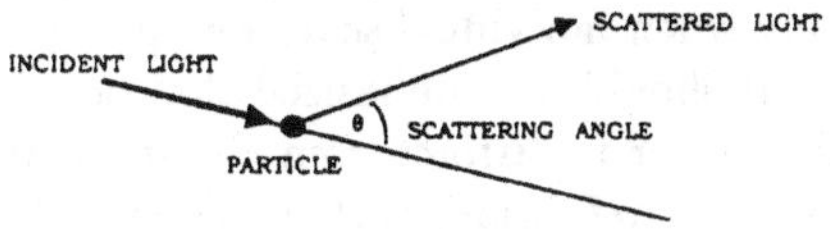

Figure 2

Particularly useful to object feature extraction in computer vision, our expression 1 for diffuse reflection allows precise characterization of the conditions under which the Lambertian model breaks down for inhomogeneous dielectrics and where our more accurate model should be used. We show that Lambert's law is valid for smooth dielectrics to within 5% only as long as both angle of incidence, ψ, and viewing angle, ϕ, are simultaneously less than 50°. This means that for applications in computer vision there are a large number of situations in which Lambert's law is significantly in error for smooth surfaces– near the occluding contour of objects under any illumination condition; for illuminations incident at greater than 50° relative to viewing there will be significant errors both near the occluding contour and over a large portion of object area bounded on one side by the shadow boundary with respect to illumination; for multiple images of a smooth dielectric object there will be significant viewpoint dependence of diffuse reflection for most object-viewpoint situations. Existing diffuse reflectance models for smooth surfaces do not account for these errors. In addition to experimentally verifying the viewpoint dependence of diffuse reflection, we show how the diffuse reflection maximum occurring between an occluding contour and the shadow boundary for a smooth dielectric illuminated by a point light source is accurately predicted because of this viewpoint dependence. For illumination incident at 90° and greater relative to viewing, a diffuse reflection maximum is not even predicted by existing models having dependence only on angle of incidence, when in fact this maximum is very prominent. Furthermore, diffuse reflection is empirically observed to be very small in the immediate vicinity of an occluding contour for *all* illumination conditions, an observation that is not supported by existing models, and yet accounted for by expression 1.

A number of papers from the optics community have studied reflectance and transmission of light from diffuse scattering within dielectric media [12], [14], [16], [1], [5]. Some of these papers have presented theories for both reflectance and transmission for arbitrary optical thicknesses of scattering media, using collimated or diffuse light sources. In relation to the optics literature, this paper studies the specific case of diffuse reflectance from a semi-infinite, plane-parallel, inhomogeneous dielectric, which is most relevant to diffuse reflection observed in computer vision (and computer graphics). We assume that individual scattering of light from inhomogeneities within the dielectric medium is isotropic. The assumption that the scatterers are isotropic stems from the observed physical characteristic of common smooth dielectric surfaces that diffuse reflectance is independent of azimuth about the surface normal with respect to viewing. Chandrasekhar [3], whose derivation we utilize for the subsurface scattering distribution, shows that azimuth symmetric distributions arise from multiple scattering only when individual scatterers produce an isotropic distribution.

It should be emphasized that non-Lambertian diffuse reflection discussed in this paper is *distinct* from the specular component arising purely from surface interface reflection, such as described in the works by Torrance and Sparrow [21], and, Beckmann and Spizzichino [2]. Only very recently has there been consideration of non-Lambertian diffuse reflection within the vision and graphics community. Healey derives an expression for diffuse reflection based on Reichman's model for the semi-infinite case of opaque dielectrics [8], [7] applying this to geometry insensitive color segmentation. Tagare and deFigueiredo [20] propose as part of their multiple lobed reflectance model for machine vision a functional approximation to the Chandrasekhar diffuse reflection Law [3] for application to photometric stereo. While we also use the Chandrasekhar diffuse reflection law for diffuse subsurface scattering, it is not nearly accurate for materials without consideration of various dielectric-air boundary effects. Oren and Nayar [15] have proposed a diffuse reflection model for very rough dielectric surfaces assuming a statistical distribution of Lambertian reflecting facets along with masking, shadowing, and, intereflection. The mechanism of using roughness elements that are assumed to be Lambertian however implies that in the limit as surface roughness gets smaller and smaller, that smooth surfaces therefore have Lambertian behavior- a phenomenon shown to be largely not true. In the Conclusion section, suggestions are made on how to possibly combine these two models to produce a unified model that accurately predicts diffuse reflection throughout the entire range of very rough to smooth surfaces.

Apart from analysis of reflected intensity distributions for diffuse reflection, Shafer [17] proposed a *dichromatic* color reflectance model for diffuse and specular components, and Wolff [22] proposed a polarization reflectance model involving the diffuse reflection component for inhomogeneous dielectrics.

2 The Physics Of Diffuse Reflection From Multiple Scattering

The analysis of diffuse reflection in this paper begins with the theory of radiative transfer developed by Chandrasekhar [3] for multiple scattering of incident light upon stellar and planetary atmospheres. The important problem of diffuse reflection and transmission from plane parallel atmospheres in astrophysics has a number of similarities with diffuse reflection from inhomogeneous dielectric materials. Incident light strikes gaseous molecules within an atmosphere whereupon some of the light is absorbed, and some of the light is scattered with an assumed intensity distribution respective to each individual molecule. Similarly, particles forming discontinuities in refractive index within a dielectric absorb and scatter light that has penetrated the surface boundary, and this process can be quantified using radiative transfer theory.

The fundamentals of single scattering theory were established by Lord Rayleigh for particles smaller than the wavelength of incident light and by Mie for spherical particles of arbitrary size [10], [11]. It is commonly assumed that the scattered light produced by a single small particle has an axial symmetric scattered energy distribution about the incident direction of light. The angular distribution of the scattered radiation is described by a ***phase function*** $P(cos\theta)$ where θ is the scattering angle of deflection away from the direction of the incident light. See Figure 2. The function $P(cos\theta)/4\pi$ represents the proportion of incident light energy flux that is scattered into a given direction per unit solid angle. The proportion of total light energy flux that is reradiated (i.e., scattered) is given by:

$$\int_{unit\ sphere} P(\cos\theta)\frac{d\omega}{4\pi} = \rho \leq 1 , \tag{2}$$

and ρ is referred to as the ***single scattering albedo.*** In nonconservative cases when $\rho < 1$ some energy is absorbed by the particle. The single scattering albedo is commonly wavelength dependent which accounts for the colored appearance of many dielectrics when illuminated by white light.

Materials for which $P(\cos\theta)$ is independent of θ allow a tractable physical analysis. Chandrasekhar's theory [3] would imply that for these materials the diffuse reflection distribution is symmetric about the surface's azimuth. Such materials include many common materials like plastics, ceramics, rubber, opaque glasses and smooth glossy paints. The details of Chandrasekhar's diffuse reflection law are explained in [24].

3 Diffuse Reflection From Smooth Dielectrics

Chandrasekhar's theory can be adapted so as to account for diffuse reflection from smooth inhomogeneous dielectric materials made up of particle inhomogeneities embedded in a medium with uniform index of refraction different from that of air. Once adapted the theory permits useful expressions that quantify the process of diffuse reflection from such materials. These expressions are summarized in this section, with the full technical derivations and details described in [24].

The first-order expression for diffuse reflection [23] quantifies reflection caused by light that penetrates into the material surface, scatters amongst sub-surface particle inhomogeneities, and then refraction back out into air. See Figure 1. This expression is given by:

$$\varrho_1 L \times [1 - F(\psi, n)] \times \cos\psi \times [1 - F(\overline{\phi}, 1/n)] d\omega \,,$$

for light incident at radiance L through solid angle $d\omega$. The functions, $F(,)$, are the Fresnel reflection coefficients [18] as functions of specular angle (external angle of incidence ψ for the first $F(,)$, and internal angle of incidence $\overline{\phi}$ for the second $F(,)$), and, index of refraction (n for light incident on the dielectric interface from air external to the dielectric for the first $F(,)$, and $1/n$ for light incident on the dielectric interface internal to the dielectric for the second $F(,)$). The internal angle $\overline{\phi}$ is related to the angle of emittance, ϕ using Snell's Law, $\overline{\phi} = \sin^{-1}(\frac{\sin\phi}{n})$. The term

$$\varrho_1 = \frac{\rho}{4\pi n^2} \frac{H_\rho(\overline{\mu}_{inc}) H_\rho(\overline{\mu}_{ref})}{\overline{\mu}_{inc} + \overline{\mu}_{ref}}, \quad \overline{\mu}_{inc} = \sqrt{1 - \frac{\sin^2\psi}{n^2}}, \quad \overline{\mu}_{ref} = \sqrt{1 - \frac{\sin^2\phi}{n^2}}$$

using the Chandrasekhar H-function is termed the ***first-order diffuse albedo*** as it represents the scaling of diffuse reflection magnitude. An Nth order approximation to the Chandrasekhar H-function [3] can be expressed;

$$H_\rho(\mu) = \frac{1}{\mu_1 \ldots \mu_n} \frac{\prod_{i=1}^N (\mu + \mu_i)}{\prod_\alpha (1 + \kappa_\alpha \mu)} \,,$$

defined in terms of the positive zeros, μ_i, of the even Legendre polynomial of order $2N$, and the positive roots, κ_α, of the associated ***characteristic equation***;

$$1 = \sum_{j=1}^N \frac{a_j \rho}{1 - \kappa^2 \mu_j^2} \,.$$

Upon refraction from inside the dielectric into air at the dielectric interface, there can be significant specular reflection back into the dielectric. This produces additional sub-surface scattering resulting in a second refraction back out into air, in turn resulting in specular reflection back into the dielectric, and so on. The contribution of Nth order diffuse reflection resulting from light that has been internally specularly reflected at the dielectric interface $N-1$ times and subsurface diffusely scattered N times (the first subsurface diffuse scattering before the first internal reflection) is given by:

$$\varrho_1 L \times [1 - F(\psi, n)] \times \cos\psi \times K^{N-1} \times [1 - F(\overline{\phi}, 1/n)] d\omega \ ,$$

where

$$K = \int_0^{\pi/2} F(\phi', 1/n)\, C_\rho(\cos\phi', \sqrt{1 - \frac{\sin^2\phi}{n^2}}) 2\pi \sin\phi' d\phi' \ ,$$

with

$$C_\rho(x, y) = \frac{\rho}{4\pi} L \frac{x}{x+y} H_\rho(x) H_\rho(y) \ ,$$

The total amount of diffuse reflection is obtained by summing all Nth order contributions resulting in the expression:

$$\varrho L \times [1 - F(\psi, n)] \times \cos\psi \times [1 - F(\overline{\phi}, 1/n)] d\omega \qquad (3)$$

$$\varrho = \sum_{i=1}^{\infty} \varrho_1 \times K^{i-1} = \frac{\varrho_1}{1-K} \ , \qquad (4)$$

and ϱ is termed the ***total diffuse albedo***. What is important to note is that the total diffuse albedo can be expressed purely in terms of physical parameters of the dielectric.

The assumptions made in the derivation of expression 3 are summarized in Table I below:

ASSUMPTION	COMMENTS
Isotropic Single Scattering	Azimuth independent diffuse reflection distribution
Uniform Distribution of Scattering Particles	Plane parallel uniformity of particles embedded in dielectric medium of uniform index of refraction. Able to apply Chandrasekhar theory.
Smooth Surface	Able to use Fresnel Coefficients, $F(,)$
Unpolarized Incident Light	Can be generalized to any incident polarization state using combination of $F_{\parallel}$ and $F_{\perp}$ terms
ρ_1 is constant	Very nearly true for all geometries to within 3%
K is constant	Very nearly true due to near isotropy of $C_\rho(,)$ for significant single scattering albedo, ρ

TABLE I

4 Experimental Results

Figure 3 shows an experimental goniometer apparatus that we built to measure diffuse reflected radiance from an optically smooth piece of white magnesium oxide ceramic, for different combinations of angles of incidence and emittance in the plane horizontal to the floor.

Figures 3 and 4 show experimental results for diffuse reflection from an optically smooth piece of white magnesium oxide ceramic. The magnesium oxide ceramic was measured with a stylus profilometer to have a variation in height profile no greater than half the wavelength of green light. The measurements were taken with a Panasonic CCD linear camera mounted at the end of a goniometer arm with the sample mounted at the center. Using a Brewster angle technique [4] the index of refraction of the ceramic was determined to be $n = 1.7$. Our diffuse reflectance model enables the empirical measurement of the single scattering albedo, ρ, by empirically determining the ratio of the strengths of the specular and diffuse components of reflection at known angles of incidence and emittance. The ratio of the specular to the diffuse reflection component was measured very near normal incidence and emittance and compared with the ratio:

$$\frac{F(0,n)}{\varrho(1-F(0,n))(1-F(0,1/n))d\omega},$$

where the total diffuse albedo ϱ is given by expression 4, and $d\omega$ is the solid angle in steradians subtended by the illuminating light source. The value of ϱ can be derived from this, and in turn, ρ can be computed by observing the graphed relationship between ρ and ϱ in Figure 5. The empirically determined specular to diffuse ratio is accounted for by a single scattering albedo, ρ, just above 0.95 which is very energy conserving.

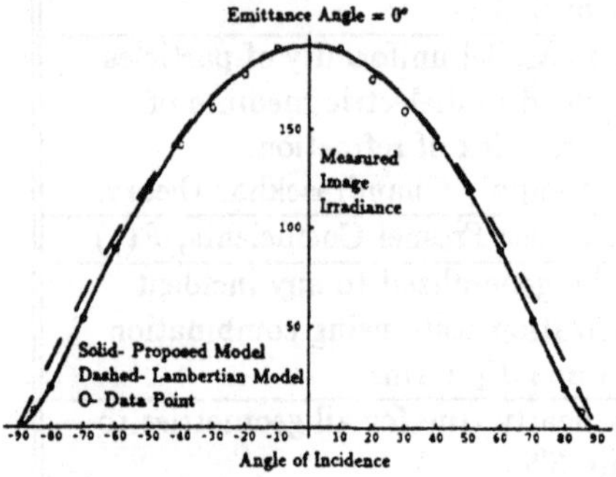

Figure 3

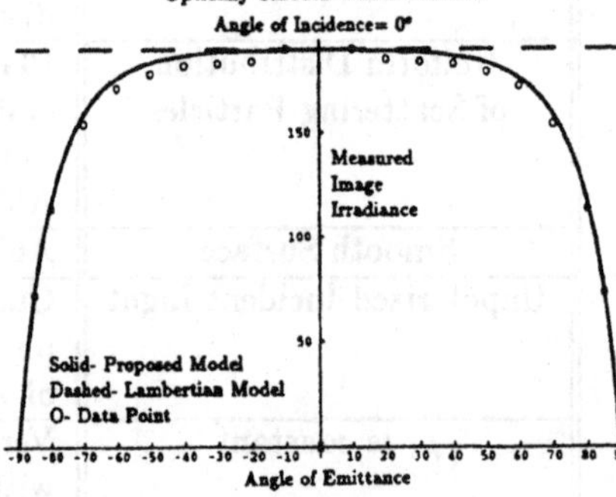

Figure 4

In Figures 3 and 4 dashed curves represent predicted Lambertian diffuse reflection, solid curves represent the diffuse reflection law of expression 3 with single scattering albedo $\rho = 0.95$ and index of refraction $n = 1.7$ (it makes little difference for $1.0 \geq \rho \geq 0.9$, $1.4 \geq n \geq 2.0$ with respect to the shape of the diffuse reflection curve). The proposed diffuse reflection law, expression 3, never deviates more than 3% from the empirical data while there exist large deviations from Lambertian behavior. The results show that the Lambertian model can be assumed to within about 5% accuracy if angle of incidence and angle of emittance are simultaneously no larger than 50°. Outside of this constraint range large deviations start to occur.

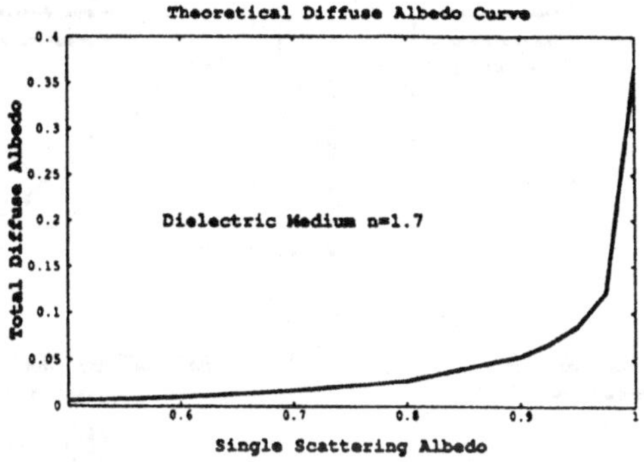

Figure 5: Graph of ϱ versus ρ using expression (4).

Figure 6 shows an ordinary scene where the Lambertian model strikingly breaks down altogether and yet is explained with high accuracy by the proposed diffuse reflection model in this paper. A ceramic coffee cup of cylindrical body shape is illuminated from the left side by an approximate point source. Starting from the left occluding contour going right the angle of incidence starts at 0° and increases. The Lambertian model predicts that image intensities should decrease going to the right. The image intensities in fact increase to a maximum intensity at about 65° surface orientation and then begin to slowly decrease. The reason is because, $\phi = 90^o - \psi$, that is the emittance angle starts at 90° at the occluding contour and decreases going right. Looking at the graph in Figure 6a diffuse reflected radiance increases sharply as angle of emittance decreases from 90°. At about surface orientation 65° ($\psi = 25^o, \phi = 65^o$) the decrease of diffuse reflected radiance with respect to increasing angle of incidence starts to overtake the increase of diffuse reflected radiance with respect to decreasing angle of emittance, and a maximum occurs. (Note the graph in Figure 6a is only for the diffuse component and does not include measurement of the specular component which occurs at relative orientation 45° in the picture of the cup.) According to Figure 6a the qualitative shape of the true diffuse reflected radiance curve (solid) is entirely different (e.g., its not even monotonic) from that for the Lambertian model (dashed). The deviation from the Lambertian model is also very high for frontal surface orientations where the angle of incidence is large. This non-Lambertian behavior of diffuse reflection occurs for a large range of lighting configurations as can be seen in Figure 6c with light source incident at 60° relative to viewing (the diffuse reflection maximum occurs at about -50° as predicted by the proposed model instead of -60° as predicted by the Lambertian model, and note the extreme drop in grey values going towards the occluding contour), and in Figure 6d with light source incident at 120° relative to viewing. For light incident at 90° or greater, Lambert's model as well as other existing diffuse reflectance models predict a monotonic diffuse reflection distribution which is empirically significantly incorrect.

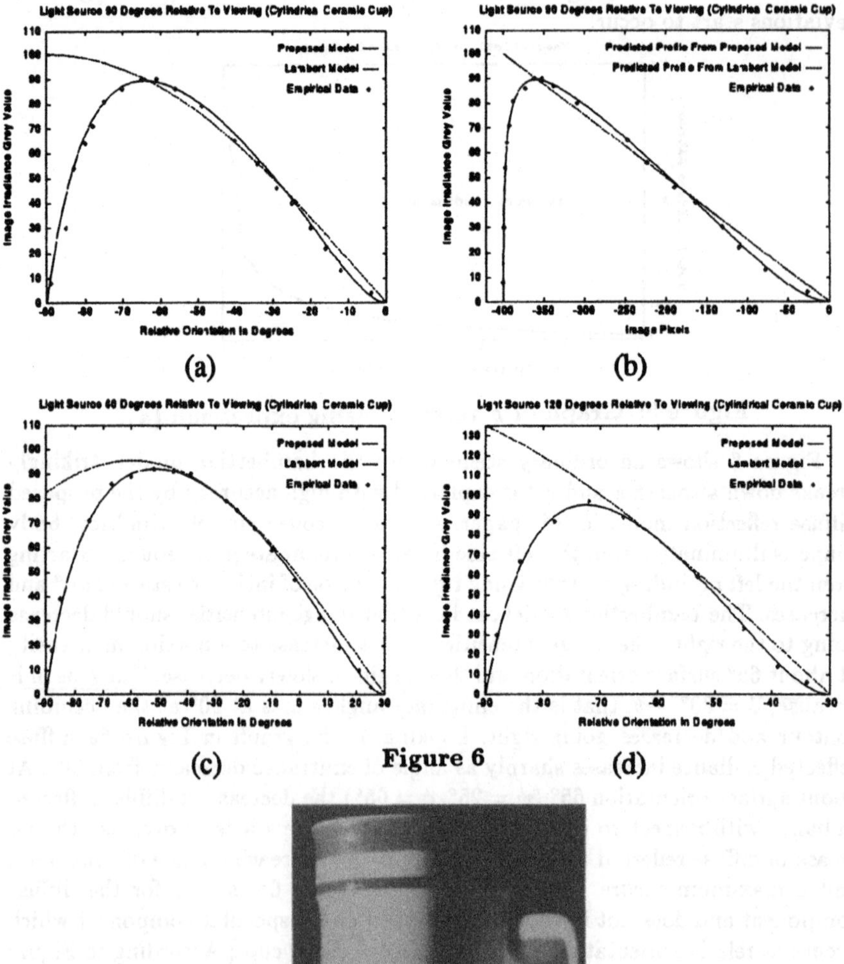

(c) **Figure 6** (d)

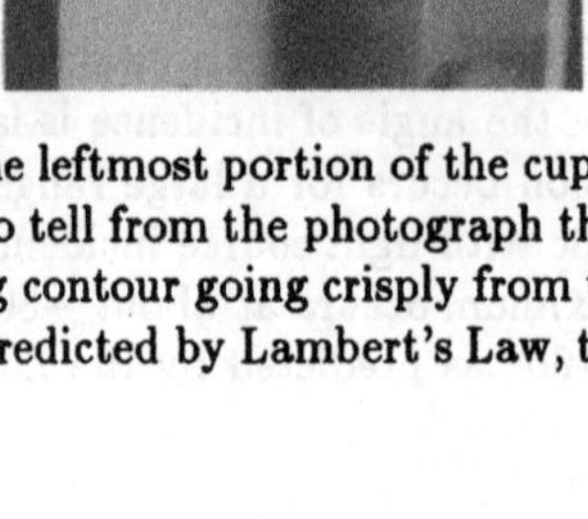

Note in Figure 6 that the leftmost portion of the cup visually looks relatively bright. It is however hard to tell from the photograph that instead of the brightness values at the occluding contour going crisply from the brightest value down to zero like a step-edge as predicted by Lambert's Law, that in fact the occluding contour edge is "blurred" over a region the size of about 10% of the radius of the cup. This is precisely predicted by our diffuse reflectance model. Figure 6b shows the brightness profile predicted by our diffuse reflection model graphed spatially across a cylinder in pixel units where the center of the cylinder is at 0 on the horizontal axis and the occluding contour is at −400 on the horizontal axis. The experimental data in Figure 6b is consistent with this profile. (Note that Lambert's law predicts a linear diffuse reflection profile across an image in this case.) Due to foreshortening, the maximum brightness value occurring at 65° relative surface orientation is about 10% of the radius of the cylinder away from the occluding contour. Between the point at which the brightness maximum occurs and the occluding contour, the brightness decreases to zero. While the brightness at the occluding contour is zero, due to the steepness of the profile shown in Figure 6b, the immediate vicinity of the occluding contour still has a relatively bright visual appearance. This is however instead of a step-edge between the occluding contour and the background. Thus occluding contour edge profiles for oblique lighting have the appearance of being blurred, and significantly displaced toward the interior of the object. With the same camera used to measure the profile in Figure 6b we have observed empirical step-edge profiles of a bright strip against a dark background that are localized to within about 2 or 3 pixels. Hence this effect at occluding contours from diffuse reflection for oblique lighting can be distinguished from typical lens blurring of step-edges as long as 10% of the radius of curvature near the occluding contour in pixel units is significantly larger than 2 or 3 pixels. Precise quantitative knowledge from our model of the brightness profile of occluding contour edges may aid in better determination of where the occluding contour is located in an image. The larger the cylinder radius (i.e., in general the larger the radius of curvature at the occluding contour) the more blurred and displaced will be the occluding contour edge defined by the profile of brightness values.

Figure 7a shows a real image of a white billiard ball illuminated by two point light sources orthogonal to viewing, one from the left side and one from the right side. Figure 7b shows a computer graphics rendering of a sphere illuminated by the same configuration of 2 point light sources assuming Lambert's diffuse reflectance law, while Figure 7c shows the same computer graphics rendering of a sphere using the diffuse reflectance law proposed in this paper. While both shadow boundaries with respect to the left and right light sources coincide along the vertically oriented great circle at the front of the sphere, there appears to be a "shadow band" of darker (i.e., smaller) intensity values about this shadow boundary due to the high fall off of diffuse reflectance at high angles of incidence near 90°. Observe that realistically this "shadow band" is in fact significantly wider in Figure 7a than predicted by the Lambert Law in Figure 7b, but more accurately predicted by the proposed diffuse reflectance law in Figure 7c.

Figures 8a, 8b, and 8c show grey level representations of isophote curves (i.e., image curves with equal intensity) corresponding respectively to Figures 7a, 7b, 7c. Lambert's Law predicts for this configuration of light sources illuminating a sphere that equal reflected radiance occurs for points forming concentric cir-

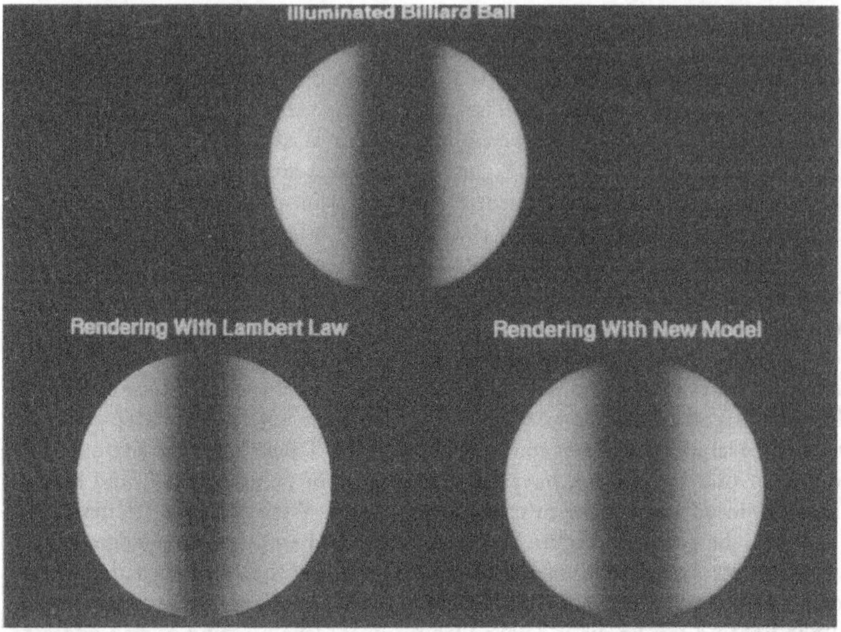

Figure 7 (Starting at top and going counterclockwise, 7a, 7b, 7c)

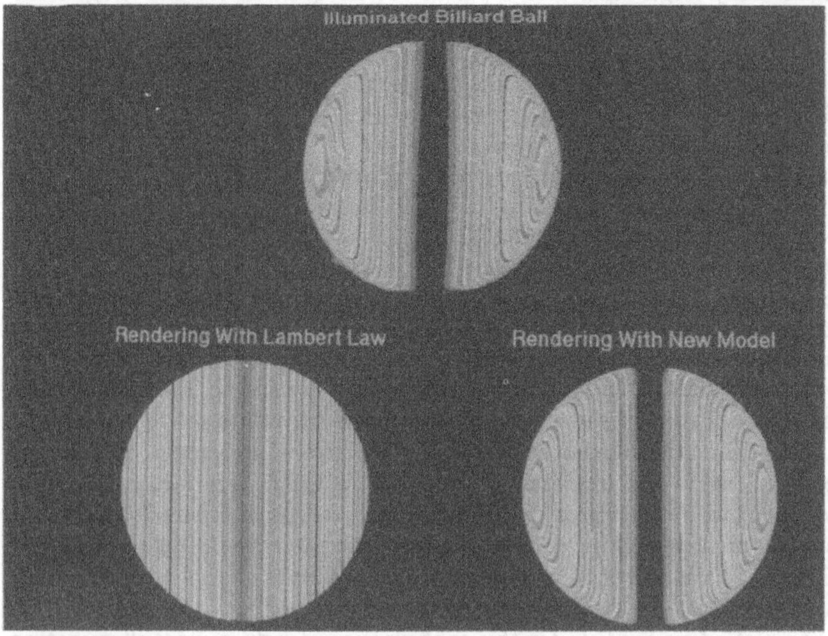

Figure 8 (Starting at top and going counterclockwise, 8a, 8b, 8c)

cles on the sphere about the left-most and right-most occluding contour points. These concentric circles of equal reflected radiance orthographically project onto straight isophote lines as depicted in Figure 8b, with maximum diffuse reflectance occurring at the left-most and right-most occluding contour points where the angle of incidence is zero. Figure 8a which is an actual depiction of the isophotes of Figure 7a shows that in fact lines of equal image intensity severely curve near the occluding contour of the sphere. Maximum diffuse reflection occurs at the center of the closed elliptical isophotes near the left-most and right-most occluding contours while diffuse reflection at the occluding contours is nearly zero, illustrating a 2-dimensional version of the effect depicted in Figure 6a. Figure 8c shows the isophotes rendered using the diffuse reflectance model proposed in this paper which are remarkably similar to the actual isophotes in Figure 8a (except for the isophotes perturbed by the specularities). Comparing Figures 8a, 8b, 8c shows very clearly how our diffuse reflectance model accurately predicts reflectance features that are significantly deviant from Lambertian behavior.

5 Conclusion

The primary result of this paper is a simple closed form expression, (equation 3 and equation 4), derived from first physical principles which accurately characterizes diffuse reflection from smooth dielectric surfaces exhibiting azimuth independent diffuse reflection about the surface normal with respect to viewing. In computer vision and computer graphics the Lambert diffuse reflection law has been almost always assumed for diffuse reflecting dielectric surfaces. Our diffuse reflection model shows that Lambert's Law is a good approximation across illuminated smooth dielectric objects when the angle of incidence and angle of emittance are simultaneously less than 50° at each object point- and that there are significant deviations from Lambertian behavior for illumination-object-viewer geometries outside of this range. For oblique lighting of objects some strikingly non-Lambertian effects have been demonstrated that occur for common smooth dielectric surfaces, that are very accurately explained by this new diffuse reflectance model. An improved algorithm for the stereo matching of intensities is described in [26] accounting for this behavior. Our proposed diffuse reflection model has the added feature that it explains the physical origin of diffuse albedo, ϱ, (equation 4), which is typically an *ad hoc* scaling coefficient. This can be used to explain the relative strengths of the specular and diffuse reflection components from smooth inhomogeneous dielectric surfaces purely in terms of the physical parameters of the surface itself as shown in [25].

The proposed model for diffuse reflectance from smooth surfaces gives significant insight into generalizing a diffuse reflection model for different levels of roughness. In particular, modeling the local geometry of rough surfaces as a collection of smooth microfacets each having the diffuse reflectance property proposed in this paper (in lieu of the Lambertian model used in [15]) makes the limiting case of diffuse reflection as surface roughness goes towards zero accurate, while simultaneously having the large potential of accurately predicting diffuse reflection from intermediate ranges of surface roughness. If closed form expressions for diffuse reflection can be obtained from such a model, this would be applicable to the entire range of rough to smooth surfaces and would be a further significant advancement.

References

1. J.R. Aronson and A.G. Emslie. Spectral reflectance and emittance of particulate materials. 2: Application and results. *Applied Optics*, 12(11):2573–2584, November 1973.
2. P. Beckmann and A. Spizzichino. *The Scattering of Electromagnetic Waves from Rough Surfaces.* Macmillan, 1963.
3. S. Chandrasekhar. *Radiative Transfer.* Dover Publications, New York, 1960.
4. D. Clarke and J.F. Grainger. *Polarized Light and Optical Measurement.* Pergamon Press, 1971.
5. A.G. Emslie and J.R. Aronson. Spectral reflectance and emittance of particulate materials. 1: Theory. *Applied Optics*, 12(11):2563–2572, November 1973.
6. W.E.L. Grimson. Binocular shading and visual surface reconstruction. *Computer Vision Graphics and Image Processing*, 28(1):19–43, 1984.
7. G. Healey. Using color for geometry-insensitive segmentation. *Journal of the Optical Society of America A*, 6(6):920–937, June 1989.
8. G. Healey and T.O. Binford. The role and use of color in a general vision system. In *Proceedings of the DARPA Image Understanding Workshop*, pages 599–613, Los Angeles, California, February 1987.
9. B.K.P. Horn and M.J. Brooks. *Shape From Shading.* MIT Press, 1989.
10. H.C. Van De Hulst. *Light Scattering by Small Particles.* John Wiley and Sons, New York, 1957.
11. M. Kerker. *The Scattering of Light and other Electromagnetic Radiation.* Academic Press, 1969.
12. P. Kubelka and F. Munk. Ein beitrag sur optik der farbanstriche. *Z. tech.*, 12:593, 1931.
13. J. H. Lambert. Photometria sive de mensura de gratibus luminis, colorum et umbrae. *Augsberg, Germany: Eberhard Klett*, 1760.
14. S. Orchard. Reflection and transmission of light by diffusing suspensions. *Journal of the Optical Society of America*, 59(12):1584–1597, December 1969.
15. M. Oren and S. Nayar. *Seeing Beyond Lambert's Law.* Proceedings of the Third European Conference on Computer Vision (ECCV 1994), Stockholm, May 1994.
16. J. Reichman. Determination of absorption and scattering coefficients for nonhomogeneous media 1: Theory. *Applied Optics*, 12(8):1811–1815, August 1973.
17. S. Shafer. Using color to separate reflection components. *Color Research and Application*, 10:210–218, 1985.
18. R. Siegal and J.R. Howell. *Thermal Radiation Heat Transfer.* McGraw-Hill, 1981.
19. G.B. Smith. Stereo integral equation. In *Proceedings of the AAAI*, pages 689–694, 1986.
20. H.D. Tagare and R.J.P. deFigueiredo. A theory of photometric stereo for a general class of reflectance maps. In *Proceedings of the IEEE conference on Computer Vision and Pattern Recognition (CVPR)*, pages 38–45, San Diego, June 1989.
21. K. Torrance and E. Sparrow. Theory for off-specular reflection from roughened surfaces. *Journal of the Optical Society of America*, 57:1105–1114, 1967.
22. L.B. Wolff. *Polarization Methods in Computer Vision.* PhD thesis, Columbia University, January 1991.

23. L.B. Wolff. Diffuse reflection. In *Proceedings of IEEE Conference on Computer Vision and Pattern Recognition (CVPR)*, pages 472–478, Urbana-Champaign Illinois, June 1992.
24. L.B. Wolff. A diffuse reflectance model for smooth dielectrics. *Journal of the Optical Society of America, (JOSA) A, Special Issue on Physics Based Machine Vision*, 11(11):2956–2968, November 1994.
25. L.B. Wolff. Relative brightness of specular and diffuse reflection. *Optical Engineering*, 33(1):285–293, January 1994.
26. L.B. Wolff and E. Angelopoulou. *3-D Stereo Using Photometric ratios.* Proceedings of the Third European Conference on Computer Vision (ECCV 1994), Stockholm, May 1994.

Image Features

Local Scale Control for Edge Detection and Blur Estimation

James H. Elder[1] and Steven W. Zucker[2]

[1] NEC Research Institute, Princeton, NJ, U.S.A.
[2] Centre for Intelligent Machines, McGill University, Montréal, Canada.

Abstract. Selecting the appropriate spatial scale for local edge analysis is a challenge for natural images, where blur scale and contrast may vary over a broad range. While previous methods for scale adaptation have required the global solution of a non-convex optimization problem [8], it is shown that knowledge of sensor properties and operator norms can be exploited to define a unique, locally-computable *minimum reliable scale* for local estimation. The resulting method for local scale control allows edges spanning a broad range of blur scales and contrasts to be reliably localized by a single system with no input parameters other than the second moment of the sensor noise. Local scale control further permits the reliable estimation of local blur scale in complex images where the conditions demanded by Fourier methods for blur estimation break down.

1 Introduction

Edge detectors are typically designed to recover step discontinuities in an image (e.g. [1–3, 11]), however the boundaries of physical structures in the world generally do not project to the image as step discontinuities, but as blurred transitions corrupted by noise (Fig. 1). This paper generalizes the detection of step discontinuities to encompass this broader, more physically realistic class of edges. We focus here on what can be computed from a *local* analysis of such luminance transitions.

Geometric models for focal and penumbral blur predict mathematically identical luminance transitions in the image, and shaded object edges can also mimic these patterns [4]. The goal of a local analysis must therefore be to reliably model luminance transitions over the broad range of conditions under which they occur, regardless of their physical origin.

To illustrate the challenge in achieving this goal, consider the scene shown in Fig. 2. Because the light source is not a point source, the contour of the cast shadow exhibits a broad variation in penumbral blur. On the left of Fig. 2 is shown the edge map generated by the Canny/Deriche edge detector [2,3], where the scale parameter has been tuned to detect the details of the mannequin. At this relatively small scale, the contour of the shadow cannot be reliably resolved and the smooth intensity gradients behind the mannequin are detected as many short, disjoint curves. On the right of Fig. 2 is shown the edge map generated by the Canny/Deriche edge detector tuned to detect the contour of the shadow. At

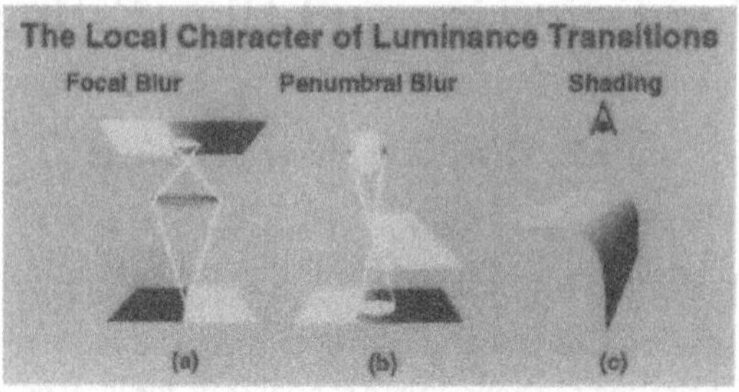

Fig. 1. Edges in the world generically project to the image as spatially blurred. From left to right: Focal blur due to finite depth-of-field; penumbral blur at the edge of a shadow; shading blur at a smoothed object edge.

this larger scale, the details of the mannequin are blurred out, and the contour of the shadow is fragmented at the section of high curvature under one arm.

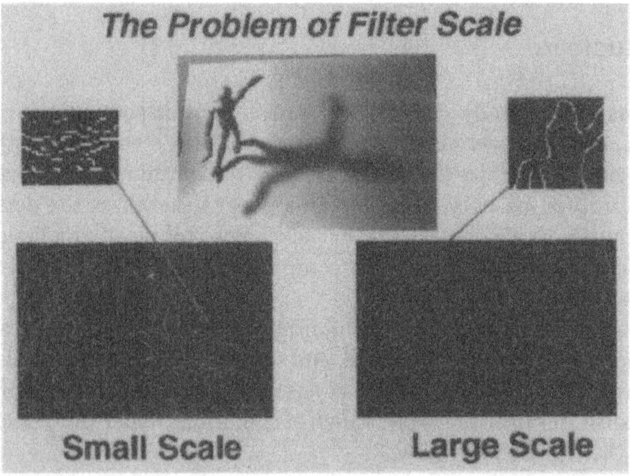

Fig. 2. Output of Canny/Deriche edge detector for small filter scale (left) and large filter scale (right).

This example suggests that to process natural images, operators of multiple scales must be employed. Of course, this conclusion has been reached by many computer vision researchers (e.g. [2,9,12,13,18]). However, the problem has been and continues to be: once a scale space has been computed, how is it used? Is there any principled way to combine information over scale, or to reason within this scale space, to produce usable assertions about the image?

2 Minimum Reliable Scale

The difficulty in reliably recovering structure from images such as Fig. 2 is that the appropriate scale for estimation varies over the image. However, while the scale of the scene structure is space-variant, the properties of the sensor are typically fixed and known in advance. Given a specific model of a luminance edge, one can use knowledge of sensor noise properties to relate the parameters of the model to a unique minimum scale at which the edge can be reliably detected. We call this unique scale the *minimum reliable scale* for the edge.

By reliable here we mean that at this scale, the likelihood of error due to sensor noise is below a standard tolerance (e.g. 5% for an entire image). This definition does not account for the danger of incorrect assertions due to the influence of scene events nearby in the image, which in any band-limited system must be an increasing function of scale. While attempts have been made by others to explicitly model this phenomenon [2], it is our view that this problem is unlikely to admit such a general solution. For example, while an ensemble of images may yield an estimate of the expected separation between edges, if a sample of the ensemble contains a fine corduroy pattern, this estimate will be of little use. The method for local scale control developed here is based on the conjecture that, given no prior knowledge of the scene being imaged, the *most* reliable scale for local estimation is the *minimum* reliable scale, as defined above.

Jeong & Kim [8] have also proposed an adaptive method for estimating a unique scale for local edge detection. They pose the problem as the minimization of a functional over the entire image. They report that their results suffered from the complicated shape of the objective function, and the resulting sensitivity of the selected scale to the initial guess. It will be shown here that such problems can be avoided; that given an appropriate model for the sensor, estimation of the minimum reliable scale can be posed as a *local* problem.

3 Modelling Edges, Blur and Sensing

An edge is modeled as a step function $Au(x) + B$ of unknown amplitude A and pedestal offset B, which, for the purposes of this discussion, will be aligned with the y-axis of the image coordinate frame. The focal or penumbral blur of this edge is modelled by a Gaussian blur kernel, $g(x, y, \sigma_b) = \frac{1}{2\pi\sigma_b{}^2} e^{-(x^2+y^2)/2\sigma_b{}^2}$ of unknown scale constant σ_b. Sensor noise $n(x, y)$ is modeled as a stationary, additive, zero-mean white noise process with standard deviation σ_n. The complete edge model is thus:

$$(A/2)(erf(x/\sqrt{2}\sigma_b) + 1) + B + n(x, y) \quad (1)$$

An example of the model is shown in Fig. 3(a).

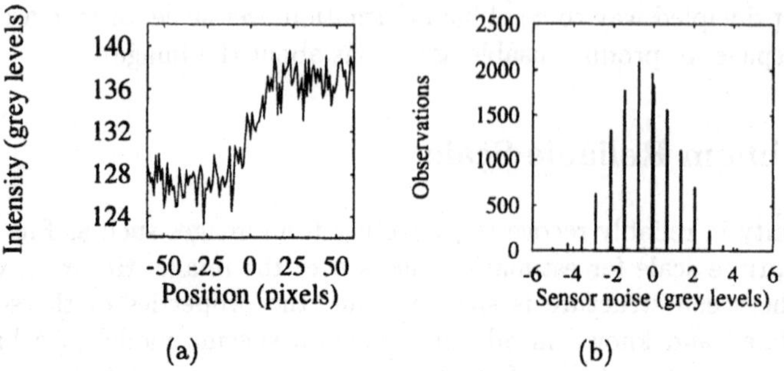

Fig. 3. a. The blurred edge model: $A = 10$ grey levels, $B = 127$ grey levels, $\sigma_b = 10$ pixels, $\sigma_n = 1.6$ grey levels. **b.** Histogram of estimated sensor noise

To estimate the sensor noise for the imaging system, a region of a defocused image of the ground plane shown in Fig. 2 was high-pass filtered with a unit-power kernel, the 2-tap filter $(1/\sqrt{2}, -1/\sqrt{2})$. The shading over this subimage varies slowly, and the defocus acts as an additional low-pass filter, so that the scene structure contributes negligible energy to the filter output. The following elementary result from the theory of random processes is now exploited [15]:

Proposition 1. *The standard deviation of a linear transformation $\mathcal{L} : \Re^n \rightarrow \Re$ of a set of i.i.d. random variables of standard deviation σ_n is the product of the L_2 norm of the linear transformation and the standard deviation of the random variables: $\sigma_{\mathcal{L}} = ||\mathcal{L}||_2 \sigma_n$.*

Thus the statistics of the unit-power filter output provide an estimate of the statistics of the sensor noise: a histogram is shown in Fig. 3(b). The standard deviation of the noise is approximately 1.6 quantization levels (for 8-bit images). The response of a unit-power operator is defined to be significant if it exceeds the mean response to white noise by $5\sigma_n$: 8 quantization levels in this case. Based on the normal model of sensor noise, the likelihood of an incorrect assertion for a single image of the size used here (256×384) is roughly 5%.

4 Local Scale Control and Gradient Estimation

A necessary condition for the local assertion of an edge is a non-zero gradient in the luminance function. The gradient can be estimated using steerable Gaussian first derivative basis filters [6,17]:

$$g_1^x(x, y, \sigma_1) = \frac{-x}{2\pi\sigma_1^4} e^{-(x^2+y^2)/2\sigma_1{}^2} \quad , \quad g_1^y(x, y, \sigma_1) = \frac{-y}{2\pi\sigma_1^4} e^{-(x^2+y^2)/2\sigma_1{}^2} ,$$

where σ_1 denotes the scale of the first derivative Gaussian estimator. The response $r_1^\theta(x, y, \sigma_1)$ of a first derivative Gaussian filter $g_1^\theta(x, y, \sigma_1)$ to an image $I(x, y)$ in an arbitrary direction θ can be computed exactly as a weighted sum of the basis filter responses:

$$r_1^\theta(x, y, \sigma_1) = \cos(\theta) r_1^x(x, y, \sigma_1) + \sin(\theta) r_1^y(x, y, \sigma_1).$$

At non-stationary points of the luminance function, $r_1^\theta(x, y, \sigma_1)$ has a unique maximum over θ, the gradient magnitude $r_1^{\theta_M}(x, y, \sigma_1)$, attained in the gradient direction $\theta_M(x, y, \sigma_1)$:

$$r_1^{\theta_M}(x, y, \sigma_1) = \sqrt{(r_1^x(x, y, \sigma_1))^2 + (r_1^y(x, y, \sigma_1))^2}$$
$$\theta_M(x, y, \sigma_1) = \arctan(r_1^y(x, y, \sigma_1)/r_1^x(x, y, \sigma_1))$$

For an image consisting of a blurred step edge along the y axis of amplitude A and blur parameter σ_b, the gradient magnitude attains its maximum on the y axis:

$$r_1^x(0, y, \sigma_1) = \frac{A}{\sqrt{2\pi(\sigma_b^2 + \sigma_1^2)}}. \tag{2}$$

To be confident that a non-zero gradient is due to the image and not to the noise, we must consider the likelihood that the response of the gradient operator could be due to noise alone. The second moment of the gradient operator can be expressed in terms of the second moments of the linear basis filter responses:

$$E[(r_1^{\theta_M}(x, y, \sigma_1))^2] = E[(r_1^x(x, y, \sigma_1))^2] + E[(r_1^y(x, y, \sigma_1))^2]$$

Using Proposition 1, we can write the second moment of the response of the gradient operator to the sensor noise alone as

$$\sqrt{E[(r_1^{\theta_M}(x, y, \sigma_1))^2]} = \sqrt{2}\sigma_n ||g_1^\theta(x, y, \sigma_1)||_2$$

Using the Cauchy inequality, this second moment can be used as a conservative estimate of the expected response of the gradient operator to the sensor noise, so that, using the chosen significance level of 5 standard deviations, it can be asserted that the gradient of the original image is non-zero if

$$r_1^{\theta_M}(x, y, \sigma_1) > 6\sqrt{E[(r_1^{\theta_M}(x, y, \sigma_1))^2]} = 6\sqrt{2}\sigma_n ||g_1^\theta(x, y, \sigma_1)||_2$$

The L_2 norm of the Gaussian first derivative operator at a point is given by $||g_1^\theta(x, y, \sigma_1)||_2 = 1/(2\sqrt{2\pi}\sigma_1^2)$ [4], thus we have the following

Definition 2. The *significance threshold function* $s_1(\sigma_1)$, tracing the threshold of reliability of the Gaussian gradient operator as a function of scale, is given by

$$s_1(\sigma_1) = 3\sigma_n/\sqrt{\pi}\sigma_1^2 \tag{3}$$

Combining Eqns. 2 and 3 and solving for σ_1, we derive the following

Proposition 3. *For a system with white sensor noise of standard deviation σ_n, and an edge of amplitude A and blur parameter σ_b, there exists a* **minimum reliable scale** $\hat{\sigma}_1$ *at which the luminance gradient can be reliably detected:*

$$\hat{\sigma}_1^2 = \frac{2}{c^2}(1 + \sqrt{1 + (c\sigma_b)^2}), \quad \text{where} \quad c = \sqrt{2}A/3\sigma_n.$$

In our experiments we attempt only to stay close to the minimum reliable scale by computing gradient estimates at octave intervals of scale, at each point using the smallest scale at which the gradient estimate exceeds the significance threshold function, i.e.

$$\hat{\sigma}_1(x, y) = \inf\{\sigma_1 : r_1^{\theta_M}(x, y, \sigma_1) > s_1(\sigma_1)\}$$

We tested this computation on the synthetic image of Fig. 4, a vertical edge blurred by a space-varying Gaussian kernel, corrupted by Gaussian i.i.d. noise. The minimum reliable scale space map for gradient estimation is shown in Fig. 4(b). Five scales were used: $\sigma_1 \in \{0.5, 1, 2, 4, 8\}$ pixels. Observe how the estimation scale increases as the strength of the gradient signal decreases.

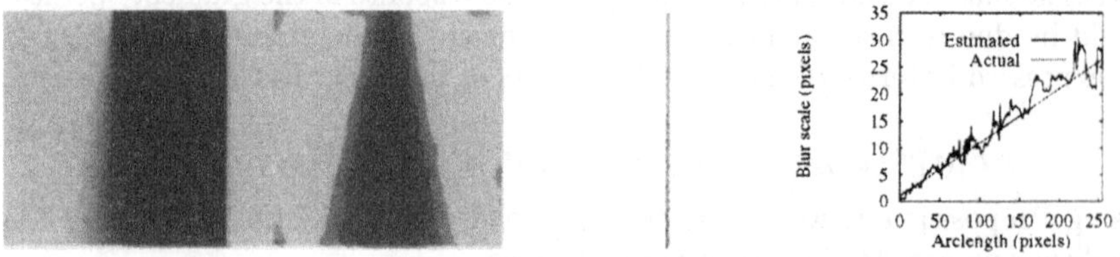

(a) (b) (c) (d)

Fig. 4. Testing the local scale control algorithm on a synthetic image. **(a)** The blur grade is linear, ranging from $\sigma_b = 1$ pixel to $\sigma_b = 26.6$ pixels. Parameters (see Section 3): $A = B = 85$ grey levels, $\sigma_b \in [1, 26.6]$ pixels, $\sigma_n = 1.6$ grey levels. **(b)** Minimum reliable scale map for gradient estimation. Larger scales are rendered in lighter grey. White indicates that no reliable estimate could be made. **(c)** Detected edge. **(d)** Estimated blur scale.

The result of this gradient computation for the image of the mannequin and shadow is shown in Fig. 5(a). While the smallest scale is reliable for the contours of the mannequin, higher scales are required to fully recover the shadow. Note that significant gradients are also detected on the smoothly shaded ground surface: gradient information alone is clearly an insufficient basis for edge detection. We now proceed to develop the complete criteria for edge selection.

(a) (b)

Fig. 5. Maps of the minimum reliable scales for gradient and second derivative estimation, respectively. Larger scales are rendered in lighter grey.

5 Edge Detection and Localization

To distinguish edges from other forms of luminance gradients, we must somehow select for the sigmoidal shape that characterizes our edge model. This information is available in the second derivative of the luminance function, which can be estimated with a steerable second derivative of Gaussian operator, $g_2^{\theta}(x, y, \sigma_2)$ [4,6,17]. Since we are interested only in the luminance variation orthogonal to the edge, at each point in the image the second derivative is steered in the direction of the gradient estimated at the minimum reliable scale. As for the gradient, one can show that near an edge there exists a unique minimum scale at which the sign of the second derivative response $r_2^{\theta}(x, y, \sigma_2)$ can be reliably determined. A second derivative map is thus obtained which describes, at each point in the image where a significant gradient could be estimated, how this gradient is changing in the gradient direction (if at all). Six scales are employed to estimate the second derivative, at octave intervals: $\sigma_2 \in \{0.5, 1, 2, 4, 8, 16\}$ pixels. The minimum reliable scale map for the mannequin image is shown in Fig. 5(b).

The importance of the second derivative in localizing blurred edges is illustrated in Fig. 6. Fig. 6(b) shows the luminance profile through the edge of the mannequin's shadow. Fig. 6(d) shows the gradient magnitude along the cross-section, and Fig. 6(c) shows the minimum reliable scales at which the gradient was estimated. Note how the scale of estimation automatically adapts as the strength of the signal varies. Although this allows the gradient to be reliably detected as non-zero over this cross-section, the response is not unimodal: there are in fact 5 maxima in the gradient along the cross section of the edge. Marking edges at extrema in the gradient function would clearly lead to multiple separate responses to this single edge.

Fig. 6(f) shows the estimated second derivative steered in the gradient direction, and Fig. 6(e) shows the minimum reliable scales for these estimates. Note again how scale automatically adapts as the signal varies in strength: larger scales are needed near the centre of the edge where the luminance function is nearly linear. Despite the rockiness of the gradient response, the adaptive second derivative response provides a unique zero-crossing to localize the edge, and strong extrema to identify its sigmoidal nature. The key here is that local estimation at the minimum reliable scale guarantees that the sign of the second derivative estimate is reliable, and hence that the zero-crossing is unique. The number of peaks in the gradient response, on the other hand, depends on the blur of the edge, and is not revealed in the response of the operator at any single point: ensuring the uniqueness of a gradient maximum is not a local problem. Thus the reliable detection and localization of blurred edges requires both gradient and second derivative information:

Definition 4. **An edge** exists at a point (x_0, y_0) if it satisfies 2 conditions:
Gradient Condition:
The gradient of the luminance function at the point is detectably non-zero:

$$\hat{\sigma}_1(x_0, y_0) = \inf\{\sigma_1 : r_1^{\theta_M}(x_0, y_0, \sigma_1) > s_1(\sigma_1)\} \neq \emptyset$$

Second Derivative Condition:
The second derivative of the luminance function in the gradient direction θ_M changes from positive to negative at the point:

$$r_2^{\theta_M}(x_+, y_+, \hat{\sigma}_2(x_+, y_+)) > 0 \quad , \quad r_2^{\theta_M}(x_-, y_-, \hat{\sigma}_2(x_-, y_-)) < 0$$

$$\begin{aligned} \textit{where} \quad x_- &= x_0 + \epsilon cos\theta_M \, , \quad & y_- &= y_0 + \epsilon sin\theta_M \\ x_+ &= x_0 - \epsilon cos\theta_M \, , \quad & y_+ &= y_0 - \epsilon sin\theta_M \end{aligned}$$

$$\begin{aligned} \textit{and} \quad \hat{\sigma}_2(x_+, y_+) &= \inf\{\sigma_2 : |r_2^{\theta_M}(x_+, y_+, \sigma_2)| > s_2(\sigma_2)\} \neq \emptyset \\ \hat{\sigma}_2(x_-, y_-) &= \inf\{\sigma_2 : |r_2^{\theta_M}(x_-, y_-, \sigma_2)| > s_2(\sigma_2)\} \neq \emptyset \end{aligned}$$

Here, ϵ is determined by the required localization precision: in this implementation, blurred edges are localized to 1 pixel precision. Note that, in general, $\hat{\sigma}_1(x, y) \neq \hat{\sigma}_2(x, y)$: the minimum reliable scales for estimating the gradient and second derivative are independent.

Roughly speaking, the above definition selects for points where the luminance function is detectably sigmoidal in shape. There is a strong connection between this definition and the machinery of logical/linear operators developed to detect in-focus edges [7].

Fig. 4(c) shows the edge points from the synthetic variable-blur image selected by Definition 4, using the local scale control algorithm. The edge is reliably and uniquely detected over a wide range of blur. Fig. 7 shows a map of the edge points in the image of the mannequin and shadow. Both the fine detail of the mannequin and the complete contour of the shadow are resolved, without spurious responses to the smooth shading gradients on the ground surface (compare with the results of the Canny/Deriche detector in Fig. 2.)

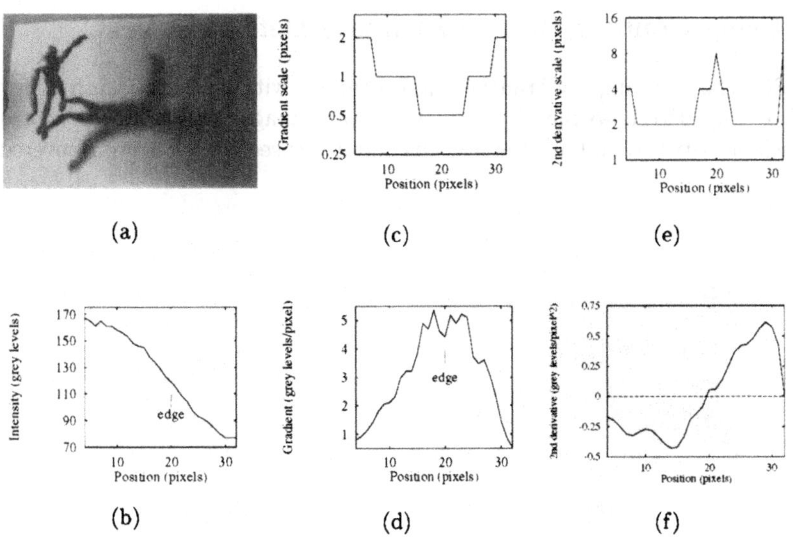

Fig. 6. **(a)** Mannequin shadow. White line indicates cross section of shadow boundary under analysis. **(b)** Cross section of luminance function across shadow boundary. **(c)** Minimum reliable scale for the gradient estimate. **(d)** Estimated gradient magnitude. **(e)** Minimum reliable scale for the second derivative estimate **(f)** Estimated directional second derivative. A unique zero-crossing localizes the edge.

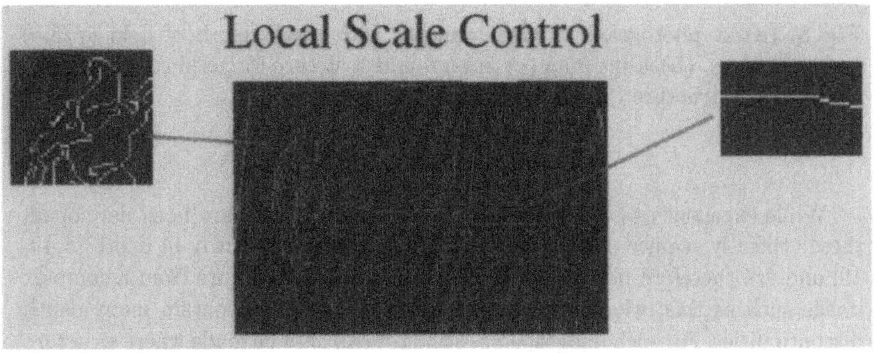

Fig. 7. Edges recovered from mannequin image using local scale control. Both the fine detail of the mannequin and the blurred boundary of the shadow are recovered.

6 Using Contour Blur to Estimate Defocus

Fig. 8(a) shows a tangle of branches, photographed with a shallow depth of field ($f/3.5$). Fig. 8(b) shows the edges detected for this image using local scale control for reliable estimation: both the in-focus and defocused branches are recovered.

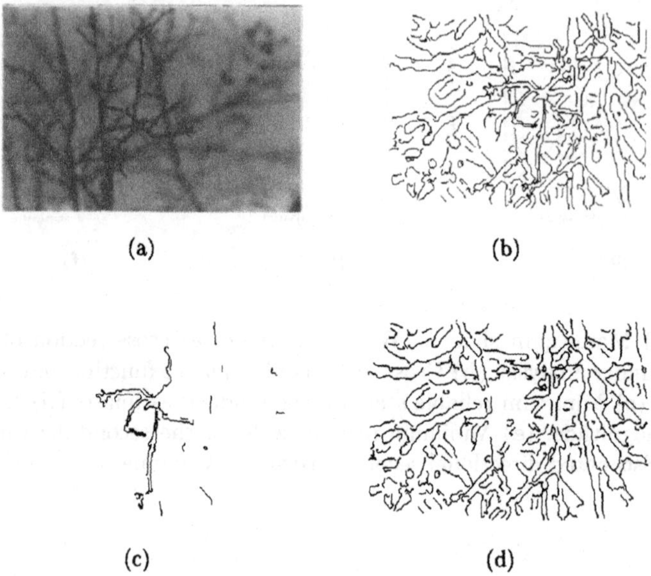

Fig. 8. **(a)** A photograph of tree branches with small depth of field ($f/3.5$) and near focus. **(b)** Edge map **(c)** Foreground structure (focused contours). **(d)** Background structure (blurred contours).

While excellent passive techniques for blur estimation have been developed, these typically require densely-textured surfaces varying slowly in depth [5, 14, 16] and are therefore not suited for recovering depth structure from a complex image such as this, where each local neighbourhood may contain many depth discontinuities. For such images, blur estimates can only be made where structure exists, and must be computed as locally as possible. The local scale control method can be extended naturally to this problem.

For the estimation of focal blur, the Gaussian kernel of our edge model represents the point spread function of the lens system of the camera employed. Without loss of generality, we again assume that the edge is oriented parallel to the y axis of the image plane. The extrema in the estimated second derivative occur at the zero-crossings x_M of the third derivative of the blurred step edge:

$$r_3^x(x_M, y_M, \sigma_2) = \frac{A}{\sqrt{2\pi(\sigma_2^2 + \sigma_b^2)}}((x_M^2/(\sigma_2^2 + \sigma_b^2) - 1)e^{-x_M^2/2(\sigma_2^2+\sigma_b^2)} = 0$$

so that $x_M = \pm\sqrt{\sigma_2^2 + \sigma_b^2}$. Defining d as the distance between second derivative extrema of opposite sign in the gradient direction, we obtain $\sigma_b = \sqrt{(d/2)^2 - \sigma_2^2}$. Thus the blur due to defocus can be estimated from the measured thickness of the contours, after compensation for the blur induced by the estimation itself.

Fig. 4(d) shows a plot of the estimated and actual blurs of the synthetic test image. While the resulting pointwise blur estimates are noisy, they provide an approximately unbiased estimate of the blur scale of the edge. This method for blur estimation was applied to the problem of segmenting the tangle of branches shown in Fig. 8. Fig. 8(c) and (d) show the extracted foreground (focused) and background (defocused) structure, respectively.

7 Space Curves from Defocus and Cast Shadows

While others have had some success in classifying contours as thin or diffuse [10, 19], this adaptive method for estimating contour blur can provide dense, accurate estimates *continuously* along image contours. As an example, consider the image of a model car (Fig. 9(a)) photographed with shallow depth of field ($f/2.5$). The lens was focused on the rear wheel of the car, so that the hood and front bumper are defocused. Fig. 9(b) shows the edges detected using local scale control. Fig. 9(c) shows a 3-D plot of one of the main contours of the car. Here the vertical axis represents the focal blur σ_b, estimated as described above, and smoothed along the contour with a Gaussian blur kernel ($\sigma = 22$ pixels). This adaptive method for blur estimation provides a continuous estimate of focal blur along the contour of the car.

This method for blur estimation can also be used to extract scene structure from penumbral blur (Fig. 9(d)). For a fixed light source, the penumbral blur is determined by the distance between the shadowing surface and the surface shadowed.[3] The results of penumbral blur estimation are shown in Fig. 9(e).

8 Conclusions

While edge detectors are typically designed to detect step discontinuities in images, physical edges in the world generally project to the image as blurred luminance transitions of unknown blur scale and contrast. In this paper, we have developed a method for adapting the spatial scale of local estimation to allow edges to be detected and characterized over this broad range of conditions.

[3] The estimation of distance from penumbral blur is more complicated than defocus, due to distortion caused by the slant of the ground surface relative to the observer and the light source. These estimates should therefore be viewed as qualitative.

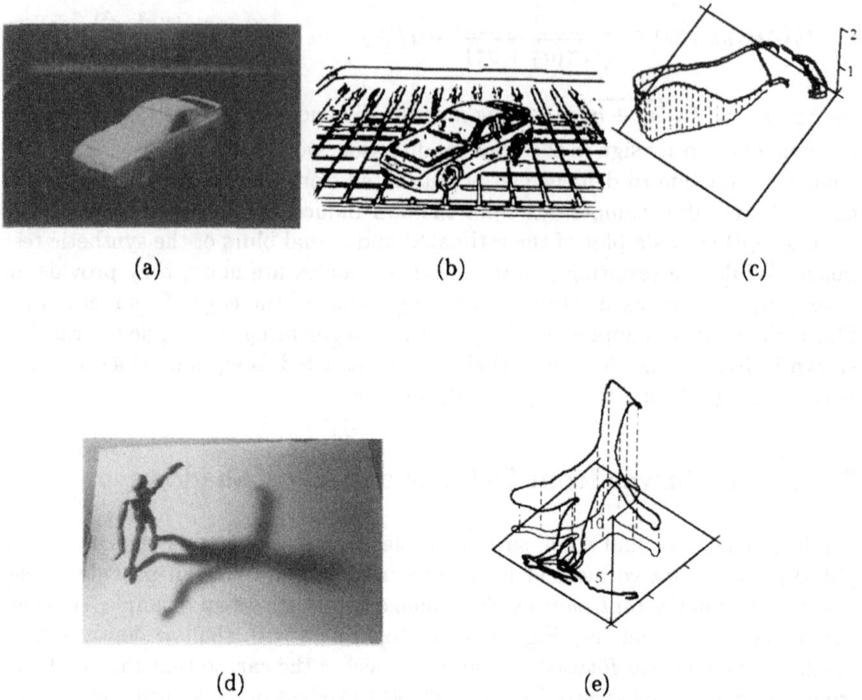

Fig. 9. **(a)** Photograph of car model with shallow depth of field ($f/2.5$). **(b)** Edges recovered using local scale control. **(c)** Space curve of contour from car. Estimated blur scale in pixels plotted on vertical axis. **(d)** Mannequin image with shadow. **(e)** Blur scale contour of mannequin shadow.

Specifically, this adaptive method reliably localizes edges while avoiding two important types of error: **(1)** multiple responses to a single transition, and **(2)** blurring of fine structure. This method for local scale control was further exploited to compute reliable estimates of local blur in complex images which do not satisfy the smoothness and texture density conditions required by traditional methods. These algorithms require no input parameters other than the second moment of the sensor noise.

References

1. A. Blake and A. Zisserman. *Visual Reconstruction.* MIT Press, Cambridge, Mass., 1987.
2. J.F. Canny. Finding edges and lines in images. Master's thesis, MIT Artificial Intelligence Laboratory, 1983.
3. R. Deriche. Using Canny's criteria to derive a recursively implemented optimal edge detector. *Int. J. Computer Vision*, 1(2):167–187, 1987.
4. J. Elder. *The visual computation of bounding contours.* PhD thesis, McGill University, Dept. of Electrical Engineering, 1995.
5. J. Ens and P. Lawrence. Investigation of methods for determining depth from focus. *IEEE Trans. Pattern Anal. Machine Intell.*, 15(2):97–108, 1993.
6. W.T. Freeman and E.H. Adelson. The design and use of steerable filters. *IEEE Trans. Pattern Anal. Machine Intell.*, 13(9):891–906, 1991.
7. L. Iverson and S.W. Zucker. Logical/linear operators for image curves. *IEEE Trans. Pattern Anal. Machine Intell.*, 17(10):982–996, 1995.
8. H. Jeong and C.I. Kim. Adaptive determination of filter scales for edge detection. *IEEE Trans. Pattern Anal. Machine Intell.*, 14(5):579–585, 1992.
9. J. Koenderink. The structure of images. *Biol. Cybern.*, 50:363–370, 1984.
10. A.F. Korn. Toward a symbolic representation of intensity changes in images. *IEEE Trans. Pattern Anal. Machine Intell.*, 10(5):610–625, 1988.
11. Y.G. Leclerc. The local structure of image discontinuities in one dimension. *IEEE Trans. Pattern Anal. Machine Intell.*, 9(3):341–355, 1987.
12. T. Lindeberg. Scale-space for discrete signals. *IEEE Trans. Pattern Anal. Machine Intell.*, 12(3):234–254, 1990.
13. D. Marr and E. Hildreth. Theory of edge detection. *Proc. R. Soc. Lond. B*, 207:187–217, 1980.
14. S.K. Nayar and N. Yasuo. Shape from focus. *IEEE Trans. Pattern Anal. Machine Intell.*, 16(8):824–831, 1994.
15. A. Papoulis. *Probability, Random Variables and Stochastic Processes.* McGraw-Hill, New York, 1965.
16. A.P. Pentland. A new sense for depth of field. *IEEE Trans. Pattern Anal. Machine Intell.*, 9(4):523–531, 1987.
17. P. Perona. Deformable kernels for early vision. *IEEE Trans. Pattern Anal. Machine Intell.*, 17(5):488–499, 1995.
18. A. Witkin. Scale space filtering. *Proc. Int. Joint Conf. on Artif. Intell.*, pages 1019–1021, 1983.
19. W. Zhang and F. Bergholm. An extension of Marr's signature based edge classification and other methods determining diffuseness and height of edges, and bar edge width. *Proc. 4th Int. Conf. on Computer Vision*, pages 183–191, 1993.

Regularization, Scale-Space, and Edge Detection Filters

Mads Nielsen[1], Luc Florack[1], and Rachid Deriche[2]

[1] DIKU, Universitetsparken 1, DK-2100 Copenhagen, Denmark
[2] INRIA, 2004 Route des Lucioles, BP 93, F-06902 Sophia Antipolis, France

Abstract. Computational vision often needs to deal with derivatives of digital images. Such derivatives are not intrinsic properties of digital data; a paradigm is required to make them well-defined. Normally, a linear filtering is applied. This can be formulated in terms of scale-space, functional minimization, or edge detection filters. The main emphasis of this paper is to connect these theories in order to gain insight in their similarities and differences. We take regularization (or functional minimization) as a starting point, and show that it boils down to Gaussian scale-space if we require scale invariance and a semi-group constraint to be satisfied. This regularization implies the minimization of a functional containing terms up to infinite order of differentiation. If the functional is truncated at second order, the Canny-Deriche filter arises.

1 Introduction

Given a digital signal in one or more dimensions, we want to define its derivatives in a well-posed way. This can be done in a distributional sense [1] by convolving the signal with a smooth test function. Instead of taking the derivatives of the signal, we take the derivatives of the convolved signal by differentiation of the smooth filter prior to convolution. In this way, the derivatives of any integrable signal are operationally defined and well-posed.

Distributional differentiation has been implemented in various conceptually different ways. In computational vision, Gaussian scale-space, regularization, or edge detection filters are typically applied. Anyway, the method must be linear since differentiation by its very definition is linear. In the following, we describe these three methods, their motivation, and discuss their relations.

Gaussian scale-space [2] can be motivated from different points of views. Koenderink [3] introduced it as a one parameter family of blurred images satisfying the causality criterion: every isophote in scale-space must be upwards convex. This requirement expresses in a precise sense that coarse scale details must have a cause at finer scales and essentially singles out the normalized Gaussian filter. Florack [4] defines a visual front-end to be linear, spatially isotropic, spatially homogeneous, separable and scale invariant, leading to the Gaussian as well. Since the Gaussian is smooth, all derivatives of the images are well-defined in Gaussian scale-space.

In Tikhonov regularization theory [5,6], we search for a differentiable function, which in some precise sense is closest to our signal. The measure of difference between the solution and the signal is a functional of the solution. In this way, the problem of well-posedness of differentiation is turned into a functional minimization problem. Using regularization we can directly specify up to which order the solution must be differentiable.

In edge detection many methods have been used for finding robust image derivatives. One method is to define a sample edge as well as criteria for an optimal detection. Canny [7] used a noisy step edge to find the optimal linear detection filter according to criteria of signal-to-noise-ratio, localization, and uniqueness of detection. Deriche [8] used these criteria on an infinite domain to find an optimal filter.

In the following sections, regularization is reviewed and the relation to Gaussian scale-space is derived. Multi-dimensional regularization under the constraint of Cartesian invariance is considered. Furthermore, it is shown that the Canny-optimal Deriche-filter corresponds to a regularization. Finally, it is shown that regularization can be implemented in an efficient way using recursive filtering without introducing further approximations.

2 Linear regularization

Regularization of a signal $g \in \mathcal{L}^2(\mathbb{R})$ can be formulated [5] as the minimization with respect to the regularized solution f of an energy functional E.

Definition 1. (Tikhonov regularization) The Tikhonov regularized solution f of the signal $g \in \mathcal{L}^2(\mathbb{R})$ minimizes the energy functional

$$E[f] \equiv \frac{1}{2}\int dx \left((f-g)^2 + \sum_{i=1}^{\infty} \lambda_i \left(\frac{\partial^i}{\partial x^i} f\right)^2 \right) \tag{1}$$

with nonnegative λ_i.

When $\lambda_n \neq 0$ and $\lambda_i = 0$ for all $i > n$ we talk about nth order regularization. The functional is convex and the minimum exists, implying a uniquely defined minimum. It can be found by linear filtering:

Proposition 2. (Regularization by convolution) *Linear convolution of the signal $g \in \mathcal{L}^2(\mathbb{R})$ by the filter h, having the Fourier transform*

$$\hat{h} = \frac{1}{\sum_{i=0}^{\infty} \lambda_i \omega^{2i}} \tag{2}$$

yields the solution of the regularization of Definition 1. By definition $\lambda_0 \equiv 1$.
Proof In the Fourier domain the energy functional (1) yields (according to Parsevals Theorem)

$$E[\hat{f}] = \frac{1}{2}\int d\omega \left((\hat{f}-\hat{g})^2 + \sum_{i=1}^{\infty} \lambda_i \omega^{2i} \hat{f}^2 \right)$$

A necessary (and sufficient) condition for E to be minimized with respect to $\hat{f}$ is that the variation of E is zero:

$$0 = \frac{\delta E}{\delta \hat{f}} = (\hat{f} - \hat{g}) + \sum_{i=1}^{\infty} \lambda_i \omega^{2i} \hat{f} \qquad \Leftrightarrow \qquad \hat{f} = \frac{1}{1 + \sum_{i=1}^{\infty} \lambda_i \omega^{2i}} \hat{g}$$

The optimal $\hat{f}$ can thus be found by a linear filtering of the initial signal g with a filter defined by the constants λ_i. Defining $\lambda_0 \equiv 1$ we obtain the linear filter h given by (2). □

Thus any regularization using only sums of quadratic terms of the derivatives of the solution can be reformulated as a linear convolution with the Fourier inverse (provided it exists) of $\hat{h}$ as given by (2). The differentiation of a regularized signal can be performed by a differentiation of the filter h prior to convolution. We make the following proposition:

Proposition 3. (Completeness of regularization) *Regularization on the form of (1) can implement any real, normalized, even, and analytic filter.*
Proof Regularization implements the linear filtering with $\hat{h}$ of (2). Since the denominator of $\hat{h}$ can be the Taylor series expansion of any analytic, even, and real function which is unity in zero, the filter $\hat{h}$ itself can be any analytic, even, and real function which is unity in zero. Hence, the spatial filter h can be any even, real, analytic, and normalized filter. □

Regularization can be formulated as a mapping of the initial function into a function space, in which all functions have the desired regularity properties. In Tikhonov regularization a square integrable function is mapped to a function within a Sobolev space of specified order. A Sobolev space of order N is the space of all functions being square integrable, and having derivatives up to order N all well-defined and square integrable.

2.1 Scale invariant regularization

For an action to be scale invariant it must be dimensionless, i.e. without units [9]. Any physical law is in general scale invariant. In order for a filter to be physically meaningful it must thus be dimensionless. In the following we use dimensional analysis to insure scale invariance of the filters. The notation $[\cdot]$ denotes "the dimension of".

A filtering h is scale invariant if it commutes with the scaling operation $S : x \mapsto x\gamma$ on the spatial domain. A scale invariant regularization other than the identity regularization cannot be constructed. A regularization parametrized by a scale parameter s so that

$$S(h(s, g)) = h(s', S(g))$$

can be constructed. Here $s' = \phi(s, S)$ is an identifiable function. We denote this scale invariant regularization.

Proposition 4. (Scale invariant regularization) *A regularization implemented by a filter $\hat{h}(\omega, s)$ is invariant to a scaling $S : x \mapsto x\gamma$ when*

$$\hat{h}(\omega, s) = \frac{1}{\sum_0^\infty \lambda_i \omega^{2i}} \qquad \textit{where} \qquad \lambda_i(s) = \frac{1}{i!} a_i s^i$$

and $a_i \in \mathbb{R}_+$ are scale independent (dimensionless) constants, $s \propto \gamma^2$, and the factorial $i!$ is included for later algebraic simplicity.

Proof If every term of the filter is dimensionless, then the filter is scale invariant. Since $[\omega] = \text{length}^{-1}$, then $[\lambda_i]$ must be $\text{length}^2 i$. This is ensured since $[s] = \text{length}^2$. □

The above construction does not limit the class of possible regularization. It expresses how the parameters of the regularization must transform if a change of units of the spatial domain is taking place.

2.2 Semi-group constraint

We now add constraints on the filter h stating that the initial signal g should not play a privileged role, but must be treated on equal footing with any reconstructed signal f. This is called a *recursivity constraint*: the regularization of a regularized signal must be expressible as a single regularization of the signal. This constraint is most intuitive when the signal g is given as a digitized signal. In this case, the digitizer has already performed a filtering of g. If g then plays a privileged role, a commitment in the image processing is made to the digitizer. We can perceive the recursivity constraint as a necessary condition to unconfound the image processing from the grid details. The recursivity constraint ensures that a filtered signal smoothed to some degree (eventually by the digitizer) can be smoothed to a higher degree by a filter from the same family [4,10]. The recursivity constraint can be expressed as the *semi-group constraint.*

Definition 5. (Semi-group property) A *semi-group* G is a set of compositions: $G \times G \to G$ which is associative.

Because regularizations can be implemented as linear filters, and linear filters are associative, so are regularizations. Regularization forms a semi-group, since it also satisfies the requirement $G \times G \to G$, which is the recursivity constraint.

If we embed h into a 1-parameter family $h(s)$, this family is a subset of all regularizations and this subset may not satisfy the semi-group constraint. We may formulate the *semi-group* property of the 1-parameter family $h(s)$ as:

Definition 6. (Semi-group property of filter family) The semi-group property of the 1-parameter family of filters $h(s)$ is

$$\forall s, t \in \mathbb{R}_+ : \; h(s \oplus_p t) = h(s) * h(t)$$

where the parameter-concatenation is the p-norm addition [3]

$$s \oplus_p t = (s^p + t^p)^{1/p}$$

[3] This is only the most general parameter concatenation when the filters must be dimensionless [14].

We call the filter-parameter the *scale parameter*, since it resembles the scale parameter in scale-space theory.

Proposition 7. (Scale invariant semi-group regularization) *If a one-parameter regularization filter family $h(s)$, $s \in \mathbb{R}_+$ is a scale invariant filter (Prop. 4) and the filter fulfills the semi-group constraint using the p-norm addition, then the filter family can be written as*

$$\hat{h}_p(\omega, t) = \frac{1}{\sum_{i=0}^{\infty} \frac{t^{ip}}{ip!} \omega^{2ip}} = e^{-(\omega^2 t)^p}$$

here given in the Fourier domain, where $t \propto s$.
Proof According to proposition 4, we can write the filter constants

$$\lambda_i(s) = \frac{1}{i!} a_i s^i$$

The semi-group property yields in conjunction with the actual form of h (given by (2))

$$\forall s, t \in \mathbb{R}_+ : \sum_{i=0}^{\infty} \lambda_i(s \oplus_p t) \omega^{2i} = \left(\sum_{i=0}^{\infty} \lambda_i(s) \omega^{2i} \right) \left(\sum_{i=0}^{\infty} \lambda_i(t) \omega^{2i} \right)$$

By separation into terms of equal power of ω:

$$\forall s, t \in \mathbb{R}_+^2 : \; \lambda_i(s \oplus_p t) = \sum_{l=0}^{i} \lambda_l(s) \lambda_{i-l}(t)$$

For clarity we now analyse the situation of $p = 1$. Choose $a_1 \equiv \tau$. By induction we find $a_i = \tau^i$, resulting in a filter of the form ($t = \tau s$)

$$\hat{h}(\omega, t) = \frac{1}{\sum_{i=0}^{\infty} \frac{t^i}{i!} \omega^{2i}} = e^{-\omega^2 t}$$

which is the well-known Gaussian. We formulate this as the result:

Result 1 (Gaussian regularization) *Scale invariance and the semi-group constraint applied to Tikhonov regularization, yields Gaussian scale-space if the 1-norm addition is used for concatenation of scales associated with the dimension of length squared.*

Applying the general p-norm scale-concatenation results in a broader class of filters: The derivation can be found in [14].

$$\hat{h}_p(\omega, t) = \frac{1}{\sum_{i=0}^{\infty} \frac{t^{ip}}{ip!} \omega^{2ip}} = e^{-(\omega^2 t)^p}$$

This is a class of filters all having the semi-group property for different scale-concatenation norms p. □

The above filters were derived from the class of regularization filters, but constitute the full class of all analytic scale invariant semi-group filters as it was derived by Pauwels et al. [11]. Here the existence of an infinitesimal generator is the starting point. Combined with scale invariance the above class of filters arises. Since the semi-group constraint is implied by the existence of an infinitesimal generator the above approach and Pauwels et al. [11] are quite similar.

All $\hat{h}_p$, $p \in \mathbb{N}$ are Fourier-invertible. For $p = 1$ it is the Gaussian, for p increasing towards infinity, it converges towards the ideal low-pass filter with cut-off frequency $\omega_c = 1/\sqrt{t}$.

2.3 Evolution equations

In the above we adjusted the weighting of the derivatives in Tikhonov regularization to implement Gaussian filtering. We can also reverse the argumentation: Gaussian scale-space can be implemented as the minimization of an energy functional (using a smoothness term of infinite order). The corresponding energy functional is

$$E[f] = \frac{1}{2}\int dx\ (f-g)^2 + \sum_{i=1}^{\infty} \frac{t^i}{i!}[\frac{\partial^i}{\partial x^i} f]^2$$

A necessary condition for f to minimize this is that the first variation yields zero. This is the Euler-Lagrange formulation and yields for the above functional

$$0 = (f-g) + \sum_{i=1}^{\infty} \frac{(-t)^i}{i!} \frac{\partial^{2i}}{\partial x^{2i}} f \equiv e^{-t\Delta} f - g$$

where Δ is the Laplacean and its exponentiation is defined by its Taylor series expansion. This differential equation is an equation using only spatial derivatives, but having the same solutions as the Heat Equation. The solution can formally (only formally since g is not necessarily differentiable) be written as

$$f = e^{t\Delta} g$$

By differentiation with respect to t, we obtain the Heat Equation:

$$\begin{cases} \frac{\partial}{\partial t} f = \Delta f \\ f|_{t=0} = g \end{cases}$$

The same derivation can be made for the other scale invariant semi-group filters under the mapping $\Delta \mapsto \Delta^p$, where Δ^p denotes the Laplacean applied p times.

2.4 Scale-space interpretation of regularization

We have seen that infinite order regularization with appropriate weights of the different orders boils down to linear scale-space. We can reformulate this result as follows: as time evolves, the signal governed by the Heat Equation travels through the minima of a one-parameter functional. In general, we would like to be able to construct a partial differential equation which travels through the minima of a given 1-parameter functional. Here, we look into the case of Tikhonov regularization. We want to find the function f minimizing the functional

$$E[f] = \int dx \left((f-g)^2 + \lambda \sum_{i=1}^{n} \lambda_i (\frac{\partial^i}{\partial x^i} f)^2 \right)$$

where $\lambda_i \geq 0$ are arbitrary. We find the Euler-Lagrange equation by setting the first variation of this to zero

$$(f-g) + \lambda \sum_{i=1}^{n} \lambda_i (-\Delta)^i f = 0$$

where $(-\Delta)^i$ denotes the Laplacean applied i times and negated if i is odd. We perceive of f as being a function of space x and evolution parameter λ and find by differentiation with respect to λ the following differential equation:

$$(1 + \lambda \sum_{i=1}^{n} \lambda_i (-\Delta)^i) \frac{\partial}{\partial \lambda} f = -\sum_{i=1}^{n} \lambda_i (-\Delta)^i f$$

This is a differential equation governing f, so that it travels through the minima of the above functional. In Fourier space it yields

$$(1 + \lambda \sum_{i=1}^{n} \lambda_i \omega^{2i}) \frac{\partial}{\partial \lambda} \hat{f} = \sum_{i=1}^{n} \omega^{2i} \hat{f}$$

In case of first order regularization (i.e. $\lambda_1 = 1$ and all other $\lambda_i = 0$) this is the Heat Equation, but with the following remapping of scale

$$d\lambda = (1 + \lambda\omega^2) ds \qquad \Leftarrow \qquad s(\lambda; \omega) = \frac{\log(1 + \lambda\omega^2)}{\omega^2}$$

In this way first order regularization can be interpreted as Gaussian scale-space with a frequency-dependent remapping of scale. Note that this remapping of scale is not Fourier invertible. Hence, a first order regularization cannot be constructed as a Gaussian scale-space with a spatially varying scale.

So far, we have discussed 1-dimensional regularization. In more dimensions mixed derivatives show up complicating the picture. However, imposing Cartesian invariance [12] (translation, rotation, and the scaling given implicit by the scaling nature of regularization) on the functional E leads to:

Proposition 8. (High-dimensional Cartesian invariant regularization) *The most general Cartesian invariant functional used for regularization in D dimensions has only one independent parameter λ_i per order of differentiation.*

The proof is given in [14]. Using this translationally and rotationally invariant regularization, we can easily generalize the results from the previous sections to higher dimensions. All what is needed is to substitute ω by $|\omega|$ in the filters. Also the semi-group property using the 1-norm addition leads trivially to the Gaussian filter.

3 Truncated smoothness operators

To be consistent with scale-space, Tikhonov regularization must be of infinite order. Nevertheless, low-order regularization is often performed [13]. We might perceive of low-order regularization as a regularization using a truncated Taylor-series of the smoothness functional. Here we list the linear filters h_z found by Fourier inversion. The notation h_z indicates the filter where all λ_i are zero except those mentioned by their index in the set z (E.g. in $h_{1,2}$ all λs are zero except λ_1 and λ_2).

$$h_1(x) = \frac{1}{\sqrt{2\lambda}} e^{\frac{-|x|}{\sqrt{\lambda}}} \tag{3}$$

$$h_2(x) = \frac{\pi}{\lambda^{1/4}} \cos\left(\frac{\sqrt{2}|x|}{\lambda^{1/4}}\right) e^{-\frac{|x|}{\sqrt{2}\lambda^{1/4}}} \tag{4}$$

In the case of mixed first and second order regularization with $\lambda = \lambda_1$ and $\lambda_2 = \lambda_1^2/2$ corresponding to the truncation to second order of the energy functional implying Gaussian filtering, we find

$$h_{1,2}(x) = \frac{\pi(|x| + \sqrt{\lambda})}{2\lambda} e^{-\frac{|x|}{\sqrt{\lambda}}}$$

We notice that the first order regularization filter is always positive, while the second order filter is similar apart from a multiplication with an oscillating term. The latter explains why second order regularization might give inadequate and oscillating results (so-called overshooting or ringing effects). The second order truncated Gaussian filter $h_{1,2}$ is always positive, and will in general give more sensible results than one without the first order term. Higher order non-oscillating regularization filters can be constructed by successive first order regularizations. An analysis of positiveness of regularization filters can be found in [14].

4 Canny optimality

Canny proposed three criteria of optimality of a feature detection filter. The feature is detected in the maxima of the linear filtering. In the case of a general feature $e(x)$ with uncorrelated Gaussian noise, the measures of signal-to-noise ratio Σ, localization Λ, and uniqueness of zero-crossing Υ are:

$$\Sigma[f] = \frac{\left|\int dx\ e(x) f(x)\right|}{\left(\int dx\ f^2(x)\right)^{1/2}}, \quad \Lambda[f] = \frac{\left|\int dx\ e'(x) f'(x)\right|}{\left(\int dx\ f'^2(x)\right)^{1/2}}, \quad \Upsilon[f] = \frac{\left(\int dx\ f'^2(x)\right)^{1/2}}{\left(\int dx\ f''^2(x)\right)^{1/2}}$$

where all integrals are taken over the real axis. We now *define* the feature to be a symmetric step edge:

$$e(x) = \int_{-\infty}^{x} dt\ \delta_0(t)$$

The symbol δ_0 denotes the Dirac delta functional. Canny tries first simultaneously to maximize Σ and Λ and finds the box filter as the optimal solution on a finite domain. To avoid the box-filter he then introduces the uniqueness measure. After this, a simultaneous optimization of all three measures using Lagrange multipliers is performed. Deriche [8] finds the optimal solution on a infinite domain. In the following we show that the uniqueness criterion can be omitted on the infinite domain leading to a conceptually simpler result, but lacking simplicity of the edges. Simplicity means they are zero-crossings of a differentiable function.

In order to find the optimal smoothing filter h to be differentiated so as to give the optimal step edge detector, we substitute $h'(x) = f(x)$ in the measures and try to find h. Assume, h is symmetric and normalized. We use one of the factors as optimality criterion and the others as constraints to find a composite functional Ψ. To obtain symmetry in this formulation we multiply all factors by an arbitrary Lagrange multiplier λ_i.

$$\Psi[h] = \int dx\ \lambda_1 e h' + \lambda_2 h'^2 + \lambda_3 e' h'' + \lambda_4 h''^2 + \lambda_5 h'''^2$$

Omitting the uniqueness constraint corresponds to $\lambda_5 = 0$. A necessary condition for an optimal filter is a zero variation:

$$\frac{\delta \Psi}{\delta h} = -\lambda_1 \delta_0 - 2\lambda_2 h^{(2)} + \lambda_3 \delta_0^{(2)} + 2\lambda_4 h^{(4)} - 2\lambda_5 h^{(6)} = 0$$

Here, a parenthesized superscript denotes the order of differentiation. By Fourier transformation we find

$$\hat{h}(\omega) = \frac{\lambda_1 + \lambda_3 \omega^2}{2(\lambda_2 \omega^2 + \lambda_4 \omega^4 + \lambda_5 \omega^6)}$$

Normalized in the spatial domain implies $\lambda_1 = 0$ and $\lambda_3 = 2\lambda_2$:

$$\hat{h}(\omega) = \frac{1}{1 + \alpha\omega^2 + \beta\omega^4}$$

where $\alpha = 2\lambda_4/\lambda_3$ and $\beta = 2\lambda_5/\lambda_3$. This is a second order regularization filter. Deriche [8] chooses $\alpha = 2\sqrt{\beta}$, and fixes in this way a one-parameter family of optimal filters. This choice corresponds well with a dimensional analysis and is the second order truncated Gaussian $h_{1,2}$.

Another way to select a one-parameter family is to select $\beta = 0$. By omitting the uniqueness criteria we find the optimal step edge detection filter to be the derivative of

$$\hat{h}(\omega) = \frac{1}{1 + \alpha\omega^2}$$

This is a first order regularization filter. It seems intuitively correct that the optimal first order filter should project into a Sobolev space of first order. We do not in general require the second derivative of the signal to exist when we consider only first order characteristics. In the spatial domain, we find the filter

$$h'(x) = \frac{\pi}{\sqrt{\alpha}} \frac{x}{|x|} e^{-|x|/\sqrt{\alpha}}$$

This is the edge detection filter proposed by Castan et al. [16]. The derivative of this filter is not well-defined, and we see that we indeed have projected into a Sobolev space of first order.

The above results can be generalized to optimal detection of higher order image features [14]. This also implies regularization but of higher order.

5 Implementation issues

The regularization using quadratic stabilizers can be implemented in various ways. It can be implemented by a gradient descent algorithm. This has the advantage that the natural boundary conditions can be implemented directly, leading to sensible results near the boundaries. The disadvantage is slow convergence (especially for large λs).

Implementation can be done by convolution in the spatial domain. In this case the boundaries can be handled by cutting off the filters and renormalizing. The computational complexity will be $O(MN\lambda^2 n)$, where the size of the image is $M \times N$ pixels and n is the order of regularization.

The filtering can also be implemented in the Fourier domain as a multiplication, using the Fast Fourier Transform (FFT). In this case the image is assumed to be cyclic, which may imply strange phenomena near the boundaries. The computational complexity is $O(MN \log M \log N)$ independently of the order of regularization and the λs.

Finally, the regularization can be implemented as a recursive filtering. In this case, the boundaries can be handled by cutting off like in the case of convolution in the spatial domain. The computational complexity is $O(MNn)$. In most practical cases, this will be the fastest implementation.

In order to deal with the recursive system, we need to reformulate the regularization on a discrete grid. We define the energy

$$E(f) \equiv \sum_x \left((f-g)^2 + \sum_{i=1}^{N} \lambda_i (d_i * f)^2 \right)$$

where d_i is the ith order difference filter. We recall the definition of the z-transform

$$\hat{h} = \sum_{x=-\infty}^{\infty} h(x) z^{-x}$$

and make the following proposition:

Proposition 9. (Recursive filtering) *The discrete regularization can be implemented by recursive filtering using no more than* $2N$ *multiplications and additions per output element, where* N *is the order of regularization.*

Proof By following the argumentation in the continuous case (and substituting the Fourier transform by the z-transform) or the discrete formulation given by

Unser et al. [17], we find that the minimization is implemented by convolution with the filter

$$\hat{h}(z) = \frac{1}{1 + \sum_{i=1}^{N} \lambda_i \hat{d}_i(z)\hat{d}_i(z^{-1})}$$

where the hat indicates the z-transform. The transform of the difference operator is given by

$$\hat{d}_i(z) = z^{-i/2}(1 - z)^i$$

implying that $\hat{d}_i(z)\hat{d}_i(z^{-1})$ is a N order polynomial in z multiplied by a N order polynomial in z^{-1}. Because the transfer function is symmetric in z and z^{-1} the $2N$ roots in the denominator will appear in pairs of z and z^{-1}. The transfer function can be decomposed as $\hat{h}(z) = \hat{h}^{+}(z)\hat{h}^{+}(z^{-1})$, where

$$\hat{h}^{+}(z) = \frac{c}{\prod_{i=1}^{N}(1 - z_i z^{-1})} = \frac{c}{1 - a_i z^{-1} - \ldots - a_N z^{-N}}$$

and z_i are the roots in the denominator ordered by length. This means that the regularization can be implemented by the forward and backward recursive filters with identical coefficients:

$$\begin{aligned} f^{+}(x) &= g(x) + a_1 f^{+}(x-1) + \ldots + a_N f^{+}(x-N) \\ f^{-}(x) &= f^{+}(x) + a_1 f^{-}(x+1) + \ldots + a_N f^{-}(x+N) \\ f(x) &= c^2 f^{-}(x) \end{aligned}$$

□

We have here proven that regularization in the discrete 1D case can be implemented (without any other approximation than the discrete implementation) as recursive filtering. The proof follows the lines of the proof given by Unser et al. [17] for the case, where a single order stabilizer is used.

6 Summary

We have presented a formulation of the Heat Equation as a minimization of a functional corresponding to a Tikhonov regularization. Furthermore, we have developed an infinite series of regularization filters all possessing the semi-group property. The simplest of these is the well-known Gaussian. The series of semi-group filters converges towards an ideal low-pass filtering, when the scale addition convention converges towards the infinity norm.

We have shown that the Canny-Deriche filter is a second order regularization filter, when we choose coefficients corresponding to the truncation of the functional leading to Gaussian scale-space. The Canny-Deriche filter is optimized for detection of zero order step edges.

Furthermore, we have shown that regularization in a higher dimensional space in its most general form is a rotation of the 1D regularization, when Cartesian invariance is imposed. This means that all of the above results can be generalized to any dimension, under the constraint that the functional is a Cartesian invariant.

The above work is a concentrate of the technical report [14], where the omitted proofs and further results can be found. We have only treated linear problems: Linear convolution, linear scale-space, linear biased diffusion. In future a major subject for studies will be to find the resemblance between non-linear biased and non-biased diffusion. Can we transform a non-linear diffusion equation [18] into a non-linear biased diffusion equation [19] (and thereby capture minimization of non-quadratic functionals) and vice versa?

References

1. L. Schwartz, "*Théorie des Distributions*", Hermann, 2nd edition, Paris, 1966.
2. A. P. Witkin, "*Scale-Space Filtering*", Proc. of IJCAI, Karlsruhe, Germany, 1983.
3. J. J. Koenderink, "*The Structure of Images*", Biol. Cybern., vol. 50, pp. 363-370, 1984.
4. L. M. J. Florack, "*The Syntactical Structure of Scalar Images*", PhD-thesis, University of Utrecht, 1993.
5. A. N. Tikhonov and V. Y. Arseninn, "*Solution of ill-posed problems*" V.H Winston & Sons, John Wiley, 1977.
6. T. Poggio, H. Voorhees and A. Yuille, "*A Regularized Solution to Edge Detection*", A. I. Memo 883, M.I.T, May 1985.
7. J. F. Canny, "*A Computational Approach to Edge Detection*", IEEE PAMI, Vol. 8, No. 6, Nov 1986.
8. R. Deriche, "*Using Canny's Criteria to Derive a Recursively Implemented Optimal Edge Detector*", Int. Jour. of Computer Vision, pp. 167-187, 1987.
9. J. Fourier, "*The Analytic Theory of Heat*", New York Dover publications, 1995, Original "*Théori Analytique de la Chaleur*", Paris, 1822.
10. T. Lindeberg, "*Scale-Space Theory in Computer Vision*", Kluwer Academic Publishers, Netherlands, 1994.
11. E. J. Pauwels, P. Fiddelaers, T. Moons, and L. J. V. Gool, "*An extended Class of Scale-Invariant and Recursive Scale-Space Filters*", Katholieke Universiteit Leuven, Tech. Rep. KUL/ESAT/MI2/9316, 1994.
12. D. Hilbert, *Ueber die vollen Invariantsystemen*, Math. Annalen, vol. 42, pp. 313–373, 1893.
13. W. E. L. Grimson, "*From Images to Surfaces*", MIT Press, Cambridge, Massachusetts, 1981.
14. M. Nielsen, L. M. J. Florack, and R. Deriche, "*Regularization and Scale-Space*", RR-2352, INRIA, France, 1994.
15. R. Deriche, "*Fast Algorithms for Low-level Vision*", IEEE PAMI, Vol. 12, No. 1, Jan 1990.
16. S. Castan, J. Zhao, and j. Shen, "*Optimal Filter for Edge Detection Methods and Results*", ECCV 90, Antibes, France, May, 1990.
17. M. Unser, A. Aldroubi, and M. Eden, "*Recursive Regularization Filters: Design, Properties and Applications*", IEEE PAMI, vol. 13, no. 3, March 1991.
18. P. Perona and J. Malik, "*Scale-Space and Edge Detection using Anisotropic Diffusion*" IEEE PAMI, vol. 12, pp. 429-439, 1990.
19. N. Nordström, "*Biased Anisotropic Diffusion - a Unified Regularization and Diffusion Approach to Edge Detection*", Image and Visual Computing, no. 8, pp. 318-327, 1990.

Direct Differential Range Estimation Using Optical Masks

Eero P. Simoncelli and Hany Farid

GRASP Laboratory
University of Pennsylvania, Philadelphia, PA 19104, USA

Abstract. We describe a novel formulation of the range recovery problem, based on computation of the differential variation in image intensities with respect to changes in camera position. The method uses a single stationary camera and a pair of calibrated optical masks to directly measure this differential quantity. The subsequent computation of the range image is simple and should be suitable for real-time implementation. We also describe a variant of this technique, based on direct measurement of the differential change in image intensities with respect to aperture size. These methods are comparable in accuracy to other single-lens ranging techniques. We demonstrate the potential of our approach with a simple example.

1 Introduction

Visual images are formed via the projection of light from the three-dimensional world onto a two-dimensional sensor. In an idealized pinhole camera, all points lying on a ray passing through the pinhole will be imaged onto the same image position. Thus, information about the distance to objects in the scene (i.e., *range*) is lost. Range information can be recovered by measuring the change in appearance of the world resulting from a change in viewing position. Traditionally, this is accomplished via simultaneous measurements with two cameras (binocular stereo), or via a sequence of measurements collected over time from a moving camera (structure from motion).

The recovery of range in these approaches frequently relies on an assumption of brightness constancy, which states that the brightness of the image of a point in the world is constant when viewed from different positions [Horn 86]. Consider the formulation of this assumption in one dimension (the extension to two dimensions is straightforward). Let $f(x; v)$ describe the intensity function measured through a pinhole camera system. The variable v corresponds to the pinhole position (along the direction perpendicular to the optical axis). The variable x parameterizes the position on the sensor. This configuration is illustrated in Figure 1. According to the assumption, the intensity function $f(x; v)$ is of the form:

$$f(x; v) = I\left(x - \tfrac{vd}{Z}\right), \tag{1}$$

where $I(x) = f(x; v)\,|_{v=0}$, d is the distance between the pinhole and the sensor and Z is the range (distance from the pinhole to a point in the world). Note

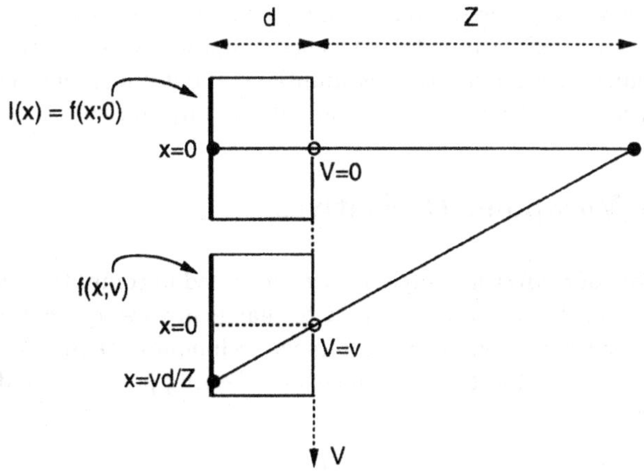

Fig. 1. Geometry for a binocular stereo system with pinhole cameras. The variable V parameterizes the position of the camera pinholes. According to the brightness constancy constraint, the intensity of a point in the world, as recorded by the two pinhole cameras, should be the same.

that this assumption will typically be violated near occlusion boundaries, where points visible from one viewpoint are invisible from another.

Several complications arise in these approaches. The degree to which the brightness constancy assumption holds will, in general, decrease with increasing camera displacement. This is due to larger occluded image regions, and increased effects of the non-Lambertianity of surface reflectances. Violations of the brightness constancy assumption lead to difficulties in matching corresponding points in the images (the so-called "correspondence problem"). Furthermore, a two-camera stereo system (or a single moving camera) requires careful calibration of relative positions, orientations, and intrinsic parameters of the camera(s).

These problems are partially alleviated in techniques utilizing a single stationary camera. A number of these techniques are based on estimation of blur or relative blur from two or more images (e.g., [Krotkov 87, Pentland 87, Subbarao 88, Xiong 93, Nayar 95]). Adelson [Adelson 92] describes an unusual method in which a lenticular array is placed over the sensor, effectively allowing the camera to capture visual images from several viewpoints in a single exposure. Dowski [Dowski 94] and Jones [Jones 93] each describe range imaging systems that use an optical attenuation mask in front of the lens. By observing local spectral information in a single image, they are able to estimate range. Both techniques rely on power spectral assumptions about the scene.

In this paper, we propose a single-camera method which avoids some of the computational and technical difficulties of the single-camera approaches discussed above. In particular, we propose a "direct" differential method for range estimation which computes the image derivative with respect to viewing position using a single stationary camera and an optical attenuation mask. We also present a variation based on the derivative with respect to aperture size (i.e., differential range-from-defocus). These approaches avoid the correspondence problem, make no spectral assumptions about the scene, are relatively straightforward to calibrate, and are computationally efficient.

2 Direct Viewpoint Derivatives

For the purpose of recovering range, we are interested in computing the change in the appearance of the world with respect to change in viewing position. It is thus natural to consider differential measurement techniques. Taking partial derivatives of $f(x;v)$ with respect to the image and viewing positions, and evaluating at $v=0$ gives:

$$\begin{aligned} I_x(x) &\equiv \frac{\partial f(x;v)}{\partial x}\Big|_{v=0} \\ &= I'(x), \end{aligned} \tag{2}$$

and

$$\begin{aligned} I_v(x) &\equiv \frac{\partial f(x;v)}{\partial v}\Big|_{v=0} \\ &= -\frac{d}{Z} I'(x), \end{aligned} \tag{3}$$

where $I'(\cdot)$ indicates the derivative of $I(\cdot)$ with respect to its argument. Combining these two expressions gives:

$$I_v(x) = -\frac{d}{Z} I_x(x). \tag{4}$$

Clearly, an estimate of the range, Z, can be computed using this equation. Note that in the case of differential binocular stereo (e.g., [Lucas 81]), the derivative with respect to viewing position, I_v, is replaced by a difference, $I_{v_1} - I_{v_2}$. A similar relationship is used in computing structure from motion (for known camera motion), where I_v is typically replaced by differences of consecutive images.

We now show a direct method for measurement of this derivative through the use of an optical attenuation mask. Consider a world consisting of a single point light source and a standard lens-based imaging system with a variable-opacity optical mask, $M(u)$, placed directly in front of the lens (left side of Figure 2). The light striking the lens is attenuated by the value of the optical mask function at that particular spatial location.[1] With such a configuration, the image of the point source will be a scaled and dilated version of the optical mask function:

$$I(x) = \frac{1}{\alpha} M\left(\frac{x}{\alpha}\right), \tag{5}$$

as illustrated in Figure 2. The scale factor, α, is a monotonic function of the distance to the point source, Z, and may be derived from the imaging geometry:

$$\alpha = 1 - \frac{d}{f} + \frac{d}{Z}, \tag{6}$$

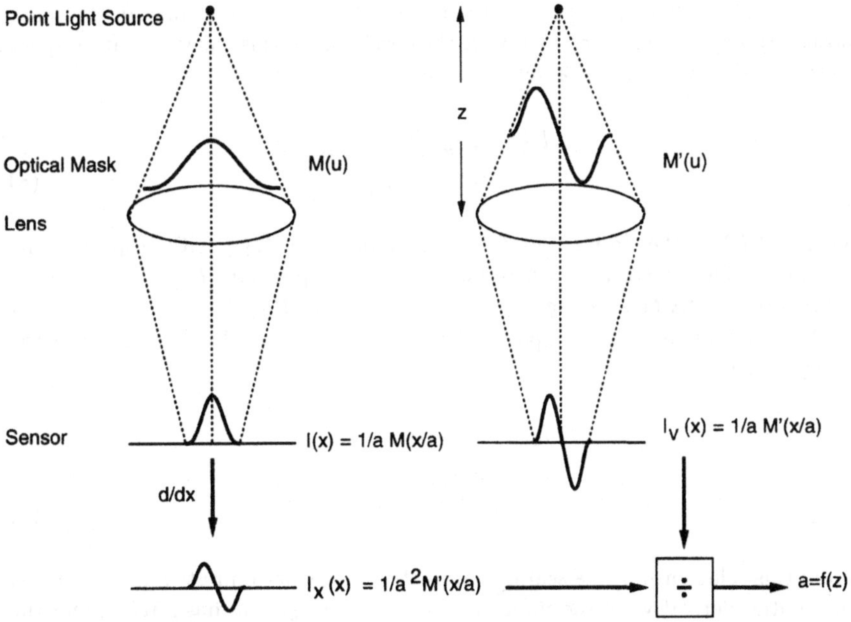

Fig. 2. Direct differential range determination for a single point source. Images of a point light source are formed using two different optical masks, corresponding to the function $M(u)$ and its derivative, $M'(u)$. In each case, the image formed is a scaled and dilated copy of the mask function (by an amount α). Computing the spatial (image) derivative of the image formed under mask $M(u)$ produces an image that is identical to the image formed under the derivative mask, $M'(u)$, except for a scale factor α. Thus, α may be estimated as the ratio of the two images. Range is computed from α using the relationship given in Equation (6).

where d is the distance between lens and sensor, and f is the focal length of the lens.

In the system shown on the left side of Figure 2, the *effective* viewpoint may be altered by translating the mask, while leaving the lens and sensor stationary.[2] The generalized intensity function, for a mask centered at position v is written as:

$$f(x; v) = \frac{1}{\alpha}\, M(\tfrac{x}{\alpha} - v), \tag{7}$$

assuming that the non-zero portion of the mask does not extend past the edge of the lens.

[1] For our purposes, we assume that the values of such a mask function are real numbers in the range [0,1].

[2] For example, consider a mask which contains a single pinhole; different views of the world are obtained by sliding the pinhole across the front of the lens.

The *differential* change in the image (with respect to a change in the mask position) may be computed by taking the derivative of this equation with respect to the mask position, v, evaluated at $v = 0$:

$$\begin{aligned} I_v(x) &\equiv \tfrac{\partial}{\partial v} f(x;v)\big|_{v=0} \\ &= -\tfrac{1}{\alpha} M'(\tfrac{x}{\alpha}), \end{aligned} \tag{8}$$

where $M'(\cdot)$ is the derivative of the mask function $M(\cdot)$ with respect to its argument. The derivative with respect to viewing position, $I_v(x)$, *may thus be computed directly by imaging with the optical mask* $M'(\cdot)$.[3]

Finally, notice that the spatial derivative of $I(x)$ is closely related to the image $I_v(x)$:

$$\begin{aligned} I_x(x) &\equiv \tfrac{\partial}{\partial x} f(x;v)\big|_{v=0} \\ &= \tfrac{1}{\alpha^2} M'(\tfrac{x}{\alpha}) \\ &= -\tfrac{1}{\alpha}\, I_v(x). \end{aligned} \tag{9}$$

From this relationship, the scaling parameter α may be computed as the ratio of the spatial derivative of the image formed through optical mask $M(u)$, and the image formed through the derivative of that optical mask, $M'(u)$. This computation is illustrated in Figure 2. The distance to the point source can subsequently be computed from α using the monotonic relationship given in Equation (6). Note that the resulting equation for distance is identical to that of Equation (4) when $d = f$ (i.e., when the camera is focused at infinity).

The necessity of the brightness constancy assumption can now be made explicit. For our system, brightness constancy means that the light emanating from a point is of constant intensity across the surface of the mask. A point light source that violates this assumption has a directionally varying light emission, $L(\cdot)$, and when imaged through the pair of optical masks will produce images of the form:

$$I(x) = \tfrac{1}{\alpha} L(\tfrac{x}{\alpha}) \cdot M(\tfrac{x}{\alpha}) \tag{10}$$

$$I_v(x) = \tfrac{1}{\alpha} L(\tfrac{x}{\alpha}) \cdot M'(\tfrac{x}{\alpha}). \tag{11}$$

As before, computing the derivative of $I(x)$ yields:

$$I_x(x) = \tfrac{1}{\alpha^2} \left(L(\tfrac{x}{\alpha}) M'(\tfrac{x}{\alpha}) + L'(\tfrac{x}{\alpha}) M(\tfrac{x}{\alpha}) \right). \tag{12}$$

Thus, if the light is not constant across the aperture (i.e., $L'(\cdot) \neq 0$) then the simple relationship between $I_v(x)$ and $I_x(x)$ (given in Equation (9)) will not hold.

---

[3] In practice, $M'(u)$ cannot be directly used as an attenuation mask, since it contains negative values. This issue is addressed in Section 4.

3 Range Estimation

Equation (9) embodies the fundamental relationship used for the direct differential computation of range of a single point light source. A more realistic world consisting of a collection of many such uniform intensity point sources imaged through an optical mask will produce an image consisting of a superposition of scaled and dilated versions of the masks. In particular, we can write an expression for the image by summing the images of the visible points, p, in the world:

$$f(x;v) = \int dx_p \, \frac{1}{\alpha_p} M\left(\frac{x-x_p}{\alpha_p} - v\right) L(x_p), \tag{13}$$

where the integral is performed over the variable x_p, the position in the sensor of a point p projected through the center of the lens. The intensity of the world point p is denoted as $L(x_p)$, and α_p is monotonically related to the distance to p (as in Equation (6)). Note again that we must assume that each point produces a uniform light intensity across the optical mask.

Again, consider the derivatives of $f(x;v)$ with respect to viewing position, v, and image position, x:

$$\begin{aligned} \frac{\partial}{\partial v} f(x;v) &= \frac{\partial}{\partial v} \int dx_p \, \frac{1}{\alpha_p} M\left(\frac{x-x_p}{\alpha_p} - v\right) L(x_p) \\ &= - \int dx_p \, \frac{1}{\alpha_p} M'\left(\frac{x-x_p}{\alpha_p} - v\right) L(x_p), \end{aligned} \tag{14}$$

and

$$\begin{aligned} \frac{\partial}{\partial x} f(x;v) &= \frac{\partial}{\partial x} \int dx_p \, \frac{1}{\alpha_p} M\left(\frac{x-x_p}{\alpha_p} - v\right) L(x_p) \\ &= \int dx_p \, \frac{1}{\alpha_p^2} M'\left(\frac{x-x_p}{\alpha_p} - v\right) L(x_p), \end{aligned} \tag{15}$$

where (as before) $M'(\cdot)$ is the derivative of $M(\cdot)$ with respect to its argument.

As in the previous section, the following two partial derivative images are defined:

$$\begin{aligned} I_v(x) &\equiv \frac{\partial}{\partial v} f(x;v) \,\big|_{v=0} \\ &= - \int dx_p \, \frac{1}{\alpha_p} M'\left(\frac{x-x_p}{\alpha_p}\right) L(x_p), \end{aligned} \tag{16}$$

and

$$\begin{aligned} I_x(x) &\equiv \frac{\partial}{\partial x} f(x;v) \,\big|_{v=0} \\ &= \int dx_p \, \frac{1}{\alpha_p^2} M'\left(\frac{x-x_p}{\alpha_p}\right) L(x_p). \end{aligned} \tag{17}$$

Equations (16) and (17) differ only in a multiplicative term of $\frac{1}{\alpha_p}$. Unfortunately, solving for α_p is nontrivial, since it is embedded in the integrand and depends on the integration variable. Consider, however, the special case where all points in the world lie on a frontal-parallel plane relative to the sensor.[4] Un-

[4] In actuality, this assumption need only be made locally.

der this condition, the scaling parameter α_p is the same for all points x_p and Equations (16) and (17) can be written as:

$$I_v(x) = \frac{1}{\alpha} \int dx_p \; M'\left(\frac{x - x_p}{\alpha}\right) L(x_p) \tag{18}$$

$$I_x(x) = \frac{1}{\alpha^2} \int dx_p \; M'\left(\frac{x - x_p}{\alpha}\right) L(x_p). \tag{19}$$

The scaling parameter, α, a function of the distance to the points in the world (Equation (6)) can then be computed as the ratio:

$$I_v(x) = -\alpha I_x(x). \tag{20}$$

As in the single-point case, this expression is identical to that of Equation (4) with $d = f$.

In order to deal with singularities (i.e., $I_x = 0$), a least-squares estimator can be used for α (as in [Lucas 81]). Specifically, we minimize $E(\alpha) = \sum_P (I_v + \alpha I_x)^2$, where the summation is performed over a small patch in the image, P. Taking the derivative with respect to α, setting equal to zero and solving for α yields the minimal solution:

$$\alpha = -\frac{\sum_P I_v I_x}{\sum_P I_x^2}. \tag{21}$$

The algorithm easily extends to a three-dimensional world: we need only consider two-dimensional masks $M(u, w)$, and the horizontal partial derivative $M'(u, w) = \partial M(u, w)/\partial u$. For a more robust implementation, the vertical partial derivative mask $\partial M(u, w)/\partial w$ may also be included. The least-squares error function becomes:

$$E(\alpha) = \sum_P (I_u + \alpha I_x)^2 + (I_w + \alpha I_y)^2. \tag{22}$$

Solving for the minimizing α gives:

$$\alpha = -\frac{\sum_P (I_u I_x + I_w I_y)}{\sum_P (I_x^2 + I_y^2)}. \tag{23}$$

4 Aperture Mask Design

Thus far, the only restriction placed on the aperture masks, $M(u, w)$ and $M'(u, w)$, is that the second be the derivative of the first. Figure 3 contains a matched pair of masks based on a two-dimensional Gaussian. In practice, the function $M'(u, w)$ has negative values and thus is not feasible for use as an optical attenuation mask. Furthermore, a positive constant cannot simply be added to $M'(u, w)$, since this will destroy the required derivative relationship between the two masks.

Due to the linearity of the imaging process, however, we can use masks that are linear combinations of the masks $M'(u, w)$ and $M(u, w)$. In particular, a

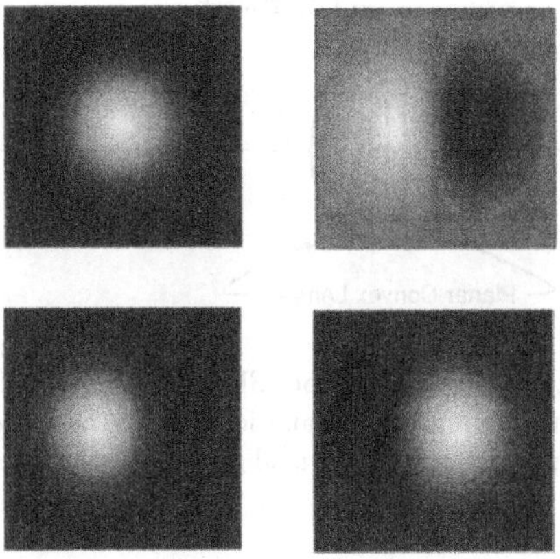

Fig. 3. Gaussian aperture masks. Top left: A two-dimensional Gaussian mask, $M(u,w)$. Top right: Gaussian partial derivative, $M'(u,w)$. Bottom row: Two non-negative aperture masks, $M_1(u,w)$ and $M_2(u,w)$. These are computed from the top masks using Equations (24) and (25).

scalar multiple of $M(u,w)$ can be added to $M'(u,w)$ in order to form a mask function that is entirely positive. The new mask, $M_1(u,w)$, shown in Figure 3, is given by:

$$M_1(u,w) = \beta M(u,w) + \gamma M'(u,w), \tag{24}$$

where β, γ are scaling constants chosen to force the function $M_1(u,w)$ to fill the range $[0,1]$. A second symmetrical mask can be formed by *subtracting* $M'(u,w)$ from $M(u,w)$:

$$M_2(u,w) = \beta M(u,w) - \gamma M'(u,w). \tag{25}$$

Note that $M_2(u,w)$ is equal to $M_1(u,w)$ rotated 180 degrees about its center, and that

$$M(u,w) = \frac{M_1(u,w)+M_2(u,w)}{2\beta} \tag{26}$$

$$M'(u,w) = \frac{M_1(u,w)-M_2(u,w)}{2\gamma}. \tag{27}$$

Again, by linearity of the imaging process, the images that would have been obtained with the masks $M(u,w)$ and $M'(u,w)$ can be recovered from images obtained with two masks $M_1(u,w)$ and $M_2(u,w)$. In particular, let $I_1(u,w)$ be the image obtained through the mask M_1, and $I_2(u,w)$ the image obtained through the mask M_2. Then:

$$I(u,w) = \frac{I_1(u,w)+I_2(u,w)}{2\beta} \tag{28}$$

$$I_v(u,w) = \frac{I_1(u,w)-I_2(u,w)}{2\gamma}, \tag{29}$$

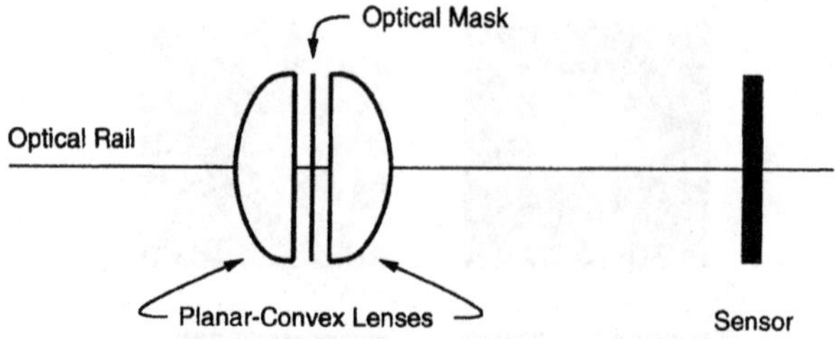

Fig. 4. Experimental camera configuration. Shown are a pair of planar-convex lenses, placed back-to-back, with an optical attenuation mask sandwiched between them. The imaging sensor is mounted on a rail that is aligned with the optical axes of the lenses.

where $I(u, w)$ and $I_v(u, w)$ are the desired quantities for estimating the range image using Equation (21).

There are still several mask design issues that need to be resolved. First, our example of Gaussian-based optical masks was somewhat arbitrary. A pair of masks should be designed from a set of optimality constraints based on derivative accuracy, effective baseline, light transmittance, etc. Once an optimal function is determined, the construction of the actual optical masks must be calibrated to include nonlinearities in the printing process (e.g., halftoning), and the effects of the intrinsic point spread function of the camera. In particular, the image of a point light source recorded by the camera with mask $M'(u, v)$ must be equal to the spatial derivative of the image recorded with mask $M(u, v)$. Finally, noise in the image measurements, $I_1(u, w)$ and $I_2(u, w)$, will be amplified by the computations in Equation (28) and Equation (29): small values of β or γ are thus undesirable.

5 Results

We have constructed a preliminary system for computing aperture derivatives and estimating range. The configuration consists of a pair of planar-convex lenses (50mm diameter, 50mm focal length), a Gaussian-based non-negative optical mask (Figure 3) printed onto a transparency from a 600 dpi laser printer, and a standard CCD sensor array (SONY XC-77R). The optical mask was sandwiched between the pair of lenses, placed back-to-back, and mounted along an optical rail in front of the CCD array (Figure 4).

A pair of images $I_1(u, w)$ and $I_2(u, w)$ are acquired by imaging with the masks M_1 and M_2. A linear combination of the images (Equations (28) and (29)) yields the required $I(u, w)$ and $I_v(u, w)$. The range image is computed from a ratio of $I_v(u, w)$ and the spatial derivative, $I_x(u, w)$, of $I(u, w)$ (Equation (21)). Spatial derivatives were computed using optimized 5×5 derivative filters [Simoncelli 94].

Illustrated in Figure 5 are results from a frontal-parallel target 250 mm from the lens, consisting of an intensity step edge. Shown are the pair of intensity images, $I_1(u, w)$ and $I_2(u, w)$, and the images $I_v(u, w)$ and $I_x(u, w)$. Shown also

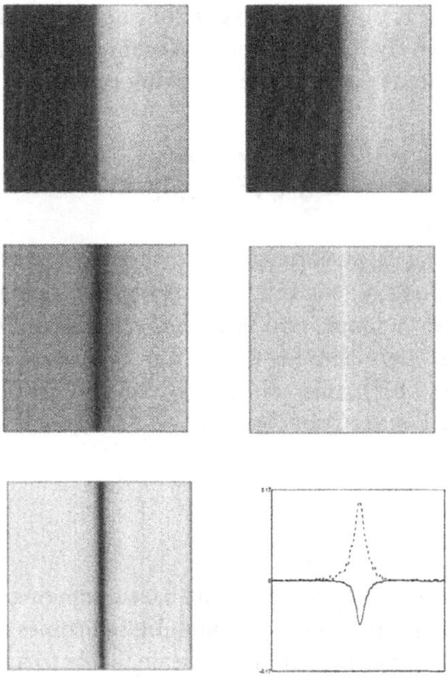

Fig. 5. Top: images obtained through masks $M_1(u, w)$ and $M_2(u, w)$. The scene is a frontal-parallel plane containing an intensity step edge placed 250 mm in front of the focal plane. Middle, left: image $I_v(x, y)$, the derivative with respect to viewing position (computed directly from $M_1(u, w)$ and $M_2(u, w)$). Middle, right: image $I_x(x, y)$, the derivative with respect to image position. Bottom, left: the image $\alpha(x, y)$. Bottom, right: slices through α-map shown at left. Also shown is a slice through another α-map for the same target placed 100mm further from the sensor (behind the focal plane).

is the computed "α-map", where α is monotonically related to the distance to the target. For comparison, the same target was translated by 100mm away from the sensor and the process repeated. Illustrated in Figure 5 are 1-D slices of the computed "α-maps". Notice that since the target moved through the focal plane, the α values are negated, that is, there is no ambiguity between objects equally spaced in front of and behind the focal plane.

6 Range from Aperture Size Derivatives

An interesting variant of the technique arises when considering a Gaussian mask, and its derivative with respect to σ:

$$G(u, w) = \frac{1}{\sigma^2} e^{-(u^2+w^2)/2\sigma^2}, \tag{30}$$

$$\begin{aligned} G_\sigma(u, w) &= \frac{\partial}{\partial \sigma} G(u, w) \\ &= -\frac{2}{\sigma^3} e^{-(u^2+w^2)/2\sigma^2} + \frac{(u^2+w^2)}{\sigma^5} e^{-(u^2+w^2)/2\sigma^2} \end{aligned} \tag{31}$$

Let $I(x,y)$ and $I_\sigma(x,y)$ be the images obtained through the masks $G(u,w)$ and $G_\sigma(u,w)$, respectively. Using the same techniques as in Section 2, it can be shown that these two images obey the following constraint:

$$\begin{aligned} I_\sigma(x,y) &= \alpha^2\sigma\left[I_{xx}(x,y) + I_{yy}(x,y)\right], \\ &= \alpha^2\sigma\nabla^2 I(x,y), \end{aligned} \tag{32}$$

where $I_{xx}(x,y)$ and $I_{yy}(x,y)$ correspond to the horizontal and vertical second partial derivatives of $I(x,y)$, and ∇^2 is the Laplace operator. As before, α is inversely proportional to range, and is given by Equation (6). This formulation provides a differential algorithm for range-from-defocus. Unlike previous formulations (e.g., [Pentland 87]), this solution avoids the artifacts arising from the computation of local Fourier transforms.

7 Discussion

An optical mask placed in front of a lens-based imaging system produces an image which is a superposition of scaled and dilated copies of the mask function. The derivative of this image is related by a scale factor to a second image created with the derivative of the first optical mask. The scale factor is monotonically related to range. This simple observation has lead us to a *direct* differential method for estimating range from a single stationary camera. In particular, the derivative with respect to viewing position is computed directly: it is simply the image formed under the derivative mask.

Two assumptions have been made in our solution to this problem. Both of these assumptions are made (although often not explicitly) in nearly every structure from stereo or motion algorithm. The first assumption is that the light emanating from each point in the scene is constant across the lens (i.e., the brightness constancy assumption). Note that this assumption will typically be violated at occlusion boundaries, because the light emanating from a partially occluded point will hit only a portion of the lens. One potential solution to this problem is to expand the function describing the light emanating from a point in a Taylor series. The coefficients of these terms may be estimated by collecting additional measurements (i.e., images) with higher-order derivative masks.

The second assumption is that of locally frontal-parallel surface orientation. This assumption was necessary in order to solve for α_p given the two image measurements described by Equations (16) and (17). Solving without this assumption is a nonlinear optimization problem (since the α_p appears inside the argument of $M(\cdot)$), which should be amenable to an iterative solution.

The accuracy of our technique has not yet been tested empirically, but we expect it to be comparable to other single-lens techniques (e.g., [Pentland 87, Adelson 92, Jones 93, Dowski 94]). As with most ranging techniques, accuracy behaves according to the rules of triangulation. In particular, errors will be proportional to the square of the range, and inversely proportional to both the focal length and baseline.[5] We have verified these relationships via simple simulations. As in many other range-imaging systems, the accuracy may be improved with the use of structured illumination.

A counterintuitive aspect of our technique is that it relies on the defocus of the image. In particular, a perfectly focused image corresponds to $\alpha = 0$, leading to a singularity in Equation (9). In practice, this may be alleviated by focusing the camera at infinity (i.e., $d = f$), thus ensuring that points at distances within the operating range of the algorithm will be sufficiently blurred.

Finally, since the calculations required for recovering range are simple (convolution followed by arithmetic combination of a pair of images), we believe it will be appropriate for real-time implementation. A computer-controlled liquid-crystal lattice could be used in place of a fixed optical attenuation mask, and could be switched back and forth between the two masks at video frame rates.

References

[Adelson 92] Adelson, E.H. and Wang, J.Y.A. Single Lens Stereo with a Plenoptic Camera. *IEEE Transactions on Pattern Analysis and Machine Intelligence*, 14(2):99–106, 1992.

[Dowski 94] Dowski, E.R. and Cathey, W.T. Single-Lens Single-Image Incoherent Passive-Ranging Systems. *Applied Optics*, 33(29):6762–6773, 1994.

[Horn 86] Horn, B.K.P. *Robot Vision*. MIT Press, Cambridge, MA, 1986.

[Jones 93] Jones, D.G. and Lamb, D.G. *Analyzing the Visual Echo: Passive 3-D Imaging with a Multiple Aperture Camera*. Technical Report CIM-93-3, Department of Electrical Engineering, McGill University, 1993.

[Krotkov 87] Krotkov, E. Focusing. *International Journal of Computer Vision*, 1:223–237, 1987.

[Lucas 81] Lucas, B.D. and Kanade, T. An iterative image registration technique with an application to stereo vision. In *Proceedings of the 7th International Joint Conference on Artificial Intelligence*, pages 674–679, Vancouver, 1981.

[Nayar 95] Nayar, S.K., Watanabe, M., and Noguchi, M. Real-Time Focus Range Sensor. In *Proceedings of the International Conference on Computer Vision*, pages 995–1001, Cambridge, MA, 1995.

[Pentland 87] Pentland, A.P. A New Sense for Depth of Field. *IEEE Transactions on Pattern Analysis and Machine Intelligence*, 9(4):523–531, 1987.

[Simoncelli 94] Simoncelli, E.P. Design of Multi-Dimensional Derivative Filters. In *First International Conference on Image Processing*, 1994.

[Subbarao 88] Subbarao, M. Parallel Depth Recovery by Changing Camera Parameters. In *Proceedings of the International Conference on Computer Vision*, pages 149–155, 1988.

[Xiong 93] Xiong, Y. and Shafer, S. Depth from Focusing and Defocusing. In *Proc. of the DARPA Image Understanding Workshop*, pages 967–976, 1993.

[5] Effective baseline in our system depends on the mask function, and is proportional to the lens diameter.

Motion

Generalised Epipolar Constraints

Kalle Åström[1] and Roberto Cipolla[2] and Peter J. Giblin[3]

[1] Dept of Mathematics, Lund University, Lund, Sweden
[2] Dept of Engineering, Univ. of Cambridge, Cambridge, UK
[3] Dept of Pure Mathematics, Univ. of Liverpool, Liverpool, UK

Abstract. The *frontier* of a curved surface is the envelope of contour generators showing the boundary, at least locally, of the visible region swept out under viewer motion. In general, the outlines of curved surfaces (apparent contours) from different viewpoints are generated by different contour generators on the surface and hence do not provide a constraint on viewer motion. Frontier points, however, have projections which correspond to a real point on the surface and can be used to constrain viewer motion by the epipolar constraint.
We show how to recover viewer motion from frontier points and generalise the ordinary epipolar constraint to deal with points, curves and apparent contours of surfaces. This is done for both continuous and discrete motion, known or unknown orientation, calibrated and uncalibrated, perspective, weak perspective and orthographic cameras. Results of an iterative scheme to recover the epipolar line structure from real image sequences using only the outlines of curved surfaces, is presented. A statistical evaluation is performed to estimate the stability of the solution. It is also shown how the full motion of the camera from a sequence of images can be obtained from the relative motion between image pairs.

1 Introduction

Structure and motion from the images of point features has attracted considerable attention and a large number of algorithms exist to recover both the spatial configuration of the points and the motion compatible with the views. Structure and motion from the outlines of curved surfaces, on the other hand, has been thought to be more difficult because of the *aperture problem*, i.e. it is not possible to get the correspondence of points between two images of the same curve.

For a smooth arbitrary curved surface an important image feature is the outline or apparent contour. This is the projection of the locus of points on the surface which separates the visible from the occluded parts (Fig. 1.A). Under perspective projection this locus – the critical set or contour generator, Σ – can be constructed as the set of points on the surface where rays through the projection centre $\mathbf{c}$ are tangent to the surface. Each viewpoint will generate a different contour generator with the contour generators 'slipping' over the visible surface under viewer motion (Fig. 1.B).

Under *known* viewer motion, the deformation of apparent contours can be used to recover the surface geometry (structure) [8, 5, 14]. This requires a spatio–

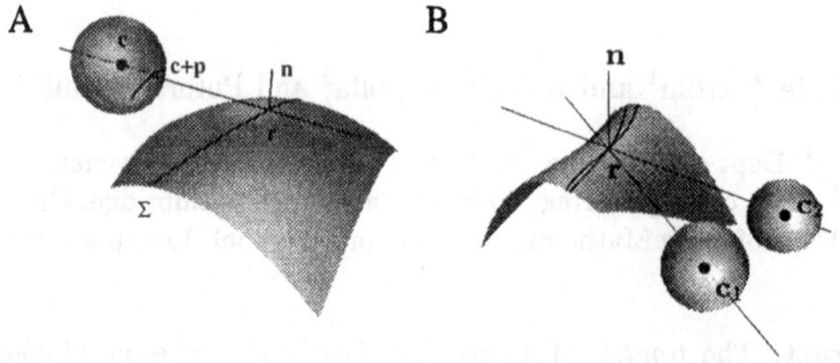

Fig. 1. Perspective projection (A): the contour generator Σ with a typical point $\mathbf{r}$, the image sphere with centre $\mathbf{c}$ and the corresponding apparent contour point $\mathbf{c}+\mathbf{p}$. Thus $\mathbf{p}$ is the unit vector joining the centre $\mathbf{c}$ to the apparent contour point. Degenerate case of epipolar parameterisation (B): The epipolar plane is a tangent plane of the surface at a frontier point. At a frontier point the contour generators from consecutive viewpoints intersect.

temporal parametrisation of the image-curve motion. The *epipolar* parametrisation is most naturally matched to the recovery of surface curvature, cf. [5].

In this paper we address the problem of recovering the viewer motion from the deformation of apparent contours. We show how frontier points can be detected in image sequences and used to recover viewer egomotion. The special case of frontier points under orthographic projection and object rotation about a single axis was considered by Rieger [13] and Giblin et al [7]. Porrill and Pollard [12], although primarily concerned with stereo calibration from 3D space curves, noted that the intersection of the two contour generators from two discrete viewpoints generated a real point visible in both images which could also be used to generate an epipolar constraint. This constraint was exploited by Carlsson [3] in the analysis of the visual motion of space-curves.

We analyse both the continuous (infinitesimal) and discrete viewer motion cases as well as considering calibrated and uncalibrated cameras, orthographic and perspective projection. We present preliminary experimental results obtained from real image sequences of curved surfaces from unknown viewpoints. An iterative technique is implemented which recovers the maximum likelihood estimate of the epipolar line structure (*fundamental and essential matrices*) from the image motion of frontier points. A statistical evaluation is also performed to estimate the stability of the solution. The work is a continuation of [4].

2 Generalised Epipolar Constraints

Consider a smooth surface M (closed, for example the boundary of a 3-dimensional object) which is viewed from a curve $\mathbf{c}(t)$ of camera centres. Each camera centre $\mathbf{c}(t)$ gives rise to a contour generator $\Gamma(t)$ consisting of points $\mathbf{r}$ in M

such that the viewline is perpendicular to the surface normal: $(\mathbf{r}-\mathbf{c}).\mathbf{n}=0$. As t changes the contour generator slides over the surface.

Now consider the camera center at two time instants $\mathbf{c}_1 = \mathbf{c}(t_1)$ and $\mathbf{c}_2 = \mathbf{c}(t_2)$, and consider all planes that go through these two camera centers which are tangent to the surface M. This pencil of planes will be called the pencil of *epipolar tangency planes* with respect to the camera motion. The image of the epipolar tangency planes is a star of lines through a point, the so called *epipole*, each line being tangent to an apparent contour. Since both pencils are the image of the same pencil of planes they must be related. Denote by $\mathbf{p}_1(s)$ the apparent contour in image 1 with curve parameter s and by $\mathbf{p}_2(s)$ corresponding apparent contour in image 2. The point $\mathbf{r}$ where the epipolar tangency point is tangent to the surface M must be on both contour generators as is illustrated in Fig. 1.B. The normal $\mathbf{n}$ to the surface at this point is orthogonal not only to $\mathbf{p}_1$ and $\mathbf{p}_2$ but also to their image tangents $(\mathbf{p}_1)'_s$ and $(\mathbf{p}_2)'_s$ and to the direction of motion $\Delta\mathbf{c} = \mathbf{c}_2 - \mathbf{c}_1$. This can be written in a compact way as the *generalised epipolar constraints*,

$$\operatorname{rank}\left[\Delta\mathbf{c}\ \mathbf{p}_1(s_1)\ (\mathbf{p}_1)'_s(s_1)\ \mathbf{p}_2(s_2)\ (\mathbf{p}_2)'_s(s_2)\right] = 2\ , \tag{1}$$

i.e. the five column vectors in the above matrix must lie in a plane perpendicular to the normal. Notice that the rank constraints involve both curve parameters s_1 and s_2 and motion parameters $\Delta\mathbf{c}$. This can be used in several ways. Firstly, once the direction of motion $\Delta\mathbf{c}$ is known or guessed the first image of the epipolar tangency point can be found by searching for the curve parameter s_1 so that $\det\left[\Delta\mathbf{c}\ \mathbf{p}_1(s_1)\ (\mathbf{p}_1)'_s(s_1)\right] = 0$, and similarly for the second image. Secondly, when the image of the epipolar tangency points has been found it can be checked whether or not $\det\left[\Delta\mathbf{c}\ \mathbf{p}_1\ \mathbf{p}_2\right] = 0$.

The above formulation of the generalised epipolar constraints depends on the specific choice of both object and image coordinate system. The motion parameters can therefore only be determined up to an unknown choice of coordinate system. The above formulation is convenient since it makes the generalisation to other camera models easy. As an example the infinitesimal version can be obtained as the limit when the time difference approaches zero.

$$\operatorname{rank}\left[\Delta\mathbf{c}\ \mathbf{p}_1\ (\mathbf{p}_1)_s\ \mathbf{p}_2\ (\mathbf{p}_2)_s\right] = \operatorname{rank}\left[\Delta\mathbf{c}/\Delta t\ \ (\mathbf{p}_1)_s\ (\mathbf{p}_2-\mathbf{p}_1)/\Delta t\ (\mathbf{p}_2)_s\right] \to$$
$$\to \operatorname{rank}\left[\mathbf{c}_t\ \mathbf{p}\ (\mathbf{p})_s\ \mathbf{p}_t\ (\mathbf{p})_s\right] = 2, \qquad \text{as } t \to 0\ .$$

Note that $\mathbf{p}$ is the orientation of the ray in the fixed reference/world frame for $\mathbb{R}^3$. For a *calibrated* camera it is determined by a spherical image position vector $\mathbf{q}$ (the orientation of the ray in a coordinate system attached to the viewer or camera) and the orientation of the camera co-ordinate system relative to the reference frame. For a moving observer the viewer coordinate system is continuously moving with respect to the reference frame. The relationship between $\mathbf{p}$ and $\mathbf{q}$ can be conveniently expressed in terms of a rotation operator, $\mathbf{R}(t)$, such that $\mathbf{p} = \mathbf{R}(t)\mathbf{q}$. The measurements in an *uncalibrated* camera, $\mathbf{w}$, are related to the spherical image position, $\mathbf{q}$, by an intrinsic calibration matrix (affine transformation), $\mathbf{A}$, such that $\mathbf{q} = \mathbf{A}(t)\mathbf{w}$. For simplicity the relationship between $\mathbf{w}$

and $\mathbf{p}$ will be expressed by a single matrix $\mathbf{S}$ representing both intrinsic calibration and orientation of the camera, i.e. $\mathbf{p} = \mathbf{R}(t)\mathbf{A}(t)\mathbf{w} = \mathbf{S}(t)\mathbf{w}$. A thorough treatment of the other cases can be found in [1]. The results are summarised in the following two tables. The direction of motion is represented with a three vector, $\Delta\mathbf{c}$, in the central projection cases and $\Delta\mathbf{k} = (\cos(\theta), \sin(\theta), 0)^T$ in the parallel projection case. In the tables $\Delta\mathbf{R}$ is a three by three rotation matrix, $\Delta\mathbf{S}$ is a general non-singular matrix, $\Delta\mathbf{B}$ is a matrix representing planar euclidean transformations and $\Delta\mathbf{C}$ is a matrix representing planar similarity transformations. The matrix $\mathbf{R}_t$ is the derivative of a rotation matrix and similarly for the other types.

Camera Model	Motion Params	Obs. d o f	Generalised Epipolar Constraints
Pure transl.	$\Delta\mathbf{c}$	2	rank $\left[\Delta\mathbf{c}\ \mathbf{p}_1\ (\mathbf{p}_1)_s\ \mathbf{p}_2\ (\mathbf{p}_2)_s\right] = 2$
Calibrated	$\Delta\mathbf{c}, \Delta\mathbf{R}$	5	rank $\left[\Delta\mathbf{c}\ \mathbf{q}_1\ (\mathbf{q}_1)_s\ \Delta\mathbf{R}\mathbf{q}_2\ \Delta\mathbf{R}(\mathbf{q}_2)_s\right] = 2$
Uncalibrated	$\Delta\mathbf{c}, \Delta\mathbf{S}$	7	rank $\left[\Delta\mathbf{c}\ \mathbf{w}_1\ (\mathbf{w}_1)_s\ \Delta\mathbf{S}\mathbf{w}_2\ \Delta\mathbf{S}(\mathbf{w}_2)_s\right] = 2$
Orthographic	$\Delta\mathbf{k}, \Delta\mathbf{B}$	3	rank $\left[\Delta\mathbf{k}\ \mathbf{m}\ (\mathbf{w}_1)_s\ (\Delta\mathbf{B}\mathbf{w}_2 - \mathbf{w}_1)\ \Delta\mathbf{B}(\mathbf{w}_2)_s\right] = 2$
Weak Persp.	$\Delta\mathbf{k}, \Delta\mathbf{C}$	4	rank $\left[\Delta\mathbf{k}\ \mathbf{m}\ (\mathbf{w}_1)_s\ (\Delta\mathbf{C}\mathbf{w}_2 - \mathbf{w}_1)\ \Delta\mathbf{C}(\mathbf{w}_2)_s\right] = 2$

Summary of relevant motion parameters, number of observable degrees of freedom and generalised epipolar constraints in the discrete case.

Camera model	Motion Params	Obs. d o f	Generalised Epipolar Constraints
Pure transl.	$\mathbf{c}_t$	2	rank $\left[\mathbf{c}_t\ \mathbf{p}\ \mathbf{p}_s\ \mathbf{p}_t\right] = 2$
Calibrated	$\mathbf{c}_t, \mathbf{R}_t$	5	rank $\left[\mathbf{c}_t\ \mathbf{q}\ \mathbf{q}_s\ \mathbf{R}_t\mathbf{q} + \mathbf{q}_t\right] = 2$
Uncalibrated	$\mathbf{c}_t, \mathbf{S}_t$	7	rank $\left[\mathbf{c}_t\ \mathbf{w}\ \mathbf{w}_s\ \mathbf{S}_t\mathbf{w} + \mathbf{w}_t\right] = 2$
Orthographic	$\mathbf{k}_t, \mathbf{B}_t$	3	rank $\left[\mathbf{k}_t\ \mathbf{m}\ \mathbf{w}_s\ \mathbf{B}_t\mathbf{w} + \mathbf{w}_t\right] = 2$
Weak Persp.	$\mathbf{k}_t, \mathbf{C}_t$	4	rank $\left[\mathbf{k}_t\ \mathbf{m}\ \mathbf{w}_s\ \mathbf{C}_t\mathbf{w} + \mathbf{w}_t\right] = 2$

Summary of relevant motion parameters, number of observable degrees of freedom and generalised epipolar constraints in the infinitesimal case.

3 Implementation

3.1 Extraction and Tracking of Apparent Contours

An important aspect in calculating motion from the deformation of apparent contours is the actual extraction and tracking of the contours. This can be achieved with B-spline snakes. In our implementation, the snake at time t_1 is used as a template to find corresponding contour in the image at the next time instant t_2. At first the snake is only allowed to move rigidly. This ensures a fast, robust, but rough positioning of the snake in the new image. The snake is then allowed to deform to match the new image. This procedure is explained in more detail in [6]. In the last steps of snake deformation we have used subpixel edge detectors that not only gives us the location of the contour but also a confidence interval in the normal direction of the curve. This is done with a new technique described in [2]. For clear well defined edges, the individual edge positions can be found with a standard deviation of roughly 1/5:th of a pixel in the normal direction. This uncertainty measure is important in estimating motion parameters later. Different frontier points are weighted according to the uncertainty in their positioning.

As a by-product of the snake type tracking, a rough guess of point correspondences are obtained. These approximate point correspondences can be used to calculate an initial estimate of motion parameters as described in the next section.

3.2 Initial Hypothesis of Motion

There are a number of different ways to obtain the initial hypothesis of the motion parameters, which are needed in order to use the generalised epipolar constraints.

Point matches can be used to estimate motion parameters using the linear eight point method [10] or non-linear methods [11].

Motion sensors: In some situations, partial knowledge of the motion can be obtained by other means, e.g. from motion sensors.

Prediction: If viewer motion is smooth it might be possible to predict motion parameters from estimates of motion history.

Simpler camera models: Approximate motion parameters can also be obtained using approximate camera models, e.g. the weak perspective model.

3.3 Maximum Likelihood Estimation

The maximum likelihood method is a natural way to estimate the motion parameters given noisy input data. It has several advantages and is relatively easy to apply. The main idea is the following. A residual, $\alpha_i = \alpha_i(m)$, as a function of motion parameters, m, is chosen. The joint distribution, $f_n(\alpha|m)$, of the residuals given motion parameters, m, is calculated. The maximum likelihood estimate, $\hat{m}$, is the motion parameter which maximises the likelihood function $L(m) = f_n(\alpha|m)$. To make this optimisation simple it is often assumed that the residuals, α_i, are independent and of Gaussian distribution with zero mean and standard deviation σ_i. This is a reasonable assumption if the images of the frontier points are not too close to each other in the image. The likelihood function is then

$$L = \prod \frac{1}{\sqrt{2\pi\sigma_i^2}} e^{-\alpha_i^2/2\sigma_i^2} \ .$$

By taking the negative logarithm maximising the likelihood is approximately the same as minimising

$$g = \sum \frac{\alpha_i^2}{\sigma_i^2} \ .$$

The estimate is the motion parameters that minimise this weighted sum of squared residuals. In our implementations minimisation is performed using either a modified Newton-Raphson method or Gauss-Newton method. For more details see [1].

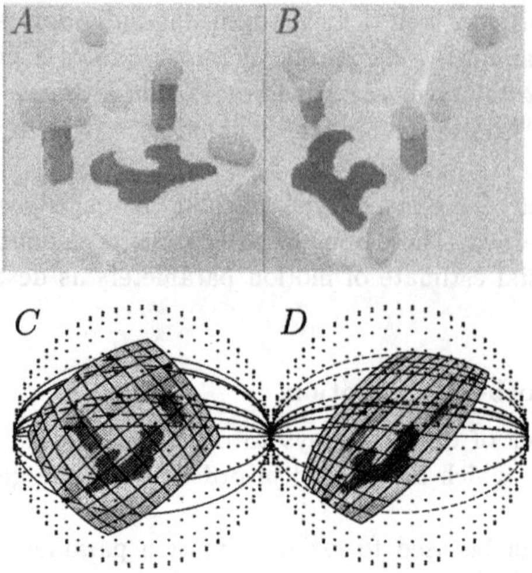

Fig. 2. Rectification of uncalibrated images. Two images (A-B) are projected onto the viewing sphere and rectified using the motion parameters (C-D). After rectification The epipolar tangency planes all intersect at the x-axis. The two sets of epipolar tangency planes should be equal. The angular difference is used as the residual.

Residuals in the discrete time, perspective projection case: After a standard rectification to parallel geometry the set of epipolar tangency planes in both images should be identical. The epipolar tangency points and the epipolar tangency planes are found using the epipolar tangency constraints

$$|\Delta \mathbf{c}\ \mathbf{w}_1\ (\mathbf{w}_1)_s| = 0\ , \tag{2}$$

$$|\Delta \mathbf{c}\ \Delta \mathbf{S} \mathbf{w}_2\ \Delta \mathbf{S}(\mathbf{w}_2)_s| = 0\ . \tag{3}$$

The angular difference, α_i, cf. Fig. 3, between the two representations of the same epipolar tangency planes, after rectification, is calculated as well as the standard deviation σ_i of this residual.

Residuals in the discrete time, parallel projection case: The orthographic and weak perspective cases are similar. After rectification the two sets of parallel epipolar tangency planes should be identical. The distance α_i, between the parallel epipolar tangency planes, is used as the residual. The residual is scaled with respect to its standard deviation σ_i. The residual variance, due to edge localisation error, is changed in these transformations. These effects are taken into account.

Residuals in the infinitesimal case: In the infinitesimal case, the direction of viewer motion $\mathbf{c}_t$ is used as an infinitesimal epipole or the focus of expansion.

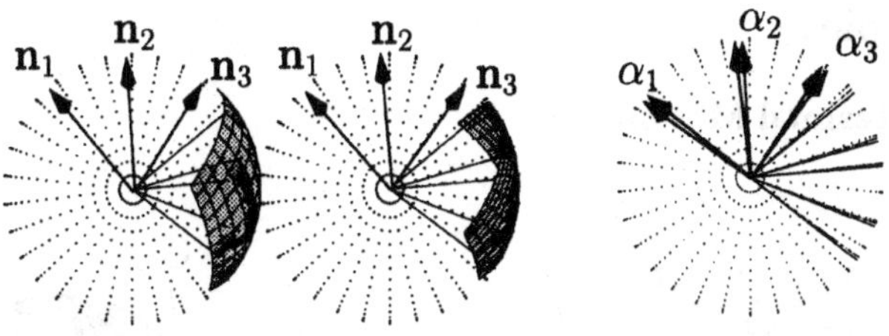

Fig. 3. Two sets of epipolar tangency planes are calculated from two images. These two sets should ideally be identical. The residual is defined as the angular difference α_i between corresponding epipolar tangency planes after rectification.

The tangency constraint is then used to find the epipolar tangency planes and corresponding frontier points. For example, in the calibrated case we have

$$\left|\mathbf{c}_t\ \mathbf{q}\ \mathbf{q}_s\right| = 0 \ .$$

Each plane defines a normal $\mathbf{n} = \mathbf{q} \times \mathbf{c}_t$. The motion constraint is then simply $\mathbf{n} \cdot (\mathbf{R}_t\mathbf{q} + \mathbf{q}_t) = 0$. It is reasonable to use

$$\alpha = \mathbf{n} \cdot (\mathbf{R}_t\mathbf{q} + \mathbf{q}_t)$$

as the residual. Errors in α are mostly due to the errors in $\mathbf{q}_t$, therefore the standard deviation, $\sigma[\alpha]$, can be approximated by $\sigma[\mathbf{n}_i \cdot \mathbf{q}_t]$, i.e. the standard deviation in estimated normal velocity. These standard deviations are obtained from the sub-pixel edge detector routines. The standard deviation is approximately constant around each frontier point, so g is in fact quadratic in $\mathbf{R}_t$ so that the minimum with respect to $\mathbf{R}_t$ can be found with linear methods. The same argument applies to the uncalibrated and parallel projection cases.

3.4 Statistical Evaluation

The maximum likelihood estimate has several good properties. One is that it is guaranteed to be asymptotically unbiased.

The residuals at the minimum can be used to estimate empirically the magnitude of edge localisation error. This can then be compared to the ones obtained from the edge detectors. The residuals can thus be used to automatically verify whether a reasonably low minimum has been found. Thus it may be possible to get out of local minima and also to remove outliers.

The second derivative matrix of g together with the variance of the scaled residuals gives us an estimate of the covariance of the estimated motion parameters, cf. [1].

4 Examples

4.1 Infinitesimal Motion

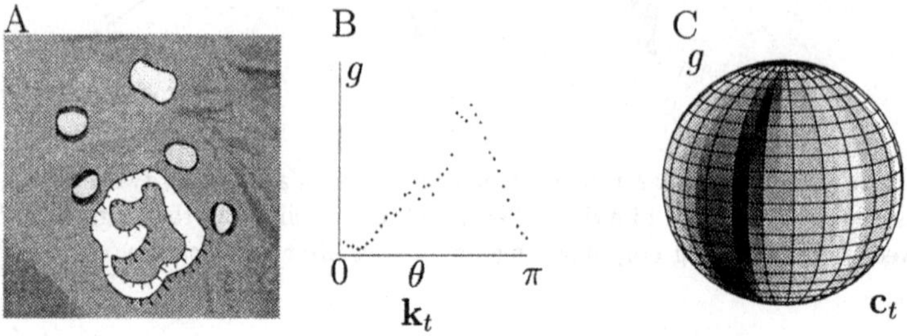

Fig. 4. Infinitesimal motion with a weak perspective camera. One image with apparent contours and the estimated normal velocity is used as input (A). The minimal error function g as a function of the tessalated line at infinity using the weak perspective camera model (B) and as a function of the sphere of directions using the uncalibrated camera model (C). Low values are dark in the figure.

One image, taken from a longer sequence, of the same scene is shown in Fig. 4.A. The deformation between this and the next image is used to estimate normal velocity. This is illustrated with small line segments.

For each of forty choices of focus of expansion $\mathbf{k}_t = [\cos(\theta), \sin(\theta), 0]^T$ with $\theta = 0, \pi/40, \ldots, 39\pi/40$, the epipolar tangency points $\mathbf{w}$ were calculated using the epipolar tangency constraint $|\mathbf{k}_t\ \mathbf{m}\ \mathbf{w}_s| = 0$. The normal velocity and its standard deviation was calculated at these points. For each $\mathbf{k}_t$ the weighted squared residual $g = \sum \alpha_i^2/\sigma_i^2$ with $\alpha = \mathbf{n} \cdot (\mathbf{B}_t\mathbf{w} + \mathbf{w}_t)$ was minimised with respect to $\mathbf{B}_t$. The minimal residual as a function of θ is shown in Fig. 4.B. Thus a rough estimate of the motion is obtained by finding the minimal g of these forty directions.

The same input was also used to illustrate the uncalibrated camera case. The same idea of tessellating the focus of expansion can be used. In the perspective projection case this means tessellating the sphere of directions. As in the previous example, for each choice of focus of expansion or direction of motion, $\mathbf{c}_t$, it is straightforward to find the epipolar tangency points. Minimising the weighted sum of squared residuals g with respect to $\mathbf{S}_t$ is a linear problem. The minimal residual as a function of $\mathbf{c}_t \in \mathbb{S}^2$ is shown in Fig. 4.C. Notice that the low values of, g, form a long dark valley on the sphere. We expect the direction of motion to be poorly located along that valley. This is confirmed by the statistical evaluation. Notice also that choosing the weak perspective model corresponds to searching this sphere along the equator only.

The minimum obtained from tessellating the sphere or the minimum obtained from the weak perspective case above can both be used as initial estimates in a

Gauss-Newton search of the minimum. This was done and about ten iterations were needed to find the minimum.

4.2 Discrete Motion, Known Rotation

Known rotation can easily be illustrated by incorporating and using a planar curve in the image. The image of a planar curve is deformed with a planar projective transformation. By detecting and aligning a planar feature in a sequence of images, *the sequence can be regarded as one of a purely translating camera.* The planar curve is then regarded as a curve on the plane at infinity and thus it has no apparent image motion. This idea has been used in [9] and is known as *projective reduction*, a generalisation of the *plane plus parallax method.*

This is illustrated in Fig. 6. Only the direction of motion $\Delta \mathbf{c}$ needs to be estimated. The sphere of possible directions can then be tessellated and the error function g can be calculated for each direction. The minimum obtained after tessellation is improved by local Newton-Raphson search (6 iterations were needed).

4.3 Discrete Motion, Uncalibrated Camera

The discrete motion case with uncalibrated camera is illustrated with a pair of images from the same sequence that was used to illustrate the infinitesimal case. The result from the infinitesimal case can be used as an initial estimates of the discrete motion parameters. A standard extrapolation is used:

$$\Delta \mathbf{S} = e^{\mathbf{S}_t \Delta t}, \qquad \Delta \mathbf{c} = \mathbf{c}_t \Delta t$$

This initial estimate is used as input in a gauss-newton search. At each iteration we use the current motion parameters to calculate the epipoles and the epipolar tangencies. Fig. 5.A and B illustrate the epipolar tangencies obtained at iteration 1. The two sets of epipolar tangency planes are then rectified using the current motion parameters, cf. Fig. 2. After rectification corresponding epipolar tangency planes should coincide. The difference, measured as the angle α is calculated. The estimated standard deviation σ of this angle due to edge localisation errors is also calculated and the weighted residual $Y = \alpha/\sigma$ is formed. The derivative of Y with respect to infinitesimal changes in motion parameters is then calculated. The weighted residual Y and its derivatives are then used to adjust the motion parameters according to the Gauss-Newton method, cf. [1]. Fig. 5 illustrates the epipolar tangencies and the rectified epipolar tangency planes of iteration 1, 4 and 10 of such a minimisation. Typically one needs a couple of iterations (6 in this example) to get close to the minima. During these iteration the error function $g = \sum Y^2$ decreases rapidly. Then a few more iterations are needed to localise the minima within machine accuracy. During these last iteration the norm of the gradient decreases rapidly, while the error function stays almost constant. This is illustrated in the table below, in which the error function g and the logarithm of the norm of the gradient of g is shown for the 10 iterations.

Iteration	1	2	3	4	5	6	7	8	9	10
g	110.5	66.54	34.24	16.45	7.695	6.482	6.476	6.476	6.476	6.476
$\log_{10}(\lvert\nabla g\rvert)$	-0.99	-1.11	-1.28	-1.48	-2.01	-3.63	-5.53	-7.62	-9.70	-9.70

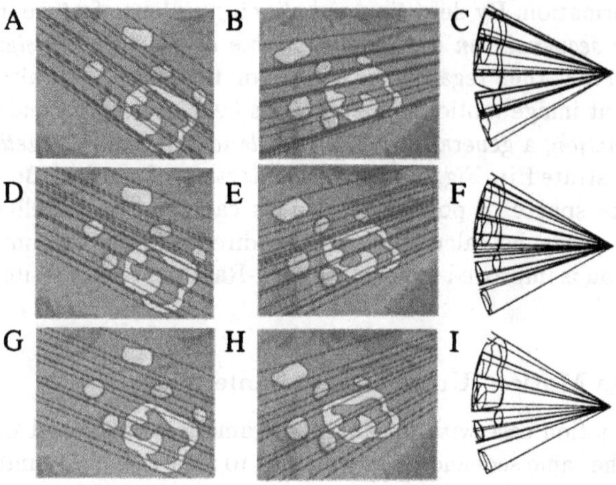

Fig. 5. Finding the uncalibrated motion parameters using the generalised epipolar constraints. Optimisation of the likelihood function. Iteration number 1 (first row), 4 (second row) and 10 (Third row). First and second columns illustrate the epipolar tangencies in the first and second image. The third column illustrate the rectified epipolar tangency planes, projected on the view-sphere and viewed along the direction of motion..

5 Extension to Image Sequences

The implementation briefly described above allows us to calculate motion parameters between pairs of images. A natural extension is to use the estimates of motion parameters for pairs of images in a sequence of images at times $(t_0, \ldots, t_n)$, to obtain the full motion of the camera. In the calibrated case the full motion of the camera as represented by camera positions $\mathbf{c}_i = \mathbf{c}(t_i)$ and camera orientation $\mathbf{R}_i = \mathbf{R}(t_i)$ must fulfill the following equations:

$$\mu_{ij}\Delta\mathbf{c}_{ij} = \mathbf{R}_i^{-1}(\mathbf{c}_j - \mathbf{c}_i)$$

$$\Delta\mathbf{R}_{ij} = \mathbf{R}_i^{-1}\mathbf{R}_j \ .$$

where $(\Delta\mathbf{c}_{ij}, \Delta\mathbf{R}_{ij})$ are the motion parameters from image i to image j. The overall coordinate system must be chosen, e.g. by choosing $\mathbf{c}_0 = 0$, $\mathbf{R}_0 = I$ and

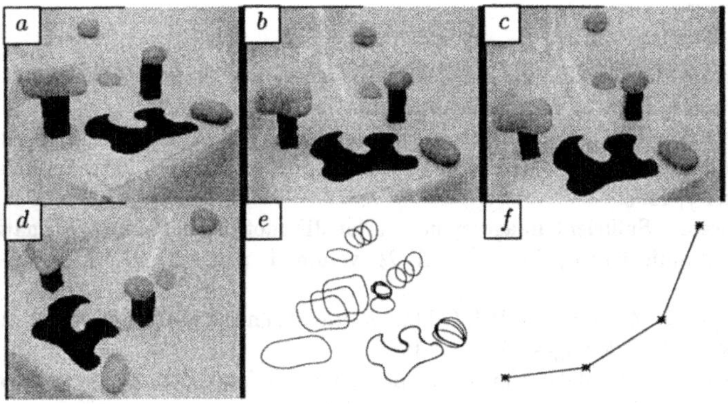

Fig. 6. The case of discrete motion with uncalibrated camera. Four images out of a longer sequence is shown in (A-D). By detecting and aligning the image of a planar feature the images can be thought of as coming from a purely translating camera. The apparent contours after alignement are shown in (E). This makes it relatively easy to extract motion parameter between each pair of images. These parameters can then be used to calculate the full motion of the camera (F).

$|\mathbf{c}_n| = 1$. The unknown scale factor μ_{ij} has to be found since it is only possible to determine $\Delta\mathbf{c}_{ij}$ up to an unknown scale factor. Similar equations apply to the other camera models. The idea is illustrated in Fig. 6. Four images of a short sequence is shown (A-D) and the camera motion are represented as the corresponding four camera positions (F).

6 Conclusions and Future Work

The apparent contour and its deformation under viewer motion is known to be a rich source of surface geometric information which can be used in visual navigation and object manipulation. Here we have shown how so called frontier points of apparent contours can be used to recover the viewer motion from the deformation of apparent contours. The epipolar constraint for points is generalised to points, curves and apparent contours, to both the continuous and discrete motion cases, to uncalibrated and calibrated cameras and to perspective and parallel camera models. An iterative method to obtain the maximum likelihood estimate of the motion parameters is presented and the problem of obtaining initial estimates is discussed. Statistical evaluation of the results are presented. These can be used to evaluate the validity of the solution but also to obtain estimates of the covariance of the estimated motion parameters. The theory is applied to real image sequences. It is also shown how motion between image pairs can be used to obtain full camera motion.

References

1. K. Åström, R. Cipolla, and P. J. Giblin. Generalised epipolar constraints. Technical report, Dept. of Mathematics, Lund University, 1996.
2. K. Åström and A. Heyden. Stochastic modelling and analysis of image acquisition and sub-pixel edge detection. Technical report, Dept. of Mathematics, Lund University, 1995.
3. S. Carlsson. Sufficient image structure for 3D motion and shape estimation. In J-O. Eklundh, editor, *Proc. ECCV'94*, volume I, pages 83–91. Springer–Verlag, 1994.
4. R. Cipolla, K. Åström, and P.J. Giblin. Motion from the frontier of curved surfaces. In *Proc. ICCV'95*, pages 269–275, 1995.
5. R. Cipolla and A. Blake. Surface shape from the deformation of apparent contours. *Int. Journal of Computer Vision*, 9(2):83–112, 1992.
6. R. Curwen and A. Blake. Dynamic contours: real-time active splines. In A. Blake and A. Yuille, editors, *Active Vision*. MIT Press, 1992.
7. P.J. Giblin, F.E. Pollick, and J.E. Rycroft. Recovery of an unknown axis or rotation from the profiles of a rotating surface. *J. Opt. Soc. America*, A11:1976–1984, 1994.
8. P.J. Giblin and R. Weiss. Reconstruction of surfaces from profiles. In *Proc. ICCV'87*, pages 136–144, London, 1987.
9. A. Heyden and K. Åström. A canonical framework for sequences of images. In *Proc. IEEE workshop on Representation of Visual Scenes, MIT, USA*, 1995.
10. H.C. Longuet-Higgins. A computer algorithm for reconstructing a scene from two projections. *Nature*, 293:133–135, 1981.
11. Q-T. Luong, R. Deriche, O. Faugeras, and T. Papadopoulo. On determining the fundamental matrix: analysis of different methods and experimental results. In *Israelian Conf. on Artificial Intelligence and Computer Vision*, Tel-Aviv, Israel, 1993. A longer version is INRIA Tech Report RR-1894.
12. J. Porrill and S.B. Pollard. Curve matching and stereo calibration. *Image and Vision Computing*, 9(1):45–50, 1991.
13. J.H. Rieger. Three dimensional motion from fixed points of a deforming profile curve. *Optics Letters*, 11:123–125, 1986.
14. R. Vaillant and O.D. Faugeras. Using extremal boundaries for 3D object modelling. *IEEE Trans. Pattern Analysis and Machine Intell.*, 14(2):157–173, 1992.

Object Models From Contour Sequences

Edmond Boyer

CRIN–CNRS / INRIA Lorraine
54506 Vandœuvre les Nancy Cedex, France.
e-mail: boyere@loria.fr

Abstract. We address the problem of building 3D object models from image sequences obtained with known camera motion. An approach based on a local reconstruction method is presented. Recovered surfaces are described as polygonal meshes. To this purpose, reconstructed points are triangulated and surface areas which are not covered by rims are detected since they may lead to false reconstructed points. Resulting meshes are then regularised in order to correct noise perturbations which affect the reconstruction. Experimental results on real data are presented.

1 Introduction

Recovering and representing three-dimensional object shape is an important task in computer vision. The derived models can be used for recognition, localisation and design automation. In the case of curved objects, rich and robust information on the shape are provided by image contours which are called *occluding contours*. The corresponding contours on the surface, the *rims*, are viewpoint dependent and defined by the fact that viewing directions at their points are tangential to the surface. In addition, it has been shown that local shape recovery from three or more occluding contours is possible given a known camera motion. Several algorithms [Cip 90, Vai 92, Sze 93] allow such a local reconstruction under the assumption of a linear camera motion. In previous works [Boy 95], we proposed an explicit solution for rim point reconstruction which is correct for any camera motion.

The approaches mentioned above are concerned with local shape estimation. However, a complete surface description is needed to build an object model. Seales and Faugeras [Sea 95] generate a polygonal surface mesh using a spline-based slicing technique. The input is a set of surface points which are recovered using the local approach of Vaillant and Faugeras [Vai 92]. Zhao and Mohr [Zha 94] attempt to recover the global surface description in a single stage by use of B-spline patches. This approach introduces a direct regularisation of the reconstructed surface, but it requires a complete *a priori* parametrisation of the surface which is usually not available. Zheng [Zhe 94] presents a global method in the case of plane camera rotations. In this work, it is shown that regions of the surface unexposed to contours are related to non-smoothness of contour distribution in the image. A detection algorithm based on this fact is proposed for plane camera rotations.

In this paper, we present a global approach for shape estimation which extends previous results on surface reconstruction from occluding contours [Boy 95] and yields robust surface model. First, we study the case of surface regions where local approximations of the reconstruction method are not valid. This the case, for example, at planar, concave parts of the surface or at surface discontinuities. This corresponds to surface areas which are unexposed to rims and where reconstruction yields 3D points which are not on the surface. We present an algorithm to detect them, the interest is to remove such *false* points from the final surface mesh. Our approach extends the result of Zheng [Zhe 94] for planar camera rotations to any camera motion.

Secondly, we propose a surface description based on a triangulation of the recovered rim points. Such a description preserves the coherence with data since no approximation functions which introduce a bias are needed to describe the surface, and it does not require any *a priori* knowledge on the surface parametrisation.

We then present an original method to regularise the reconstructed surface. The idea is to optimise point positions by minimising an energy function which takes into account the surface area.

Finally, significant results on a real object are shown which prove the reliability of the method.

2 Preliminaries

In this section, we summarise some elementary notions to be used along this paper.

We assume that the imaging system is based on the pinhole model (i.e. perspective projection). Therefore, the vector position of a point P on the object surface can be written

$$r = C + \lambda T, \tag{1}$$

where C is the camera centre position, T the unit viewing direction and λ the depth of the point P along the viewing direction. For a given camera position there is a locus of points on the surface S where the normal N is perpendicular to the viewing direction. This set of points is called the *rim* and its projection onto the image plane is called the *occluding contour.*

An essential property of an occluding contour is that the normal to the surface on the corresponding rim can be computed from image information T and τ:

$$N = \frac{T \wedge \tau}{|T \wedge \tau|}, \tag{2}$$

where τ is the tangent to the occluding contour. This property leads to the following expression for the depth at a rim point [Cip 90]:

$$\lambda = \frac{C_t.N}{T_t.N}, \quad (3)$$

where t denotes time (suffix denoting derivative). This formula is defined for continuous observations of the object surface. In the case of discrete information, depth can be computed only by approximation using successive contours. This can be done by using at least three occluding contours and locally approximating the surface up to order two. Such an approximation leads to a linear estimation of rim point depth. For further information about the local reconstruction method, the reader is referred to [Boy 95].

3 Detection of regions unexposed to contours

In the process of reconstruction from contours, parts of the observed surface [1] may not be covered by rims, depending on the local form of the object surface and the camera motion. When the camera turns around such a region, rims *jump* from one extremity to the other and a non-smoothness appears in the spatio-temporal surface. These unexposed regions correspond to part of the observed surface which are concave, parabolic, planar [Zhe 94] or self-occluded and they can not be recovered using only contour information. Moreover, the local model used to compute surface point positions is not valid in these areas. Therefore, they should be detected in order to build a surface model. In this section, we propose a method to detect them.

3.1 Continuous observations

In the case of continuous observations, we will say that a surface point P is exposed to contours if there exists a camera centre position $C(t)$ in the sequence for which the viewing line at P does not pierce the observed surface but is tangent to it.

A particular case corresponds to points which belong to exposed parts of the surface and where the viewing line is tangent to the surface at more than one point (including itself). Such *critical* points constitute the limit of surface areas exposed to contours.

From a quantitative point of view, we can see that at critical points P, the viewing direction T and the normal to the surface N are continuous functions[2] of t but not the depth λ; indeed the observed surface point jump from the first

[1] From now on, we will suppose that the term *observed surface* means the union of all object surfaces which are present in the scene.

[2] t usually refers to the time, however, a more rigourous definition is that t parametrises the camera motion

point of tangency P^- to the second P^+. If we denote by λ^- and λ^+ depths at P^- and P^+, and by using the depth formula 3 we have:

$$\lambda^- = \frac{C_t{}^-.N}{T_t{}^-.N}, \quad \lambda^+ = \frac{C_t{}^+.N}{T_t{}^+.N},$$

where $C_t{}^-, T_t{}^-$ denote the left derivatives according to t and $C_t{}^+, T_t{}^+$ denote the right derivatives. If we suppose that t parametrises the camera motion $C(t)$, then C_t is defined at P_c and:

$$T_t{}^+.N = \frac{C_t.N}{\lambda^+}, \quad T_t{}^-.N = \frac{C_t.N}{\lambda^-}.$$

This shows that the viewing direction T is not differentiable according to the camera motion parameter t at critical points. Therefore, regions unexposed to contours correspond to non-smoothness of the spatio-temporal surface [Gib 94, Fau 93] defined by $T(s,t)$, where s parametrises the occluding contours. Thus, they may be detected by considering the continuity of T_t along epipolar curves. This result generalises the one obtained in [Zhe 94] for planar camera motions to the case of perspective projections and any camera motions.

3.2 Discrete observations

In the case of discrete observations, the depth at a rim point is computed by a second order approximation [Boy 95]. Such a reconstruction can lead to *false* surface points if the viewing lines which are used jump over a surface area unexposed to contours. To detect these *false* points, we can estimate λ^- and λ^+. However, this requires more information than the three occluding contours used in the second order approximation. Another approach consists in estimating $T_t{}^-$ and $T_t{}^+$ at the reconstructed point. Such values should be equal for a point exposed to contours. Thus, we can detect false points by computing:

$$\sigma = \frac{T_t{}^+.N}{T_t{}^-.N}, \tag{4}$$

This can be done by first order approximation and, consequently, three successive occluding contours.

We denote by $C(t_1)$, $C(t_2)$ and $C(t_3)$ three successive camera positions at time t_1, t_2 and t_3 and by $T(t_1)$, $T(t_2)$ and $T(t_3)$ three viewing directions which are epipolar correspondents ($T(t_1)$ with $T(t_2)$, and $T(t_2)$ with $T(t_3)$). At the rim point $P(t_2)$ on the viewing line define by $C(t_2)$ and $T(t_2)$, we have at order 1:

$$T_t^- = \frac{(T(t_2) - T(t_1))}{(t_2 - t_1)}, \quad T_t^+ = \frac{(T(t_3) - T(t_2))}{(t_3 - t_2)},$$

and if we suppose that $||C_t||$ is constant between the three camera positions, which is not a constraint in the discrete case, we can write:

$$(t_2 - t_1) = \frac{||(C(t_2) - C(t_1))||}{||C_t||}, \quad (t_3 - t_2) = \frac{||(C(t_3) - C(t_2))||}{||C_t||},$$

where $|| \, ||$ denotes the euclidian norm. Finally:

$$\sigma = \left(\frac{T(t_3).N(t_2)}{||(C(t_2) - C(t_1))||} / \frac{-T(t_1).N(t_2)}{||(C(t_3) - C(t_2))||} \right), \tag{5}$$

where $N(t_2)$ is the normal to the surface at $P(t_2)$. Due to the approximation σ will not necessarily be equal to one at a point exposed to contour but close to this value. Therefore, surface areas which correspond to false reconstructed points will be detected by use of a threshold. Hence, reconstructed points which verify:

$$|1 - \sigma| < thresh, \tag{6}$$

where $| \, |$ denotes the absolute value, will be considered as surface points exposed to contours and points which do not satisfy this condition will be considered as false points occluding an area unexposed to contour. Note that in the case of discrete observations (i.e., occluding contours at discrete times) the fact that the reconstructed points are false or not depend on *a priori* information on the surface. Indeed, *tresh* corresponds to the limit below which a planar surface region will be considered as a concavity. This value should therefore be set according to the application.

4 Triangulation

The result of the reconstruction is a set of 3D contour points. A parametric description of this set of points is required for a global surface representation. Such a description is used to build functions which approximate or model the object surface. This can be either a parametrisation of the surface or first, a triangulation of the rim points. In this section, we briefly discuss such a description and we present our approach.

If a parametrisation of the surface is available then an approximation can be computed using, for example, spline functions [dB 78]. However, without any *a priori* information on the object surface, it might be difficult to compute such a parametrisation. Another approach consists in triangulating the reconstructed points. The object surface can then be represented as a set of connected polygonal facets. The advantage is that only topological information are required: the neighbours of each vertex of the triangulation.

An *optimal* triangulation in the 2D case is the Delaunay triangulation which maximises the minimum angle of the resulting mesh. The generalisation to the 3D case leads to the tetrahedrisation of the set of points which is a volume. This method is therefore not well adapted to the case of rim points; indeed rims describe the object surface or only part of it and thus, they do not necessarily define a volume. In addition, the 3D points are organised in contours and hence, the resulting triangulation should conserve this information.

Consequently, our approach consists in constructing a triangular mesh which respect the adjacency of successive rims:

- two 3D points which are not on the same rim can be connected if and only if they are on two consecutive rims.

Thus, the problem of triangulating the set of rim points is reduced to one of triangulating each pair of consecutive rims in the sequence. This leads to the following condition for a triangular facet:

- A facet is defined by two points on one rim and one point on the next or the previous rim.

This condition is not sufficient to define a unique triangulation of a rim pair and an additional criterion must be used to isolate one set of facets. The triangulation of a rim pair corresponds therefore to the problem of finding a minimum cost path in a directed graph, in which the vertices correspond to the set of all possible connections between the points of the rim pair [Fuc 77]. An solution can be found in n^2 operations where n is the number of points on a rim.

The additional criteria that can be used are: the sum of facet areas, the sum of edge lengths or the sum of angles. We choose the sum of facet areas, thus the resulting surface minimises the total surface area.

The triangulation algorithm is applied to the whole set of reconstructed points including false points. Then, all the triangular facets which contains one false point (according to 6) are removed from the surface mesh.

5 Regularisation

The resulting triangular mesh approximates the part of the surface which was covered by the observed rims. However, since the 3D reconstruction process is very sensitive, the reconstructed surface may present perturbations such as folds. This is due to different reasons including:

- the noise which is present in the acquisition system,
- camera calibration errors,
- contour tracking errors,

In order to correct these defaults, positions of mesh vertices are optimised by minimising a functional E:

$$E = E_{dist} + \alpha\, E_{reg},$$

where E_{dist} controls the fitness to the data and E_{reg} the smoothness of the reconstructed surface. In this section, we precise the original energy functions that are used.

5.1 Distance energy

In the context of a reconstruction from occluding contours, data consist of image positions $\{p_{i,j}\}$ of mesh vertices, where i is the image number and j is the point number on the occluding contour. Thus the fidelity to the data can be characterised by the distance between the image point and the projection of the corresponding mesh vertex $P_{i,j}$ onto the image plane. Hence:

$$E_{dist}(P_{i,j}) = \sum_{i,j} |M_i P_{i,j} - p_{i,j}|^2,$$

where $\{M_i\}$ are the calibration matrices (i.e., perspective projection matrices) of the different image planes. This expression is consistent with the fact that original data are viewing lines and not 3D reconstructed points. In the optimisation procedure, surface point displacements are therefore not limited to a closed neighbourhood of the reconstructed point, but to a closed neighbourhood of the corresponding viewing line.

5.2 Regularising energy

In order to optimise surface point positions and thus, smooth the reconstructed surface we introduce a regularising energy. Classically, such energies are based on curvatures or, equivalently, second derivatives of surface point position function [Pog 85, Ter 86]. To this aim, derivatives can be approximated by finite differences [Wel 94] or discrete curvatures can be computed [Hen 92]. However, in the case of a triangular mesh, resulting functionals may be strongly non-linear and thus, difficult to minimise. Furthermore, errors such as surface folds may not be corrected by considering discrete curvatures.

We therefore introduce a term which is based on the triangle areas. Hence, the regularising energy is given by:

$$E_{reg}(\Gamma_{k\in\{1,N_t\}}) = \sum_{k=1}^{N_t} S(\Gamma_k)^2,$$

where N_t is the number of triangles and $S(\Gamma_k)$ the area of a triangle. This energy, and its derivatives, are easy to compute. Consequently, it can be minimised using a classical optimisation method.

Finally, the total energy can be written:

$$E = \sum_{i,j} |M_i P_{i,j} - p_{i,j}|^2 + \alpha \sum_{k=1}^{N_t} S(\Gamma_k)^2. \qquad (7)$$

Only positions of points which do not belong to the surface boundaries are optimised. The parameter α controls the trade-off between fidelity to the data and variation of the sum of squared triangle areas and should therefore be set by the user.

6 Experimental results

We present here results for a real image sequence of a jug (see figure 1(a)(b)). This sequence was taken using a rotating turntable and the occluding contours were tracked using snakes [Ber 94, Kas 88]. The result of the reconstruction is shown in figures 1(c)(d). In these figures, the triangular mesh corresponds to the hull of the object surface defined by the observed contours. Note that due to the contour tracking, the flower which appears on the jug yields a fold in the triangular mesh (see figure 1(c)). In figure 2(a) the triangular facets which contains one or more false point (according to the detection algorithm) are shown. These facets were removed from the final surface as shown in figures 2(b). Note that the fold corresponding to the flower has been corrected (see figure 2(b)). In figures 2(c)(d) the final surface was rendered using a ray tracer and projected in two images of the sequence. This is done by use of the perspective projection matrices computed during a preliminary calibration step. It shows that the resulting surface is coherent with the original one.

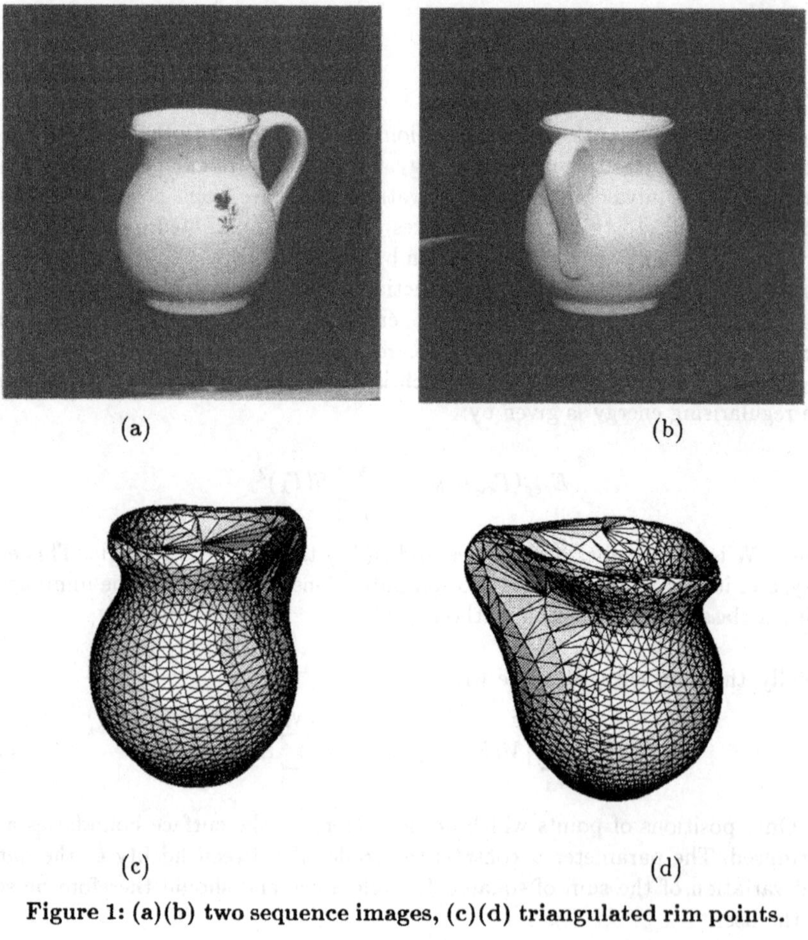

(a) (b) (c) (d)

Figure 1: (a)(b) two sequence images, (c)(d) triangulated rim points.

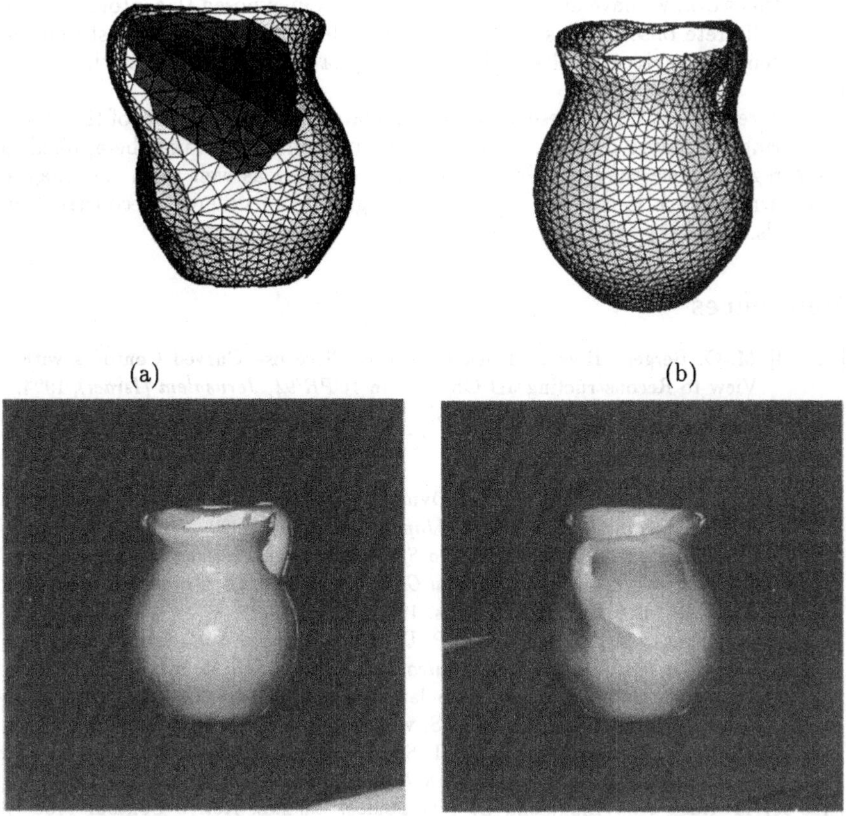

(c) (d)

Figure 2: (a) facets which are detected as unexposed to contours, (b) final mesh surface, (c)(d) projections of the final surface in the original image.

7 Conclusion

We have described a reconstruction procedure that produces a smooth surface from image sequences. Resulting polygonal meshes are obtained by triangulating reconstructed rim points. Such approach is well adapted to the shape from contour problem since it does not require any *a priori* informations on the observed surface. Thus, it allows partial as well as complete surface descriptions. In addition, mesh point positions may be optimised by considering the regularity of the surface. We have proposed a regularising term which is based on triangle areas. This term as well as its derivatives are easy to compute and allows reconstruction perturbations such as folds to be corrected. This make it possible to build object models.

In this work, we have also studied surface areas unexposed to contours. In the case of discrete observations, such regions may leads to false points estimation. We extended previous results and we proposed a criterion to detect these false points.
The interest of detecting these areas is also related to improvements of the global recovering procedure. Indeed, another reconstruction method may be applied to these regions once they are clearly determined in 3D as well as in the images. Our current work is concern with such integration of different reconstruction methods.

References

[Ber 94] M.-O. Berger. How to Track Efficiently Piecewise Curved Contours with a View to Reconstructing 3D Objects. In *ICPR'94, Jerusalem (Israel)*, 1994.

[Boy 95] E. Boyer and M.-O. Berger. 3D Surface Reconstruction Using Occluding Contours. In *CAIP'95, Prague (Czech Republic)*, September 1995. LNCS, volume 970.

[Cip 90] R. Cipolla and A. Blake. The Dynamic Analysis of Apparent Contours. In IEEE, editor, *ICCV'90, Osaka (Japan)*, December 1990.

[dB 78] C. de Boor. *A Practical Guide to Splines.* Springer-Verlag, 1978.

[Fau 93] O. Faugeras. *Three-Dimensional Computer Vision: A Geometric Viewpoint.* Artificial Intelligence. MIT Press, 1993.

[Fuc 77] H. Fuchs, Z.M. Kedem, and S.P. Uselton. Optimal Surface Reconstruction from Planar Contours. *Communications of the ACM*, 20(10), 1977.

[Gib 94] P.J. Giblin and R.S. Weiss. Epipolar Fields on Surfaces. In *ECCV'94, (Stockholm, Sweden)*, May 1994. LNCS, volume 801.

[Hen 92] Henry P. Moreton and Carlo H. Séquin. Functional Optimization for Fair Surface Design. In *Computer Graphics (Proceedings Siggraph)*, July 1992.

[Kas 88] M. Kass, A. Witkin, and D. Terzopoulos. Snakes: Active Contour Models. *International Journal of Computer Vision*, 1: 321–331, 1988.

[Pog 85] T. Poggio, V. Torre, and C. Koch. Computational Vision and Regularization theory. *Nature*, pages 314–319, 1985.

[Sea 95] W.B. Seales and O.D. Faugeras. Building Three-Dimensional Object Models From Image Sequences. *CVIU*, 61(3): 308–324, 1995.

[Sze 93] R. Szeliski and R. Weiss. Robust Shape Recovery from Occluding Contours Using a Linear Smoother. In *CVPR'93, New York (USA)*, 1993.

[Ter 86] D. Terzopoulos. Regularizarion of Inverse Visual Problems Involving Discontinuities. *IEEE Transactions on PAMI*, 8: 413–424, 1986.

[Vai 92] R. Vaillant and O. Faugeras. Using Extremal Boundaries for 3-D Object Modeling. *IEEE Transactions on PAMI*, 14(2): 157–173, February 1992.

[Wel 94] William Welch and Andrew Witkin. Free-Form Shape Design Using Triangulated Surfaces. In *Computer Graphics (Proceedings Siggraph)*, July 1994.

[Zha 94] C. Zhao and R. Mohr. Relative 3D Regularized B-Spline Surface Reconstruction Through Image Sequences. In *ECCV'94, (Stockholm, Sweden)*, May 1994. LNCS, volume 801.

[Zhe 94] Jiang Yu Zheng. Acquiring 3-D Models from Sequences of Contours. *IEEE Transactions on PAMI*, 16(2): 163–178, 1994.

Directions of Motion Fields Are Hardly Ever Ambiguous

Tomas Brodsky, Cornelia Fermüller, and Yiannis Aloimonos

Computer Vision Laboratory, Center for Automation Research, Dept. of Computer Science and Institute of Advanced Computer Studies, University of Maryland, College Park, MD 20742-3275

Abstract. Recent literature [7, 10, 11, 9, 13, 17] provides a number of results regarding uniqueness aspects of motion fields and exact image displacements due to 3-D rigid motion. Here instead of the full motion field we consider only the direction of the motion field due to a rigid motion and ask what can we say about the three-dimensional motion information contained in it. This paper provides a geometric analysis of this question based solely on the fact that the depth of the surfaces in view is positive (i.e. that the surface in view is in front of the camera). With this analysis we thus offer a theoretical foundation for image constraints employing only the sign of flow in various directions and provide a solid basis for their utilization in addressing 3D dynamic vision problems.

For two different rigid motions (with instantaneous translational and rotational velocities $(\mathbf{t}_1, \omega_1)$ and $(\mathbf{t}_2, \omega_2)$) to yield the same direction of the flow, the surfaces in view must satisfy certain inequality and equality constraints, called critical surface constraints. A complete description of image areas where the constraints cannot be satisfied is derived and it is shown that if the imaging surface is a whole sphere, any two motions with different translation and rotation axes can be distinguished using only the direction of the flow. In the case where the imaging surface is a hemisphere or a plane, it is shown that two motions could give rise to the same direction of the flow if $(\mathbf{t}_1 \times \mathbf{t}_2) \cdot (\omega_1 \times \omega_2) = 0$ and several additional constraints are satisfied. For this to occur, the surfaces in view must satisfy all the critical surface constraints; thus at some points only a single depth value is allowed. Similar results are obtained for the case of multiple motions. Consequently, directions of motion fields are hardly ever ambiguous.

1 Introduction and Motivation

The basis of the majority of visual motion studies has been the motion field, i.e., the projection of the velocities of 3D scene points on the image. Classical

The support of the Advanced Research Projects Agency (ARPA Order No. 8459) and the U.S. Army Topographic Engineering Center under Contract DACA76-92-C-0009, the Office of Naval Research under Contract N00014-93-1-0257, National Science Foundation under Grant IRI-90-57934, , and the Austrian "Fonds zur Förderung der wissenschaftlichen Forschung", project No. S 7003, is gratefully acknowledged.

results on the uniqueness of motion fields [7, 10, 11] as well as displacement fields [9, 13, 17] have formed the foundation of most research on rigid motion analysis that addressed the 3D motion problem by first approximating the motion field through the optical flow and then interpreting the optical flow to obtain 3D motion and structure [15, 8, 16, 18, 14].

The difficulties involved in the estimation of optical flow have recently given rise to a small number of studies considering as input to the visual motion interpretation process some partial optical flow information. In particular the projection of the optical flow on the gradient direction, the so-called normal flow [6, 12], and the projections of the flow on different directions [1, 3, 4] have been utilized. In [3] constraints on the sign of the projection of the flow on various directions were presented. These constraints on the sign of the flow were derived using only the rigid motion model, with the only constraint on the scene being that the depth in view has to be positive at every point—the so-called "depth-positivity" constraint. In the sequel we are led naturally to the question of what these constraints, or more generally any constraint on the sign of the flow, can possibly tell us about three-dimensional motion and the structure of the scene in view. Thus we would like to investigate the amount of information in the sign of the projection of the flow. Since knowing the sign of the projection of a motion vector in all directions is equivalent to knowing the direction of the motion vector, our question amounts to studying the relationship between the directions of 2D motion vectors and 3D rigid motion.

The 2D motion field on the imaging surface is the projection of the 3D motion field of the scene points moving relative to that surface. We use a coordinate system $OXYZ$ fixed to the camera. The center of projection is at the origin and the image is formed on a sphere with radius 1. A scene point $\mathbf{R}$ is projected onto an image point $\mathbf{r} = \mathbf{R}/|\mathbf{R}|$, where $|\mathbf{R}|$ is the norm of the vector $\mathbf{R}$.

Suppose the observer is moving rigidly with instantaneous translation $\mathbf{t} = (U, V, W)$ and instantaneous rotation $\boldsymbol{\omega} = (\alpha, \beta, \gamma)$. The well-known equation for the motion field at point $\mathbf{r}$ is

$$\dot{\mathbf{r}} = v_{\mathrm{tr}}(\mathbf{r}) + v_{\mathrm{rot}}(\mathbf{r}) = \frac{1}{|\mathbf{R}|}((\mathbf{t}\cdot\mathbf{r})\mathbf{r} - \mathbf{t}) - \boldsymbol{\omega}\times\mathbf{r} = -\frac{1}{|\mathbf{R}|}(\mathbf{r}\times(\mathbf{t}\times\mathbf{r})) - \boldsymbol{\omega}\times\mathbf{r} \ .$$

The first term $v_{\mathrm{tr}}(\mathbf{r})$ corresponds to the translational component which depends on the depth $Z = |\mathbf{R}|$, the distance of $\mathbf{R}$ to the center of projection. The direction of $v_{\mathrm{tr}}(\mathbf{r})$ is along great circles (longitudes) pointing away from the Focus of Expansion ($\mathbf{t}$) and towards the Focus of Contraction ($-\mathbf{t}$). The second term $v_{\mathrm{rot}}(\mathbf{r})$ corresponds to the rotational component which is independent of depth. Its direction is along latitudes around the axis of rotation (counterclockwise around $\boldsymbol{\omega}$ and clockwise around $-\boldsymbol{\omega}$).

As can be seen, even from exact optical flow, without additional constraints there is an ambiguity in the computation of shape and translation. It is not possible to disentangle the effects of $\mathbf{t}$ and $|\mathbf{R}|$, and thus we can only derive the direction of translation. If we only consider the sign of optical flow, in addition we are also restricted in the computation of the rotational parameters. If we multiply $\boldsymbol{\omega}$ by a positive constant, leave $\mathbf{t}$ fixed, but multiply $\frac{1}{|\mathbf{R}|}$ by the same positive constant, the sign of the flow is not affected. Thus from the direction of the flow we can at most compute the axis of rotation and, as discussed before, the axis of translation.

Hereafter, for the sake of brevity, we will refer to the motion field also as the flow field or simply flow, and to the direction of the motion field as the directional flow field or simply directional flow.

In this paper, by pursuing a theoretical investigation of the amount of information present in directional flow fields, we demonstrate the power of the qualitative image measurements already used empirically, and justify their utilization in global constraints for three-dimensional dynamic vision problems. We study here the theoretical question, investigating uniqueness issues in a noise-free flow field. In practical situations, of course, inaccurate estimation of image displacement directions might influence the result of 3D motion estimation, but this is not the issue we are concerned with here.

The organization of this paper is as follows: In Sect. 2 we develop the preliminaries: Given two rigid motions, we study what the constraints are on the surfaces in view for the two motion fields to have the same direction at every point. In Sect. 3, using these constraints, we study conditions under which two rigid flow fields could have the same direction at every point on a hemisphere. In Sect. 4, we investigate the ambiguity of multiple motions. Section 5 summarizes the results.

2 Critical Surface Constraints

Let us assume that two different rigid motions $(\mathbf{t}_1, \boldsymbol{\omega}_1)$ and $(\mathbf{t}_2, \boldsymbol{\omega}_2)$ yield the same directional flow at every point in the image. To simplify the explanations, we assume $\mathbf{t}_1 \times \mathbf{t}_2 \neq 0$ and $\boldsymbol{\omega}_1 \times \boldsymbol{\omega}_2 \neq 0$. The special cases are dealt with in Sect. 3.3.

Since from the direction of flow we can only recover the directions of the translation and rotation axes, we assume all four vectors $\mathbf{t}_1$, $\mathbf{t}_2$, $\boldsymbol{\omega}_1$ and $\boldsymbol{\omega}_2$ to be of unit length. Let $Z_1(\mathbf{r})$ and $Z_2(\mathbf{r})$ be the functions, mapping points $\mathbf{r}$ on the image into the real numbers, that represent the depths of the surfaces in view corresponding to the two motions. In the future we will refer to Z_1 and Z_2 as the two depth maps. We assume that the two depths are positive, and allow Z_1 or Z_2 to be infinitely large. Thus we assume $1/Z_1 \geq 0$ and $1/Z_2 \geq 0$.

2.1 Notation

We denote $f_\omega(\mathbf{r}) = [\boldsymbol{\omega}_1\, \boldsymbol{\omega}_2\, \mathbf{r}]$, $f_t(\mathbf{r}) = [\mathbf{t}_1\, \mathbf{t}_2\, \mathbf{r}]$, and $g_{ij}(\mathbf{r}) = (\boldsymbol{\omega}_i \times \mathbf{r}) \cdot (\mathbf{t}_j \times \mathbf{r})$, where $[\mathbf{abc}] = (\mathbf{a} \times \mathbf{b}) \cdot \mathbf{c}$ denotes the triple product of vectors $\mathbf{a}$, $\mathbf{b}$ and $\mathbf{c}$.

These functions have a simple geometric meaning. Function $f_\omega(\mathbf{r})$ is zero for points $\mathbf{r}$ lying on a geodesic passing through $\boldsymbol{\omega}_1$ and $\boldsymbol{\omega}_2$. The geodesic is the locus of points $\mathbf{r}$ where $v_{\mathrm{rot}_1}(\mathbf{r})$, the rotational component of the first motion, is parallel to $v_{\mathrm{rot}_2}(\mathbf{r})$, the rotational component of the second motion. Similarly, the geodesic passing through $\mathbf{t}_1$ and $\mathbf{t}_2$ is the locus of points where $f_{\mathbf{t}}(\mathbf{r}) = 0$ and $v_{\mathrm{tr}_1}(\mathbf{r})$ is parallel to $v_{\mathrm{tr}_2}(\mathbf{r})$.

Equation $g_{ij}(\mathbf{r}) = 0$ defines a second order contour consisting of two closed curves on the sphere, the so-called zero motion contour of motion $(\mathbf{t}_j, \boldsymbol{\omega}_i)$ (see Fig. 1). It is the locus of points where $v_{\mathrm{rot}_i}(\mathbf{r})$ is parallel to $v_{\mathrm{tr}_j}(\mathbf{r})$, and therefore the locus of points where the flow due to the motion $(\mathbf{t}_j, \boldsymbol{\omega}_i)$ could be zero. Throughout the paper the functions $f_i(\mathbf{r})$, $g_{ij}(\mathbf{r})$ and the curves defined by their zero crossings will play very important roles.

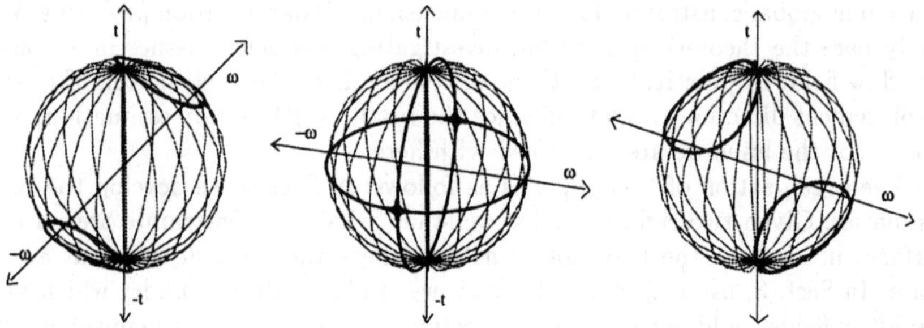

Fig. 1. The zero motion contour (the locus of points $\mathbf{r}$ where $\dot{\mathbf{r}}$ could be zero) consists of two closed curves on the sphere. Three possible configurations are (a) $(\omega \cdot \mathbf{t}) > 0$, (b) $(\omega \cdot \mathbf{t}) = 0$, and (c) $(\omega \cdot \mathbf{t}) < 0$.

To simplify the notation we will usually drop $\mathbf{r}$ and write only f_i and g_{ij} where the index i in f_i can take values $\mathbf{t}$ and $\boldsymbol{\omega}$. By simple vector manipulation, we can prove a useful relationship between f_i and g_{ij}

$$g_{11}g_{22} = f_{\mathbf{t}}f_{\omega} + g_{12}g_{21} \tag{1}$$

2.2 Conditions for Ambiguity

The two motions can give rise to the same directional flow fields if at any point $\mathbf{r}$ there exists $\mu > 0$ such that

$$-\frac{1}{Z_1}(\mathbf{r} \times (\mathbf{t}_1 \times \mathbf{r})) - \boldsymbol{\omega}_1 \times \mathbf{r} = \mu(-\frac{1}{Z_2}(\mathbf{r} \times (\mathbf{t}_2 \times \mathbf{r})) - \boldsymbol{\omega}_2 \times \mathbf{r}) \tag{2}$$

Projecting the vector equation (2) on directions $\mathbf{t}_2 \times \mathbf{r}$ and $\mathbf{r} \times (\boldsymbol{\omega}_2 \times \mathbf{r})$ and using $\mu > 0$, we obtain two constraints for Z_1:

$$\mathrm{sgn}(\frac{1}{Z_1}f_{\mathbf{t}} + g_{12}) = \mathrm{sgn}(g_{22}) \tag{3}$$

$$\mathrm{sgn}(\frac{1}{Z_1}g_{21} - f_{\omega}) = \mathrm{sgn}(\frac{1}{Z_2}g_{22}) \tag{4}$$

where sgn(·) denotes the sign function.

We define $s_1 = -g_{12}/f_t$ and $s'_1 = f_\omega/g_{21}$. At any point, f_i and g_{ij} are constant, so (3) and (4) provide simple constraints on $1/Z_1$. We call them the s_1-constraint and the s'_1-constraint respectively.

We see that the depth Z_1 has a relationship to the surfaces $1/s_1(\mathbf{r})$ and $1/s'_1(\mathbf{r})$. To express the surfaces in scene coordinates $\mathbf{R}$, we substitute in the above equations $Z(\mathbf{r})\mathbf{r} = \mathbf{R}$ and obtain

$$(\mathbf{t}_1 \times \mathbf{t}_2) \cdot \mathbf{R} + (\boldsymbol{\omega}_1 \times \mathbf{R}) \cdot (\mathbf{t}_2 \times \mathbf{R}) = 0 \tag{5}$$

and

$$(\boldsymbol{\omega}_2 \times \mathbf{R}) \cdot (\mathbf{t}_1 \times \mathbf{R}) - ((\boldsymbol{\omega}_1 \times \boldsymbol{\omega}_2) \cdot \mathbf{R})\mathbf{R}^2 = 0 \tag{6}$$

Thus Z_1 is constrained by a second order surface through (5) and by a third order surface through (6). At some points it has to be inside the first surface and at some points it has to be outside the first surface. In addition, at some points it has to be inside the second surface and at some points it has to be outside the second surface. Figure 2 provides a pictorial description of the two surfaces constraining Z_1.

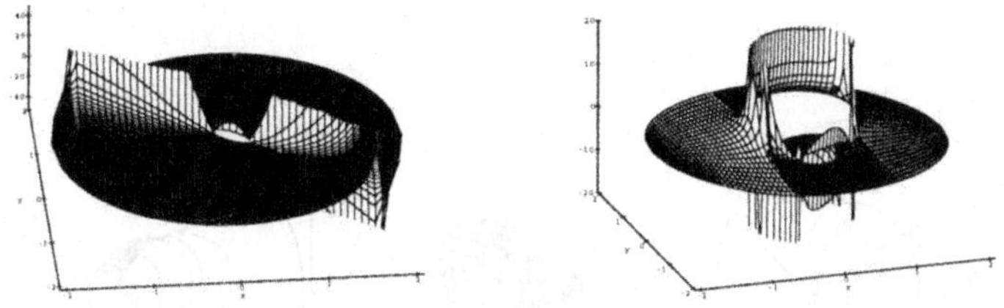

(a) (b)

Fig. 2. Two rigid motions $(\mathbf{t}_1, \omega_1)$, $(\mathbf{t}_2, \omega_2)$ constrain the possible depth Z_1 of the first surface by a second and a third order surface. The particular surfaces shown in the coordinate system of the imaging sphere, projected stereographically, correspond to the motion configuration of Fig. 3.

We can repeat the above derivation for the depth map Z_2. Projecting (2) on vectors $\mathbf{t}_1 \times \mathbf{r}$ and $\mathbf{r} \times (\boldsymbol{\omega}_1 \times \mathbf{r})$, we obtain

$$\mathrm{sgn}(g_{11}) = \mathrm{sgn}(-\frac{1}{Z_2} f_t + g_{21}) \tag{7}$$

$$\mathrm{sgn}(\frac{1}{Z_1} g_{11}) = \mathrm{sgn}(\frac{1}{Z_2} g_{12} + f_\omega) \tag{8}$$

We define $s_2 = g_{21}/f_t$, $s'_2 = -f_\omega/g_{12}$. Equations (7) and (8) provide constraints on $1/Z_2$, and we thus call them the s_2-constraint and the s'_2-constraint.

2.3 Interpretation of Surface Constraints

At each point we have a s_1-constraint (depending on signs of f_t, g_{12}, and g_{22}), a s'_1-constraint (depending on signs of g_{21}, f_ω, and g_{22}), and an additional constraint $1/Z_1 \geq 0$. If the constraints cannot be satisfied, the two flows at this point cannot have the same direction and we say that we have a *contradictory point.*

The existence of a solution for $1/Z_1$ depends also on the sign of $s_1 - s'_1$. Using (1), we can write $s_1 - s'_1 = -(g_{11}\, g_{22})/(f_t\, g_{21})$. So we see that $\mathrm{sgn}(s_1 - s'_1)$ can change only at points where at least one of f_i, g_{ij} is zero.

At points where $g_{22} = 0$ we have the s_1-constraint $1/Z_1 = s_1$, the s'_1-constraint $1/Z_1 = s'_1$, and also $s_1 = s'_1$; thus at these points there is a unique value of Z_1 satisfying the constraints.

The analysis for Z_2 yields the same results. In summary, the curves $f_i(\mathbf{r}) = 0$ and $g_{ij}(\mathbf{r}) = 0$ separate the sphere into a number of areas. Each of the areas is either contradictory (i.e., containing only contradictory points), or ambiguous (i.e., containing points where the two motion vectors can have the same direction). Two different rigid motions can produce ambiguous directions of flow if the image contains only points from ambiguous areas. There are also two scene surfaces constraining depth Z_1 and two surfaces constraining depth Z_2. If the depths do not satisfy the constraints, the two flows are not ambiguous.

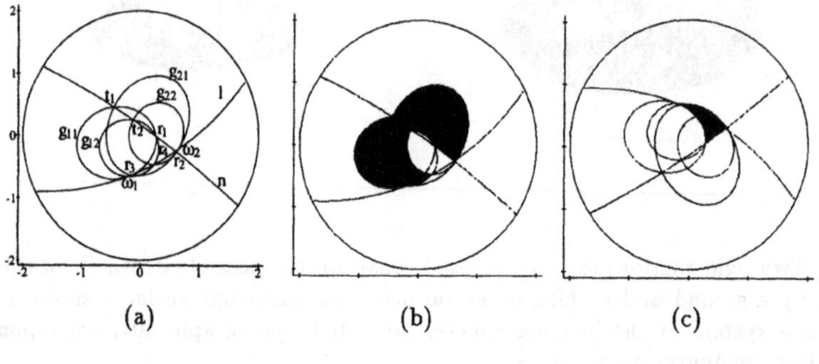

Fig. 3. (a) Separation of the sphere through curves $f_i = 0$ and $g_{ij} = 0$. (b), (c) Contradictory areas for both halves of the sphere.

2.4 Contradictory Points

Since we are interested in contradictory areas, we investigate only a general case, i.e. assume $f_i \neq 0$ and $g_{ij} \neq 0$. Special cases are discussed separately.

If $1/Z_1 < s_1 < 0$, or $1/Z_1 \leq s'_1 < 0$, there is no solution for Z_1. This happens under conditions C_1 and C_2:

$$\mathrm{sgn}(f_t) = \mathrm{sgn}(g_{12}) = -\mathrm{sgn}(g_{22}) \tag{9}$$

$$\mathrm{sgn}(f_\omega) = -\mathrm{sgn}(g_{21}) = \mathrm{sgn}(g_{22}) \tag{10}$$

The similar conditions C_3 and C_4 for Z_2 are

$$\mathrm{sgn}(f_t) = -\mathrm{sgn}(g_{21}) = \mathrm{sgn}(g_{11}) \tag{11}$$

$$\mathrm{sgn}(f_\omega) = \mathrm{sgn}(g_{12}) = -\mathrm{sgn}(g_{11}) \tag{12}$$

We call these four constraints (C_1–C_4) Contradictory Point conditions, or CP-conditions for short. We can show that a point (where $f_i \neq 0$ and $g_{ij} \neq 0$) is contradictory if and only if at least one of the four conditions is satisfied.

2.5 Antipodal Pairs of Points

Let us now consider point $\mathbf{r}$ and its antipodal point $-\mathbf{r}$ such that $f_i(\mathbf{r}) \neq 0$ and $g_{ij}(\mathbf{r}) \neq 0$. Clearly, $f_i(-\mathbf{r}) = -f_i(\mathbf{r})$, but $g_{ij}(-\mathbf{r}) = g_{ij}(\mathbf{r})$.

Both point $\mathbf{r}$ and point $-\mathbf{r}$ are ambiguous only if the CP-conditions are not satisfied. This occurs when

$$\mathrm{sgn}(g_{11}(\mathbf{r})) = \mathrm{sgn}(g_{12}(\mathbf{r})) = \mathrm{sgn}(g_{21}(\mathbf{r})) = \mathrm{sgn}(g_{22}(\mathbf{r})) \tag{13}$$

Let the imaging surface be a whole sphere. The two motions are different, thus there always exist areas on the sphere where condition (13) is not satisfied. Since for any point the image contains also its antipodal point, clearly some areas in the image are contradictory.

3 The Geometry of the Depth-Positivity Constraint

In the previous section we have shown that any two rigid motions can be distinguished using directional flow if the image is a whole sphere. From now on, we assume that the image is a half of the sphere. Let the image hemisphere be bounded by equator q and let $\mathbf{n}_0$ be a unit vector normal to the plane of q such that point $\mathbf{n}_0$ lies in the image.

3.1 Half Sphere Image: The General Case

Let us assume that $(\boldsymbol{\omega}_1 \times \boldsymbol{\omega}_2) \cdot (\mathbf{t}_1 \times \mathbf{t}_2) \neq 0$. We show that under this condition the two rigid motions cannot produce motion fields with the same direction everywhere in the image.

We project point $\boldsymbol{\omega}_1$ on the geodesic n connecting $\mathbf{t}_1$ and $\mathbf{t}_2$ to obtain point $\mathbf{r}_1 = (\mathbf{t}_1 \times \mathbf{t}_2) \times (\boldsymbol{\omega}_1 \times (\mathbf{t}_1 \times \mathbf{t}_2))$. It can be shown that point $\mathbf{r}_1$ is always contradictory and also one of the areas around $\mathbf{r}_1$ is contradictory, because (12) or (11) is satisfied.

Just as we projected $\boldsymbol{\omega}_1$ on geodesic n connecting $\mathbf{t}_1$ and $\mathbf{t}_2$ to obtain $\mathbf{r}_1$, we project $\boldsymbol{\omega}_2$ on n to obtain $\mathbf{r}_2$, and we project $\mathbf{t}_1$ and $\mathbf{t}_2$ on geodesic l, connecting $\boldsymbol{\omega}_1$ and $\boldsymbol{\omega}_2$, to obtain $\mathbf{r}_3$ and $\mathbf{r}_4$ (see Fig. 3). Again, each of $\mathbf{r}_i$ (and one of its neighboring areas) is contradictory, since one of the CP-constraints must be valid.

3.2 Half Sphere Image: The Ambiguous Case

We now consider the case $(\mathbf{t}_1 \times \mathbf{t}_2) \cdot (\boldsymbol{\omega}_1 \times \boldsymbol{\omega}_2) = 0$. However, we still assume $\mathbf{t}_1 \times \mathbf{t}_2 \neq 0$ and $\boldsymbol{\omega}_1 \times \boldsymbol{\omega}_2 \neq 0$. Then contour $f_t = 0$ is perpendicular to $f_\omega = 0$, and $\mathbf{r}_1 = \mathbf{r}_2 = \mathbf{r}_3 = \mathbf{r}_4$.

The motions can be ambiguous only if contours g_{ij} do not intersect the equator q (because of (13)). We obtain a condition

$$l = [\mathbf{t}_j \boldsymbol{\omega}_i \mathbf{n}_0]^2 - 4(\boldsymbol{\omega}_i \cdot \mathbf{n}_0)(\mathbf{t}_j \cdot \mathbf{n}_0)(\boldsymbol{\omega}_i \cdot \mathbf{t}_j) < 0. \qquad (14)$$

By considering all the possible sign combinations for terms f_i and g_{ij} around $\mathbf{r}_1$, it can be verified that the motions are contradictory unless

$$\mathrm{sgn}(\mathbf{t}_1 \cdot \mathbf{n}_0) = \mathrm{sgn}(\mathbf{t}_2 \cdot \mathbf{n}_0) \qquad (15)$$

$$\mathrm{sgn}(\boldsymbol{\omega}_1 \cdot \mathbf{n}_0) = \mathrm{sgn}(\boldsymbol{\omega}_2 \cdot \mathbf{n}_0) \qquad (16)$$

$$\mathrm{sgn}(((\boldsymbol{\omega}_1 \times \boldsymbol{\omega}_2) \times (\mathbf{t}_1 \times \mathbf{t}_2)) \cdot \mathbf{n}_0) = \mathrm{sgn}(\mathbf{t}_1 \cdot \mathbf{n}_0)\,\mathrm{sgn}(\boldsymbol{\omega}_1 \cdot \mathbf{n}_0) \qquad (17)$$

In summary, two rigid motions can be ambiguous on one hemisphere if vector $(\mathbf{t}_1 \times \mathbf{t}_2)$ is perpendicular to vector $(\boldsymbol{\omega}_1 \times \boldsymbol{\omega}_2)$ and all conditions (14), (15), (16), and (17) are satisfied. In addition, as shown in Sect. 2, the two surfaces in view are constrained by a second and a third order surface. Figure 4 gives an example of such a contradictory configuration.

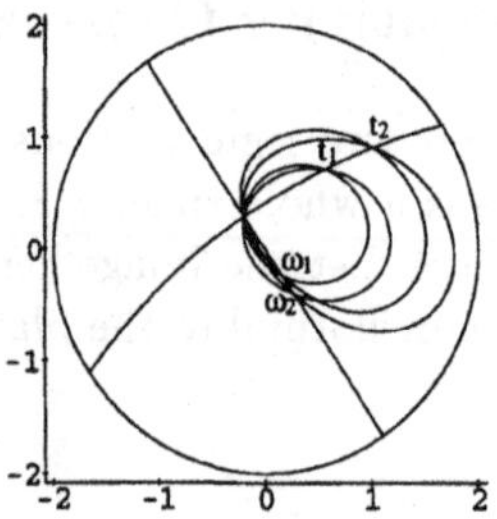

(a)

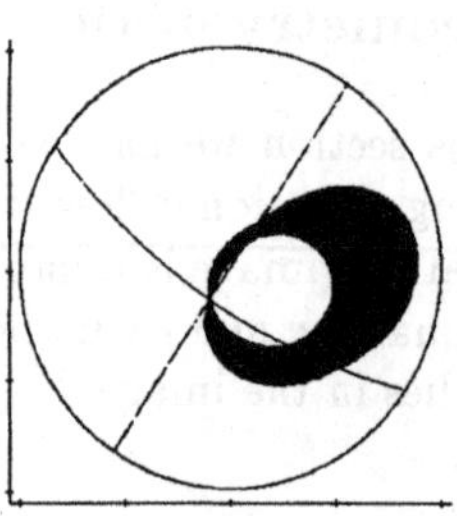

(b)

Fig. 4. Both halves of the sphere showing two rigid motions for which there do not exist contradictory areas in one hemisphere. (a) Hemisphere containing only ambiguous areas. (b) Contradictory areas on the other hemisphere.

3.3 Special Cases

For the special cases it can be shown that if $\mathbf{t}_1 \times \mathbf{t}_2 = 0$ or $\boldsymbol{\omega}_1 \times \boldsymbol{\omega}_2 = 0$, ambiguity cannot occur. A detailed analysis is given in [2].

4 Ambiguity of More than Two Motions

In [2], the analysis has been extended to multiple 3-D motions. Considering three rigid motions, conditions similar to the CP-conditions have been developed. Checking their validity at the intersections of the various zero-motion contours and geodesics through the points $\mathbf{t}_i$ and $\boldsymbol{\omega}_i$, it has been shown that three or more rigid motions $(\mathbf{t}_i, \boldsymbol{\omega}_i)$ could give rise to the same directional flow only if all the $\mathbf{t}_i$ lie on a geodesic n and all the $\boldsymbol{\omega}_i$ lie on a geodesic l perpendicular to n. Also, all the zero-motion contours $g_{ii} = 0$ must intersect in the same points. For an illustration see Fig. 5.

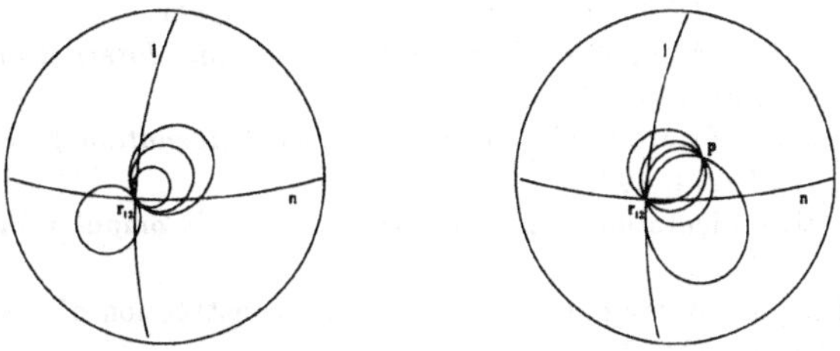

(a) (b)

Fig. 5. Two possible configurations of multiple ambiguous motions. (a) All the zero motion contours are tangent at point $\mathbf{r}_{12}$. (b) All the zero motion contours cross at one point $\mathbf{p} \neq \mathbf{r}_{12}$. There can be an additional ambiguous motion with a degenerate zero-motion contour containing only point $\mathbf{p}$.

5 Conclusions

In this paper we have analyzed the amount of information inherent in the directions of rigid flow fields. We have shown that in almost all cases there is enough information to determine up to a multiplicative constant both the 3D-rotational and 3D-translational motion from a hemispherical image. Ambiguities can result only if the surfaces in view satisfy certain inequality and equality constraints. Furthermore, for two 3D motions to be compatible the two translation vectors must lie on a geodesic perpendicular to the geodesic through the two rotation vectors. With this analysis we have also shown that visual motion analysis does not necessarily require the intermediate computation of optical flow or exact correspondence. Instead, many dynamic vision problems might be solved with the use of more qualitative (and thus robust) flow estimates if appropriate global constraints are found.

References

1. N. Ancona and T. Poggio. Optical flow from 1-D correlation: Application to a simple time-to-crash detector. *International Journal of Computer Vision: Special Issue on Qualitative Vision*, Y. Aloimonos (Ed.), 14:131–146, 1995.
2. T. Brodsky, C. Fermüller and Y. Aloimonos. Directions of motion fields are hardly ever ambiguous. Technical Report CAR-TR-780, Center for Automation Research, University of Maryland, 1995.
3. C. Fermüller. Passive navigation as a pattern recognition problem. *International Journal of Computer Vision: Special Issue on Qualitative Vision*, Y. Aloimonos (Ed.), 14:147–158, 1995.
4. C. Fermüller and Y. Aloimonos. Direct Perception of Three-Dimensional Motion from Patterns of Visual Motion. *Science*, 270:1973–1976, 1995.
5. C. Fermüller and Y. Aloimonos. On the geometry of visual correspondence. *International Journal of Computer Vision*, 1995. To appear.
6. C. Fermüller and Y. Aloimonos. Qualitative egomotion. *International Journal of Computer Vision*, 15:7–29, 1995.
7. B. Horn. Motion fields are hardly ever ambiguous. *International Journal of Computer Vision*, 1:259–274, 1987.
8. B. Horn. Relative orientation. *International Journal of Computer Vision*, 4:59–78, 1990.
9. H. Longuet-Higgins. A computer algorithm for reconstruction of a scene from two projections. *Nature*, 293:133–135, 1981.
10. S. Maybank. *Theory of Reconstruction from Image Motion*. Springer, Berlin, 1993.
11. S. Negahdaripour. Critical surface pairs and triplets. *International Journal of Computer Vision*, 3:293–312, 1989.
12. S. Negahdaripour and B. Horn. Direct passive navigation. *IEEE Transactions on Pattern Analysis and Machine Intelligence*, 9:163–176, 1987.
13. M. Spetsakis and J. Aloimonos. Structure from motion using line correspondences. *International Journal of Computer Vision*, 1:171–183, 1990.
14. M. Spetsakis and J. Aloimonos. A unified theory of structure from motion. In *Proc. ARPA Image Understanding Workshop*, Pittsburgh, PA, pages 271–283, 1990.
15. M. Spetsakis and J. Aloimonos. Optimal Visual Motion Estimation. *IEEE Transactions on PAMI*, 14:959–964, 1992.
16. M. Tistarelli and G. Sandini. Dynamic aspects in active vision. *CVGIP: Image Understanding: Special Issue on Purposive, Qualitative, Active Vision*, Y. Aloimonos (Ed.), 56:108–129, 1992.
17. R. Tsai and T. Huang. Uniqueness and estimation of three-dimensional motion parameters of rigid objects with curved surfaces. *IEEE Transactions on Pattern Analysis and Machine Intelligence*, 6:13–27, 1984.
18. T. Vieville and O. D. Faugeras. Robust and fast computation of edge characteristics in image sequences. *International Journal of Computer Vision*, 13:153–179, 1994.

Euclidean Reconstruction: from Paraperspective to Perspective *

Stéphane Christy and Radu Horaud

GRAVIR-IMAG & INRIA Rhône-Alpes
46, avenue Félix Viallet 38031 Grenoble FRANCE

Abstract. In this paper we describe a method to perform Euclidean reconstruction with a perspective camera model. It incrementally performs reconstruction with a paraperspective camera in order to converge towards a perspective model. With respect to other methods that compute shape and motion from a sequence of images with a calibrated perspective camera, this method converges in a few iterations, is computationally efficient, and does not suffer from the non linear nature of the problem. Moreover, the behaviour of the algorithm may be simply explained and analysed, which is an advantage over classical non linear optimization approaches. With respect to 3-D reconstruction using an approximated camera model, our method solves for the sign (reversal) ambiguity in a very simple way and provides much more accurate reconstruction results.

1 Introduction and background

The problem of computing 3-D shape and motion from a long sequence of images has received a lot of attention for the last few years. Previous approaches attempting to solve this problem fall into several categories, whether the camera is calibrated or not, and/or whether a projective or an affine model is being used. With a calibrated camera one may compute Euclidean shape up to a scale factor using either a perspective model [8], or a linear model [9], [10], [6]. With an uncalibrated camera the recovered shape is defined up to a projective transformation or up to an affine transformation [4]. One can therefore address the problem of either Euclidean, affine, or projective shape reconstruction. In this paper we are interested in Euclidean shape reconstruction with a calibrated camera. In that case, one may use either a perspective camera model or an affine approximation – orthographic projection, weak perspective, or paraperspective.

The perspective model has associated with it, in general, non linear reconstruction techniques. This naturally leads to non-linear minimization methods which require some form of initialization [8], [4]. If the initial "guess" is too far-away from the true solution then the minimization process is either very slow or it converges to a wrong solution. Affine models lead, in general, to linear resolution methods [9], [10], [6], but the solution is defined only up to a sign (reversal)

* This work has been supported by "Société Aérospatiale" and by DRET.

ambiguity, i.e., there are two possible solutions. Moreover, an affine solution is just an approximation of the true solution.

One way to combine the perspective and affine models could be to use the linear (affine) solution in order to initialize the non-linear minimization process associated with perspective. However, there are several drawbacks with such an approach. First, such a resolution technique does not take into account the simple link that exists between the perspective model and its linear approximations. Second, there is no mathematical evidence that a non-linear least-squares minimization method is "well" initialized by a solution that is obtained linearly. Third, there are two solutions associated with the affine model and it is not clear which one to choose.

The perspective projection can be modelled by a projective transformation from the 3-D projective space to the 2-D projective plane. Weak perspective and paraperspective are the most common affine approximations of perspective. Weak perspective may well be viewed as a zero-order approximation: $1/(1+\varepsilon) \approx 1$. Paraperspective is a first order approximation of full perspective: $1/(1+\varepsilon) \approx 1-\varepsilon$. Recently, in [3] a method has been proposed for determining the pose of a 3-D shape with respect to a single view by iteratively improving the pose computed with a weak perspective camera model to converge, at the limit, to a pose estimation using a perspective camera model. At our knowledge, the method cited above, i.e., [3] is among one of the first computational paradigms that link linear techniques (associated with affine camera models) with a perspective model. In [5] an extension of this paradigm to paraperspective is proposed. The authors show that the *iterative paraperspective* pose algorithm has better convergence properties than the *iterative weak perspective* one.

In this paper we describe a new Euclidean reconstruction method that makes use of affine reconstruction in an iterative manner such that this iterative process converges, at the limit, to a set of 3-D Euclidean shape and motion parameters that are consistent with a perspective model. The novelty of the method that we propose is twofold: (i) it extends the iterative pose determination algorithms described in [3] and in [5] to deal with the problem of shape and motion from multiple views and (ii) it is a generalization to perspective of the factorization methods [9], [6] and of the affine-invariant methods [10]. More precisely, the *affine-iterative reconstruction* method that we propose here has a number of interesting features:

- It solves the sign (or reversal) ambiguity that is inherent with affine reconstruction;
- It is fast because it converges in a few iterations (3 to 5 iterations), each iteration involving simple linear algebra computations;
- We show that the quality of the Euclidean reconstruction obtained with our method is only weakly influenced by camera calibration errors;
- It allows the use of either weak perspective [1] or paraperspective camera models (paraperspective in this paper) which are used iteratively, and
- It can be combined with almost any affine shape and motion algorithm. In particular we show how our method can be combined with the factorization method [9], [6].

2 Camera models

Let us consider a pin hole camera model. We denote by P_i a 3-D point with Euclidean coordinates X_i, Y_i, and Z_i in a frame that is attached to the object – the object frame. The origin of this frame may well be the object point P_0. An object point P_i projects onto the image in p_i with image coordinates u_i and v_i and we have ($\mathbf{P}_i$ is the vector $\overrightarrow{P_0P_i}$ from point P_0 to point P_i):

$$\begin{pmatrix} su_i \\ sv_i \\ s \end{pmatrix} = \begin{pmatrix} \alpha_u & 0 & u_c & 0 \\ 0 & \alpha_v & v_c & 0 \\ 0 & 0 & 1 & 0 \end{pmatrix} \begin{pmatrix} \mathbf{i}^T & t_x \\ \mathbf{j}^T & t_y \\ \mathbf{k}^T & t_z \\ \mathbf{0} & 1 \end{pmatrix} \begin{pmatrix} X_i \\ Y_i \\ Z_i \\ 1 \end{pmatrix}$$

The first matrix describes the projective transformation between the 3-D camera frame and the image plane. The second matrix describes the rigid transformation (rotation and translation) between the object frame and the camera frame.

From now on we will be assuming that the intrinsic camera parameters are known and therefore we can compute camera coordinates from image coordinates: $x_i = (u_i - u_c)/\alpha_u$ and $y_i = (v_i - v_c)/\alpha_v$

In these equations α_u and α_v are the vertical and horizontal scale factors and u_c and v_c are the image coordinates of the intersection of the optical axis with the image plane.

The relationship between object points and camera points can be written as:

$$x_i = (\mathbf{i} \cdot \mathbf{P}_i + t_x)/(\mathbf{k} \cdot \mathbf{P}_i + t_z) \tag{1}$$

$$y_i = (\mathbf{j} \cdot \mathbf{P}_i + t_y)/(\mathbf{k} \cdot \mathbf{P}_i + t_z) \tag{2}$$

We divide both the numerator and the denominator of eqs. (1) and (2) by t_z. We introduce the following notations:

- $\mathbf{I} = \mathbf{i}/t_z$ is the first row of the rotation matrix scaled by the z-component of the translation vector;
- $\mathbf{J} = \mathbf{j}/t_z$ is the second row of the rotation matrix scaled by the z-component of the translation vector;
- $x_0 = t_x/t_z$ and $y_0 = t_y/t_z$ are the camera coordinates of p_0 which is the projection of P_0 – the origin of the object frame, and
- We denote by ε_i the following ratio:

$$\varepsilon_i = \mathbf{k} \cdot \mathbf{P}_i/t_z \tag{3}$$

We may now rewrite the perspective equations as:

$$x_i = (\mathbf{I} \cdot \mathbf{P}_i + x_0)/(1 + \varepsilon_i) \tag{4}$$

$$y_i = (\mathbf{J} \cdot \mathbf{P}_i + y_0)/(1 + \varepsilon_i) \tag{5}$$

Whenever the object is at some distance from the camera, the ε_i are small compared to 1. We may therefore introduce the paraperspective model as a first order approximation of the perspective equations, Figure 1. Indeed, with the approximation: $1/(1+\varepsilon_i) \approx 1-\varepsilon_i \ \forall i, i \in \{1...n\}$ we obtain x_i^p and y_i^p which are the coordinates of the paraperspective projection of P_i:

$$x_i \approx (\mathbf{I}\cdot\mathbf{P}_i + x_0)(1-\varepsilon_i) \approx \mathbf{I}\cdot\mathbf{P}_i + x_0 - x_0\varepsilon_i = \frac{\mathbf{i}\cdot\mathbf{P}_i}{t_z} + x_0 - x_0\frac{\mathbf{k}\cdot\mathbf{P}_i}{t_z} = x_i^p \quad (6)$$

where the term in $1/t_z^2$ was neglected. There is a similar expression for y_i^p.

Finally, the paraperspective equations are:

$$x_i^p - x_0 = \frac{\mathbf{i} - x_0\,\mathbf{k}}{t_z}\cdot\mathbf{P}_i \quad (7)$$

$$y_i^p - y_0 = \frac{\mathbf{j} - y_0\,\mathbf{k}}{t_z}\cdot\mathbf{P}_i \quad (8)$$

3 Reconstruction with a perspective camera

Let us consider again the perspective equations (4) and (5). These equations may be also written as (there is a similar expression for y_i):

$$x_i(1+\varepsilon_i) - x_0 - x_0\underbrace{\frac{1}{t_z}\mathbf{k}\cdot\mathbf{P}_i}_{\varepsilon_i} = \frac{1}{t_z}\mathbf{i}\cdot\mathbf{P}_i - x_0\frac{1}{t_z}\mathbf{k}\cdot\mathbf{P}_i$$

These equations can be written more compactly as:

$$(x_i - x_0)(1+\varepsilon_i) = \mathbf{I}_p\cdot\mathbf{P}_i \quad (9)$$

$$(y_i - y_0)(1+\varepsilon_i) = \mathbf{J}_p\cdot\mathbf{P}_i \quad (10)$$

with: $\mathbf{I}_p = \frac{\mathbf{i}-x_0\,\mathbf{k}}{t_z}$ and $\mathbf{J}_p = \frac{\mathbf{j}-y_0\,\mathbf{k}}{t_z}$ Equations (9) and (10) can be interpreted in two different ways: (i) we can consider x_i and y_i as the perspective projection of P_i or (ii) we can consider $x_i(1+\varepsilon_i) - x_0\varepsilon_i$ and $y_i(1+\varepsilon_i) - y_0\varepsilon_i$ as the paraperspective projection of P_i.

The basic idea of our method is to estimate values for ε_i *incrementally* such that one can compute the paraperspective projections of the 3-D points from the perspective projections which are the *true image measurements*. Therefore, the perspective reconstruction problem is reduced to the problem of iterative paraperspective reconstruction.

Let us consider now k views of the same scene points. We assume that image-to-image correspondences have already been established. Equations (9) and (10) can be written as:

$$\mathbf{s}_{ij} = A_j\mathbf{P}_i \quad (11)$$

In this formula the subscript i stands for the ith point and the subscript j for the jth image. The 2-vector $\mathbf{s}_{ij}$ is equal to:

$$\mathbf{s}_{ij} = \begin{pmatrix} (x_{ij} - x_{0j})(1+\varepsilon_{ij}) \\ (y_{ij} - y_{0j})(1+\varepsilon_{ij}) \end{pmatrix} \quad (12)$$

In these equations ε_{ij} (see eq. (3)) is defined for each point and for each image:

$$\varepsilon_{ij} = \mathbf{k}_j\cdot\mathbf{P}_i/t_{z_j} \quad (13)$$

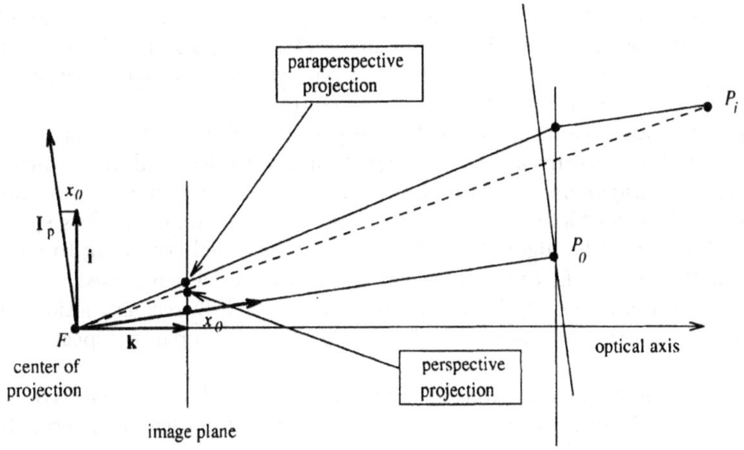

Fig. 1. This figure shows the principle of projection with a paraperspective camera model. We consider the plane through P_0 parallel to the image plane. A 3-D point P_i first projects onto this plane along the direction of FP_0 and then is projected onto the image along a line passing through F. Notice that the two vectors $\mathbf{I}_p$ and $\mathbf{J}_p$ are orthogonal to the direction of projection FP_0 ($\mathbf{I}_p$ only is depicted here).

The reconstruction problem is now the problem of solving simultaneously $2 \times n \times k$ equations of the form of eq. (11). We introduce a method that solves these equations by affine iterations. More precisely, this method can be summarized by the following algorithm:

1. $\forall i,\ i \in \{1...n\}$ and $\forall j,\ j \in \{1...k\}$ set: $\varepsilon_{ij} = 0$ (initialisation);
2. Update the values of $\mathbf{s}_{ij}$ according with eq. (12) and using the newly computed values for ε_{ij};
3. Perform an Euclidean reconstruction with a paraperspective camera;
4. $\forall i,\ i \in \{1...n\}$ and $\forall j,\ j \in \{1...k\}$ estimate new values for ε_{ij} according with eq.(13);
5. Check the values of ε_{ij}:
   if $\forall(i,j)$ the values of ε_{ij} just estimated at this iteration are identical with the values estimated at the previous iteration, *then* stop;
   else go to step 2.

The most important step of this algorithm is step 4: estimate new values for ε_{ij}. This computation can be made explicit if one considers into some more detail step 3 of the algorithm which can be further decomposed into: (i) Affine reconstruction and (ii) Euclidean reconstruction.

The problem of affine reconstruction is the problem of determining both A_j and $\mathbf{P}_i$, for all j and for all i. It is well known that affine reconstruction determines shape and motion up to a 3-D affine transformation. Indeed, for any 3×3 invertible matrix T we have: $A_j\mathbf{P}_i = A_jTT^{-1}\mathbf{P}_i$. In order to convert affine shape and motion into Euclidean shape and motion one needs to consider some Euclidean constraints associated either with the motion of the camera or with the shape being viewed by the camera. Since we deal here with a calibrated camera, we may well use rigid motion constraints in conjunction with paraperspective

[6]. See [7] for the case of an uncalibrated affine camera. Therefore, step 3 of the algorithm provides both Euclidean shape ($\mathbf{P}_1 ... \mathbf{P}_n$) and Euclidean motion. Based on the parameters of the Euclidean shape and motion thus computed one can estimate ε_{ij} for all i and for all j using eq. (13) – step 4.

The above algorithm can be easily interpreted as follows. The first iteration of the algorithm performs a 3-D reconstruction using the initial image measurements and a paraperspective camera model. This first reconstruction allows an estimation of values for the ε_{ij}'s which in turn allow the image vectors $\mathbf{s}_{ij}$ to be *modified* (step 2 of the algorithm). The $\mathbf{s}_{ij}$'s are modified according to eq. (12) such that they better fit the approximated camera model being used.

The next iterations of the algorithm perform a 3-D reconstruction using (i) image vectors that are incrementally modified and (ii) a paraperspective camera model.

At convergence, the equations (11) are equivalent with the perspective equations (9), (10). In other terms, this algorithm solves for Euclidean reconstruction with a perspective camera by iterations of Euclidean reconstruction with a paraperspective camera. Therefore, before we proceed further in order to understand some important features of this iterative algorithm, it is necessary to have insights into the problem of Euclidean reconstruction with a paraperspective camera.

The iterative algorithm outlined in this paper is best illustrated on Figure 2. At the first iteration, the algorithm considers the *true* perspective projections of P_i and attempts to reconstruct the 3-D points as if they were projected in the image using paraperspective. At the second iteration the algorithm considers modified image point positions. At the last iteration, the image point positions were modified such that they fit the paraperspective projections.

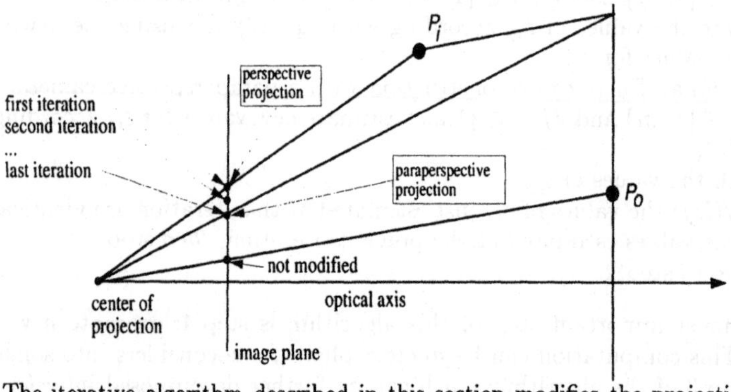

Fig. 2. The iterative algorithm described in this section modifies the projection of a 3-D point from true perspective to paraperspective (see text).

4 Reconstruction with a paraperspective camera

In this section we develop step 3 of the algorithm outlined in the previous section. Methods that use a linear camera model provide a 3-D affine reconstruction if at least 2 views of 4 non-coplanar points are available and if the motion is not a pure translation. However, 3 views are necessary in order to convert this affine reconstruction into an Euclidean one. While the affine-invariant method allows a more direct analysis of the problem, [10] the factorization method is more convenient from a practical point of view.

The factorization method, [9] computes shape and motion simultaneously by performing a singular value decomposition of the $2k \times n$ matrix σ which is formed by concatenating eq. (11) for all i and j: $\sigma = AS = O_1 \Sigma O_2$. Affine shape and motion, i.e., the $2k \times 3$ matrix A and the $3 \times n$ matrix S, can be computed only if the rank of the *measurement* matrix σ is equal to 3.

The rank of σ is equal to the rank of the $n \times n$ diagonal matrix Σ. Even if the rank condition stated above is satisfied, the rank of Σ may be greater than 3 because of numerical instability due to noise. Tomasi & Kanade [9] suggested to solve the rank problem by truncating the matrix Σ such that only the 3 largest diagonal values are considered. They claim that this truncation amounts to removing noise present in the measurement matrix. Therefore, one can write the singular value decomposition of the measurement matrix as $\sigma = O_1' \Sigma' O_2' + O_1'' \Sigma'' O_2''$ where Σ' is a 3×3 diagonal matrix containing the 3 largest diagonal terms of Σ.

Finally, affine shape and affine motion are given by $S = (\Sigma')^{1/2} O_2'$ and $A = O_1' (\Sigma')^{1/2}$.

4.1 From affine to Euclidean

Obviously, the factorization method described above does not provide a unique decomposition of the measurement matrix σ. The method that we describe here for recovering Euclidean shape and motion with a paraperspective camera is an alternative approach to the method described in [6].

One has to determine now Euclidean shape and motion by combining the affine reconstruction method just described and the Euclidean constraints available with the camera model being used. As already mentioned, one has to determine a 3×3 invertible matrix T such that the affine shape S becomes Euclidean: $\left(\mathbf{P}_1 \ldots \mathbf{P}_n\right) = T^{-1}\left(\mathbf{S}_1 \ldots \mathbf{S}_n\right)$ and the affine motion becomes rigid: $\left(R_1 \ldots R_k\right)^T = \left(A_1 \ldots A_k\right)^T T$. Indeed, in order to avoid confusion we denote by S and A affine shape and affine motion and by P and R Euclidean shape and rigid motion. The matrices R_j are given by:

$$R_j = \begin{pmatrix} \mathbf{I}_{p_j} \\ \mathbf{J}_{p_j} \end{pmatrix}$$

The Euclidean constraints allowing the computation of T are the following [6]:

$$||\mathbf{I}_{p_j}||^2/(1+x_{0_j}^2) = ||\mathbf{J}_{p_j}||^2/(1+y_{0_j}^2)$$

and

$$\mathbf{I}_{p_j} \cdot \mathbf{J}_{p_j} = x_{0_j} y_{0_j}/2 \left(||\mathbf{I}_{p_j}||^2/(1+x_{0_j}^2) + ||\mathbf{J}_{p_j}||^2/(1+y_{0_j}^2)\right)$$

We denote by $\mathbf{a}_j$ and $\mathbf{b}_j$ the row vectors of matrix A_j. Using the constraints above, for k images one obtains $2k$ constraints for the matrix T:

$$\mathbf{a}_j^T T T^T \mathbf{a}_j/(1+x_{0_j}^2) - \mathbf{b}_j^T T T^T \mathbf{b}_j/(1+y_{0_j}^2) = 0 \tag{14}$$

$$\mathbf{a}_j^T T T^T \mathbf{b}_j = x_{0_j} y_{0_j}/2 \left(\mathbf{a}_j^T T T^T \mathbf{a}_j/(1+x_{0_j}^2) + \mathbf{b}_j^T T T^T \mathbf{b}_j/(1+y_{0_j}^2)\right) \tag{15}$$

These constraints are homogeneous and non linear in the coefficients of T. In order to avoid the trivial null solution the scale factor must be fixed in advance. For example, one may choose $||\mathbf{I}_{p_1}||^2 = 1 + x_{0_1}^2$ or $||\mathbf{J}_{p_1}||^2 = 1 + y_{0_1}^2$. Hence we obtain one additional constraint such as:

$$\mathbf{a}_1^T T T^T \mathbf{a}_1 = 1 + x_{0_1}^2 \tag{16}$$

These constraints are non linear in the coefficients of T. With the substitution $Q = TT^T$ equations (14), (15), and (16) become linear and there are 6 unknowns because, by definition, Q is a 3×3 symmetric positive matrix. Since we have $2k+1$ independent equations and 6 unknowns, at least 3 views are necessary to estimate Q. Finally T can be derived from Q using a factorization of Q. As it will be explained later in section 5 there is an ambiguity associated with the factorization of the symmetric semi-definite positive matrix Q and this ambiguity is the origin of the reversal ambiguity associated with any affine camera model.

Next we determine the parameters of the Euclidean motion by taking explicitly into account the paraperspective camera model. The method presented below is an alternative to the method proposed in [6] and it is equivalent to the problem of computing pose with a paraperspective camera [5].

First we determine the translation vector. From the formulae above we have:

$$t_{z_j} = 1/2 \left((\sqrt{1+x_{0_j}^2})/(||\mathbf{I}_{p_j}||) + (\sqrt{1+y_{0_j}^2})/(||\mathbf{J}_{p_j}||) \right)$$

and $t_{x_j} = x_{0_j} t_{z_j}$, $t_{y_j} = y_{0_j} t_{z_j}$.

Second, we derive the three orthogonal unit vectors $\mathbf{i}_j$, $\mathbf{j}_j$, and $\mathbf{k}_j$ as follows. $\mathbf{I}_p$ and $\mathbf{J}_p$ may be written as:

$$\mathbf{i}_j = t_{z_j}\, \mathbf{I}_{p_j} + x_{0_j}\, \mathbf{k}_j \tag{17}$$

$$\mathbf{j}_j = t_{z_j}\, \mathbf{J}_{p_j} + y_{0_j}\, \mathbf{k}_j \tag{18}$$

The third vector, $\mathbf{k}_j$ is the cross-product of these two vectors $\mathbf{k}_j = \mathbf{i}_j \times \mathbf{j}_j$. Let's, for convenience, drop the subscript j. We obtain for $\mathbf{k}$:

$$\mathbf{k} = t_z^2\, \mathbf{I}_p \times \mathbf{J}_p + t_z y_0\, \mathbf{I}_p \times \mathbf{k} - t_z x_0\, \mathbf{J}_p \times \mathbf{k}$$

Let $S(\mathbf{v})$ be the skew-symmetric matrix associated with a 3-vector $\mathbf{v}$, and $I_{3\times 3}$ be the identity matrix. The previous expression can now be written as follows:

$$\underbrace{(I_{3\times 3} - t_z y_0\, S(\mathbf{I}_p) + t_z x_0\, S(\mathbf{J}_p))}_{B}\ \mathbf{k} = t_z^2\, \mathbf{I}_p \times \mathbf{J}_p \tag{19}$$

This equation allows us to compute $\mathbf{k}$, provided that the linear system above has full rank. Indeed, one may notice that the 3×3 matrix B is of the form:

$$B = \begin{pmatrix} 1 & c & -b \\ -c & 1 & a \\ b & -a & 1 \end{pmatrix}$$

Its determinant is strictly positive and therefore, B has full rank and one can easily determine $\mathbf{k}_j$ using eq. (19) and $\mathbf{i}_j$ and $\mathbf{j}_j$ using eqs. (17) and (18). As a consequence, it is possible to compute the rigid motion between each camera position and the 3-D scene, i.e., $\mathbf{i}_j$, $\mathbf{j}_j$, $\mathbf{k}_j$, t_{x_j}, t_{y_j}, and t_{z_j} and to estimate ε_{ij} for each image and for each point (eq. (13)).

5 Solving the reversal ambiguity

The algorithm outlined in Section 3 solves for Euclidean reconstruction with a perspective camera by iterations of an Euclidean reconstruction method with a paraperspective camera. In this section we show how this iterative algorithm has to be modified in order to solve the reversal ambiguity problem which is inherent with any affine camera model. Indeed, let us consider again the affine shape and motion recovery method outlined in the previous section. A key step of this method consists of computing a transformation T that converts affine structure into Euclidean structure. This transformation must be computed by decomposition of a symmetric semi-definite positive matrix Q: $Q = TT^T$. There are at least two ways to determine T:

1. Q can be written as $Q = ODO^T$, where O is an orthogonal matrix containing the eigenvectors of Q and D is a diagonal matrix containing the eigenvalues of Q. Since the eigenvalues of a symmetric semi-definite positive matrix are all real and positive, one may write Q as: $Q = (OD^{1/2})(OD^{1/2})^T = KK^T$.
2. Alternatively one may use the Cholesky decomposition of Q: $Q = LL^T$ where L is a lower triangular matrix.

Let H be a non singular matrix such that $L = KH$ and we have:

$$Q = LL^T = KHH^TK^T = KK^T \tag{20}$$

We conclude that H is necessarily an orthogonal matrix. The orthogonality of H is also claimed in [11] but without any formal proof. Therefore H represents either a rotation or a mirror transformation (its determinant is either $+1$ or -1) and there are two classes of shapes that are possible:

- a *direct* shape which is defined up to a rotation and
- a *reverse* shape which is obtained from the direct shape by applying a mirror transformation.

Since shape is defined up to rotation and without loss of generality we choose the mirror transformation to be $-I$ where I is the identity matrix. Therefore the affine shape and motion equation can be written as: $\sigma = AS = (-A)(-S)$. Because of this reversal ambiguity, there are two solutions for the ε_{ij}'s at each iteration of the reconstruction algorithm described above.

The vectors $\mathbf{k}_j$ are computed using eq. (19). This equation may use either $\mathbf{I}_p$ and $\mathbf{J}_p$ (the first solution) or $-\mathbf{I}_p$ and $-\mathbf{J}_p$ (the second solution). Therefore we obtain two distinct solutions, that is, $\mathbf{k}_j^1$ and $\mathbf{k}_j^2$. The two solutions for ε_{ij} correspond to $\mathbf{k}_j^1$ and $\mathbf{P}_i$ and to $\mathbf{k}_j^2$ and $-\mathbf{P}_i$: $\varepsilon_{ij}^{1,2} = \pm\mathbf{k}_j^{1,2} \cdot \mathbf{P}_i/t_{z_j}$.

At each iteration of the perspective reconstruction algorithm two values for ε_{ij} are thus estimated. Therefore, after N iterations there will be 2^N possible solutions. All these solutions are not, however, necessarily consistent with the image data and a simple verification technique allows to check this consistency and to avoid the explosion of the number of solutions. Finally, a unique solution is obtained.

The first iteration of the algorithm makes available two solutions – a "positive" shape S and a "negative" shape $(-S)$ – that are both considered. At the

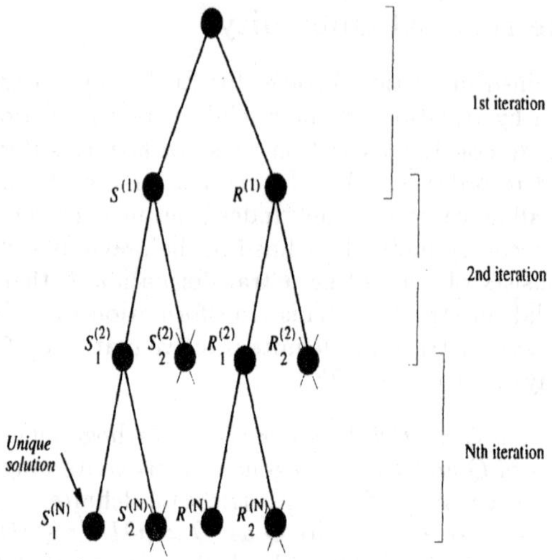

Fig. 3. A strategy for selecting a unique solution (see text).

next iterations of the algorithm two shapes are maintained: one shape consistent with S and another shape consistent with $-S$. Therefore, at convergence, one obtains two solutions, each one of these solutions being consistent with one or the other of the initial shapes. Finally, the solution that best fits the image data is selected as the unique solution. This solution selection process is best illustrated on Figure 3.

6 Experimental results and discussion

In this section we describe two types of experiments: (i) experiments with synthetic data which allow us to study both the accuracy of the 3-D reconstruction and the behaviour of the iterative algorithm, and (ii) experiments with real data.

Let us consider some synthetic data. We designate by D the distance between the center of these data and the camera center of projection divided by the size of the data's diameter – D is therefore a relative distance. Hence, D is approximatively equal to $1/\overline{\varepsilon_{ij}}$ where $\overline{\varepsilon_{ij}}$ is the average value of ε_{ij} for all i and j. For a fixed value of D we consider 10 camera motions, each motion being composed of 15 images. Each such motion is farther characterized by a translation vector and a rotation axis and angle. The directions of the translation vector and rotation axis are randomly chosen. The angle of rotation between two images is equal to 2^0. Moreover, the image data obtained by projecting this object onto the image plane is perturbed by adding Gaussian noise with a standard deviation equal to 1.

The accuracy of the reconstruction is measured by the difference between the theoretical 3-D points and the reconstructed 3-D points. We compute the mean and the maximum values of these differences over all motions at a fixed relative distance D. Figure 4 summarizes the results.

Finally, we consider one experiment performed with real images: A sequence of 13 images of a wood piece with 10 tracked points (figure 5); In all these

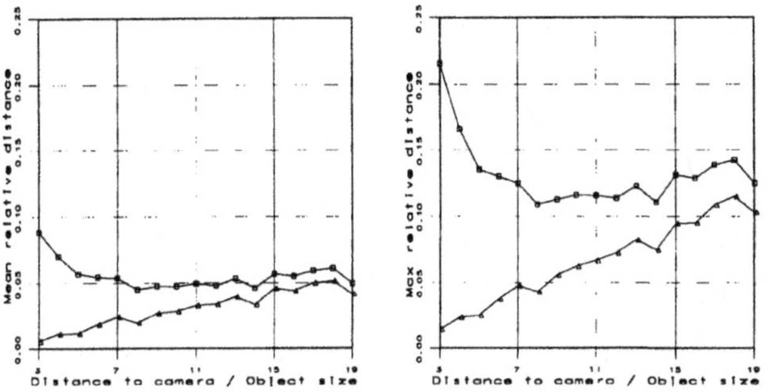

Fig. 4. The behaviour of the factorization method (small squares) is compared with the behaviour of the iterative method described in this paper (small triangles) as a function of the relative distance between the object and the camera (see text). The left side shows the mean value of the distance between object points and reconstructed points and the right side shows the maximum value of this distance.

experiments the camera center was fixed to $u_c = v_c = 256$ and the horizontal and vertical scale factors were fixed to $\alpha_u = 1500$ and $\alpha_v = 1000$.

In this paper we described a method for solving the Euclidean reconstruction problem with a perspective camera by incrementally performing Euclidean reconstruction with a paraperspective camera model. The method converges, on an average, in 5 iterations, is computationally efficient, and it produces accurate results even in the presence of image noise and/or camera calibration errors. The method may well be viewed as a generalization to perspective of shape and motion computation using factorization and/or affine-invariant methods. It is well known that with a linear camera model, shape and motion can be recovered only up to a sign (reversal) ambiguity. The method that we propose in this paper solves for this ambiguity and produces a unique solution even if the camera is at some distance from the scene.

Although the experimental results show that there are little convergence problems, we have been unable to study the convergence of the algorithm from a theoretical point of view. We studied its convergence based on some numerical and practical considerations which allow one to determine in advance the optimal experimental setup under which convergence can be guaranteed [2]. In the future we plan to study more thoroughly the convergence of this type of algorithms.

References

1. S. Christy and R. Horaud. A quasi linear reconstruction method from multiple perspective views. In *Proc. IROS*, pages 374–380, Pittsburgh, Pennsylvania, August 1995.
2. S. Christy and R. Horaud. Euclidean shape and motion from multiple perspective views by affine iterations. *IEEE PAMI*, 1996.
3. D. F. DeMenthon and L. S. Davis. Model-based object pose in 25 lines of code. *Int. J. Comp. Vis.*, 15(1/2):123–141, 1995.
4. R. I. Hartley. Euclidean reconstruction from uncalibrated views. In Mundy, Zisserman, & Forsyth, editors, *Appl. of Inv. in Comp. Vis.*, pages 237–256. Springer Verlag, 1994.

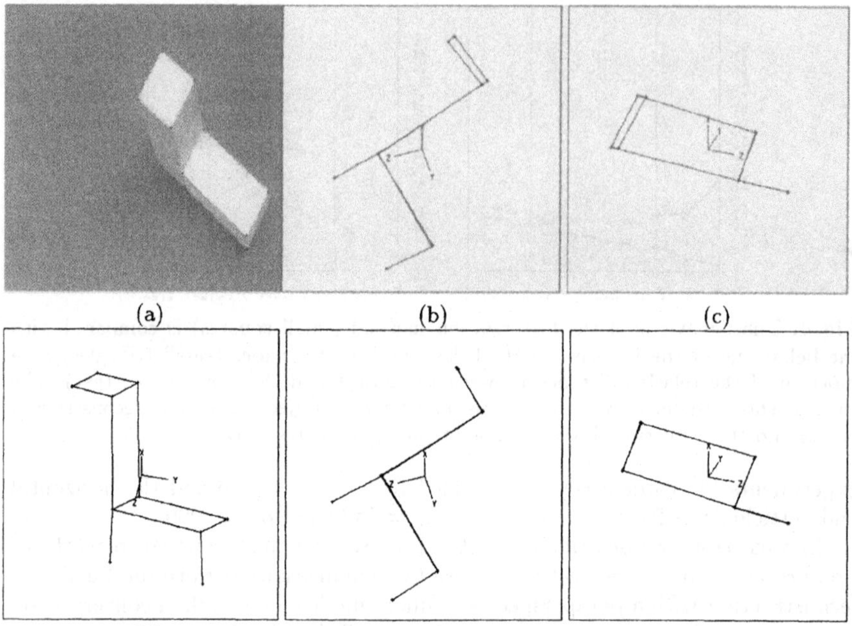

Fig. 5. This figure shows one image (a) out of a sequence of 13 images where only 10 points were tracked and reconstructed. The first row (b) and (c) shows the result of reconstruction using the factorization method with a paraperspective model, while the second row (d), (e), and (f) shows the result of reconstruction with the iterative method and a perspective model. In this example the iterative algorithm converged in 4 iterations.

5. R. Horaud, S. Christy, F. Dornaika, and B. Lamiroy. Object pose: Links between paraperspective and perspective. In *5th ICCV*, pages 426–433, Cambridge, Mass., June 1995.
6. C. J. Poelman and T. Kanade. A paraperspective factorization method for shape and motion recovery. In *3rd ECCV*, pages 97–108. Stockholm, Sweden, May 1994.
7. L. Quan and R. Mohr. Self-calibration of an affine camera from multiple views. In *CAIP'95*, pages 448–455, Prague, September 1995.
8. R. Szelinski and S. B. Kang. Recovering 3-D shape and motion from image streams using non-linear least squares. Tech. Rep. CRL 93/3, Digital - Cambr. Res. Lab., March 1993.
9. C. Tomasi and T. Kanade. Shape and motion from image streams under orthography: a factorization method. *Int. J. Comp. Vis.*, 9(2):137–154, November 1992.
10. D. Weinshall. Model-based invariants for 3-d vision. *Int. J. Comp. Vis.*, 10(1):27–42, February 1993.
11. D. Weinshall and C. Tomasi. Linear and incremental acquisition of invariant shape models from image sequences. *IEEE PAMI*, 17(5):512–517, May 1995.

Optical Flow and Phase Portrait Methods for Environmental Satellite Image Sequences

Isaac COHEN, Isabelle HERLIN

AIR Project,
INRIA, Rocquencourt
B.P. 105, 78153 Le Chesnay CEDEX, France.
Email Isaac.Cohen@inria.fr, Isabelle.Herlin@inria.fr

Abstract. We present in this paper a motion computation and interpretation framework for oceanographic satellite images. This framework is based on the use of a non quadratic regularization technique in optical flow computation that preserves flow discontinuities. We also show that using an appropriate tessellation of the image according to an estimate of the motion field can improve optical flow accuracy and yields more reliable flows. This method defines a non uniform multiresolution scheme that refines mesh resolution only in the neighborhood of moving structures. The second part of the paper deals with the interpretation of the obtained displacement field. We use a phase portrait model with a new formulation of the approximation of an oriented flow field. This allows us to consider arbitrary polynomial phase portrait models for characterizing salient flow features. This new framework is used for processing oceanographic and atmospheric image sequences and presents an alternative to the very complex physical modelling techniques.

Keywords: Optical flow, Non quadratic regularization, Finite element method, Adaptive mesh, Phase portrait, Flow pattern classification, Ocean circulation.

1 Introduction

Oceanographic and atmospheric images obtained from environmental satellite platforms present a new challenge for geosciences and computer vision. The wide ranges of remote sensors allow to characterize natural phenomena through different physical measurements. For example Sea Surface Temperature (SST), Altimetry and Ocean color can be used simultaneously for characterizing vortex structures in the ocean. A major advantage of environmental remote sensing is the regular sampling of the measurements and their availability. These regular temporal and spatial data samplings allow to characterize the short range evolution of atmospheric and oceanographic processes with image sequence processing.

In this paper we focus on dynamic satellite image analysis. The purpose of this paper is to derive a complete framework for processing large oceanographic and atmospheric image sequences in order to detect global displacements (oceanographic streams, clouds motion, ...), or to localize particular structures like vortices and fronts. These characterizations will help in initializing particular processes in a global monitoring system.

Processing such an image sequence raise some specific problems. Indeed, computing an apparent motion field to characterize short range evolution must take into account discontinuities of the motion field that occur near SST temperature fronts and clouds' boundaries. For this purpose we make use of a new regularization method for solving the optical flow constraint equation which involves a non quadratic regularization allowing flow discontinuities. Furthermore, an important issue is the reduction of the numerical complexity due to the large size of the images (typically 2048×1024). The minimization model is handled through a finite element method allowing the use of a non uniform domain tessellation. This tessellation is obtained with a mesh subdivision model allowing to obtain, locally, a finer mesh resolution near moving structures. These structures are detected and localized by studying the norm of the estimated motion field along the direction of the image gradient. Using such a non uniform domain tessellation yields an accurate optical flow near moving structures and a lower numerical complexity.

The second part of the paper deals with the interpretation of the obtained displacement field. A velocity field can be studied from two different view points. First, it is a motion field and therefore it can be studied according to the analysis of fluid motion. On the other hand, it is a vector field whose topology is described by its critical points and its salient features. This approach is more suitable in our context since characterizing salient flow features will help us in locating interesting structures (like vortices and fronts) which represent physical phenomena appearing in SST and atmospheric image sequences. We present in this paper a new approach for approximating an orientation field and characterizing the stationary points of the trajectories obtained from an arbitrary polynomial phase portrait. Furthermore the model is always linear, independently of the polynomial representation.

2 Non Quadratic Optical Flow Computation

The differential techniques used for the computation of optical flow are based on the image flow constraint equation:

$$\frac{dI(x,y,t)}{dt} = I_x u + I_y v + I_t = 0, \tag{1}$$

where the subscripts x, y and t represent the partial derivatives. This equation, based on the assumption that the image grey level remains constant, relates the temporal and spatial changes of the image grey level $I(x,y,t)$ at a point (x,y) in the image to the velocity (u,v) at that point [10]. Equation (1) is not sufficient for computing the image velocity (u,v) at each point since the two velocity components are constrained by only one equation; this is the aperture problem. Therefore, most of the techniques use a regularity constraint that restrains the space of admissible solutions of equation (1) ([1] and references therein). This regularity constraint is generally quadratic and therefore enforces the optical flow field to be continuous and smooth. But, true discontinuities can occur in the optical flow and they are generally located on the boundary between two surfaces representing two objects with different movements. This type of discontinuity occurs for example on temperature front in SST images, and cloud boundary in atmospheric images. Recovering this discontinuity is necessary for further analysis of oceanographic and atmospheric images. Indeed, locating and tracking of temperature fronts in oceanographic images represent an accurate estimation of the oceanic surface circulation. These fronts are defined as regions where the temperature variation is high. An accurate computation of optical flow components near these regions must take into account the flow field discontinuity along the temperature front. For this purpose, we defined a non quadratic regularization scheme preserving flow discontinuities while insuring a unique solution of equation (1). Several authors [2, 3] proposed non-quadratic schemes for the motion field regularity constraint. These schemes are based on the Graduated Non Convexity method, and allow a computation of optical flow components that preserves discontinuity but can not handle non uniform tessellation of the image which is an important issue when dealing with very large image sequences. The proposed method makes use of the L^1 norm (defined by $|u|_1 = \int |u|$) for the regularization constraint. The advantage of this norm is that the variation of expressions like $|u|_1$ produces singular distributions as coefficients (e.g. Dirac functions). This property allows to preserve sharp signals as well as discontinuities in the space of L^1 functions. Such a property can be used to constrain the set of admissible solutions of Eq. (1). Considering the space of functions with bounded variation, i.e.: $BV_1 = \left\{ f = (f_1, f_2) \text{ such that } \int_\Omega |\nabla f_1| + |\nabla f_2|\, dxdy < +\infty \right\}$, the optical flow problem can be stated as the minimization of the functional:

$$\int_\Omega \sqrt{u_x^2 + u_y^2} + \sqrt{v_x^2 + v_y^2} + (I_x u + I_y v + I_t)^2\, dxdy, \tag{2}$$

where u_x and u_y (resp. v_x and v_y) represent the partial derivatives of u (resp. v) with respect to x and y.

This minimization problem (2) can also be viewed as a constrained minimization problem where we search for a vector flow field in the space of functions with bounded variations BV_1, with the constraint of satisfying the optical flow stationarity equation (1).

The solution of the minimization problem (2) is obtained through the Euler - Lagrange equations:

$$\begin{cases} \mathcal{D}u + (uI_x^2 + vI_xI_y + I_xI_t) = 0 \\ \mathcal{D}v + (uI_xI_y + vI_y^2 + I_yI_t) = 0 \end{cases} \tag{3}$$

where $\mathcal{D}$ is the nonlinear operator defined by:

$$\mathcal{D}f = -\frac{\partial}{\partial x}\left(\frac{f_x}{\sqrt{f_x^2 + f_y^2}}\right) - \frac{\partial}{\partial y}\left(\frac{f_y}{\sqrt{f_x^2 + f_y^2}}\right). \tag{4}$$

Equation (3) is nonlinear and therefore must be processed in a particular way. An efficient method for solving this kind of nonlinear partial differential equations is to consider the associated evolution equations, or equivalently, the gradient descent method [8, 15]. This time-dependent approach means that we solve the evolution equation:

$$\begin{cases} \frac{\partial u}{\partial t} + \mathcal{D}u + (uI_x^2 + vI_xI_y + I_xI_t) = 0 \\ \frac{\partial v}{\partial t} + \mathcal{D}v + (uI_xI_y + vI_y^2 + I_yI_t) = 0 \end{cases} \tag{5}$$

and a stationary solution characterizes a solution of Eq. (3).

This evolution equation is solved through a finite element method allowing the use of arbitrary tessellations of the image domain by taking into account an index characterizing image motion in order to reduce the numerical complexity of the algorithm and increase its accuracy near moving structures (see Section 3).

3 Mesh Subdivision

Computing an optical flow field over an image sequence using a classical approach leads to the solution of a large set of linear equations in the case of the quadratic regularizer or to an iterative solution for the non quadratic case. In both approaches space discretization (i.e. image tessellation) is an important issue since it defines the accuracy of the solution and the numerical complexity of the algorithm. In this section we propose a selective multi-resolution approach. This method defines a new approach for coarse to fine grid generation. It allows to increment locally the resolution of the grid according to the studied problem. The advantage of such a method is its lowest numerical complexity and its higher accuracy. Each added node refines the grid in a region of interest and increases the numerical accuracy of the solved problem [4].

The method is based on a recursive subdivision of a given tessellation of the domain. This tessellation must fulfill the *conform triangulation* requirement of the FEM scheme [4], *i.e.*: *any face of any n-simplex $\mathcal{T}_1$ in the triangulation is either a subset of the boundary of the domain, or a face of another n-simplex $\mathcal{T}_2$ in the triangulation.* This requirement restrains the type of n-simplex that can be used for an automatic non-uniform cell subdivision. We consider only triangular cells which are well adapted for domain triangulation and allow to derive a simple recursive subdivision scheme.

Given a triangulation $\mathcal{T} = \bigcup_i \mathcal{T}_i$ of the image domain, we have to refine it only near moving structures. These structures are characterized through $w_\perp$ representing the norm of the estimated motion field along the direction of the image gradient. Such a characterization is based on the image spatio-temporal gradients and does not require the knowledge of the optical flow field.

Let I represents the image brightness and $\overrightarrow{\nabla I}$ its gradient vector field, then the optical flow equation (1) can be rewritten as [18]: $\frac{dI}{dt} = \frac{\partial I}{\partial t} + ||\nabla I|| w_\perp$ where $w_\perp$ is the norm of the component $\overrightarrow{w_\perp}$ of the motion field $\overrightarrow{w} = (u, v)$ along the direction of $\overrightarrow{\nabla I}$. If the flow constraint equation is satisfied (*i.e.* $\frac{dI}{dt} = 0$) and $||\nabla I|| \neq 0$, we obtain:

$$\overrightarrow{w_\perp} = -\frac{\partial I/\partial t}{||\nabla I||} \cdot \frac{\nabla I}{||\nabla I||} \tag{6}$$

representing the component of the motion field along the direction of the gradient in term of the partial derivatives of I. Although $w_\perp$ does not always characterize image motion due to the aperture problem, it allows to locate moving points. Indeed, $w_\perp$ is high near moving points and becomes null near stationary points.

The definition of $w_\perp$ gives the theoretical proof of the motion measure D defined by Irani *et al* [11] and used by several authors [12, 17]:

$$D(x,y,t) = \frac{\sum_{(x_i,y_i)\in W} |I(x_i,y_i,t+1) - I(x_i,y_i,t)|\,|\nabla I(x_i,y_i,t)|}{\sum_{(x_i,y_i)\in W} |\nabla I(x_i,y_i,t)|^2 + C} \tag{7}$$

where W is a small neighborhood and C a constant to avoid numerical instabilities. This motion measure, defined as *residual motion*, is a particular form of $w_\perp$ where the numerator and the denominator are summed over a small neighborhood.

The subdivision scheme is based on a splitting strategy. We start with a coarse tessellation of the image and split each cell $\mathcal{T}$ of the triangulation according to the norm $w_\perp$ summed all over the cell: $w_\perp^{\mathcal{T}}$. A cell is subdivided while $w_\perp^{\mathcal{T}}$ is greater than a given threshold and while its area is greater than another threshold. Furthermore, each subdivided cell must satisfy the *conform triangulation* requirement.

4 Experimental Results

We have experimented our method on the "Hamburg Taxi Sequence". In this street scene, there were four moving objects at different speeds. We first construct an adaptive mesh from the spatio-temporal gradient according to the algorithm described in section 3. We set the motion threshold, *i.e.* $w_\perp^{\mathcal{T}}$, to one pixel/frame and the cell area threshold to 15 pixels. The obtained mesh and the optical flow field are displayed in figure 1. We can observe that the mesh resolution is finer near moving objects allowing an important reduction of the algorithmic complexity, since we deal with 499 nodes and 956 triangles yielding an optical flow numerical accuracy lower than 0.5 pixel near moving structures. To achieve such an accuracy with a classical rectangular mesh, one has to consider 2560 nodes. The complete processing (*i.e.* the image gradients and the solution of Eq. (5)) takes 20 seconds on a Alpha 3000/500 workstation.

Our main objective is to derive the surface ocean circulation from a sequence of Sea Surface Temperature measurements. These infra-red measurements are the most reliable for surface motion estimation and are daily available through the NOAA satellites. These measurements are corrupted by the presence of clouds, consequently, the data usually considered are composite images obtained by considering at each image pixel the maximal value during a period of time. Figure 2 shows a SST image which concerns the confluence region near Argentina coasts where a combined stream is formed by the Falkland northward flowing current (cold water), and the southward flowing Brazilian current (hot water). The discontinuities are located in the regions where these two streams are combined (top of figure 2). The obtained displacement field and mesh used are also displayed.

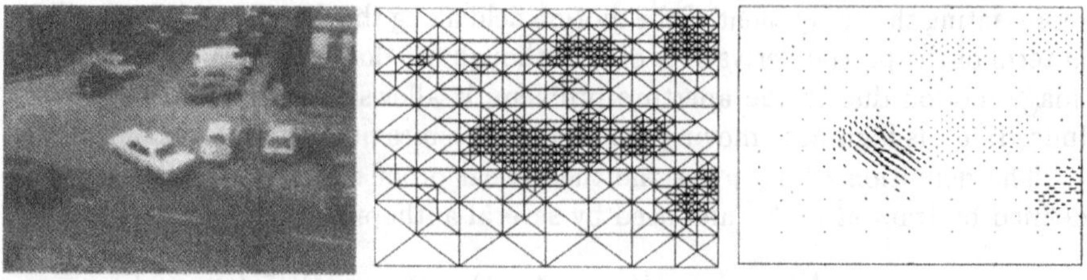

Fig. 1. Example of the use of an adaptive mesh to increase the numerical accuracy of the computed flow field while reducing the algorithmic complexity of the method. We display a frame of the Hamburg Taxi sequence, the associated mesh and the optical flow field.

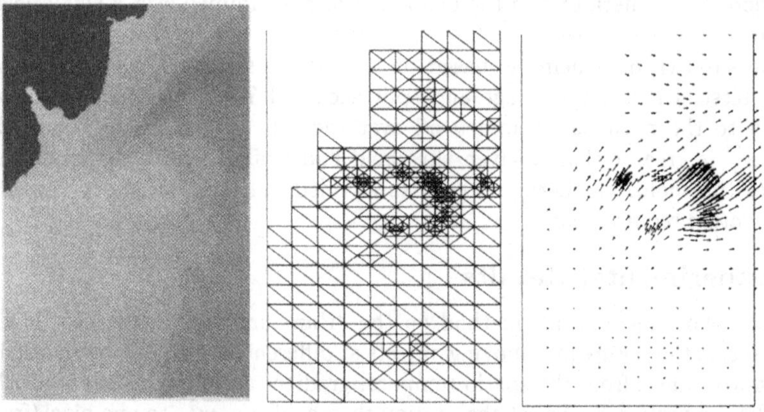

Fig. 2. An illustration of the adaptive mesh approach for computing optical flow components from an image sequence. This figure displays a frame of the SST image sequence of confluence region near Argentina coasts. In this region discontinuities are located in the regions where the streams are combined.

5 Estimation of a Phase Portrait

The computation of optical flow yields a quantitative measure of the flow field on each image point. By processing sea surface temperature and atmospheric image sequences we are also interested by the nature of motion since some phenomena like vortices are characterized through some specific patterns of the motion field. The extraction of higher level descriptors from a flow field is naturally crucial when studying fluid motion and vector analysis.

A classical approach is to use qualitative differential equations to characterize the orientation flow field by considering it as the velocity field of a particle in a dynamic system. Consider a particle governed by the model:

$$g(x,y) = \begin{cases} \frac{dX}{dt} = P(x,y) \\ \frac{dY}{dt} = Q(x,y) \end{cases} \tag{8}$$

where $P(x,y)$ and $Q(x,y)$ are continuously differentiable functions. The particle trajectories are defined by the curves $(\varphi(t), \psi(t)), t > 0$, satisfying: $(\varphi\prime(t), \psi\prime(t)) = (P(\varphi(t), \psi(t)), Q(\varphi(t), \psi(t)))$.

Modeling the orientation flow field by a dynamic system allows to characterize the flow field through the particles trajectories and their stationary points.

Different works were led on linear phase portrait models and their use for characterizing oriented texture fields [14, 16, 19]. The main drawback of a linear phase portrait is that it can handle only one critical point. Ford and Strickland [6] proposed a nonlinear phase portrait model allowing multiple critical point, but this model is computationally expensive and can not be generalized to arbitrary polynomials. This paper propose a new approach for approximating an orientation field and characterizing the stationary points of the trajectories obtained from an arbitrary polynomial phase portrait. Furthermore this model is always linear, independently of the polynomial representation.

Given two vectors A and B, a distance between them may be given by [9]:

$$dist(A,B) = |A|\,|B|\,|sin(\theta)| \tag{9}$$

where θ is the angle subtended by the oriented segments. This distance represents also the area of the triangle formed by these oriented segments. Several authors [7, 14] used this measure to recover the six parameters of a linear phase model by minimizing locally the functional:

$$S_1 = \frac{1}{2} \sum_{i,j \in W} |A_{ij}|^2 \, |B_{ij}|^2 \, |sin(\theta_{1ij} - \theta_{2ij})|^2 \,, \tag{10}$$

where W is the local region over-which the linear approximation is searched, θ_1 represents the linear model and θ_2 the vector field to approximate.

The use the of Levenberg-Marquardt method for minimizing such a non quadratic functional, leads to a slow convergence rate [14]. In the following, we make use of another definition of the area subtended by two oriented segments that leads to the minimization of a quadratic functional. This other definition of the distance is obtained from the cross product of the vectors A and B:

$$dist(A,B) = \frac{1}{2} |A \times B| \,. \tag{11}$$

Hence, considering a polynomial phase portrait model $g(x,y)$ defined by Eq. (8), where P and $Q \in Q_n(\mathbb{R}^2) = \left\{p, \text{ st } p(x,y) = \sum_{i,j \leq n} a_{ij} x^i y^j \right\}$, we fit the model to the given orientation field $f^t = (f_1, f_2)$ (obtained from the optical flow field framework) by minimizing locally:

$$S_2(g) = \frac{1}{2} \sum_{i,j \in W} |f \times g|^2 \,, \tag{12}$$

where W is a neighborhood of the image point (i,j).

This criterion is easier to handle since it is quadratic. In [5] we have proved that recovering the coefficients of the two polynomials P and $Q \in Q_n(I\!R^2)$ by minimizing S_2 amounts to an eigenvalue problem.

This new formulation of the phase portrait similarity measure allows to derive a linear algorithm characterizing arbitrary polynomial portraits.

6 Flow Pattern Classification

Analyzing a flow pattern consists in deriving a symbolic description from the model (8) fitted to the given local orientation field. In the case of optical flow, it would give a displacement information while in the case of fluid dynamic it will characterize the fluid velocity and the coherent structures of the flow. For example, the dynamic of the clouds can be tracked with infrared measurements, one can use such an approach in order to obtain a qualitative description of the atmospheric circulation without using a complex physical model of the underlying phenomena (see Figure 3).

The symbolic description of the trajectory of the phase portrait model was extensively used in the case of a linear model. In the general case (*i.e.* arbitrary polynomials) a classification of the stationary points can be obtained from the linearization of the polynomial model [13] but this assumes a location of critical points (i.e. to solve an arbitrary polynomial system). Instead we use the Index of a vector field to locate and characterize the stationary points. Let $g = (P, Q)$ be a vector field defined over a Jordan curve J in the Euclidean plane, with no critical point on J. The index of g over J is proportional to the angular variation of the vector $g(M)$ (applied at $M \in J$) as M describes J. For the system (8), the index over an oriented Jordan curve J is given by:

$$\text{Index(J)} = \frac{1}{2\pi} \oint_J d\left(\arctan \frac{Q}{P}\right) = \frac{1}{2\pi} \oint_J \frac{PdQ - QdP}{P^2 + Q^2}. \tag{13}$$

A classification of the flow field g can be obtained by computing the index over a small circle surrounding an isolated critical point. Since the computation of the stationary points of system (8) leads to the problem of solving a system of polynomials with arbitrary degree, we choose to compute locally the index of $g = (P, Q)$ over the whole flow field: At each point we consider a circle contained in the centered window W and we use the following classification:

- The index of a focus, a center or a node is equal to $+1$,
- The index of a saddle point is -1.

Although this characterization is compendious, it characterizes the most important structures in a fluid flow field: the stationary points. The index measure, computed over all the flow, allows to obtain the critical points locations without computing the roots of the system (8). Once we locate these points we may use the linearization technique to obtain a complete description of the flow field in the neighborhood of its stationary points.

Figure 3 illustrates the complete framework for processing an infrared image sequence (Meteosat) of Europe (Courtesy of LMD). A frame of the sequence is

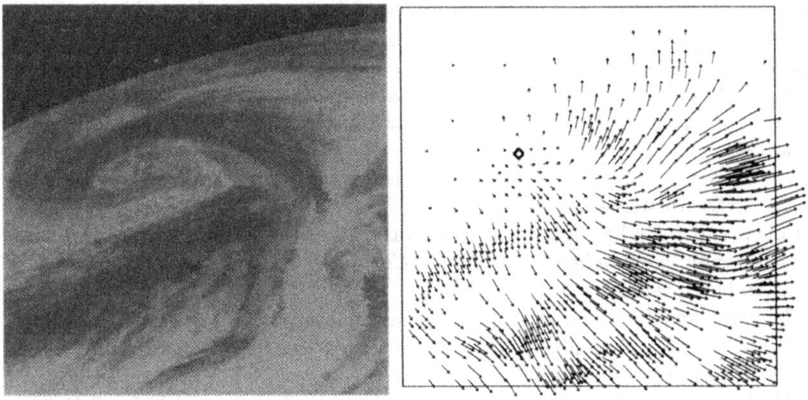

Fig. 3. A frame of the infrared image sequence, and the plot of the computed optical flow characterizing clouds motion. The black quadrangle represents the vortex detected with the index approach. In this case, the given flow field (*i.e.* the optical flow) was approximated with a $Q_2(\mathbb{R}^2)$ polynomial.

represented on the left part of the figure. The processing is done in two steps: A first step consists in computing the apparent motion field and a second one the characterizing the flow patterns. The right part of figure 3, shows the optical flow obtained on a given frame. One can easily localize the vortex on the upper left corner of the figure from the flow structure. An approximation of this orientation field with a $Q_2(\mathbb{R}^2)$ polynomial phase portrait model and using the flow pattern classification described in section 6 gives an accurate localization of the vortex.

7 Conclusion

This research was done within the applicative context of environmental dynamic satellite images. We tried to solve some problem arising in processing a sequence of images representing the evolution of a physical phenomenon. This computer vision approach represents an alternative to the complex modeling of the underlying physical processes.

We proposed a two stages framework allowing the processing of environmental image sequences. The first stage concerns the efficient computation of an optical flow field that preserves flow discontinuities. The later one is concerned with the interpretation of the obtained displacement field.

We are currently studying the comparison between the computer vision approach and the classical method used by oceanographic and atmospheric researchers which deal with more elaborated models.

References

1. J.L. Barron, D.J. Fleet, and S.S. Beauchemin. Performance of optical flow techniques. *International Journal of Computer Vision*, 12(1):43–77, February 1994.

2. M. J. Black. Recursive non-linear estimation of discontinuous flow fields. In *Third European Conference on Computer Vision*, pages 138–145, Sweden, May 1994. Springer-Verlag.
3. M.J. Black and P. Anandan. Robust dynamic motion estimation over time. In *IEEE Proceedings of Computer Vision and Pattern Recognition*, pages 296–302, June 1991.
4. P. G. Ciarlet. *The finite element methods for elliptic problems.* NORTH-HOLLAND, Amsterdam, 1987.
5. I. Cohen and I. Herlin. A motion computation and interpretation framework for oceanographic satellite images. In *IEEE, Computer Vision Symposium*, pages 13–18, Florida, November 1995.
6. R. M. Ford and R. N. Strickland. Nonlinear phase portrait models for oriented textures. In *IEEE Proceedings of Computer Vision and Pattern Recognition*, pages 644–645, New-York, June 1993.
7. R. M. Ford, R. N. Strickland, and B. A. Thomas. Image models for 2-D flow visualization and compression. *CVGIP: Graphical models and Image Processing*, 56(1):75–93, January 1994.
8. R. Glowinski. *Numerical Methods for Nonlinear Variational Problems.* Springer-Verlag, New-York, 1984. Springer Series in Computational Physics.
9. G. H. Golub and C. F. Van Loan. *Matrix Computations.* The Johns Hopkins University Press, London, second edition, 1989.
10. B.K.P. Horn and G. Schunck. Determining optical flow. *Artificial Intelligence*, 17:185–203, 1981.
11. M. Irani, B. Rousso, and S. Peleg. Detecting and tracking multiple moving objects using temporal integration. In *Proceedings of the Second European Conference on Computer Vision 1992*, pages 282–287, May 1992.
12. J.R. Muller, P. Anandan, and J.R. Bergen. Adaptive-complexity registration of images. In *IEEE Proceedings of Computer Vision and Pattern Recognition*, pages 953–957, 1994.
13. V. V. Nemytskii and V. V. Stepanov. *Qualitative Theory of Differential Equations.* Dover Publications, New York, 1989.
14. A. R. Rao and R. C. Jain. Computerized flow field analysis: Oriented textures fields. *IEEE Transactions on Pattern Analysis and Machine Intelligence*, 14(7):693–709, July 1992.
15. J.B. Rosen. The gradient projection method for nonlinear programming. Part I: Linear constraints. *J. Soc. Indust. Appl. Math.*, 8(1):181–217, March 1960.
16. C.-F. Shu and R.C. Jain. Vector field analysis for oriented patterns. In *IEEE Proceedings of Computer Vision and Pattern Recognition*, pages 673–676, Urbana Champaign, Illinois, June 1992.
17. R. Szeliski and H.Y. Shum. Motion estimation with quadtree splines. Technical report, DEC Cambridge Research Lab, March 1995.
18. A. Verri and T. Poggio. Motion field and optical flow: Qualitative properties. *IEEE Transactions on Pattern Analysis and Machine Intelligence*, 11(5):490–498, May 1989.
19. J. Zhong, T.S. Huang, and R.J. Adrian. Salient structure analysis of fluid flow. In *IEEE Proceedings of Computer Vision and Pattern Recognition*, pages 310–315, Seattle, Washington, June 1994.

Refinement of Optical Flow Estimation and Detection of Motion Edges

Andrea Giachetti and Vincent Torre

Dip. Fisica Università di Genova, Via Dodecaneso 33, 16146 Genova
Tel: 39-10-3536311 Fax: 39-10-3536311
E-mail: giach@ge.infm.it

Abstract. In this paper a multi-scale method for the estimation of optical flow and a simple technique for the extraction of *motion edges* from an image sequence are presented. The proposed method is based on a differential neighborhood-sampling technique combined with a multi-scale approach and flow filtering techniques. The multi-scale approach is introduced to overcome the aliasing problem in the computation of spatial and temporal derivatives. The flow filtering is useful near motion boundaries to preserve discontinuities. A ***residual function***, which is a confidence measure of the least-squares fit used to compute the optical flow, is introduced and used to filter the flow and to detect motion boundaries. These boundaries, that we call *motion edges* are extracted by searching for the directional maxima of the map obtained by thinning this residual function. The proposed method has been tested in a variety of conditions. The results obtained with test images show that the proposed approach is an improvement of previous techniques available in the literature.

Introduction

The computation of the optical flow with a precise detection of image locations where the motion field presents discontinuities is a relevant and hard problem of Computer Vision. Because the gray level evolution at one pixel is not sufficient to determine the flow at its location, all the methods used to compute the image motion must use information coming from the surrounding pixels, making assumptions of constancy or smoothness of the motion in these points. *Neighborhood-sampling* optical flow estimators [2, 3, 5] provide a simple way to roughly identify the image locations where motion discontinuities occur. These methods assume that the flow is constant or linear over a small window centered in the considered point, writing for each pixel of the window an equation relating the velocity components to the image derivatives (usually the gray-level constancy constraint $dE/dt = 0$ [1]) and solving the obtained overconstrained system with standard least-square techniques. If the flow is discontinuous in the window, a low confidence measure of the least-square fit is found and tests on the quality of the fit can be used for the detection of motion discontinuities. Nagel [6], suggests the use of stochastic tests on parameters measured by a flow estimator to detect the motion boundaries. A confidence measure of the fit can also be used to improve the accuracy of the computed flow near the discontinuities as proposed by Bartolini et al. [8]. However, to detect motion discontinuities and improve the accuracy of the flow, other sources of error should not affect the confidence measure. The flow can be wrong for at least other three reasons: if the brightness constancy equation is not satisfied, i.e. $dE/dt \neq 0$; if the approximation of the derivatives with finite differences is not good; if the texture in the window is poor or ambiguous (aperture problem). The analysis of these errors suggested us to develop a refined optical flow algorithm based on a multiscale approach [11, 9, 10] combined with edge-preserving flow filters [8]. We

also realized a new algorithm for the localization of "motion edges" based on a search for the local directional maxima of a refined *residual function* which is a local measure of the error of the multi-scale estimate. When the shape of the motion boundaries is available because of some a priori information, a better detection of motion boundaries can be obtained by a parametric estimation of shape templates through the detected motion edges.

1 The basic optical flow algorithm

Fitting the derivatives to the constant local velocity model - Let us assume that that the condition $dE/dt = E_x u + E_y v + E_t = 0$ linking the optical flow components $\mathbf{v} = (u, v)$ to the derivatives of the grey level E: E_x, E_y, E_t holds true over a window centered in the considered point. In detail, given the spatio-temporal window W with $n \times n \times m$ pixels centered in the point we have:

$$\tilde{E}_x(i,j,k)u(x,y,t) + \tilde{E}_y(i,j,k)v(x,y,t) = -\tilde{E}_t(i,j,k) \qquad (i,j,k \in W) \quad (1)$$

where $\tilde{E}_x, \tilde{E}_y, \tilde{E}_t$ are the estimated derivatives. Applying the standard least-squares technique we obtain as the best estimate $\tilde{\mathbf{v}}$ the one minimizing the quantity: $\sum_{i,j,k \in W}(\tilde{E}_t(i,j,k) + \tilde{E}_x(i,j,k)u(x,y,t) + \tilde{E}_y(i,j,k)v(x,y,t))^2$,

that is: $\tilde{\mathbf{v}} = (\mathbf{B}^T\mathbf{B})^{-1}\mathbf{B}^T\mathbf{Y}$ with $\mathbf{B} = \begin{pmatrix} \tilde{E}_x(\mathbf{x}_1) & \tilde{E}_y(\mathbf{x}_1) \\ ... & ... \\ \tilde{E}_x(\mathbf{x}_n) & \tilde{E}_y(\mathbf{x}_n)) \end{pmatrix}$.

$\mathbf{B}$ is the "coefficient matrix", where $(\mathbf{x}_1 ... \mathbf{x}_n)$ are the coordinates of the pixels inside the window, while $\mathbf{Y} = (-\tilde{E}_t(x_1), \ ... \ , -\tilde{E}_t(x_n))$. is the "data" vector. Supposing that eqn. (1) holds true, the estimated optical flow is close to the true image velocity only if: 1) the system is well conditioned; 2) the assumption of constant $\mathbf{v}$ holds true. Condition 2 suggests the use of small windows, so as to reduce the number of wrong estimates due to the presence of discontinuities inside the window. In this case, however, the equations of the system are heavily correlated due to the effect of the derivative mask and of image smoothing[10]. In order to have a well conditioned system larger windows must be used; to reduce the computational weight, as the data from neighboring points are strongly correlated, it is possible to eliminate some equations by subsampling the windows as suggested in Fig. 1. Large windows are not sufficient to guarantee a correct estimate. In fact, in order to compute a reliable flow, spatial derivatives must be nonzero in more than one point and the ratio E_x/E_y should not be constant. This can be ensured by requiring a nonzero determinant of the matrix $\mathbf{B}^T\mathbf{B}$. To eliminate wrong estimates it is necessary to have a threshold d_T on the value of *Det* $(\mathbf{B}^T\mathbf{B})$. A threshold on the conditioning number c_T, defined as the absolute value of the ratio between the minimum and the maximum eigenvalue of the matrix avoids numerical instability of the inversion of $(\mathbf{B}^T\mathbf{B})$ [7].

Accuracy of the fit - The quality of the fit can be evaluated by analyzing the value of the *residual function*:

$$Q(W(\mathbf{x})) = \sum_{i,j,k \in W} (\tilde{E}_t(i,j,k) + \tilde{E}_x(i,j,k)\tilde{u}(x,y,t) + \tilde{E}_y(i,j,k)\tilde{v}(x,y,t))^2/N(W) \quad (2)$$

where N(W) is the number of the pixel of the window used for the estimation. High values of this function indicate a bad fit of the model and an inaccurate flow field. This is the case when there are points in the window where the gray level constancy is false or where the estimation of the derivatives is wrong or if there is a sharp motion discontinuity inside the window. In the following we will always

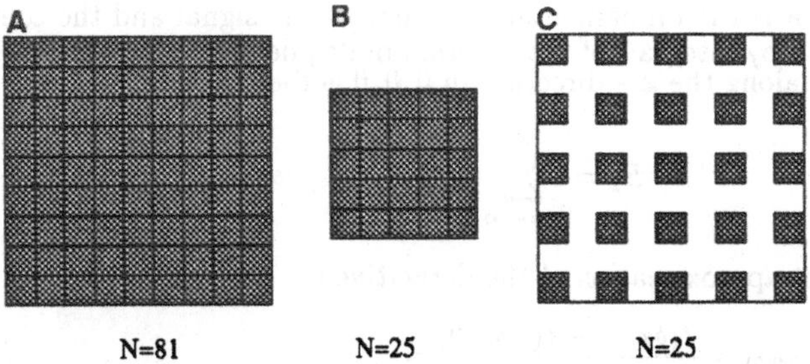

Fig. 1: Advantages of using sub-sampled windows. A: $9 \times 9(\times 1)$ window: the size is almost always sufficient to have a well conditioned system, but the number of equation is high. B: 5×5 window: the computation is speeded up, but the conditioning of the system is often poor. C: 9×9 subsampled window: the number of equations is the same as in B, but the system is almost always well conditioned.

consider condition (1) to hold. If this is true and the estimates of the derivatives are correct, Q can be then used for the location of motion discontinuities. In order to do this it is necessary to analyze carefully the problems in computing derivative estimation.

2 Errors in computing derivatives

Two main kinds of errors affect derivatives computed from finite difference or polynomial fitting: errors due to poor texture, and errors due to the spatio-temporal sampling steps.

Poor texture - It is evident that, whenever a spatial derivative is almost zero, eqn. (1) becomes meaningless because of the large relative errors. Let us consider, the three point approximation of the derivatives: $\tilde{E}_x(x,y) = [E(x+\delta x, y) - E(x-\delta x, y)]/2\delta x$ and suppose that the first order approximation of the signal is good so that the use of the finite difference does not introduce error. The only error will be due to the uncertainty on the values of E. Let us suppose that this uncertainty is limited by the quantity δE; then the error on the derivative will be $\delta E/\delta x$. The relative error is then:

$$\delta E_x / E_x = \delta E / (\tilde{E}_x \delta x) \tag{3}$$

that is high when local variations of E are negligible. A similar analysis show that the second order derivatives computed with the standard 5-points mask are much more sensitive to noise: in fact the relative error is $16\delta E/3\tilde{E}_{xx}\delta^2 x$ and the average value of E_{xx} is usually lower than the average value of E_x.

Spatio-temporal sampling / Aliasing - The recovery of the derivatives from a sampled signal is not always possible, because the knowledge about the original signal is not complete due to sampling.
To see the effect of high frequencies on the derivatives of our sampled image we can compare the estimated derivatives in the point $(0,0,0)$ (we choose this point to simplify the formulas, but it would be the same for a general location) with their true values. We consider a neighborhood I of $(0,0,0)$ defined by $|x-ut| < \lambda$, $|y-vt| < \lambda$, $|t| < \tau$, so that, assuming $\mathbf{v}$ constant in I, the signal can be written as [9]: $E(x,y,t) = f(x-ut, y-vt)$, where f can be expressed by the Fourier series:

$$f(x-ut, y-vt) = \sum_{n=-\infty}^{\infty} \sum_{m=-\infty}^{\infty} A_{nm} \exp(j(n\omega(x-ut) + m\omega(y-vt))) \tag{4}$$

where $\omega = 2\pi/\lambda$ is the fundamental frequency of the signal and the coefficients A_{nm} are defined by integrals of the function multiplied by exponentials as usual. The derivative along the $x-$ direction in $0,0,0$ is then:

$$E_x = \sum_{n=-\infty}^{\infty} \sum_{m=-\infty}^{\infty} A_{nm} jn\omega \tag{5}$$

The three point approximation of the derivative is:

$$\begin{aligned} \tilde{E}_x|_{(0,0,0)} &= \frac{f(\delta x, 0) - f(-\delta x, 0)}{2} = \\ &= \sum_{n,m=-\infty}^{\infty} A_{nm} \frac{e^{jn\omega\delta x} - e^{-jn\omega\delta x}}{2\delta x} = \sum_{n,m=-\infty}^{\infty} A_{nm} j \frac{\sin(n\omega\delta x)}{\delta x} \end{aligned} \tag{6}$$

The difference between the estimated and the true value is:

$$E_x - \tilde{E}_x = \sum_{n=-\infty}^{\infty} \sum_{m=-\infty}^{\infty} A_{nm} j(n\omega\delta x - \sin(n\omega\delta x)) \tag{7}$$

We immediately notice that the estimated derivative is correct only if all the coefficients A_{nm} corresponding to frequencies $n\omega << 1/\delta x$ are negligible. The approximation of the derivative is then good only if for the maximum frequency with relevant energy, $N\omega$, we have: $\sin(N\omega\delta x) \simeq N\omega\delta x$ *i.e.* $N\omega\delta x << 1$
When many frequencies over this limit present relevant coefficients, the relative error may be in the order of the unity. The analysis of the spatial derivative suggests to remove frequencies higher than $1/\delta x$ to avoid large errors. With the same method used before, it is possible to see that difference between the true value and the 3-point estimate of the temporal derivative in (0,0,0) is:

$$E_t - \tilde{E}_t = \sum_{n=-\infty}^{\infty} \sum_{m=-\infty}^{\infty} A_{nm} j \frac{n\omega u\delta t + m\omega v\delta t - \sin(n\omega u\delta t + m\omega v\delta t)}{\delta t} \tag{8}$$

The estimate is correct only if for the highest frequencies $N\omega, M\omega$ with relevant coefficients A_{NM}, we have: $\sin(N\omega u\delta t + M\omega v\delta t) \simeq \omega(Nu + Mv)\delta t$
To eliminate all the frequencies introducing large errors in the estimates, it is therefore necessary to use a low-pass filter with a cutoff frequency chosen according to the largest value of $\mathbf{v}$ to be detected.

3 Multi-scale techniques

A good solution to the problem of aliasing without the loss of all the high-frequencies information is the *multi-scale* approach [9, 11]. Multi-scale techniques are usually based on the "pyramid" decomposition of the images. Each of the original $l_x \times l_y$ images is smoothed and then subsampled keeping only even or odd pixels. The new $l_x/2 \times l_y/2$ images can then be decomposed in the same way until the coarsest resolution $l_x/2^N \times l_y/2^N$ is obtained. Moving from the finer $(l_x \times l_y)$ to the coarser $(l_x/2 \times l_y/2)$ image removes higher frequencies. To avoid the poor estimate due to the derivative errors, it is necessary to stop the decomposition at a resolution where the motion is in the order of 1 pixel/frame. The coarsest resolution, however, should be sufficiently fine to keep the assumption of flow constancy in the windows well approximated. We always used two or three resolutions in the experiments, enough to compute the largest motions

without introducing large errors. In the multi-scale flow computation, the integration of the information relative to the different scales is performed in a *coarse to fine* framework. In [9] the flow is computed at finer scales only where the one computed at the coarse scale is considered wrong. We have chosen a different solution, consisting of computing the flow at the coarser resolution and then compute at the finer resolution corrective terms, i.e. differences between the flow and the integer approximation of the previously computed flow, with a scheme recently introduced by Xu [11]. Let us denote with $\mathbf{V}^N = (U^N, V^N)$ the integer approximation of the flow $\mathbf{v}^N$ computed at the resolution $l_x/2^N \times l_y/2^N$. If at the finer resolution $l_x/2^{N-1} \times l_y/2^{N-1}$, we compute the *shifted derivatives* defined as follow:

$$E_t'^{N-1} = \frac{E(x+U_x^N, y+V_x^N, t+1) - E(x-U_x^N, y-V_x^N, t-1)}{2} \tag{9}$$

in all the points of a window $W(\mathbf{x})$ centered in the point $\mathbf{x}$ of interest; from the overconstrained system

$$E_x^{N-1}(\mathbf{x})c_x + E_y^{N-1}(\mathbf{x})c_y + E_t'^{N-1}(\mathbf{x}) = 0 \qquad (\mathbf{x} \in W(\mathbf{x})) \tag{10}$$

it is possible to compute the *corrective terms* c_x, c_y, representing the additional term to be added to the integer values U^N, V^N. If the residual function $Q^{N-1}(\mathbf{x})$ of this fit is less than the residual fit $Q^N(\mathbf{x})$ computed estimating the value of $\mathbf{v}^N$, we consider as the best estimate of the flow in $\mathbf{x}$ at the scale $N-1$, the quantity described by:

$$u^{N-1} = U^N + c_x^{N-1} \qquad v^{N-1} = V^N + c_y^{N-1} \tag{11}$$

and we replace the value of the residual Q stored in memory with the new value Q^{N-1}. In this way information coming from finer scales is used without introducing further error. The coarse to fine velocity propagation is not correct if a a pixel of the coarser scale correspond to a motion discontinuity. Near these points it is necessary to improve the quality of the estimation before computing the corrective terms, and use some special procedure for the flow correction (see Section 5).

4 Discontinuities

Detection of discontinuities - When a discontinuity occurs in the window (or close to a window) used for the estimation of $\mathbf{v}(\mathbf{x})$, the result obtained is wrong for two reasons: firstly, the assumption of constant $\mathbf{v}$ inside the window fails; secondly, the derivative estimates obtained where the discontinuity is located inside a derivative mask are not reliable. Supposing to have eliminated the effects of the errors in derivative computation described before, we can identify the regions where a sharp motion discontinuity occurs, with the pixels where the the residual function $Q(W(\mathbf{X}))$ presents high values. $Q(W(\mathbf{X}))$ is, in this case, approximately zero when the whole window is inside a coherently moving region, nonzero when a discontinuity crosses the window and presents its directional maxima where a discontinuity passes near the center of the window. We can roughly identify these maxima with the "motion edges".

Flow filtering near discontinuities: the multi-window approach - The residual function can also be used to improve the flow accuracy near the discontinuities, by using a non linear filter. A similar approach has been successfully introduced by Bartolini et al. [8]. Their algorithm is very simple: for each pixel location **x** 9 windows containing the point and shifted as shown in Fig. 2 A are considered. The estimates obtained in the points $x - mask * i, y - mask * j$ with $i, j = -1, 0, 1$ and the corresponding average squared residuals are considered (*mask* is equal to half of the mask dimension). The correct flow estimated in **x** is considered to be the one obtained in the window with the lowest residual function. In this way a window containing the central point and with uniform motion inside is searched in the eight directions and the quality of the estimates is improved.

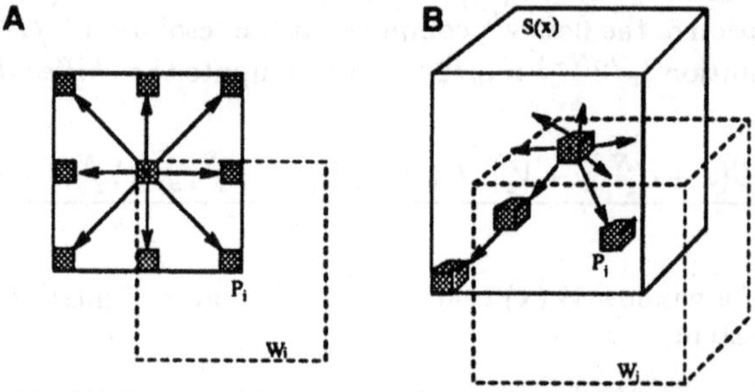

Fig. 2: A: The multi-window filter of Bartolini et al. It is assumed as velocity estimate in the central point the one obtained in the window centered in one of the nine points where the residual function is lower. B: Generalization of the residual filter: the minimum of the residual function is searched in a 3D (x,y,t) search space S including pixels where the flow estimate depends on the gray value in the central point. The best velocity estimate is the one obtained where the residual function is minimum.

The generalized residual filter - This filtering procedure can be generalized with a search for the minimum value of the residual on a set of points of the neighborhood of **x** corresponding to masks containing **x** or where the velocity estimate depends on $E(\mathbf{x})$ because of the extension of the derivative mask. It is important to extend the search for continuity for our filter even in the temporal direction, to compensate for the effects of temporal discontinuities of the motion. In our filter the search for the minimum of the residual function is performed in a search space $S(\mathbf{x})$ of $9 \times 9 \times 3$ pixels (the mask dimension used is $9 \times 9 \times 1$). The filtered flow estimate $\mathbf{v}_{rf}(\mathbf{x})$ in the point **x** is therefore chosen as the one obtained in the point of $S(\mathbf{x})$ where the residual function is minimum.

Regularization preserving the edges - We introduced also a flow regularization preserving the sharp variations of the flow near the discontinuities. First the average $\mathbf{v}_{avg}$ of the flow vectors over a set of points of the neighborhood of **x** is computed. This set is composed by pixels $\mathbf{x}'$ where the residual function is smaller than a threshold and the difference between the computed flow $\mathbf{v}(\mathbf{x}')$ and the output of the residual filter $\mathbf{v}_{rf}(\mathbf{x})$ is less than 1 *pixel*/*frame*. The regularized estimate of $\mathbf{v}_{reg}(x)$ is given by: $\mathbf{v}_{reg}(x) = (\mathbf{v}_{rf} + \mathbf{v}_{avg})/2$

Thinning of the residual map and edge detection - The final multi-scale residual map $Q(\mathbf{x})$ identifies the presence of discontinuities, but its localization is poor because of the window size. It is possible, however, to obtain a thinning of this map during the residual filtering, by assuming as new residual value $Q'(\mathbf{x})$ at each pixel location the minimum value of the residual in the search space $S(\mathbf{x})$. Motion discontinuities are still revealed because the size of the region around **x** where the flow estimate depends on $E(\mathbf{x})$ is larger than the size of the search space. Our algorithm for the detection of *motion edges* is therefore based on the search for local maxima along $x-$ and $y-$ directions of the map $Q'(\mathbf{x})$, larger than a defined threshold.

5 Description of the complete multi scale algorithm

The above analysis suggested us to design a multi-scale algorithm for the detection of motion edge and flow estimate, consisting of the following steps:

1. The image sequence is filtered and decomposed in a Gaussian pyramid.
2. The optical flow is computed at the coarsest scale over $9 \times 9 \times 1$ (subsampled) windows, rejecting equations obtained where spatial derivatives (computed with a 5 point mask) are below a threshold and rejecting estimates obtained where the determinant and the conditioning number of the matrix $\mathbf{B}^T\mathbf{B}$ are less than the thresholds d_T and c_T and the residual function is above the threshold r_T.
3. The residual filter is applied to the flow.
4. The regularizing filter is applied to the flow;
5. Flow corrections are computed at the finer scale. If the residual function computed at the coarser scale had a high value, several corrective terms are computed shifting the derivatives not only of $(int)v(x, y, t)$, but also of $(int)v(x+i, y+j, t+k)$ with $i, j, k = -1, 0, 1$. The shift corresponding to the minimum value of the residual function at the finer scale is kept.
6. The corrected flow at the finer scale is kept if its residual is lower than the one computed at the coarser scale, then the residual map Q is corrected with the new value. If the residual is higher, the flow computed at the coarser scale is still considered the best estimate.
7. Filtering is applied to the corrected flow and the procedure is repeated until the finer scale is reached.
8. When the highest resolution is reached, the thinned residual map is generated as described in Section 4 and the motion edges are extracted from this map.

6 Experimental results

We have tested our algorithms on calibrated sequences (with known true displacements). To evaluate the accuracy of the computed flows we used the angular distance between the computed $\mathbf{v} = (u, v)$ and the real $\mathbf{v}' = (u', v')$ displacements, defined by Barron [4] as: $dist(\mathbf{v}, \mathbf{v}') = \arccos\left(\frac{uu'+vv'+1}{\sqrt{(|\mathbf{v}|+1)(|\mathbf{v}'|+1)}}\right)$

Enhancement of flow accuracy - The first series of experiments aimed at testing the proposed optical flow algorithm. We have measured the improvement in the flow accuracy on a synthetic sequence where a circle with radius equal to 100 pixels translates with velocity $(1, 0.5)$, over a background translating with velocity $(0, -0.5)$, and on the Marbled Block sequence created by Michael Otte (Karlsruhe), a calibrated sequence where the camera is moved while the white block is translating (a reduced version of $256 \times 256 pixels$). All the parameters are known, so the correct map of the image displacements is known. In the first case, using the single scale algorithm with 9×9 windows and gaussian smoothing with $\sigma = 1.5$, the average angular distance was 4.59^o degrees (100% density). With the two scale algorithm (N=2) the distance was reduced to 3.97^o. Introducing the residual filter at each scale the distance became 3.52^o and adding the regularizing filter 2.92^o. For the marbled block sequences the average angular distance from the true displacements was 7.77^o (100% density) with the basic estimator, 7.45^o introducing the two scale decomposition, 6.65^o and 5.33^o with the residual filter and both the residual and regularizing filters. Another test has been done to demonstrate the possibility of speeding up the computation by window subsampling. Table 1 shows the average angular differences from the true motion of flows computed on the synthetic sequence of Fig. 4 in different cases. Reducing the number of equations by subsampling does not lead to a deterioration of the flow.

Window	dist	std.dev.	dens.
9×9 (81 eqns)	2.63°	6.71°	82.6
5×5 (25 eqns)	4.33°	7.49°	69.2
9×9 sub. (25 eqns)	2.86°	7.32°	77.1
9×9 sub. (9 eqns)	4.23°	8.33°	65.3

Table 1: If the number of equations is reduced by reducing the window size, the estimate gets poorer, while subsampling a window of the same size, the equations are reduced with no relevant effects on flow accuracy.

Quantitative comparison with other techniques - We have compared the average angular differences between our flows and the corresponding true displacements on some test sequences. The Translating Tree and Diverging Tree sequences, created by David Fleet, are synthetic sequences simulating the translation perpendicular and parallel to the optical axis of the camera of a slanted planar surface. In these sequences no motion discontinuities are present, but the presence of large motions (~ 2 pixel/frame) shows the utility of the multi-scale approach: the accuracy of our results is satisfactory compared with the values reported in[4, 10, 11] (Table 2). The "Yosemite Valley" sequence (Fig. 3 A), created by Lynn Quam, is a challenging test for optical flow algorithms, because it presents large displacements, occlusions and regions with poor texture. The flow obtained with our algorithm (Fig. 3 B) is more accurate than those obtained with other techniques (Table 2).

Algorithm	Translating Tree			Diverging Tree			Yosemite		
	ang.d.	std.dev.	dens.	ang.d.	std. dev.	dens.	ang.d.	std.dev.	dens.
Horn & Schunck	38.72°	27.67°	100	12.02°	11.72°	100	32.43°	30.28°	100
Heeger	4.53°	2.41°	57.8	4.49°	3.10°	74.2	10.51°	12.11°	15.2
Anandan	4.54°	2.98°	100	7.64°	4.96°	100	15.84°	13.46°	100
Lucas & Kanade	0.66°	0.67°	39.8	1.94°	2.06°	48.2	4.10°	9.58°	35.1
Fleet & Jepson	0.32°	0.38°	74.5	0.99°	0.78°	61.0	4.25°	11.34°	34.1
Weber & Malik	0.49°	0.35°	96.8	3.18°	2.50°	88.6	4.31°	8.66°	64.2
Xu	1.89°	4.23°	95.7	4.11°	6.56°	93.7	9.93°	11.03°	99.8
Our Impl.(raw)	0.51°	0.46°	95.0	3.97°	2.60°	95.0	4.01°	7.12°	70.9
Our Impl.(reg.)	0.25°	0.23°	95.0	2.07°	1.37°	95.0	2.82°	6.98°	70.9

Table 2: Quantitative comparison of the flow accuracy obtained on the sequences Translating Tree, Divering Tree and Yosemite with our algorithm and with other techniques.

Detection of motion edges - The final series of experiments aimed at detecting motion edges through the search for the directional maxima of the thinned residual function. Fig. 4 A shows the residual function obtained from the computation of the single - scale flow on the synthetic sequence described before. High values of the residual function do not correspond only to motion discontinuities due to the error in derivatives computation. With the multi-scale approach, the residual map clearly presents its highest values near the discontinuities (Fig. 4 B). Applying the residual function thinning as described in Section 5, a clear localization of the boundaries of the moving region is found (Fig. 4 C). The "motion edges" extracted from this map are finally superimposed to the original image in Fig. 4 D showing their proximity to the true boundaries of the moving region. In the case of the *Marbled Block* sequence, it is much more difficult to estimate the position of the motion boundaries, because flow differences between different objects are not so relevant. The refined residual map obtained from the reduced 256×256 sequence with our multi-scale algorithm is shown in Fig. 4E. In spite of the difficulties, the boundaries of the nearest column and of the marbled block are clearly revealed (Fig. 4 F).

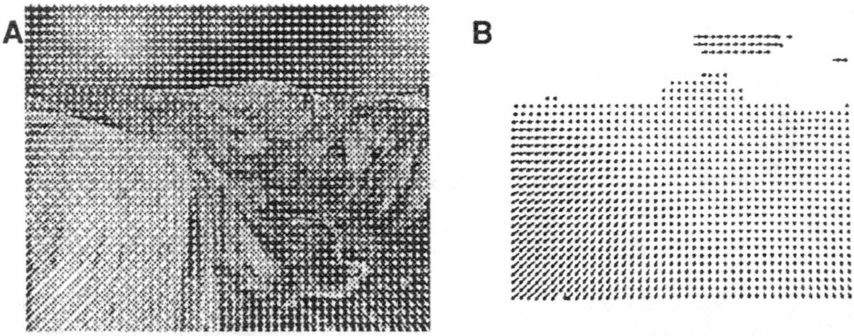

Fig. 3: Results obtained on the Yosemite Valley sequence. A: Sequence image with the true motion superimposed. B: Optical flow computed with our algorithm.

Use of shape models - The *motion edges* extracted with our technique are close to the boundaries of the moving regions, but are inevitably irregular and incomplete due to the texture properties and to errors. In order to refine the detection of these boundaries it is therefore necessary to introduce other information and other processing. If templates of the moving objects are known, a possible solution is a parametric fitting of the model through the detected edges (Fig. 5 A,B).

7 Conclusions

Our work shows that motion edges can be extracted and the quality of the optical flow estimation can be improved without the introduction of complex techniques. In a similar way also motion discontinuities can be satisfactorily revealed.

References

1. B.K.P.Horn & B.G.Schunck, "Determining optical flow," Art. Int. **17,** 185–203 (1981).
2. B. Lucas and T. Kanade, "An iterative image registration technique with an application to stereo vision" Proc. DARPA Image Und. Workshop 121–130 (1981).
3. M. Campani and A. Verri, "Motion analysis from first order properties of optical flow," CVGIP - Image Understanding **56,** 1:90–107 (1992).
4. J.L. Barron, D.J. Fleet and S.S. Beauchemin, "Performance of optical flow techniques" Int. J. Comp. Vision **12**, 1:43–77 (1994).
5. M. Otte and H.H. Nagel, "Optical flow estimation: advances and comparisons" Proc. Third ECCV, **14**, 51–60 (1994).
6. H. H. Nagel, "Optical Flow Estimation and the Interaction Between Measurement Errors at Adjacent Pixel Positions" Int. J. Comp. Vision **15**:3, 271–288 (1995).
7. E.De Micheli, S.Uras & V.Torre, "The accuracy of the computation of optical flow and of the recovery of motion parameters," IEEE Trans. PAMI **15,** 5:434–447 (1993).
8. F.Bartolini, V.Cappellini, C.Colombo, A.Mecocci, "Multiwindow least-square approach to the estimation of optical flow with discontinuities" Opt. Eng., **32,** 6: 1250–1256, (1993).
9. R. Battiti, E. Amaldi and C. Koch, "Computing Optical Flow Across Multiple Scales: An Adaptive Coarse to Fine Strategy", Int. J. Comp. Vision **6**:2, 133–145 (1991).
10. J. Weber and J. Malik "Robust Computation of Optical Flow in a Multi-Scale Differential Framework" Int. J. Comp. Vision **14,** 1:67–81 (1995).
11. S. Xu "Motion and Optical Flow in Robot Vision" Linköping Studies in Science and Technology - Thesis No. 442 (1994).

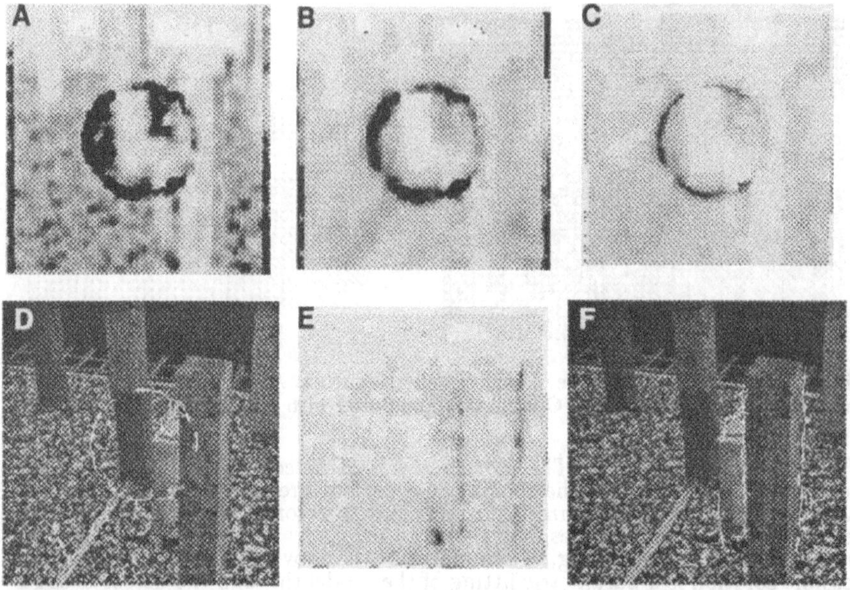

Fig. 4: Extraction of "motion edges". A: Residual function obtained from the single-scale flow computation on a synthetic with the marbled block texture. A central circle and the background are differently translating. B: Residual function obtained from the multi-scale algorithm on the same sequence. C: "Thinned" residual map. D: "Motion edges" superimposed to the image. E: Thinned multi-scale residual map obtained on the 256 × 256 Marbled Block sequence. F: "Motion edges" superimposed to the image.

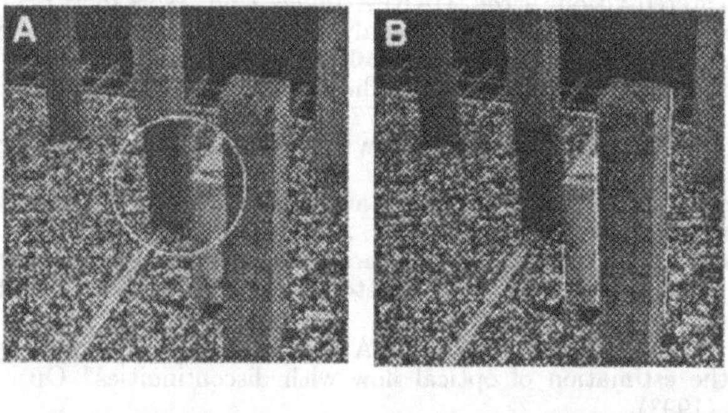

Fig. 5: A: Precise detection of the moving circle obtained from the search for the best circumference approximating the edges of Fig. 4D. Circles with radius between 20 and 80 pixels, with at least half of the points close to an edge and with a local maximum in the number of points close to edges were extracted. The parameters of the circle (center in 130, 129, radius 50 pixels) were correctly estimated. B: Improved detection of the boundaries obtained from the search for the nearly vertical segments of at least 15 pixels better approximating the edges of 4F.

Reliable Extraction of the Camera Motion using Constraints on the Epipole

Jonathan Lawn and Roberto Cipolla

Department of Engineering, University of Cambridge

Abstract. An accurate estimate of the epipole (direction of camera translation) is necessary if image motion is to be decomposed into rotational and translational components, which give the camera rotation and feature depths respectively. In this paper we introduce the Linearised Subspace Method to find direct constraints on the epipole which are independent of camera rotation and scene structure. We present methods to compute reliable constraints and their uncertainties from image motion. We show how erroneous constraints due to errors in tracking can be rejected and how the valid constraints should be combined to form accurate estimates of the direction of translation. Experimental results show these methods lead to improvements in the recovery of camera motion and that the uncertainty estimates are accurate and useful in detecting degenerate scene structure or camera motions.

1 Introduction

In the structure from motion problem image motion is used to recover unknown camera motion and scene structure. The Fundamental Matrix [?] is commonly used to encode the camera motion (up to a projective transformation) and the rigidity or epipolar constraint. It is independent of scene structure; can be estimated from point correspondences and has successfully been exploited in segmentation and outlier rejection algorithms [?]. In principle the Fundamental matrix can be decomposed to find the epipole (direction of translation) and hence the camera motion and scene structure up to a projective transformation. In practice this is very ill-conditioned and leads to inaccurate estimates of the epipole.

An accurate estimate of the epipole is, however, an essential component of the structure-from-motion problem if motion and scene structure is to be recovered. Finding the epipole accurately removes much of the uncertainty from the calculation of the rotational component of visual motion, and therefore also improves the calculation of the feature depths [?, ?], and the rejection of outliers caused by spurious or mis-tracked features.

This paper proposes a novel method – the Linearised Subspace Method– of finding linear constraints on the epipole from small regions of a full perspective image. Perturbation analysis is used to produce estimates of the uncertainty in these constraints and reliable constraints are combined to produce an accurate estimate of the epipole even in the presence of outliers and degenerate scene structure and motion. Experimental results are presented.

2 Constraints on the Epipole

If the parameters of the motion of the camera between frames $k-1$ and k (in the co-ordinate frame of k) are a rotation, $\mathbf{R}$, and a translation, $\boldsymbol{T}$, then the motion of each point is given by the rigid body transformation:

$$\boldsymbol{X}_i' = \mathbf{R}(\boldsymbol{X}_i + \boldsymbol{T}) \tag{1}$$

where $\boldsymbol{X}_i'$ and $\boldsymbol{X}_i$ are the vectors to the stationary feature point, i, in the coordinate frames of the frames $k-1$ and k respectively. Points in the images are represented by the 3-dimensional vectors $\boldsymbol{P}_i'$ and $\boldsymbol{P}_i$, where $\boldsymbol{X}_i = r_i \boldsymbol{P}_i$ and r_i is the unknown scalar depth.

The Subspace method [7] for obtaining a constraint on the position of the epipole is based on cancelling the rotational component of motion in a similar way to motion parallax [12]. A discrete time/view version of the Subspace method can be obtained by rearranging the rigid motion equation (1) into the product relationship below.

$$\boldsymbol{T}.(\boldsymbol{P}_i' \times \boldsymbol{P}_i) = \rho_i \boldsymbol{T}.\left(\boldsymbol{P}_i' \times (R^{-1}\boldsymbol{P}_i')\right) \tag{2}$$

Approximating $\rho_i = \frac{r_i'}{r_i}$ as a constant, $\bar{\rho}$, *which is intrinsic to the continuous time formulation used originally* [7], and taking a weighted sum across a set of points:

$$\boldsymbol{T}.\sum_i c_i(\boldsymbol{P}_i' \times \boldsymbol{P}_i) \approx \bar{\rho}\boldsymbol{T}.\sum_i c_i\left(\boldsymbol{P}_i' \times (R^{-1}\boldsymbol{P}_i')\right) \tag{3}$$

The weights, c_i, can be chosen to cancel the second order terms of $\boldsymbol{P}_i'$ by making

$$\sum_i c_i \boldsymbol{P}_i' \boldsymbol{P}_i'^T = 0 \tag{4}$$

and then the right hand side of Equation 3 will equal zero, and therefore

$$\boldsymbol{T}.\sum_i c_i(\boldsymbol{P}_i' \times \boldsymbol{P}_i) \approx 0 \tag{5}$$

for any camera rotation, R. *The linear sum cancels the rotational component of the visual motion, just as in Affine Motion Parallax [1, 10], giving a constraint on the direction of translation.* The constraint on the weights, c_i, in Equation 4 can be rewritten as

$$\begin{bmatrix} & p_{xi}'p_{xi}' & \\ \cdots & p_{yi}'p_{xi}' & \cdots \\ & p_{zi}'p_{xi}' & \\ & p_{yi}'p_{yi}' & \\ \cdots & p_{zi}'p_{yi}' & \cdots \\ & p_{zi}'p_{zi}' & \end{bmatrix} \begin{pmatrix} c_1 \\ c_2 \\ \vdots \end{pmatrix} = [\phi_1\ \phi_2\ \cdots]\,\boldsymbol{c} = \boldsymbol{\Phi}\boldsymbol{c} = \boldsymbol{0} \tag{6}$$

where $P'_i = [p'_{xi}\, p'_{yi}\, p'_{zi}]^T$, and Φ is a $6 \times n$ matrix with columns, ϕ_i made up of the 6 independent elements of the matrix $P'_i P'^T_i$. Given enough points ($n > 6$), a subspace of possible weights, $\mathbf{C}$, can be found.

$$\Phi\,[c_1\; c_2\; \cdots] = \Phi\mathbf{C} = 0 \tag{7}$$

This space is invariant to affine deformations of the image [7]. It is therefore possible to write (from Equation 5)

$$T^T \mathbf{MC} = 0 \tag{8}$$

where the $3 \times n$ matrix $\mathbf{M}$ is defined as

$$\mathbf{M} = \left[\, (P'_1 \times P_1)\; (P'_2 \times P_2)\; \cdots \right] \tag{9}$$

Since $\mathbf{M}$ and $\mathbf{C}$ can be computed from measurements in the image, each column of the matrix $\mathbf{MC}$ provides a constraint on the direction of translation and the epipole. These constraints are in fact equivalent to epipolar lines.

2.1 The Linearised Subspace Method

The accuracy and sensitivity to noise of the Subspace algorithm can be improved by looking at constraints from *small* regions of the image. If small image regions are used, then the equation for each weight vector, c_a (Equation 4) is approximately linear [11].

$$\sum_i \hat{c}_i P'_i = \mathbf{0} \tag{10}$$

(Small here means in a region of radius less than 0.1 radians approximately, so that the first order variations from the mean image position are significantly greater than the second order terms.) Only four points are now needed to extract a constraint instead of seven using the Heeger and Jepson Subspace method. This has advantages for outlier rejection (see Section 5). The linearisation may also improve the stability of the solution to noise on the image motion measurements.

Each region should only produce one constraint because under the weak perspective assumption for the small field of view in each neighbourhood the constraints will have the same direction. In the original algorithm [7], an arbitrary weight vector was chosen. Here we choose the constraint with the least uncertainty (as computed below). This takes into account image localisation errors and degeneracy in the structure being viewed.

Though each constraint, $\hat{c}_a$, (generated from four or more points) must come from a small region, different constraints can be generated from different small image regions in a full perspective image and combined to estimate the epipole. *This method is very similar to the Affine Motion Parallax algorithm* [10], *generalised for any number of feature points* (≥ 4). It has the same geometric intuitiveness: combine the visual motions of a number of features from a small image region in a linear sum, dependent on their positions only, to cancel the rotational component of the motion.

Figure 1(a) shows the image motion from a sequence of views of Trinity College. The image displacement vectors were automatically obtained by tracking "corner" features [5]. The Linearised Subspace method was used to compute constraints on the epipole by looking at 14 image regions. These are displayed in 1(b).

3 Uncertainty Analysis

For real, noisy data the epipole constraints will not be exact, and therefore constraints from more than two image regions will not intersect. A method is needed that will provide an estimate of the accuracy of each constraint. This will allow *excessively uncertain* constraints to be rejected; *optimal combination* of the good constraints to provide the least uncertain result possible; *degeneracy* of the structure or motion to be detected by measuring the significance of components quantitatively, and *outliers* (due to independent motion and incorrect visual motion measurements) to be detected.

The feature measurement errors on the image plane (which may be an output of the feature detector, or be considered constant and determined in advance) will determine the uncertainty in the constraint. The propagation of the image measurement errors to compute the uncertainty in the constraint is in fact directly linked to the camera motion and 3D structure of the points being viewed.

The approach to propagating the errors through the calculations is to assume that each vector or scalar (eg. $\boldsymbol{b}$) has a small additive gaussian uncertainty which can be represented by a covariance matrix ($\mathcal{V}[\boldsymbol{b}]$) [8]. These error predictions are then propagated through the calculations by linearising around each result algebraically. Differentiating about each input variable ($\boldsymbol{a}$) gives the Jacobian ($J_{\boldsymbol{a}}(\boldsymbol{b})$) and this shows how the input covariances contribute to the output covariance.

The constraint vector, $\mathbf{M}\boldsymbol{c}$, will have two elements to its uncertainty: $\delta\mathbf{M}\ \boldsymbol{c}$ and $\mathbf{M}\ \delta\boldsymbol{c}$. The contribution of both can be calculated. However, we would expect the "velocity component" ($\delta\mathbf{M}$) to be more uncertain than the component dependent on image position. We therefore expect the first term, $\delta\mathbf{M}\ \boldsymbol{c}$, to be the more significant. Experiments validate this [11]. We therefore calculate the variance on the constraint to be:

$$\mathcal{V}[\mathbf{M}\boldsymbol{c}] \approx \sum_i c_i^2\ \mathcal{V}[\boldsymbol{m}_i] = \sum_i c_i^2\ \mathcal{V}[\boldsymbol{P}_i \times \boldsymbol{P}'_i] \tag{11}$$

$$= \sum_i c_i^2 \left(\boldsymbol{P}'_{i\times} \mathcal{V}[\boldsymbol{P}_i]\, \boldsymbol{P}'^{t}_{i\times} + \boldsymbol{P}_{i\times} \mathcal{V}[\boldsymbol{P}'_i]\, \boldsymbol{P}^{T}_{i\times} \right) \tag{12}$$

where $\boldsymbol{a}_\times$ is the skew-symmetric matrix that produces the vector product with $\boldsymbol{a}$. Figures 1(c) and (d) show two constraints and their respective uncertainties. The uncertainty on the constraints depends on 3D structure of the points as well as the image localisation errors. Nearly planar scene structure results in a very uncertain constraint (Figure 1(d)).

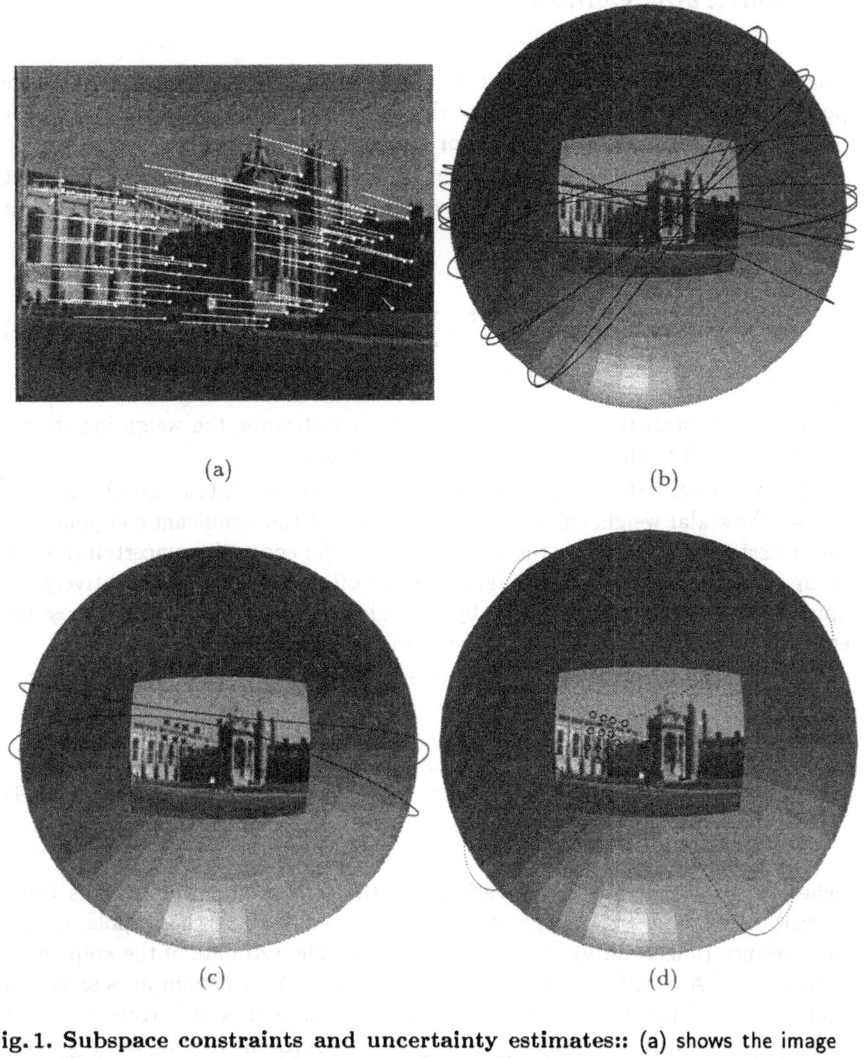

Fig. 1. Subspace constraints and uncertainty estimates:: (a) shows the image feature displacements measured between two frames. The features were automatically extracted and matched and approximately 6 are incorrect correspondences. The translation of the camera was along the path just visible in the bottom right corner, but the image motion is mostly due to rotation. (b) shows the set of constraints obtained from 14 image regions by the Linearised Subspace Method. (c) shows the 95% uncertainty band and the image points contributing to the constraint. (d) shows an excessively uncertain constraint arising from distant points which are approximately on a plane

4 Constraint Combination

The problem of finding the epipole now becomes finding the direction or image point which is the best intersection of the set of uncertain constraints. This must be done after "rogue" measurement or outliers are rejected.

The uncertainty of the epipole estimate can be minimised by weighted least squares estimation. Generalising the constraints as unit vectors $\{\boldsymbol{n}_i\}$, we must minimize:

$$\sum_i w_i(e.\boldsymbol{n}_i)^2 = \boldsymbol{e}^T \left(\sum_i w_i \boldsymbol{n}_i \boldsymbol{n}_i^T \right) \boldsymbol{e} = \boldsymbol{e}^T \mathbf{A} \boldsymbol{e} \tag{13}$$

Assuming the statistical independence of the constraints, the weighting should be proportional to the inverse of the scalar variance [9].

An initial estimate of the epipole (eg. from an unweighted estimate) is needed so that the scalar weight can be determined only by the significant component of the uncertainty (parallel to the epipole). Since the constraint uncertainties are often highly anisotropic, many iterations are often necessary. Alternatively, an approximate weight can be found by averaging the uncertainty for all possible epipoles (for example, using the trace of the variance matrix).

An attractive feature of this form of epipole estimation is that the weighted outer product sum, $\mathbf{A}$,

$$\mathbf{A} = \sum_i \mathbf{A}_i = \sum_i \frac{1}{\sigma_i^2} \boldsymbol{n}_i \boldsymbol{n}_i^T \tag{14}$$

(where $\sigma_i^2 = \boldsymbol{e}_0^T \mathcal{V}[\boldsymbol{n}_i]\, \boldsymbol{e}_0$ is the variance of the constraint normal in the (estimated) direction of the epipole, $\boldsymbol{e}_0$) encodes the estimate of the epipole, $\boldsymbol{e}$, and the variance (uncertainty) of the estimate, $\mathcal{V}[\boldsymbol{e}]$. The estimate of the epipole, $\boldsymbol{e}$, minimises $\boldsymbol{e}^T \mathbf{A} \boldsymbol{e}$ while $\mathcal{V}[\boldsymbol{e}]$ is the pseudo-inverse of $\mathbf{A}$. An example is shown in Figure 3 (c). Before being able to combine constraints it is important incorrect meaurement(outliers) of visual motion due to failures in tracking are rejected.

5 Outlier Rejection

There are many reasons why some of the visual motion measurements will be incorrect and should not be used in the structure-from-motion calculation. Incorrect visual motion measurements arise from poor matching or tracking of features and independently moving objects. They can be removed by trying to form a solution from as many measurements as possible without using any that do not consistent with the solution found.

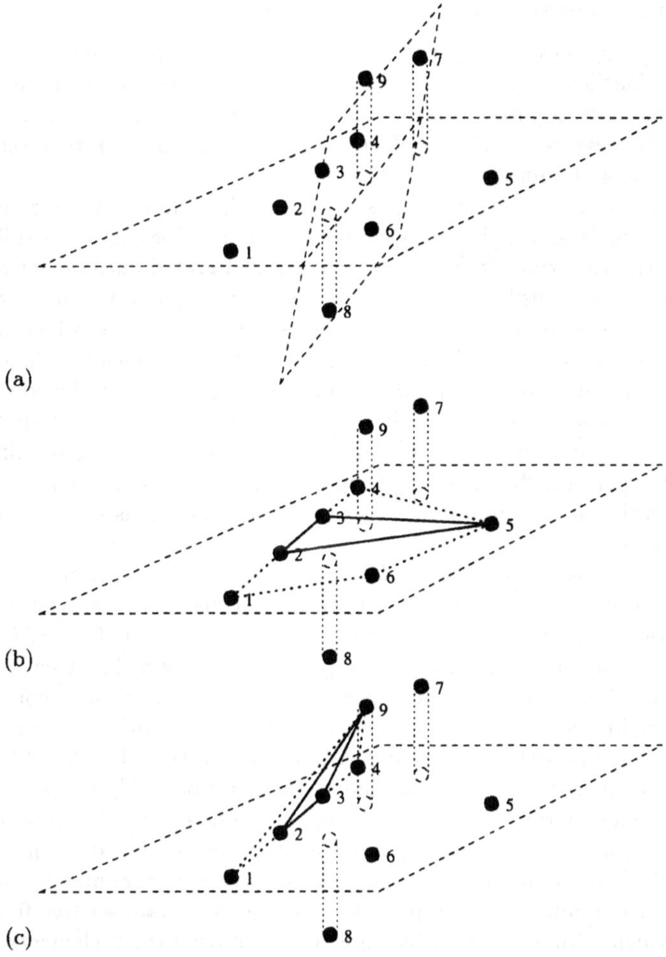

Fig. 2. Example rejection problem: Because the rejection algorithms considered in this section are generally applicable, a low-dimensional example problem will be used to introduce them: find a plane that fits a large number of the 3D points in a set. Each picture above shows the six inliers (1-6), the three outliers (7-9), and the plane. **(a) Poor initialisation using whole set:** Outliers are often of greater magnitude than true measurements, and can therefore greatly influence any initialisation using the whole set (as shown). In this case, take-one-out cross-validation would also fail because we have three significant outliers. **(b) Good solution from random initialisation set:** If the assembling of the plane estimate is initialised by three correct points (eg. 2, 3 and 5 as shown) then the other points can also be found. An average of four random trials (approximately) will be needed to find the plane. **(c) False solution from degenerate structure:** Solutions can also appear to be found when the verification comes from degenerate structure. Points 1 and 4 appear to be verifying the plane found from 2, 3 and 9, because they form a degenerate structure (a line) with all but one of these points. This effect should not cause trouble when a good solution can be found, as it will receive better verification (as in (a)). However, when the scene is truly degenerate, this can cause an (often strong) false positive.

5.1 Approaches to Rejection of Outliers

The aim of every rejection algorithm is to find a self-consistent set of "elements" with as few outliers as possible. However there are a large number of algorithms that can be used. Figure 2 presents an example problem of finding a plane of points in the presence of outliers, which will be used to demonstrate the strengths and weaknesses of a number of algorithms.

One method for rejecting outliers from any algorithm, is to start with the complete set of data, and then to reject the subset that does not *agree* sufficiently with the estimate formed from the set. This can be iterated until the set is stable. Though this seems sensible, large outliers can sufficiently bias the initial result so that they are never removed (see Figure 2a). A better approach, which was used in earlier papers on Affine Motion Parallax [10], is to disassemble the estimate using *take-one-out cross-validation* [15]. Each element is removed in turn and is checked against an estimate formed by the others. The constraint that agrees least is removed and the process repeated, until all constraints agree sufficiently. Though this performs better, the computational cost is high, and a few outliers can sufficiently bias the initial result so that removing one is still not sufficient (see Figure 2a).

A different method is to *assemble* the estimate, starting from an estimate from a minimal set, and combining only those constraints that agree. If too few constraints agree then a new initial estimate is needed. RANSAC [3] is a popular method that generalises this approach (see Figure 2b). Generally, this will be more efficient than the Hough methods. The problem here comes when the initial estimate is uncertain. If the uncertainty of the initial solution is great, then it is probably best to find a new initialisation, rather than build from this one. Even if the uncertainty in reasonable, there are probably still a number of erroneous elements that could be included, and different results may be found depending on how many of the agreeing elements are combined at once.

Each of the above methods can be iterated or two or more may be cascaded. Here we use a minimal basis to provide an initial candidate set free from gross outliers, which is then depleted by iteratively removing those elements that do not agree with a combined estimate.

5.2 Alternative Stages of Rejection

Most structure-from-motion algorithms that use outlier rejection, attempt to form a full projective egomotion solution and then use the epipolar constraint to remove outliers [14]. However, if the egomotion determination is broken into stages, as it has been in the algorithms presented here (finding the constraints and then the epipole before the egomotion), then the rejection can also be split into stages. Here, in addition to checking if the individual feature motions comply with the projective egomotion found for the set, we can also test if the individual features in small regions agree with the constraint found from that region, and if the constraints agree with one another. In this section we analyse and then test each stage in turn, in an attempt to determine if any advantage is gained from this approach.

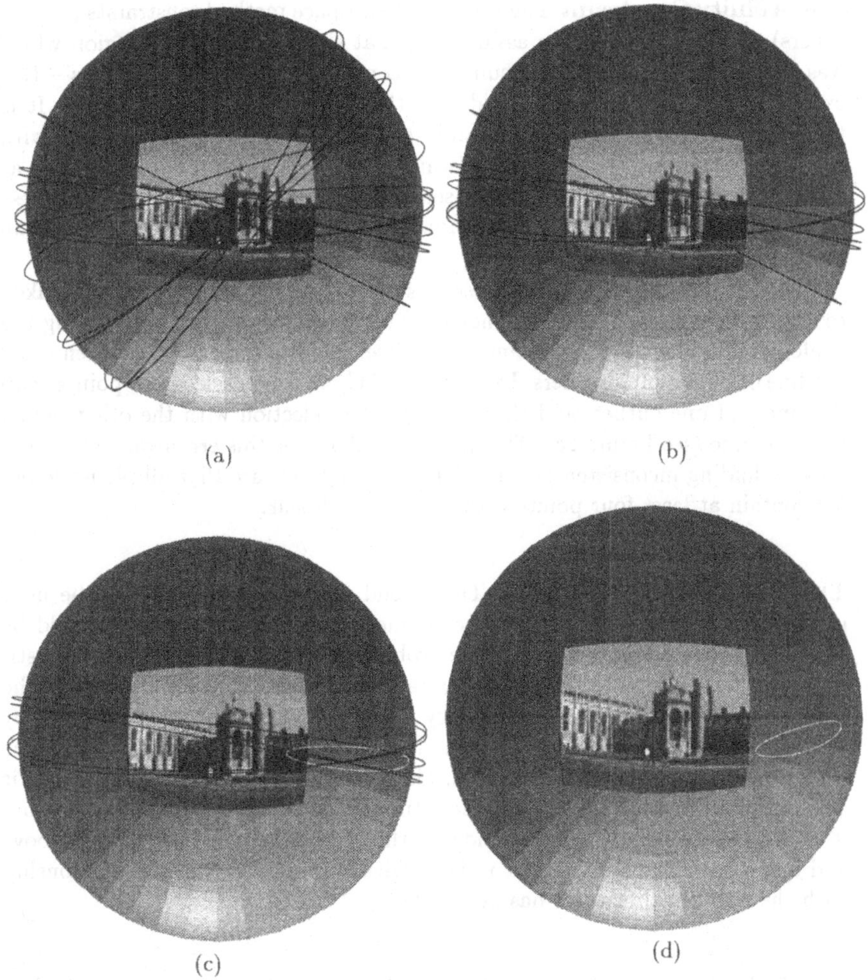

Fig. 3. The outlier rejection constraints: This figure gives a limited demonstration of the Consistency constraint and excessive error rejection for the features in Figure 5. (a) shows all the constraints found, (b) shows those with a reasonably small uncertainty (as in Figure ??) and (c) shows all but one being accepted as agreeing with an epipole estimate, which is also shown as its 95% certainty ellipse. (d) shows the epipole ellipse given by the Fundamental Matrix calculated with the inliers and constrained to the epipole from (c).

The Affinity Constraint The Linearised Subspace method constraints (amongst others) are formed using the assumption that each comes from a region which was deforming affinely. This assumption constrains the feature motions – this can be seen more easily by considering the Affine Fundamental Matrix. It is therefore possible to test whether all of the features in the region comply with this assumption, and to reject those which do not. This could in fact be used as a perceptual grouping stage for any set of feature motions detected, regardless of the structure-from-motion algorithm that will use them, so long as the small motion approximation applies.

Though the algorithm initially seems simple to formulate, care must be taken to ensure that a few outliers are not masking a degenerate situation. Using the analogy of finding a plane of points in 3-space, the degenerate case is when there is a line of points plus outliers. Using RANSAC, we might select two points from the line and one outlier, and then validate this selection with the other points from the line (see Figure 2c). Though we could accept this erroneous set (in the hope of finding inconsistencies later) it is better to ensure that all planes found are contain at least four points with no three collinear.

The Consistency Constraint The second rejection stage that can be used is when the epipole constraints are combined: the constraint vectors should be coplanar if they are accurate, and the problem becomes finding the epipole estimate which is normal to the most constraint estimates. This is simpler than the Affinity Constraint, because degeneracy is more obvious, but is complicated by the tendency with all these constraint algorithms for correct and erroneous line constraints to pass through the image. This produces a bias in the solution but it also produces more candidate epipole estimates (intersections). For this reason, it is highly beneficial if the rejection methods discussed in the sections above and below are efficient. Another problem with this stage is that the relationship with the individual features has been lost.

The Epipolar Line Constraint Once the epipole has been estimated, then the projective "rotation" matrix can be found, and the rotational component of image motion predicted for each feature. The remaining image motion should be the translational component which will be towards (or away from) the epipole but have unknown magnitude (due to unknown feature depth). Any remainder not in this direction is evidence of an incompatibility. The effect should be similar to the Affinity Constraint above.

This is the rejection method most commonly used as the Fundamental Matrix finds the epipolar lines directly [14]. Here we must ensure that the Fundamental Matrix complies with the epipole found. It should therefore be formed from only the inliers from the previous stages, and should then be forced to have the same epipole. This is achieved by projecting the vectors that make up the Fundamental Matrix estimate onto a plane perpendicular to the epipole found (ie. using $\frac{1}{e^T e}\left(e^T e - e e^T\right) \mathbf{F}$, where e is the epipole), giving the nearest complying matrix to that measured (by the Frobenius norm). This matrix can also be decomposed to give the camera motion estimate.

The Positive Depth Constraint The previous stages all effectively used the epipolar constraint, each time updating the epipolar line to include more evidence. Another useful though often ignored constraint is that all objects must be in front of the camera, and therefore all have translational motions in the same sense: either all towards or all away from the epipole. This can be applied in this scheme with the Epipolar Line Constraint. However, an egomotion estimate is needed to determine the rotational component of the visual motion, and therefore the Fundamental Matrix alone is not sufficient.

5.3 Using Uncertainty Estimates in Rejection

Uncertainty estimates can be used to provide intelligent bounds on the rejection or acceptance in the algorithms above: the χ^2 test on the Mahalanobis distance provides a threshold value, the confidence bound of which can be set [4]. This is the strictest use we have for the uncertainty estimates, as we are now accepting or rejecting on their absolute values, and not just weighting according to their relative values. If the uncertainty estimates are considered inaccurate then a "safety factor" can be used to reduce either false positives or negatives. However this does not detract from the advantages of having a threshold value that reflects the uncertainty in the element and the proposed solution.

6 Conclusions

Figures 4 and 5 show two contrasting situations. In the first, the epipole is approximate in the centre of the image. Most existing algorithms would have little difficulty in localising this epipole. There are a number of incorrect feature motion measurements, but both the (linear, non-iterative) Fundamental Matrix and the Linearised Subspace methods give good results. In the second however, the epipole is well right of the image. Here the Fundamental Matrix estimate is wrong, again placing the epipole in the image. The Linearised Subspace method gives an accurate result though, demonstrating our belief that epipole constraints can be both more accurate and more robust. The uncertainty predicted for the epipole estimate is high, but in both cases, applying this constraint to the Fundamental Matrix will ensure that the decomposition is correct. Another benefits of our approach is the uncertainty estimages are accurate.

Future work will concentrate on how best to select the small regions to give good coverage without excessive computation, and whether rigorous optimal combination is computationally viable. We conclude that the Fundamental Matrix benefits considerably from more direct estimates of the epipole. We have presented a novel epipole constraint algorithm and shown how to combine the constraints. We have also shown that outlier rejection is more reliable if it is performed at each stage of the calculation.

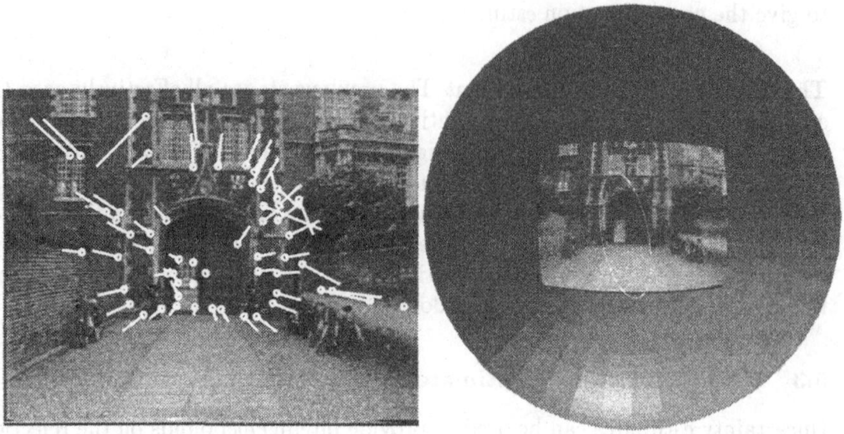

(a) (b)

Fig. 4. Example needing considerable rejection: (a) shows the flow measured between two frames using approximate calibration. This motion, towards the centre of the doors with little rotation, is easier than the previous motion for most algorithms. However, a considerable proportion of features have incorrect correspondences. (b) shows the epipole estimated after using the Linearised Subspace method and the rejection methods outlined.

References

1. R. Cipolla, Y. Okamoto, and Y. Kuno. Robust structure from motion using motion parallax. In *Proc. 4th International Conf. on Computer Vision*, pages 374–382, 1993.
2. O.D. Faugeras, Q.-T. Luong, and S.J. Maybank. Camera self-calibration: Theory and experiments. In G. Sandini, editor, *Proc. 2rd European Conf. on Computer Vision*, pages 321–334. Springer–Verlag, 1992.
3. M. A. Fischler and R. C. Bolles. Random sample consensus: A paradigm for model fitting with applications to image analysis and automated cartography. *Communications of the ACM*, 24(6):381–395, 1981.
4. G.H. Golub and C.F. van Loan. *Matrix Computations.* North Oxford Academic, 1986.
5. C.G. Harris. Geometry from visual motion. In A. Blake and A. Yuille, editors, *Active Vision*, chapter 16. MIT Press, 1992.
6. R.I. Hartley and P. Sturm. Triangulation. In *Proceedings of the DARPA Image Understanding Workshop*, pages 957–966, 1994.
7. A.D. Jepson and D.E. Heeger. Linear subspace methods for recovering translational direction. In L. Harris and M. Jenkin, editors, *Spatial vision in humans and robots.* Cambridge University Press, 1993. Also, University of Toronto, Department of Computer Science, Technical Report: RBCV-TR-92-40, April 1992.
8. K. Kanatani. Computational projective geometry. *Computer Vision, Graphics and Image Processing (Image Understanding)*, 54(3):333–348, 1991.

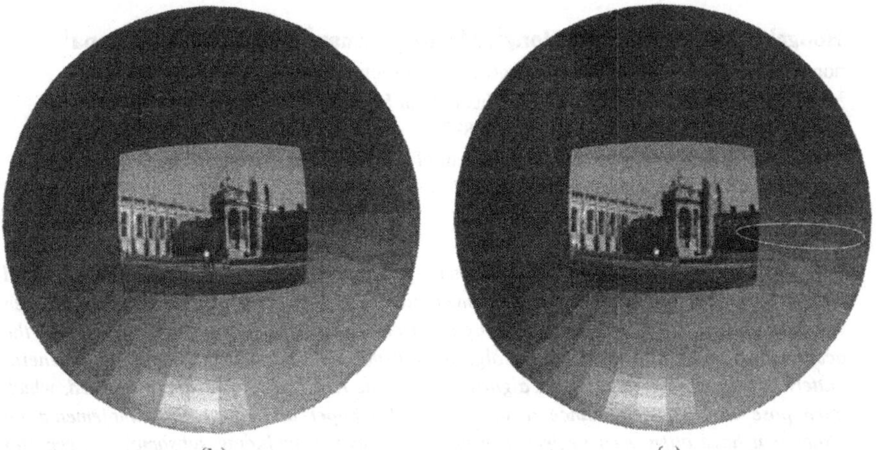

(b) (c)

Fig. 5. Example with challenging motion: (a) shows the flow measured between two frames using approximate calibration. Half a dozen instances of poor feature correspondences are visible, mostly in the top left corner. The translation was along the path just visible in the bottom right corner, but the image motion is mostly due to rotation. (b) shows the epipole and uncertainty found by the linear Fundamental Matrix method and rejection, which is assigned a low uncertainty but is incorrect, and (c) the same using the Linearised Subspace method and the rejection methods outlined, which is very close to the actual epipole.

9. K. Kanatani. Renormalization for unbiased estimation. In *Proc. 4th International Conf. on Computer Vision*, pages 599–606, 1993.
10. J.M. Lawn and R. Cipolla. Robust egomotion estimation from affine motion-parallax. In *Proc. 3rd European Conf. on Computer Vision*, pages I.205–210, 1994.
11. J.M. Lawn and R. Cipolla. Reliable extraction of the camera motion using constraints on the epipole. Technical Report CUED/F-INFENG/TR.240, University of Cambridge, 1995.
12. H.C. Longuet-Higgins and K. Prazdny. The interpretation of a moving retinal image. *Proc. of the Royal Society of London, Series B*, 208:385–397, 1980.
13. C. Rothwell, G. Csurka, and O. Faugeras. A comparison of projective reconstruction methods for pairs of views. In *Proc. 5th International Conf. on Computer Vision*, pages 932–937, 1995.
14. P.H.S. Torr. *Motion Segmentation and Outlier Detection*. PhD thesis, Department of Engineering Science, University of Oxford, 1995.
15. D.H. Wolpert. Combining generalizers using partitions of the learning set. In L.Nadel and D.Stein, editors, *1992 Lectures in Complex Systems*. Addison-Wesley, 1993.

Accuracy vs. Efficiency Trade-offs in Optical Flow Algorithms

Hongche Liu[†‡], Tsai-Hong Hong[‡], Martin Herman[‡], and Rama Chellappa[†]
hongche@etak.com, hongt@cme.nist.gov, herman@cme.nist.gov, rama@cfar.umd.edu
[‡]Intelligent Systems Division, National Institute of Standards and Technology (NIST)
Blg. 220, Rm B124, Gaithersburg, MD 20899
[†]Center for Automation Research/Department of Electrical Engineering,
University of Maryland, College Park 20742

Abstract

There have been two thrusts in the development of optical flow algorithms. One has emphasized higher accuracy; the other faster implementation. These two thrusts, however, have been independently pursued, without addressing the accuracy vs. efficiency trade-offs. Although the accuracy-efficiency characteristic is algorithm dependent, an understanding of a general pattern is crucial in evaluating an algorithm as far as real world tasks are concerned, which often pose various performance requirements. This paper addresses many implementation issues that have often been neglected in previous research, including subsampling, temporal filtering of the output stream, algorithms' flexibility and robustness, etc. Their impacts on accuracy and/or efficiency are emphasized. We present a critical survey of different approaches toward the goal of higher performance and present experimental studies on accuracy vs. efficiency trade-offs. The goal of this paper is to bridge the gap between the accuracy and the efficiency-oriented approaches.

1. Introduction

Whether the results of motion estimation are used in robot navigation, object tracking, or some other applications, one of the most compelling requirements for an algorithm to be effective is adequate speed. No matter how accurate an algorithm may be, it is not useful unless it can output the results within the necessary response time for a given task. On the other hand, no matter how fast an algorithm runs, it is useless unless it computes motion sufficiently accurately and precisely for subsequent interpretations.

Both accuracy and efficiency are important as far as real world applications are concerned. However, recent motion research has taken two approaches in opposite directions. One neglects all considerations of efficiency to achieve the highest accuracy possible. The other trades off accuracy for speed as required by a task. These two criteria could span a whole spectrum of different algorithms, ranging from very accurate but slow to very fast but highly inaccurate. Most existing motion algorithms are clustered at either end of the spectrum. Applications which need a certain combination of speed and accuracy may not find a good solution among these motion algorithms. To evaluate an algorithm for practical applications, we propose a 2-dimensional scale where one of the coordinates is accuracy and the other is time efficiency. In this scale, an algorithm that allows different parameter settings generates an accuracy-efficiency (AE) curve, which will assist users in understanding its operating range (accuracy-efficiency trade-offs) in order to optimize the performance.

In evaluating the accuracy-efficiency trade-offs, we also consider implementation issues such as subsampling, temporal filtering and their effects on both accuracy and speed.

Since Barron, et al. have published a detailed report [4] regarding accuracy aspects of optical flow algorithms, we start here with a survey of real-time implementations.

2. Previous Work on Real-Time Implementations

Regarding the issue of speed, there is a prevailing argument in most motion estimation literature that with more advanced hardware in the near future, the techniques could be implemented to run at frame rate [4] . In a recent report, many existing algorithms' speeds (computing optical flow for the diverging trees sequence) are compared and compiled in a tabular form[5] . We use the data from this table and calculate the time (in years) it may take for these algorithms to achieve frame rate, assuming computing power doubles every year [28] . This result is displayed in Table 1. Note that some

Table 1: Existing algorithms' speed and expected time to achieve frame rate

Techniques	Horn	Uras	Anandan	Lucas	Fleet	Bober
Execution time (min:sec) (from [5])	8:00	0:38	8:12	0:23	30:02	8:10
Approximate execution time on HyperSparc 10	2:00	0:10	2:03	0:06	6:00	2:03
Expected time to achieve frame rate (years)	12	8	12	7	14	12

algorithms can take up to 14 years (from when the table was compiled) to achieve frame rates. This would drastically limit the potential of such algorithms for many practical applications over the next decade.

There have been numerous attempts to realize fast motion algorithms. There are two major approaches: the hardware approach and the algorithmic approach. They are summarized in Table 2 and Table 3 given below and elaborated in the following paragraphs.

Table 2: Real-time motion estimation algorithms—hardware approach

Category	Type	Difficulties
Parallel computers	Connection machine [6] [26] [37] [39] , Parsytec transputer [32] and hybrid pyramidal vision machine (AIS-4000 and CSA transputer)[11]	high cost, weight and power consumption
Image processing hardware	PIPE [1] [9] [29] [35] , Datacube [25] and PRISM-3 [27]	low precision
Dedicated VLSI chips	Vision Chips: gradient method [23] [34] , correspondence method [10] [33] and biological receptive field design [12] [22]	low resolution
	Non-Vision Chips: analog neural networks [17] digital block matching technique [3] [15] [38]	coarsely quantized estimates

Table 3: Real-time motion estimation algorithms—algorithmic approach

Technique	Algorithms	Difficulties
Sparse feature motion	tracking [1] [20] [24] , computing time-to-contact (and hence obstacle avoidance)[9] and segmentation [32]	requirement of temporal filtering
Special constraints	constraint on motion velocity [8] , constraint on projection [40]	constraint on input images
Efficient algorithm	1-D spatial search [2] , separable filter design (Liu's algorithm[18])	requirement of careful algorithm design

The hardware approach uses specialized hardware to achieve real-time performance. There have been three categories of specialized hardware employed for motion estimation: parallel computers, specialized image processing hardware and dedicated VLSI chips. The hardware approach generally suffers from high cost and low precision.

The most popular algorithmic method is to compute sparse feature motion. Recent advances in this approach have enabled versatile applications including tracking [1] [20] [24] , computing time-to-contact (and hence obstacle avoidance)[9] and even segmentation [32] , which were believed to be better handled with dense data. However, in order to interpret the scenes with only sparse features, these algorithms need to use extensive temporal information (e.g., recursive least squares, Kalman filtering), which is time-consuming. Therefore, either the speed is not satisfactory[1] or they need to run on special hardware to achieve high speed[9] [20] [32] .

Another method is to constrain the motion estimation to a more tractable problem. For example, images can be subsampled so that the maximum velocity is constrained to be less than 1 pixel per frame. Therefore, a correlation method [8] can simply perform temporal matching in linear time instead of spatial searching in quadratic time. Another technique is to use a different projection. For example, any 3-D vertical line appears as a radial line in a conic projected image. This fact has been exploited for real-time navigation[40] .

Another elegant idea is to work on efficient design and implementation of the flow estimation algorithm. The main goal of this approach is to reduce the computational complexity. Suppose the image size is S and the maximum motion velocity is V. Traditional correlation algorithms performing spatial search or gradient-based algorithms using 2-D filters have the complexity $O(V^2S)$. Several recent spatio-temporal filter based methods [13] [14] even have $O(V^3S)$ complexity. However, some recent algorithms have achieved $O(VS)$ complexity. These include a correlation-based algorithm that uses 1-D spatial search [2] and a gradient-based algorithm [18] that exploits filter separability. These algorithm are so efficient that they achieve satisfactory rate on general purpose workstations or microcomputers.

3. Accuracy vs. Efficiency Trade-offs

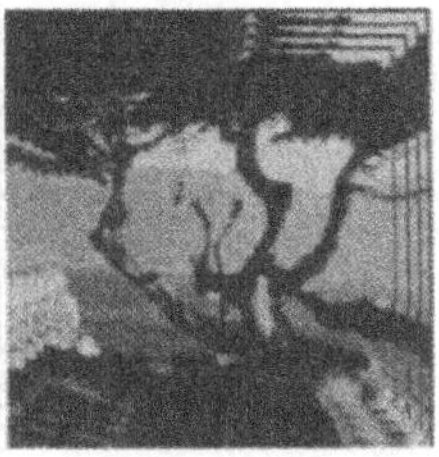

Fig 1. Diverging tree sequence

Although real-time is often used to mean video frame rates, in this paper, real-time is loosely defined as sufficiently fast for interactions with humans, robots or imaging hardware. The following subsections discuss the issues that are only of interest when one is concerned about both accuracy and speed. All the experiments illustrating our discussions are done on the diverging tree sequence[4] .

3.1 Accuracy-efficiency curve

If a motion algorithm is intended to be applied in a real-world task, the overall performance, including accuracy and efficiency, should be evaluated. Analogous to the use of electronic devices (e.g., transistor), without the knowledge of an algorithm's full operating range and characteristics, one may fail to use it in its optimal condition. Using accuracy (or error) as one coordinate and efficiency (or execution time) as the other, we propose the use of a 2-D accuracy-efficiency (AE) curve to characterize an algorithm's performance. This curve is generated by setting parameters in the algorithm to different values.

For correlation methods, the template window size and the search window size are common parameters. For gradient methods, the (smoothing or differentiation) filter size is a common parameter. More complex algorithms may have other parameters to consider. The important thing is to characterize them in a quantitative way.

For optical flow, accuracy has been extensively researched in Barron, et al.[4] . We will use the error measure in [4] , that is, the angle error between computed flow $(u_c, v_c, 1)$ and the ground truth flow $(u, v, 1)$, as one quantitative criterion. For efficiency, we use throughput (number of output frames per unit time) or its reciprocal (execution time per output frame) as the other quantitative criterion.

In the 2-D performance diagram depicted below (Fig 2.), the x axis represents the angle error; the y axis represents the execution time. A point in the performance diagram corresponds to a certain parameter setting. The closer the performance point is to the origin (small error and low execution time), the better the algorithm is. An algorithm with different parameter settings spans a curve, usually of negative slope. The distance from the origin to the AE curve represents the algorithm's AE performance. In Fig 2, there are two AE curves and several points*. It can be seen that some algorithms (e.g., Fleet & Jepson[13]) may be very accurate but very slow while some

* The implementations of all algorithms except Liu, et al. and Camus are provided by Barron [4] . Some of the algorithms produce different density, we simply project the error by extrapolation. In Liu's curve, the filter size used range from 5x5x5 to 17x17x11. In Camus's curve the template size ranges from 7x7x2 to 7x7x10. The execution time for all algorithms is the approximate elapsed time running on the same machine (80MHz HyperSparc 10 board).

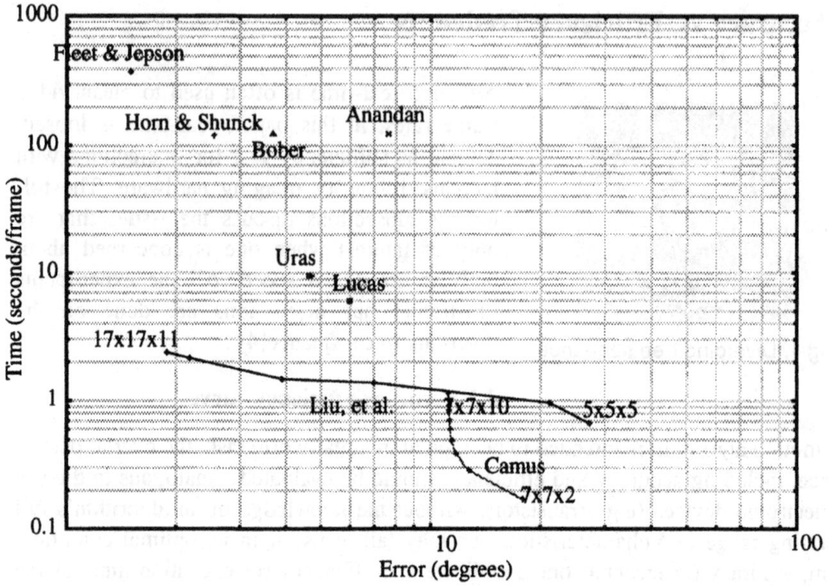

Fig 2. 2-D performance diagram

algorithms (e.g., Camus[8]) may be very fast but not very accurate. In terms of AE performance, Liu, et al.'s algorithm [18] is most flexible and effective because the curve is closest to the origin. It is also interesting to find that Liu, et al.'s curve [18] is relatively horizontal and Camus's [8] curve is relatively vertical they intersect each other.

The AE curve is also useful in understanding the effect and cost of certain types of processing. For example, Fig 3 shows the effect and cost of using different orders of image derivatives in Liu, et al.'s algorithm. The trade-off is clearer in this example. Using only up to second order derivatives saves 50% in time while sacrificing 85% in accuracy.

3.2 Subsampling effect

The computational complexity of an optical flow algorithm is usually proportional to the image size. However, an application may not need the full resolution of the digitized image. An intuitive idea to improve the speed is to subsample the images. Subsampling an image runs the risk of undersampling below the Nyquist frequency, resulting in aliasing, which can confuse motion algorithms. To avoid aliasing, the spatial sampling distance must be smaller than the scale of image texture and the temporal sampling period must be shorter than the scale of time. That is, the image intensity pattern must evidence a phase shift that is small enough to avoid phase ambiguity[30] .

Subsampling should be avoided on an image sequence with high spatial frequency and large motion. The aliasing problem can be dealt with by smoothing (low-pass filtering)

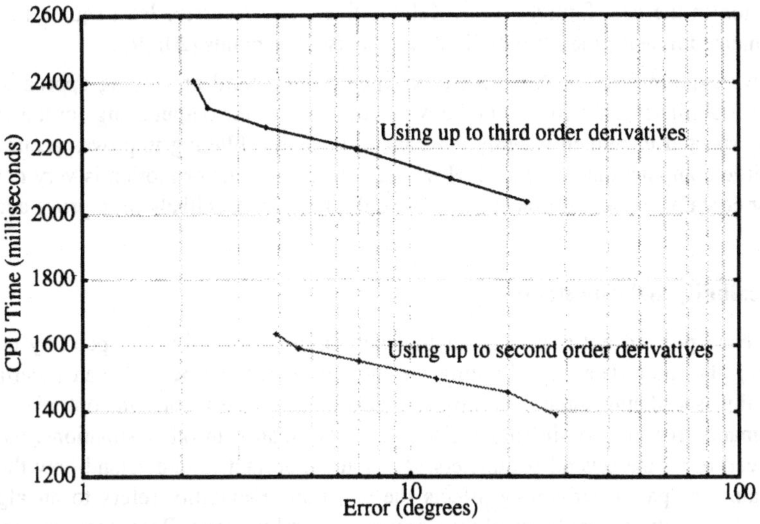

Fig 3. The effect and cost of using different order of derivatives.

before subsampling. However, since smoothing is done on the original images and the computational cost is proportional to the original image size, the advantage of subsampling is lost.

Aliasing is not the only problem in subsampling. Object size in subsampled images is reduced quadratically but the object boundaries are reduced linearly (in terms of number of pixels). Hence the density of motion boundaries are higher. This is detrimental to optical flow algorithms in general.

In short, subsampling can improve efficiency but needs a careful treatment.

3.3 Temporal processing of the output

Most general purpose optical flow algorithms are still very inefficient and often operate on short "canned" sequences of images. For long image sequences, it is natural to consider the possibility of temporally integrating the output stream to achieve better accuracy.

Temporal filtering is often used in situations where noisy output from the previous stage cannot be reliably interpreted. Successful applications of temporal filtering on the output requires a model (e.g., Gaussian with known variance) for noise (Kalman filtering) or a model (e.g., quadratic function) for the underlying parameters (recursive least squares). Therefore, these methods are often task specific.

A general purpose Kalman filter has been proposed in [31] where the noise model is tied closely to the framework of the method. In this scheme, an update stage requires point-to-point local warping using the previous optical flow field (as opposed to global warping using a polynomial function) in order to perform predictive filtering. It is computationally very expensive and therefore has little prospect of real-time implementation in the near future. So far, Kalman filters and recursive least square filters implemented in real-time are only limited to sparse data points [20] [9] .

We have experimented with a simple, inexpensive temporal processing approach—exponential filtering. We found out that when the noise in the output is high or the output is expected to remain roughly constant, exponential filtering improves accuracy with little computational overhead. However, when the scene or motion is very complex or contains numerous objects, exponential filtering is less likely to improve accuracy.

3.4 Flexibility and robustness

It has been pointed out that some motion algorithms achieve higher speed by constraining the input data, e.g., limiting the motion velocity to be less than a certain value, thus sacrificing some flexibility and robustness. Some algorithms optimize the performance for limited situations. For good performance in other situations, users may need to retune several parameters. It is thus important to understand how these constraints or parameter tuning affects the accuracy. Flexibility refers to an algorithm's capability to handle widely varying scenes and motions. Robustness refers to resistance to noise. These two criteria prescribe an algorithm's applicability to general tasks.

To evaluate an algorithm's flexibility, we have conducted the following simple experiment. A new image sequence is generated by taking every other frame of the diverging trees sequence. The motion in the new sequence will be twice as large as that in the original sequence. We then run algorithms on the new sequence using the same parameter setting as on the original sequence and compare the errors in the two outputs. A flexible algorithm should yield similarly accurate results, so we observe performance variation rather than absolute accuracy here. Fig 4 illustrates the results. The algorithms' performance variation for these two sequences ranges from 16% (Liu, et al.) to 75% (Horn & Shunck). Fleet and Jepson's algorithm, which has been very accurate, failed to generate nonzero density on the new sequence so is not plotted.

To evaluate an algorithm's noise sensitivity or robustness, we have generated a new diverging tree sequence by adding Gaussian noise of increasing variance and observed the algorithms' performance degradation. Fig 4 illustrates the algorithms' noise sensitivity. Some algorithms (Lucas & Kanade, Liu, et al., Anandan, Camus) have linear noise sensitivity with respect to noise magnitude; some (Fleet and Jepson) show quadratic noise sensitivity.

3.5 Output density

With most algorithms, some thresholding is done to eliminate unreliable data and it is hoped that the density is adequate for the subsequent applications. In addition, the threshold value is often chosen arbitrarily (by users who are not experts on the algorithms) without regard to the characteristics of the algorithms. The important characteristics that should be considered are flexibility and robustness to noise. If an algorithm is accurate but not flexible and not robust to noise, then it is better off gener-

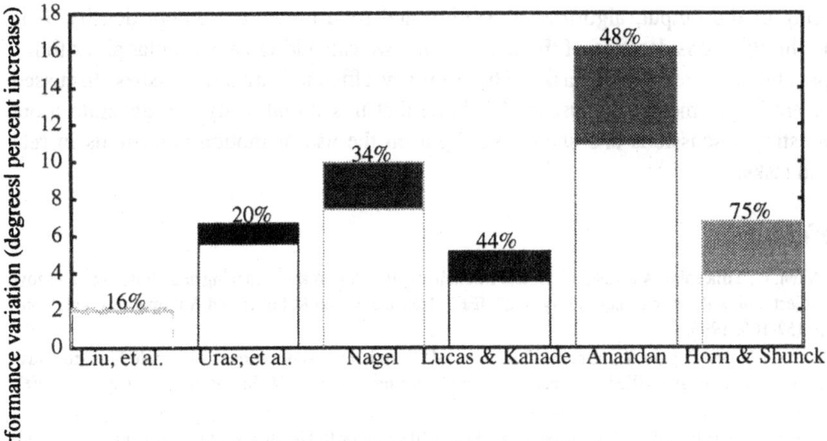

Fig 4.Algorithms's flexibility in handling different motion.

Noise sensitivity analysis on diverging trees sequence

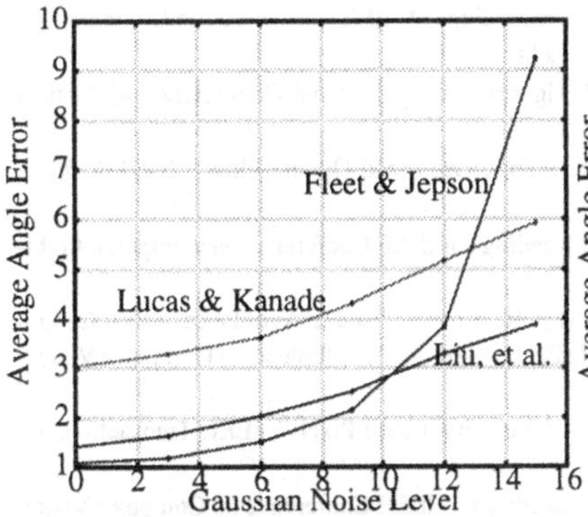

Fig 4.1 Noise sensitivity for 50% density data

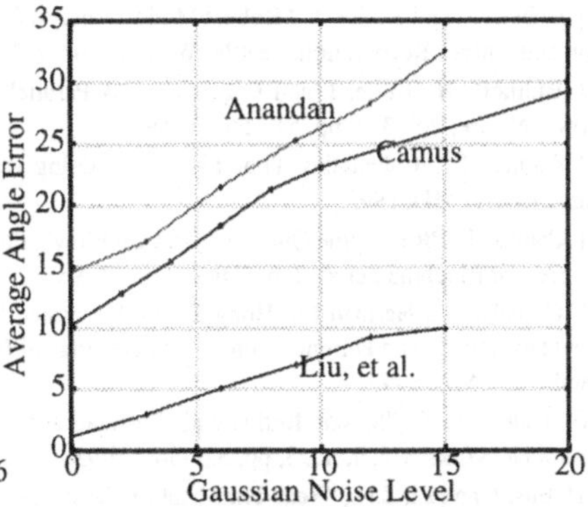

Fig 8.2 Noise sensitivity for 100% density data.

ating a sparse field because the more data it outputs, the more likely it will contain noisy data. However, the output density should really be determined by the requirements of the subsequent applications. Although dense flow field is ideal, selecting the right density of sufficiently accurate output is perhaps a more practical approach.

Specific experiments in [18] [19] using NASA sequence have shown output density as well as accuracy is the decisive factor in obstacle avoidance.

4. Conclusion

Motion research has typically focused on only accuracy or only speed. We have reviewed many different approaches to achieving higher accuracy or speed and pointed out their difficulties in real world applications. We also have raised the issues of accuracy-efficiency trade-offs resulting from subsampling effects, temporal processing of the output, algorithm flexibility and robustness, and output density. It is only through consideration of these issues that we can address a particular algorithm's applicability to real world tasks. The accuracy-efficiency trade-off issues discussed here are by no means exhaustive. We hope that this initial study can generate more interesting discussions and shed some light on the use of motion algorithms in real world tasks.

References

[1] Allen, P., Timcenko, A., Yoshimi, B. and Michelman, P., "Automated Tracking and Grasping of a Moving Object with a Robotic Hand-Eye System", IEEE Transactions on Robotics and Automations, vol. 9, no. 2, pp. 152-165, 1993.

[2] Ancona, N. and Poggio, T., "Optical Flow from 1-D Correlation: Applications to a Simple Time-to-Crash Detector", Proceedings of IEEE International Conference on Computer Vision, Berlin, Germany, pp. 209-214, 1993.

[3] Artieri, A. and Jutand, F., "A Versatile and Powerful Chip for Real-Time Motion Estimation", Proceedings of IEEE International Conference on Acoustics, Speech, and Signal Processing, Glasgow, UK, pp. 2453-2456, 1989.

[4] Barron, J. L., Fleet, D. J. and Beauchemin, S. S., "Performance of Optical Flow Techniques", International Journal of Computer Vision, vol. 12, no. 1, pp. 43-77, 1994.

[5] Bober, M. and Kittler, J., "Robust Motion Analysis", Proceedings of IEEE Conference on Computer Vision and Pattern Recognition, Seattle, WA, pp. 947-952, 1994.

[6] Bülthoff, H., Little, J. and Poggio, T., "A Parallel Algorithm for Real-Time Computation of Optical Flow", Nature, vol. 337, pp. 549-553, 1989.

[7] Camus, T., "Calculating Time-to-Contact Using Real-Time Quantized Optical Flow", NISTIR-5609, Gaithersburg, MD, 1995.

[8] Camus, T., "Real-Time Quantized Optical Flow", Proceedings of IEEE Conference on Computer Architectures for Machine Perception, Como, Italy, 1995.

[9] Coombs, D., Herman M., Hong T. and Nashman, M., "Real-time Obstacle Avoidance Using Central Flow Divergence and Peripheral Flow", Proceedings of IEEE International Conference on Computer Vision, Cambridge, MA, 1995.

[10] Delbruck, T., "Silicon Retina with Correlation-Based Velocity-Tuned Pixels", IEEE Transactions on Neural Networks, vol. 4, no. 3, pp. 529-541, 1993.

[11] Ens, J. and Li, Z.N., "Real-Time Motion Stereo", Proceedings of IEEE Conference on Computer Vision and Pattern Recognition, pp.130-135, New York, NY, 1993.

[12] Etienne-Cummings, R., Fernando, S.A., Van der Spiegel, J. and Mueller, P., "Real-Time 2D Analog Motion Detector VLSI Circuit", Proceedings of IEEE International Joint Conference on Neural Networks, vol. 4, pp. 426-431, New York, NY, 1992.

[13] Fleet, D.J. and Jepson, A.L., "Computation of Component Image Velocity from Local Phase Information", International Journal of Computer Vision, vol. 5, no.1, pp. 77-104, 1990.

[14] Heeger, D. J., "Optical Flow Using Spatiotemporal Filters", International Journal of Computer Vision, vol. 1, no. 4, pp. 279-302, 1988.

[15] Inoue, H., Tachikawa, T. and Inaba, M., "Robot Vision System with a Correlation Chip for Real-Time Tracking, Optical Flow and Depth Map Generation", Proceedings IEEE Conference on Robotics and Automation, Nice, France, vol. 2, pp. 1621-1626, 1992.

[16] Koch, C., Marroquin, J. and Yuille, A., "Analog 'Neural' Networks in Early Vision", Proceedings of the National Academy of Sciences, vol. 83, pp. 4263-4267, 1986.

[17] Lee, J.C., Sheu, B. J., Fang, W.C. and Chellappa, R., "VLSI neuroprocessors for Video Motion Detection", IEEE Transactions on Neural Networks, vol. 4, no. 2, pp. 178-191, 1993.

[18] Liu, H., Hong, T., Herman, M. and Chellappa, R., "A General Motion Model and Spatio-temporal Filters for Computing Optical Flow", University of Maryland TR -3365, November, 1994; NIST-IR 5539,

Gaithersburg MD, January, 1995, to appear in International Journal of Computer Vision.

[19] Liu, H., "A General Motion Model and Spatio-temporal Filters for 3-D Motion Interpretations", PH.D. Dissertation, University of Maryland, September, 1995.

[20] Matteucci, P., Regazzani, C.S. and Foresti, G.L., "Real-Time Approach to 3-D Object Tracking in Complex Scenes", Electronics Letters, vol. 30, no. 6, pp. 475-477, 1994.

[21] Mioni, A., "VLSI Vision Chips Homepage", http://www.eleceng.adelaide.edu.au/Groups/GAAS/Bugeye/visionchips.

[22] Mioni, A., Bouzerdoum, A., Yakovleff, A., Abbott, D., Kim, O., Eshraghian, K. and Bogner, R.E., "An Analog Implementation of Early Visual Processing in Insects", Proceedings International Symposium on VLSI Technology, Systems, and Applications, pp. 283-287, 1993.

[23] Moore, A. and Koch C., "A Multiplication Based Analog Motion Detection Chip", Proceedings of the SPIE, vol. 1473, Visual Information Processing: From Neurons to Chips, pp. 66--75, 1991.

[24] Nashman, M., Rippey, W., Hong, T.-H. and Herman, M., "An Integrated Vision Touch-Probe System for Dimensional Inspection Tasks", NISTIR 5678, National Institute of Standards and Technology, 1995.

[25] Nelson, R., "Qualitative Detection of Motion by a Moving Observer", Proceedings of the IEEE Conference on Computer Vision and Pattern Recognition, Lahaina, HI, pp. 173-178, 1991.

[26] Nesi, P., DelBimbo, A.D. and Ben-Tzvi, D. "A Robust Algorithm for Optical Flow Estimation", Computer Vision and Image Understanding, vol. 62, no. 1, pp. 59-68, 1995.

[27] Nishihara, H.K., "Real-Time Stereo- and Motion-Based Figure Ground Discrimination and Tracking Using LOG Sign Correlation", Conference Record of the Twenty-Seventh Asilomar Conference on Signals, Systems and Computers, vol. 1, pp. 95-100, 1993.

[28] Patterson, D. and Hennessy, J., "Computer Organization and Design: the Hardware/Software Interface", Morgan Kaufman, San Mateo, CA 1994.

[29] Rangachar, R., Hong, T.-H., Herman, M., and Luck, R., "Three Dimensional Reconstruction from Optical Flow Using Temporal Integration", Proceedings of SPIE Advances in Intelligent Robotic Systems: Intelligent Robots and Computer Vision, Boston, MA, 1990.

[30] Schunck, B., "Image Flow Segmentation and Estimation by Constraint Line Clustering", IEEE Transactions on Pattern Analysis and Machine Intelligence, vol. 11, no.10, pp. 1010-1027, 1989.

[31] Singh, A., "Optical Flow Computation: A Unified Perspective", IEEE Computer Society Press, 1991.

[32] Smith, S.M., "ASSET-2 Real-Time Motion Segmentation and Object Tracking", Proceedings of the Fifth International Conference on Computer Vision, pp. 237-244, Cambridge, MA, 1995.

[33] Tanner J. and Mead C., "A Correlating Optical Motion Detector", MIT Advanced Research in VLSI, pp. 57--64, 1984.

[34] Tanner, J. and Mead, C., "An Integrated Analog Optical Motion Sensor", VLSI Signal Processing II, R.W. Brodersen and H.S. Moscovitz, Eds., pp. 59-87, New York, 1988.

[35] Waxman, A.M., Wu J. and Bergholm F. "Convected Activation Profiles and Receptive Fields for Real Time Measurement of Short Range Visual Motion", Proceedings of IEEE Conference on Computer Vision and Pattern Recognition, Ann Arbor, MI, pp. 717-723, 1988.

[36] Weber, J. and Malik, J., "Robust Computation of Optical Flow in a Multi-Scale Differential Framework", Proceedings of the Fourth International Conference on Computer Vision, Berlin, Germany, 1993.

[37] Weber, J. and Malik, J., "Robust Computation of Optical Flow in a Multi-Scale Differential Framework", International Journal of Computer Vision, vol. 14, pp. 67-81, 1995.

[38] Weiss, P. and Christensson, B., "Real Time Implementation of Subpixel Motion Estimation for Broadcast Applications", IEE Colloquim on Applications of Motion Compensation, London, UK, no. 128, pp. 7/1-3, 1990.

[39] Woodfill, J. and Zabin, R., "An Algorithm for Real-Time Tracking of Non-Rigid Objects", Proceedings of the Ninth National Conference on Artificial Intelligence, pp. 718-723, 1991.

[40] Yagi, Y., Kawato, S. and Tsuji, S., "Real-Time Omnidirectional Image Sensor (COPIS) for Vision-Guided Navigation", IEEE Transactions on Robotics and Automation, vol. 10, no. 1, pp. 11-22, 1994.

Rigorous Bounds for Two–Frame Structure from Motion

J Oliensis

NEC Research Institute, Princeton NJ 08540, USA

Abstract. We analyze the problem of recovering rotation from two image frames, deriving an exact bound on the error size. With the single weak requirement that the average translational image displacements are smaller than the field of view, we demonstrate rigorously and validate experimentally that the error is small. These results form part of our correctness proof for a recently developed algorithm for recovering structure and motion from a multiple image sequence. In addition, we argue and demonstrate experimentally that in the complementary domain when the translation is large the whole motion can typically be recovered robustly, assuming the 3D points vary significantly in their depths.

1 Introduction

We have demonstrated in [12, 13] a new algorithm for structure from motion (SFM) which, in the appropriate domain, provably reconstructs structure and motion correctly. In the current paper, we present a part of the correctness proof for this algorithm, deriving tight bounds and estimates on the error in recovering rotation between two image frames when the translation is moderate. We show in [9] that these bounds typically give good estimates of the rotation error.

It has traditionally been believed that recovering rotation is difficult. We prove here that this is not true when the translation is moderate or small. Intuitively, our result is straightforward: if the translational displacements of image points are not large, then merely aligning the fields of view (FOV) of two images gives reasonable rotation estimates. We make this intuition precise in this paper.

We also argue and show experimentally that in the complementary large translation domain, if we additionally assume that perspective effects are important, then the complete motion and structure typically can be determined reliably. Experimentally, it seems that the domain where "small translation" techniques reliably determine the rotation overlaps the domain where "large translation" techniques work.

Our experimental work on the large translation domain suports our previous claim [13, 12] that this domain is essentially an easy one: any SFM algorithm, including the notoriously unreliable "8–point" [6] algorithm, typically works well in this domain. In addition, we demonstrate that often most of the error in recovering the Euclidean structure is due to a single structure component.

2 Rotation Error Bound

In this section we derive an explicit bound on the error in recovering the relative rotation between two noisy images **neglecting a nonzero translation between the two images**. The derived bound goes beyond a first order estimate; it does not require that the translation and noise be infinitesimally small.

Let the first image be represented by unit rays $\mathbf{p}_i$ and the second by unit rays $\mathbf{p}'_i$. Let there be N_p points in both images. We assume that the rotation is determined by finding R minimizing the least square error for the unit rays

$$\sum_i |\mathbf{p}'_i - \mathrm{R}\mathbf{p}|^2, \tag{1}$$

which is equivalent to maximizing

$$\sum_i \mathbf{p}'^t_i \; \mathrm{R}\mathbf{p}_i. \tag{2}$$

The rotation minimizing (1) can be found by standard techniques, e.g., using the SVD. (1) is a least square error in the image ray orientations rather than in the image plane coordinates (as is usual). For moderate FOV this makes little difference, and the ray–based error may actually better represent typical noise distributions. Probably, the rotations computed from (1) will be close to those computed by any standard algorithm and the bounds derived here will be relevant to such algorithms.

Let R_T represent the true rotation between the two images. Define $\mathrm{R} \equiv \mathrm{R}_E\mathrm{R}_T$ (R_E is the error rotation) and $\bar{\mathbf{p}}_i \equiv \mathrm{R}_T\mathbf{p}_i$. R_E is the rotation maximizing

$$\sum_i \mathbf{p}'^t_i \; \mathrm{R}_E\bar{\mathbf{p}}_i. \tag{3}$$

Let the unit vector $\hat{\omega}$ define the axis and θ the angle of rotation for R_E. $|\theta|$ measures the size of the error in the recovered rotation. We will derive a bound on $|\theta|$ starting from (3).

Proposition 1. *Define the* 3×3 *symmetric matrix* M *by*

$$M \equiv \frac{1}{N_p}(\mathbf{1}_3 \sum_i \mathbf{p}'^t_i \; \bar{\mathbf{p}}_i - \frac{1}{2}(\mathbf{p}'_i \; \bar{\mathbf{p}}^t_i + \bar{\mathbf{p}}_i \; \mathbf{p}'^t_i)), \tag{4}$$

where $\mathbf{1}_3$ *is the identity matrix, and define the 3 vector*

$$\mathbf{V} \equiv \frac{1}{N_p}\sum_i \mathbf{p}'_i \times \bar{\mathbf{p}}_i = \frac{1}{N_p}\sum_i (\mathbf{p}'_i - \bar{\mathbf{p}}_i) \times \bar{\mathbf{p}}_i. \tag{5}$$

Assume M *is positive definite. Then* $|\theta| \leq \left|M^{-1}\mathbf{V}\right|$.

Remark. As we discuss below, M will be positive definite when the translation and noise are sufficiently small. The noise is usually insignificant.

Proof. (3) can be decomposed as

$$\sum_i \mathbf{p}_i'^t \, R_E \bar{\mathbf{p}}_i = \sum_i (\hat{\omega}^t \mathbf{p}_i')(\hat{\omega}^t \bar{\mathbf{p}}_i)$$
$$+ \sum_i (\hat{\omega} \times \mathbf{p}_i')^t \, (\cos\theta(\hat{\omega} \times \bar{\mathbf{p}}_i) - \sin\theta(\hat{\omega} \times (\hat{\omega} \times \bar{\mathbf{p}}_i))), \quad (6)$$

where the first term on the right hand side involves the rotation only through $\hat{\omega}$ and is θ–independent. The coefficient of $\cos(\theta)$ in the second term is proportional to $\hat{\omega}^t M \hat{\omega}$ and is nonvanishing since M is positive definite. The θ_ω giving the maximum can be solved for explicitly:

$$\tan(\theta_\omega) = \frac{\sum_i (\hat{\omega} \times \mathbf{p}_i')^t \, \bar{\mathbf{p}}_i}{\sum_i (\hat{\omega} \times \mathbf{p}_i')^t \, (\hat{\omega} \times \bar{\mathbf{p}}_i)} \equiv \frac{A(\hat{\omega})}{B(\hat{\omega})}, \quad (7)$$

where

$$A(\hat{\omega}) \equiv \frac{1}{N_p} \sum_i (\hat{\omega} \times \mathbf{p}_i')^t \, \bar{\mathbf{p}}_i, \qquad B(\hat{\omega}) \equiv \frac{1}{N_p} \sum_i (\hat{\omega} \times \mathbf{p}_i')^t \, (\hat{\omega} \times \bar{\mathbf{p}}_i). \quad (8)$$

Substituting $\tan(\theta_\omega)$ back into (6) yields $N_p(A^2(\hat{\omega}) + B^2(\hat{\omega}))^{1/2}$ for the second term at the maximum. The first term in (6) can be rewritten as

$$\sum_i (\hat{\omega}^t \mathbf{p}_i')(\hat{\omega}^t \bar{\mathbf{p}}_i) = \sum_i \mathbf{p}_i'^t \, \bar{\mathbf{p}}_i - N_p B(\hat{\omega}). \quad (9)$$

Thus after the substitution for $\tan(\theta_\omega)$, (3) becomes up to rotation independent terms and an irrelevant factor of N_p

$$E(\hat{\omega}) \equiv \sqrt{A^2(\hat{\omega}) + B^2(\hat{\omega})} - B(\hat{\omega}) = B(\sqrt{1 + (A/B)^2} - 1). \quad (10)$$

We also have $B(\hat{\omega}) = \hat{\omega}^t M \, \hat{\omega}$ and $A(\hat{\omega}) = \mathbf{V}^t \, \hat{\omega}$. where M, $\mathbf{V}$ are the matrix and vector defined in the statement of the theorem.

Define $X(\hat{\omega}) \equiv (1/2) \, A^2(\hat{\omega})/B(\hat{\omega})$. As we discuss below, when the translation and noise are small $|A/B|$ will in general also be small and $E(\hat{\omega}) \approx X(\hat{\omega})$. Thus, as a first step toward maximizing E, we consider maximizing $X(\hat{\omega})$.

Lemma 2. *Let $\hat{\omega}_M$ be the value of the unit vector $\hat{\omega}$ maximizing $X(\hat{\omega})$. Assume M is positive definite. Then the ratio $|A(\hat{\omega}_M)/B(\hat{\omega}_M)| = |M^{-1}\mathbf{V}|$, and $\theta_M \equiv \tan^{-1}(A(\hat{\omega}_M)/B(\hat{\omega}_M))$ satisfies $|\theta_M| \leq |M^{-1}\mathbf{V}|$.*

Proof. Since M is real symmetric and positive definite, we can write $M = M^{1/2}M^{1/2}$ where $M^{1/2}$ is real symmetric and positive definite. Let $\boldsymbol{\omega}' \equiv M^{1/2} \, \hat{\omega}$ and $V' \equiv M^{-1/2}V$. Maximizing X over unit vectors $\hat{\omega}$ is equivalent to maximizing $(\hat{\omega}' \cdot \mathbf{V}')^2$, where $\hat{\omega}' \equiv \omega'/|\omega'|$, over the ellipsoid $\boldsymbol{\omega}'^t M^{-1} \boldsymbol{\omega}' = 1$. Clearly, the maximum occurs at one of the two points on the ellipsoid where $\boldsymbol{\omega}'$ is parallel to V', corresponding to a value for $\hat{\omega}$ of $\hat{\omega}_M \sim M^{-1/2}\mathbf{V}' = M^{-1}\mathbf{V}$. Substituting this value into the expressions for $A(\hat{\omega}_M)$, $B(\hat{\omega}_M)$ yields $|A(\hat{\omega}_M)/B(\hat{\omega}_M)| = |M^{-1}\mathbf{V}|$ and since $|\theta_M| \leq |\tan(\theta_M)|$, it follows that $|\theta_M| \leq |M^{-1}\mathbf{V}|$.

We now return to the maximization of the exact expression $E(\hat{\omega})$. In terms of $X(\hat{\omega})$ and $\rho(\hat{\omega}) \equiv A(\hat{\omega})/B(\hat{\omega})$,

$$E(\hat{\omega}) = \frac{2X(\hat{\omega})}{\rho^2(\hat{\omega})}(\sqrt{1+\rho^2(\hat{\omega})}-1) \equiv X(\hat{\omega})\ K(\rho^2(\hat{\omega})), \tag{11}$$

where $K(x) \equiv 2(\sqrt{1+x}-1)/x$ is a monotonically decreasing function for $x \geq 0$. Recall that $\hat{\omega}_M \sim M^{-1}\mathbf{V}$ gives the maximum value X_M of $X(\hat{\omega})$. Let $\rho_M \equiv \rho(\hat{\omega}_M)$; the lemma showed that $|\rho_M| = |M^{-1}\mathbf{V}|$. The only way to achieve a value $E(\hat{\omega})$ larger than $E(\hat{\omega}_M) = X_M K(\rho_M^2)$ is via a value of $\hat{\omega}$ such that $K(\rho^2(\hat{\omega})) > K(\rho_M^2)$. But since $K(x)$ is monotonic decreasing this implies $|\rho(\hat{\omega})| < |\rho_M|$. Since $|\theta(\hat{\omega})| \leq |\tan(\theta(\hat{\omega}))| = |\rho(\hat{\omega})|$ we have the desired result: the error in the rotation is bounded by $|\theta| < |M^{-1}\mathbf{V}|$.

2.1 Estimates on the Rotation Bound

In the remainder of this section we discuss the expected magnitude of the derived bound and the conditions under which it is valid—that is, the conditions under which M is positive definite. More detail can be found in [11], where we focus especially on how the rotation error is affected by the FOV size.

Let $\delta\mathbf{p}_i \equiv \mathbf{p}'_i - \bar{\mathbf{p}}_i$. Recall that $\delta\mathbf{p}_i$ reflects the displacement due to noise and translation only since the rotation has been compensated exactly in the definition of $\bar{\mathbf{p}}_i$. We derive crude but simple bounds on the eigenvalues of M and on $|M^{-1}\mathbf{V}|$ in terms of the $\delta\mathbf{p}_i$. For extensions see [11].

Using the notation $\langle x\rangle \equiv \sum_i x_i/N_p$, where i varies over all N_p points,

$$|\mathbf{V}| = \frac{1}{N_p}|\sum_i(\delta\mathbf{p}_i \times \bar{\mathbf{p}})| \ \leq\ \langle|\delta\mathbf{p}|\rangle, \tag{12}$$

$$B(\hat{\omega}) = \frac{1}{N_p}\left(\sum_i |\ \hat{\omega} \times \bar{\mathbf{p}}_i|^2 + (\ \hat{\omega} \times \delta\mathbf{p}_i)^t(\ \hat{\omega} \times \bar{\mathbf{p}}_i)\right). \tag{13}$$

In (13), the second term is also bounded by $\langle|\delta\mathbf{p}|\rangle$,

$$\left|\frac{1}{N_p}\sum_i(\ \hat{\omega} \times \delta\mathbf{p}_i)^t(\ \hat{\omega} \times \bar{\mathbf{p}}_i)\right| \leq \frac{1}{N_p}\sum_i|\delta\mathbf{p}_i \cdot \bar{\mathbf{p}}_i| \leq \langle|\delta\mathbf{p}|\rangle, \tag{14}$$

while the first term is positive for arbitrary $\hat{\omega}$ (assuming the image points $\bar{\mathbf{p}}_i$ are not all collinear). We denote the first term of (13) by $B_0(\hat{\omega})$:

$$B_0(\hat{\omega}) = \frac{1}{N_p}\sum_i|\ \hat{\omega} \times \bar{\mathbf{p}}_i|^2 = \ \hat{\omega}^t M_0\ \hat{\omega}, \tag{15}$$

where

$$M_0 \equiv \mathbf{1}_3 - \frac{1}{N_p}\sum_i \bar{\mathbf{p}}_i\bar{\mathbf{p}}_i^t. \tag{16}$$

Denote the eigenvalues of M_0 by e_i and those of M by e'_i, ordered from greatest to least. e_3 depends just on the initial observed image points and is typically of order 1 if the FOV is moderate or large [11]. Elementary linear algebra implies

$$B(\hat{\omega}) \equiv \hat{\omega}^t M \,\hat{\omega} \;\geq e'_3 \geq \; e_3 - \langle|\delta\mathbf{p}|\,\rangle. \tag{17}$$

If $\langle|\delta\mathbf{p}|\rangle$ is sufficiently small, $e_3 - \langle|\delta\mathbf{p}|\rangle > 0$ and $B(\hat{\omega}) > 0$ for arbitrary $\hat{\omega}$, implying that M is positive definite. Moreover, $e_3 - \langle|\delta\mathbf{p}|\rangle > 0$ implies

$$|\theta| \leq |M^{-1}\mathbf{V}| \leq \frac{|\mathbf{V}|}{e'_3} \leq \frac{\langle|\delta\mathbf{p}|\,\rangle}{e_3 - \langle|\delta\mathbf{p}|\,\rangle}. \tag{18}$$

When $e_3 \approx 1$ and $\langle|\delta\mathbf{p}|\rangle \ll 1$, the rotation error (in radians) is bounded approximately by $\langle|\delta\mathbf{p}|\rangle$, the average displacement of an image point due to the translation or noise. Recall that $\delta\mathbf{p}_i$ is a difference of unit vectors; $\langle|\delta\mathbf{p}|\rangle \approx 1$ would imply a large average translational shift in the image of about 45°.

2.2 An Improved Estimate

(18) is a pessimistic bound on $|M^{-1}\mathbf{V}|$ since the overlap of $\mathbf{V}$ with the least eigendirection of M—the one corresponding to e'_3—is typically small. We can improve this bound based on the propositions below (proofs omitted). Propositions similar to these appear in [2] and [16] but we obtain slightly better bounds since they apply to the **least** eigenvalues.

Definition 3. Let H', H be $N \times N$ symmetric matrices, and let H be positive definite. Define the perturbation $\Delta H \equiv H' - H$. Let e'_i, e_i be the eigenvalues respectively of H' and H, ordered from greatest to least, and let $\mathbf{E}'_i$ and $\mathbf{E}_i$ be the corresponding unit eigenvectors. Define $P_{\perp N}$ to be the projection operator $P_{\perp N} \equiv \mathbf{1}_N - \mathbf{E}_N\mathbf{E}^t_N$, where $\mathbf{1}_N$ is the $N \times N$ identity matrix. Define

$$D \equiv P_{\perp N}\,(H' - \mathbf{1}_N(e_N + \Delta H_{NN}))\,P_{\perp N}, \tag{19}$$

where $\Delta H_{NN} \equiv \mathbf{E}^t_N\; \Delta H\; \mathbf{E}_N$, and also define $\Delta H_{\perp\perp} \equiv P_{\perp N} \Delta H P_{\perp N}$. Let a be a lower bound on the eigenvalues of the matrix D (excluding the zero eigenvalue associated with $\mathbf{E}_N$) and let $b \equiv |P_{\perp N} \Delta H \mathbf{E}_N|$. Finally, define

$$B_2(\xi) = \frac{\sqrt{1+4\xi^2}-1}{2} \leq \xi^2.$$

Proposition 4 Eigenvalue bound. *Assume that* $a > 0$. *Then the perturbation of the least eigenvalue is bounded by*

$$\Delta H_{NN} \geq e'_N - e_N \geq \Delta H_{NN} - aB_2(b/a). \tag{20}$$

Definition 5. With the definitions above and assuming that $\mathbf{E}'_i \cdot \mathbf{E}_i \neq 0$, define $\delta\mathbf{E}_i$ by

$$\mathbf{E}'_i \equiv \frac{\mathbf{E}_i + \delta\mathbf{E}_i}{|\mathbf{E}_i + \delta\mathbf{E}_i|}, \qquad\qquad \mathbf{E}_i \cdot \delta\mathbf{E}_i = 0.$$

Proposition 6 Eigenvector bound. *Assume that* $a > 0$. *Then*

$$|\delta \mathbf{E}_N| \le \frac{2b}{a}.$$

Based on these propositions, a detailed theoretical analysis [11] shows that, even for moderate to small FOV,

1. M is positive definite and the derived bound is valid unless the average image displacement due to the translation or noise is comparable to the FOV.
2. Larger rotation errors can be expected for translations parallel to the image plane and for 3D points lying on a plane approximately perpendicular to the image plane.
3. The rotation error due to neglecting the translation is bounded roughly by the ratio of the average translational image displacement to the FOV.

2.3 Experiments

We have experimentally verified these theoretical expectations. The results of our synthetic experiments appear in Table 1, where each entry summarizes 100 trials. For each trial, feature points were selected randomly within the specified FOV (shown in the Table), and values of $|\mathbf{T}|/r_i$ (the inverse radial depth scaled by the translation magnitude) were chosen randomly varying uniformly from 0 up to the input maximum value. (Note that using the uniform distribution for the **inverse** depths gives a stringent test of our claims. Typically, the radial **depths** r_i are distributed uniformly; assuming this instead for the inverse depths causes the 3D points to cluster at small depths, with probability density $P(r_i) \sim 1/r_i^2$.) A second image was generated from the first using the specified translation and randomly computed structure; then Gaussian image noise with 2 pixel standard deviation was added to each image point in the second image.

The maximum computed rotation error over all trials and all cases is 23.4° $\sim$ .41 radians. In our structure from motion algorithm [13, 12], the first order effects of rotation error are eliminated. Thus the effect of rotation error on the estimates of structure and motion are in the worst case scaled by $.41^2 \approx 17\%$. As shown in the last column, expecially at moderate to large FOV the upper bound B is a very good estimate of the actual rotation error $|\theta|$. It can be shown that a lower bound on $|\theta|$ exists which at such FOV is very close to the upper bound. Since for 3D planar scenes the rotation errors are expected to be relatively large, we also ran experiments explicitly for this case. The results were similar to those reported in the Table.

We also report results on two real image sequences: the rocket field sequence [1, 17, 10] and the PUMA sequence [14, 15] (Figure 7b). The rocket sequence consists of nine images obtained by approximately forward motion in an outdoor environment. The maximum translation magnitude, between the first and ninth image, was 7.4 feet, while the depth of the nearest point relative to the first (farthest) camera position was 17.8 feet. The field of view was about 45°. We

N_{pts}	FOV°	T	$\max(\|T\|/r_i)$	range($\|\theta\|$)	range(B)	$\bar{m}, \sigma(B)$	range($\|\theta\|/B$)
20	90	$(1\,0\,0)$	.3	5.5°–10.8°	5.5–11.0	8.4°, 1.2°	.984–.997
20	40	$(1\,0\,0)$	.3	6.3–12.5	6.3–13.3	9.3, 1.4	.94–.996
20	20	$(1\,0\,0)$	.3	5.6–19.0	5.7–28.3	11.5, 4.0	.62–.995
20	20	$(1\,0\,1)$	.3	4.5–13.7	4.5–18.1	8.8, 2.4	.75–.995
20	20	$(1\,0\,2)$	.3	3.3–10.9	3.3–11.6	5.6, 1.5	.92–.998
20	20	$-(1\,0\,2)$	.3	2.3–7.7	2.3–8.1	4.1, 1.2	.93–.9995
20	40	$(1\,0\,0)$	.5	9.7–18.1	9.8–22.9	15.3, 2.6	.78–.989
20	30	$(1\,0\,0)$	.5	9.5–23.4	9.6–45.0	17.8, 5.3	.39–.988
20	30	$(1\,0\,1)$	.5	7.9–20.5	7.9–31.0	15.0, 3.5	.66–.992
20	30	$-(1\,0\,1)$	.5	6.0–14.9	6.0–16.6	9.4, 1.9	.88–.995
40	30	$(1\,0\,1)$	.5	10.2–17.6	10.3–21.0	14.1, 2.1	.81–.988
40	30	$(1\,0\,0)$	.5	10.8–17.8	10.9–23.5	15.7, 2.6	.73–.986
20	30	$(1\,0\,2)$	.5	5.8–16.4	5.8–20.1	10.2, 2.1	.81–.994

Table 1. Measured rotation error computed by SVD and bound $B \equiv M^{-1}\mathbf{V}$. Each entry summarizes the results of 100 trials.

computed the rotations neglecting the translation. The rotation was recovered with errors increasing nearly linearly with the translation size from .33° to 4.3°.

The PUMA sequence consists of 16 images of 32 points, with predominantly rotational motion of the camera at the rate of about 4° per image. The maximum translation was 1.5 feet, while the depth of the nearest point to the first camera position was 13.4 feet. The maximum rotation from the first image was 60.5°. The FOV was about 40°. Neglecting the translation, the rotation was recovered with error increasing almost perfectly linearly with the translation magnitude. The error varied from .37° to a maximum of 4.3°.

3 Large Translations

We now turn to the case when the translation is of the same order or large compared to the distances of the 3D points. We argue and demonstrate experimentally that if the feature points vary significantly in depth then typically both the rotation and translation can be recovered accurately[1]. The structure can also be recovered accurately except for very distant points.

The reasons why the structure and motion can be recovered robustly for large translation from 2 images are straightforward. Most importantly, because of the significant depth variation, it is easy to distinguish translations from rotations. For a rotation, the image displacements depend smoothly on the image point positions, while for translations the displacements for near and far points are very different and typically uncorrelated with image position. A large translation amplifies these differences allowing easy disambiguation. This leads to accurate recovery of the entire motion since it is well known that the rotation/translation

[1] Unless the 3D points lie close to a critical surface.

ambiguity is the main error source. With accurate motion, the structure computation is also typically robust; the nonlinearities in computing the depth are minimized because for large translations image rays from the two images tend to intersect at large angles.

We conducted a set of 100 synthetic experiments to test these claim. In each trial, 20 3D points were chosen randomly in a 40° FOV, with depths varying uniformly from 2 to 40 in units of the translation size. The translation was always in the x–direction (translation parallel to the image plane is traditionally believed to be a difficult case), and the rotation was chosen randomly up to a maximum of about 25°. The image noise was Gaussian with a σ of one pixel. The results are displayed in the figures. The angular error in the translation direction as computed by the "8-point" algorithm is shown in Figure 1a, while an improved error as computed by Levenberg–Marquardt starting from the result of the "8–point" algorithm is shown in Figure 1b. Even though the "8–point" algorithm is notorious for its unreliability, it reconstructs the translation direction accurately when the translation is large. Similar results for the rotation error (in degrees) are shown in Figures 2a and 2b.

Figures 3a and b show 3D histograms of the number of points reconstructed for given values of the depth and depth error; the results for the percentage depth error appear in Figure 4a[2]. Clearly the error is small for depths less than about 20 units.

The results for the depths are actually better than indicated. It can be shown that the structure component that is typically least accurately recovered is the constant component of the **inverse** depths (e.g., [5, 8, 7, 9]). In Figures 4b—6a, this component has been corrected to the ground truth for each of the 100 trials and the histogram shows the residual depth error. In Figures 6b and 7a, the fraction of points of a given depth with depth errors greater than 10% are shown before and after correction of the constant component.

Finally, we computed the rotation directly assuming that the translation is zero as in the earlier sections. Since the minimum 3D depth in these experiments is 2, the "large translation" domain considered here actually overlaps the "small translation" domain previously discussed, where the maximum inverse depth was .5. The maximum and average rotation errors obtained were 12° and 6.4°. These experiments can be considered successfully as either large translation or small translation.

We have also obtained similar results for translations in the directions (1 0 1), (1 0 2), and (0 0 1). In these cases the single constant component accounts for less of the error (as expected [9])) but the motion is recovered more accurately.

Lastly, we ran trials similar to those above but using the ground truth structure obtained from the real PUMA image sequence (Figure 7b and [14, 15]). The results for the "8–point" algorithm are shown in Table 2, where each entry summarizes 500 trials. 1 pixel uniform noise was added to each image point, and

[2] Note that in these histograms the bins corresponding to the largest depth error also include all hits with errors larger than the nominal value.

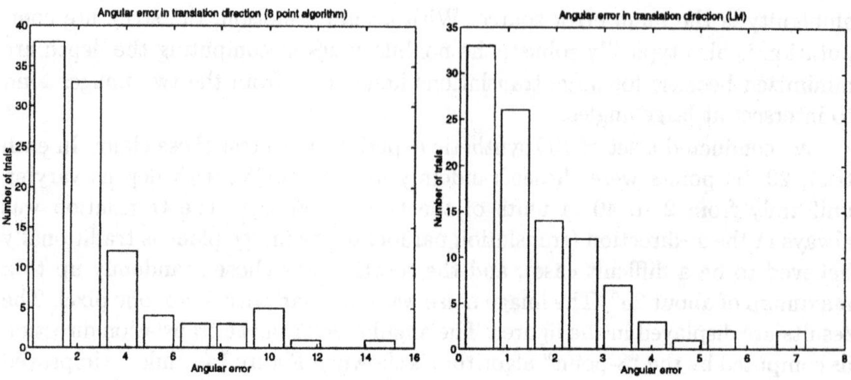

Fig. 1. a, b

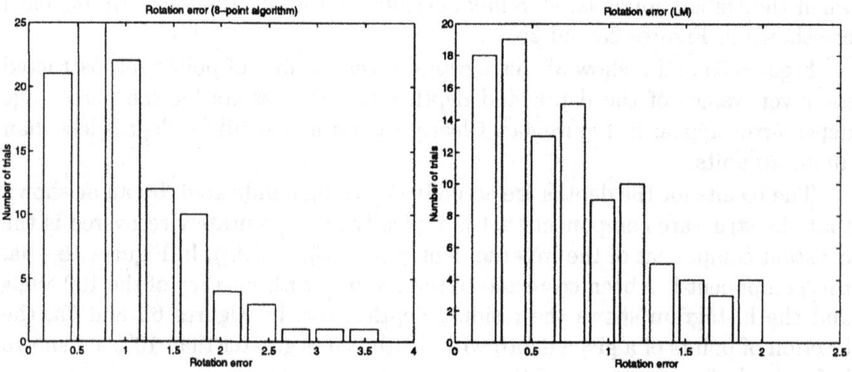

Fig. 2. a, b

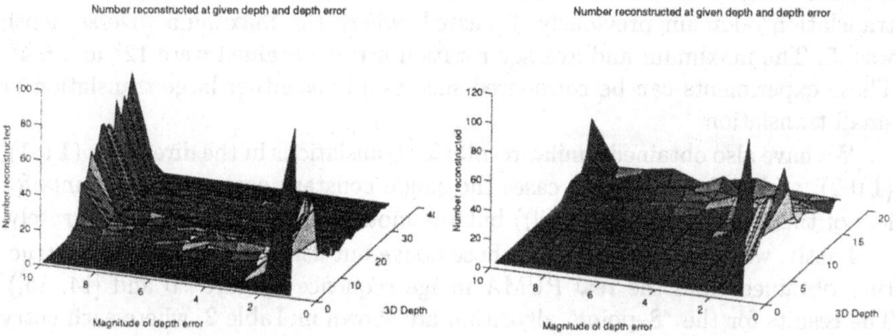

Fig. 3. a, b

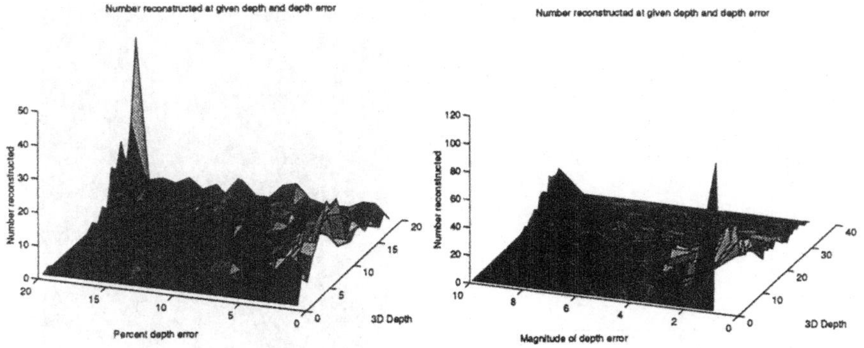

Fig. 4. a, b

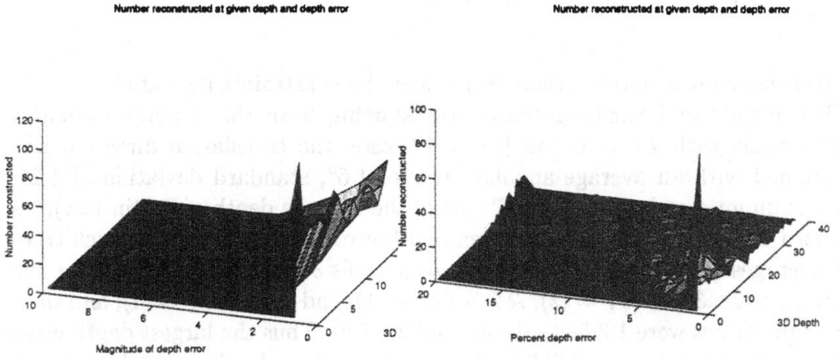

Fig. 5. a, b

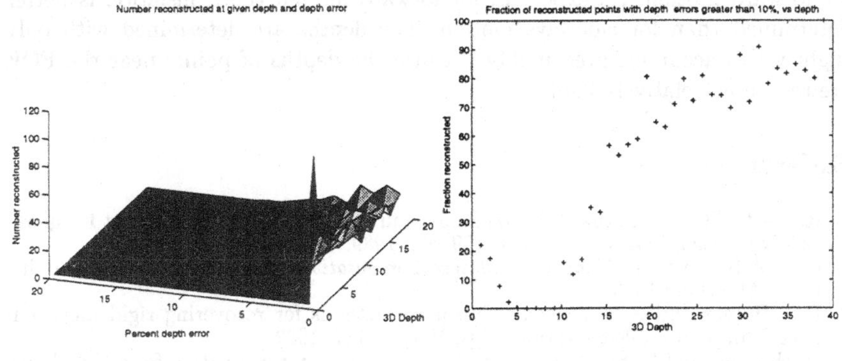

Fig. 6. a, b

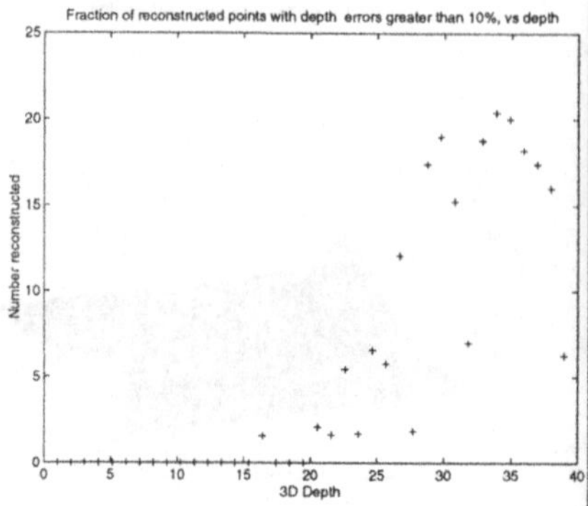

Fig. 7. a, b

the translation was selected randomly under the constraints indicated.

The results of Levenberg–Marquardt starting from the 8–point estimates on 250 trials with $Tz = 0$ and $|T| = 12$ were: the translation direction was determined with an average angular error of 1.6°, standard deviation of 1.2°, and maximum angular error of 8.6°. Also, the average depth error (in feet), its standard deviation, and the maximum depth error were recorded for each trial. The averages (and standard deviations over trials σ_T) of these results over the 250 trials were .3 feet ($\sigma_T = .2$), .2 feet ($\sigma_T = .1$), and .7 feet ($\sigma_T = .4$), and their maximum values were 1.3 feet, .6 feet, and 2.5 feet. Thus the largest depth error for any feature point over all 250 trials was 2.5 feet. Clearly, these results support our claim that the depth can be recovered accurately from just two image frames when the translational displacement is large. These experiments were repeated for forward motion. As expected, for forward motion the translation is better determined than for sideways motion. The depths are determined with only slightly less accuracy, presumably because the depths of points near the FOE are recovered relatively badly.

References

1. R. Dutta, R. Manmatha, L.R. Williams, and E.M. Riseman, "A data set for quantitative motion analysis," *CVPR*, 159-164, 1989.
2. G. Golub and C. F. Van Loan, *Matrix Computations*, John Hopkins Press, Baltimore, Maryland,1983.
3. D.J. Heeger and A.D. Jepson, "Subspace methods for recovering rigid motion I: Algorithm and implementation," **IJCV** 7, 95-117, 1992.
4. R. Hummel and V. Sundareswaran, "Motion parameter estimation from global flow field data," **PAMI** 15, 459-476, 1993.
5. A.D. Jepson and D.J. Heeger, "Subspace methods for recovering rigid motion II: Theory," University of Toronto Technical Report RBCV-TR-90-36, 1990.
6. H. C. Longuet–Higgins, "A computer algorithm for reconstructing a scene from two projections," *Nature*, 293:133–135, 1981.
7. S. Maybank, *Theory of Reconstruction from Image Motion*, Springer, Berlin, 1992.

Table 2. Two Frame Motion Recovery for Large Translation Using the 8–point algorithm. The angular error in determining the translation direction is displayed (in degrees) for different magnitudes of the input 3D translation. The mean of the angular error, its standard deviation, the maximum angular error, and the number of trials with errors respectively greater than 10° and 20° are shown. The angular error quoted is $\min(|\Delta\theta|, |180° - \Delta\theta|)$ since both T and $-T$ are correct solutions.

	Baseline (ft)	4	6	8	10	12	14	16
	Mean($\lvert\Delta\theta\rvert$)	41	14	7	6	5	5	4
	$\sigma_{\lvert\Delta\theta\rvert}$	25	12	6	5	4	3	3
$T_z = 0$	max($\lvert\Delta\theta\rvert$)	90	81	32	29	20	19	14
	$N_{>10}$	437	253	130	84	45	37	27
	$N_{>20}$	370	114	17	5	1	0	0
	Mean($\lvert\Delta\theta\rvert$)	22	7	4	3	2		
	$\sigma_{\lvert\Delta\theta\rvert}$	18	6	4	4	2		
Forward T	max($\lvert\Delta\theta\rvert$)	89	42	23	20	19		
	$N_{>10}$	369	106	39	21	5		
	$N_{>20}$	191	19	3	1	0		
	Mean($\lvert\Delta\theta\rvert$)	32	18	12	9	8	7	7
	$\sigma_{\lvert\Delta\theta\rvert}$	16	10	7	5	5	4	4
Backward T	max($\lvert\Delta\theta\rvert$)	88	86	42	35	25	22	22
	$N_{>10}$	476	411	287	210	171	123	112
	$N_{>20}$	393	174	66	22	14	3	6

8. S. Maybank, "A Theoretical Study of Optical Flow," Doctoral Dissertation, University of London, 1987.
9. J. Oliensis, in preparation.
10. J. Oliensis, "Provably Correct Algorithms for Multiframe Structure from Motion: Constant Translation Direction," *ECCV* 1996.
11. J. Oliensis, "Rigorous Bounds for Two–Frame Structure from Motion," NEC TR October 1995, http://www.neci.nj.nec.com/homepages/oliensis.html.
12. J. Oliensis, "A Linear Solution for Multiframe Structure from Motion," *IUW*, 1225–1231, November 1994.
13. J. Oliensis, "Multiframe Structure from Motion in Perspective," *Workshop on the Representations of Visual Scenes*, 77–84, June 1995.
14. H. S. Sawhney, J. Oliensis, and A. R. Hanson, A "Description and Reconstruction from Image Trajectories of Rotational Motion", *ICCV*, 494-498, 1990.
15. H.S. Sawhney and A.R. Hanson, "Comparative results of some motion algorithms on real image sequences," *IUW*, 307-313, 1990.
16. G.W. Stewart and J. G. Sun, *Matrix Perturbation Theory*, Academic Press, Boston, 1990.
17. J. Inigo Thomas, A. Hanson, and J. Oliensis, "Refining 3D reconstructions: A theoretical and experimental study of the effect of cross–correlations", **CVGIP:IU**, Vol. 60, 359–370, 1994.

The Rank 4 Constraint in Multiple (≥ 3) View Geometry

Amnon Shashua[1] *and Shai Avidan*[2]

[1] Technion — Israel Institute of Technology, CS department, Haifa 32000, Israel
[2] Hebrew University, Institute of Computer Science, Jerusalem 91904, Israel

Abstract. It has been established that certain trilinear froms of three perspective views give rise to a tensor of 27 intrinsic coefficients [8]. Further investigations have shown the existence of quadlinear forms across four views with the negative result that further views would not add any new constraints [3, 12, 5]. We show in this paper first general results on any number of views. Rather than seeking new constraints (which we know now is not possible) we seek connections across trilinear tensors of triplets of views. Two main results are shown: (i) trilinear tensors across $m > 3$ views are embedded in a low dimensional linear subspace, (ii) given two views, all the induced homography matrices are embedded in a four-dimensional linear subspace. The two results, separately and combined, offer new possibilities of handling the consistency across multiple views in a linear manner (via factorization), some of which are further detailed in this paper.

Keywords: Trilinearity, 3D recovery from 2D views, Matching Constraints, Projective Structure, Algebraic and Projective Geometry.

1 Introduction

The algebraic and geometric relations across multiple perspective views is a recent and growing interest which is relevant to a number of topics including (i) issues of 3D reconstruction from 2D data, (ii) representations of visual scenes from video data, (iii) image synthesis and animation, and (iv) visual recognition and indexing.

Typical to these topics is the question about the limitations and possibilities of going from two-dimensional (2D) measurements of point matches (correspondences) across two or more views to properties of the three-dimensional (3D) object or scene. Since the relationship between the 3D world and the 2D image space combines together 3D shape parameters, camera viewing parameters and 2D image measurements, the question of limitations and possibilities, in its widest scope, is about (i) 2D constraints across multiple views (matching constraints), (ii) characterizations of the space of all images of a particular object (indexing functions). In other words, one seeks to best represent, in terms of efficiency, compactness, flexibility and scope of use, two kinds of manifolds: (i) the manifold of image and viewing parameters (invariance to shape), and (ii) the manifold of image and object parameters (invariance to viewing parameters).

It has been established that certain trilinear forms of three perspective views give rise to a tensor of 27 intrinsic coefficients [8]. Further work on the properties of the "trilinear tensor" with relevancy to 3D reconstruction was described in [10, 4]. Other investigations have established the existence of quadlinear forms (with total of 81 coefficients) across four views with the negative result that further views would not add any new constraints [3, 12, 5]. Also, adopting the representation put forward by [3], dual trilinear tensors were established by [2, 13].

In this paper we extend the investigation to any number $m > 3$ views. There are two motivations to this work. First, is the practical aspect — if any additional view over the fourth view is redundant, what is the best and most efficient way of capturing that redundancy (in a linear manner)? Second, the existence of the quad-linearities is somewhat unsettling because the number of coefficients has risen from 27 to 81, whereas a view adds only 12 parameters. In other words, the quad-linearities may be too redundant a representation of the constraints over four views.

Our line of approach is to investigate the space of *all* trilinear tensors and to look for rank deficiencies in that space. Any finding of that sort is extremely useful because it readily allows a statistical way of putting together many views, simply by means of factorization. Moreover, a finding of that nature promises progress on the task of novel-view synthesis from model images ("image-based rendering") because a rank deficiency implies that trilinear tensors are related together by linear combinations — which is a necessary property for synthesizing tensors from a number of model tensors.

Two main results are shown: (i) trilinear tensors across $m > 3$ views are embedded in a low dimensional linear subspace, (ii) given two views, all the induced homography matrices are embedded in a four-dimensional linear subspace. The two results, separately and combined, offer new possibilities of handling multiple views in a linear manner (via factorization), some of which are further detailed in this paper.

2 Preliminaries About the Trilinear Tensor

Let P be a point in 3D projective space projecting onto p, p', p'' three views ψ, ψ', ψ'' represented by the two dimensional projective space. The relationship between the 3D and the 2D spaces is represented by the 3×4 matrices, $[I, 0]$, $[A, v']$ and $[B, v'']$, i.e.,

$$p = [I, 0]P$$
$$p' \cong [A, v']P$$
$$p'' \cong [B, v'']P$$

We may adopt the convention that $p = (x, y, 1)^{\top}$, $p' = (x', y', 1)^{\top}$ and $p'' = (x'', y'', 1)^{\top}$, and therefore $P = [x, y, 1, \rho]$. The coordinates $(x, y), (x'y'), (x'', y'')$

are matching points (with respect to some arbitrary image origin — say the geometric center of each image plane). The 3×3 matrices A and B are 2D collineations (homography matrices) from ψ to ψ' and ψ'', respectively, induced by the plane $\rho = 0$. The vectors v' and v'' are the epipolar points (the projection of the first camera center onto views ψ' and ψ'', respectively). The trilinear tensor is an array of 27 entries:

$$\alpha_i^{jk} = v'^k b_i^j - v''^j a_i^k. \qquad i, j, k = 1, 2, 3 \tag{1}$$

where superscripts denote contravariant indices (representing points in the 2D plane, like v') and subscripts denote covariant indices (representing lines in the 2D plane, like the rows of A). Thus, a_i^k is the element of the k'th row and i'th column of A, and v'^k is the k'th element of v'. The tensor α_i^{jk} forms the set of coefficients of certain trilinear forms that vanish on any corresponding triplet p, p', p'' and which have the following form: let s_k^l be the matrix,

$$s = \begin{bmatrix} 1 & 0 & -x' \\ 0 & 1 & -y' \end{bmatrix}$$

and, similarly, let r_j^m be the matrix,

$$r = \begin{bmatrix} 1 & 0 & -x'' \\ 0 & 1 & -y'' \end{bmatrix}$$

Then, the tensorial equations are:

$$s_k^l r_j^m p^i \alpha_i^{jk} = 0, \tag{2}$$

with the standard summation convention that an index that appears as a subscript and superscript is summed over (known as a contraction). For further details on this derivation, see Appendix A. Hence, we have four trilinear equations (note that $l, m = 1, 2$). In more explicit form, these functions (referred to as "trilinearities") are:

$$x''\alpha_i^{13}p^i - x''x'\alpha_i^{33}p^i + x'\alpha_i^{31}p^i - \alpha_i^{11}p^i = 0,$$
$$y''\alpha_i^{13}p^i - y''x'\alpha_i^{33}p^i + x'\alpha_i^{32}p^i - \alpha_i^{12}p^i = 0,$$
$$x''\alpha_i^{23}p^i - x''y'\alpha_i^{33}p^i + y'\alpha_i^{31}p^i - \alpha_i^{21}p^i = 0,$$
$$y''\alpha_i^{23}p^i - y''y'\alpha_i^{33}p^i + y'\alpha_i^{32}p^i - \alpha_i^{22}p^i = 0.$$

Since every corresponding triplet p, p', p'' contributes four linearly independent equations, then seven corresponding points across the three views uniquely determine (up to scale) the tensor α_i^{jk}. More details and applications can be found in [8, 9]. Also worth noting is that these trilinear equations are an extension of the three equations derived by [11] under the context of unifying line and point geometry.

Furthermore, for any arbitrary (covariant) vector $\rho_j = (\rho_1, \rho_2, \rho_3)$, the matrix $\rho_j \alpha_i^{jk}$ (recall the summation convention, i.e., $\rho_j \alpha_i^{jk} = \rho_1 \alpha_i^{1k} + \rho_2 \alpha_i^{2k} +$

$\rho_3 \alpha_i^{3k}$, which is a matrix) is not just any matrix, it is a 2D ***homography*** (a 2D collineation) from image 1 to image 2 via some plane, whose parameters are determined by ρ_j (the vector ρ_j is in direction of the normal to the plane in a coordinate system whose origin is the first camera center and its axes are aligned with the axes of the third camera coordinate system). Therefore, if we take ρ_j to be $(1,0,0), (0,1,0)$ and $(0,0,1)$, we obtain three homography matrices, which we will denote by E_1, E_2, E_3, respectively. In other words, the elements of the tensor α_i^{jk} are ***rearranged*** to comprise three matrices E_1, E_2, E_3 ($E_1 = \alpha_i^{1k}$, for example). For example, the "fundamental" matrix F between ψ and ψ' can be linearly determined from the tensor by: ${E_j}^{\top} F + F^{\top} E_j = 0$, which yields 18 linear equations of rank 8 for F. More details can be found in [10].

3 Tensors and Rank 12

Consider the following arrangement: we are given views $\psi_1, \psi_2, ... \psi_{m+2}$, $m \geq 1$. For each (ordered) triplet of views there exists a unique trilinear tensor. Consider m triplets of views each containing ψ_1, ψ_2, i.e., the triplets $< \psi_1, \psi_2, \psi_i >$, $i = 3, ..., m+2$. Consider each of the tensors as a vector of 27 components and concatenate all these vectors as columns of a $27 \times m$ matrix. The question is what is the rank of this matrix when $m \geq 27$? Clearly, if the rank is smaller than 27 we obtain a line of attack on the task of putting together many views. The motivation for considering this arrangement is that a view adds only 12 parameters (up to scale). It may be the case that the redundancy of representing an additional view with 27 numbers (a column vector in the $27 \times m$ matrix), instead of 12, comes to bear only at a non-linear level — in which case it will not affect the rank of the system above. Therefore, a rank deficiency implies an important property of a collection of tensors.

We arrange each tensor as a 27 column vector as follows:

$$\begin{pmatrix} E_1 = \alpha_i^{1k} \\ E_2 = \alpha_i^{2k} \\ E_3 = \alpha_i^{3k} \end{pmatrix}$$

where E_j (a 3×3 matrix) is arranged column-wise as a 9 vector. To simplify notation of indices, let $[B_i, v^{(i)}]$ denote the camera transformation matrices for view $\psi_3, ...$, and $[A, v']$ the camera transformation matrix for view ψ_2. In the next formula, $A, B_i, v', v^{(i)}$ all appear as column vectors: A is a arranged as a 9-vector column wise, i.e., $(a_1^1, a_2^1, a_3^1, a_1^2, ..., a_3^3)$, and likewise $B_i^{\top}$. It is simply a matter of observation to verify that the following holds:

$$\begin{bmatrix} \\ \\ \\ \\ \end{bmatrix}_{27\times m} = \begin{bmatrix} v' & 0 & \cdots & 0 \\ 0 & v' & \cdots & 0 \\ \cdot & 0 & \cdots & \\ \cdot & \cdot & \cdots & \cdot \\ \cdot & \cdot & \cdots & \cdot \\ \cdot & \cdot & \cdots & \cdot \\ 0 & 0 & \cdots & v' \end{bmatrix}_{27\times 9} \begin{bmatrix} B_1^T \cdots B_m^T \end{bmatrix}_{9\times m} - \begin{bmatrix} A & 0 & 0 \\ \\ 0 & A & 0 \\ \\ 0 & 0 & A \end{bmatrix}_{27\times 3} \begin{bmatrix} v^{(1)} \cdots v^{(m)} \end{bmatrix}_{3\times m}$$

$$= \begin{bmatrix} v' & 0 & \cdots & 0 & A & 0 & 0 \\ 0 & v' & \cdots & 0 & & & \\ \cdot & 0 & \cdots & & & & \\ \cdot & \cdot & \cdots & \cdot & 0 & A & 0 \\ \cdot & \cdot & \cdots & \cdot & & & \\ \cdot & \cdot & \cdots & \cdot & & & \\ 0 & 0 & \cdots & v' & 0 & 0 & A \end{bmatrix}_{27\times 12} \begin{bmatrix} B_1^T \cdots B_m^T \\ \\ v^{(1)} \cdots v^{(m)} \end{bmatrix}_{12\times m}$$

Thus, we have proved the following theorem:

Theorem 1 (Rank 12) *All trilinear tensors live in a manifold of $\mathcal{P}^{26}$. The space of all trilinear tensors with two of the views fixed, is a 12'th dimensional linear sub-space of $\mathcal{R}^{27}$.*

From the factorization principle above we see that each additional view adds, linearly, only 12 parameters — as expected. Moreover, these 12 parameters constitute the camera transformation matrix associated with the new view. Next we show that the linear subspace of all tensors with two views fixed is *closed*, i.e., any linear combination of such tensors produces an admissible tensor describing the configuration of the two fixed cameras and some third camera position.

Theorem 2 *The linear subspace containing tensors of views $<\psi_1, \psi_2, \psi>$, for all views ψ, is closed.*

The proof is detailed in Appendix B. In particular, consider the two tensors $<\psi_1, \psi_2, \psi_i>$ and $<\psi_1, \psi_2, \psi_j>$ as two points on the (non-linear) manifold of *all* tensors; the inifinite line passing through the two points *lives* inside the manifold — which implies that the manifold is decomposed into ruled surfaces. This property is the basis for novel-view synthesis which we will touch upon later in the paper.

4 Collineations and Rank 4

Consider a similar exercise done with homography matrices between two fixed views. Given some plane in space projecting onto views ψ and ψ', the corresponding image points are mapped to each other by a collineation (homography matrix), $Ap \cong p'$ for all matching pairs p, p'. Since the homography matrix A depends on the orientation and location of the planar object, we obtain a *family* of homography matrices when we consider all possible planes. It is also known, that given a homography matrix A of *some* plane, then all other homography matrices can be described by,

$$\lambda A + v'n^{\mathsf{T}}.$$

Consider homography matrices $A_1, A_2, ..., A_k$ each as a column vector in a $9\times k$ matrix. We ask again, what is the rank of the system? It would be convenient if it were 4, because each additional homography matrix represents a plane, and a plane is determined by 4 parameters. Let $A_i = \lambda_i A + v'n_i{}^{\mathsf{T}}$. The following can be verified by inspection:

$$\begin{bmatrix} \\ \\ \end{bmatrix}_{9\times k} = \begin{bmatrix} \\ \lambda_1 A \cdots \lambda_k A \\ \\ \end{bmatrix}_{9\times k} + \begin{bmatrix} v' & 0 & 0 \\ 0 & v' & 0 \\ 0 & 0 & v' \end{bmatrix}_{9\times 3} \begin{bmatrix} n_1 \cdots n_k \end{bmatrix}_{3\times k}$$

$$= \begin{bmatrix} & v' & 0 & 0 \\ A & 0 & v' & 0 \\ & 0 & 0 & v' \end{bmatrix}_{9\times 4} \begin{bmatrix} \lambda_1 \cdots \lambda_k \\ n_1 \cdots n_k \end{bmatrix}_{4\times k}$$

We have thus proven the following result:

Theorem 3 (Collineations, Rank 4) *The space of all homography matrices between two fixed views is embedded in a 4 dimensional linear subspace of* $\mathcal{P}^8$.

5 Tensors and Rank 4

We recall from Section 2 that the tensor α_i^{jk} can be contracted into three homography matrices, associated with three distinct planes, between ψ_1 and ψ_2. Hence, consider the same situation as before where we have the tensors of the triplets $< \psi_1, \psi_2, \psi_i >$, $i = 3, ..., m+2$. But now, instead of arranging each tensor as a 27 column vector, we arrange it in a 9×3 block, where each column is the homography α_i^{jk}, $j = 1, 2, 3$. We obtain a $9 \times 3m$ matrix,

$$M = \begin{bmatrix} \\ \alpha_i^{1k} \; \alpha_i^{2k} \; \alpha_i^{3k} \cdots \\ \\ \end{bmatrix}_{9\times 3m}$$

From Theorem 3 we know that its rank must be 4. Therefore, we have the following result:

Theorem 4 (Tensors and Rank 4) *The space of all trilinear tensors with two of the views fixed can be decomposed into three separate linear subspaces, each of dimension 4, of $\mathcal{R}^{27}$.*

Four columns of M span all tensors $< \psi_1, \psi_2, \psi >$, for any view ψ. Instead of being spanned as a 27 vector it is spanned three times each as a 9 vector. Thus, the new tensor is determined by 12 coefficients (of the linear combinations of the 4 columns). As a consequence, each additional tensor would require only 6 matching points, instead of 7:

Corollary 1 (Tensor-and-a-third) *A tensor of views $< \psi_1, \psi_2, \psi_3 >$ and "third" of the tensor $< \psi_1, \psi_2, \psi_4 >$, linearly span, with 12 coefficients, all tensors $< \psi_1, \psi_2, \psi >$ (over all views ψ). Each such tensor can be recovered using 6 matching points with ψ_1 and ψ_2.*

We can use instead a single tensor to linearly span all tensors $< \psi_1, \psi_2, \psi >$ (for all views ψ) by recovering the fundamental matrix F from the tensor (see Section 2). The homography matrix $[v']_x F$ (see [6]), where $[v']_x$ is the skew-symmetric matrix associated with vector products, can replace the "third" tensor of above (note that $[v']_x F$ is of rank 2, therefore is not linearly spanned by the three homography matrices provided from the tensor — unless the three camera centers are collinear). We have therefore the following result:

Corollary 2 (Tensor + F) *The tensor of views $< \psi_1, \psi_2, \psi_3 >$ and the epipolar constraint (matrix F and epipole v') together linearly span, with 12 coefficients, all other tensors $< \psi_1, \psi_2, \psi >$ (running over all views ψ). Each additional view ψ contributes linearly 12 parameters and its tensor with ψ_1, ψ_2 can be determined linearly using 6 matching points.*

6 Applications and Experimental Results

The results presented in this paper have three areas of application (that have been identified so far). First, is the obvious application of enforcing numerical consistency across two or more tensors. This can done as follows.

Assume we are given views $\psi_1, \psi_2, ...\psi_{m+2}$ and consider the tensors of the triplets $< \psi_1, \psi_2, \psi_i >$, $i = 3, ..., m+2$. Arrange the tensors into a $9 \times 3m$ matrix M as described in the previous section. Perform an SVD and keep only the 4 largest singular values. We have thus obtained a new matrix $\hat{M}$ that enforces the rank 4 constraint, which in turn, enforces the consistency across all the tensors. Separate the tensors of interest from $\hat{M}$ (each tensor still occupies a 9×3 block of M).

Another variant on this application is to recover the four principle components of M, as follows. Given M as above, perform a principle component analysis and obtain the four principle components $A_1, ..., A_4$ (each is a 9-vector). These vectors encode the geometry between ψ_1, ψ_2 alone (represent four homography

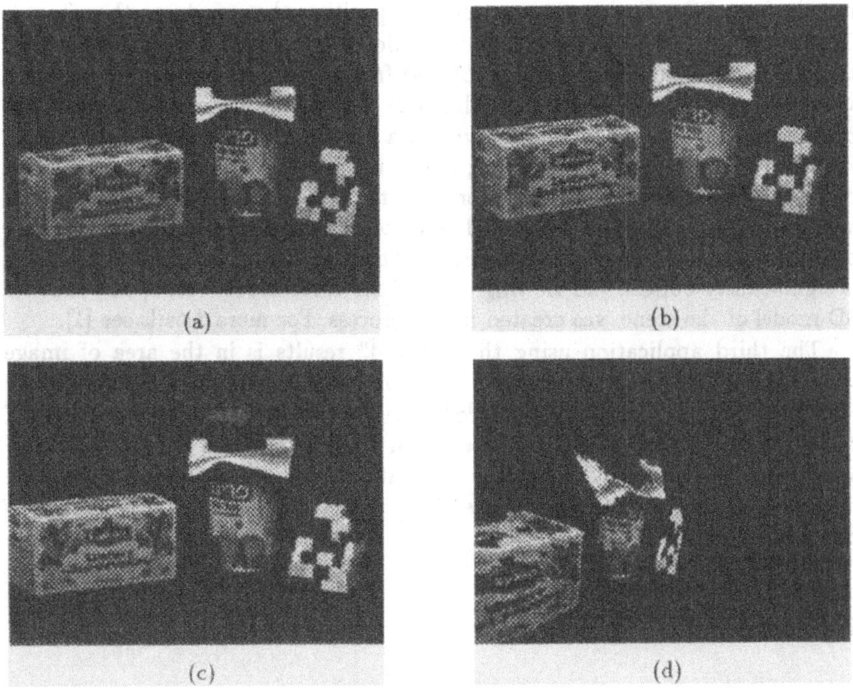

Fig. 1. An example of image synthesis using optic-flow and a tensor. (a) (b) and (c) are the original three images. (d) is a synthetic image created by the tensor of the three model images, the user-specified virtual camera motion, and the (dense) correspondence between images (a) and (b). Note that the virtual view is significantly outside the viewing cone of the original model views. For more details see [1].

matrices). Hence, we can use them to recover F from the (over-determined) linear system of 24 equations $A_j{}^{\top} F + F^{\top} A_j = 0$, $j = 1, ..., 4$.

The second area of application is in novel-view synthesis (image-based rendering). Theorem 4 can be rewritten as follows:

$$\begin{pmatrix} \hat{E}_1 \\ \hat{E}_2 \\ \hat{E}_3 \end{pmatrix} = \begin{bmatrix} l_{11} & \cdots & \cdot \\ \cdot & \cdots & \cdot \\ \cdot & \cdots & l_{34} \end{bmatrix} \begin{bmatrix} E_1 \\ E_2 \\ E_3 \\ A \end{bmatrix} \tag{3}$$

where E_1, E_2, E_3 are the three homography matrices comprising the tensor of views $< \psi_1, \psi_2, \psi_3 >$, A is some arbitrary homography matrix from ψ_1 to ψ_2 not spanned by E_1, E_2, E_3; the matrix L is a 3×4 matrix comprising the 12 coefficients necessary to span a new tensor of the views $< \psi_1, \psi_2, \psi >$, for some view ψ, arranged as $\hat{E}_1, \hat{E}_2, \hat{E}_3$ as a 3×9 matrix, i.e., each matrix $\hat{E}_j$ is a 9-vector arranged column-wise. Furthermore, the matrix L is a camera transformation matrix from 3D space to the new camera position that produces the view ψ. This result implies that in principle we could be in a position to ***synthesize*** new tensors on demand (by specifying L) and from those new tensors to reproject views ψ_1, ψ_2 and create views ψ. This idea was taken further in [1] where the possibility of creating synthetic movies of a 3D scene was demonstrated. Fig. 1 shows three model views of a scene, and a new synthesized image of the scene created by synthesizing a new tensor from the tensor of the three model views and then reprojecting the two model views on the top row onto the new viewing position using the new synthesized tensor. Note that the synthesized virtual view is significantly outside the viewing cone of the three model views, and that no 3D model of the scene was created in the process. For more details see [1].

The third application using the "rank 4" results is in the area of image stabilization. In [7] it is shown that Theorem 3 can be used as a building block for recovering the small-angle approximation of the rotational component of camera motion *directly* and in closed-form from a tensor of three views. The significance of this result is that the translational component of camera motion need not be recovered in the process, and that the process is fairly robust in practice.

7 Summary

This paper has presented a new approach for investigating the inter-relationship among a collection of four or more views. The approach is based on searching for rank deficiencies in the space of all trilinear tensors.

We have shown that families of trilinear tensors are embedded in a low dimensional linear subspace of tensor space (the manifold where all tensors live). First, this result enables a factorization approach to enforce consistency among many views (via consistency among the tensors). We showed, for instance, that one can use the factorization principle to obtain the fundamental matrix of two views from any number of views. Secondly, the theorems and their corollaries provide a tight bound on the contribution of additional views over three views.

We view these results as forming "building blocks" for future applications, not necessarily as forming an application on their own. We have pointed out two areas of application where these results have already proven fruitful — the first is in the area of "image-based rendering" where one is interested in synthesizing novel views of a 3D scene without necesserily creating a 3D model of the scene, and the second application is the area of video sequence stabilization.

A Deriving the Trilinear Tensor

The trilinear equations were first derived in [8] together with the equation of the tensor. The derivation presented here is more compact and more details can be found in [9].

The camera transformation between images ψ and ψ' is represented by $p' \cong [A, v']P$ where $P = (x, y, 1, \rho)^{\top}$. Let s_k^l be the matrix,

$$s = \begin{bmatrix} 1 & 0 & -x' \\ 0 & 1 & -y' \end{bmatrix}$$

It can be verified by inspection that $p' \cong [A, v']P$ can be represented by the following two equations:

$$\rho s_k^l v'^k + p^i s_k^l a_i^k = 0, \tag{4}$$

with the standard summation convention that an index that appears as a subscript and superscript is summed over (known as a contraction). Note that we have two equations because $l = 1, 2$ is a free index.

Similarly, the camera transformation between views ψ and ψ'' is $p'' \cong [B, v'']P$. Likewise, let r_j^m be the matrix,

$$r = \begin{bmatrix} 1 & 0 & -x'' \\ 0 & 1 & -y'' \end{bmatrix}$$

And likewise,

$$\rho r_j^m v''^j + p^i r_j^m b_i^j = 0, \tag{5}$$

Note that k and j are dummy indices (are summed over) in equations 4 and 5, respectively. We used different dummy indices because now we are about to eliminate ρ and combine the two equations together. Likewise, l, m are free indices, therefore in the combination they must be separate indices. We eliminate ρ and after some rearrangement and grouping we obtain:

$$s_k^l r_j^m p^i (v'^k b_i^j - v''^j a_i^k) = 0,$$

and the term in parenthesis is the trilinear tensor.

B The linear subspace of tensors is closed under linear combinations

Theorem 2: *The linear subspace containing tensors of views* $<\psi_1, \psi_2, \psi>$, *for all views* ψ, *is closed.*

Proof. We need to show that for any views ψ_3, ψ_4, the linear combination of tensors $<\psi_1, \psi_2, \psi_3>$ and $<\psi_1, \psi_2, \psi_4>$ produces a tensor $<\psi_1, \psi_2, \psi>$ for some view ψ. Let $[B_3; v^{(3)}]$ and $[B_4; v^{(4)}]$ be the camera transformation matrices associated with views ψ_3 and ψ_4, respectively. From Theorem 1 it is clear that the linear combination will produce a tensor $<\psi_1, \psi_2, \psi>$ where the camera transformation matrix associated with view ψ is $[aB_3 + bB_4; av^{(3)} + bv^{(4)}]$ where a, b are the coefficients of the linear combination. The subtle point in this argument is that although the homography matrix $aB_3 + bB_4$ does not correspond to the same plane (the plane $\rho = 0$, see Section 2) associated with B_3 and B_4, it can be transformed into such a homography matrix by correspondingly transforming the projective representation of the 3D scene. In other words, the projective representation of the scene can undergo a projective transformation (which effectively translates the scene along the optical axes of the first view) which is interchangeable with the camera motion from view to view. The interchangeability point is effectively contained in the arguments of [3] about the geometric content of each trilinearity. □

References

1. S. Avidan and A. Shashua. Novel view synthesis in tensor space. CS report, CIS-9602, Technion, January 1996. Also http://www.cs.huji.ac.il/ avidan/render/render.html.
2. S. Carlsson. Duality of reconstruction and positioning from projective views. In *Proceedings of the workshop on Scene Representations*, Cambridge, MA., June 1995.
3. O.D. Faugeras and B. Mourrain. On the geometry and algebra of the point and line correspondences between N images. In *Proceedings of the International Conference on Computer Vision*, Cambridge, MA, June 1995.
4. R. Hartley. A linear method for reconstruction from lines and points. In *Proceedings of the International Conference on Computer Vision*, pages 882–887, Cambridge, MA, June 1995.
5. A. Heyden. Reconstruction from image sequences by means of relative depths. In *Proceedings of the International Conference on Computer Vision*, pages 1058–1063, Cambridge, MA, June 1995.
6. Q.T. Luong and T. Vieville. Canonic representations for the geometries of multiple projective views. In *Proceedings of the European Conference on Computer Vision*, pages 589–599, Stockholm, Sweden, May 1994. Springer Verlag, LNCS 800.
7. B. Rousso, S. Avidan, A. Shashua, and S. Peleg. Robust recovery of camera rotation from three frames. In *Proceedings of the ARPA Image Understanding Workshop*. Palm Springs, CA, February 1996.
8. A. Shashua. Algebraic functions for recognition. *IEEE Transactions on Pattern Analysis and Machine Intelligence*, 17(8):779–789, 1995.
9. A. Shashua and P. Anandan. The generalized trilinear constraints and the uncertainty tensor. In *Proceedings of the ARPA Image Understanding Workshop*, Palm Springs, CA, February 1996.
10. A. Shashua and M. Werman. On the trilinear tensor of three perspective views and its underlying geometry. In *Proceedings of the International Conference on Computer Vision*, June 1995.
11. M.E. Spetsakis and J. Aloimonos. A unified theory of structure from motion. In *Proceedings of the ARPA Image Understanding Workshop*, 1990.
12. B. Triggs. Matching constraints and the joint image. In *Proceedings of the International Conference on Computer Vision*, pages 338–343, Cambridge, MA, June 1995.
13. D. Weinshall, M. Werman, and A. Shashua. Shape tensors for efficient and learnable indexing. In *Proceedings of the workshop on Scene Representations*, Cambridge, MA., June 1995.

Using Singular Displacements for Uncalibrated Monocular Visual Systems

T. Viéville and D. Lingrand

INRIA, Sophia, BP93, 06902 Valbonne, France.
tel : +33 93 65 76 88 fax : +33 93 65 78 45
email : vthierry@sophia.inria.fr

Abstract. In the present paper, we review and complete the equations and the formalism which allow to achieve a minimal parameterization of the retinal displacement for a monocular visual system without calibration.
Considering the emergence of active visual systems for which we can not consider that the calibration parameters are either known or fixed, we develop an alternative strategy using the fact that certain class of special displacements induces enough equations to evaluate the calibration parameters, so that we can recover the affine or Euclidean structure of the scene when needed.
A synthesis of what can be recovered for singular displacements in terms of camera calibration, scene geometry and kinematics is proposed. We give, for the different levels of calibration, an exhaustive list of the geometric and kinematic information which can be recovered. Following a strategy based on special kind of displacements, such as fixed axis rotations or pure translations for instance, we describe how to detect this particular classes of displacement.

Key-Words Structure and Motion, Singular Displacements, Self-Calibration

1 Introduction

The analysis of motion in the case of an uncalibrated monocular image sequence has already been developed by several authors, considering point and/or line correspondences or correspondences between planar patches and using either a discrete or a continuous representation of the rigid displacement between two or more frames.

These studies are motivated by the fact that *we must not consider an active visual system is calibrated* [6]. However, it has been demonstrated that, in the general case, it is not possible to self-calibrate the camera when zooming or modifying the intrinsic calibration parameters.

Considering this fact, the key idea of the present study is that **several singular displacements induce enough equations to evaluate the calibration parameters.**

For instance, fixed axis rotations of known angles or pure rotations [5] allow to estimate the calibration parameters, their uncertainty and, for a given kind of displacement, which parameters are optimally estimated, so that active visual strategies can be developed. On the other hand, pure translations do not allow to calibrate the Euclidean geometry of the scene [8], but its affine geometry [11].

Collecting all this information and considering a suitable statistical framework as in [1], it is then possible to infer which kind of displacement will increase at most the information (usually represented by the inverse of a covariance matrix) on the scene geometry, object kinematics and calibration parameters.

This is the goal of this paper.

In order to attain this objective, we are first going to review the theory of motion when no calibration: equations, parameterization of motion, etc...

We then are going to propose a synthesis of what can be recovered in terms of scene geometry and kinematics when calibration is not given as an input: describe the different forms of calibration, the different levels of calibration and give an exhaustive list of the different geometric and kinematic information to be recovered, depending the chosen geometry.

2 Reviewing the theory of motion when no calibration.

Notations. We write vectors and matrices using bold letters, matrices being written with capital letters. The duals of vectors are represented as the transpose of a vector and scalars in italic. The notation $\mathbf{x} \wedge \mathbf{y} = \tilde{\mathbf{x}}\mathbf{y}$ corresponds to the cross-product, the dot-product being written as $\mathbf{x}^T\mathbf{y}$. $\tilde{\mathbf{x}}$ is a 3×3 skew-symmetric matrix[1] . The identity matrix is written **I**. Geometric objects such as points, lines, planes are written with capital letters in 3D, and small letters in 2D. We represent the components of a matrix or a vector using superscripts from 0 to 2, e.g.: $\mathbf{x} = (x^0, x^1, x^2)^T$.
We write $\mathbf{a} \equiv \mathbf{b}$ if $\mathbf{a}$ is equal to $\mathbf{b}$ up to a scale factor, i.e. $\exists k, \mathbf{a} = k\,\mathbf{b}$.

2.1 Setting the equations

Camera model and frame of reference. We use *the standard pinhole model* for a camera, assuming the camera performs a perfect perspective transform with center C (the camera optic center) at a distance f (the focal length) of the retinal plane. The pinhole model can still be used for a zoom lens if the object-to-image distance is not considered as fixed.

All coordinates are related to an affine frame of reference $\mathcal{R} = (C, \mathbf{x}, \mathbf{y}, \mathbf{z})$ *attached to the retina*, $\mathbf{z}$ being aligned with the optical axis, $\mathbf{x}$ and $\mathbf{y}$ being aligned with the horizontal and vertical axe in the image. The retinal plane is thus perpendicular to the optical axis Cz,as shown in figure 1.

[1] Remember that a 3×3 skew-symmetric matrix has 3 parameters and can always be represented by the crossproduct of a vector, i.e. is of the form $\tilde{\mathbf{x}}$ for some $\mathbf{x}$.

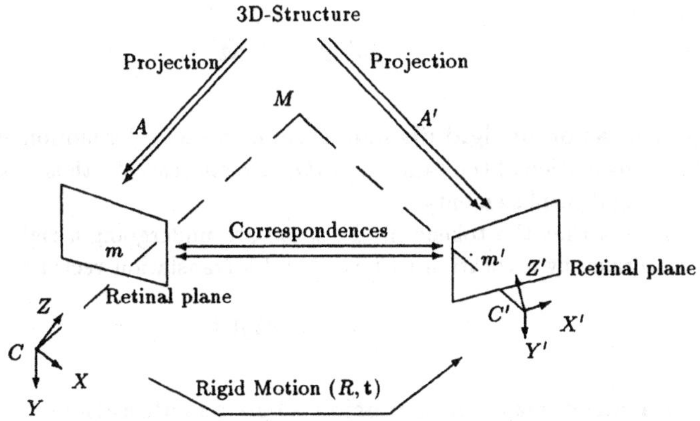

Fig. 1. Elements used in the definition of the problem

Using points as primitives. We represent a 3D-point M by the vector $\mathbf{M} = \mathbf{CM} = (X, Y, Z)^T$ using Euclidean coordinates. Points in the retina, with pixel coordinates coordinates (u, v) will be represented as homogeneous 3-D vectors: $\lambda\ \mathbf{m} = \lambda\ \mathbf{Cm} = \lambda\ (u, v, 1)^T$, corresponding to lines of a given direction passing through the optical center (2-D projective space).

Other primitives will be represented using set of points. This will be discussed in the sequel.

A suitable model of the intrinsic parameters of the camera. In this study, *we do not assume the system is calibrated.* However, we are in a specific situation because we have chosen a "canonical" frame attached to the retina. Therefore, we consider only the matrix of the intrinsic parameters (called A-matrix) in the projection and write:

$$Z\ \mathbf{m} = \mathbf{A}\ \mathbf{M} \quad , \quad \mathbf{A} = \begin{pmatrix} f & 0 & u_0 \\ 0 & f & v_0 \\ 0 & 0 & 1 \end{pmatrix} \tag{1}$$

A complete review can be found in [1].

In the present model, (u_0, v_0) is the principal point, and f the focal length; following [8], we assume that **we know the ratio between the horizontal and vertical focal length** and that we assume that **the two retinal coordinates are orthogonal.** It has been shown experimentally that these assumptions are valid for standard cameras [8] and also for high-level visual sensors [10]. Using this simple model will allow us to improve the obtained results.

We also assume that the intrinsic parameters are different for each camera position, as during a zoom. In the consecutive frame $\mathcal{R}' = (C', \mathbf{x}', \mathbf{y}', \mathbf{z}')$ we write:

$$Z' \mathbf{m}' = \mathbf{A}'\mathbf{M}' \tag{2}$$

Representation of rigid displacements. We consider motion of rigid objects and the ego-motion of the camera, *in the discrete case.* We thus represent motion through rigid displacements.

It means that the tokens in the scene are undergoing a rigid displacement parameterized by a rotation matrix R and a translation vector $\mathbf{t}$:

$$\mathbf{M}' = \mathbf{R}\,\mathbf{M} + \mathbf{t} \tag{3}$$

2.2 Parameterization of motion when no calibration.

The goal of the parameterization of motion is the following: given a set of points in correspondence between two views, i.e. a set of matches $\{m.m'\}$ we want to analyse all constraints which relate the two points i.e find the equations of the form $\forall\{m.m'\}, f(m, m') = 0$. In particular, we would like to predict the location of a point given its correspondent, i.e. a relation of the form $\forall\{m.m'\}, m' = g(m)$. Having such parameterization allows to exact all information available from the retinal displacements, which is measured through the set of matches.

The Qs-representation and the F-matrix. Considering the 2D correspondences between two points m and m' in two different frames, we obtain, combining equations (1),(2) and (3):

$$Z' \mathbf{m}' = Z \underbrace{\mathbf{A}'\mathbf{R}\mathbf{A}^{-1}}_{\mathbf{H}_\infty} \mathbf{m} + \underbrace{\mathbf{A}'\mathbf{t}}_{\mathbf{s}} \tag{4}$$

where the Q-matrix $\mathbf{H}_\infty$ corresponds to the "uncalibrated rotational component of the rigid displacement", or more geometrically *the collineation of the plane at infinity,* while the s-vector corresponds to the "uncalibrated translational component of the rigid displacement", also called "focus of expansion" by some authors, and more geometrically *the epipole.* These notations have been introduced in [8] to analyse the motion of points and lines in the general case.

If we eliminate Z and Z' in equation (4) (by taking the cross-product with $\mathbf{s}$ and multiplying by $\mathbf{m}'^T$) we obtain:

$$\mathbf{m}'^T \underbrace{[\tilde{\mathbf{s}}\,\mathbf{H}_\infty]}_{\mathbf{F}} \mathbf{m} = 0 \tag{5}$$

The matrix $\mathbf{F} = \tilde{\mathbf{s}}\,\mathbf{H}_\infty$ is the *Fundamental matrix* and is also called the "essential matrix in the uncalibrated case". If we consider that the only information available is related to the retinal correspondences between points, without any knowledge about the depths Z, equation (5) is the only equation that can be derived [8].

Considering a set of matches related by equation (3) the equation (5) is well defined if and only if (i) $\mathbf{s} \neq 0$ and (ii) there is no linear relations between all m' and m. The degenerated cases occur only if the translation is zero, or if all points belong to the same plane [8][2]. This particular case will be analysed in detail.

As discussed in [12] an efficient criterion is the average retinal Euclidean distances between each point and its epipolar line. The following symmetric least-square sum is minimized:

$$\mathbf{F}_\bullet = argmin_{\mathbf{F}} \underbrace{\left[\sum_{\{\mathbf{m}\}} w_{\mathbf{m}} \underbrace{\left[d(\mathbf{m}', \mathbf{F}\mathbf{m})^2 + d(\mathbf{m}, \mathbf{F}^{\mathrm{T}}\mathbf{m}')^2 \right]}_{f_{\mathbf{m}}(\mathbf{F})^2} \right] / \left[2 \sum_{\{\mathbf{m}\}} w_{\mathbf{m}} \right]}_{[\epsilon_{\mathbf{F}}(\mathbf{F})]^2} \qquad (6)$$

where $w_{\mathbf{m}}$ is a weighted corresponding to the precision of the match, in fact the inverse of the variance of the precision of the match. The quantity $w_{\mathbf{m}}$ is given in *pixel*$^{-2}$, while $\epsilon_{\mathbf{F}}(\mathbf{F})$, the *average distance to the epipolar*, is in pixel.

A camera for which F has been computed is called *a weak calibrated camera.*

The vector $\mathbf{s}$ is defined as the basis vector of the kernel of $\mathbf{F}^T$.

The case of a pure rotation, and the planar case. As pointed out previously, in the case of a pure rotation or if the set of points belongs to a unique planar structure, we cannot estimate the F-matrix because all points in one view are related to points in the other view by a relation of the form:

$$\mathbf{m}' \equiv \mathbf{H}\,\mathbf{m} \qquad (7)$$

which corresponds to two equations for each match.

There, if the matrix $\mathbf{F}$ is undefined, we still can estimate the matrix $\mathbf{H}$ as in [8], using $\mathbf{H} = \mathbf{I}$ as initial value.

Following the same method as for the F-matrix, an efficient criterion is to minimize the residual disparity again, as in [8] and obtain $\mathbf{H}$ through:

$$\mathbf{H}_\bullet = argmin_{\mathbf{H}} \underbrace{\left[\sum_{\{\mathbf{m}\}} w_{\mathbf{m}} \underbrace{\left\| \mathbf{m}' - \frac{\mathbf{H}\,\mathbf{m}}{((\mathbf{h}^2)^T\,\mathbf{m})} \right\|^2}_{f_{\mathbf{m}}(\mathbf{H})^2} \right] / \left[\sum_{\{\mathbf{m}\}} w_{\mathbf{m}} \right]}_{[\epsilon_{\mathbf{H}}(\mathbf{H})]^2} \qquad (8)$$

[2] From algebraic point of view, equation (5) has singular solutions if and only if there exist a linear relation between $\mathbf{m}$ and $\mathbf{m}'$, i.e. a relation of the form $\mathbf{m}' = \mathbf{H}\,\mathbf{m}$. This situation corresponds to the case were the points are related by a collineation, i.e. correspond to a planar structure as reviewed in the sequel.

where we write $\mathbf{H} = (\mathbf{h}^0, \mathbf{h}^1, \mathbf{h}^2)$ in order to have a compact notation[3].

The error $\epsilon_{\mathbf{H}}(\mathbf{H})$, given in pixel, will be called ***residual disparity after motion reduction*** in the sequel.

Reciprocally, as soon as the points belongs to at least two planes, we can defined a F-matrix [3]. This degenerated situation thus only exists in the case of an unique plane.

3 Using specific displacements for motion analysis.

Let us now discuss situations for which the F-matrix or the H-matrix have a particular form. Considering a robotic system, it is very often that a displacement is not a general displacement but a constrained motion such as a pure translation, a fixed axis rotation, etc... as illustrated in figure 2.

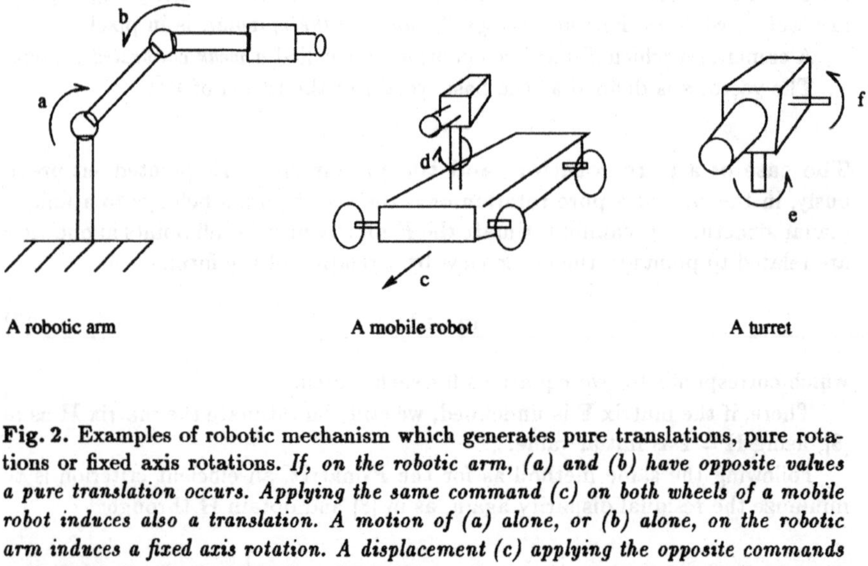

Fig. 2. **Examples of robotic mechanism which generates pure translations, pure rotations or fixed axis rotations.** ***If, on the robotic arm, (a) and (b) have opposite values a pure translation occurs. Applying the same command (c) on both wheels of a mobile robot induces also a translation. A motion of (a) alone, or (b) alone, on the robotic arm induces a fixed axis rotation. A displacement (c) applying the opposite commands on both wheels of a mobile robot also induces a fixed axis rotation. Turrets for camera can induce pure rotations in pan (e) or tilt (f) around the optical center.***

[3] The relation $\mathbf{m}' = \frac{\mathbf{H}\,\mathbf{m}}{((\mathbf{h}^2)^T\,\mathbf{m})}$ is a vectorial form for :

$$\begin{cases} u' = \frac{H^{00}\,u + H^{01}\,v + H^{02}}{H^{20}\,u + H^{21}\,v + H^{22}} \\ v' = \frac{H^{10}\,u + H^{11}\,v + H^{12}}{H^{20}\,u + H^{21}\,v + H^{22}} \\ 1 = 1 \end{cases}$$

Moreover we can make the following assumptions about this kind of hardware [6]:

- The displacements are reproducible.
- We can measure the angles of the rotation. For technical reasons we do not assume the same thing for translations (it is not sure that we can estimate the norm of the linear translation of a zoom for instance [6] while the precision of the translation of a mobile robot is not very high).
- All extrinsic parameters are unknown, and equations are expected to provide unstable estimates of them [1].

These particular constraints are far from having negative properties. On the contrary, they induce additional equations which help solving the reconstruction or calibration problem.

Furthermore, the estimation of the displacement are easier in these cases, because we have to evaluate less parameters.

However, the system must be able to recognize if the displacement corresponds to such a particular case, so that we must characterize the situation in each case.

Finally, the different class of displacements might have several implications on the perception strategy, which is also to be discussed.

The parameterization of all these different kind of displacements have been given in [7] and will not be reported here.

4 Defining a hierarchical motion module

4.1 Combining different models of displacements

Following the previous discussion, when we estimate a rigid displacement, we consider several cases, depending on the nature of the displacement. Collecting all constraints proposed in [7], we can describe the following set of models:

Considering a rigid structure, the following class of displacements can be identified, N is the number of parameters:

Class of Displacement	Parameterization or constraint	Information Recovered	N
Pure rotation	$\mathbf{F} = 0$	$\mathbf{H}_\infty, \mathbf{t} = 0$	0
Z-axis pure translation	$\mathbf{F} = \begin{pmatrix} 0 & -1 & 0 \\ 1 & 0 & 0 \\ 0 & 0 & 0 \end{pmatrix}$	$\mathbf{H}_\infty = \mathbf{R} = \mathbf{I}, \mathbf{t} \equiv \mathbf{z}$	0
Pure retinal translation	$\mathbf{F} = \begin{pmatrix} 0 & 0 & a \\ 0 & 0 & b \\ -a & -b & 0 \end{pmatrix}, \lVert\mathbf{F}\rVert = 1$	$\mathbf{H}_\infty = \mathbf{R} = \mathbf{I}, \mathbf{t} \equiv \begin{pmatrix} \cos(\theta) \\ \sin(\theta) \\ 0 \end{pmatrix}$ $\theta = atan(\frac{s_1}{s_0})$	1
Pure translation	$\mathbf{F} = \begin{pmatrix} 0 & c & a \\ -c & 0 & b \\ -a & -b & 0 \end{pmatrix}, \lVert\mathbf{F}\rVert = 1$	$\mathbf{H}_\infty = \mathbf{R} = \mathbf{I}$	2
Retinal displacement	$\mathbf{F} = \begin{pmatrix} 0 & 0 & a \\ 0 & 0 & b \\ c & d & e \end{pmatrix}, \lVert\mathbf{F}\rVert = 1$	$\mathbf{R}, \mathbf{t}/\lVert\mathbf{t}\rVert, eq(\mathbf{a})$	4
Zoom displacement	$\mathbf{F} = \begin{pmatrix} 0 & f & a \\ -f & 0 & b \\ c & d & e \end{pmatrix}, \begin{matrix} cb - ad = 0 \\ \lVert\mathbf{F}\rVert = 1 \end{matrix}$	$\mathbf{R} = \mathbf{I}, \mathbf{t}/\lVert\mathbf{t}\rVert, eq(\mathbf{a})$	4
Fixed axis rotation	$det(\mathbf{F} + \mathbf{F}^T) = 0,\ det(\mathbf{F}) = 0,\ \lVert\mathbf{F}\rVert = 1$	$eq(\mathbf{a})$	6
Retinal translation	$det(\mathbf{F}) = 0, s^2 = 0, \lVert\mathbf{F}\rVert=1$	$\mathbf{t} \equiv \begin{pmatrix} \cos(\theta) \\ \sin(\theta) \\ 0 \end{pmatrix}, eq(\mathbf{a})(Kruppa)$ $\theta = atan(\frac{s_1}{s_0})$	6
General rigid displacement	$det(\mathbf{F}) = 0, \lVert\mathbf{F}\rVert=1$	$eq(\mathbf{a})(Kruppa)$	7

where $eq(\mathbf{a})$ means that we obtain equations about the intrinsic calibration parameters, these equations being either linear equations or the quartic Kruppa equations, as specified. In these cases, it is not possible to maintain an estimation of all calibration parameters.

In the planar case, we have:

Class of Displacement	Parameterization or constraint	Information Recovered	Number of Parameters
Stationary structure	$\mathbf{H} = \mathbf{I}$	$\mathbf{R} = \mathbf{I}, \mathbf{t} = 0, \mathbf{a} = \mathbf{a}'$	0
Constant retinal displacement	$\mathbf{H} = \begin{pmatrix} 0 & 0 & a \\ 0 & 0 & b \\ 0 & 0 & 1 \end{pmatrix}$	$\mathbf{R} = I, \mathbf{t} \equiv (a, b, 0), \mathbf{a} = \mathbf{a}', \mathbf{n} \equiv \mathbf{z}$	2
Retinal planar zoom	$\mathbf{H} = \begin{pmatrix} c & 0 & a \\ 0 & c & b \\ 0 & 0 & 1 \end{pmatrix}$	eq(a)	4
Retinal planar rotation	$\mathbf{H} = \begin{pmatrix} c & d & a \\ -d & c & b \\ 0 & 0 & 1 \end{pmatrix}$	R, eq(a)	4
Pure planar retinal translation	$\mathbf{H} = \mathbf{I} + \mathbf{s}\nu^T, s^2 = 1$	$\mathbf{R} = \mathbf{I}, \mathbf{s}/\lVert\mathbf{s}\rVert, \nu$	5
Pure planar translation	$\mathbf{H} = \mathbf{I} + \mathbf{s}\nu^T, s^2 = 0$	$\mathbf{R} = \mathbf{I}, \mathbf{s}/\lVert\mathbf{s}\rVert, \nu$	5
Retinal planar displacement	$\mathbf{H} = \begin{pmatrix} a & b & c \\ d & e & f \\ 0 & 0 & 1 \end{pmatrix}$	R, t/‖t‖, n, eq(a)	6
General planar displacement	$\mathbf{H} = \begin{pmatrix} a & b & c \\ d & e & f \\ g & h & 1 \end{pmatrix}$		8

In fact some other variants have also been introduced in order to have alternative models with very few parameters. For instance a model with zero parameters, corresponding to a collineation equal to the identity, i.e. a stationary structure is introduced. This allows to have a simple model assuming that points are not moving.

Furthermore, this set of model has a very interesting structure, i.e. some models are generalizations of others. This allows to take as best model the first model, starting from the bottom, which statistical significance is smaller that every models immediatly higher in the hierarchy.

4.2 Experimental results

An example with real data. We consider a sequence of 16 imagesfor which the displacement is an approximative retinal translation, with some erroneous rotation because of the actual set-up. A retinal displacement is thus expected.

The early-vision module has provided matches between 44 points and the errors are given in table (1).

displacement	number of outliers	residual error
stationary structure	0	17.9633
pure retinal translation	0	17.4948
planar retinal displacement	11	6.52437
retinal displacement (F22 = 0)	8	1.95843
retinal displacement (F22 = 1)	8	0.735489
pure translation	0	14.7264
zoom displacement (s2 = 0; s'2 = 0; s0.s'0 != 0)	0	11.677
zoom displacement linear (s2 = 1; s'2 = 1)	4	0.887803
general rigid displacement	11	0.84458

Table 1. Table of residues for the real scene.

The model is correctly estimated also in this real case, which thus allow us to conclude on the validity of the proposed mechanism.

5 Conclusion

In the present paper we have reviewed and completed the description of a general framework which allows not only to estimate a minimal parameterization of the rigid displacement between two frames, but also to determine several particular cases which occur in practice and have important advantages with respect to the calibration problem. This is true for several standard displacement, except a zoom displacement which seems to be a singular case, for the proposed model.

The statistical framework to implement these equations has been already described in [8] and has been applied here to the estimation of collineations from a minimal parameterization. This paper however generalizes the set of models to general rigid displacements, and proposes a complete analysis of the underlying rigid displacement in each case.

Similar attempts to use degenerated models of parameterization of motion have been already issued in the past [8, 4, 9]. However, we collect here new results about the Euclidean representation associated to each parameterization.

Furthermore, the implementation also integrates two new aspects: (i) clustering data and (ii) testing different models to represent the data.

Finally, this work tries to develop -with a certain degree of completeness- all the different singularities which occur for a rigid displacement and which can be detected without calibration. A practical motion module has been developed and successfully experimented.

A step further, we will use this hierarchical approach to not only parameterize the retinal displacement but also analyse the calibration of the visual system and recover the scene structure. A preliminary study has been issued [2] for retinal displacements.

References

1. O. Faugeras. *Three-dimensional Computer Vision: a geometric viewpoint.* MIT Press, Boston, 1993.
2. D. Lingrand and T. Viéville. Dynamic foveal 3d sensing using affine models. Technical Report RR-2687, INRIA, 1995.
3. T. Luong. *Matrice Fondamentale et Calibration Visuelle sur l'Environnement.* PhD thesis, Université de Paris-Sud, Orsay, 1992.
4. P. Torr, A. Zisserman, and S. Maybank. Robust detection of degenerated configurations for the fundamental matrix. In *5th International Conference on Computer Vision*, pages 1037–1042, 1995.
5. T. Viéville. Autocalibration of visual sensor parameters on a robotic head. *Image and Vision Computing*, 12, 1994.
6. T. Viéville, E. Clergue, R. Enciso, and H. Mathieu. Experimentating with 3D vision on a robotic head. *Robotics and Autonomous Systems*, 14(1), 1995.
7. T. Viéville and D. Lingrand. Using singular displacements for uncalibrated monocular visual systems. Technical Report RR-2678, INRIA, 1995.
8. T. Viéville, C. Zeller, and L. Robert. Using collineations to compute motion and structure in an uncalibrated image sequence. *International Journal of Computer Vision*, 1995. To appear.
9. C. Wiles and M. Brady. Closing the loop on multiple motion. In *5th International Conference on Computer Vision*, pages 308–313, 1995.
10. R. Willson. *Modeling and Calibration of Automated Zoom Lenses.* PhD thesis, Department of Electical and Computer Engineering, Carnegie Mellon University, 1994.
11. C. Zeller and O. Faugeras. Applications of non-metric vision to some visual guided tasks. In *The 12th Int. Conf. on Pattern Recognition*, pages 132–136, 1994.
12. Z. Zhang, R. Deriche, Q.-T. Luong, and O. Faugeras. A robust approach to image matching: Recovery of the epipolar geometry. In *Proc. International Symposium of Young Investigators on Information\Computer\Control*, pages 7–28, Beijing, China, Feb. 1994.

Acknowledgments. We are especially thankful to **O.D. Faugeras** for some powerful ideas which are at the origin of this work. This work is partially achieved under
Esprit Project P8878/REALISE.

Duality of Multi-Point and Multi-Frame Geometry: Fundamental Shape Matrices and Tensors*

Daphna Weinshall[1], Michael Werman[1] and Amnon Shashua[2]

[1] Inst. of Computer Science, Hebrew University, Jerusalem 91904, Israel
{werman,daphna}@cs.huji.ac.il
[2] Dept. of Computer Science, Technion, 32000 Haifa, Israel

Abstract. We provide a complete analysis of the geometry of N points in 1 image, employing a formalism in which multi-frame and multi-point geometries appear in symmetry: points and projections are interchangeable. We derive bilinear equations for 6 points, trilinear equations for 7 points, and quadrilinear equations for 8 points. The new equations are used to design new algorithms for the reconstruction of projective shape from many frames. Shape is represented by shape descriptors, which are sufficient for object recognition, and for the simulation of new images of the object. We further propose a linear shape reconstruction scheme which uses all the available data - all points and all frames - simultaneously. Unlike previous approaches, the equations developed here lead to *direct* and *linear* computation of shape, without going through the cameras' geometry.

1 Introduction

The geometry of multiple primitives in multiple frames, where a $3D$ model consisting of many primitives is projected to a sequences of images via unknown cameras, has two inherent unknowns: the camera geometry, and the shape geometry. We do not know in advance the parameters of the projection from $3D$ to $2D$ in each image, and we do not know the parameters of the $3D$ shape (position) of each point in the model. Depending on the application at hand, we may want to compute shape, or camera geometry, or both. For example, tracking requires the knowledge of camera geometry, whereas object recognition requires the knowledge of shape.

Very often, the sequence of computations had been argued to be the following: First, compute the camera geometry (by camera geometry we refer here to both explicit representations, e.g., rotation and translation matrices, or implicit representations, e.g., the fundamental matrix). Second, compute the shape using the known projection geometry. This order of events makes particular sense

* This research was supported by the Israeli Ministry of Science under Grant 032.7568, and by the U.S. Office of Naval Research under Grant N00014-93-1-1202, R&T Project Code 4424341—01.

when the first task that one needs to solve is the tracking of features across frames.

But the reverse order of computation is also sensible: first compute shape, then compute camera geometry if necessary. This is the logical order when only shape is needed, for example, in a pure recognition task. If this is the case, it makes sense to compute the shape first, rather than rely in the shape computation on an otherwise unnecessary intermediate computation (the computation of camera geometry), which may not always be reliable.

This line of reasoning lead us to consider representations of shape which differ from the traditional Cartesian coordinates (whether in a Euclidean, affine, or projective basis). We seek shape representations that can be computed from images directly and robustly, and which include sufficient detail to unambiguously identify and simulate new images of the same shape. These are necessary requirements for a shape representation to be "good". Thus we propose below new shape descriptors, describing the shape of 6-8 points, which can be directly and linearly computed from at least 2-4 images. The computation of these shape descriptors does not require the computation of camera parameters (camera calibration, see also [13]).

Most of the literature on the subject followed the first path, namely, computing camera calibration first using techniques derived from multi-frame geometry (see, e.g., [6, 7, 10, 4]). Much less is known about multi-point geometry under perspective projection with uncalibrated cameras, and this gap is filled by our paper. We present here a complete analysis of the multi-point geometry, describing relations between the projections of many points in an image (each relation has a dual relation in multi-camera geometry). This analysis provides the foundation for a computation where shape is computed first, and camera geometry second. A similar analysis was presented by Carlsson in [1] (see also [15]).

More specifically, in Section 2 we show dual results to the ones obtained by computing camera geometry first. We first observe that a projection matrix from a $3D$ projective world to a $2D$ projective image is really a point in $\mathcal{P}^3$ (a geometrical interpretation of this point is given in [1]). We then observe that the relations between models, projections, and images can be written in a symmetrical form where models and projections are interchangeable. Using these observations, for every known relation between images and projection matrices we can derive a dual relation between images and models.

We use the dual results to develop new algorithms for the direct computation of shape, without first computing camera geometry. In particular, we describe a linear algorithm to compute the fundamental shape matrix of 6 points from at least 4 images, a linear algorithm to compute the fundamental shape tensor of 7 points from at least 3 images, and a linear algorithm to compute the fundamental shape tensor of 8 points from at least 2 images. Experiments with real images are described in Section 3. In Section 4 we show how to enhance the shape computation to include many points simultaneously. The computed shape descriptions are sufficient for the identification of novel images of the same object, and for the prediction of new images, as described in Section 5.

2 Algebraic and Geometrical derivation of Results

In this section we derive bilinear, trilinear, and quadrilinear relations using a formalism analogous (but dual) to [3]. We show duality between multi-frame and multi-point geometries: every relation between the vector coordinates of one point in many images has an almost identical equivalent here, a relation between the vector coordinates of many points in 1 image. This duality follows from the employment of a symmetrical formalism to describe multi-point and multi-camera geometry, where $3D$ coordinates of points and projection parameters are interchangeable (see Section 2.2).

2.1 Problem definition

Consider a model which includes more than 5 $3D$ points in $\mathcal{P}^3$ (all 5 point sets are projectively equivalent). W.l.o.g. (assuming uncalibrated cameras) we choose a coordinate system where the first 5 points are the standard projective basis, and $\mathbf{M}_i = [X_i, Y_i, Z_i, W_i]$ denotes the coordinates of the (i+5) point. Thus the shape matrix of the model $\mathtt{M}$, a $4 \times n$ matrix, is:

$$\mathtt{M} = \begin{pmatrix} 1 & 0 & 0 & 0 & 1 & X_1 & \dots & X_n \\ 0 & 1 & 0 & 0 & 1 & Y_1 & \dots & Y_n \\ 0 & 0 & 1 & 0 & 1 & Z_1 & \dots & Z_n \\ 0 & 0 & 0 & 1 & 1 & W_1 & \dots & W_n \end{pmatrix} \tag{1}$$

Similarly and again w.l.o.g, we choose the image coordinates of the first 4 points to be the standard projective basis in $\mathcal{P}^2$. Let $\mathbf{m}_i = [a_i, b_i, c_i] \in \mathcal{P}^2$ denote the vector of homogeneous coordinates of the $(i+5)$ point in the image. The projected shape matrix $\mathtt{m}$, a $3 \times n$ matrix, is:

$$\mathtt{m} = \begin{pmatrix} 1 & 0 & 0 & 1 & a_0 & \dots & a_n \\ 0 & 1 & 0 & 1 & b_0 & \dots & b_n \\ 0 & 0 & 1 & 1 & c_0 & \dots & c_n \end{pmatrix}$$

Since the image shape matrix $\mathtt{m}$ is a projection of the shape matrix $\mathtt{M}$, there exists a 3×4 projection matrix $\mathtt{P}$ such that the following equality holds in $\mathcal{P}^2$:

$$\mathtt{P} \cdot \mathtt{M} = \mathtt{m} \tag{2}$$

Given our particular selection of projective bases, and using the fact that the first 4 points are transformed from a basis in $\mathcal{P}^3$ to a basis in $\mathcal{P}^2$, the projection matrix $\mathtt{P}$ is of the form [2]:

$$\mathtt{P} = \begin{bmatrix} \alpha & 0 & 0 & \delta \\ 0 & \beta & 0 & \delta \\ 0 & 0 & \gamma & \delta \end{bmatrix}$$

We define a corresponding projection vector in $\mathcal{P}^3$ (a geometrical interpretation of this vector can be found in [1]):

$$\mathbf{p} = (\alpha \quad \beta \quad \gamma \quad \delta)$$

2.2 Multi-frame and multi-point geometry

Using Eq. (2) with the $5+i$ model point $\mathbf{M}_i$ and the j-th frame obtained by the projection camera $\mathbf{p}_j$, producing the image point $(x_{ji}, y_{ji}, 1)$ gives:

$$x_{ji} = \frac{\alpha_j X_i + \delta_j W_i}{\gamma_j Z_i + \delta_j W_i}, \quad y_{ji} = \frac{\beta_j Y_i + \delta_j W_i}{\gamma_j Z_i + \delta_j W_i}$$

Clearly these equations are completely symmetrical with respect to $\mathbf{M}_i$ and $\mathbf{p}_j$: if we interchange the 2 vectors, we will get exactly the same image point. With k frames and $n+4$ points, we get the $2k \times n$ measurements matrix $\mathtt{W}$ whose ji element is x_{ji} for $j \leq k$, and $y_{(j-k)i}$ for $k < j \leq 2k$ (cf. [11, 14]). Now if we read $\mathtt{W}$ by columns, the i-th column gives us multi-camera geometry; if we read it by rows, the j-th and $(j+k)-th$ rows give us multi-point geometry in a single image.

We use this symmetry to obtain dual relations to those obtained for multi-camera geometry, by reading the data-matrix by rows instead of by columns; there are the following role changes:

- The solution vector in $\mathcal{P}^3$ is a projection operator $\mathbf{p}_j$ in multi-point geometry, and a $3D$ point $\mathbf{M}_i$ in multi-camera geometry.
- In multi-point geometry the data vectors are the 2 row-vectors $[x_{1j}, \ldots, x_{nj}]$ and $[y_{1j}, \ldots, y_{nj}]$ - the image coordinates in the j-th frame of all the points. In multi-frame geometry, the data vector is the column-vector $[x_{i1}, \ldots, x_{ik}, y_{i1}, \ldots, y_{ik}]$ - the trajectory of the $i+5$ point in k frames.

2.3 Multi-point geometry

From now on we fix the frame and ignore the subscript j. Every $3D$ point $\mathbf{M}_i = [X_i, Y_i, Z_i, W_i]$ from the 5th on ($i \geq 0$), which is projected to an image point $[a_i, b_i, c_i]$, defines 2 constraints on the projection matrix $\mathtt{P}$. We write these as linear homogeneous equations constraining the projection vector $\mathbf{p}$, $\forall\, i \geq 0$:

$$\begin{bmatrix} c_i X_i & 0 & -a_i Z_i & c_i W_i - a_i W_i \\ 0 & c_i Y_i & -b_i Z_i & c_i W_i - b_i W_i \end{bmatrix} \cdot \mathbf{p} = 0 \tag{3}$$

(note that $X_0 = Y_0 = Z_0 = W_0 = 1$). Using these equations, we obtain relations between models and images.

Given $n+4$ points, including the 4 image basis points and n additional points, (3) expands to $2n$ linear equations for $\mathbf{p}$. The matrix representing this over-constrained linear system is $2n \times 4$. Since the linear system is homogenous and has a non-trivial solution $\mathbf{p}$, the rank of this matrix must be smaller than 4. Thus the determinant of every subset of 4 rows of this matrix must be 0. This gives us $\binom{2n}{4}$ constraints on $\mathbf{p}$.

Bilinear equations Given 6 points (or $n = 2$), the constraints matrix is 4×4 and of rank 3; thus we have 1 constraint – the determinant of the matrix must be 0. This gives us the following equation:

$$\begin{array}{l} (-a_0\, b_1 + a_0\, c_1)(W_1\, X_1 - Y_1\, Z_1) \;+\; (a_1\, b_0 - b_0\, c_1)(W_1\, Y_1 - Y_1\, Z_1) \;+ \\ (-a_1\, c_0 + b_1\, c_0)(W_1\, Z_1 - Y_1\, Z_1) \;+ \\ (-a_0\, c_1 + b_0\, c_1)(X_1\, Y_1 - Y_1\, Z_1) \;+\; (a_0\, b_1 - b_1\, c_0)(X_1\, Z_1 - Y_1\, Z_1) \;=\; 0 \end{array} \tag{4}$$

We propose the following shape vector as a shape descriptor, "good" in the sense discussed in the introduction, and fully describing the projective shape of the 6 points. In other words, the projective coordinates of the 6th point can be computed from this vector (if the shape representation is needed for some purpose other then recognition or the generation of new images)[3]:

$$\mathbf{V6} \;=\; [W_1\, X_1,\; W_1\, Y_1,\; W_1\, Z_1,\; X_1\, Y_1,\; X_1\, Z_1] - Y_1\, Z_1 \mathbf{1}_5 \tag{5}$$

This representation can be computed from at least 4 pictures (or directly from the $3D$ model).

We can rewrite (4) in a matrix form, in a dual way to the use of the fundamental matrix to describe the epipolar geometry:

$$\mathbf{m}_0^T G_{01} \mathbf{m}_1 \;=\; [a_0, b_0, c_0] \begin{bmatrix} 0 & X_1\, Z_1 - W_1\, X_1 & -X_1\, Y_1 + W_1\, X_1 \\ W_1\, Y_1 - Y_1\, Z_1 & 0 & X_1\, Y_1 - W_1\, Y_1 \\ Y_1\, Z_1 - W_1\, Z_1 & -X_1\, Z_1 + W_1\, Z_1 & 0 \end{bmatrix} \begin{bmatrix} a_1 \\ b_1 \\ c_1 \end{bmatrix} \;=\; 0 \tag{6}$$

Whereas the fundamental matrix depends on the camera calibration, the fundamental shape matrix G_{01} here depends on the $3D$ shape of the object. It has only 4 degrees of freedom in it (up to scale), since its elements sum to 0.

Trilinear equations Given 7 points (or $n = 3$), the constraints matrix is 6×4, giving us $\binom{6}{4} = 15$ constraints. Using algebraic tools, we show that there are only 7 independent equations: 3 bilinear equations involving subsets of 6 points, and 4 new trilinear equations. The 4 new constraints give us the following set of equations:

$$\begin{bmatrix} 0 & 0 & -a_0 c_1 c_2 + b_0 c_1 c_2 & 0 \\ -a_0 a_2 c_1 + a_2 c_0 c_1 & 0 & a_0 b_2 c_1 - b_2 c_0 c_1 & 0 \\ a_0 a_2 c_1 - a_0 c_1 c_2 & 0 & -a_0 b_2 c_1 + a_0 c_1 c_2 & 0 \\ 0 & -a_0 c_1 c_2 + b_0 c_1 c_2 & 0 & 0 \\ 0 & -a_2 b_0 c_1 + a_2 c_0 c_1 & 0 & -b_0 b_2 c_1 + b_2 c_0 c_1 \\ 0 & a_2 b_0 c_1 - b_0 c_1 c_2 & 0 & b_0 b_2 c_1 - b_0 c_1 c_2 \\ a_0 a_1 c_2 - a_1 c_0 c_2 & a_0 b_1 c_2 - b_1 c_0 c_2 & 0 & 0 \\ 0 & 0 & -a_1 b_0 c_2 + a_1 c_0 c_2 & b_0 b_1 c_2 - b_1 c_0 c_2 \\ -a_1 a_2 c_0 + a_1 c_0 c_2 & -a_2 b_1 c_0 + b_1 c_0 c_2 & a_1 b_2 c_0 - a_1 c_0 c_2 & -b_1 b_2 c_0 + b_1 c_0 c_2 \\ -a_0 a_1 c_2 + a_0 c_1 c_2 & -a_0 b_1 c_2 + a_0 c_1 c_2 & 0 & 0 \\ 0 & 0 & a_1 b_0 c_2 - b_0 c_1 c_2 & -b_0 b_1 c_2 + b_0 c_1 c_2 \end{bmatrix}^T \begin{pmatrix} X_1 Y_2 - W_1 Z_2 \\ X_1 Z_2 - W_1 Z_2 \\ X_1 W_2 - W_1 Z_2 \\ Y_1 X_2 - W_1 Z_2 \\ Y_1 Z_2 - W_1 Z_2 \\ Y_1 W_2 - W_1 Z_2 \\ Z_1 X_2 - W_1 Z_2 \\ Z_1 Y_2 - W_1 Z_2 \\ Z_1 W_2 - W_1 Z_2 \\ W_1 X_2 - W_1 Z_2 \\ W_1 Y_2 - W_1 Z_2 \end{pmatrix} = 0$$

[3] $\mathbf{1}_n$ below deontes the vector of length n, whose elements are all 1.

The "good" shape descriptor of the 7 points is the following shape vector:

$$\begin{aligned}\mathbf{V7} = & [X_1 Y_2,\ X_1 Z_2,\ X_1 W_2,\ Y_1 X_2,\ Y_1 Z_2,\ Y_1 W_2,\ Z_1 X_2,\ Z_1 Y_2, \\ & Z_1 W_2,\ W_1 X_2,\ W_1 Y_2] - W_1 Z_2 \mathbf{1}_{11}\end{aligned} \tag{7}$$

As in multi-camera geometry, we can write each new trilinear constraint in a tensor form, using a $3 \times 3 \times 3$ fundamental shape tensor T. More specifically, we have:

$$\sum_{i,j,k} \mathsf{T}_{ijk} (\mathbf{m}_0)_i (\mathbf{m}_1)_j (\mathbf{m}_2)_k = 0 \tag{8}$$

Each of the 4 constraints has a different shape tensor. For example, the tensor of the first constraint is:

$$T = \begin{bmatrix} 0\,0 & Z_1 X_2 - W_1 X_2 \\ 0\,0 & 0 \\ 0\,0 & -X_1 W_2 + W_1 X_2 \end{bmatrix} \begin{bmatrix} 0\,0\,0 \\ 0\,0\,0 \\ 0\,0\,0 \end{bmatrix} . \begin{bmatrix} 0\,0 & -Z_1 X_2 + Z_1 W_2 \\ 0\,0 & 0 \\ 0\,0 & 0 \end{bmatrix}$$

However, there are only 11 unknowns in all 4 tensors, and writing the constraints as above allows us to compute the shape vectors from 3 images or more.

Quadrilinear equations With 8 points we have $\binom{8}{4}$= 70 constraints. Using algebraic tools, we show that there are only 15 independent equations: 3 bilinear equations involving subsets of 6 points, and 12 trilinear equations involving 7 points. However, there are new quadrilinear equations that define 22 independent constraints on a new 41-dimensional shape descriptor.

$$\begin{aligned}\mathbf{V8} = & [W_1 W_2 Z_3,\ W_1 X_2 Z_3,\ W_1 X_2 W_3,\ W_1 Y_2 X_3,\ Z_1 W_2 Y_3,\ Z_1 W_2 Z_3,\ Z_1 W_2 W_3,\ W_1 X_2 Y_3,\ Z_1 X_2 W_3, \\ & Z_1 Y_2 X_3,\ W_1 Z_2 W_3,\ W_1 W_2 X_3,\ W_1 W_2 Y_3,\ W_1 Y_2 Z_3,\ W_1 Y_2 W_3,\ W_1 Z_2 X_3,\ W_1 Z_2 Y_3,\ W_1 Z_2 Z_3, \\ & Y_1 Z_2 X_3,\ Y_1 Z_2 Z_3,\ Y_1 Z_2 W_3,\ Y_1 W_2 X_3,\ Y_1 W_2 Z_3,\ Y_1 W_2 W_3,\ Z_1 X_2 Y_3,\ Z_1 X_2 Z_3,\ Y_1 X_2 Z_3, \\ & Y_1 X_2 W_3,\ Z_1 Z_2 Y_3,\ Z_1 Z_2 W_3,\ Z_1 W_2 X_3,\ Z_1 Y_2 Z_3,\ Z_1 Y_2 W_3,\ Z_1 Z_2 X_3,\ X_1 Y_2 W_3,\ X_1 Z_2 Y_3, \\ & X_1 Z_2 Z_3,\ X_1 Z_2 W_3,\ X_1 W_2 Y_3,\ X_1 W_2 Z_3,\ X_1 W_2 W_3] - X_1 Y_2 Z_3 \mathbf{1}_{41}\end{aligned} \tag{9}$$

The derivation of the equations constraining this shape descriptor are omitted. Using more than 8 points does not lead to any new equation.

3 Experiments

Using a sequence of real images, where features had been automatically detected and tracked, we compute the projective shape of the tracked points using the following procedure:

Initially, we choose an arbitrary basis of 5 points. For each additional point:

1. The corresponding shape vector **V6** is computed in 2 ways:

Linear computation: all the available frames are used to solve an over-determined linear system of equations, where each frame provides the single constraint given in (4).

Non-linear computation: we use 3 frames only: the first, middle, and last, and the following non-linear constraint on the elements of $\mathbf{V6}$: $\mathbf{V6}_1\mathbf{V6}_2\mathbf{V6}_5 - \mathbf{V6}_1\mathbf{V6}_3\mathbf{V6}_4 + \mathbf{V6}_2\mathbf{V6}_3\mathbf{V6}_4 - \mathbf{V6}_2\mathbf{V6}_3\mathbf{V6}_5 - \mathbf{V6}_2\mathbf{V6}_4\mathbf{V6}_5 + \mathbf{V6}_3\mathbf{V6}_4\mathbf{V6}_5 = 0$

(where $\mathbf{V6}_i$ denotes the i-th component of the shape descriptor $\mathbf{V6}$.) A similar non-linear constraint was derived in [8]. Optimally, the non-linear computation should use the solution of the linear system defined by all the frames, and project the result onto the surface defined by the equation above. The non-linear computation normally gives 3 solutions.

2. From the shape vector $\mathbf{V6}$, the homogeneous coordinates of the point are computed as follows:

$$\frac{X_1}{W_1} = \frac{\mathbf{V6}_4 - \mathbf{V6}_5}{\mathbf{V6}_2 - \mathbf{V6}_3}, \quad \frac{Y_1}{W_1} = \frac{\mathbf{V6}_4}{\mathbf{V6}_1 - \mathbf{V6}_3}, \quad \frac{Z_1}{W_1} = \frac{\mathbf{V6}_5}{\mathbf{V6}_1 - \mathbf{V6}_2}$$

3. Finally, in order to compare the results with the real $3D$ shape of the points, the projective homogeneous coordinates are multiplied by the actual $3D$ coordinates of the projective basis points, to obtain the equivalent Euclidean representation.

We used a sequence obtained from the 1991 motion workshop. It includes 16 images of a robotic laboratory, obtained by rotating a robot arm 120° (one frame is shown in Fig. 1a). 32 corner-like points were tracked. The depth values of the points in the first frame ranged from 13 to 33 feet; moreover, a wide-lens camera was used, causing distortions at the periphery which were not compensated for. (See a more detailed description in [9] Fig. 4, or [5] Fig. 3.)

We computed the shape of the 32 points as described above, using all the 16 frames in the linear computation, and using the non-linear computation. The real $3D$ coordinates of about half the points, the corresponding linearly reconstructed $3D$ coordinates, and the best reconstructed $3D$ coordinates (among the 4 solutions provided by the linear and non-linear computations), are shown below. We also give the median relative error (where the error at each point is divided by the distance of the point from the origin), computed over all points:

real shape:

$$\begin{bmatrix} -0.3 & -1.7 & -0.3 & 1.8 & 5.3 & 9.9 & 3.2 & -2.3 & 1.5 & -0.6 & 0.5 & 1.5 & -0.5 \\ -4 & -2.6 & 4.4 & 6.3 & 4.2 & -1.6 & -2.8 & -2 & 5 & 3 & 2 & 0.9 & 1 \\ 16.4 & 17.1 & 19.7 & 20 & 25.3 & 29.8 & 31.6 & 15.1 & 21.7 & 21.5 & 21.6 & 21 & 21.6 \end{bmatrix}$$

linear computation:

$$\begin{bmatrix} -0.3 & -1.8 & -0.6 & 0.9 & 3.8 & 8.9 & -0.8 & -2.4 & 0.7 & -0.6 & 0.3 & 1.3 & -0.4 \\ -3.6 & -1.3 & 5 & 6.2 & 4.4 & -1.5 & 0.7 & -1.3 & 4.8 & 2.6 & 1.8 & 0.4 & 0.7 \\ 15.8 & 14.7 & 21.5 & 24.9 & 27.5 & 30.1 & 9.7 & 13.4 & 23.5 & 20.2 & 21.1 & 21.1 & 19.2 \end{bmatrix}$$

median relative error: 12%

a) b)

Fig. 1. a) One frame from the lab sequence. b) One frame from the box sequence.

best of linear and non-linear computations:

$$\begin{bmatrix} -0.3 & -0.8 & -0.4 & 1.3 & 3.8 & 8.9 & 3.2 & -2.4 & 1.4 & -0.6 & 0.2 & 1.3 & -0.4 \\ -3.6 & -2.4 & 4.6 & 7.9 & 4.4 & -1.5 & -2.2 & -1.3 & 5.3 & 3.3 & 1.9 & 0.4 & 0.7 \\ 15.8 & 18.6 & 19.4 & 18 & 27.5 & 30.1 & 31.9 & 13.6 & 21.3 & 20.9 & 21.3 & 21.1 & 19.2 \end{bmatrix}$$

median relative error: 5.5%

4 Shape from many points and many frames

W.l.o.g., consider the trilinear shape descriptor **V7** defined in (7).

Lemma 1 rank 4. *Given* $6+n$ *points, the* $11 \times n$ *matrix whose i-th column is the shape vector* **V7** *of the points* $< 1, 2, \ldots, 6, 6+i >$, $i = 1, \ldots, n$, *is of rank 4.*

In other words, we first select 6 fixed points and recover the shape vectors of the sets $< 1, \ldots, 6, 7 >, < 1, \ldots, 6, 8 >, \ldots, < 1, \ldots, 6, 6+n >$. We then concatenate the shape vectors into a $11 \times n$ matrix denoted $\mathtt{W}$. Our claim is that the resulting matrix $\mathtt{W}$ is of rank 4 (instead of 11).

More specifically, let $\mathbf{V7}^i$ denote the shape vector of the set of points $< 1, \ldots, 6, 6+i >$. Let $\mathtt{W} = [\,\mathbf{V7}^1 \;\; \ldots \;\; \mathbf{V7}^n\,]$. It follows from (7) that:

$$\mathtt{W} = \begin{bmatrix} 0 & 0 & 0 & Y_1 & 0 & 0 & Z_1 & 0 & 0 & W_1 & 0 \\ X_1 & 0 & 0 & 0 & 0 & 0 & 0 & Z_1 & 0 & 0 & W_1 \\ -W_1 & X_1 - W_1 & -W_1 & -W_1 & Y_1 - W_1 & -W_1 & -W_1 & -W_1 & -W_1 & -W_1 & -W_1 \\ 0 & 0 & X_1 & 0 & 0 & Y_1 & 0 & 0 & Z_1 & 0 & 0 \end{bmatrix}^T \begin{bmatrix} X_2 \ldots X_n \\ Y_2 \ldots Y_n \\ Z_2 \ldots Z_n \\ W_2 \ldots W_n \end{bmatrix}$$

Thus W is of rank 4.

This result gives us the following algorithm for the reconstruction of shape using many views and many points, for an object with $n+6$ points:

1. Choose a subset of 6 "good" points (an algorithm on how to choose good basis points in described in [14]).
2. For every additional point M_i, $i \geq 7$:
   Using all available frames (but at least 3), compute the shape vector $\mathbf{V7}^i$ of the set of 7 points $< 1,2,3,4,5,6,6+i >$
3. Define the $11 \times n$ matrix $\mathtt{W}$, whose i-th column is the shape vector $\mathbf{V7}^i$. From Result 1, the rank of $\mathtt{W}$ is 4. Let $\tilde{\mathtt{W}}$ denote the matrix of rank 4 which is the closest (in least squares) to $\mathtt{W}$; $\tilde{\mathtt{W}}$ is computed using SVD factorization of $\mathtt{W}$ (see Section 4), with the decomposition: $\mathtt{W} \approx \tilde{\mathtt{W}} = \mathtt{U} \cdot \mathtt{V}$, where $\mathtt{U}$ is a 11×4 matrix, and $\mathtt{V}$ is a $4 \times n$ matrix.
4. Notice that $\tilde{\mathtt{W}} = \mathtt{UTT}^{-1}\mathtt{V}$ for every non-singular 4×4 matrix $\mathtt{T}$. Compute $\mathtt{T}$ such that

$$\mathrm{UT} = \begin{bmatrix} 0 & 0 & 0 & Y_1 & 0 & 0 & Z_1 & 0 & 0 & W_1 & 0 \\ X_1 & 0 & 0 & 0 & 0 & 0 & 0 & Z_1 & 0 & 0 & W_1 \\ -W_1 & X_1 - W_1 & -W_1 & -W_1 & Y_1 - W_1 & -W_1 & -W_1 & -W_1 & -W_1 & -W_1 & -W_1 \\ 0 & 0 & X_1 & 0 & 0 & Y_1 & 0 & 0 & Z_1 & 0 & 0 \end{bmatrix}^T \tag{10}$$

   for some constants X_1, Y_1, Z_1, W_1. This defines a homogeneous linear system of equations, which can be solved using, e.g., SVD decomposition. Note that this system can only be solved in a least-squares sense, as there are 40 equations with only 16 unknowns (the elements of $\mathtt{T}$).
5. (a) $\mathtt{T}^{-1}\mathtt{V}$ is the shape matrix of points $7, \ldots, 6+n$.
   (b) the coordinates of point 6 are obtained from $\mathtt{UT}$ and (10).

When using this algorithm with real images, our first results have been very sensitive to noise. Clearly the use of robust statistics (or outlier removal), in the solution of the linear system that defines the trilinear shape vector, is necessary.

5 Simulation of new images:

Rather than compute projective shape, the shape vectors described above can be used directly to simulate new (feasible) images of the object. It follows from Section 2.3 that 5 $3D$ points can be projected to any location in the image. Moreover, there is only 1 constraint on the location of the 6th point. Thus we start by choosing a random location for the first 5 points, and the first coordinate of the 6th point. The location of the remaining points can now be computed using the shape vectors:

The 6th point: we compute the shape vector **V6** of the first 6 points, from which we obtain the fundamental shape matrix G_{01}. We plug into (6) G_{01}, the coordinates of the 5th point in the new frame, and 1 constraint on the coordinates of the 6th point. This gives us a linear equation with 1 unknown, and we solve for the unknown coordinate of the 6th point.

The remaining points: for each point P_i, $i = 7..n$, we compute the shape vectors $\mathbf{V7}^l$, $l = 1..4$, which describe the shape of the first 6 points and P_i. From each shape vector we compute the trilinear shape tensor T_l. We plug each T_l, the coordinates of the 5th point in the new frame, and the coordinates of the 6th point computed in the previous stage, into (8). Thus the 4 trilinear constraints give us 4 homogeneous linear equations with 3 unknowns, the coordinates of the i-th point. For each point P_i we solve this system using SVD, thus obtaining the coordinates of all the points.

To test this algorithm we used a box sequence, which includes 8 images of a rectangular chequered box rotating around a fixed axis (one frame is shown in Fig. 1b). 40 corner-like points on the box were tracked. The depth values of the points in the first frame ranged from 550 to 700 mms, and were given. (See a more detailed description of the sequence in [9] Fig. 5, or [5] Fig. 2.)

We rotated the box by up to $\pm 60^o$, translated it in $\mathcal{R}^3$ by up to ± 100 mms, and then projected it with uncalibrated perspective projection, to obtain new images of the box. The new images differed markedly from the original 8 images used for the computation of the shape vectors. We selected a "good" basis of 5 points, using the procedure described in [14]. In each image, we transformed 4 of the basis points to the non-standard basis of $\mathcal{P}^2$: $[1,0,1]$, $[0,1,1]$, $[0,0,1]$, $[1,1,1]$.

We used the image coordinates of the 5th point, and the x coordinate of the 6th point, to compute the shape of the remaining points as described above. In a typical image, in which the size of the projected box was 83×68 pixels, the median prediction error was 0.32 pixels; the mean prediction error was 0.46. The mean error could get larger in some simulated images, when large errors occurred in outlier points. The error at each point was computed in the image, by the Euclidean distance between the real point and its predicted location, using the original (metric) coordinate system of the image.

6 Discussion

When looking at data streams containing sequences of points, we wish to use all the available data: all frames, all points, or all frames and all points. Under weak perspective projection: [12] showed how to use all the points and 1.5 frames to linearly compute affine structure, [13] showed how to use all the frames of 4 points to linearly compute Euclidean structure, [11] used all the data to linearly compute affine shape and camera orientation, and [14] used all the data to linearly compute Euclidean shape.

The situation under perspective projection is more complex: [6, 7, 10, 4] (among others) showed how to linearly compute the camera calibration from 2-4 frames using all the points. Here (as well as in [1]) we showed how to linearly compute the projective shape of 6-8 points from all the frames. We also showed a 2-step algorithm to linearly compute the projective shape of all the points from all the frames. This does not yet accomplish the simplicity and robustness demonstrated by the algorithms which work under the weak perspective approximation.

References

1. S. Carlsson. Duality of reconstruction and positioning from projective views. In *IEEE Workshop on Representations of Visual Scenes*, Cambridge, Mass, 1995.
2. O. Faugeras. What can be seen in three dimensions with an uncalibrated stereo rig? In *Proceedings of the 2nd European Conference on Computer Vision*, pages 563–578, Santa Margherita Ligure, Italy, 1992. Springer-Verlag.
3. O. Faugeras and B. Mourrain. On the geometry and algebra of the point and line correspondences between N images. In *Proceeding of the Europe-China Workshop on geometrical modeling and invariant for computer vision*. Xidian University Press, 1995.
4. R. Hartley. Lines and points in three views - an integrated approach. In *Proceedings Image Understanding Workshop*, pages 1009–10016, San Mateo, CA, 1994. Morgan Kaufmann Publishers, Inc.
5. R. Kumar and A. R. Hanson. Sensitivity of the pose refinement problem to accurate estimation of camera parameters. In *Proceedings of the 3rd International Conference on Computer Vision*, pages 365–369, Osaka, Japan, 1990. IEEE, Washington, DC.
6. H. C. Longuet-Higgins. A computer algorithm for reconstructing a scene from two projections. *Nature*, 293:133–135, 1981.
7. Q.-T. Luong and O. Faugeras. The fundamental matrix: theory, algorithms, and stability analysis. *International Journal of Computer Vision*, 1995. in press.
8. L. Quan. Invariants of 6 points from 3 uncalibrated images. In *Proceedings of the 3rd European Conference on Computer Vision*, pages 459–470, Stockholdm, Sweden, 1994. Springer-Verlag.
9. H. S. Sawhney, J. Oliensis, and A. R. Hanson. Description and reconstruction from image trajectories of rotational motion. In *Proceedings of the 3rd International Conference on Computer Vision*, pages 494–498, Osaka, Japan, 1990. IEEE, Washington, DC.
10. A. Shashua and M. Werman. Trilinearity of three perspective views and its associated tensor. In *Proceedings of the 5th International Conference on Computer Vision*, Cambridge, MA, 1995. IEEE, Washington, DC.
11. C. Tomasi and T. Kanade. Shape and motion from image streams under orthography: a factorization method. *International Journal of Computer Vision*, 9(2):137–154, 1992.
12. S. Ullman and R. Basri. Recognition by linear combinations of models. *IEEE Transactions on Pattern Analysis and Machine Intelligence*, 13(10):992–1006, 1991.
13. D. Weinshall. Model-based invariants for 3D vision. *International Journal of Computer Vision*, 10(1):27–42, 1993.
14. D. Weinshall and C. Tomasi. Linear and incremental acquisition of invariant shape models from image sequences. *IEEE Transactions on Pattern Analysis and Machine Intelligence*, 17(5):512–517, 1995.
15. D. Weinshall, M. Werman, and A. Shashua. Shape tensors for efficient and learnable indexing. In *Proceedings of the IEEE Workshop on Representations of Visual Scenes*, Cambridge, Mass, 1995.

On the Appropriateness of Camera Models

Charles Wiles[1,2] and Michael Brady[1]

[1] Robotics Research Group, Department of Engineering Science, University of Oxford, Parks Road, Oxford, OX1 3PJ, UK.
[2] TOSHIBA, Kansai Research Laboratories, Osaka, Japan.

Abstract. Distinct camera models for the computation of structure from motion (sfm) can be arranged in a hierarchy of uncalibrated camera models. Degeneracies mean that the selection of the most appropriate camera from the hierarchy is key. We show how accuracy of fit to the data; efficiency of computation; and clarity of interpretation enable us to compute a measure of the *appropriateness* of a particular camera model for an observed trajectory set and thus automatically select the most appropriate model from our hierarchy of camera models. An elaboration of the idea, that we call the *combined appropriateness* allows us to determine a suitable frame at which to switch between camera models.

1 Introduction

A number of different camera models have been suggested for the computation of structure from motion (sfm). The appropriateness of the particular model depends on a number of factors. Usually, known imaging conditions and the required clarity in the computed structure are used to choose the camera model in advance. However, for many tasks (for example motion segmentation), it is neither possible to know in advance what the imaging conditions will be nor is the clarity of the computed structure important. Moreover, under degenerate motion or degenerate structure, particular algorithms for particular models may fail due to degeneracies in the solution. In this paper we propose a new measure called "*appropriateness*", which can be determined directly from observed image trajectories and is robust to gross outliers. The measure can be used to automatically select the most appropriate model from a hierarchy of camera models on the basis of *accuracy*, *clarity* and *efficiency*. The hierarchy of models allows degeneracy to be dealt with by explicitly employing reduced models. Moreover, a measure of *combined appropriateness* [4] enables automatic model switching when different camera models are appropriate for different parts of the sequence.

Previous research has tackled only the forward problem of choosing a camera model when the process of image formation is known. The far more difficult inverse problem of choosing the appropriate camera model to compute structure given only the observed image trajectories has not been tackled. This paper is an initial foray into the problem.

2 Rigidity

A camera projects a 3D world point $\mathbf{X} = (X, Y, Z, 1)^T$ to a 2D image point $\mathbf{x} = (x, y, 1)^T$. The *projective* camera model [1] can be written

$$\lambda \begin{bmatrix} x \\ y \\ 1 \end{bmatrix} = \begin{bmatrix} P_{11} & P_{12} & P_{13} & P_{14} \\ P_{21} & P_{22} & P_{23} & P_{24} \\ P_{31} & P_{32} & P_{33} & P_{34} \end{bmatrix} \begin{bmatrix} X \\ Y \\ Z \\ 1 \end{bmatrix} = \mathbf{P} \begin{bmatrix} X \\ Y \\ Z \\ 1 \end{bmatrix}, \quad (1)$$

This formulation places no restriction on the frame in which $\mathbf{X}$ is defined, but the *rigidity constraint* can be enforced by fixing the frame to the object in such a way that $\mathbf{X}$ is constant for all time. Equation 1 can be rewritten $\lambda \mathbf{x}_i(j) = \mathbf{P}(j)\mathbf{X}_i$, where $\mathbf{x}_i(j)$ is the image coordinate of the ith point in the jth pose frame $\mathbf{P}(j)$.

The observed point trajectories, $\mathbf{z}_i(j) = (x_{ij}, y_{ij})^T$, will generally be perturbed by measurement noise and a solution for $\mathbf{P}(j)$ and $\mathbf{X}_i$ is obtained by minimising the sum of the squared error in the image plane. Thus $2nk$ observations are used to estimate $11k$ pose parameters and $3n$ structure parameters. The computed structure will be unique up to an arbitrary 4×4 projective transformation or *collineation*, $\mathbf{H}$. There are 15 independent elements in $\mathbf{H}^{-1}$, so 15 of the pose parameters can be fixed arbitrarily when solving for sfm. Thus the total number of degrees of freedom in the system for the projective camera model is, $dof_p = 2nk - 11k - 3n + 15$. For a unique solution to be found dof must be positive for all values of n and k (with k greater than 1). Thus $n_{min} \geq 7$ and $k_{min} \geq 2$.

3 Degeneracy

Under certain imaging conditions and camera motions, the solution to sfm may be underconstrained or ill-conditioned and thus *degeneracy* must be studied. The causes of degeneracy for the sfm problem fall into four categories: (i) Degenerate structure (critical surfaces); (ii) degenerate motion; (iii) degenerate spatial positioning of features; and (iv) poor preconditioning. Degenerate *structure* occurs when, for example, the sfm problem is solved using the 3D projective camera model and the object being observed is in fact planar. Degenerate *motion* occurs, for example, when using the 3D affine camera model [3] and there is no rotation about a shallow object: effectively only one view of the object is seen. Degenerate *spatial positioning* occurs when the features on an object are *by chance* poorly positioned. For example, two features "close together" may give only one feature's worth of information. Alternatively, when trying to fit the planar projective camera model to four points, there will be a degeneracy in the solution if any three points are collinear. *Poor preconditioning* leads to numerical conditioning problems when using the projective camera models [2].

When degenerate structure or motion are the cause of degeneracy, switching to a reduced camera model allows the cause of degeneracy to be removed explicitly and this has lead to our *hierarchy of camera models* (see Table 1). Not only is it necessary to use reduced models when analysing real scenes, but it is also desirable, as the reduced models generally can be computed more efficiently and can be more easily interpreted (if for example, the planar projective camera model accurately models the data, then the observed object is known to be a plane). In essence when two camera models both accurately model the same data then the reduced model will be the more "appropriate".

In short, by using ***the simplest applicable camera model that accurately models the data*** we may expect the following gains: (i) removal of degeneracy and improved numerical conditioning (*accuracy*); (ii) greater computational efficiency (*efficiency*); and (iii) clearer interpretation (*clarity*). We are now faced with two questions, (a) what set of camera models should we choose; and (b) how should we switch between them?

The models in Table 1 are listed with the most complex model, the projective camera model, at the top of our hierarchy. We choose to use only uncalibrated camera models, since calibrated camera models are computationally expensive and often fail to converge to an accurate solution. The *image aspectation* model is applicable when the object is shallow and there is no rotation of the camera around the object such that only one view of the object is ever seen (degenerate motion). This situation occurs frequently in road based scenes, for example when following a car on a highway. Though the image aspectation model is not necessary in theory—it is a special case of the planar affine camera model—it proves valuable in practice, since the structure is interpreted as being either 3D *or* planar. The *planar affine* and *planar projective* models are necessary for dealing with degenerate structure. For example, when only one side of an object is visible (when being overtaken by a lorry, for example) the solutions to the 3D *affine* and 3D *projective* models are underconstrained. The planar projective model is exploited further when finding the ground plane.

4 Camera model "appropriateness"

The affine camera can be thought of as an uncalibrated version of the weak perspective camera in the same way as the projective camera model can be thought of as an uncalibrated version of the perspective camera model. We refer to shallow scenes viewed with a long focal length camera as meeting *affine imaging conditions* and deep scenes or scenes viewed with a short focal length camera as meeting *projective imaging conditions*.

We propose the following heuristic metric for computing the "appropriateness" of a model,

$$\Xi = \eta^{\rho_\eta} \kappa^{\rho_\kappa} \nu^{\rho_\nu} \quad , \tag{2}$$

where η, κ and ν are measures of efficiency, clarity and accuracy respectively and ρ_η, ρ_κ and ρ_ν are their weightings in the equation. We call η, κ and ν the Ξ*-measures* and each takes a lowest value of 0 (extremely inappropriate) and a highest value of 1 (proper). The weights also take values between 0 and 1 and are defined according to the application.

3D Projective	$\mathbf{P}_p = \begin{bmatrix} P_{11} & P_{12} & P_{13} & P_{14} \\ P_{21} & P_{22} & P_{23} & P_{24} \\ P_{31} & P_{32} & P_{33} & P_{34} \end{bmatrix}$ $\lambda \begin{bmatrix} x \\ y \\ 1 \end{bmatrix} = \begin{bmatrix} P_{11} & P_{12} & P_{13} & P_{14} \\ P_{21} & P_{22} & P_{23} & P_{24} \\ P_{31} & P_{32} & P_{33} & P_{34} \end{bmatrix} \begin{bmatrix} X \\ Y \\ Z \\ 1 \end{bmatrix}$	– Deep object and any motion. – $doa_p = 15$.
3D Affine	$\mathbf{P}_a = \begin{bmatrix} P_{11} & P_{12} & P_{13} & P_{14} \\ P_{21} & P_{22} & P_{23} & P_{24} \\ 0 & 0 & 0 & 1 \end{bmatrix}$ $\begin{bmatrix} x \\ y \end{bmatrix} = \begin{bmatrix} P_{11} & P_{12} & P_{13} & P_{14} \\ P_{21} & P_{22} & P_{23} & P_{24} \end{bmatrix} \begin{bmatrix} X \\ Y \\ Z \\ 1 \end{bmatrix}$	– Shallow object and shallow motion containing rotation of the camera around the object. – $doa_a = 12$.
Planar Projective	$\mathbf{P}_{pp} = \begin{bmatrix} P_{11} & P_{12} & 0 & P_{14} \\ P_{21} & P_{22} & 0 & P_{24} \\ P_{31} & P_{32} & 0 & P_{34} \end{bmatrix}$ $\lambda \begin{bmatrix} x \\ y \\ 1 \end{bmatrix} = \begin{bmatrix} P_{11} & P_{12} & P_{14} \\ P_{21} & P_{22} & P_{24} \\ P_{31} & P_{32} & P_{34} \end{bmatrix} \begin{bmatrix} X \\ Y \\ 1 \end{bmatrix}$	– Deep plane and any motion. – Points at infinity and any motion. – $doa_{pp} = 8$.
Planar Affine	$\mathbf{P}_{pa} = \begin{bmatrix} P_{11} & P_{12} & 0 & P_{14} \\ P_{21} & P_{22} & 0 & P_{24} \\ 0 & 0 & 0 & 1 \end{bmatrix}$ $\begin{bmatrix} x \\ y \end{bmatrix} = \begin{bmatrix} P_{11} & P_{12} & P_{14} \\ P_{21} & P_{22} & P_{24} \end{bmatrix} \begin{bmatrix} X \\ Y \\ 1 \end{bmatrix}$	– Shallow plane and shallow motion containing rotation of the camera around the plane. – $doa_{pa} = 6$.
Image Aspectation	$\mathbf{P}_{ia} = \begin{bmatrix} C & S & 0 & P_{14} \\ -kS & C & 0 & P_{24} \\ 0 & 0 & 0 & 1 \end{bmatrix}$ $\begin{bmatrix} x \\ y \end{bmatrix} = \begin{bmatrix} C & S & P_{14} \\ -kS & C & P_{24} \end{bmatrix} \begin{bmatrix} X \\ Y \\ 1 \end{bmatrix}$	– Shallow plane and shallow motion with no rotation of the camera around the plane. – Shallow object and shallow motion with no rotation of the camera around the object. – $doa_{ia} = 5$.

Table 1. *A hierarchy of uncalibrated camera models for analysing arbitrary scenes containing unknown independent motions.*

The *efficiency measure* η depends on the particular algorithm used to implement the model. In general it will be a function of the number of observations. For iterative methods of solution it will also depend greatly on the degenerate nature of the data since as the data becomes more degenerate the accuracy of the initial estimate and the speed of descent will both be poor. However by explicitly incorporating degenerate models into our selection procedure we remove this as a factor in the efficiency measure.

In practice the efficiency measure is only useful in the recursive update of structure and motion (in a batch process, once a solution has been obtained the efficiency of its computation is no longer relevant). In our subsequent analysis we deal only with batch methods and so set $\rho_\eta = 0$.

The *clarity measure* κ can be defined by considering the ambiguities in the structure solutions. We choose the following definition: $\kappa = doa_{min}/doa_m$, $\rho_\kappa = 1$, where doa_m is the number of degrees of freedom in the structure solution for a particular model m and doa_{min} is the known minimum value of doa_m for all the models in the hierarchy. The division by doa_{min} scales the clarity measure such that the maximum possible clarity score is $\kappa = 1$.

The ideal *accuracy measure* ν would be to find how accurately the computed structure fits the real structure taking into account the known ambiguity in the solution. This is achieved by finding the sum of the squared error between the real structure and the estimated structure when it is transformed into the frame of the real structure. The transformation can be computed by finding the elements of $\hat{\mathbf{H}}$ (where $\mathbf{H}$ takes the relevant form for the known structure ambiguity) that minimise the sum of the squared structure error,

$$J_E = \sum_{i=1}^{n} |\, \mathbf{X}_i^E - \hat{\mathbf{X}}_i^E \,|^2 ,$$

where $\mathbf{X}^E$ is the actual Euclidean structure and $\hat{\mathbf{X}}^E$ is the computed structure transformed into the Euclidean frame,

$$\hat{\mathbf{X}}_i^E = \begin{bmatrix} 1\,0\,0\,0 \\ 0\,1\,0\,0 \\ 0\,0\,1\,0 \end{bmatrix} \frac{\hat{\mathbf{H}}\mathbf{X}_i}{\hat{\mathbf{H}}_3^\top \mathbf{X}_i} , \qquad \mathbf{H} = \begin{bmatrix} \mathbf{H}_1^\top \\ \mathbf{H}_2^\top \\ \mathbf{H}_3^\top \\ \mathbf{H}_4^\top \end{bmatrix} .$$

We then define $\nu = 1/(1+J_E/\sigma^E)$, $\rho_\nu = 1$, where σ^E is the expected standard deviation in the structure error in the Euclidean frame and must be determined in advance. This measure is not robust to outliers in the computed structure and a more robust measure would be to compute

$$J'_E = \sum_{i=1}^{n} f_{\sqcup}\left(\frac{|\, \mathbf{X}_i^E - \hat{\mathbf{X}}_i^E \,|}{3\sigma^E} \right) , \quad f_{\sqcup}(x) = \begin{cases} 0 \text{ if } -1 < x < 1 \\ 1 \text{ otherwise.} \end{cases} ,$$

and $\nu = 1 - \frac{J'_E}{n}$. Unfortunately, $\mathbf{X}_i^E$ are not known and the prior computation of σ^E is difficult. Indeed, the standard deviation in the structure error will depend on the nature of the relevant motion of the camera and object, and is likely to be anisotropic and different for different strcuture vectors. In essence, the accuracy measure cannot, in general, be computed in this way. However, it is possible to compute the sum of the squared image error (indeed this is what we minimise when solving sfm), but the sum of the squared image error bears no direct relation to the sum of the squared structure error. However, the error between each observation and the projected structure estimate either supports or does not support the hypothesis that each structure estimate is accurate. It is thus possible to compute the *robust image error*,

$$J' = \sum_{j=1}^{k}\sum_{i=1}^{n} f_{\sqcup}\left(\frac{|\,\mathbf{z}_i(j) - \hat{\mathbf{z}}_i(j)\,|}{3\sigma}\right) , \tag{3}$$

where σ is the standard deviation of the expected image noise and must be determined in advance. J' is simply a count of the number of observed trajectory points that fail to agree with the projected structure estimates. The accuracy is then computed $\nu = 1 - (J'/nk)$.

In theory, the robust image error will be in the range 0 to $nk-(n_{min}(k-1)+n)$ and not 0 to nk. This is because in the computation of sfm all n points can be made to agree in at least one frame and at least n_{min} points (where the vlaue of n_{min} depends on the camera model) can be made to agree in the remaining $k-1$ frames. Thus, for small data sets or short time sequences, the accuracy measure can be refined to be $\nu = 1-(J'/dof'_m(n,k))$, where we call $dof'_m(n,k) = nk - (n_{min,m}(k-1)+n)$ the *robust degrees of freedom*. In practice it is only necessary to use the refined measure for very small data sets.

The Ξ-equation that we use to measure camera model appropriateness is thus,

$$\Xi = \kappa\nu = \left(\frac{doa_{min}}{doa_m}\right)\left(1 - \frac{1}{nk}\sum_{j=1}^{k}\sum_{i=1}^{n} f_{\sqcup}\left(\frac{|\,\mathbf{z}_i(j) - \hat{\mathbf{z}}_i(j)\,|}{3\sigma}\right)\right) , \tag{4}$$

where $\hat{\mathbf{z}}_i(j)$ are a function of the model, m.

5 Results

In order to test our measure of appropriateness we selected five trajectory sequences each containing structure, motion and imaging conditions well suited to one of the five camera models. The trajectories were automatically generated by the algorithm described by Wiles and Brady [4, 5] and presegmented so that they are known to be consistent with a rigid motion. We then computed sfm on each test set using each of the five uncalibrated models above and computed the appropriateness, Ξ. The results are shown in Table 2(a) and the ideal values are shown in Table 2(b). The ideal values are those that would be generated with perfect data such that the accuracy measure equals one for models above the diagonal. The measured values will be less than or equal to the ideal values since a small percentage of outliers will cause the accuracy measure to be less than one.

Sequence	Image Aspectation $doa_{ia} = 5$	Planar Affine $doa_{pa} = 6$	Planar Projective $doa_{pp} = 8$	3D Affine $doa_a = 12$	3D Projective $doa_p = 15$
"ambulance" jeep	0.662	0.584	0.438	0.336	0.167
"buggy" front	0.104	0.557	0.004	0.384	0.039
"basement" floor	0.199	0.336	0.528	0.273	0.292
"buggy"	0.000	0.160	0.176	0.337	0.113
"basement"	0.196	0.149	0.141	0.167	0.294

(a)

Imaging conditions appropriate for ...	Image Aspectation $doa_{ia} = 5$	Planar Affine $doa_{pa} = 6$	Planar Projective $doa_{pp} = 8$	3D Affine $doa_a = 12$	3D Projective $doa_p = 15$
Image Aspectation	1.0	0.83	0.62	0.42	0.33
Planar Affine	-	0.83	0.62	0.42	0.33
Planar Projective	-	-	0.62	-	0.33
3D Affine	-	-	-	0.42	0.33
3D Projective	-	-	-	-	0.33

(b)

Table 2. *Camera model appropriateness values, Ξ. (a) Results for a number of sequences. (b) Ideal camera model appropriateness values. The dashes may be any value lower than the diagonal value on that row.*

The "ambulance" sequence consists of the camera following a jeep down a road at a roughly constant distance and an ambulance overtaking the camera on the right. The jeep is a shallow object about which the camera does not rotate and thus the image aspectation camera model should be most appropriate for this object. All the models give accurate solutions for sfm, but the image aspectation camera model has the highest value of Ξ due to its greater clarity. The resulting planar structure is shown in Figure 1(b). The figure was generated by mapping texture from the image onto the computed corner structure.

The "buggy" sequence consists of a model buggy rotating on a turntable. It is viewed from a distance with a camera with a narrow field of view creating a shallow three-dimensional scene. Thus, the affine camera model should be most appropriate. Both the projective and affine camera models should give accurate results, but the projective camera fails to find a solution (due to degenerate imaging conditions). Hence the affine camera model has the highest value of Ξ. The resulting structure is shown in Figure 2(c).

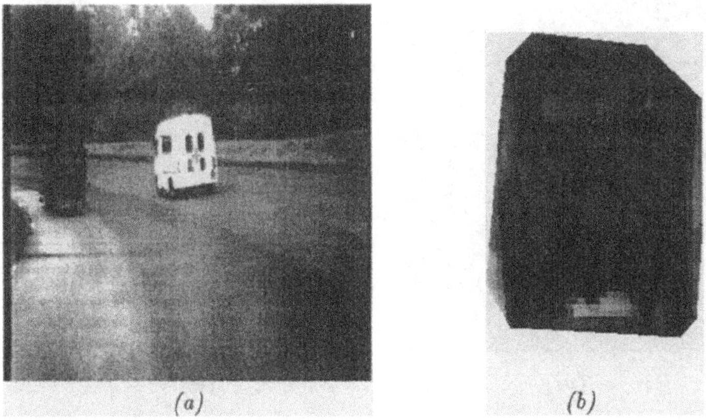

Fig. 1. *Planar affine structure computed from the "ambulance" sequence. (a) The image from the last frame. (b) the planar affine structure computed by applying the image aspectation camera model. The X and Y axes of the computed structure are perpendicular and lie in the plane of the paper.*

The subset of the trajectories corresponding to the front of the "buggy" forms a shallow planar scene. Thus, the planar affine camera model should be the most appropriate, and they are. All models except the image aspectation camera model should generate accurate solutions, but again the projective models fail to find a solution (due to degenerate imaging conditions). The two affine models had equal accuracy measures, but the planar model had greater clarity in its solution. Hence the planar affine camera model has the highest value of Ξ.

The "basement" sequence consists of a camera translating along the optical axis through a deep three-dimensional scene. Thus, the projective camera model should be most appropriate. Indeed, only the projective camera model gives an accurate solution and hence it has the highest value of Ξ. The resulting structure is shown in Figure 2(d).

The subset of the trajectories corresponding to the basement floor forms a deep planar scene. Thus, the planar projective camera model should be the most appropriate. Both the planar projective and projective camera models accurately model the observed trajectories. The sum of the squared image errors is a factor of two smaller for the full projective model due to the greater degrees of freedom in the solution. This demonstrates that, for the purpose of computing appropriateness, the sum of the squared image errors is a poor accuracy measure. However, the robust image error is approximately equal for the two models indicating that on accuracy alone the two models have equal appropriateness. The clarity measure for the planar model is twice as good as the three-dimensional model and hence the planar projective camera model has the highest value of Ξ. The resulting structure is shown in Figure 3(b).

(a)

(b)

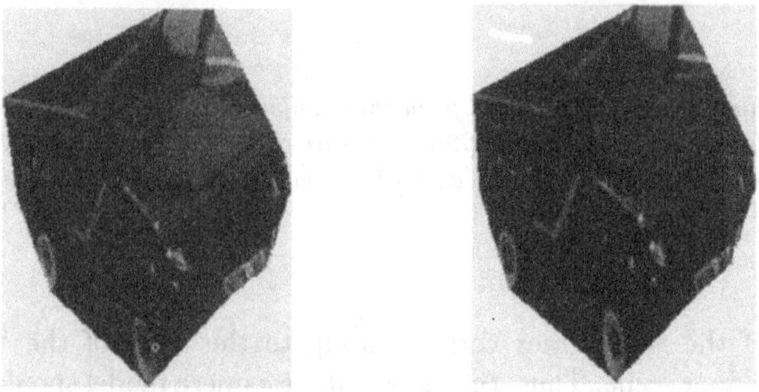

(d)

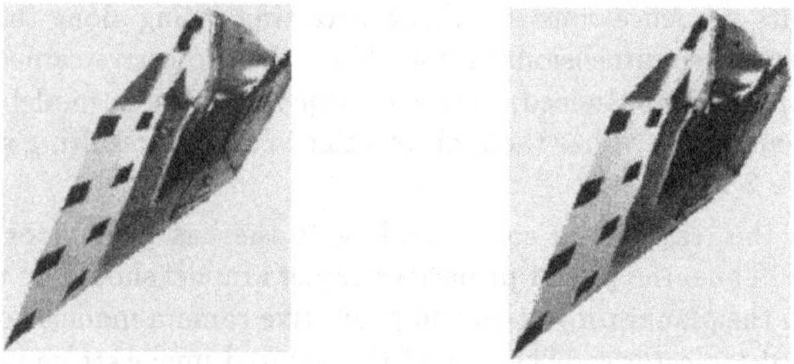

(c)

Fig. 2. 3D ***structure computed from the "buggy" and "basement" sequences. (a) & (b) Images from the last frames. (c) & (d) Cross-eyed stereo pairs showing the structure computed by applying the affine and projective camera models to the sequences respectively.***

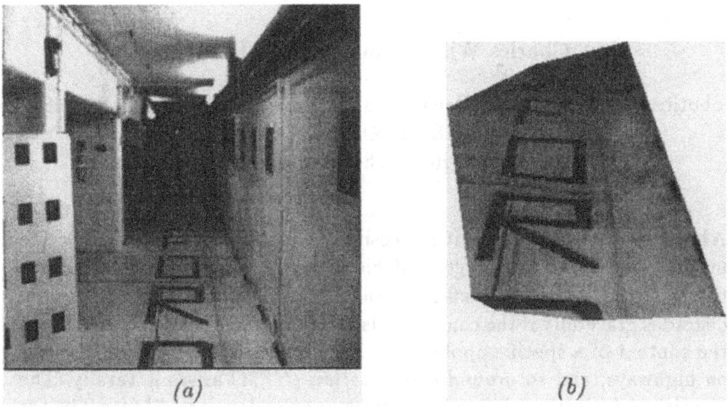

Fig. 3. *Planar projective structure computed from the "basement" sequence floor. (a) The image from the last frame. (b) the planar projective structure computed by applying the planar projective camera model. The X and Y axes of the computed structure are perpendicular and lie in the plane of the paper.*

References

1. O. Faugeras. *Three-Dimensional Computer Vision.* The MIT Press, 1993.
2. R.I. Hartley. In defense of the 8-point algorihtm. In *Proc. 5th International Conference on Computer Vision*, pages 1064–1070, June 1995.
3. J.L. Mundy and A. Zisserman, editors. *Geometric invariance in computer vision.* The MIT Press, 1992.
4. C.S. Wiles. *Closing the loop on multiple motions.* PhD thesis, Dept. Engineering Science, University of Oxford, Michaelmas 1995.
5. C.S. Wiles and J.M. Brady. Closing the loop on multiple motions. In *Proc. 5th International Conference on Computer Vision*, pages 308–313, June 1995.

Ground Plane Motion Camera Models

Charles Wiles[1,2] and Michael Brady[1]

[1] Robotics Research Group, Department of Engineering Science, University of Oxford, Parks Road, Oxford, OX1 3PJ, UK.
[2] TOSHIBA, Kansai Research Laboratories, Osaka, Japan.

Abstract. We show that it is possible to build an application-specific motion constraint into a general hierarchy of camera models so that the resulting algorithm is efficient and well conditioned; and the system degrades gracefully if the constraint is violated. The current work arose in the context of a specific application, namely *smart convoying* of vehicles on highways, and so *ground plane motion (GPM)* arises naturally. The algorithm for estimating motion from monocular, uncalibrated image sequences under ground plane motion involves the computation of three kinds of structure α, β and γ for which algorithms are presented. Typical results on real data are shown.

1 Introduction

Current approaches to process image sequences generated by vehicles moving on the ground plane directly enforce application-specific constraints in their algorithms. The resulting algorithms are often effective as long as the assumptions and constraints around which the systems are built hold, but the moment these do not do so such systems may fail catastrophically. Moreover, such algorithms tend to be unportable and can be used in only a single problem domain. Rather than building application-specific constraints such as ground plane motion into the primary stages of scene analysis we propose that primary analysis should be general, obtaining a basic understanding of the scene, then further constraints should be tested for and enforced to obtain a more detailed understanding. This process conforms to the principle of *graceful degradation*, whereby on failure to obtain a full solution, a partial solution is still recovered.

A hierarchy of camera models has been proposed by Wiles and Brady [6] that is completely general, and which has been applied to numerous quite general motion sequences. However, like all general methods, it raises the question of whether or not one can exploit application-specific constraints to enhance and improve the analysis of particular kinds of motion and scene. We show that the hierarchy of camera models can indeed be modified so that it only allows the "recognition" of motions that are consistent with ground plane motion of the camera with respect to the world.

We begin by describing the *ground plane motion constraint* and then describe the projective GPM camera model. Structure from motion (sfm) can be efficiently computed using the GPM camera models, by first computing sfm with general models and then switching into the GPM forms. In doing so it is possible to validate the ground plane motion constraint and recover projective structure if the constraint does not hold. Three types of structure are introduced, α, β and γ. In all three, verticals in the real world are recovered as verticals in the computed structure and the plane in which the camera moves is recovered as perpendicular to the verticals. **γ-structure** is the most ambiguous. The remaining degrees of freedom are completely general and the structural ambiguity a subgroup of the projective group. **β-structure** requires the ground plane to be identified. The structure is recovered up to an arbitrary planar affine transformation parallel to the ground plane and a scale factor perpendicular to it. The structural ambiguity is a subgroup of the affine group. **α-structure** is the least ambiguous. The structure is recovered up to an arbitrary planar similarity transformation parallel to the ground plane and a scale factor perpendicular to it. The resulting structure is as close to Euclidean as is possible under ground plane motion. Detatils of the computation of α, β and γ-structure and of planar and affine GPM camera models can be found in [5].

2 General camera models

The *projective* camera model [1, 2, 3, 4] can be written in terms of a 3×4 projection matrix $\mathbf{P}$. It is convenient to decompose $\mathbf{P}$ as follows:

$$\mathbf{P} = \mathbf{K}\mathbf{I}_p\mathbf{G} = \begin{bmatrix} K_{11} & K_{12} & K_{13} \\ K_{21} & K_{22} & K_{23} \\ 0 & 0 & K_{33} \end{bmatrix} \begin{bmatrix} 1 & 0 & 0 & 0 \\ 0 & 1 & 0 & 0 \\ 0 & 0 & 1 & 0 \end{bmatrix} \begin{bmatrix} G_{11} & G_{12} & G_{13} & G_{14} \\ G_{21} & G_{22} & G_{23} & G_{24} \\ G_{31} & G_{32} & G_{33} & G_{34} \\ G_{41} & G_{42} & G_{43} & G_{44} \end{bmatrix} . \quad (1)$$

The matrix $\mathbf{G}$ represents a projective transformation in $\mathcal{P}^3$ such that structure defined in an arbitrary frame is transformed into the camera frame where the X axis is aligned with the x-axis of the image plane and the Z-axis with the optical axis of the camera. *Camera centred* coordinates are defined $\mathbf{X}^c = \mathbf{G}\mathbf{X}$ and $\mathbf{G}$ encodes the pose or *extrinsic* parameters of the camera.

The *perspective* camera model enforces special forms on the matrices $\mathbf{G}$ and $\mathbf{K}$. First, world coordinates are assumed to undergo a *Euclidean* transformation into the camera frame and hence the extrinsic matrix takes the form,

$$\mathbf{G}_e = \begin{bmatrix} \mathbf{R} & \mathbf{T} \\ \mathbf{0}^\top & 1 \end{bmatrix} , \quad \mathbf{R} = \begin{bmatrix} \mathbf{R}_1^\top \\ \mathbf{R}_2^\top \\ \mathbf{R}_3^\top \end{bmatrix} ,$$

where $\mathbf{R}$ is a 3×3 rotation matrix and $\mathbf{T} = (T_x, T_y, T_z)^T$ is a translation vector giving the coordinates of the origin of the object centred frame in the camera frame. $\mathbf{G}$ now has only 6 degrees of freedom.

Second, the intrinsic matrix $\mathbf{K}$ can be written:

$$\mathbf{K} = \mathbf{K}_e = \begin{bmatrix} f_x & 0 & o_x \\ 0 & f_y & o_y \\ 0 & 0 & 1 \end{bmatrix} . \quad (2)$$

3 Ground plane motion

The *ground plane motion constraint (GPM)* holds if the following three conditions are met: (i) the ground is locally planar over the path which the camera traverses; (ii) the camera is rigidly fixed to a vehicle so that it undergoes all translations and rotations in a plane parallel to the ground plane; and (iii) the camera is aligned with the ground plane such that the x axis in the image plane is parallel to the ground plane ($\mathbf{x} \cdot \hat{\mathbf{n}}^c = 0$) and the tilt of the optical axis with respect to the ground plane remains constant ($\mathbf{Z}^c \cdot \hat{\mathbf{n}}^c = -\sin\alpha$.), where $\hat{\mathbf{n}}^c$ is the unit normal of the ground plane in the camera frame.

When the GPM constraint holds, the matrix $\mathbf{G}_e$ can be decomposed such that,

$$\mathbf{G}_{e_g} = \mathbf{G}_\alpha \mathbf{G}_{YZ} \mathbf{G}_g = \begin{bmatrix} 1 & 0 & 0 & 0 \\ 0 & c_\alpha & s_\alpha & 0 \\ 0 & -s_\alpha & c_\alpha & 0 \\ 0 & 0 & 0 & 1 \end{bmatrix} \begin{bmatrix} 1 & 0 & 0 & 0 \\ 0 & 0 & 1 & 0 \\ 0 & 1 & 0 & 0 \\ 0 & 0 & 0 & 1 \end{bmatrix} \begin{bmatrix} c_\beta & s_\beta & 0 & T_x \\ -s_\beta & c_\beta & 0 & T_z \\ 0 & 0 & 1 & 0 \\ 0 & 0 & 0 & 1 \end{bmatrix} . \quad (3)$$

$\mathbf{G}_g$ is the ground plane motion allowing rotation and translation in the XY-plane only ($c_\beta = \cos\beta$, $s_\beta = \sin\beta$, and β is the time varying pan angle). $\mathbf{G}_{YZ}$ swaps the Y and Z axis since in camera centred coordinates the Y axis is perpendicular to the ground plane. $\mathbf{G}_\alpha$ is a rotation representing the tilt of the camera ($c_\alpha = \cos\alpha$, $s_\alpha = \sin\alpha$, and α is the fixed tilt angle). The result in the third ground plane motion constraint above can be verified to be $\mathbf{Z}^c \cdot \hat{\mathbf{n}}^c = \mathbf{Z}^c \cdot (\mathbf{G}_{e_g} \hat{\mathbf{n}}) = -\sin\alpha$.

When $\mathbf{G}_{e_g}$ is used as the extrinsic camera matrix in Equation 1, the result is the *perspective GPM camera model*,

$$\mathbf{P}_{e_g} = \mathbf{K}_e \mathbf{I}_p \mathbf{G}_{e_g} = \begin{bmatrix} f_g c_\beta - o_x s_\beta & f_g s_\beta + o_x c_\beta & o_x t & f_g T_x + o_x T_z \\ o_h(-s_\beta) & o_h c_\beta & 1 & o_h T_z \\ -s_\beta & c_\beta & t & T_z \end{bmatrix} , \quad (4)$$

where $o_h = o_y + f_y t_\alpha$, $f_g = f_x / c_\alpha$, and $t = -t_\alpha / f_h$, where $t_\alpha = \tan\alpha$ and $f_h = f_y - o_y t_\alpha$. Note that when the tilt angle, α, is zero, then $o_h = o_y$, $f_g = f_x$ and $t = 0$. The three elements of the third column are independent of the other nine and are compounded by the Z parameter in the computed structure. *Thus it is only possible to recover Euclidean structure up to an arbitrary overall scaling and arbitrary stretch in the Z axis using the perspective GPM camera model.* The third column has been divided through by f_h to reflect this. Indeed the computed structure can only be computed up to an arbitrary transformation of the form

$$\mathbf{H}_\alpha = \begin{bmatrix} c_\beta & s_\beta & 0 & T_x \\ -s_\beta & c_\beta & 0 & T_z \\ 0 & 0 & H_{33} & 0 \\ 0 & 0 & 0 & H_{44} \end{bmatrix} \quad (5)$$

and we call such structure **α-structure** (see Table 1).

Structure	Form that H takes	Structure ambiguity
Projective structure	$\mathbf{H}_p = \begin{bmatrix} H_{11} & H_{12} & H_{13} & H_{14} \\ H_{21} & H_{22} & H_{23} & H_{24} \\ H_{31} & H_{32} & H_{33} & H_{34} \\ H_{41} & H_{42} & H_{43} & H_{44} \end{bmatrix}$	Any projective transformation.
Affine structure	$\mathbf{H}_a = \begin{bmatrix} H_{11} & H_{12} & H_{13} & H_{14} \\ H_{21} & H_{22} & H_{23} & H_{24} \\ H_{31} & H_{32} & H_{33} & H_{34} \\ 0 & 0 & 0 & H_{44} \end{bmatrix}$	Any affine transformation.
γ-structure	$\mathbf{H}_\gamma = \begin{bmatrix} H_{11} & H_{12} & 0 & H_{14} \\ H_{21} & H_{22} & 0 & H_{24} \\ 0 & 0 & H_{33} & 0 \\ H_{41} & H_{42} & 0 & H_{44} \end{bmatrix}$	Any projective transformation such that verticals remain perpendicular to the XY-plane.
β-structure	$\mathbf{H}_\beta = \begin{bmatrix} H_{11} & H_{12} & 0 & H_{14} \\ H_{21} & H_{22} & 0 & H_{24} \\ 0 & 0 & H_{33} & 0 \\ 0 & 0 & 0 & H_{44} \end{bmatrix}$	Arbitrary scaling perpendicular to ground plane and affine transformation parallel to ground plane.
α-structure	$\mathbf{H}_\alpha = \begin{bmatrix} c_\beta & s_\beta & 0 & T_x \\ -s_\beta & c_\beta & 0 & T_z \\ 0 & 0 & H_{33} & 0 \\ 0 & 0 & 0 & H_{44} \end{bmatrix}$	Arbitrary scaling perpendicular to ground plane and similarity transformation parallel to ground plane.

Table 1. *Summary of structural ambiguities.*

The perspective GPM camera model can be decomposed so that

$$\mathbf{P}_{e_g} = \mathbf{K}_{e_g}\mathbf{I}_p\mathbf{G}_g = \begin{bmatrix} f_g & o_x & o_x t \\ 0 & o_h & 1 \\ 0 & 1 & t \end{bmatrix} \mathbf{I}_p \begin{bmatrix} c_\beta & s_\beta & 0 & T_x \\ -s_\beta & c_\beta & 0 & T_z \\ 0 & 0 & 1 & 0 \\ 0 & 0 & 0 & 1 \end{bmatrix} , \qquad (6)$$

where $\mathbf{K}_{e_g}$, the *GPM intrinsic matrix*, contains all the fixed parameters of the camera model and $\mathbf{G}_g$ contains all the time varying parameters. f_g, o_x, o_h, and t can be considered to be the calibration parameters for the perspective GPM camera model. When solving the sfm problem with the perspective GPM camera model, $2nk$ observations are used to estimate $3k+4$ pose parameters and $3n$ structure parameters. There are 5 degrees of ambiguity in the computed α-structure which means that the total number of degrees of freedom in the system are given by $dof_{e_g} = 2nk - 3k - 3n + 1$.

4 The projective GPM camera model

It is useful to generalise Equation 4 to form the *projective GPM* camera model,

$$\lambda \begin{bmatrix} x \\ y \\ 1 \end{bmatrix} = \mathbf{P}_{p_g}\mathbf{X} = \begin{bmatrix} P_{11} & P_{12} & a & P_{14} \\ oP_{31} & oP_{32} & 1 & oP_{34} \\ P_{31} & P_{32} & b & P_{34} \end{bmatrix} \begin{bmatrix} X \\ Y \\ Z \\ 1 \end{bmatrix} , \tag{7}$$

which has six time varying parameters, P_{ij}, and three fixed parameters o, a and b. The time varying parameters have the full six degrees of freedom since we have fixed the scale factor of the matrix by enforcing $P_{23} = 1$.

We compute sfm using the projective GPM camera model first using the full projective camera model and then transforming the pose matrices into the correct form for the projective GPM camera model using a *model switching matrix*. This two stage process is particularly useful since it is possible to test whether or not the ground plane motion constraint is violated. The structure computed using the projective GPM camera model is related to the real Euclidean structure by a 4×4 homography, $\mathbf{H}$, so that $\mathbf{P}_p(j) = \mathbf{P}_e(j)\mathbf{H}$, where we have dropped the g subscript on the pose matrices. Multiplying column i of $\mathbf{H}$ by rows 2 and 3 of $\mathbf{P}_e(j)$ results in the following sets of equations for $i = 1, 2, 4$:

$$oP_{3i}(j) = -o_h s_\beta(j) H_{1i} + o_h c_\beta(j) H_{2i} + H_{3i} + o_h T_z(j) H_{4i} ,$$
$$P_{3i}(j) = -s_\beta(j) H_{1i} + c_\beta(j) H_{2i} + tH_{3i} + T_z(j) H_{4i} ,$$

and eliminating P_{3i} gives

$$(o_h - o)(-s_\beta(j) H_{1i} + c_\beta(j) H_{2i} + T_z(j) H_{4i}) + (1 - ot) H_{3i} = 0 .$$

The parameters o and H can be thought of as defining a plane in a three dimensional space and the parameters $s_\beta(j)$, $c_\beta(j)$ and $T_z(j)$ can be thought of as the coordinates of points in the same space. In general these points will describe the surface of a cylinder and not a plane. Since at least one of H must be non-zero the equation can only be true for all j if $o = o_h$ and $H_{3i} = 0$.

Multiplying column 3 of $\mathbf{H}$ by rows 2 and 3 of $\mathbf{P}_e(j)$ results in the following set of equations:

$$1 = -o_h s_\beta(j) H_{13} + o_h c_\beta(j) H_{23} + H_{33} + o_h T_z(j) H_{43} ,$$
$$b = -s_\beta(j) H_{13} + c_\beta(j) H_{23} + tH_{33} + T_z(j) H_{43} .$$

This can only be true for all j if $H_{13} = H_{23} = H_{43} = 0$, and $b = t$. Similarly $a = o_x t$.

So $\mathbf{H}$ must have the form

$$\mathbf{H}_\gamma = \begin{bmatrix} H_{11} & H_{12} & 0 & H_{14} \\ H_{21} & H_{22} & 0 & H_{24} \\ 0 & 0 & H_{33} & 0 \\ H_{41} & H_{42} & 0 & H_{44} \end{bmatrix} . \tag{8}$$

Thus, when solving the sfm problem with the projective GPM camera model, $2nk$ observations are used to estimate $6k + 3$ pose parameters and $3n$ structure parameters. There are 9 degrees of ambiguity in the computed structure which means that the total number of degrees of freedom in the system are, $dof_{p_g} = 2nk - 6k - 3n + 6$.

This interpretation of $\mathbf{H}_\gamma$ is that all points with the same X and Y coordinates in the Euclidean frame will have the same X and Y coordinates in the computed projective frame. The physical meaning of this result is that verticals in the real world will remain vertical in the computed structure. We call such structure *γ-structure* (see Table 1).

If the ground plane is identified, the structure computed using the projective GPM camera model can be transformed to remove the projective distortions. We introduce the term **β-structure** to describe the ambiguity in the computed structure. Assume that we have computed sfm so that we have a set of pose matrices in the form of the projective GPM camera model, $\mathbf{P}_{p_g}(j)$, and structure, $\mathbf{X}_i$, that minimises the sum of the squared errors in the image plane. The computed structure is related to the actual Euclidean structure, $\mathbf{X}_i^e$, by the transformation, $\mathbf{X}_i^e = \mathbf{H}_\gamma^{-1}\mathbf{X}_i$.

$\mathbf{H}_\gamma^{-1}$ can be decomposed as follows:

$$\mathbf{H}_\gamma^{-1} = \mathbf{H}_\beta \mathbf{H}_\lambda = \begin{bmatrix} H'_{11} & H'_{12} & 0 & H'_{14} \\ H'_{21} & H'_{22} & 0 & H'_{24} \\ 0 & 0 & H'_{33} & 0 \\ 0 & 0 & 0 & H'_{44} \end{bmatrix} \begin{bmatrix} 1 & 0 & 0 & 0 \\ 0 & 1 & 0 & 0 \\ 0 & 0 & 1 & 0 \\ H_{41} & H_{42} & 0 & H_{44} \end{bmatrix} , \tag{9}$$

where $\mathbf{H}_\beta$ is an affine transformation in the XY-plane only and $\mathbf{H}_\lambda$ is a matrix that removes the projective distortion.

Points in the ground plane (or any plane parallel to the ground plane) will be transformed under $\mathbf{H}_\lambda$,

$$\begin{bmatrix} 1 & 0 & 0 & 0 \\ 0 & 1 & 0 & 0 \\ 0 & 0 & 1 & 0 \\ H_{41} & H_{42} & 0 & H_{44} \end{bmatrix} \begin{bmatrix} X_i \\ Y_i \\ Z_i \\ 1 \end{bmatrix} = \begin{bmatrix} X'_i \\ Y'_i \\ h \\ 1 \end{bmatrix} ,$$

so that the Z-coordinate becomes equal to h, the height of the camera from the ground plane, for all such points. If h is unknown, it can be arbitrarily set equal to 1.

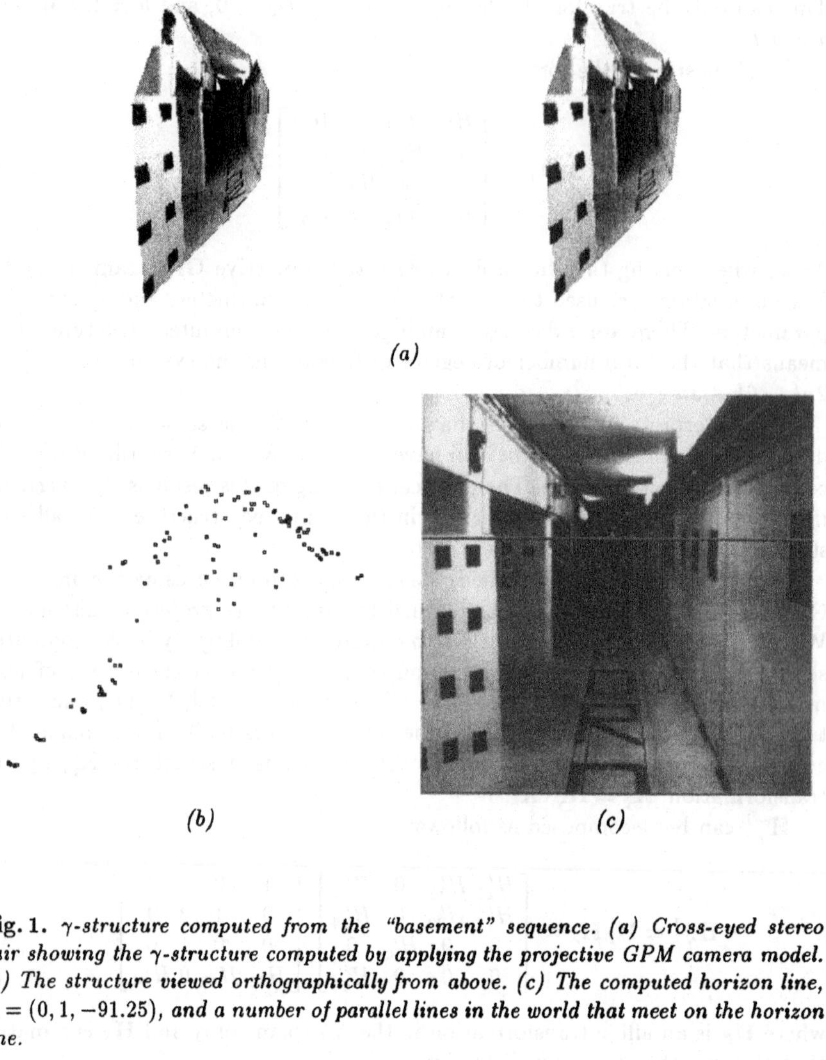

Fig. 1. *γ-structure computed from the "basement" sequence. (a) Cross-eyed stereo pair showing the γ-structure computed by applying the projective GPM camera model. (b) The structure viewed orthographically from above. (c) The computed horizon line,* $\mathbf{l}_h = (0, 1, -91.25)$, *and a number of parallel lines in the world that meet on the horizon line.*

The resulting structure is related to the actual Euclidean structure by a purely affine transformation parallel to the ground plane and a scale factor perpendicular to the ground plane as described by the matrix $\mathbf{H}_\beta$. We call this structure *β-structure* (see Table 1) and it combines both the features of γ-structure (the third row and third column of $\mathbf{H}_\beta$ have the form $(0, 0, H, 0)$) and affine structure (the bottom row of $\mathbf{H}_\beta$ equals $(0, 0, 0, H)$).

Fig. 2. *β-structure computed from the "basement" sequence. (a) Cross-eyed stereo pair showing the β-structure computed when three points on the ground plane are identified in the structure computed by the projective GPM camera model (Figure 1).*

5 Results

The projective GPM camera model was applied to trajectories computed from a "basement" sequence. The results are shown in Figure 1. All verticals in the real world remain vertical in the computed γ-structure and the plane parallel to the ground plane in which the camera lies is perpendicular to these verticals. Finally, Figure 1(c) shows the position of the computed horizon line, $\mathbf{l}_h = (0, 1, -91.25)$, and shows that a number of lines that are parallel in the real world and parallel to the ground plane do indeed meet at a vanishing point on the horizon line.

Figure 2 shows the structure computed when three points in the ground plane are identified and the structure transformed so that these points have the same value of Z. The computed β-structure can be seen to be related to the real structure by an affine transform parallel to the ground plane and a scaling perpendicular to the ground plane.

A quantitative measure of the accuracy of the computed β-structure was obtained by transforming a subset of features from the computed β-structure into the Euclidean frame of the *measured* Euclidean structure. The affine transformation that minimised the sum of the squared structure errors in the Euclidean frame was found to be

$$\mathbf{H}_a = \begin{bmatrix} 185.4 & -243.9 & -4.9 & 45.2 \\ -0.4 & 488.2 & 0.8 & -331.2 \\ 0.3 & 3.2 & -122.3 & 119.7 \\ 0 & 0 & 0 & 1 \end{bmatrix}$$

The measured Euclidean structure, transformed β-structure and individual structure errors are given in Table 2 and shown graphically in Figure 3. The resulting

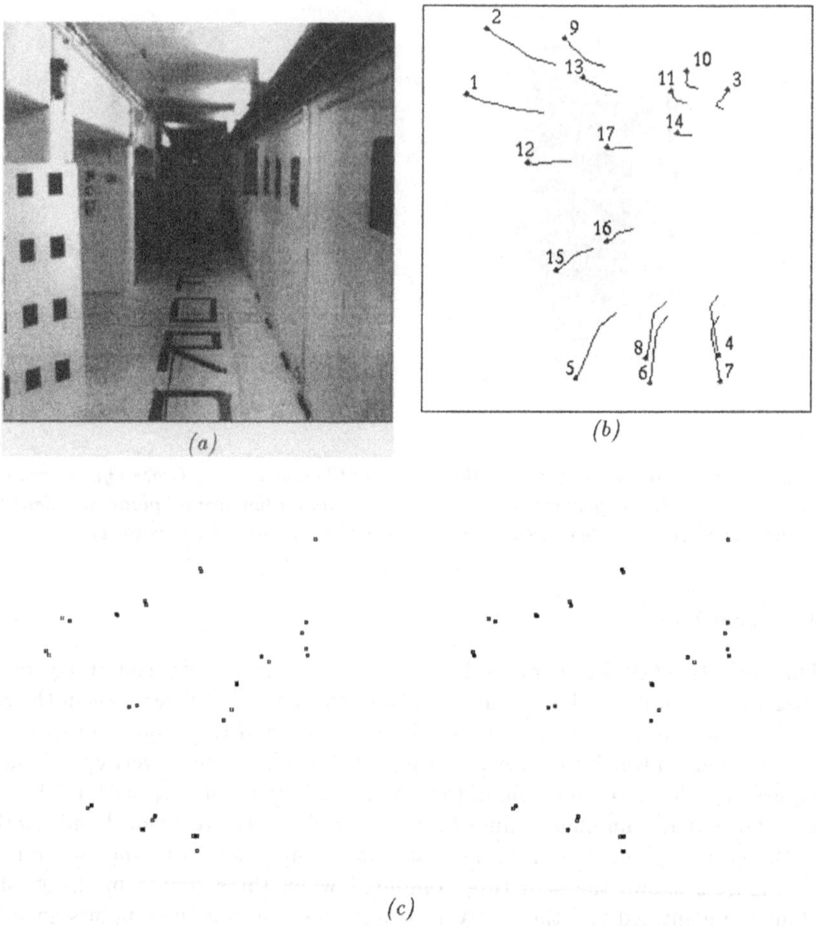

Fig. 3. ***β*-structure errors from the "basement" sequence. (a) The image from the last frame. (b) A subset of the computed trajectories (the labels correspond to those in Table 2). (c) A cross-eyed stereo pair showing the measured Euclidean structure and superimposed the *β*-structure transformed into the Euclidean frame under an affine transformation.**

root mean square structure error was found to be 12.5 cm and the measured dimensions of the basement corridor being 186×891×196 cm.

The elements of $\mathbf{H}_a$ are in good agreement with the expected form for the structural ambiguity in β-structure except for the element in the third row and fourth column which corresponds to the height of the camera at 120cm above the ground plane. The individual structure errors are relatively small with the

Label	Measured Euclidean structure (X,Y,Z) (cm±3)	Transformed β-structure (X,Y,Z) (cm)	Error (X,Y,Z) (cm)
1	(-66, -35, 148)	(-72, -32, 149)	(6, -3, -1)
2	(-66, -20, 183)	(-59, -23, 183)	(-7, 3, -0)
3	(120, -20, 155)	(120, -42, 151)	(0, 22, 4)
4	(115, -10, 18)	(118, -6, 16)	(-3, -4, 2)
5	(0, 0, 0)	(-1, 8, -1)	(1, -8, 1)
6	(60, 0, 0)	(61, -3, -0)	(-1, 3, 0)
7	(120, 0, 0)	(120, 1, -0)	(-0, -1, 0)
8	(60, 31, 0)	(60, 35, 1)	(0, -4, -1)
9	(-19, 82, 196)	(-14, 75, 196)	(-5, 7, -0)
10	(120, 95, 176)	(112, 96, 179)	(8, -1, -3)
11	(120, 235, 176)	(115, 266, 181)	(5, -31, -5)
12	(-100, 265, 115)	(-105, 282, 112)	(5, -17, 3)
13	(-19, 270, 196)	(-15, 265, 196)	(-4, 5, 0)
14	(120, 290, 145)	(124, 281, 143)	(-4, 9, 2)
15	(-66, 320, 0)	(-69, 307, 3)	(3, 13, -3)
16	(0, 530, 0)	(2, 527, 2)	(-2, 3, -2)
17	(-11, 856, 140)	(-11, 851, 137)	(-0, 5, 3)

Table 2. *β-structure errors from the "basement" sequence. The computed β-structure was transformed into the frame of the measured Euclidean structure under an affine transformation. The labels of the trajectories correspond to those shown in Figure 3(b) and the errors are displayed graphically in Figure 3(c).*

error being greatest in the Y-axis corresponding to the difficulty in accurately estimating the coordinate parallel to the optical axis of the camera.

References

1. O. Faugeras. *Three-Dimensional Computer Vision.* The MIT Press, 1993.
2. R. Mohr, F. Veillon, and L. Quan. Relative 3D reconstruction using multiple uncalibrated images. *Proc. Conference Computer Vision and Pattern Recognition*, pages 543–548, 1993.
3. J.L. Mundy and A. Zisserman, editors. *Geometric invariance in computer vision.* The MIT Press, 1992.
4. R. Szeliski and S.B. Kang. Recovering 3D shape and motion from image streams using non-linear least squares. DEC technical report 93/3, DEC, 1993.
5. C.S. Wiles. *Closing the loop on multiple motions.* PhD thesis, Dept. Engineering Science, University of Oxford, Michaelmas 1995.
6. C.S. Wiles and J.M. Brady. Closing the loop on multiple motions. In *Proc. 5th International Conference on Computer Vision*, pages 308–313, June 1995.

Medical Applications

Snakes and Splines for Tracking Non-Rigid Heart Motion

Amir A. Amini, Rupert W. Curwen, and John C. Gore

333 Cedar St., P.O. Box 208042, Yale University, New Haven, CT 06520
E-mail: amini@minerva.cis.yale.edu

Abstract. MRI is unique in its ability to non-invasively and selectively alter tissue magnetization, and create tagged patterns within a deforming body such as the heart muscle. The resulting patterns (radial or SPAMM patterns) define a time-varying curvilinear coordinate system on the tissue, which we track with B-snakes and coupled B-snake grids. The B-snakes are optimized by a dynamic programming algorithm operating on B-spline control points in discrete pixel space. Coupled B-snake optimization based on an extension of dynamic programming to two dimensions, and gradient descent are proposed. Novel spline warps are also proposed which can warp an area in the plane such that two embedded snake grids obtained from two SPAMM frames are brought into registration, interpolating a dense displacement vector field. The reconstructed vector field adheres to the known displacement information at the intersections, forces corresponding snakes to be warped into one another, and for all other points in the plane, where no information is available, a second order continuous vector field is interpolated.

1 Introduction

In the past, much of the work in image sequence processing has dealt with motion analysis of rigidly moving objects. Non-rigidity however occurs abundantly in motion of both solids and fluids: motion of trees, muscular motion of faces, and non-rigid movement and pumping motion of the left-ventricle (LV) of the heart, as well as fluid motion are all non-rigid [10, 14, 1].

MRI is an excellent imaging technique for measuring non-rigid tissue motion and deformation. MR imaging provides depiction of cardiac anatomy and provides dynamic images with reasonable time resolution that in principle can be used to track the movement of individual segments of myocardium[1] or other structures. In common with other imaging modalities, however, MR images are recorded as selected "snap-shots" at discrete intervals through-out the cardiac cycle that are registered relative to a coordinate system external to the body. Conventional MR or CT imaging therefore can not be used to infer the actual trajectories of individual tissue elements, as such images can only provide geometric information about anatomical object boundaries. Furthermore, since motion of the LV is non-rigid, it is not possible to determine the trajectory of individual

[1] heart muscle

tissue points from boundary information alone, limiting any motion or deformation measurement scheme. To overcome such limitations, tagging methods [19] and phase-contrast [16] measurements of motion have been developed. Phase-contrast techniques provide velocity information about the deforming structure and must be integrated to yield approximate displacements.[2] Tagging techniques allow for direct measurement of displacements at specifically labeled tissue locations.

Leon Axel developed the SPAMM technique in which a striped pattern of altered magnetization is placed within the myocardial tissue [3]. An alternate method described by Zerhouni and co-workers applies a radial array of thin striped tags over a short axis view of the heart [19]. For analysis of tagged images, Prince and his collaborators used optical flow techniques [8], McVeigh et al. used least squares to best localize individual points along a radial tag line assuming known LV boundaries [9], Young and Axel applied FEM models to fit known displacements at tag intersections [18], and Park, Metaxas, and Axel applied volumetric ellipsoids to SPAMM analysis and extracted aspect ratios of their model [15].

Deformable models are powerful tools which can be used for localization as well as tracking of image features [4, 5, 7, 11, 12]. For tagged MR images, several groups have applied snakes for tracking tag lines [2, 18, 17]. In this paper, our efforts are summarized and B-spline snakes in 1D (curves) as well as 2D (grids) are applied to MR tag localization and tracking. A new dynamic programming (DP) algorithm for optimization of B-snakes is proposed. In order to construct an energy field for localization of radial tag lines, we use normalized correlations of simulated image profiles with the data. For SPAMM tags, coupled B-snake grids are used which interact via shared control points. The energy function for coupled B-snake optimization uses the image intensity information along grid lines, as has been used by [2, 18], but sum-of-squared-differences (SSD) of pixel windows at snake intersections are also used in the grid optimization. Optimizing snake grids with standard DP is not possible. For this reason, an extension of DP to two dimensions is discussed where interaction of horizontal and vertical grid lines is allowed. Since this technique is not practical, we have developed a gradient descent algorithm for optimizing spline grids. For SPAMM images, techniques are also discussed for measuring strain as an index of non-rigid deformation from the coupled snake analysis.

Finally, we develop a new class of image warps for bringing two successive snake grids into registration, interpolating a dense displacement vector field. The vector field adheres to the known displacement information at snake intersections, forces corresponding snakes to be warped into one another, and for all other points in the plane, where no information is available, a second order continuous vector field is interpolated.

---

[2] In essence, in the Taylor expansion, $x(t) = x_0 + vt + \frac{1}{2}at^2 + \cdots$, phase-contrast MRI provides v. See [13] for methods for analysis of PC MRI data.

2 Optimization of B-Snakes with Dynamic Programming

B-spline curves are suitable for representing a variety of industrial and anatomical shapes [7, 12, 4, 5]. The advantages of B-spline representations are: (1) They are smooth, continuous parametric curves which can represent open or closed curves. For our application, due to parametric continuity, B-splines will allow for sub-pixel localization of tags, (2) B-splines are completely specified by few control points, and (3) Individual movement of control points will only affect their shape locally. In medical imaging, local tissue deformations can easily be captured by movement of individual control points without affecting static portions of the curve.

2.1 Snake Optimization

To localize each radial tag line, our approach is to minimize the following expression along a quadric B-spline, $\alpha(u)$:

$$E_{total} = -\{\int \rho^n(\alpha(u))du + \rho_e^n(\alpha(0)) + \rho_e^n(\alpha(u_{max}))\} \tag{1}$$

In equation (1), the first term maximizes ρ^n along the length of the snake, and second and third terms attract the snake to the endpoint of tag lines by maximizing ρ_e^n. The limits of summation above include all points on the spline except the two endpoints, $\alpha(0)$ and $\alpha(u_{max})$. Due to the few control points needed to represent tag deformations, we have been able to obtain good results without any additional smoothness constraints on B-splines. In the more general case, the derivative smoothness constraints can also be included.

The discrete form of E_{total}, for a quadric spline may be written as:

$$E_{total} = E_0(p_0, p_1, p_2) + \cdots + E_{N-3}(p_{N-3}, p_{N-2}, p_{N-1}) \tag{2}$$

where p_i is a B-spline control point, and E_i is the sum energy of one B-spline span. DP may be used to optimize the curve in the space of spline control points using the following recurrence

$$S_i(p_i, p_{i+1}) = \min_{p_{i-1}}\{E_{i-1}(p_{i-1}, p_i, p_{i+1}) + S_{i-1}(p_{i-1}, p_i)\} \tag{3}$$

for $i \geq 2$, and $S_1(p_1, p_2) = \min_{p_0} E_0(p_0, p_1, p_2)$. The constructed table, S_i, is called the optimal value function. The window of possible choices for p_{i-1} is typically a 3×3 or a 5×5 pixel area. But, non-square rectangular search areas may be used as well. In general, for an order k B-spline, S_i is a function of k control points, and E_i is a function of $k+1$ control points.[3] Figure 1 shows results from localization and tracking of a cardiac sequence.

[3] Note that the minimization yields the optimal open spline. For other applications, given an external energy field, it may be necessary to optimize a closed snake. In such a case, one performs M applications of the recurrence, where M is the number of possible pixel choices for the endpoint p_i, and for each optimization fixes the *end point* to be one of the M choices, repeating for all M possibilities, and finally choosing the minimum.

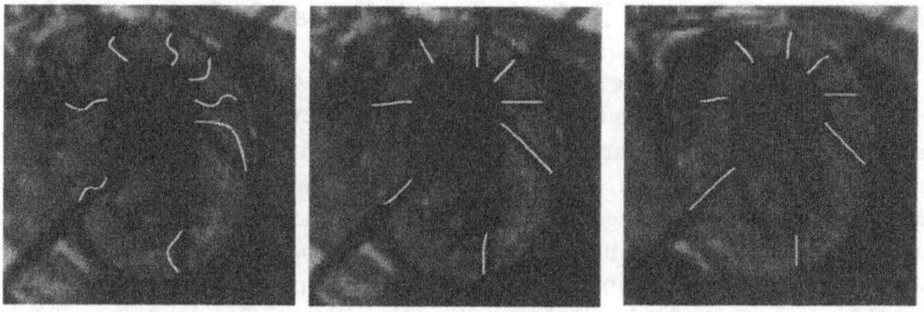

Fig. 1. Results from tracking with B-snakes. Two frames from a sequence are shown. The initial, unoptimized placement of B-snakes is shown on the left. The first image is repeated in the middle, displaying localized tags.

2.2 Snake Energy Field

In a normal MR image, the image intensity is obtained by measuring the NMR signal corresponding to each tissue location. The detected signal is a function of tissue relaxation parameters, as well as the local proton density, and depends on the choice of applied radio-frequency (RF) pulse sequence. In a radial tagged MR image, a series of spatially selective RF pulses is applied prior to a conventional imaging sequence voiding the NMR signal at selective locations, and creating tag patterns within the deforming tissue.

In order to create an external energy field for snake optimization, we simulate time-dependent tag profiles using physics of image formation (see [9]) and concatenate a series of profiles along the vertical axis to create a template (g_1), which subsequently is correlated with the data. Let the image be represented by g_2. The normalized correlation, $\rho(\delta_x, \delta_y)$, between the template g_1 and the image g_2 satisfies $0 \leq \rho \leq 1$, with $\rho = 1$ when g_1 is a constant multiple of g_2. In order to increase the discrimination power of the technique, the energy field is set to $-\rho^n(\delta_x, \delta_y)$, where n is a positive integer less than 10. The higher the value of n, the more discriminating against inexact template matches the energy function becomes, in the limit only *accepting* exact matches. Endpoint energy fields are termed ρ_e, and are obtained from correlating endpoint templates with the tag data.

There is a trade-off between the spatial resolution of this energy field and noise from approximate matches; the degree of trade-off controlled by the template size. The larger the template size, the lower the resolution will be, but the energy field will be more robust to noise and inexact profile matches. The simulated correlation template is successively rotated to create kernels along other orientations. [4]

[4] (0,45,90,135) degrees for 4 radial tag orientations

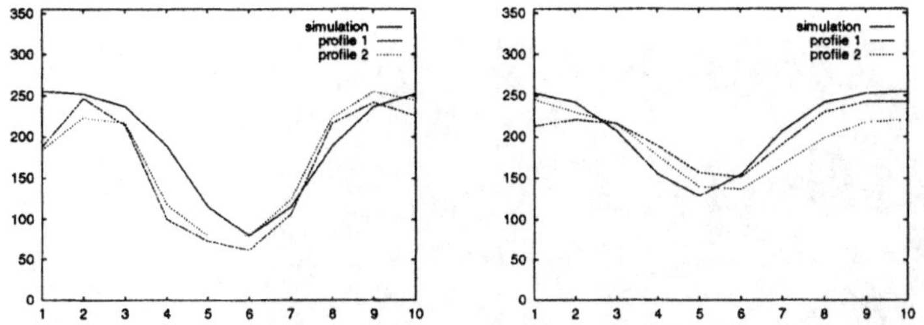

Fig. 2. Comparison of simulated profiles and actual profiles taken from different images.

3 Coupled B-Snake Grids

Coupled snake grids are a sequence of spatially ordered snakes, represented by B-spline curves, which respond to image forces, and track non-rigid tissue deformations from SPAMM data. The spline grids are constructed by having the horizontal and vertical grid lines share control points. By moving a spline control point, the corresponding vertical and horizontal snakes deform. This representation is reasonable since the point of intersection of two tag lines is physically the same material point, and furthermore tissues are connected.

We define a MN spline grid by $(M \times N)-4$ control points which we represent by the set

$$\{\{p_{12}, p_{13}, \cdots, p_{1,N-1}\}, \{p_{21}, p_{22}, \cdots, p_{2,N}\}, \cdots, \{p_{M,2}, p_{M,3}, \cdots, p_{M,N-1}\}\} \quad (4)$$

where p_{ij} is the spline control point at row i and column j.

3.1 Grid Optimization with Gradient Descent

To detect and localize SPAMM tag lines, we optimize grid locations by finding the minimum intensity points in the image, as tag lines are darker than surrounding tissues. However, there is an additional energy term present in our formulation which takes account of the local 2D structure of image intensity values at tag intersections. Although we can not specify an exact correspondence for points along a tag line, we do know the exact correspondence for points at tag intersections. This is the familiar statement of aperture problem in image sequence analysis. The way to incorporate this familiar knowledge into our algorithm is by use of the SSD function in (5). The energy function which we minimize is

$$\mathcal{E}(p_{12}, \cdots, p_{M,N-1}) = \lambda_1 \sum_k \int I(\alpha_k(u))du + \lambda_2 \sum_{ij} SSD(v_{ij}) \quad (5)$$

where v_{ij} denotes the intersection point of horizontal and vertical snake curves, and λ_1 and λ_2 are pre-set constants. The SSD function determines the sum-of-squared-differences of pixels in a window around point v_{ij} in the current frame

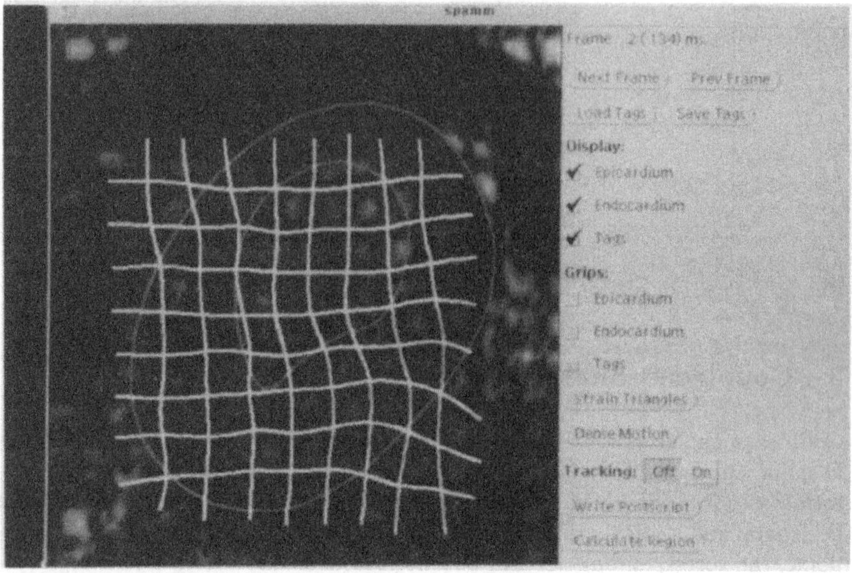

Fig. 3. Locating tag lines in SPAMM images. An intermediate frame in a sequence of tagged images is shown. The LV contours in the figure are hand-drawn B-spline curves. Note that the portions of the snake grids lying within the ventricular blood and in the liver do not contribute to the energy of the snake grid.

with a window around the corresponding B-snake grid intersection in the next frame. In order to minimize the discretized version of $\mathcal{E}$, in each iterative step, we compute the gradient of $\mathcal{E}$ with respect to p_{ij}, perform a line search in the $\nabla\mathcal{E}$ direction, move the control points to the minimum location, and continue the procedure until the change in energy is less than a small number, defining convergence. In practice, we have an additional constraint in computing the energy function: we only use the intersections, and points on the snake grid which lie on the heart tissue. Results from grid optimization for localization and tracking of a spline grid with gradient descent is shown in figure 3.

3.2 Grid Optimization with Dynamic Programming

In this section, we describe a DP formulation for grid optimization. We note that optimizing an interacting grid with standard DP is not possible due to the dependence of horizontal and vertical snakes on one another.

As stated in section 2, the DP optimal value function for 1D splines is a function of spline control points. To apply 2D DP to this problem, we define the optimal value function, $S : \mathcal{R} \times \mathcal{F} \longrightarrow \mathcal{R}$ with $\mathcal{R}$ denoting the real line and $\mathcal{F}$ being a space of one dimensional splines

$$S(x, V_k(x,.)) = \min_j \{S(x+\Delta x, V_j(x+\Delta x,.)) + \mathcal{E}_1(V_k(x,.), V_j(x+\Delta x,.)) + \mathcal{E}_2(V_j(x+\Delta x,.))\} \quad (6)$$

where x is a real number indexing location along the horizontal axis of the image where tag lines are located. Δx represents vertical tag separation, and V represents a realization of a spline. The class of all admissible vertical splines is $\mathcal{F}$. Note that

$$\mathcal{E}_1(V_j(x+\Delta x,.), V_k(x,.)) = \sum_i \int -I(H_i(u))du \tag{7}$$

where i indexes horizontal snakes between x and $x+\Delta x$, H_i is snake number i, and $1 \leq i \leq m$ with m being the number of horizontal snakes. $\mathcal{E}_2(V_j(x+\Delta x,.))$ is the energy of spline number j at location $x+\Delta x$

$$\mathcal{E}_2(V_j(x+\Delta x,.)) = -\int I(V_j(x+\Delta x, u))du \tag{8}$$

We have defined a recursive form for obtaining optimal grids with the new DP algorithm. Finally, we note that this algorithm is not practical, since we need to know all the possible deformations of a vertical or horizontal snake before hand. Moreover, the stated formulation will only apply to linear splines (to further generalize to non-linear splines, $\mathcal{E}_1$ will depend not only on splines at x and $x+\Delta x$, but also on other neighboring splines). On the positive side, if all the possible deformations were known, the 2D DP algorithm would guarantee the global optimality of the grid.

3.3 Tissue Strain from Snake Interactions

Strain is a measure of local deformation of a line element due to tissue motion and is independent of the rigid motion. To compute the local 2D strain for a given triangle, correspondence of 3 vertices with a later time is sufficient. With this information known, an affine map F is completely determined. Under the local affine motion assumption for the LV, strain in the direction of vector $\mathbf{x}$ can be expressed as

$$\Sigma = \frac{1}{2}\left(\frac{|F\mathbf{x}|^2}{\mathbf{x}^T\mathbf{x}} - 1\right) \tag{9}$$

Two directions within each such triangle are of particular interest, namely, the directions of principal strain, representing the maximum and minimum stretch within a triangle. The results of strain analysis performed on a typical tagged image is shown in figure 4 with the corresponding maximum and minimum principal strain directions and values.

4 Smooth Warps

Tracking tissue deformations with SPAMM using snake grids provides 2D displacement information at tag intersections and 1D displacement information along other 1D snake points. In this section, we describe *smooth warps* which

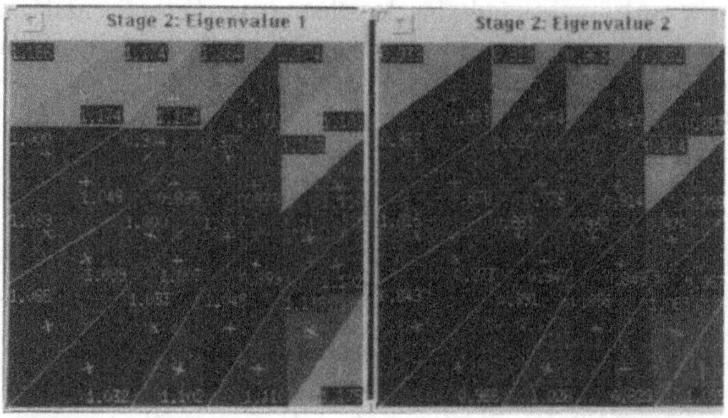

Fig. 4. Strain measurement made from two SPAMM frames.

reconstruct a dense displacement vector field using the optimized snakes and the available displacement information, assuming only 2D motion (as is roughly the case towards the apical end of the heart). We note that this analysis can be extended to 3D.

To proceed more formally, the continuity constraint is the bending energy of a thin-plate which is applied to the x and y component of the displacement field $(u(x,y), v(x,y))$:

$$\int\int u_{xx}^2 + 2u_{xy}^2 + u_{yy}^2 dxdy \tag{10}$$

$$\int\int v_{xx}^2 + 2v_{xy}^2 + v_{yy}^2 dxdy \tag{11}$$

These constraints serve as the smoothness constraints on the reconstructed vector field, characterizing approximating thin-plate splines [6].

With the intersection springs in place,

$$\sum(u - d^u)^2 + (v - d^v)^2 \tag{12}$$

is also to be minimized. In (12), d^u and d^v are the x and y components of displacement at tag intersections. The form of the intersection spring constraints is similar to depth constraints in surface reconstruction from stereo, and has also been used in a similar spirit by Young et al. [18].

Assuming 2D tissue motion, a further physical constraint is necessary: any point on a snake in one frame must be displaced to lie on its corresponding snake in all subsequent frames. This constraint is enforced by introducing a sliding spring. One endpoint of the spring is fixed on a grid line in the first frame, and its other endpoint is allowed to slide along the corresponding snake in the second frame, as a function of iterations. We minimize

$$\sum \{(x+u-\bar{x})^2 + (y+v-\bar{y})^2\} \tag{13}$$

along 1D snake points. In the above equation, (x, y) are the coordinates of a point on the snake in the current frame, and $(\bar{x}, \bar{y})$ is the closest point to $(x+u, y+v)$ on the corresponding snake in the second frame. Adding (10), (11), (12), and (13), and deriving the Euler-Lagrange equations will yield a pair of equations involving partial derivatives of $(\bar{x}, \bar{y})$: [5]

$$\begin{aligned} \lambda_1 \nabla^4 u + \lambda_2(u - d^u) + \lambda_3 \{(u + x - \bar{x})(1 - \bar{x}_u) + (v + y - \bar{y})(-\bar{y}_u)\} &= 0 \\ \lambda_1 \nabla^4 v + \lambda_2(v - d^v) + \lambda_3 \{(v + y - \bar{y})(1 - \bar{y}_v) + (u + x - \bar{x})(-\bar{x}_v)\} &= 0 \end{aligned} \tag{14}$$

where $\lambda_1 \geq 0$ everywhere, $\lambda_2 \geq 0$ at tag intersections, and $\lambda_3 \geq 0$ at all 1D snake points. We now make two approximations. For vertical grids, the x-coordinates of curves only vary slightly, and as the grid lines are spatially continuous, $\bar{x}_u$ is expected to be small. Furthermore, for vertical grids $\bar{y}$ changes minutely as a function of u, so that $\bar{y}_u \approx 0$. For horizontal grids, the y coordinates of curves also vary slightly along the length of grid lines, and since these are spatially continuous curves, $\bar{y}_u$ is expected to be small. Note that these approximations will hold under smooth local deformations, as is expected in the myocardial tissue. Only $\bar{x}_u$ for horizontal grids, and $\bar{y}_v$ for vertical grids is expected to vary more significantly. Though the effect of these terms is to modulate λ_3, for completeness, we include these derivatives in our equations:

$$\begin{aligned} \lambda_1 \nabla^4 u + \lambda_2(u - d^u) + \lambda_3(u + x - \bar{x})(1 - T_{hor}\bar{x}_u) &= 0 \\ \lambda_1 \nabla^4 v + \lambda_2(v - d^v) + \lambda_3(v + y - \bar{y})(1 - T_{ver}\bar{y}_v) &= 0 \end{aligned} \tag{15}$$

The variables T_{hor} and T_{ver} are predicates equal to one if the snake point of interest lies on a horizontal, or a vertical grid line. An iterative solution to (15) from finite differences has been adopted which converges in 400-500 iterations for two SPAMM frames.

The results of applying (15) to two snake grids in figure 7 is shown in figure 5 with $\lambda_1 = 8.0$, $\lambda_2 = 1.0$, and $\lambda_3 = 10.0$. At iteration 1, $NZP = 1008$, $TPS = 12429.06$, $IE = 432.06$, and $PE = 2742.26$. At convergence, $NZP = 298$, $TPS = 68.52$, $IE = 1.39$, and $PE = 56.22$. Where IE is given by equation (12) , and PE is given by equation (13). NZP is the number of non-zero points in a 2D scratch array where all snake coordinates on both snake grids are exclusive-ored together. The thin-plate spline measure (TPS) is the sum of the two integral measures in equations (10), and (11). These measures were computed for (intersection and grid) points within the myocardium. Note that PE does not include the intersection points.

Utilizing the dense displacement vector field recovered with the method outlined in this section, in addition to strain, differential measures describing local

[5] Note that in practice, $(\bar{x}, \bar{y})$ is smoothed by local averaging

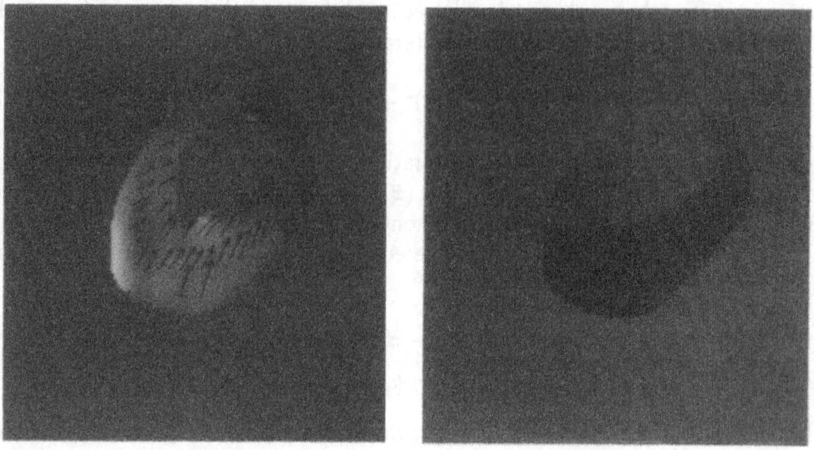

Fig. 5. Reconstruction of dense u and v components of displacement from 2 snake grids in figure 3. The vector field is superimposed on the u, and v reconstructions for the final results (iteration 500).

rotations and expansions may be obtained. Tissue expansion or contraction in an arbitrary area may be computed by applying Gauss's theorem, and local tissue rotations may be computed by applying Stoke's theorem. As a final note, the dense displacement field suggested here can only be an approximation to the true displacement vector field. However, as the tag lines become closer, the approximation becomes very accurate.

5 Conclusions

In conclusion, we have described new computational algorithms suitable for analysis of both radial and SPAMM tagged data. We described a new DP algorithm which given an external energy field, can optimize B-snakes. We have argued that in comparison to other forms of parametrization, use of B-splines for representing curves has several advantages, including subpixel accuracy for tag localization and parametric continuity, as well as the need to only optimize the location of few control points in order to determine the location of a complete tag line.

A different aspect of our work involves reconstruction and interpolation of dense displacement vector fields directly from tracked snake grids. To this end, we presented smooth warps which warp an area in the plane such that two embedded grids of curves are brought into registration.

References

1. A. A. Amini. A scalar function formulation for optical flow. In *European Conference on Computer Vision*, Stockholm, Sweden, May 1994.
2. A. A. Amini and et al. Energy-minimizing deformable grids for tracking tagged MR cardiac images. In *Computers in Cardiology*, pages 651–654, 1992.
3. L. Axel, R. Goncalves, and D. Bloomgarden. Regional heart wall motion: Two-dimensional analysis and functional imaging with MR imaging. *Radiology*, 183:745–750, 1992.
4. B. Bascle and R. Deriche. Stereo matching, reconstruction, and refinement of 3d curves using deformable contours. In *International Conference on Computer Vision*, 1993.
5. A. Blake, R. Curwen, and A. Zisserman. A framework for spatio-temporal control in the tracking of visual contours. *International Journal of Computer Vision*, 11(2):127–145, 1993.
6. F. Bookstein. Principal warps: Thin-plate splines and the decomposition of deformations. *IEEE Transactions on PAMI*, 1989.
7. A. Gueziec. Surface representation with deformable splines: Using decoupled variables. *IEEE Computational Science and Engineering*, pages 69–80, Spring 1995.
8. S. Gupta and J. Prince. On variable brightness optical flow for tagged MRI. In *Information Processing in Medical Imaging (IPMI)*, pages 323–334, 1995.
9. M. Guttman, J. Prince, and E. McVeigh. Tag and contour detection in tagged MR images of the left ventricle. *IEEE-TMI*, 13(1):74–88, 1994.
10. T. Huang. Modeling, analysis, and visualization of nonrigid object motion. In *International Conference on Pattern Recognition*, 1990.
11. M. Kass, A. Witkin, and D. Terzopoulos. Snakes: Active contour models. *International Journal of Computer Vision*, 1(4):321–331, 1988.
12. S. Menet, P. Saint-Marc, and G. Medioni. B-snakes: Implementation and application to stereo. In *International Conference on Computer Vision*, pages 720–726, 1990.
13. F. Meyer, T. Constable, A. Sinusas, and J. Duncan. Tracking myocardial deformations using spatially constrained velocities. In *IPMI*, pages 177–188, 1995.
14. C. Nastar and N. Ayache. Non-rigid motion analysis in medical images: A physically based approach. In *IPMI*, pages 17–32, 1993.
15. J. Park, D. Metaxas, and L. Axel. Volumetric deformable models with parameter functions: A new approach to the 3d motion analysis of the LV from MRI-SPAMM. In *International Conference on Computer Vision*, pages 700–705, 1995.
16. N. Pelc, R. Herfkens, A. Shimakawa, and D. Enzmann. Phase contrast cine magnetic resonance imaging. *Magnetic Resonance Quarterly*, 7(4):229–254, 1991.
17. D. Reynard, A. Blake, A. Azzawi, P. Styles, and G. Radda. Computer tracking of tagged 1H MR images for motion analysis. In *Proc. of CVRMed*, 1995.
18. A. Young, D. Kraitchman, and L. Axel. Deformable models for tagged MR images: Reconstruction of two- and three-dimensional heart wall motion. In *IEEE Workshop on Biomedical Image Analysis*, pages 317–323, Seattle, WA, June 1994.
19. E. Zerhouni, D. Parish, W. Rogers, A. Yang, and E. Shapiro. Human heart: Tagging with MR imaging – a method for noninvasive assessment of myocardial motion. *Radiology*, 169:59–63, 1988.

Local Quantitative Measurements for Cardiac Motion Analysis

Serge Benayoun[1], Dany Kharitonsky[1]
Avraham Zilberman[2] and Shmuel Peleg[1]

[1] Institute of Computer Science, The Hebrew University, Jerusalem 91904, Israel
[2] Cardiology Department, Bikur Holim Hospital, Jerusalem

Abstract. We design for this work a new practical tool for computation of non-rigid motion in sequences of 2D heart images. The implementation of our approach allows us to integrate several constraints in the computation of motion : optical flow, matching of different kinds of shape-based landmarks and regularity assumption. Based on the determination of spatio-temporal trajectories, we next propose several measurements to analyze quantitatively the local motion of the left ventricle wall. Some experimental results on cardiac images issued from clinical cases illustrate our approach.

1 Introduction

Cardiac motion analysis has received these last years a great attention from the computer vision community [2, 3, 5, 9, 11, 12]. Since it is a non-rigid organ, the recovery of quantitative parameters featuring heart deformations is a very difficult problem. New medical imagery modalities like Computed Tomography (Fast-CT) or Magnetic Resonance Imagery (MRI) produce now 3D sequences of the beating heart, the second method having the major advantage to be non invasive, while requiring post-synchronization to reconstruct a complete cardiac cycle. More sophisticated imagery techniques like Tagged MRI [1] or Phase Velocity MRI [6] give now direct physical information about the motion. But, until now, these techniques are not often clinically used for a number of reasons. First, the cost of such machines is relatively high. Secondly, sophisticated post-processing tools are necessary to take advantage of these data, like 3D visualization or 3D motion analysis. These tools are not widespread at the moment.

On the other side, cardiologists very often use modalities like Cardiac Ultrasonography, Doppler echocardiography, Angiography or Ventriculography which produce real-time sequences of heart cross-sections or 2D projections. It is clear that the heart deforms in the 3D space and consequently, an accurate study of its deformations must be done with 3D acquisitons. Nevertheless, at this time, it seems that computer vision techniques can be very helpful for cardiologists to give local quantitative measurements on 2D data. These local measurements, combined with global cardiac parameters and the "visual" experience of heart specialists, could help them to improve evaluation, comparison and classification of sequence data. For this study, we designed a simple and fast tool for computing motion in sequences of $2D$ images.

In our approach we track closed curves which are representative of the anatomical deformations through the time sequence. Several works have already been done on this subject, in particular *Cohen et al.* [5] who match closed 2D *snakes* by minimizing a couple of energies. One energy measures the difference of curvatures between matched points. The second measures the regularity of the correspondence function. For high curvature points, they privilege the first energy. *Mc Eachen et al.* [9] use parametric deformable curves. The matching is also based on curvature with a particular attention to high curvature points, but also integrates information provided by Phase Velocity MRI. Our approach has the following particularities : (i) we perform matching between iso-intensity curves, (ii) we use two kinds of shape-based landmarks : curvature extrema and curvature zero-crossings. This approach can be generalized to 3D images and we plan to do so in the future.

This paper is organized as follows. We first present the method we used to compute motion field between 2D curves (section 2). Based on the determination of spatio-temporal trajectories, we next propose different kinds of local quantitative measurements for assessing myocardial function (section 3). Finally, we illustrate our study with some experimental results issued from clinical cases (section 4).

2 Motion Computation

In this section, we present the different steps of the method for computing motion fields. We first consider a particular curve on each frame (section 2.1), we next extract shape-based landmarks on these curves (section 2.2) and we finally use these landmarks to perform curve matching (section 2.3).

2.1 Isolines

In medical imaging the grey level (or *intensity*) of a pixel is often representative of the tissue which is imaged. These curves are generally called *iso-intensity curves* or *isolines*. Actually, an isoline may be not everywhere significant and it may be necessary to consider only its relevant parts.

Considering a time sequence of images $(I_t)_{t=1,N}$, we first smooth them with a recursive gaussian filter [10]. The width of the filter (about 4 pixels) was chosen in order to preserve accuracy while having smooth curves. Next we extract an isoline (L_t) on each image. We assume that the set $(L_t)_{t=1,N}$ represent the evolution of the same physical boundary during time. The intensity of isolines is chosen manually and is the same for the whole sequence. By construction, these curves are closed, ordered and defined in the real 2D space, i.e. they are defined at a sub-pixel resolution. The problem now amounts to compute the local motion between two successive isolines.

The intensity of the pixel is also often used to do tracking in a time sequence. Optical flow is now a well-known technique for recovering motion, it is based on the assumption that the intensity of a particular point is constant. First used on sequences of 2D images like in [7], it has been next generalized for 3D images [12]. Thus, by matching isolines with same iso-intensity, one integrates implicitely the optical flow assumption.

2.2 Shape-based Landmarks

The most common way for recovering curve nonrigid motion consists in tracking some particular points in the curve. Theses points are generally called *landmarks*. A landmark may have several characteristics. One of them is that they must be trackable in a time sequence. Another is that they must well characterize the overall motion. In fact, it means that one may compute the motion just by propagating the values computed on these landmarks. There are different kinds of landmarks : implanted markers, physical landmarks and shape-based landmarks.

Implanted markers are distinctive points physically put in the organ and can thus be tracked easily during the motion [9]. Unfortunately this method is only used for research purpose and can not be clinically generalized.

Tagged MRI is a new technique which provides physical landmarks during the cardiac cycle. It generates a magnetization grid superimposed on the MRI data. Next this grid moves with the heart tissue as the heart moves [1]. Then one has just to compute the displacement of the grid nodes (or intersect points between tag lines and anatomical structures) to obtain the right motion of some points. This method has been successfully used for example to track a volumetric deformable model of the left ventricle [11].

Shape-based landmarks are directly computed from the images. Unlike preceeding categories, these landmarks have not always anatomically meaningfull although some works proved their relevance for tracking cardiac motion [9]. In our case, we work on 2D curves and we use two kinds of landmarks : (i) *curvature extrema points* which have been widely used to track left ventricle wall in ultra-sound images [5] or MRI cross-sections [9] (ii) *curvature zero-crossing points* which also are trackable and representative of the anatomical shape.

The estimation of curvature along the isoline is based on the computation of angles between three successive points in the curve. One may compute this curvature at different scales by considering different distances between these successive points. We used here a scale of 20 pixels.

2.3 Curves Matching

The landmarks matching procedure uses the assumption that time resolution is sufficiently fine. This assumption is common for more sophisticated techniques as well. We use a two-step method :

- For each landmark detected in an isoline, we search the closest landmark in the following isoline. If this landmark is sufficiently close and if the difference of curvature of this couple of points is sufficiently small, we accept the match. In this first step, the distance between points is computed in the 2D space.

To add robustness, we keep only *symmetric* matches, i.e. for which the result is identical if we consider the curves in reverse order.
- In the second step, the distance is computed along the curves, i.e. we consider the difference of arclength. For computing arclength, the closest match obtained in the first step gives us the two *starting points*.

This method runs successively with curvature extrema points and curvature zero-crossing points.

In order to obtain a dense motion field between the two curves, we perform linear interpolation between each couple of adjacent displacement vectors given by the landmarks matching algorithm. Next we smooth the resulting motion field by iterative local averaging. All these computations are done at a sub-pixel resolution. We have also taken into account the *starting point problem* by considering at each step of our process the 2D curves as closed *loops* [5].

3 Local Quantitative Measurements

In order to detect cardiac diseases, it is important to make the distinction between global and local quantitative measurements. For example, *ejection fraction* which measures the relative variation of the heart volume between end-of-diastole and end-of-systole is a global parameter. Its normal value is comprised between 0.5 and 0.8. If the computed value is for example less than 0.5, it means that the heart is diseased but it can not help to determine accurately the location of the pathology. Another example concerns mathematical analysis of motion fields with modal analysis which gives a compact but global description of the field [3]. In this work we try to determine local quantitative measurements because they can prove more useful for the cardiologist to locate and quantify a pathology.

3.1 Spatio-temporal Trajectories

trajectories computation To consider a single point motion, we define a *spatio-temporal trajectory* as a list of points $(P_t)_{t=1,N}$ for which each point P_t belongs to the isoline (L_t) in a time sequence and represents the position of the same physical point P a time t. Like for motion estimation, the trajectory is computed at a sub-pixel resolution.

periodicity constraint It is logical to compute periodic trajectories because of the periodic nature of the heart beats. The method we used is described in [6]. We first compute two trajectories, one in the forward direction and the other in the backward direction. Next we compute an averaged trajectory by combining these two trajectories. The averaged points are computed along the curves. The weighting of each trajectory depends on the temporal position of the point. If it is far from the starting point, the weight will be low and vice versa. Thus, by construction, the averaged trajectory will be periodic.

temporal smoothing The periodicity constraint integrates the temporal dimension in the motion computation. However it is desirable to add an other temporal constraint in order to further regularize the spatio-temporal trajectory. Such a result can be obtained just by smoothing the curvature along the trajectory. One convenient way to do this consists in minimizing the trajectory length. In order to preserve the original shape of the trajectory, the smoothing is performed only if the local length of the trajectory can be substantially decreased, it means for most unregular pathes. This processing is iterated until the trajectory does not change anymore. Figure 1 shows the trajectory of a point with all the isolines.

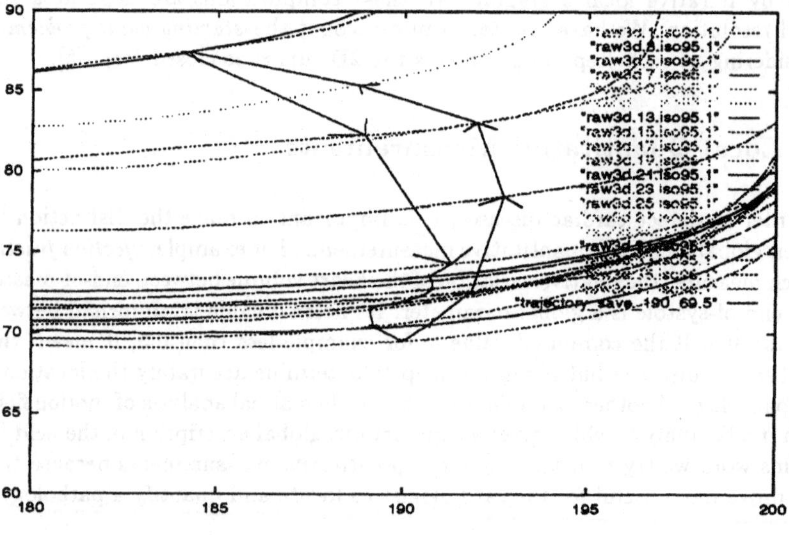

Fig. 1. Spatio-temporal trajectory : this result integrates motion field computation, periodicity constraint and temporal smoothing

trajectory kynesis Velocity of a point $\frac{\partial P}{\partial t}$ and its acceleration $\frac{\partial^2 P}{\partial t^2}$ may be useful for cardiologists in particular to check the regularity of the motion which helps to quantify the local physical characteristics of the heart muscle.

3.2 Segments Time Evolution

If we consider the trajectories of two neighbor points along the isoline, they describe the time evolution of the left ventricular wall segment between them. Thus we may calculate during time some measurements to characterize it : (i) *length* which gives an index of local elasticity of the curve segment (ii) ***bending energy*** which is equal to the sum of squared curvature along the segment, gives information about its local deformability.

4 Experimental Results

4.1 Ventriculography with Catheterization

Angiography is an imaging technique which consists in introducing an X-ray contrast agent in the arteries with a catheter. By this mean, one can inspect the interior of the arteries, and, in pathologic cases, detect artherosclerotic or fibromuscular stenoses inside an artery which can lead to disturbances in blood flow. In some cases, the little balloon at the top of the catheter may be next used to open stenotic coronary blood vessels [4].

Ventriculography is a similar examination but, this time, the catheter is directly introduced inside the left ventricle. So, with the contrast agent, it becomes possible to evaluate the motion of the left ventricle wall. Figure 2 shows an image produced by this examination.

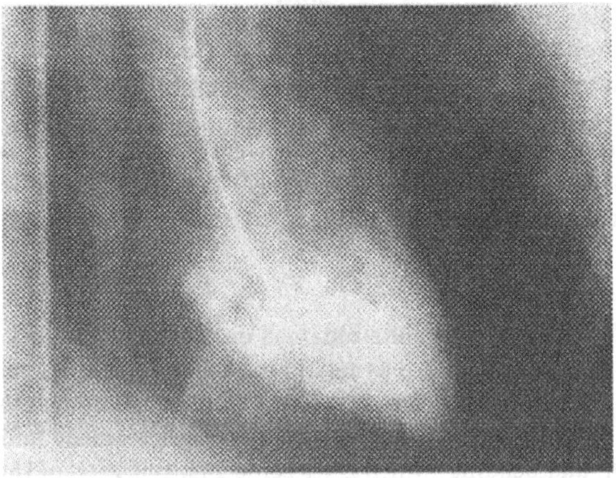

Fig. 2. Ventriculography with catheterization data (courtesy of Bikur Holim Hospital)

Generally, distinction is made between four kinds of kynesis [8] : (i) normal kynesis (ii) hypokynesis : the motion is less than normal (iii) akynesis : some parts of the wall do not move (iv) diskynesis : some parts of the wall do not move in the normal direction. An accurate evaluation of the wall motion can point to a certain problem in the heart muscle or in the coronary arteries.

4.2 Clinical Cases Analysis

We did experiments on clinical cases provided by the cardiology department of the *Bikur Holim* hospital, at Jerusalem. Ventriculography is done routinely in this hospital. Nevertheless, our processing requires some preliminar conditions : (i) the quantity of injected contrast product must be sufficient in order that all the ventricle wall may be apparent (ii) the structures appearing in the image background may not perturb the isoline extraction first stage. Under these conditions, we perform the following chain of computations for each sequence :

- compute all the motion fields
- consider the relevant part of the isoline and divide it into N segments of equal arclength
- for each segment, compute the following features
  - **Midpoint Motion** : trajectory length (TOT), maximal displacement between two frames (MAX), minimal displacement (MIN), mean displacement (MEAN), standard deviation (SD) and amplitude (AMP) which corresponds to the maximal distance between two positions in the trajectory.
  - **Segment Length** : maximal length (MAX), minimal length (MIN), length fraction (FRAC = $\frac{MAX-MIN}{MAX}$), mean length (MEAN) and standard deviation (SD).
  - **Segment Bending** : maximal bending (MAX), minimal bending (MIN), bending fraction (FRAC) , mean bending (MEAN) and standard deviation (SD).

  These parameters are based on the currently used cardiological techniques for the evaluation of the heart muscle [4] and the doctors demands. Other characteristic measurements of the heart muscle function are being considered.

CASE 1 We present now results obtained on a clinical case. By a cardiologist diagnosis, it is a case of normal kynesis. For the first example, we considered 10 segments. Figure 3 presents the evolution of the segments length and bending during time. Table 1 presents the evolution of each segment midpoint motion, overall length, and bending, which gives some quantification of the heart muscle behavior at that segment.

CASE 2 By a cardiologist diagnosis, this second example is a case of slight hypokynesis near the heart apex. Figure 3 presents the evolution of the segments length and bending during time. Table 2 presents the same features as table 1 computed on each segment. The hypokynesis is confirmed by the low amplitude

Table 1. Features computed on the segments (CASE 1)

Segment Number		*1*	*2*	*3*	*4*	*5*	*6*	*7*	*8*	*9*	*10*
Midpoint Trajectory	*TOT*	79.5	100.8	107.3	163.6	154.9	177.3	130.5	104.0	92.0	98.7
	MAX	8.9	17.4	17.4	23.1	21.6	24.8	30.0	28.1	23.4	19.1
	MIN	0.4	0.2	0.3	0.2	0.7	1.0	0.3	0.5	0.5	0.1
	MEAN	4.7	5.9	6.3	9.6	9.1	10.4	7.7	6.1	5.4	5.8
	SD	2.4	3.7	4.5	5.7	5.2	5.9	7.0	5.9	5.1	4.8
	AMP	30.5	43.4	48.8	65.1	72.0	82.6	58.0	43.4	34.6	34.0
Length	*MAX*	56.2	50.6	67.8	59.0	59.2	55.8	57.2	48.4	55.5	47.5
	MIN	28.5	30.1	32.9	39.5	32.0	21.1	18.3	23.5	29.6	25.9
	FRAC	0.5	0.4	0.5	0.3	0.5	0.6	0.7	0.5	0.5	0.5
	MEAN	41.8	38.6	44.8	45.8	46.4	42.1	40.3	38.0	41.3	33.7
	SD	8.0	6.2	9.9	6.3	9.0	8.9	12.1	9.6	7.3	7.0
Bending	*MAX*	16.4	7.7	9.0	11.4	1.2	24.3	13.2	2.4	1.2	5.7
	MIN	6.0	0.1	0.5	0.0	0.0	12.9	2.1	0.2	0.1	0.0
	FRAC	0.6	1.0	0.9	1.0	1.0	0.5	0.8	0.9	0.9	1.0
	MEAN	11.7	1.8	2.8	3.3	0.5	18.7	5.5	0.9	0.4	2.1
	SD	3.1	2.0	2.6	3.4	0.3	3.6	3.3	0.6	0.3	2.0

in segments 4, 5, 6, 7 (near the heart apex) relative to case 1. We are currently working on a statistical model of heart kynesis with a large number of cases.

Conclusion

We have presented a fast and practical tool for computing motion field on sequence of 2D cardiac data. Based on the determination of spatio-temporal trajectories, this tool allowed us to compute local quantitative measurements representative of the heart activity during the whole cardiac cycle. These measurements may be useful for cardiologists to evaluate more accurately some cardiovascular diseases, like coronary arteries disease. The next step will consist of doing clinical validation on a large number of cases. We hope to present such results at the time of the conference. Future work includes definition of standard local parameters in order to reduce the heart deformations to a minimal number of representative values. Another direction concerns the developpement of the user-interface for a practical use. Finally we intend to extend these measurements for other kinds of modalities, like cardiac ultrasonography and also sequences of 3D data.

Acknowledgments

This work was supported by Silicon Graphics Biomedical Ltd., Jerusalem.

Table 2. Features computed on the segments (CASE 2)

Segment Number		1	2	3	4	5	6	7	8	9
Midpoint Trajectory	TOT	160.4	146.0	112.1	108.1	107.7	81.7	85.1	165.9	209.8
	MAX	37.9	33.2	25.0	27.6	27.9	22.9	25.2	40.3	46.6
	MIN	1.0	0.4	1.1	0.6	0.2	0.6	0.8	0.1	0.3
	MEAN	10.7	9.7	7.5	7.2	7.2	5.4	5.7	11.1	14.0
	SD	8.5	8.1	5.9	6.0	6.3	5.0	5.9	11.4	13.5
	AMP	70.5	68.6	47.5	41.9	41.7	32.3	37.4	50.0	64.6
Length	MAX	62.2	55.6	44.9	47.0	57.1	44.8	57.9	43.5	43.5
	MIN	43.5	31.6	22.1	14.4	28.4	28.8	27.4	13.0	16.9
	FRAC	0.3	0.4	0.5	0.7	0.5	0.4	0.5	0.7	0.6
	MEAN	53.2	40.6	34.6	31.9	43.1	37.3	37.6	31.0	31.3
	SD	6.4	5.6	8.5	9.4	8.2	5.7	8.5	10.7	9.1
Bending	MAX	5.8	20.5	4.0	16.9	18.7	1.8	1.6	3.2	5.0
	MIN	0.2	2.3	0.4	1.6	3.0	0.1	0.1	0.3	0.2
	FRAC	1.0	0.9	0.9	0.9	0.8	0.9	0.9	0.9	1.0
	MEAN	2.5	13.4	2.3	11.1	8.4	0.7	0.6	1.8	2.0
	SD	1.5	6.5	1.0	4.2	4.5	0.5	0.5	0.9	1.5

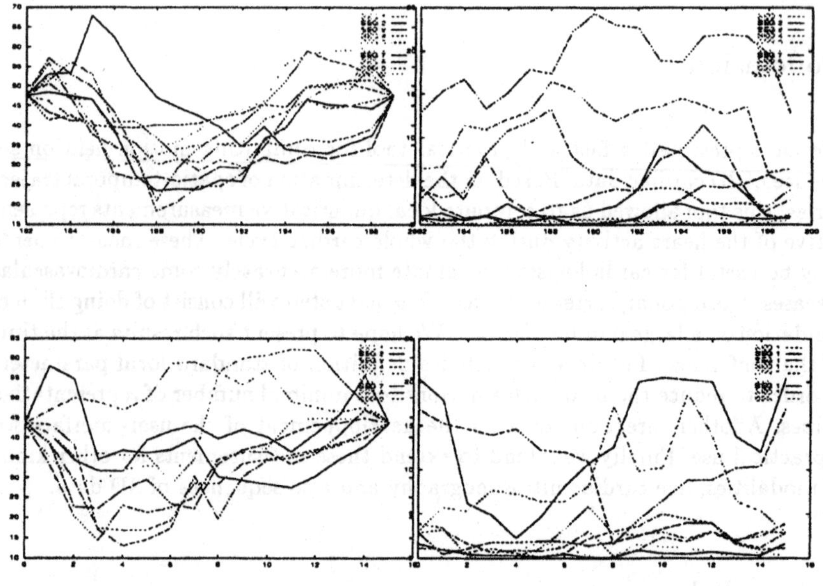

Fig. 3. Evolution of length (left) and bending (right) for each segment. On up, CASE 1 and on bottom, CASE 2

References

1. H. Azhari, J.L. Weiss, W.J. Rogers, C.O. Siu, and E.P. Shapiro. A noninvasive comparative study of myocardial strains in ischemic canine hearts using tagged mri in 3-d. *American Journal of Physiology*, 268:1918–1926, 1995.
2. E. Bardinet, L.D. Cohen, and N. Ayache. Superquadrics and free-form deformations : a global model to fit and track 3d medical data. In *Proceedings of the First International Conference on Computer Vision, Virtual Reality and Robotics in Medicine (CVRMed'95)*, Nice, France, April 1995.
3. S. Benayoun, C. Nastar, and N. Ayache. Dense non-rigid motion estimation in sequences of 3d images using differential constraints. In *Proceedings of the First International Conference on Computer Vision, Virtual Reality and Robotics in Medicine (CVRMed'95)*, Nice, France, April 1995.
4. E. Braunwald. *Heart Disease : A Textbook of Cardiovascular Medicine*, volume 1. W.B. Saunders, 1992. Fourth Editon.
5. Isaac Cohen, Nicholas Ayache, and Patrick Sulger. Tracking points on deformable objects using curvature information. In *European Conference on Computer Vision*, pages 458–466, Santa Margherita Ligure, Italy, May 1992.
6. R.T. Constable, K.M. Rath, A.J. Sinusas, and J.C. Gore. Development and evaluation of tracking algorithms for cardiac wall motion analysis using phase velocity mr imaging. *Magnetic Resonance in Medicine*, 32:33–42, 1994.
7. S. Gong and M. Brady. Parallel computation of optical flow. In *European Conference on Computer Vision*, pages 124–133, Antibes, France, April 1990.
8. Harrison. *Principles of Internal Medicine.* Mc Graw Hill.
9. J.C. McEachen, F.G. Meyer, R.T. Constable, A. Nehorai, and J.S. Duncan. A recursive filter for phase velocity assisted shape-based tracking of cardiac non-rigid motion. In *IEEE International Conference on Computer Vision*, pages 653–658, Cambridge, Massachusetts, June 1995.
10. O. Monga, R. Lengagne, and R. Deriche. Extraction of the zero-crossings of the curvature derivative in volumic 3d medical images : a multi-scale approach. In *IEEE Conference on Computer Vision and Pattern Recognition*, Seattle, June 1994.
11. J. Park, D. Metaxas, and L. Axel. Volumetric deformable models with parameter functions: a new approach to the 3D motion analysis of the lv from mri-spamm. In *IEEE International Conference on Computer Vision*, pages 700–705, Cambridge, Massachusetts, June 1995.
12. S.M. Song, R.M. Leahy, D.P. Boyd, B.H. Brundage, and S. Napel. Determining cardiac velocity fields and intraventricular pressure distribution from a sequence of ultrafast ct cardiac images. *IEEE Transactions on Medical Imaging*, 13(2), June 1994.

Application of Model Based Image Interpretation Methods to Diabetic Neuropathy

M.J. Byrne and J. Graham
Dept. of Medical Biophysics,
University of Manchester, Oxford Road,
Manchester, M13 9PT, UK.
Email: mjb@sv1.smb.man.ac.uk

Abstract

We present two applications of model based computer vision methods to measurement of image features significant in the diagnosis of diabetic neuropathy. The first involves the location of the boundaries of nerve fascicles in light microscope images. The second involves the segmentation of capillary cell regions using electron microscope images. In each case the boundaries required are of arbitrary shape and characterised by local texture or changes in textured regions.

The fascicular boundary is located using an Active Contour Model responding to a texture measure based on edge directionality. A start position for the model is automatically generated. The capillary segmentation is performed using a region–based snake responding to a weighted combination of texture measures followed by a local boundary refinement using dynamic programming. These methods show that application of various types of Active Contour Model, accompanied by appropriate starting cues, or followed by local refinements, can locate robustly positioned and intuitively correct boundaries in these images. The aim of the work is the automation of diagnostic measurements currently performed manually. We discuss the implications of automated analysis for procedures in quantitative histology.

1. INTRODUCTION

There are various symptoms and side–effects of the disease diabetes. Important among these are the effects on the nervous system. The most frequent pattern of involvement of diabetes on the nervous system is a peripheral symmetric neuropathy (non–traumatic disorder of the peripheral nerves) of the lower extremities affecting both motor and sensory functions[14].

The major effect is degeneration of the insulating myelin sheath that surrounds each nerve fibre leading to deterioration in motor and sensory functions. Biopsies are taken from patients and images obtained using light and electron microscopy. Number densities and size distributions of myelineated nerve fibres are obtained from these images. Currently these fibre measurements are made manually[2] which is a time consuming process. We have developed automated methods for nerve fibre detection in both light and electron microscope images. We do not present these methods here, although we make use of the results in section 3.3.

Nerve fibres are grouped together to form fascicles (fig.1).

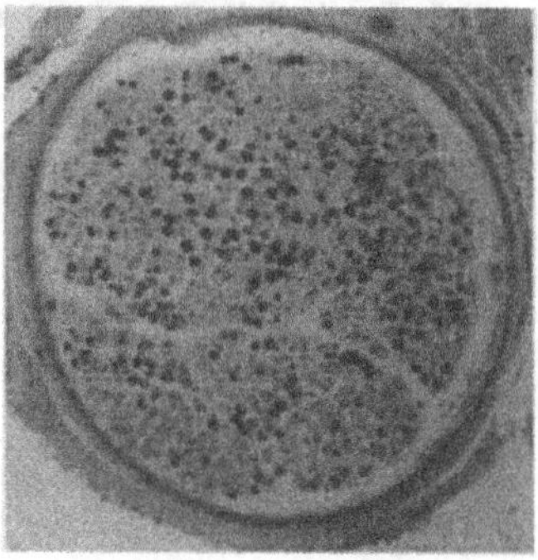

Fig.1. Example of nerve fibre fascicle

It is also necessary to locate the fascicle boundary so that an accurate measure of the fascicular area, and hence accurate fibre number densities can be obtained. In section 3 we describe a method for automatically locating the fascicular boundary using an Active Contour Model responding to an appropriate image measure with a starting cue generated as a result of the automated fibre detection.

Another symptom of interest is microangiopathy (disease of small blood vessels) with effects manifested within the nerve fibres themselves and in capillaries in the endoneurium (the interstitial connective tissue in peripheral nerves separating individual nerve fibres). The two main effects are;

i) A thickening of the basement membrane or an accumulation of basement membrane material in the capillaries (visible as an apparent contraction of the luminal area), and
ii) A proliferation of endothelial cell material together with basement membrane thickening. This manifests itself in arteries, arterioles and occasionally venules.

The structure of two endoenuerial capillaries as they appear in electron micrographs is shown in fig.2. Fig.2a is a healthy example whilst fig.2b displays a degree of diabetic neuropathy and shows these effects.

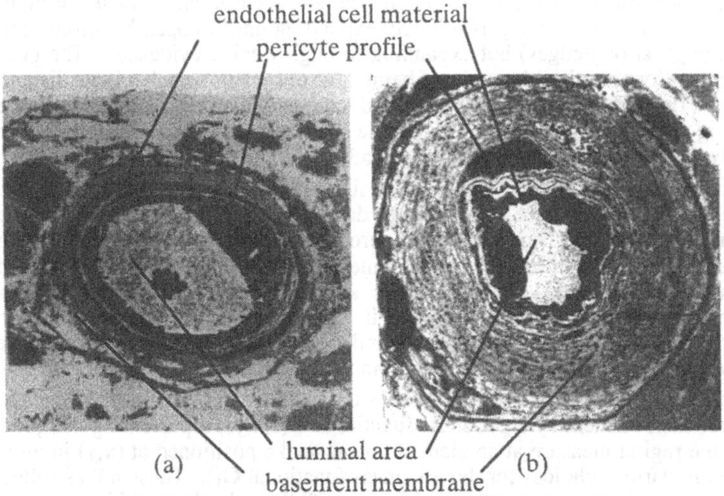

Fig.2. Electron Microscope images of endoneurail capillaries (a) normal and (b) showing neuropathy

Currently the various regions are delineated by hand[13,16]. In section 4 we will describe methods of automated segmentation of the three capillary regions using Active Contour Models followed by local boundary refinement using dynamic programming.

2. MODEL BASED METHODS

Both the fascicle and capillary images are complex. The image evidence defining the required regions involves several parameters which vary from image to image. Often the evidence is poor or missing requiring local interpolation of the data. Location and segmentation of the desired regions in both sets of images therefore require the use of some form of model based method. Statistically based methods such as Point Distribution Models (PDMs) have been sucessfully applied, as part of a constrained image search strategy[5] (Active Shape Models), to the location of poorly defined boundaries in noisy images[6]. PDMs rely on the ability to label a consistent set of landmark points representing the boundary shape in a set of training images in order to construct a statistical model of the expected boundary shape and its allowed degree of variation. In both problems described in this paper it is not possible to identify a set of landmark points that are consistent from image to image. The lack of correspondence among training images introduces such a large degree of variability between different examples that statistics gathered on shape and appearance impose little constraint on the model. As a result, for the problems presented in this paper, Active Shape Models have not provided a useful approach to image segmentation.

2.1. SNAKES

Active Contour Models (snakes)[10] do not rely on a description of the expected region shape to constrain image search but instead impose internal constraints using a quasi–physical model to control the spacing and curvature of boundary elements. The snake combines these internal constraints with external "forces" derived from the image evidence and iteritively re–positions itself to achieve a minimum energy configuration.

In the absence of any external image forces a snake reaches equilibrium by collapsing either to a single point or line, as dictated by the internal constraints . Furthermore if the snake search is initiated too far away from the desired contour the snake will fail to converge to the correct solution. Cohen[4] added an extra inflationary term causing the snake to behave as a balloon. The inflationary term takes the form of an isotropic pressure potential resulting in an outward pressure force acting along the normal to each snake element. The balloon is inflated, expanding until trapped by strong image evidence (e.g. strong edges) but expanding through weaker evidence. The pressurised snake therefore has the advantage of being able to start its search a large distance away from the desired contour but has the disadvantage that an image feature must produce a strong response in order to overcome the snake's internal pressure force. Hence it is not always possible to segment an image adequately using a pressurised snake.

An adaptation proposed by Ivins and Porrill[9] is the statistical snake, an active region model linking the pressure term to image data within the region enclosed by the snake. An initial seed region is defined either through user interaction or through some form of cue generation. Within this region the means and variances for a suitable set of image measures are determined. These measures should be capable of distinguishing between the region of interest and those around it. The snake is allowed to expand from the boundary of the seed region until the boundary elements encounter pixels whose image measures differ significantly from those in the seed region.

An energy term for the region measure is obtained by multiplying the local change in area for each element by a goodness functional **G(I(x,y))** representing the goodness of fit of the region measure at an element of the snake positioned at **(x,y)** in an image **I**. There are various choices for the goodness functional **G()**. These are as follows for a region having a mean response μ and a range of allowed values within k standard deviations of μ.

Unary Pressure: The goodness functional **G()** is set to unity for pixels with image measures within the range specified by the seed region and zero for pixels outside this range.

Binary Pressure: The goodness functional **G()** is set to +1 for pixels with image measures within the range specified by the seed region and −1 for those with measures outside the range. When a snake element encounters pixels outside the seed region's range the direction of expansion at that element is reversed.

Linear Pressure. A normalised linear pressure term allows the model to reach equilibrium when its boundary elements encounter pixels at the statistical limits where the goodness functional and hence the pressure force evaluate to zero.

Several image features may be combined by use of a Mahalanobis pressure term.

3. FASCICULAR BOUNDARY LOCATION

To obtain images with sufficient resolution for fibre detection using light microscopy a magnification of 40 times is required. At this magnification several fields are required to represent an entire fascicle. As a result the fibre detection is carried out on a mosaic of images connected using cross–correlation. The composite images produced typically have dimensions of 1000–2000 x 1000–2000 pixels.

Application of active contour models to boundary detection requires the choice of a suitable image force and a method of generating a suitable starting position for the model.

3.1 CHOICE OF IMAGE FORCE

The choice of image force is determined by the fascicular boundary structure which consists of a series of closely spaced, fairly parallel lines (fig.3a,b).

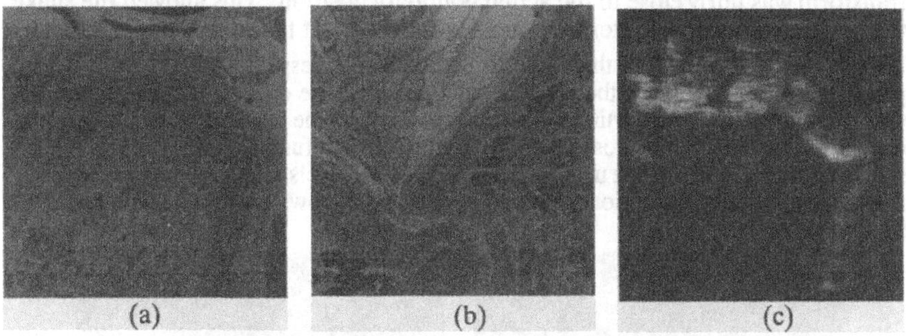

Fig. 3. Examples of fascicle boundary structure(a,b) and result of direction algorithm (c)

However the contrast between these lines and the background is often poor and simple measures based on edge magnitude or contrast fail. To achieve robust detection we have used a texture measure based on edge frequency and edge directionality.

Image Force Algorithm: The algorithm implemented makes uses of the response of a Canny edge detector[3].

The Canny response along the fascicular boundary to consists of a number of parallel edges. Generally the response over the remainder of the image is low, except around nerve fibres, where it shows very little local directionality.

The image feature used to generate the snake's image force is the modal value of the direction of the Canny output within a local neighbourhood. The following algorithm generates an intrinsic image based on this feature.

1. Apply Canny Operator
2. Threshold Canny output to retain only salient edges
3. Quantise edge directions to 16 values
4. Calculate modal direction value within local neighbourhood
5. Retain number of responses at modal direction as pixel output

A local neighbourhood half–width of 10 pixels was empirically found to give the most robust boundary response. An example of the algorithm's output is given in fig.3c.

3.2. GENERATION OF A START POSITION

A starting cue for the snake can be obtained by making use of myelineated nerve fibres detected by our automated method. The fascicle boundary lies in a region surrounding that containing the nerve fibres. In most cases the boundary is not a great distance from the fibre region. The limit of the fibre region is calculated using the distance transform[8] of the image containing the detected nerve fibres.

The outer boundary of the fibre region is determined by thresholding the distance transform at a range of increasing values until a single isolated contour is obtained. A single contour is typically obtained at a distance just greater than the maximum distance between adjacent nerve fibres. This contour, after being smoothed using morphological closing, is used as the starting position of the snake. A degree of smoothing is required since the contour produced by the distance transform is extremely jagged.

3.3. RESULT OF FASCICLE SNAKE

A snake comprised of 61 elements, based on an algorithm by Williams and Shah[17] using the image force described in section.3.1, was applied to a series of sample images. In most examples the starting contour generated by the distance transform was fairly close to the actual boundary position. This allowed the snake to stabilise within 10−15 iterations achieving a close fit to the true boundary.

To assess the robustness of the boundary location with respect to the snake's starting point, an eroded version of the contour produced from the distance transform was obtained. This produced a starting point lying well within the fascicular boundary resulting in the snake having to cross regions of potentially confusing image evidence. Fig.4 shows an example of a snake using a cue eroded by 50 pixels. Fig.4a shows the detected fibres with the eroded position superimposed. Fig.4b shows the final position reached by the snake.

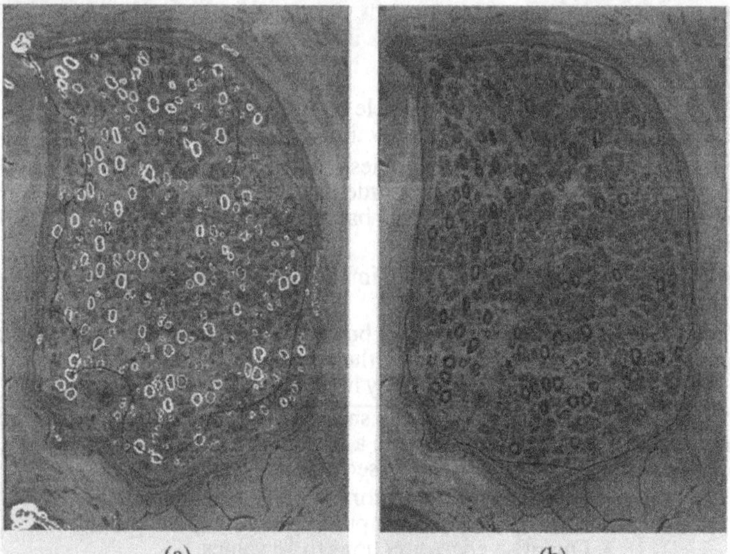

(a) (b)

Fig.4. Result of snake starting from eroded cue

While fig.4 demonstrates the robustness of the texture measure generating the image force, the greater search space and number of iterations required from the use of a distant starting point result in a substantial loss of efficiency. The advantage of having a starting point close to the true boundary position is in the generation of a rapid solution. In some cases exhibiting a high level of neuropathy the limit of the fibre region may be further away from the fascicle boundary requiring the use of an extended search space. The distance between the fascicle boundary and the limit of the fibre region may be a useful diagnostic measure in such cases.

4. SEGMENTATION OF CAPILLARY CELL REGIONS

The method used to segment the areas of interest is determined by the appearance of the three areas. The lumen is generally light in colour and is usually flat or shows only light texture. The endothelial cell material is dark in colour and shows a high degree of structure. The basement membrane area is lighter in appearance and generally shows less structure with greater directionality than the endothelial cell area. At a coarse scale, across examples, the three regions can be characterised by specific textures. At a finer scale the region boundaries are characterised by a great deal of detailed structure.

4.1. APPROACH

A two stage strategy has been implemented. An initial approximation to the boundary is obtained by application of a region–based (statistical) snake starting from a user supplied seed region. The use of a region–based snake is appropriate since it makes use of the consistent texture within a region to locate the boundary. The only restriction on the snake's starting point is that it lies inside the region to be segmented. The approximate boundary obtained in this way is refined using a higher resolution search method based on dynamic programming which takes advantage of the approximation produced by the region–based snake. The dynamic programming method searches pixel by pixel near the approximate boundary using a measure based on the texture contrast between regions. This measure is capable of greater sensitivity to texture changes when applied close to the true boundary.

4.2. CHOICE OF IMAGE FORCE FOR REGION BASED SNAKE

An image measure is required that can distinguish between the various regions of interest in the capillary. A selection of texture measures was used with discriminant analysis applied to produce a weighted combination capable of producing good classification between either lumen and endothelial cell material or endothelial cell material and basement membrane area. The discriminant analysis is carried out separately for the two boundaries of interest producing a separate set of weights for each. The image measures used were:

- Local average luminance: the average grey level within a local neighbourhood.
- Gradient: [15] a measure of gradient as a function of distance between pixels using the distance dependent texture description function g(d) computed for a user specified distance d.
- Smoothness: A measure of the number of pixels within a local neighbourhood that lie within a specified grey level range of the central pixel value.
- Entropy: to distinguish between regions with little or no texture and regions with some degree of semi–random structure.
- Laws Texture Filters: Six combinations of Laws [11] texture filters were used. These are a well known set of 1D filters which can be combined to represent 2D texture primitives.

These measures were performed on the various regions of interest for a series of training images. A discriminant analysis was carried out on the measures obtained to produce a classification between the lumen and endothelial cell area and then the basement membrane and endothelial cell area. This produced a weighted combination of the region measures. A binary goodness functional based on the results of the discriminant analysis was used as the image force for the region based snake.

4.3. REFINEMENT USING DYNAMIC PROGRAMMING

Dynamic programming as a search technique has been applied to a variety of problems in machine vision[1]. An advantage of dynamic programming is that is always guaranteed to find the optimal path for a given objective function. It also compares well to other techniques such as heuristic search algorithms which depend critically upon the quality of the forward cost estimate. Its advantage as a refinement method is that the cost function is based on local measures, in contrast to the global energy function of the snake methods. Lutkin[12] has shown that dynamic programming can be an effective method of assessing local image evidence based on an existing model boundary.

In order to refine the result of the region based snake, dynamic programming was applied to a "straightened" image constructed from single pixel spaced normals to the snake's boundary approximation. Each pixel on the straightened boundary corresponds to a node in the search graph[12]. The search then proceeds through the graph finding the best route as dictated by the cost function:

$$cost_i = \alpha\ bound_i + \beta\,(1 - (\theta_i - \theta_{i+1}))$$

where $bound_i$ is the normalised boundary response at node i and θ_i and θ_{i+1} are the angles at nodes i and i+1, and α and β are weighting constants. The cost function is designed to respond to the image evidence whilst maintaining a degree of compatibility with the initial boundary approximation. The compatibility constraint at a transition between nodes is a measure of the angle between the path from one node to the next and the direction of the approximate boundary. Since the dynamic programming is applied to a "straightened" version of the region based boundary approximation the second term in the cost function constrains the refined solution to remain close to this initial approximation.

Choice of Image Measure for Dynamic Programming: The image measure used is the difference in texture between two circular regions centred along the normal to the estimated boundary position. As in section 4.2, the texture measure is a weighted combination of the region measures, the weights being determined by discriminant analysis. At a true boundary point this difference between region responses should be maximised.

4.4. RESULTS

For the segmentation of the lumen from the endothelial cell area the starting point for the snake was within the lumen. For the boundary between the basement membrane and the endothelial cell area the region based snake was positioned in basement membrane and allowed to contract inwards towards the boundary. For these experiments the endothelial cell area was not used as a starting point since its structure is less consistent than that of the other two regions.

Results for the location of both boundaries are shown below. Fig.5 shows the region based snake applied to segmentation of the lumen endothelial cell area boundary. This boundary is very distinct and the region–based snake has produced a good approximation to the actual boundary position. Fig.6 shows an enlarged section of the capillary shown in fig.5 showing application of the boundary based refinement to the result of the region based snake.

Fig.7 shows the region based snake applied to locating the boundary between the basement membrane and the endothelial cell area. This boundary is less consistent than than the lumen endothelial boundary. In most places the boundary is distinct but in other places the image evidence is poor with the boundary appearing non–existent in some places. Fig.8 shows an enlarged version of fig.7 showing the detailed improvement achieved by the dynamic programming refinement.

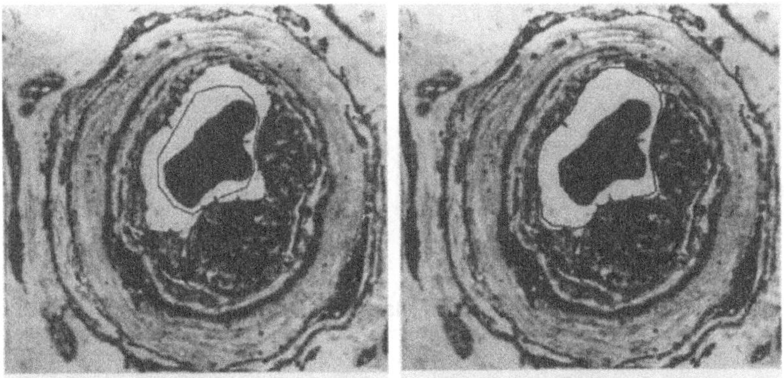

(a) (b)

Fig.5. Segmentation of lumen/endothelial cell area boundary
a)Starting position b)Result of region based snake

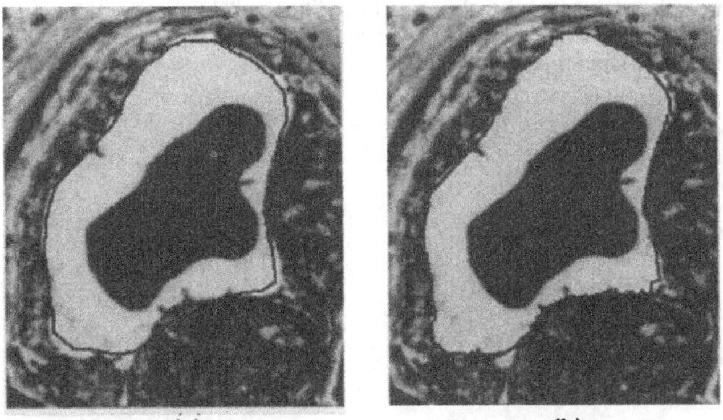

(a) (b)

Fig.6. Comparison of a)region–based snake result and
b)refinement due to dynamic programming

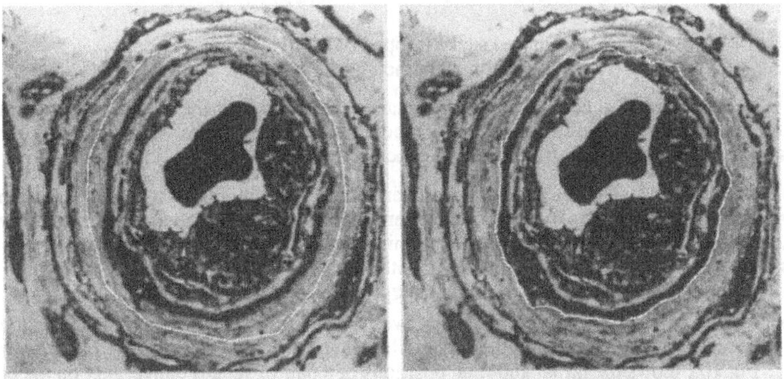

(a) (b)

Fig.7. Application of region based snake to location of basement membrane/endo-
boundary. (a) starting point (b) result of region based snake.

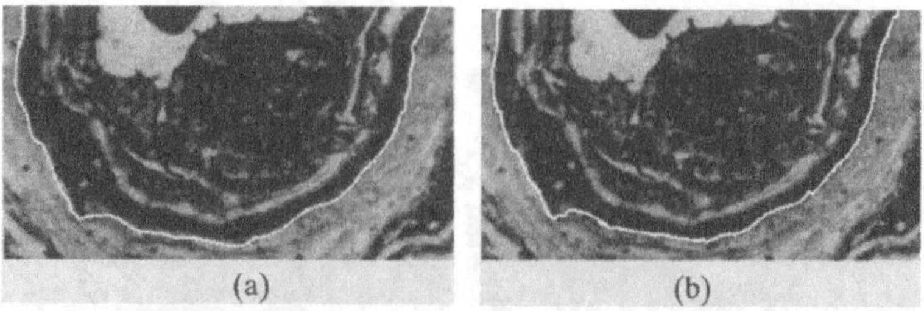

Fig.8. Comparison of a)region–based snake and b)dynamic programming results.

4.5. APPLICATION OF PAIRED SNAKES

A problem encountered when segmenting the endothelial cell area from the basement membrane is the presence of small regions within the basement membrane showing similar texture to that of the endothelial cell area. Although they are small in comparison to the main body of the endothelial cell area, the region based snake can still be "trapped" by these regions (fig.9a). A snake placed within the endothelial cell area can encounter similar problems due to the inconsistent structure of this area (fig.9b).

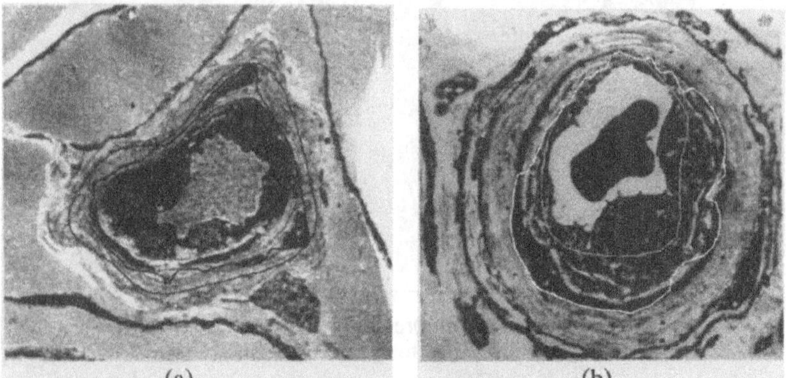

Fig.9. Region based snake trapped by confusing local evidence
a)Starting from basement membrane b)Starting from endothelial cell area

This is an example of a general problem with snakes and arises from the fact that the only internal constraints on the snake are associated with smoothness. There is no way of preventing a smooth yet incorrect solution arising from confusing local evidence.

Problems with confusing evidence can be overcome by combining two or more independent assessments of the available evidence. We attempt to obtain two independent views of the evidence through the use of a pair of snakes running simultaneously from differing starting positions. In the absence of conflicting evidence both snakes would be expected to arrive at the same answer. Points where there is disagreement suggest the need for further analysis.

The snakes of section 4.4 were augmented by two further snakes initialised within the endothelial cell area, one contracting towards the lumen, the other expanding towards the basement membrane. Fig.10a shows the initial and final positions of a pair of snakes converging on the boundary between the basement membrane and the endothelial cell area. In many places the snakes arrive at an identical position. As in the example shown in fig.9a the outer snake has been trapped by the outlying regions around the endothelial area. However in these places the inner snake has generally arrived at a satisfactory solution

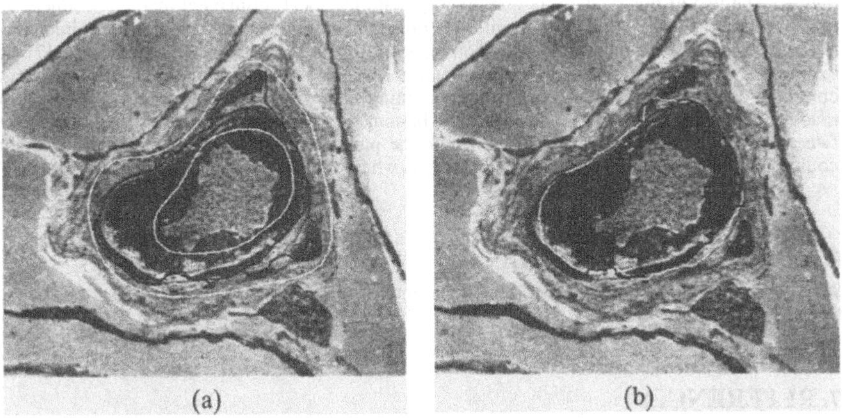

Fig.10. a)Start(white) and end(black) positions of two region based snakes
b) Average snake position(white) and result(black) of application of dynamic programming

Our initial approach to resolving the conflicting evidence has been to use the average position of the two snakes as the model boundary for dynamic programming refinement. Fig.10b shows that the result of the refinement to be an acceptable representation of the region boundary.

5. CONCLUSIONS AND DISCUSSION

The overall aim of the work presented is the development on an automated system for measurement of diabetic neuropathy encompassing image capture and automated nerve fibre detection as well as the two applications discussed in this paper. The purpose of the fibre detection and fascicular boundary measurements is perform a study of the effects of diabetes on the number densities and size distributions of myelineated nerve fibres. This requires measurement to be made on samples from a large number of patients. Thus a suitably efficient and reliable automated system to replace the need for manual measurements is extremely desirable. Details of the diagnostic utility of the methods will be published elsewhere. The purpose of this paper is to describe the computer vision methods applied.

Fibre detection and fascicular boundary location is a fully automated process. Snake based methods have been shown to successfully locate the fascicular boundary using a starting cue generated from the results of automated fibre detection. The accuracy of boundary location is not sensitive to the effectiveness of this cue, but the fact that the starting point generated is generally close to the final position has a beneficial effect on the method's efficiency.

The capillary segmentation is intended to be as nearly automated as possible and is intended to replace manual delineation of region boundaries. The results shown in this paper have have been based on manually positioned starting points. It may be possible to generate image–based cues for this application as in the case of the fascicular boundary location. The generation of such cues has not been investigated as yet but a possible candidate might use the (usually) uniformly light luminal area.

Technically the achievement of this work has been the segmentation of structurally complex images. A principled approach to characterising texture boundaries has been taken based on trainable image features. These have been shown to be robust when used in conjunction with appropriate forms of Active Contour Model. Two problems which arise from from the use of snakes have proved particularly relevant to this work. Firstly the use of a global energy function means that they do not respond readily to detailed local boundary structure. Secondly the reliance on smoothness as the constraint on snake shape leads to a lack of shape specificity. In our case this means that confusing image evidence can lead to an incorrect solution. The first problem has been addressed through the use of an additional boundary refinement phase which takes locally detailed structure into account by using dynamic programming. The second problem has been addressed through the use of paired snakes to obtain two different views of the image evidence. The use of this strategy in combination with the local refinement method produces adequate results. Further experiments will determine whether greater robustness is required. This could be achieved by more rigourous combination of the evidence. Gunn and Nixon[7], for example, have adopted an approach using paired snakes coupled together to encourage them to converge. Alternatively the dynamic programming refinement could weight the contribution made to its cost function by the approximate boundary according to the level of agreement between the paired snakes. At positions where the paired snakes disagree the local refinement could be allowed more freedom than at positions where the snakes are in agreement.

6. ACKNOWLEDGEMENTS

We would like to thank Dr. R.A. Malik of the Manchester Royal Infirmary for his assistance, and for providing the images used in this work.

7. REFERENCES

1. **A.A.Amir,** Using Dynamic Programming for Solving Variational Problems in Vision,*IEEE Trans. PAMI, 12(9),* 1990, pp.855–867.
2. **S.T. Britland, R.J. Young, A.K. Sharma, B.F. Clarke,** Acute and Remitting Painful Diabetic Neuropathy: A Comparison of Peripheral Nerve Fibre Neuropathy, *Pain, 48,* 1992, pp.316–370.
3. **J. Canny,** A Computational Approach to Edge Detection, *IEEE Trans. PAMI, 8(6),* 1986, pp679–698
4. **L.D. Cohen,** On Active Contours and Balloons, *CVGIP, Image Understanding, 53(2),* 1991, pp21–218
5. **T.F. Cootes, C.J. Taylor, D.H. Cooper, J. Graham,** Active Shape Models: Their Training and Application. *Computer Vision and Image Understanding 61,* 1995, pp.38–59.
6. **T.F. Cootes, A. Hill, C.J. Taylor, J. Haslam.** Use of Active Shape Models for Locating Structure in Medical Images, *Proc. IPMI (13),* 1993, pp33–47.
7. **S.R. Gunn, M.S. Nixon,** A Model–Based Dual Active Contour, *Proc. BMVC 94,* BMVC Press 1994, pp305–314.
8. **RM. Haralick, L.G. Shapiro.** Computer and Robot Vision Vol.1 *Addison–Wesley* 1992 pp221–223.
9. **J. Ivins, J. Porrill,** Statistical Snakes: Active Region Models, *Proc. BMVC 94,* BMVC Press 1994, pp377–386.
10. **M. Kass, A. Witkin, D. Terzopoulos.** Snakes: Active Contour Models. *Proc. 1st Intl. Conf. Computer Vision.* 1987, pp259–266.
11. **K.I. Laws.** Rapid Texture Identification. *SPIE Conf. on Image Processing for Missile Guidance.* vol.238 1980, pp376–380.
12. **J.P. Lutkin.** Interactive Segmentation of Medical Images. *Msc Thesis Manchester University,* 1994.
13. **R.A. Malik, S. Tesfaye, S.D. Thompson, A. Veves, A.K. Sharma, A.J.M. Boulton, J.D. Ward.** Microangiopathy in Human Diabetic Neuropathy: Relationship Between Capillary Abnormalities and the Severity of Neuropathy. *Diabetologia, 30,* 1989, pp.92–102.
14. **R.A. Malik, P.G. Newrick, A.K. Sharma, A. Jennings, A.K. Ah–See, A.J.M. Boulton, J.D. Ward.** Endoneurial Localisation of Microvascular Damage in Human Diabetic Neuropathy, *Diabetologia 36,* 1993 pp454–459.
15. **R. Sutton, E. Hall.** Texture Measures for Automatic Segmentation of Pulmonary Diseases, *IEEE Trans. Comp. c–21.* 1972, pp667–678.
16. **R.G. Tilton, P.L. Hoffman, C. Kilo, J.R. Williamson.** Pericyte Degeneration and Basement Membrane Thickening in Skeletal Capillaries of Human Diabetes, *Diabetes 30,* 1981, pp.326–334.
17. **D.J. Williams, M. Shah.** A Fast Algorithm for Active Contours and Curvature Estimation. *CVGIP: Image Undrstanding vol.55(1)* 1992, pp14–26.

Global Alignment of MR Images Using a Scale Based Hierarchical Model

S. Fletcher, A. Bulpitt & D. Hogg

School of Computer Studies
University of Leeds, Leeds LS2 9JT, UK

Abstract. This paper proposes a novel automated method for global alignment of three dimensional MR images. The matching algorithm employed is closely related to a common constraint based tree searching algorithm [1], but uses a novel multi-resolution encoding of the search space to improve the search time and permit searching of curved surfaces. The algorithm uses the shape index defined by Koenderink [2] which provides the very useful property of invariance to uniform scale. The surfaces of the objects are extracted from the MR images automatically using a 3D deformable model [3]. An intelligent mechanism is used for selecting unusual surface features that are common to both objects.

1 Overview

The global alignment of three dimensional objects is used widely in medical imaging to perform comparative and composite analysis of medical images. Global alignment can also be used to match images to labelled reference images (anatomical maps) in order to label each region of the image resulting in the segmentation of the image [4].

This paper presents a novel automated approach to the global matching stage of a system used to match 3D MR images of the human head. The system uses the surface characteristics of the objects in the images during the matching process to avoid the need for external markers placed on or about the patient which in general require special scanning protocols or the interaction of an expert user [5]. Our approach uses the shape index defined by Koenderink [2] which has the very useful property of invariance to uniform scale. A 3D deformable model developed by Bulpitt & Efford [3] is used to extract the surfaces to be aligned from the MR images. Such models provide good segmentation tools for MR images as they provide some measure of immunity to image noise and missing data. The surface characteristics used in the matching algorithm are easily computed from the model and have been chosen to provide invariance to both scale and rotation.

The matching algorithm employed is a constraint based matching algorithm based on an interpretation tree search [1]. The algorithm uses a depth first search to match the image data to a hierarchal model of the object [6]. The use of a hierarchical model structures the search space and greatly reduces the computational cost of the matching process [7].

The following section describes the surface characteristics used by the matching process and illustrates their use as constraints in the matching process. The details of the matching procedure and hierarchical model are given in Section 3. Section 4 presents the main results of tests on the performance the system. These results are then summarised in Section 5.

2 Surface Information

In order to produce an effective matching algorithm the surface properties used for the matching process must be invariant to rotation, translation and scale of the objects to be matched. Koenderink [2] defines two surface properties which may be used to describe the shape of a surface, the shape index and curvedness values. Both of these properties define good discriminating measures for object matching (see Section 2.1).

The method used to compute the surface properties is based on the approach of Fua and Sander [8]. The triangular mesh produced by the deformable model allows the computation of the surface properties to be more efficient than methods which use the voxel data directly.

2.1 Surface Properties as Constraints

The matching algorithm described in the next section will find plausible transformations for two data sets by matching surface points, however, the search time of the algorithm can be greatly reduced by selecting points that are uncommon in each image but are present in both. In our application we chose points that have an unusual shape index or a particularly high curvature. Figure 1 illustrates the distribution of surface characteristics taken from six head data sets. The histograms show how each head has a similar distribution of shape index and

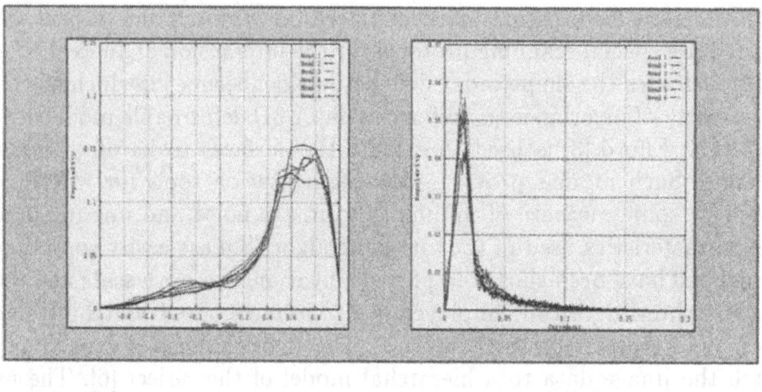

Fig. 1. Distributions of the surface point characteristics.

curvedness values. The majority of the points on surface are shown to be convex ($S \sim 0.7$) and have a low curvedness value (~ 0.02). This is as expected, as these points correspond to the back and the top of the head. Figures 2 and 3 show how the shape index and curvedness values are distributed and structured over the head. For many areas, particularly in the region of the face, similar contours can be found for each individual. The figures show how areas of high curvature on the head are typically associated with regions of rapid shape change. These

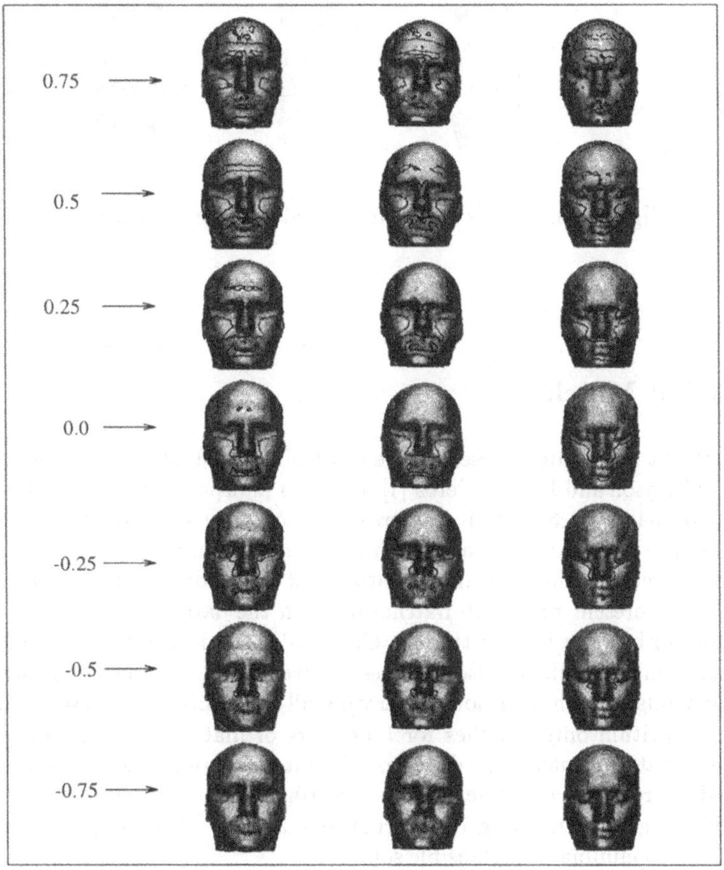

Fig. 2. Shape index contours plotted onto head surfaces.

observations provide the basis for selecting points for matching. Firstly points of high curvedness should be used. These points indicate strong surface features such as the eyes and nose and there are also fewer of these points on the surface of the head, (Figure 1). This means that using a value of high curvedness will greatly reduce the search time of the matching algorithm. The value of shape index should then be chosen to correspond to points of high curvature. Figure 2 shows how shape index values at or around 0.0 are found within these regions.

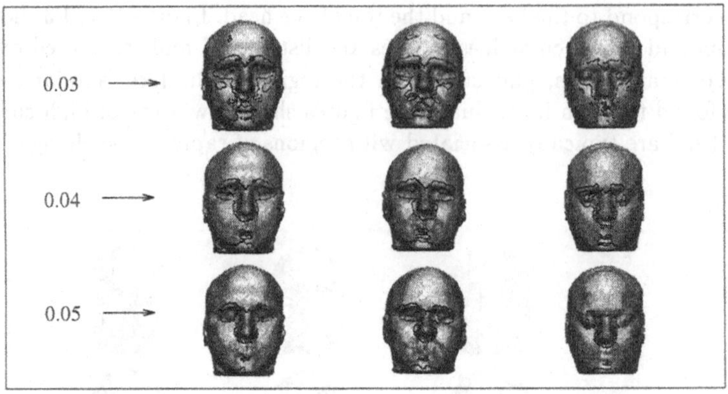

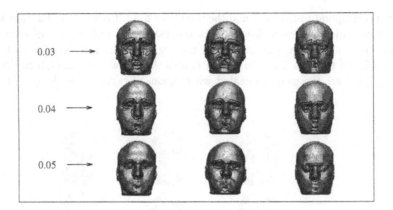

Fig. 3. Curvedness contours plotted onto head surfaces.

3 Global Matching

The matching algorithm is based on the constraint based search technique developed by Grimson and Lozano-Perez [1], using an interpretation tree to structure the search space. Each node in the tree represents a partial match between a model feature and a data feature. Each node branches into more nodes, one for each match between the next model feature and all the data features. The leaves on the tree represent potential matches between the data and the model, many of which will be false interpretations. Grimson's technique eliminates the need to evaluate many of these false matches by *pruning* the tree using constraints, to identify false matches without generating all the pairings associated with it.

Our algorithm only searches for four pairs of matched model and surface points so that the maximum depth of the interpretation tree is always four. Although more points could be used, four pairings lead to a single linear problem for estimating the similarity transformations and still allows effective use of constraints to eliminate implausible solutions.

3.1 Hierarchical Model

To cope with curved surfaces and to eliminate unproductive parts of the search space, we employ a *hierarchical* model. The top level of the model contains a very coarse description of the overall model surface. Lower levels contain increasingly more detailed descriptions of regions of the model surface. At the bottom level, the surface is described in full, as a set of surface points. This produces a multi-resolution description that is particularly suited to curved surfaces.

The hierarchical model is constructed as follows:

1. The model surface is processed to yield a collection of surface elements. Each element has associated with it: its location in 3D space, the normal of the model surface at this point, the shape index and curvedness of the surface at this point. This provides a detailed description of the model surface.
2. A triangular approximation to a sphere of unit radius is placed at the centre of the model. The first level approximation to the sphere is an octahedron which consists of 8 triangles. Successive levels of approximation to the sphere are generated using a recursive subdivision technique that splits each of the existing triangles into four new triangles. Once the new vertices are generated, they are pushed out to the surface of the sphere. Assuming a sphere of radius 1 centred at the origin, this is achieved by normalising the vertex coordinates. For the model used here, the subdivision is performed 5 times generating 8192 triangles at the highest level.
3. A vector is created from the centre of the sphere to each model surface point. This vector intersects with one of the triangles which approximate the sphere. For each triangle, the average location of the model points whose vector passes through the triangle, along with the average of each property for these points (i.e. shape index, curvedness and normal direction) are stored to form the bottom level of the model hierarchy.
4. Because the sphere approximation is generated using a recursive subdivision technique, each triangle is related to three others by a single triangle in the previous level of recursion. The next level of the model hierarchy is therefore created by averaging information contained in the bottom level triangles. This produces a set of model parts representing larger areas of the original model surface. There are four times fewer model parts at this level than at the bottom level.
5. The averaging procedure is repeated for all the remaining model levels. Each model parts is created by averaging the information contained in four model parts from the previous level. This continues until the second level is reached, which contains eight model parts. To generate the top level model part, all eight of the second level model parts are averaged. This is equivalent to averaging all the original model surface point information and therefore represents a very coarse and concise description of the complete surface.

3.2 Constraints

Given any four data points, an exhaustive search over all the bottom level leaf model parts to find a set of compatible points is computationally too expensive. In order to reduce this search a constraint based matching technique over the interpretation tree is used. Implausible nodes are pruned from the tree if the partial or full match described by the node does not satisfy the following constraints:

Order: The order of points must be the same between the model parts and the data points, otherwise symmetrical objects may generate a match that is a flipped version of the correct match. The order of points is established by constructing a plane between three of the four points and testing on which side of the plane the forth point lies.

Distance: The distance constraint measures the distance between equivalent pairs of points from the surfaces. For rigid transformations this can be used to ensure that the distances are equal. Otherwise, the constraint can be used to limit variations in the scale of the surfaces to be matched.

Normal Direction: The surface normal is invariant to translation and uniform scale of the data set, but is not invariant to rotation. The normal constraint therefore uses the angle between the normals of two points on a surface rather than their absolute values. Due to local variations in the shape of two individuals heads this angle may vary slightly. A model parameter, the *binary constraint parameter* is therefore used to allow a small degree of difference. If it is known that the two surfaces are orientated in the same manner, the normal direction for a point on one surface should be similar to the normal direction of an equivalent point on the second surface. In this case the normal directions may also be constrained using the *unary constraint parameter.*

Curvedness and Shape Index: Both the curvedness and shape index values are invariant to the rotation and translation of the surface. Therefore for rigid matching both the shape index and the curvedness can be used as unary constraints. The curvedness value however is not invariant to scale and can only be used if bounds are known on the allowable scale change.

4 Results

The system has been tested on six data sets of "normal" healthy adults. The tests performed investigate the effect of different parameters of the matching algorithm on both its efficiency and on the quality of the matches produced. The efficiency of the algorithm is compared with that of Grimson and Lozano-Perez [1] –hereafter referred to as the GLP algorithm. The results presented here provide an overview of these tests. The full results and details of these tests can be found in [6]. The efficiency of the matching algorithm is measured in terms of the number of interpretation tree nodes visited and rejected during a single search. This measure is directly related to the constraints used to prune the search tree. When the constraints are very tight the number of false interpretations will be small and consequently the search will be very efficient. When the constraints are relaxed, the converse is true since more false nodes need to be examined.

The quality of the match produced is measured as the average distance between the two surfaces and is found be to dependent on the number of sets of points used. If only one set of points is chosen, this set could be unique to one data set as a result of noise and may therefore not be able to produce a good match. By running the algorithm multiple times the probability of the results being affected by noise is reduced and confidence in the final match is increased. Typically more than 10 sets of points did not produce a significant improvement in the quality of the match produced.

4.1 Constraints

The quality and efficiency measures described above are ultimately affected by the algorithm's parameters which control the effectiveness of the constraints. Tests have therefore been performed on the algorithm to show the effect of each of the constraints on the algorithm's performance.

There are six constraint parameters used in the matching algorithm. For each test five of the parameters are held constant whilst the other is varied within a set range. The parameters values used are as follows:

Parameter	Fixed	Range
shape index	0.0	-1.0 to 1.0
change in shape index	0.0	0.0 to 0.5
curvedness	0.05	0.0 to 0.1
unary angle	$\frac{\pi}{5}$	0.0 to π
binary angle	$\frac{\pi}{5}$	0.0 to π
scale difference	0.2	0.0 to 0.6

Each experiment uses an N-1 testing method, where each model is matched with the five remaining data sets and the information gathered is then averaged. The information logged for each run includes: the final cost of the fit (which is the average distance between the two surfaces); the total number of nodes visited; the number of nodes rejected by each constraint; and the number of runs which failed to produce a valid transformation.

Shape index: The results produced for points chosen with different values of shape index are shown in Figure 4. Graph A shows how both algorithms are more efficient for high and low values of shape index, although, our algorithm searches less nodes than the GLP algorithm, especially near the mid-range shape index values. Graph C indicates that for these values, the shape information is unreliable, causing the number of failed matches to increase. This suggests that a shape index value closer to zero should be chosen. Graph B shows how the quality of the matches produced is stable for values of shape index greater than -0.2. Curvedness: Figure 5 shows the results for points selected with different values of curvedness. Graph A shows how the number of nodes searched rapidly decreases as the curvedness increases. This is as expected as there are fewer points of high curvedness on the surface of the head (Figure 3). The GLP algorithm is again found to search more nodes than the multi-resolution method. Graph C shows the high values of curvedness also produce more failed matches as these few points do not always correspond to similar points on different individuals. Graph B shows the curvedness value has no significant effect on the quality of match produced.

For the remaining constraints: Change in shape index, Unary angle, Binary angle and Scale difference, the performance of the matching algorithm was as expected. The efficiency of the algorithm was highest when then constraints were very tight and decreased rapidly as the constraints were weakened. Each constraint was found to have little effect on the quality of the match produced and as expected, failed matches are more likely to occur for the tightest constraint values.

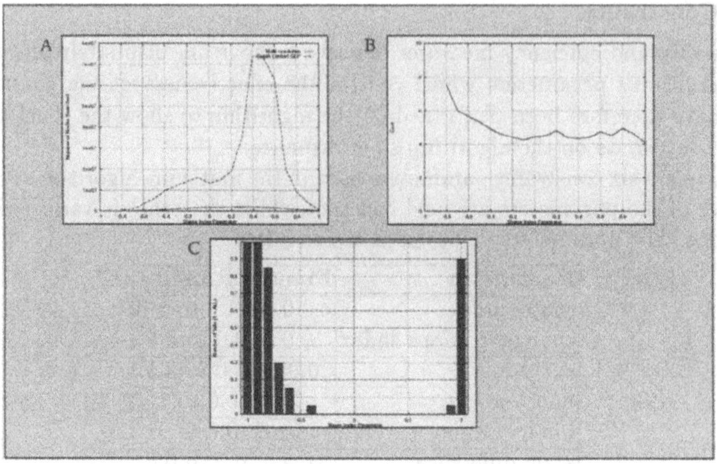

Fig. 4. Shape index parameter information. (A) The number of nodes searched. (B) The final cost (C) The number of failed matches.

4.2 Constraint Effectiveness

The matching algorithm described in Section 3 operates at a number of layers – one for each layer of the model hierarchy. At each layer a number of interpretation trees are created and searched. Two factors govern the total number of nodes searched in all of the interpretation trees of a given layer. The first is the number of leaf nodes in the previous layer. The second is the ability of the constraints to prune nodes from each tree in the current layer. To examine the effectiveness of each constraint at each layer of the search, the algorithm was executed for each model/data set combination using the fixed parameters and the number of nodes rejected by each constraint at each layer of the search was accumulated. This information was then averaged to provide the typical number of nodes rejected by the constraints. Figure 6 illustrates the results of this test. Each graph contains information about the number of nodes rejected for a given constraint. The constraints are numbered 1..5 in the order they are applied as follows:

1. Shape index and curvedness constraint.
2. Unary angle constraint.
3. Binary angle constraint.
4. Point order (orientation) constraint.
5. Distance constraint.

Except for the first layer the most effective constraint is the first one. The reason that this constraint is not so effective in the first layer is due to the coarse description of each model part. The unary angle constraint is the second most effective and it is not surprising that when this constraint is weakened the number of nodes searched increases rapidly. The point order and distance constraint are

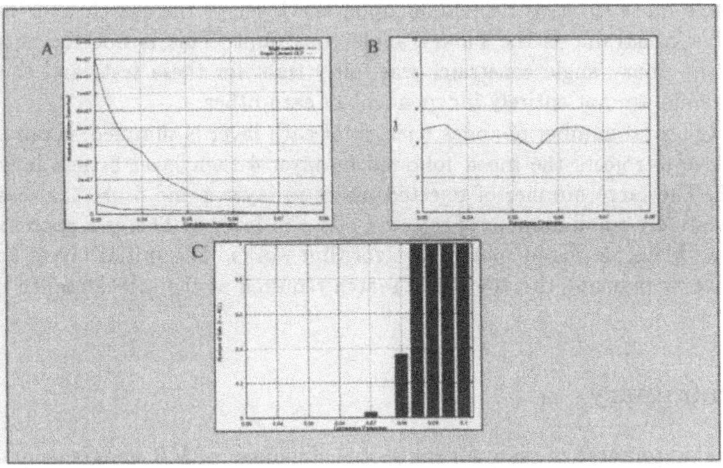

Fig. 5. Curvedness parameter information. (A) The number of nodes searched. (B) The final cost (C) The number of failed matches.

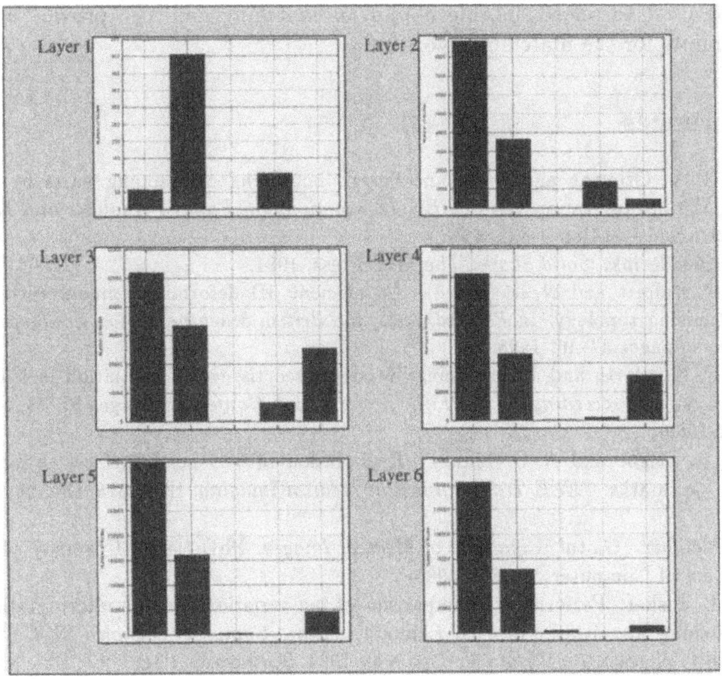

Fig. 6. The effectiveness of each constraint.

the next most effective depending upon the layer of the search and the least effective constraint is the binary angle constraint. This is not too surprising since the unary angle constraint was fairly tight for these tests and these two constraints are not entirely independent of each other.

If the total number of nodes rejected at each layer is summed, it can be seen that level 5 rejects the most, followed by layer 4 then layer 6 then layers 3, 2 and 1. The large number of rejected nodes in layer 4 and 5 are the reason for the relatively small number of rejected nodes in layer 6. This is a good example of how this scale based matching algorithm works. The initial layers are very effective at reducing the amount of search required at the most detailed bottom layer.

5 Summary

A novel automated method for the global alignment of MR images using a scale based hierarchical model has been presented. The paper shows how this multi-resolution approach structures the search space such that large proportions of the search tree can be removed during the initial stages of the search, thereby greatly reducing the overall search time.

The results show how the shape index and curvedness values of the surface can be used to select suitable points for matching and also provide effective constraints for the matching algorithm.

References

1. W. E. L. Grimson and T. Lozano-Perez. Localizing overlapping parts by searching the interpretation tree. *IEEE Transactions on Pattern Analysis and Machine Intelligence*, 9(4):469–482, 1987.
2. J. Koenderink. *Solid Shape*. The MIT Press, 1991.
3. A. J. Bulpitt and N. D. Efford. An efficient 3D deformable model with a self-optimising topology. In *Proceedings of the British Machine Vision Conference*, volume 1, pages 37–46, 1995.
4. P. H. Mowforth and J. Zhengping. Model based tissue differentiation in MR brain images. In *Proceedings of the Fifth Alvey Vision Conference*, pages 67–73, September 1989.
5. W. R. Fright and A. D. Linney. Registration of 3-D head surfaces using multiple landmarks. *IEEE Transactions on Medical Imaging*, 12(3):515–520, September 1993.
6. S. Fletcher. *Global Alignment of Medical Images*. PhD thesis, Unversity of Leeds, School of Computer Studies, 1995.
7. R. B. Fisher. Performance comparison of ten variations on the interpretation-tree matching algorithm. In J-O. Eklundh, editor, *Computer Vision - ECCV94*, volume 1, pages 507–512, Stockholm, May 1994. Springer Verlag.
8. P. Fua and P. Sander. Reconstructing surfaces from unstructured 3D points. In *Image Understanding Workshop*, San Diego, California, January 1992.

Reconstruction of Blood Vessel Networks from X-Ray Projections and a Vascular Catalogue

Peter Hall[1], Milton Ngan, and Peter Andreae
Department of Computer Science
PO Box 600
Victoria University of Wellington
Wellington
New Zealand
Email: {peter | milton | pondy}@comp.vuw.ac.nz

Abstract

Reconstruction of blood vessel networks from their x–ray projections is a challenging problem. This is because the correspondence problem must be solved for vessels which appear to twist, turn and overlap. Moreover, vessel networks vary in both branching structure and vessel shape. The extent of these variations is not catalogued in the clinical literature, or elsewhere. We have built a working system that accepts a few x–rays – separated by an angle of about ninety degrees – and reconstructs a three dimensional model of the vessel network. This task is impossible unless a priori information is used; how this information is represented is widely regarded as a key issue. Our representation makes a contribution by building and using a catalogue of anatomy that explicitly accounts for the wide variations in branching structure and shape. It is extensible in the sense that new information can be added to it at any time, and it is task independent in the sense that it can be used in many applications. We demonstrate its use in the problem of reconstructing vasculature from angiograms. Our reconstruction algorithm seeks to explain angiograms in terms of the vascular model. It can reconstruct vessel structures, even when vessels appear highly tangled, are missing, or extra vessels are present. The ability to recover complicated structure is the contribution made by our reconstruction method.

Keywords

Blood vessel networks, reconstruction, representation, x–rays

1 Introduction

Our task is to build a three–dimensional (3D) model of blood vessels in the brain (cerebral vasculature). An angiogram is an image of vasculature; biplane x–ray angiograms are perspective projections separated by an angle of about 90°. For brevity we refer to biplane x–ray angiograms as angiograms. The clinical motivation (see [6]) is to locate any arterio-venous malformations (AVM). We have previously presented an algorithm for automatic segmentation and 3D reconstruction of AVM's [4]. Here we focus on a method for reconstructing the structural elements of vasculature in 3D.

Angiograms separated by an angle of about 4° may be reconstructed using stereopsis. This is because well known techniques exist for solving the correspondence problem. We work with biplane angiograms for two reasons: (1) clinicians routinely acquire them; and (2) reconstructions from them are far more spatially accurate [2]. We face an under-determined problem that is solvable only with *a priori* information.

---

1. Author now at University of Wales Cardiff, Cardiff CF2 3XF, UK. peter@cs.cf.ac.uk

Systems are made individual by the way they represent the *a priori* information. It is widely held that a suitable representation is a key issue in solving the correspondence problem. This is because vasculature exhibit wide inter–individual variances of both branching structure (topology) and individual vessel shape (geometry). Moreover, the extent of these variations are not recorded in the clinical literature or elsewhere.

Previous approaches to the problem have used expert systems. Stansfield [13] provided three levels of rules that enabled coronary arteries (around the heart) to be segmented and identified. The low level rules relate to image processing methods for segementing pieces of vessels as trapeziums. Medium level rules combine these into strands. High level rules recognise strands as particular blood vessels. This type of scheme has been used by others [11, 14]. Delaere *et al.* [1] uses slightly different rules and, in addition, a Boolean function that predicts valid structures as seen in angiograms. The Boolean function comprises statements of the form 'this vessel has this clinical label', and evaluates to TRUE when a valid set of labels is used. We note that all of these systems are characterised by a fixed rule base that is used to describe the appearance of vasculature in angiograms acquired from specific points of view; the rules relate to angiograms and not to vasculature. New rules are needed for each vascular sub–system and for each point of view. Garreau *et al.*[3] provide an exception by modelling vasculature in three–dimensions. They do not appear to make the geometry explicit, although possible variances in topology are accounted for. However, the representation is fixed – so the system cannot "learn" from new instances of vasculature. None of these systems seem able to cope with the complexity of arteries in the brain, the cerebral vasculature, where inter–individual differences are most manifest and the vessels are smaller and more "tortuous" in their paths. All the systems described so far seem able to support only the task of reconstruction from angiograms; they are task–dependent.

Our representation is a fully three–dimensional representation of a collection of 3D models in which both topology and geometry are explicitly represented. It is a Vascular Catalogue (VC) in which each entry is an anatomical model of an individual vascular system. The content of the VC is incrementally updatable, so the VC can "learn" new instances. This is an important property since the full extent of vascular variations are not recorded in the literature. Inter–individual variances in both topology and geometry are made explicit and are maintained throughout the learning process. The VC uses manually reconstructed instances to initially learn the common vasculatures. As the VC learns more of the common structures, the need for the manually reconstructed instances becomes less. Because our VC contains an anatomical model it exhibits a high degree of task–independence. We have used it for simulating x–ray angiograms [8] and combining information from images and text [7]. The VC representation is such that it can be used to reconstruct other types of vasculature, for example coronary arteries.

We describe the VC and how to use it for reconstruction from x–ray angiograms. With regard to that task it offers two advantages: 1. it does not require different rule bases for different sub–systems; 2. it can reconstruct using angiograms from any point of view. The reconstruction process is a search for the model that best explains all input angiograms simultaneously, before recovering 3D information. The VC provides the *a priori* information needed to solve the correspondence problem. We do not consider methods for processing angiograms to extract vessels from their background, nor do we consider reconstruction of geometry once correspondence is solved, both issues are addressed in the literature. We focus on the difficult problem of solving correspondence.

2 The Vascular Catalogue

The VC is constructed from instances, each a valid anatomical model of a vascular system drawn from a library, L. It separates topological information from geometrical information. It stores the information in a compact form that allows easy access and generalisation. The VC contains a labelled graph, $G^+ = (N^+, A^+)$, that we call the *proto–graph*; this is built from L. Individual vasculature are also represented as a graph, $G_i = (N_i, A_i)$. Each instance, G_i, represents the topology of a particular vasculature. In all the graphs, nodes correspond to places where vessels divide (*furcations*) and arcs correspond to *vessel fragments* that connect furcations. Vessels as recognised by clinicians [12] are paths, or possibly small subgraphs, in a graph. The manner in which an individual graph corresponds to a real vasculature is shown in Figure 1.

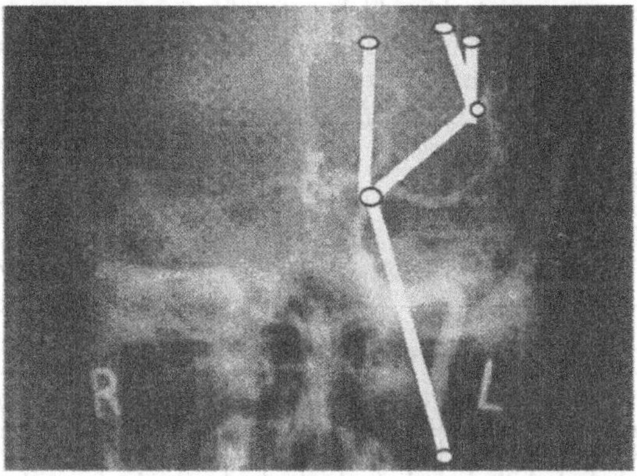

Figure 1. A real angiogram with part of a graph model drawn over it.

The labels of the nodes N^+ and the arcs A^+ contain information relating to topology and geometry, and their variances. These are generalised versions of the labels in each of the N_i and A_i, which contain only geometric information specifying the shape of a furcation or vessel fragment. The set N^+ is formed by merging elements in each of the N_i. Similarly, A^+ is formed by merging elements in each of the A_i. For example, vessel fragments with a sufficiently close geometrical similarity all map into a single element of A^+. Hence the merging operation results in a compact form for the proto–graph since many nodes (or arcs) may map to one. Unfortunately topological information may be destroyed by the merger, how this is avoided is explained next.

2.1 Topology

Topological information about the instance graphs can be maintained at a node (or arc) by a list of the graphs in L that contain the node (or arc). We can pull these labels off the nodes and arcs and use them as terms in a Boolean valued function that we call the *discrimination proposition*. This discriminates subgraphs of G^+ that belong to L and those that do not. The discrimination proposition is most easily represented by a matrix in which each column contains the label that came from a node or arc, and each row is an individual graph in L. A VC contains a proto–graph and a discrimination proposition. An example VC is depicted in Figure 2.

Rows of the matrix specify those topologies that are in L. We can add new instances to the proto–graph and simultaneously maintain the discrimination proposition (see [5] for details). This means that topological information in the VC can be incrementally updated. Updating the VC requires looking for the maximal common–subgraph (MCS) between the new instance and the current proto–graph. Searching for the MCS is an NP–complete problem in general. We are fortunate in having nodes and arcs that are richly labelled with geometric information, which makes our problem tractable. When searching for the MCS we ignore the discrimination proposition and match against the whole proto–graph. Once a match is found we check the common–subgraph, using the discrimination proposition, to decide whether the new instance is already in the library. If it is not we are able to decide which of its parts come from which instances in the library, and which of its parts are new. These operations are independent of the source of the new instance and is, therefore, independent of reconstruction from x–ray angiograms. However, we can decide which parts (of a new instance) came from which individuals (in the library). This is useful for reconstruction tasks because it provides information needed to generalise from examples.

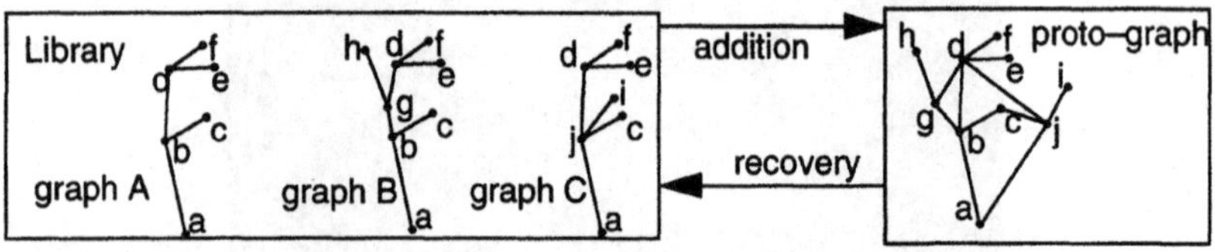

discrimination proposition: columns are labels on proto–graph.

	a	b	c	d	e	f	g	h	i	j	ab	aj	bc	bd	bg	cj	de	df	dg	dj	gh	ij
graph A	1	1	1	1	1	1	0	0	0	0	1	0	1	1	0	0	1	1	0	0	0	0
graph B	1	1	1	1	1	1	1	1	0	0	1	0	1	0	1	0	1	1	1	0	1	0
graph C	1	0	1	1	1	1	0	0	1	1	0	1	0	0	0	1	1	1	0	1	0	1

Figure 2. Example vascular catalogue with discrimination proposition.

2.2 Geometry

Geometric information is held in labels and specifies the 3D shape of the vasculature. The shape of each furcation is represented by a list of the direction of the outgoing vessel fragments relative to the incoming parent. Vessel fragments dominate the geometry of a vasculature and so we concentrate upon their description. Each vessel fragments a generalised cone whose medial–axis is a piecewise cubic curve and whose cross–section is circular. The vessel fragments in a set $V = \{ v_i : i = 1 \ldots n \}$ each correspond to a particular real vessel and all map to the same element of A^+. Each vessel fragment, v_i, is represented as a sequence of point, tangent, radius triplets; $v_i = < (\mathbf{p}_{ij}, \mathbf{t}_{ij}, r_{ij}) : j = 0 \ldots M >$, M is constant over all elements of V. The points, $\mathbf{p}_{ij}$, are separated by a distance D/M, along the medial axis; D is the length of the medial axis. Thus we obtain M+1 triplets, with the *jth* at position jD/M. To describe variances in geometry, we assume: 1. the sequence of triplets in each v_i is consistently ordered, and 2. each component in the point, tangent, and radius triplet is an independent random variable. Triplets of equal index, j, correspond by definition and are generalised by computing mean and normal distribution of each component. In this way we produce a generalised vessel–fragment for the set V. Before computing any statistical measure, we translate each vessel fragment so that its first point lies at the origin and its principal axes align to a standard frame.

The labels in elements of A^+ are generalised vessel–fragments. The representation is incrementally updatable. An important property of the generalised vessel fragment is that it includes vessel instances not yet seen: as in the topological case, we have generalised from examples. The representation can be projected onto a viewplane, which is a useful property for reconstruction, which is discussed next.

3 Reconstruction

The reconstruction module accepts angiograms as input and produces as output a reconstruction that accounts for the input angiograms. The role of the VC within the reconstruction process is to provide a model that is used to determine correspondence between angiograms. For example, if a furcation in one angiogram and a furcation in another angiogram have each been identified as corresponding to the same furcation in the proto-graph, then we can deduce that they correspond to each other. Correspondences such as this are a prerequisite to recovery of a vascular model in 3D. The description below concentrates on the central issue of how the VC is used to find such correspondences.

During reconstruction it is helpful to represent each angiogram as a graph that we call an *α–graph.* The α–graphs are obtained from the angiograms via segmentation. Angiograms are usually acquired from more than one point of view and we write $\alpha_i = (X_i, S_i)$ for the *ith* angiogram. In such a graph, each node in X_i corresponds either to a furcation in the vasculature or to a *crossing*. Crossings are artefacts of projection created by the apparent intersection of vessels. The arcs in the graph (elements of S_i) represent fractions of a vessel segment (*slivers*) that connect furcations and crossings as seen in an angiogram. This construction is shown in Figure 3. The set of correspondences between the α–graphs and the proto-graph generate a subgraph in G^+. This subgraph can be tested to determine whether it belongs to the vascular library, L, an operation discussed previously.

Reconstruction is now a search for the maximal, consistent set of mappings from nodes in the VC to nodes and arcs in each α–graph, and from arcs in the VC to nodes and arcs in each α–graph. A node in the VC will usually correspond to a node in the α-graph. Under certain circumstances, such as the node being hidden behind a vessel fragment, the node may correspond to a localised region of a sliver. An arc in the VC may map to a sequence of nodes and arcs (*sliver path*) in an α-graph because vessel fragments (arcs in the VC) may appear to cross in angiograms. In addition, more than one vessel fragment may account for a single sliver. This may occur in situations where the model has more pieces than the vasculature being reconstructed. There are a large number of possible ways correspondences can occur. We have found it is useful to record these correspondences in a *match space*, as explained next.

3.1 Match Space

The set of consistent, maximal mappings from G^+ to the set of n α–graphs, $\{\alpha_i\}$, exists in a match space with dimension $|\{\alpha_i\}|+1$. One axis of this match space comprises all elements in the proto–graph, $M^+ = N^+ \cup A^+ \cup \{\emptyset\}$; $\emptyset$ is the null element. The other axes each comprise elements in a particular α–graph, $M_i = X_i \cup S_i \cup \{\emptyset\}$. The order of these elements is of no consequence. Points within the volume of match space represent assertions of the form CR(h, $h_1, \ldots, h_n$) where $h \in M^+$ and each $h_i \in M_i$. We assign these points a value in the range [0,1] to represent the degree to which the correspondence has been validated. The value -1 is used to assert nothing is known about the correspondence. The null element is required on each axis so that we can explicitly represent those cases where an element has no correspondence. Of most interest are the planes of match space that are generated by pairs of axes – we must find correspondences on these planes. In particular, the correspondences in the planes defined by pairs of distinct axes M_i, M_j are of interest because these are needed to recover geometry in 3D by back–projection. The plane defined by M^+ and one of the M_i records correspondences between the proto–graph and the *ith* α-graph. Pairs of such planes are used to deduce the correspondences on the M_i, M_j plane.

To make these remarks a little more concrete, consider the example shown in Figure 3, which is a mock up of the experiments reported later in this paper. A model (the proto–graph without the vessel fragment BF), has been projected onto a biplane angiogram and α–graphs formed for each. We have supposed that the correspondences between the proto–graph and each of the α–graphs have been found – typically by matching the projected proto–graph (including the vessel–fragment BF). This projection is not shown on the diagram which is intended to be an illustration of a biplane angiogram. We have also supposed that B projects to somewhere in sliver 12 in α–graph$_1$ and to somewhere in sliver bb in α–graph$_2$; this helped fix the correspondences. Three planes of match space can be seen. Two planes record correspondences between the proto–graph and an α-graph. The third plane records correspondences that were deduced between the α-graphs by using the other two planes.

The example shows simple correspondences, such as CR(A, 1), on the plane defined by the proto–graph and α-graph$_1$. A situation in which a node of the proto–graph may correspond to a localised region on a sliver is also shown in the example: node C is hidden in α–graph$_2$, which we record as CR(C, de). Multiple matches can be recorded; CR(AB, 12), CR(B, 12), and CR(BD, 12) in the proto–graph / α–graph$_1$ plane shows how many elements from the proto–graph account for a single sliver. In the same plane we see CR(CD, 23_2), CR(CD, 3), and CR(CD, 35), showing that vessel fragment CD accounts for many elements in the α-graph. For brevity we shall indulge in abuse of notation and write CR(AB–B–BD, 12) or CR(CD, 23_2–3–35) for multiple correspondences. This notation has the advantage of clearly denoting paths in the proto–graph (which may be vessels as recognised by clinicians) and sliver paths. Once the set of correspondences has been constructed from the proto–graph to each α-graph, we can deduce how elements of the α-graphs correspond. Simple cases exist: CR(A, a) and CR(A, 1) lead us to conclude CR(a,1), for example. A more complex example is CR(CD, cd_1–d–de) which pairs with CR(CD, 23_2–3–35). We write these corresponding sliver paths as CR(cd_1–d–de, 23_2–3–35). In the plane defined by the α–graphs this is recorded in each point of the Cartesian product $\{23_2, 3, 35\} \times \{cd_1, d, de\}$. In Figure 3 such Cartesian products are highlighted by connecting the elements; without further geometric constraints this is the best we can do. We can deduce CR(cd_2–d–de, 23_1–3–34), which has elements in common with the previous example – this is a result of the physical overlap. The most complex example arises from the set of correspondences CR(AB, ab–bb–bb), CR(B, bb), CR(BB, bb–b–bc), and CR(AB-B-BD, 12); which can be used to deduce CR(12, ab-b-bb-b-bc). Match space contains the correspondences we desire and also allows us to build structures within it that limit the search and record its passage. The example demonstrates the role of the VC in solving the correspondence problem. However, determining the correspondences between the model and an α–graph has not been discussed so far. We address that issue next.

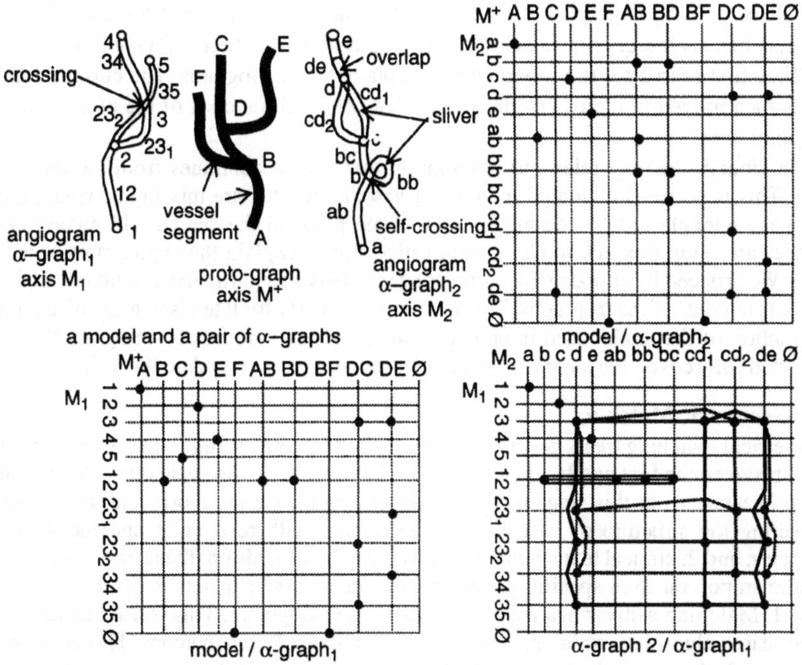

Figure 3. A model, its α-graphs, and three planes in match space

3.2 Search Algorithm

To find correspondences between the proto–graph and an α–graph we traverse the proto-graph in breadth-first order, accounting for the largest most important vessels first. By choosing such a strategy we are taking advantage of our knowledge of vascular structures. The search is initiated by manual selection of a node–arc–node sequence in the proto–graph. We assume the model is properly aligned with the angiograms and proceed by searching for correspondences in the α–graphs as follows:

1. Traverse the proto–graph, pulling off node pairs and connecting arc. This provides us with a connected sequence of elements from M^+, and sliver paths we suppose must exhibit similar connectivity. The first furcation in this sequence is usually assumed to match nodes in each of the α–graphs. In the proto–graph and each α–graph we call this the current node. We discuss the case where we relax this assumption below.
2. Hypothesise that the vessel fragment, h, exists as a sliver path in each of the α–graphs. We test this by searching in each of the α–graphs; the search is initiated with a set of mutually independent slivers. Each such sliver is a potential start for the sliver path and must be connected to the current node. The sliver path is constructed by building paths from each starting sliver, and pruning when some geometric tolerance is exceeded; beam search is used. Geometric matching is explained below.
3. The result of step 2 is a sequence $CR(h, h_{ij} \ldots h_{ik})$ for the ith α-graph. If such a

exists for every α-graph, the correspondences are recorded in match space. If more than one such sequence exists for all α-graphs then only best is retained; we use a best–first search for the subgraph that explains the angiograms. The current node is then moved to the end of the vessel fragment and traversal of the proto–graph continues.

As mentioned, we may relax our assumption that search continues from the current node. This is so we can "jump" sections of vasculature that are missing or which are otherwise unexplained by the model. If at some point in the search all geometrical matches are poor, then the model cannot satisfactorily explain the angiograms at that point. We proceed by supposing a correspondence between particular candidates in the VC and elements of the α–graphs. This is a useful property for it tends to maximise both the fraction of the model used in an explanation and the amount of angiogram that is explained. Moreover, the supposition can be used as a basis to update the VC.

3.3 Geometric Matching

Geometrical matching treats generalised vessel–fragments as straight lines by using the chord between the first and last point in the triplet sequence. We chose this geometrical match because: 1. at this stage of system development, we are more concerned with recovering the structure of vasculature than we are with recovering specific vessel geometry; and 2. clinical texts indicate that where a vessel ends up (the territory it feeds) is more important than specific geometry. The geometrical match is defined in two parts. 1. Matching a sliver against a projected vessel–fragment. This is used in stage 2 of the search algorithm explained above. 2. Accumulate of such matches. This is in line with stage 3 of our search. An individual sliver is matched to a vessel–fragment using

$$m = \begin{cases} \left(\dfrac{|\mathbf{S}|}{|\mathbf{F}|}\right)\cos\theta + \left(1 - \dfrac{|\mathbf{F}_1 - \mathbf{S}_1|}{r}\right) & \text{if } |\mathbf{S}| > |\mathbf{F}| \\ \left(2 - \dfrac{|\mathbf{S}|}{|\mathbf{F}|}\right)\cos\theta + \left(1 - \dfrac{|\mathbf{F}_1 - \mathbf{S}_1|}{r}\right) & \text{otherwise} \end{cases}$$

in which $\mathbf{F}_1$ is a point on the projected vessel–fragment; $\mathbf{F}_2$ is the end–point of the vessel–fragment; $\mathbf{F} = \mathbf{F}_2 - \mathbf{F}_1$; $\mathbf{S}_1$ is the starting point of the sliver, and $\mathbf{S}$ is the sliver chord. $\cos\theta$ is the dot product of unit vectors $\mathbf{S}/|\mathbf{S}|$ and $\mathbf{F}/|\mathbf{F}|$. r is a positive constant which measures vessel width. Initially, $\mathbf{F}_1$ lies at the start of the vessel–fragment and moves over it as the sliver and chord are matched. The measure is designed to penalise large slivers should they turn away from the expected direction, and any sliver that is too distant from the vessel–fragment. The positive constant, r, penalises distance. We form a complete measure of how N such slivers, $\mathbf{S}_i$, match a vessel–fragment by a weighted sum of these matches. Despite the crudeness of this measure, our initial results have been highly satisfactory and vindicate our approach. In particular, we have been able to recover complicated structures, even when vessels appear to be highly tangled, when vessels are missing, or when extra vessels are present.

4 Results

We have tested our method with biplane angiograms produced by our simulator [8]. This uses the same vascular catalogue as the reconstruction. Our test method was as follows:

1. Extract a model from the VC and modify it by: (a) by adding new pieces; (b) removing old pieces.

2. Use the simulator to produce both x–ray angiograms and an equivalent α–graph for each. In this we assume that the problem of segmenting slivers and crossings has been solved; an assumption justified by reference to the literature that performs a similar task [9, 10].
3. Reconstruct a vascular model using the α–graphs and the VC. Geometric form can be "borrowed" from the VC rather than recovered by back projection. This is because, here, we are more concerned with recapturing structure. We are developing mechanisms to recover geometry from the correspondences found in the match space.
4. Compare the reconstructed vascular model to the original from stage 1

Figure 4 shows an x–ray simulation of complete cerebral vasculature; this is our test VC. The α-graph representation and the reconstructed model are also shown in the same figure. This example shows we can reconstruct complicated structures. Notice that the reconstruction process has been able to "unravel" highly tangled blood vessels, including vessels that appear to overlap along their entire length from some points of view. Our comparison of the reconstructed model to the original showed the structures to be isomorphic. We can reconstruct sub–vasculature too; the image in Figure 5 shows a simulated x–ray and reconstruction of the left carotid system, which is a sub–vasculature of that in Figure 4. We found a perfect structural match between the reconstructed model and the original. This experiment was repeated with the exception that pieces of vasculature were removed. The simulated x–ray of the original, and the reconstructed model, are shown in Figure 6. The experiment was repeated again, this time with pieces added. The simulated x–ray of the original, and the reconstructed model, are again shown in Figure 6. Each time we found a perfect structural match between the original and the reconstructed model. The only reason the extra vessel does not appear in the reconstructed model is that, currently, we "borrow" reconstructed geometry from the model – and are unable to do so for pieces not in the VC.

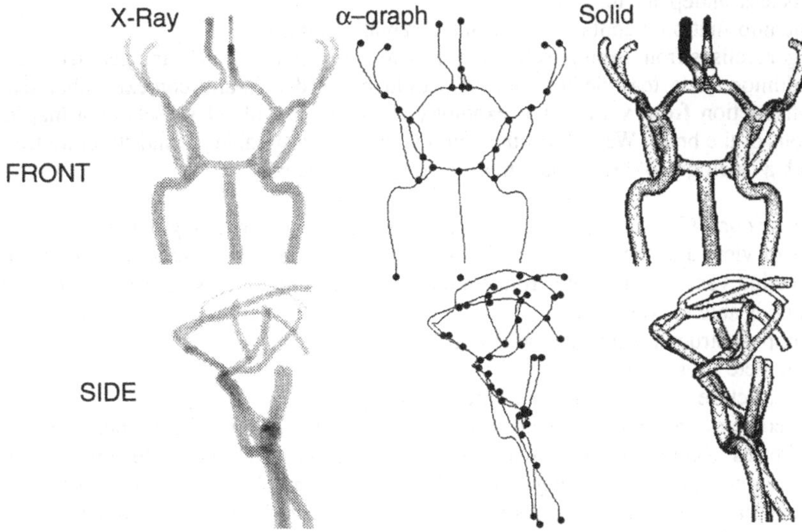

Figure 4. A complete VC as x–ray, α–graph, and reconstructed model.

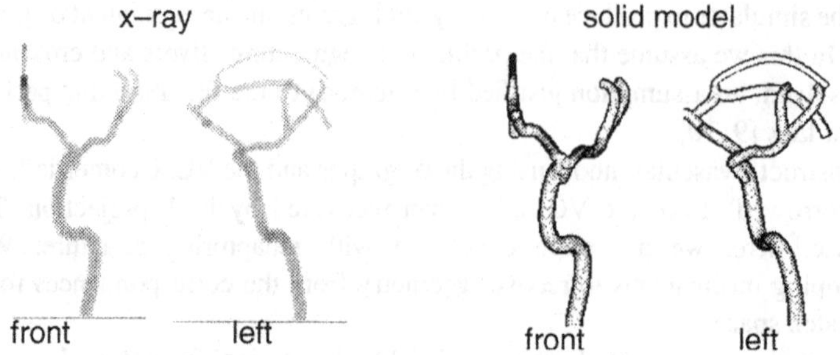

Figure 5. X-ray and reconstructed model of left carotid sub–vasculature.

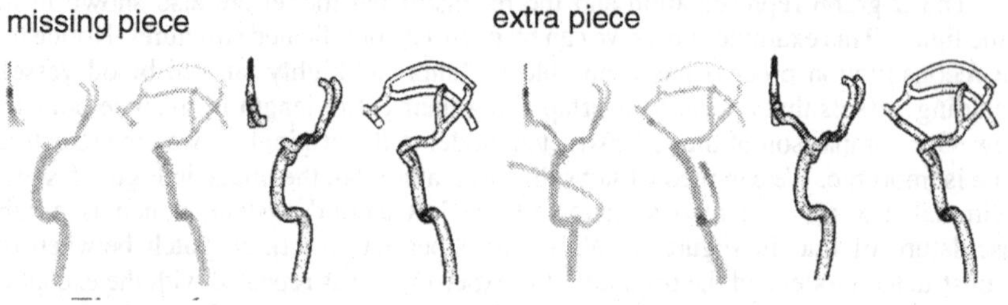

Figure 6. Reconstruction with a piece missing and with an extra piece.

5 Conclusion

We have introduced a Vascular Catalogue to model a collection of vasculature. It offers a number of improvements over previous systems:

- Structural and geometrical variations are made explicit.
- It is task–independent.
- The information it contains can be incrementally updated.
- It generalises from examples to create possibly unseen topologies and geometries.

In addition, it is feasible that our VC could be used in many contexts other than reconstruction from x–rays. For example, clinicians use blood vessels as a map of regions of the brain. We suggest that our VC provides a suitable ground structure from which a catalogue of brain anatomies could be constructed.

Because our VC describes anatomy it can be used to reconstruct angiograms from any point of view and there is no restriction on their number. We have shown its use in recovering vascular structure from simulated biplane angiograms. Our reconstruction method is able to cope with:

- Complex structures that appear to overlap.
- Complete substructures.
- Substructures with missing or additional parts.

Our geometric match is crude but surprisingly effective. This is an important result for it indicates specific geometry may not be as important as structural features. Future work is aimed at: (1) predicting structural features; (2) developing a more sophisticated geometrical match; and (3) segmenting the angiograms rather than assuming it.

6 References

[1] D. Delaere, C. Smets, P. Suetens, and G. Marchal, "A knowledge-based system for the angiographic reconstruction of blood vessels from two angiographic projections," In *Proc. of the North Sea Conference on Biomedical Engineering,* Nov. 1990.

[2] D. Delaere. (personal communication).

[3] M. Garreau, J.L. Coatrieux, R. Collerec, and C. Chardenon, "A knowledge based approach for reconstruction and labelling of vascular networks from biplane angiographic projections," *IEEE Trans Med Imag*, 10(2), 1991, pp 122-131.

[4] P.M. Hall, J. Brady, A.H. Watt, L. Walton, and U. Bergvall, "Segmenting and reconstruction vascular lesions from biplane angiogram projections," In *Proc. Digital Image Computing Techniques and Applications*, Sydney, Australia, 1993, pp 802-809.

[5] P.M. Hall, and J.J. McGregor, "A graph based model of a collection of physical vasculature." *Proc. Digital Image Computing Techniques and Applications,* 1993, Sydney, Australia, pp 414-421.

[6] P.M. Hall, R. Feltham, and T. Fitzjohn, "Automated analysis of x-ray angiograms," In *Proc. New Zealand Computer Society Conference*, Wellington, New Zealand, Aug. 1995, pp 253–260.

[7] P.M. Hall, and P. Mc Kevitt, "Automatic Interpretation of Vasculature," *AI Review,* 10(3-4), 1996, (forthcoming).

[8] P.M. Hall, "Simulating angiography," *Mathematical Modelling and Scientific Computing*, 6, 1996, (forthcoming).

[9] K.R. Hoffmann, D. Kunio, H.P. Chan, L. Fencil, H. Fujita, and A. Muraki, "Computer reproduction of the vasculature using an automated tracking method," In *Proc. SPIE 767 Medical Imaging*, 1987, pp 449–453.

[10] T.V. Nguyen, and J. Sklansky, "Reconstructing the 3–D medial axes of coronary arteries in single–view cineangiograms," IEEE Trans. Med. Imag. 13(1), March 1994, pp 61–73

[11] S.T. Rake, and L.D.R. Smith, "The interpretation of x-ray angiograms using a blackboard control architecture," *Proc. Computer Assisted Radiology* 1987, pp 681-685.

[12] G. Salamon, and Y.P. Huang, "*A radiological anatomy of the brain,*" Springer-Verlag, Berlin. 1976.

[13] S.A. Stansfield, "ANGY: A rule base expert system for automatic segmentation of coronary vessels from digital subtracted angiograms," *IEEE Transactions on Pattern Analysis and Machine Intelligence*. 8(2), 1986, pp 188-199.

[14] P. Suetens, J. van Cleynenbreugel, F. Fierens, C. Smets, A. Oosterlink, and G. Wilms, "An expert system for blood vessel segmentation on subtraction angiogram," *SPIE 767 Medical Imaging*, 1987, pp 454-459.

Tracking (2)

Combining Multiple Motion Estimates for Vehicle Tracking

Sylvia Gil, Ruggero Milanese and Thierry Pun

University of Geneva
Computer Science Department
24, rue Général Dufour,
1211 Genêve-4, Switzerland
E-mail: {*gil, milanese, pun*}*@cui.unige.ch*

Abstract. In this paper, the problem of combining estimates provided by multiple models is considered, with application to vehicle tracking. Two tracking systems, based on the bounding-box and on the 2-D pattern of the targets, provide individual motion parameters estimates to the combining method, which in turn produces a global estimate. Two methods are proposed to combine the estimates of these tracking systems: one is based on their covariance matrix, while the other one employs a Kalman filter model. Results are provided on three image sequences taken under different viewpoints, weather conditions and varying vehicle/road contrasts. Two evaluations are made. First, the performances of individual and global estimates are compared. Second, the two global estimates are compared and the superiority of the second method is assessed over the first one.

1. Introduction

In order to model a physical process, multiple models can often be designed, each of which is specialized on a particular aspect of the problem. This gives rise to the issue of combining multiple estimates, which is often done in two ways: by selecting the model providing the best estimate on a particular instance of the process (also called *hard-switch* method), or by constructing a combined estimate which weights the results of the individual ones and adaptively allows smooth transitions between them (also called *soft-switch* method [1]). The availability of multiple models can be useful for a robust vehicle tracking system which has to process data acquired under a large variety of conditions. For instance, weather and light conditions can produce drastic changes in the contrast between the road and the vehicles, introducing road reflectance and irregularities (rain), shadows (sunny), or the presence of vehicle lights at night. Let us assume that several tracking techniques are available. Any given input sequence $I(x, y, t)$ can be partitioned along the *time* dimension into various classes, representing characteristic illumination/weather conditions. Each estimator will generally perform best on some of these partitions. Thus, for each data partition or class, several estimators can provide meaningful information for the tracking task. Since it is not feasible to automatically estimate at each time the "best" estimator, an unsupervised combination of the tracking systems is needed.

In this work, we first describe two individual tracking systems (section 2), each of which independently estimates object motion parameters based on different visual features. The first feature used to represent the target is contour-based, consisting of the coordinates of the bounding rectangle obtained from the convex hull approximation of a moving-object's profile. The second feature is region-based, consisting of the 2-D pattern of the object. Then, two unsupervised methods for combining individual estimates of the two independent estimators are presented in section 3. Both combining methods take into account the instantaneous performance of the individual estimators. The first one computes a set of instantaneous weighting coefficients, used to combine the individual outputs into a global estimate. The second one provides in addition to the global estimate a confidence measure, which in turn is used to gate its own update. The performances (section 4) of the global estimates are first compared with the individual ones, then, a comparison is made between the two combining methods, in order to favor one. Results have been obtained on three image sequences with different time/weather/illumination conditions, acquired from different cameras. Finally, conclusions are reported in section 5.

2. Two Independent Tracking Systems

Traffic surveillance on urban and highway scenes has been widely studied in the past five years. One of the most popular methods, called model-based tracking, uses a 3-D model of a vehicle and is structured in two steps: (i) computation of scale, position and 3-D orientation of the vehicle (pose recovery), and (ii) tracking of the vehicle by fitting the model in subsequent frames by means of maximum-a-posteriori techniques [2] or Kalman filters [3] [4]. The vehicle model being quite detailed (3-D model), model-based tracking provides an accurate estimate of the vehicles 3-D position, which might not be needed for most applications, especially for highway surveillance. A simplified model of the vehicle is proposed in [5] by means of a polygon with fixed number of vertices, enclosing the convex hull of some vehicle's features. This approach dramatically reduces the model complexity. In [5] Kalman filters are used in order to estimate a vehicle's position and its motion using an affine model, which allows for translation and rotation. Although the method has shown good results, the fixed number of polygon vertices allows little variations on the objects shape. Some improvements on this point are proposed in [6] through the use of B-cubic splines, instead of polygons. In this case, a Kalman filter is used in order to track the curve in subsequent frames with a search strategy guided by the local contrast of the target in the image, i.e. with no use of motion information. In the context of traffic scenes, especially in the case of highways, vehicle's motion may be a powerful cue to direct the search for the target's position in subsequent frames. Another system that combines active contour models with Kalman filtering has been presented in [7]. In this case, the use of separate filters for the vehicle position and other motion parameters (affine model: translation and scale), has been shown to provide better results.

2.1 The Features Choice

Several choices are possible for the target's features, such as its color, its contour or a pattern defining its spatial layout. The major tendencies in the existing literature correspond to representations of the target's contour and to its description as a region. Both representations have advantages and drawbacks. Contour-based approaches [8] are fast, since they are based on the (efficient) detection of spatio-temporal gradients. Their major drawback is that, they may not have a physical meaning. Indeed, contour extraction depends on the local intensity variation between an object and the background, so that changes in their relative intensity may cause the appearance/disappearance of a contour. This type of features is thus reliable only when the contrast between the target and the background is sufficiently high, and constant in time. On the other hand, region-based approaches [9] represent the target through a 2D-pattern; they are quite accurate and do not depend on the background. Their drawbacks are the high computing time required for their manipulation (such as pattern matching [10]) and the sensitivity of pattern matching techniques to changes in scale and rotation. Contour- and region-based approaches thus appear to be complementary.

In this paper, both types of representations have been implemented. The contour-based feature is based on the bounding rectangle of the convex polygon and is represented through its center of gravity computed through its two characteristic corners (upper-left and lower-right). The region-based feature is the spatial pattern of the target, which is stored in a rectangular window. For both types of features, the tracked position remains the same: the center of the bounding box which is also the center of the 2D-pattern of the vehicle. An affine motion model (translation and scale changes) is used and ruled by Kalman filters.

2.2 The Kalman Filters

The two tracking systems are based on similar Kalman filter equations described in [3][7]. For both systems, the Kalman filter is used to track each visual feature of the moving target. For each feature we use a state vector, $\underline{x}_k$ to represent its position x_k, y_k and instantaneous motion parameters u_k, v_k, s_k. An affine motion model is used, which takes into account the translations along the x and y axes: u_k, v_k, as well as the scaling factor s_k representing the shrinkage/magnification of the target as it moves away or gets closer from/to the camera: $\hat{x}_k = (x_k, y_k, u_k, v_k, s_k)^T$.

The measurements $\underline{z}_k$ are the features positions, as computed from the k-th image frame. Therefore, the correspondence between the measurements and the position state vector is given by Equation (1), where $\underline{w}_k$ is the measurement error and where the *observation matrix* H_k for a given feature at a given time is defined in Equation (2).

$$\underline{z}_k = H\hat{\underline{x}}_k(-) + \underline{w}_k, \tag{1}$$

$$H_k = \begin{bmatrix} 1 & 0 & 1 & 0 & \hat{x}_k(-)-x_{ck} \\ 0 & 1 & 0 & 1 & \hat{y}_k(-)-y_{ck} \end{bmatrix}. \tag{2}$$

The last column of the matrix H_k represents the vector joining the center of gravity of the target $(x_{ck}, y_{ck})^T$ to one of the corners of the bounding rectangle. A change in this vector indicates a change in the target's scale, and allows the estimation of the scale factor s_k. The tracking system is decoupled into two inter-dependent subsystems composed of the position coordinates of the target on the one hand, and of the velocity parameters on the other hand. This decomposition allows for a dimensionality reduction of the system which is advantageous for low rank matrix inversion and also for the association of a covariance matrix to each one of the sub-system state variables.

2.3 The Features Measurement

The first feature to be considered is the bounding rectangle of the moving vehicles. Its computation is based on the convex hull approximation of the targets profile [7]. The convexity assumption is not restrictive in the case of vehicles because in most projections their profiles are pretty compact. It also considerably simplifies the matching step required by the tracking procedure, since it allows to by-pass problems such as contour regularization [11] [13]. Furthermore, an extensive literature is available describing efficient methods for convex hull computation [14]. The measurement of the convex hull can be summarized as follows. At each step *k*, a search window is obtained by translating the previous bounding rectangle, according to the predicted motion and to a tolerance margin for safety. The spatial and the temporal gradients are computed inside the search window, and points where the gradient exceeds two fixed thresholds respectively, are used for the convex hull computation[10] (cf. Figure 1). It is then straightforward to derive its bounding rectangle, whose center is the tracked feature. This feature leads to a considerable information compression and avoids the problem of tracking vectors of varying size (variable number of vertices).

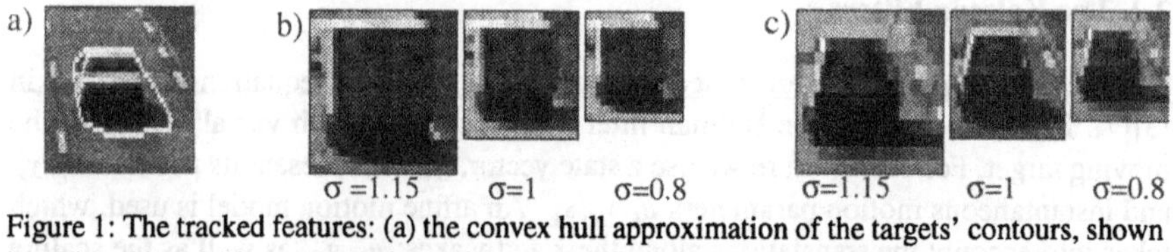

Figure 1: The tracked features: (a) the convex hull approximation of the targets' contours, shown in white; (b) and (c) representations of the target 2D patterns, with different scaling factors.

During the computation of the convex hull, some errors may occur. For instance, those introduced by an incorrect estimation of the predicted motion. In this case, the computed convex hull may exclude part of the target. Another source of errors degrading the measurement step is the fact of processing odd and even image fields (image parity change) resulting from an error in the image video sampling. This produces artificial temporal gradients at locations of a high spatial gradients, causing a deterioration in the shape of the convex hull.

The second feature representing the tracked objects is its 2-D pattern, stored as a gray-level mask. Given the target mask at a given frame, two scaled versions are computed by bilinear interpolation: $M_\sigma(x, y)$. These two masks provide an approximation of the object's appearance, caused by the objects's approaching or getting further away from the camera (positive or negative scale parameter). For each target, three scaled patterns are thus available for matching, corresponding to the scaling parameters $\sigma = \{0.8, 1, 1.15\}$ (cf. Figure 1). The measurement is performed by the correlation of the masks $M_\sigma(x, y)$ and a window of interest in the next image frame. For each scaled mask $M_\sigma(x, y)$, a recognition rate R_σ is computed based on the correlation peak value, and the autocorrelation of the scaled masks:

$$R_\sigma = \max_{u,v}\left(\frac{\sum_u\sum_v (I(x+u, y+v, t+1) - \overline{I(x+u, y+v, t+1)}) \cdot (M_\sigma(u, v) - \overline{M_\sigma(u, v)})}{\sum_m\sum_n (M_\sigma(m, n) - \overline{M_\sigma(m, n)})^2}\right) \quad (3)$$

where $I(x, y, t+1)$ is the frame used for the correlation, and the sign '$\overline{\ }$' denotes the average value. The value of R_σ quantifies the similarity between each scaled pattern and the target in the next frame. The locations of the highest peaks of the correlation surfaces (obtained with the masks $M_\sigma(x, y)$ yielding the highest value of R_σ) are retained as measurement candidates. When a scaled mask $M_\sigma(x, y)$, $\sigma \neq 1$ yielding the highest recognition rate has been selected for a sufficient number of consecutive frames, the 2-D pattern of the target is updated, its new scaled versions are computed, and the scale parameter of the motion model is updated. The mask update is achieved by copying into the new 2-D pattern window (with a new size), the values of image $I(x, y, t+1)$ around the selected correlation peak. As the size of the targets 2D pattern gets smaller, its correlation generally becomes unstable, giving rise to wrong measurements. In order to prevent these problems, a minimum size for the target is fixed, below which shrinkage of the 2D pattern is prevented. One of the advantages of this region-based tracking system is its immunity to incorrect temporal sampling (image parity change). The dynamic equations defining the tracking process require an initialization step, performed by a motion detection system described in [15].

3. Combining Estimates

Independently from motion tracking applications, estimates combination techniques are widely used in domains such as forecasting (see [16] [12] for a survey), statistics and neural network for problems such as regression, classification and time-series prediction, which require robustness to noise and the capability to cope with missing features. The most popular combination techniques are probably the "winner take all" and the averaging schemes. Despite its simplicity, averaging can provide interesting results [17]. For instance, the averaged output of a set of n unbiased and uncorrelated estimators, perturbed with uncorrelated noise, yields a mean squared error which is n times smaller than the mean squared error produced by individual estimators. In next section an overview of other more sophisticated combining techniques is presented.

3.1 Related Work

Although estimate combination is an intuitively attractive idea, some considerations must be made in order to quantify its actual efficiency. The work described in [18] considers the consequences of combining a set of individual estimates into a global one, in terms of global bias and variance for regression. Here, the computation of weighting factors multiplying the estimators' outputs is not addressed, the focus being on the performances that can be reached by combining estimates. Some general conclusions can be stated. The bias depends exclusively on single-estimator properties. The variance is composed of two additive terms: the first one depends exclusively on each estimator and thus grows as the number of estimators increases. The second term depends on the covariance of the different estimators. Thus, uncorrelated estimators should be used in order to decrease the variance.

An interesting method for combining estimates, called *generalized ensemble method* (GEM) was originally introduced for regression problems [17]. It consists of a linear combination of a population of n individual estimators weighted by normalized factors α_i. It is shown that weakly correlated estimators provide large weighting factors. The performance of the GEM saturates as the number of individual regression estimators increases, that is, when individual estimators start violating the uncorrelation assumption. Therefore a small enough number of independent estimates generally provides the best performances of the GEM. Another formulation of the problem leads to similar conclusions [19]. In combining methods, it is useful to introduce the notion of *ambiguity*, which quantifies the disagreement of an ensemble of estimators, i.e. how a single estimator's output differs from the averaged output of all estimators [20]. It has been shown that the largest the ambiguity, the smaller the quadratic ensemble error. As it has already been pointed out, it is important to choose estimators which do not agree (i.e. are not correlated) in order to increase the ambiguity term and so decrease the ensemble error.

The fusion of data issued from individual sensors through a formalism called *hierarchical estimation* is presented in [21]. Its goal is to merge n local estimates produced by n different sensors (or "local agents"), into a single global estimate. The assumption underlying this formulation is the linearity of the modeled dynamic process. The proposed architecture is shown in Figure 2.

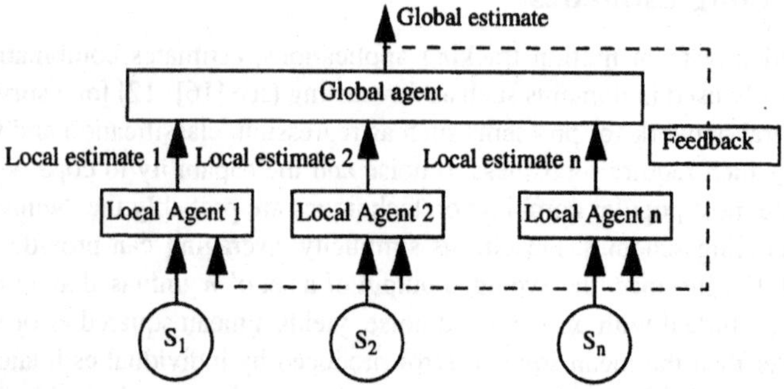

Figure 2: The fusion system is composed of n sensors, n local agents and one global agent. The feedback switch can be set *on* or *off*.

Global and individual (local) estimators are all ruled by Kalman filters. Local equations update is performed according to the customary Kalman equations, whereas the global equations update is achieved by the integration of local information. Therefore, the global equations are updated according to the performances of individual estimators, that is inversely proportional to their respective measurement error. This method for computing the global estimate has two advantages: it builds a global covariance matrix which provides a confidence measure, and it prevents from updating the estimate when individual measurement errors are too large. Several methods for computing normalized *weighting functions* depending on the input data have been presented in [1]. Here we only report the *variance-based* method. The main idea is to use the estimators which are the most *certain* of their estimation. Under the assumption that individual estimators are uncorrelated and unbiased, it can be shown that the combined estimator is also unbiased and has the smallest variance, if the weighting functions are inversely proportional to the variance of the individual estimator.

3.2 Our Contribution

In consideration of these previous works, the following points should be emphasized: (i) it is important to combine only uncorrelated estimators, and (ii) a small number of estimators must be selected, in order not to increase the global bias and variance. Thus, a reduced number of independent estimators is considered. In particular, the two independent tracking systems based on different visual features (cf. section 2) are used as individual estimators. Two combining methods based on the literature are tested.

Method 1: The Co-variance Method. The first combination method applied to our problem is derived from the method introduced in [17], where the i-th constant weighting factors α_i is obtained from the covariance matrix C of the misfit functions m_i:

$$\alpha_i = \sum_j C_{ij}^{-1} / \sum_k \sum_j C_{kj}^{-1} \tag{4}$$

$$C_{ij} = E[m_i(x) \cdot m_j(x)] \tag{5}$$

$$m_i(x) = f(x) - f_i(x) \tag{6}$$

where $f(x)$ is the target function and $f_i(x)$ is the i-th estimation. However, in the context of object tracking, target functions are not available and therefore the misfit functions cannot be computed. An alternative error function able to quantify the adequacy of the estimate to the target function is thus required. The measurement error, i.e. the error between the estimate prediction and its measurement, is proposed as the alternative error function:

$$\underline{w}_k = \underline{z}_k - \hat{\underline{x}}_k(+) . \tag{7}$$

This approximation is valid as long as the measurement is close to the target function and it becomes biased as soon as the measurement is not exact. In this case, our misfit functions will be worst-case estimates of the original misfit functions and it will be best-case when both, the measurement and the prediction, are equally biased. The advantage is that the error function of the individual estimates is computed at each time k, yielding an instantaneous covariance matrix of the misfit functions, and providing the instantaneous weights.

Method 2: The Hierarchical Estimation Method. The second combination method applied to our tracking problem is derived from [21]. Local agents provide measurements at each time t, modeled by the following equation:

$$z_i(t) = H_i \cdot x(t) + v_i(t) \qquad i = 1, ..., n \tag{8}$$

where $z_i(t)$ is the measurement vector of agent i, $x(t)$ is the state vector to be estimated, $v_i(t)$ is the zero-mean Gaussian measurement noise, with covariance matrix $R_i(t)$, and H_i is a known measurement matrix. The fusion agent, collecting all individual measurement equations yields a global observation equation:

$$z(t) = H \cdot x(t) + v(t) \tag{9}$$

where:

$$z(t) = [z_1^T(t), ..., z_n^T(t)]^T \tag{10}$$

$$H = [H_1^T, ..., H_n^T]^T \tag{11}$$

$$v(t) = [v_1^T(t), ..., v_n^T(t)]^T \tag{12}$$

The individual estimate and covariance matrix are *updated* according to the customary Kalman equations [3], whereas the global estimate $\hat{x}$ and the global covariance matrix P are updated by integrations of local agents information:

$$P^{-1}(t|t) = P^{-1}(t|t-1) + \sum_{i=1}^{n} [P_i^{-1}(t|t) - P_i^{-1}(t|t-1)] \tag{13}$$

$$P^{-1}(t|t) \cdot \hat{x}(t|t) = P^{-1}(t|t-1) \cdot \hat{x}(t|t-1) + \sum_{i=1}^{n} [P_i^{-1}(t|t) \cdot \hat{x}_i(t|t) - P_i^{-1}(t|t-1) \cdot \hat{x}_i(t|t-1)]$$

4. Experiments

In this section the tracking performances of the individual and global estimators are analyzed. The two methods described in section 3 are used to combine estimates. In order to accurately quantify the position error of individual and global estimates, the "true" position of the vehicles at each frame has been manually extracted for each vehicle for all image sequences. In some cases, the decision of assigning a true position to a vehicle is not obvious, especially when vehicle shadows appear or disappear from one frame to the next. Two comparisons are made. The first one is between the position error of the individual methods vs. the global one, in order to quantify the gain introduced by using global estimates. The second comparison is between the two combining methods, in order to make a selection between them.

In order to compute the global position estimate, two individual estimates issued from the tracking systems are available: the bounding rectangle, and the 2-D pattern of the target. For the combining method 2, the covariance matrix of the position needs to reflect the confidence of the bounding rectangle center-of-gravity position. A worst-case approach is used by choosing the position covariance matrix of the corner presenting the largest trace (highest uncertainty). For the motion parameters, three individual estimates are available: two provided by the corners of the bounding rectangle, and one originated by the correlation peak.

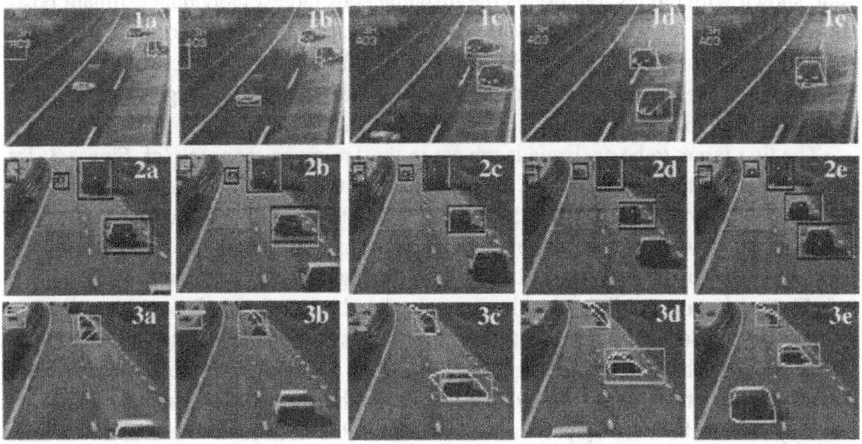

Figure 3: Three image sequences used in the experiments. In the first and third rows, the convex hull measurement is presented in white, while in the second row the correlation peak appears as a single white dot. The global prediction is represented (only for the second row) by a black rectangle. For all sequences, position predictions are reported by gray rectangles.

4.1 Input Image Sequences

The results obtained from three different image sequences are presented. They have been recorded at different locations, with different cameras, each of them with its own relative position to the road (cf. Figure 3). Of the three sequences, the latter two have been taken on a sunny day, which produces strong vehicles shadows. The first sequence has also been recorded with sunshine, but immediately after a rainstorm. The road thus appears partially wet and shadows are noticeable only where the road has dried. Moreover, shadows due to neighboring trees darken parts of the road. It should be noticed that the contrast between the cars and the road changes significantly between these three sequences.

In the third sequence (frame 3c), a wrong parity image occurs. This produces a noticeable deformation of the convex hull due to the artificially introduced strong spatio-temporal gradients on the white lane marks. Objects moving far away from the camera present weak gradients which produce an unstable convex hull shape. This effect can be observed in the third row of Figure 3, especially for the group of cars close to the top of the image. Finally, two consecutive frames of the third sequence are missing. Although this is not noticeable in Figure 3, the effect of this loss appears as a high error peak in the results (cf. Figure 4, i) and j), frame #18).

4.2 Comparing Global and Individual Estimates

The comparison of the error of the individual vs. global estimates are shown through diagrams where three error plots are superposed: those of the two individual methods

$e_k^1 = \hat{x}_k^1(+) - x_k$, $e_i^2 = \hat{x}_k^2(+) - x_k$ and the combined one $e_k^{\text{comb}} = \hat{x}_k^{\text{comb}}(+) - x_k$, where x_k are the vector of the hand-entered "true" vehicle position at time k. The first plot, shown with circles, represents the instantaneous largest individual estimate position error: $e_k^{\max} = \max(e_k^1, e_k^2)$, while the second curve (crosses) shows the instantaneous lowest position error: $e_k^{\min} = \min(e_k^1, e_k^2)$. Finally, the last curve (diamonds, solid line) represents the global estimate position error. For each image sequence, two such diagrams will be shown: the left one for method 1 and the right one for method 2.

The first target to be analyzed is the car in the 1st image sequence (cf. Figure 3, 1st row), located at the top the image. In this sequence the difficulty of dealing with vehicle shadows is manifest. During the initialization step, performed on a darker part of the road, shadows were not visible and so were excluded from the vehicle's mask. Shadows then appear in frames 12 and 13 producing a significant increase of the error (cf. Figure 4, (a) and (b)). In this case, the hand-entered "true" positions did not include car shadows. The gradient-based tracking system is highly sensitive to these morphological changes of the target, and produces an increase in the estimated scale factor and an enlargement of the prediction of the target width and height. This strongly affects the tracking process and points out the need for two independent scaling factors. On these critical frames, the convex hull measurement (including shadow) is considerably different from its prediction (no shadow, cf. Figure 4 (a) and (b)). For this reason, the measurement error of this tracked feature is high and its associated weights remain small, compared to those of the correlation-based method. It can be seen that both combining methods yield a global estimate that is close to the best individual estimate, which is the correlation-based one.

Let's now analyze the second image sequence (Figure 3, 2nd row) by focusing on the small car, the one closer to the truck. Individual estimators provide robust tracking performance, despite extremely small frame-to-frame displacements, thanks to the high vehicle/road contrast. However, its individual estimators present large errors, relative to the small target size (≅ 10 pixels). The estimate of the combining method 1 is often between the two individual ones, indicating that individual measurement errors are of the same order (cf. Figure 4 (c) and (d)). The estimate of the combining method 2 (cf. Figure 4, (d)), does not follow any of the individual estimates, indicating that the individual measurement errors are large. Its error is almost always smaller than even the best individual one. Let's now consider the truck. Individual errors are very small, relative to the targets size (≅ 27 pixels). Since the 2-D pattern is so large, the correlation-based method is very reliable. Some errors are introduced by the image crop when the truck slowly exits from the upper border of the image (cf. Figure 4 (e) and (f)). It can be seen that both global predictors perform well.

Let's analyze the performances on two vehicles of the third image sequence (Figure 3, 3rd row). The first object to be analyzed (Figure 4 (g) and (h)) is the group of cars at the top of the image. For the correlation-based tracking system, errors are due to a wrong mask initialization which includes part of the road. For the other tracking system, the cause of the large errors are the weak spatio-temporal gradients. Similarly to the previous experiments, when both individual measurement errors are equally high, the estimate of the combining method 1 falls in between the individual ones. For the combining method 2, these large measurement errors prevent from updating the global estimate, which thus provides almost always smaller error than both individual ones.

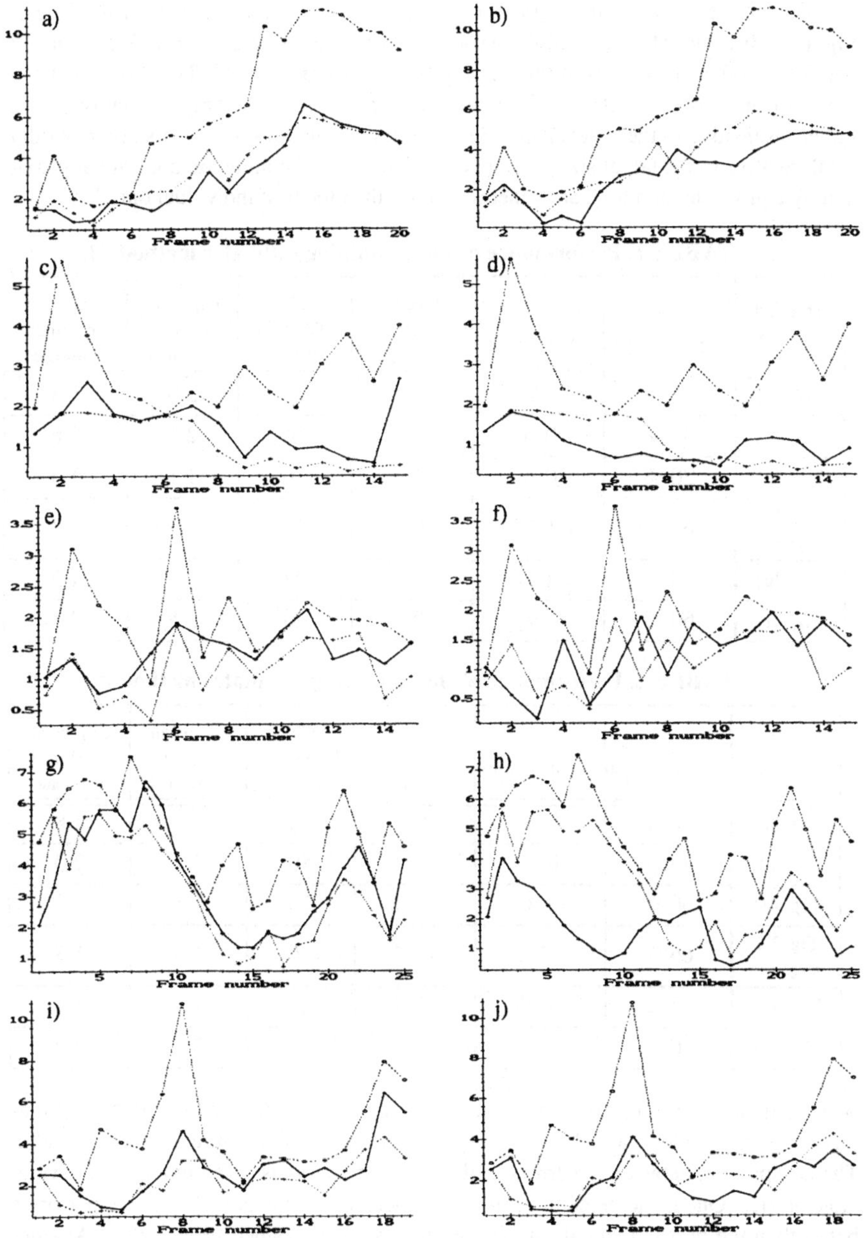

Figure 4: Position error diagrams: largest individual (–⊖–), lowest individual (+), and global (–◇–) with solid line. First sequence: (a), (b) car at the top of the image. Second sequence: (c), (d) small car; (e), (f) truck. Third sequence: (g), (h) group of small cars; (i), (j) single car. Left: combining method 1; right: combining method 2.

The vehicle described in Figure 4 (i) and (j) is the large clear car on the right lane of Figure 3, 3rd row. The effect of the wrong temporal sampling is clear in frames nos. 7 and 8, as well as the effects of the missing frames at frame no. 18. These two peaks in the error are mostly limited to the gradient-based tracking system, correlation being rather insensitive to these accidents. Besides these critical frames, errors are generally small relative to the target size (≅ 50 pixels). The two combining methods are approximately equivalent, and both are generally better than the best individual one.

TABLE 1. Performance of the combining estimates method #1.

Tracked vehicle	% $e_k^{comb} \le e_k^{min}$	% $e_k^{comb} \ge e_k^{max}$	% better than average	approx. target size (pixels)	avrg. error of best individual	avrg. error combined estimate
Fig. 4a)	16 %	8 %	64 %	20	3.0	3.3
Fig. 4c)	31 %	0 %	63 %	50	2.1	2.6
Fig. 4e)	20 %	0 %	80 %	10	1.1	1.5
Fig. 3.2 large car	60 %	6.66 %	80 %	36	1.5	1.3
Fig. 4g)	40 %	20 %	53 %	27	1.1	1.2
Fig. 4i)	60 %	0 %	90 %	20	3.5	3.3

TABLE 2. Performance of the combining estimates method #2.

Tracked vehicle	% $e_k^{comb} \le e_k^{min}$	% $e_k^{comb} \ge e_k^{max}$	% better than average	approx. target size (pixels)	avrg. error of best individual	avrg. error combined estimate
Fig. 4b)	88 %	0 %	96 %	20	3.0	1.8
Fig. 4d)	73 %	0 %	89 %	50	2.1	2.0
Fig. 4f)	60 %	0 %	100 %	10	1.1	1.0
Fig. 3.2 large car	66 %	6.7 %	93 %	36	1.5	1.3
Fig. 4h)	13 %	26 %	53 %	27	1.1	1.3
Fig. 4j)	70 %	0 %	95 %	20	3.5	3.0

A more quantitative comparison of the performances of individual vs. global estimators is given in tables 1 and 2, which respectively concern combining methods 1 and 2. The comparison between the errors of different estimators is given in terms of several table entries. One is the percentage of occurrences where the global estimate performs better than the best individual estimate or, in error terms, when $e_k^{comb} \le e_k^{min}$. Another index counts the percentage of occurrences in which the global estimates is worse than the worst individual estimates ($e_k^{comb} \ge e_k^{max}$). Finally, we compare the global estimates with what could be the estimates of a trivial integration method, i.e. the simple average of the individual estimates ($e_k^{comb} \le (e_k^1 + e_k^2)/2$). The two last columns show the average of the position error, over multiple frames. In order to have an indication of relative importance of these errors, the vehicle size (in pixels) is reported with each set of measures.

In terms of average errors, by looking at table 1, it appears that the estimate of the 1st combining method, although it remains within reasonable bounds, only outperforms the best individual estimate for one vehicle. The second combining method, however, has much better performance, consistently better than the lowest individual error. In the only case where this is not true, the individual error is already small (1.1 pixels), for a target size of approximately 27 pixels. It appears that the global estimate provided by the combining method 1 performs well only when the features measurement does not present major difficulties for neither individual tracking systems, as predicted in section 3. Apart from this specific weakness, global estimates of both methods are in general sensitive to the following sources of errors: (i) wrong mask initialization; (ii) inaccurate mask update; and (iii) consistently wrong feature measurements in both individual tracking methods (only for the combining method 1).

5. Conclusions

In this paper, we introduced the problem of combining multiple models of a dynamic process, and presented an application to target tracking using multiple motion estimators. Two independent tracking systems are used: one is based on the bounding-box of the moving object, the other one uses the object's 2-D pattern. Both individual tracking systems provide motion parameters estimates which are then combined into a global one. Several classes of combining methods presented in the literature are reviewed. Tracking performances are evaluated on 6 vehicles, from three different outdoors image sequences. To precisely compute each individual and global estimate error, the "real" vehicle positions at each frames have been hand entered.

The first method is based on a linear combination of the individual estimates, whose weights are inversely proportional to the covariance matrix coefficients. When the two individual methods both provide good estimates, then the results of this combining method represent an improvement, yielding smaller position error than each individual one. When neither individual method performs correctly, however, then combining method does not introduce any improvements. This is due to the practical need to replace the error function, usually computed through a training set, by the measurement error, which is a pessimistic estimate when the measurement is not exact, and optimistic when both measurement and prediction are equally wrong. Overall, this combining method has been shown to be superior to the averaging technique. However, only in some cases does this method outperform the best individual estimate.

The second combining method integrates the estimates of the individual methods using a Kalman filtering approach. In this case, the combined estimates clearly outperform both the averaging and the best individual estimate. In five cases out of six, the error of the combined method was significantly reduced, while in the sixth case, the individual estimates were already very good. The performances of this combining method can be further improved by avoiding errors due to wrong mask initialization, and improper mask updates. A comparison between the two proposed combining methods clearly shows the superiority of the second, Kalman filter-based one, both in terms of average position error, and in the number of frames where the results outperform the best instantaneous individual method. These result are encouraging.

Acknowledgments We would like to thank researchers from the connectionnist and vision mailing list, Prof. J. Malik (U. Berkeley), Dr. D. Koller, IRISA and IRBSA for their help, encouragement and discussions, and for providing image sequences.

References

1. V. Tresp, M. Taniguchi, "Combining Estimators Using Non-Constant Weighting Functions", to appear in Proc. of the Neural Information Processing Systems, 1995.
2. D. Koller, K. Daniilidis, H.H. Nagel, Model-Based Object Tracking in Monocular Image Sequences of Road Traffic Scenes, *Int. Jour. of Computer Vision*, Vol. 3, pp. 257-281, 1993.
3. A. Gelb, Applied Optimal Estimation, The MIT Press, MA, and London, UK, 1974.
4. K. Baker, G. D. Sullivan, Performance Assessment of Model-based Tracking, *Proc. of the IEEE Workshop on Applications of Computer Vision*, pp. 28-35, Palm Springs, CA, 1992.
5. F. Meyer, P. Bouthémy, Region-Based Tracking in an Image Sequence, *Proceedings of the European Conference on Computer Vision*, pp. 476-484, S. Margarita-Ligure, Italy, 1992.
6. A. Blake, R. Curwen, A. Zisserman, A Framework for Spatiotemporal Control in the Tracking of Visual Contours, *Int. Journal of Computer Vision*, Vol. 11, pp. 127-147, 1993.
7. D. Koller, J. Weber, J. Malik, Robust Multiple Car Tracking with Occlusion Reasoning, *Proceedings of the Third European Conference on Computer Vision*, Vol. 1, pp. 189-199, Stockholm, Sweden, 1994.
8. O. Faugeras, Three-Dimensional Computer Vision: A Geometric Viewpoint, The MIT Press, London, UK, 1993.
9. J. Shi, C. Tomasi, Good Features to Track, *IEEE Conf. on Computer Vision and Pattern Recognition*, pp. 593-600, Seattle, USA, 1994.
10. S. Gil, R. Milanese, T. Pun, "Comparing Features for Target Tracking in Traffic Scenes", to appear in *Pattern Recognition*.
11. D. Geiger, J.A. Vlontzos, Matching Elastic contours, *IEEE Conf. on Computer Vision and Pattern Recognition*, pp. 602-604, New York, June 1993.
12. International Journal of Forecasting, special issue on combining forecasts, Vol. 4(4), 1989.
13. F. Leymarie, D. Levine, Tracking Deformable Objects in the Plane Using an Active Contour Model, *IEEE Trans. on Pattern Analysis and Machine Intelligence*, Vol. 15, 1993.
14. F. P. Preparata, M. I. Shamos, Computational Geometry, Springer-Verlag, 1985.
15. S. Gil, T. Pun, Non-linear Multiresolution Relaxation for Alerting, *Proceedings of the European Conference on Circuit Theory and Design*, pp. 1639-1644, Davos, CH, 1993.
16. C.V. Granger, "Combining forecasts - twenty years later", Journal of Forecasting, Vol. 8, pp. 167-173, 1989.
17. M.P. Perrone and L.N. Cooper, "When networks disagree: Ensemble methods for hybrid Neural Networks", in "Neural Networks for Speech and Processing", Editor R.J. Mammone, Chapman-Hall, 1993.
18. R. Meir, "Bias, variance and the combination of estimators; the case of linear least squares", Preprint, Technion, Heifa, Israel, 1994.
19. S. Hashem, "Optimal Linear Combinations of Neural Networks", Ph.D. Thesis, Purdue University, December 1993.
20. A. Krogh, J. Vedelsby, "Neural Network Ensemble, Cross Validation and Active Learning", to appear in Proc. of the Neural Information Processing Systems, 1995.
21. Y. Bar-Shalom, "Multitarget Multisensor Tracking: Advanced Applications", Artech, 1990.

Visual Surveillance Monitoring and Watching

Richard Howarth and Hilary Buxton

School of Cognitive and Computing Sciences,
University of Sussex,
Falmer, Brighton BN1 9QH, UK

Abstract. This paper describes the development of computational understanding for surveillance of moving objects and their interactions in real world situations. Understanding the activity of moving objects starts by tracking objects in an image sequence, but this is just the beginning. The objective of this work is to go further and form conceptual descriptions that capture the dynamic interactions of objects in a meaningful way. The computational approach uses results from the VIEWS project [1]. The issues concerned with extending computational vision to address high-level vision are described in the context of a surveillance system. In this paper we describe two systems: a passive architecture based on "event reasoning" which is the identification of behavioural primitives, their selection and composition; and an active architecture based on "task-level control" which is the guidance of the system to comply with a given surveillance task.

1 Introduction

Until recently it has been rare to find issues of control connected with computer vision (but see for example Rimey and Brown [26] and Howarth [17]). The focus tends to be on techniques that extract information from images rather than on identifying visual behaviours appropriate for visual tasks and how these operate. Ballard's landmark paper [2] identified how these two approaches could be integrated in what he called "animate vision". In this paper, we describe some of the advantages obtained by reinterpreting a pipelined, passive vision system under a more active vision approach. We use surveillance of wide-area dynamic scenes as our problem domain.

Our surveillance problem has the following simplifications that make visual understanding more tractable: we use a fixed camera that observes the activity of rigid objects in a structured domain. Examples include: a road traffic scene where the main interest is the road vehicles, and airport holding areas where we are interested in the activities of the various special vehicles that unload and service the passenger aeroplanes. We call this single viewpoint of the fixed camera the "official-observer". From this camera input we wish to obtain a description of the activity taking place in the dynamic wide-area scene, and then an understanding of the dynamic and improvised interactions of the scene objects.

[1] Thanks to project partners on ESPRIT EP2152 (VIEWS) and to EPSRC GR/K08772 for continued funding of this work.

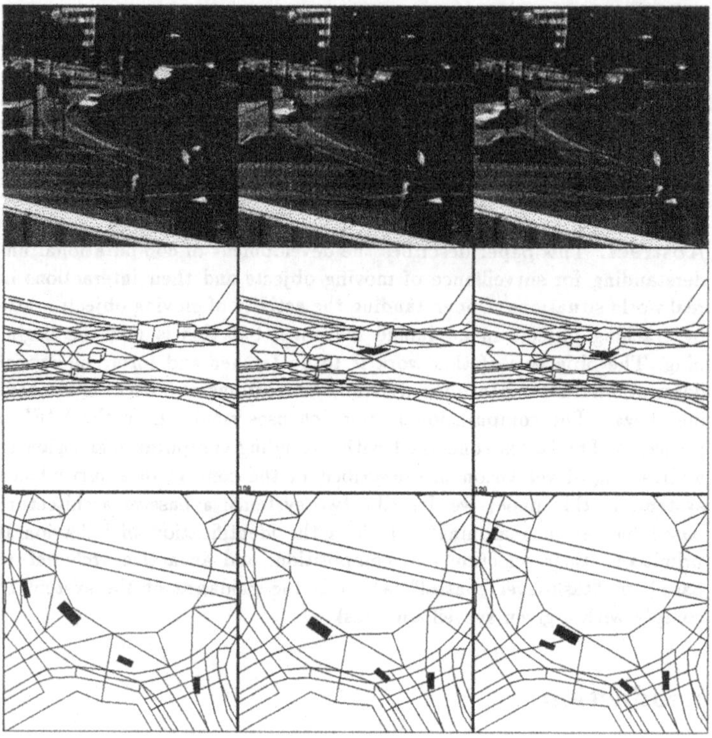

Fig. 1. Three images showing typical vehicle activity on the roundabout and how the 3D pose descriptions can be transformed to a ground-plane view.

There are constraints on the official-observer's interpretation of the objects in the scene: we only see the objects that are in the camera's field-of-view; we do not know each participant's goal (typically something like "go to place X"); and what we see is mostly reactive behaviour (rather than deeply planned).

To illustrate the difference between the passive and active approaches we will describe two systems that represent different formulations of the surveillance problem. The first is called HIVIS-MONITOR, embodying the initial design. The second system, called HIVIS-WATCHER, is a response to the problems encountered while developing our initial system. To demonstrate the different behaviour of the two systems we will use examples drawn from the road traffic domain. Here (figure 1) we illustrate this with three image frames selected from a sequence taken at a German roundabout. In this part of the sequence a number of episodic behaviours are unfolding: one vehicle leaves the roundabout; another is in an entry lane to the roundabout; also towards the rear of the image a car begins to overtake a lorry. Below the image frames we provide an illustration of the poseboxes, which are results from a model-matcher (see [9, 31] for details).

2 HIVIS-MONITOR

In our first system, HIVIS-MONITOR, we adopt a pipelined approach that reflects the general flow of data from images to conceptual descriptions. The visual processing centres around three components: extracting spatio-temporal primitives from the stream of compact encodings produced by low- and intermediate-level visual processing, detecting events from these primitives, and composing the events to form episodic sequences which are stored in an evolving database. This database is extended as new events continue, begin or end the various episodes under construction. The problem of matching behaviours to a user question is left to the query-based component that interrogates the database. At first sight, this seems an admirable system design, allowing the parallel, separate development of the perceptual processing component and the behavioural one. However, as we will see, the passive, data-driven flow of the processing causes problems for visual control.

2.1 Script-based approach

To describe the behaviour of participants in the scene we are using an ontology based upon that described by Nagel [23] to describe events, which captures the common sense notions of the terms being used. Neumann [24] provides an example set of events used in his NAOS system. Our use of events and episodes is similar in some respects to that described by Schank and Abelson [27], who compose events into scripts to describe typical behaviour of a customer at a restaurant such as entering, going to a table, ordering, eating and paying. This provides a hierarchical layering from events to episodes, and then to more complex script-like behaviours. This hierarchical decomposition and relationships between the behavioural elements can be used to define a grammar where events are terminal symbols in the language to be parsed. This approach could use syntactic methods such as attributed grammars as described by Frost [12] and Clark [6], or the island parsing described by Corrall and Hill [8].

HIVIS-MONITOR is data-driven and follows the script-based approach, constructing an interpretation of object behaviour in an evolving database that holds entries for the history of each individual object and the interactions between them. This approach reflects the flow of data from image to conceptual descriptions. Maintaining a history of the behaviour that takes place in the scene involves noting the event primitives that have been detected and then using an ongoing interpretation process to see how these events fit together. The input given to the database consists of the events and activities associated with a particular property. In addition to the functions that compute these values there are further functions that update the temporal structure by beginning, extending or ending the continuity of the value/signal for each property. To identify an episode we use a filter that matches the necessary property values.

2.2 Spatio-temporal representation

The spatial representation used by HIVIS-MONITOR is based on a model of space developed by Fleck [10, 11] for representing digitised spaces for both edge detection and stereo matching. Also, in her thesis [11], she describes how this representation can be used for qualitative reasoning and for modelling natural language semantics. The spatial and temporal representation Fleck uses and calls "cellular topology" is based on the mathematical foundation of combinatorial topology. Cellular topology uses cells to structure the underlying space and is augmented here by adding a metric (see also [16, 18]). It is to this underlying spatial representation that we can attach information about the world.

The stream of posebox data supplied by the model-matcher describes the space swept out in time by each object's path to form what we call a "conduit". The conduit is used to provide an approximation of the time at which a region is exited or entered. To do this, we extrapolate the space-time description between updates. Once we have generated the conduits, we have the problem of interpreting what they mean. If they intersect then there is a likely collision or near miss, but intersections of conduits is unusual. Other tests can be made possible by removing a pertinent dimension and testing to see if the components of the reduced model overlap, in the test for `following` behaviour we tested for an overlap with some time delay. Overtaking can be identified by ignoring the spatial dimension parallel to the objects direction of motion however, this spatial dimension should really be the 2D manifold that fits the space curve of each object's path. Mapping the conduits into one of these manifolds to perform such as test is difficult, although in principle it should be possible.

2.3 General features

We claim that HIVIS-MONITOR demonstrates typical traits of the class of traditional AI approaches we have called "script-based". In general, all script-based systems will have the following features: Maximal detail is derived from the input data. This approach obtains a description of all objects and all interactions, over the whole scene, for all the episodes it has been designed to detect; Representation is extracted first and the results are placed in an evolving database that is used to construct more abstract descriptions using hindsight. Single object reasoning is performed with ease using this approach. Simple implementation can be achieved using standard AI techniques. It is quite likely that better implementations could be developed that fulfill the script-based approach [2] but there would still be limitations.

2.4 Limitations

HIVIS-MONITOR has the following limitations:

[2] Achievements of the project are illustrated by the video [3] of the ESPRIT project VIEWS, and by Corrall and Hill [7, 8], King et al. [21] and Toal and Buxton [28]

- It is passive in its processing, operating a simple control policy, that is, not affected by changes in the perceived data.
- It is not real-time because the construction of the results database is an off-line process, and does not send feedback to any form of intermediate-level visual processing. This means that there is a problem getting timely recognition of perceived object activity.
- Unbounded storage is required because any pieces of data contained in the results database might be needed later either to compose some more abstract description or to be accessed by the user to answer a query. Since we do not retract what we have seen or the episodes that we have identified, the database structure is monotonically increasing in size.
- Multiple object reasoning is difficult within the global coordinate system used to express pose positions. A solution to this is needed because contextual knowledge is not enough to analyse the interactions, although it does provide a context for interpretations.
- The computation performed by HIVIS-MONITOR is mainly dependent upon the number of objects in the input data, i.e., it is *data-dependent*.
- This model is inflexible because it only deals with known episodes. Within the constraints of the predicates provided (language primitives that describe events and activities), new behavioural models can be added. However, defining new predicates may be difficult.
- The addition of new operators increases the number of tests performed on all the objects in the scene. For a single object operator there is a $O(n)$ increase, for most binary object operators there is a $O(n^2)$ increase, and for most multiple object operators the increase is polynomial with a maximum of $O(n^n)$, where n is the number of objects in the scene.
- The behavioural decomposition does not take into consideration the temporal context in which the events have occurred, which contributes to the process of interpretation. It is possible that the selection of the "correct" episode description is not possible due to only seeing part of an episode.

2.5 Discussion

From these features and limitations we can identify the following key problems: computation is performed to obtain results that may never be required; and as the database of results increases in size, the performance of the system will degrade. It might be possible to address these by extending the script-based approach however, we will not take this evolutionary route. Instead we will investigate a more situated approach. This new approach differs greatly from the passive, data-driven script-based approach and requires a complete reformulation of the problem to obtain an active, task-driven situated solution.

3 Reassessment

To begin this reformulation we first consider the use of more local forms of reasoning in terms of the frame-of-reference of the perceived objects, the spatial arrangements of these objects and the use of contextual indexing from knowledge about the environment. In HIVIS-MONITOR a global extrinsic coordinate system was assumed. By taking a global view we comply with a commonly held Western view of how to represent space in a map-like way as opposed to the egocentric approach described by Hutchins [20] as being used by the Micronesians to perform navigation. The absolute coordinate system also fits well with the concept of the optic-array (see Gibson [13] and Kosslyn et al. [22] for details), if we can consider placing a grid over the ground-plane to be analogical to the optic-array of the perceiver. This representation would allow reasoning to be performed that does not need full object recognition with spatial relationships represented in terms of the optic-array's absolute coordinates (in some respects this is like the video-game world used by Agre and Chapman [1] where the "winner-takes-all" recognition mechanism (see Chapman [5] and Tsotsos [29]) allows objects and their positions to be identified by key properties, such as, colour and roundedness).

In contrast to this global viewpoint, when reasoning about the behaviour of each scene object it would be useful if the representation of the properties related to each object could be described in its own relative coordinate system. However, this involves recognising each object to the extent that an intrinsic-front can be identified together with its spatial extent. This requirement places the need for a more sophisticated understanding of how the image data present in the optic-array relates to how objects exist in the environment. In our surveillance problem we can obtain the pose-positions of the scene objects via model-matching making local reasoning attractive, although its extra cost in terms of the complexity of intermediate-level vision should be noted. The **local-form** is representation and reasoning that uses the intrinsic frame-of-reference of a perceived object (exocentric with respect to the observer). The **global-form** is representation and reasoning that uses the perceiver's frame-of-reference, which operates over the whole field-of-view (egocentric with respect to the observer). The global-form is not a public-world since it, like the local-form, only exists for the perceiver. We are not dealing with representing a shared world in terms of each participant. The suitability of each HIVIS-system is detailed in table 1, indicating the extent of the reformulation for the surveillance problem.

HIVIS-MONITOR would be useful for off-line query of behaviour, whereas in HIVIS-WATCHER, by asking the question first, we remove the importance of the results database because we are no longer providing a query-based system. This removes the need to remember everything and solves the problem of the monotonically increasing database because in HIVIS-MONITOR it is difficult to know when something can be forgotten. The development of a more situated approach in HIVIS-WATCHER is part of the adoption of a more local viewpoint that uses a deictic representation of space and time. In some applications, using HIVIS-MONITOR and processing all scene objects might be necessary however, in cases where it is not, the HIVIS-MONITOR approach is ungainly. In the surveillance problem where we are inherently concerned with the "here-and-now" (the evolving contexts of both observer and scene objects), it is important

HIVIS-MONITOR	HIVIS-WATCHER	illuminates
off-line/pipelined	on-line	immediacy
structured	purposive	approaches
global	local	viewpoint
maximal detail	sufficient detail	investigation
passive	active	control
unlimited resources	limited resources	complexity
extract representation first	ask question first	timeliness
answer question from representation data	answer question from scene data	memory cost
data dependent	task dependent	propagation

Table 1. This table summarises the comparison between the two HIVIS-based systems, with the illuminates column describing what each row is about.

to form a consistent, task relevant interpretation of this observed behaviour. By taking a deictic approach in HIVIS-WATCHER we don't name and describe every object, and we register only information about objects relevant to task. By doing this the information registered is then proportional to properties of interest and not the number of objects in the world.

4 HIVIS-WATCHER

In HIVIS-WATCHER we remove the reliance on the pipelined flow of data and instead use feedback to control the behaviour of the system. By making the perceptual processing and behavioural interpretation in the HIVIS-systems more tightly coupled we provide a more active control that can direct the processing performed by the system to those elements that are relevant to the current surveillance task. Deictic representation plays an important role in this framework because it supports attentional processing with emphasis placed on the behaviour of the perceiver as it interprets the activity of the scene objects rather than just representing the behaviour of the scene objects on their own.

4.1 Situated approach

Background details to the situated approach is given in Howarth [17] and its role in perceptual processing is further described in [4]. In HIVIS-WATCHER we have three separate elements: the "virtual-world" which holds data about the world, the "peripheral-system" which operators that access the world, the "central-system" which controls system behaviour. The peripheral-system is based on Ullman's [30] visual routine processor following the approach described by Agre and Chapman [5]. Horswill [15] describes a real-time implementation of such a visual routine processor. Both HIVIS-systems employ event detection operators however, the key difference is that in the HIVIS-WATCHER peripheral-system operators are not run all the time, they are only run when selected by the task-level control system.

We have separated the operators in the peripheral-system into preattentive ones that are global, simple, and of low-cost and attentive ones which are applied to a single object and are more complex. The preattentive operators are used to guide application of attentive ones. Example preattentive operators include gross-change-in-motion which is described below, and mutual-proximity which is described in Howarth and Buxton [19]. The motivation behind the preattentive and attentional cues chosen here, was their potential usefulness on low-level data such as the identification of possible objects from clustering flow-vectors (see Gong and Buxton [14] where knowledge about a known ground plane is used to develop expectations of likely object motion). Once we have these coarse descriptions, and if they comply with the preattentive cue, then they would become candidates for further attentional processing such as model-matching (or some other form of object-recognition) to obtain aspects about the object. Basically, once we have found where something interesting is, we then try and work out what it is.

There are two types of marker in HIVIS-WATCHER. The agent type are all used by the same cluster of rules that identify events concerning changes in velocity, type-of-spatial-region-occupied, relative-position-of-other-local-objects, etc.. These rules represent the observer's understanding of typical-object-behaviour. The kernel type are each run by different sets of rules to fulfill some specific purpose, for example, the `*stationary-marker*` is allocated to any object that has recently stopped moving, by a perceiver-routine that is interested in objects that have just stopped.

4.2 An implementation of perceiver routines

To illustrate how perceiver routines work we will describe the routines associated with the official-observer looking for the presence of giveway behaviour. As mentioned above, the preattentive cue identifies any gross-change-in-motion (i.e., instances where an object changes state between `stationary` and `moving`. We can use this to initiate detection when, for example, a vehicle stops at a junction. Because this task of looking for giveway behaviour requires the identification of a number of distinct stages that involve different scene participants. The perceptual task of the official-observer involves three important entities: the first two correspond to the two roles in the giveway episode and are denoted by Stationary for ***the-stationary-vehicle***, and Blocker for ***the-vehicle-that-Stationary-is-giving-way-to***; and the third is denoted CA for ***the-conflict-area*** (a special region). When the two roles of Stationary and Blocker have been found, an area of mutual conflict, CA, can be identified (the space infront of Stationary and through which Blocker will pass). This area links Stationary to its cause. All that remains is to determine that Stationary is giving way to approaching traffic, and exhibits no other plausible behaviour (e.g., broken down, parked).

We separate the giveway episode into five routines that use region-based-prediction and perceiver level coordination. These routines are:

- Notice-stopping-object, which on completion generates **event-gw1**. The gross change in motion from moving to stationary allocates an agent and prompts the question "why is vehicle stationary?".
- Look-for-path-blocker, which on completion generates **event-gw2**. To be blocking it does not need to be physically in the way, it can also block by having "right-of-way" such that its path *will* block.
- Work-out-conflict-area, which on completion generates **event-gw3**. Having predicted the paths of Stationary and Blocker above, intersect them to find the mutually shared conflict area, CA.
- Watch-for-enter-conflict-area, which on completion generates **event-gw4**. In order to determine whether Stationary gives way, wait until Blocker has passed through CA.
- Notice-starts-to-move, which on completion generates **enter-gw5**. We then observe if Stationary moves. The gross change in motion from stationary to moving reallocates an agent to Stationary.

The five routines given above order and, as a continuous sequence, describe a temporal sequence of perceiver activity that identifies a giveway episode.

4.3 Results

Here we compare the effect of using two different tasks: "look for likely overtaking behaviour" and "look for likely giveway behaviour", to illustrate how HIVIS-WATCHER is able to solve the problem of identifying overtaking encountered in HIVIS-MONITOR. Also, we see how changing the observation task given to HIVIS-WATCHER noticeably alters the performance of the system (results are displayed in figure 2– figure 4).

Overtaking The purpose of this example is to show that HIVIS-WATCHER can pick out a pair of vehicles that are performing an overtaking episode. To do this we will use the policy "attend to likely overtaking and ignore likely following". The missing entries are because of the occlusion where no mutually proximate objects are visible to the observer. The vehicle shapes given in outline denote uninteresting peripheral objects, the number near each vehicle is its index reference (or buffer slot number), and the vehicle outlines that have marker shapes "attached" to them are selected objects that have been allocated an agent.

During frames 96 and 108 one of the vehicles occludes the other from the camera. The camera's field-of-view which affects the contents of the frame updates because we are dependent upon what is visible from the camera position not what is visible from the overhead view. By frame 132 overtaking is positively identified. A comparison between this policy and the similar one for "attend to likely following and ignore likely overtaking", together with more implementation details, is given in Howarth and Buxton [19].

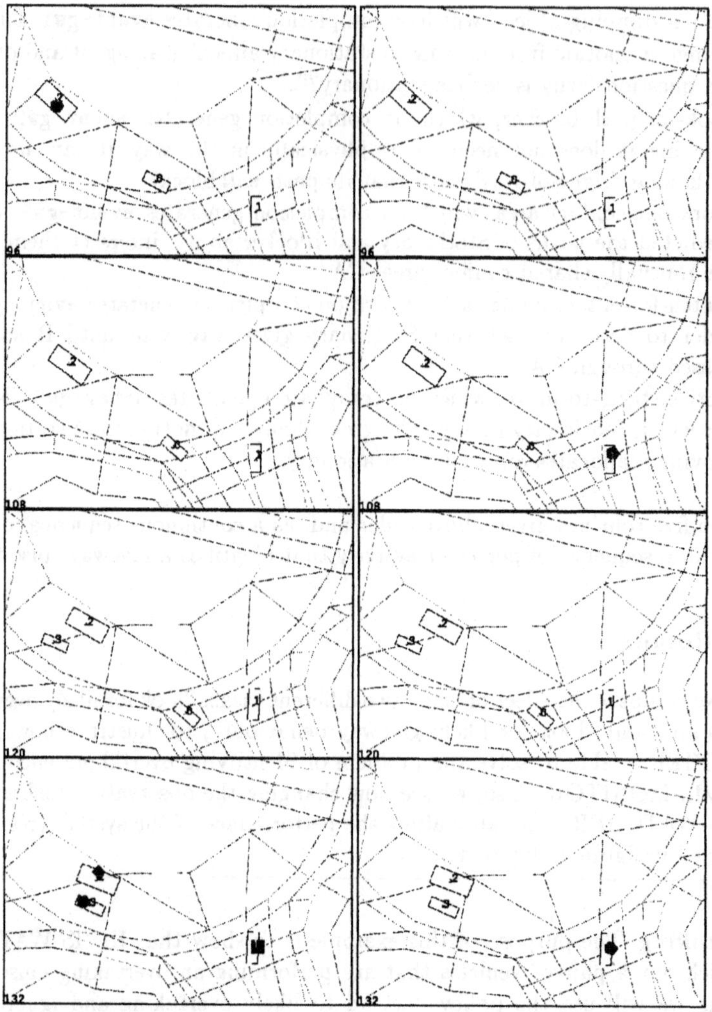

Fig. 2. Part 1 (overtaking policy left, giveway policy right).

Giveway To illustrate the need for local and global viewpoints we use the policy "look for likely giveway behaviour". HIVIS-WATCHER uses three attentional markers to perform the giveway detection routine, and the events correspond to the five routines described in section 4.2. The frames 108, 132–156 describe the allocation of **`*agent2*`** cued by gross-change-in-motion. At frame 120 the vehicle moved again, before the motion-prior was altered from **`moving`** to **`stationary`** by the agency operator **`change-motion-prior!`**. The value of motion-prior has changed by frame 168 because the object ceases to have an interesting motion property. Frame 192 shows the results from the region path predictions that

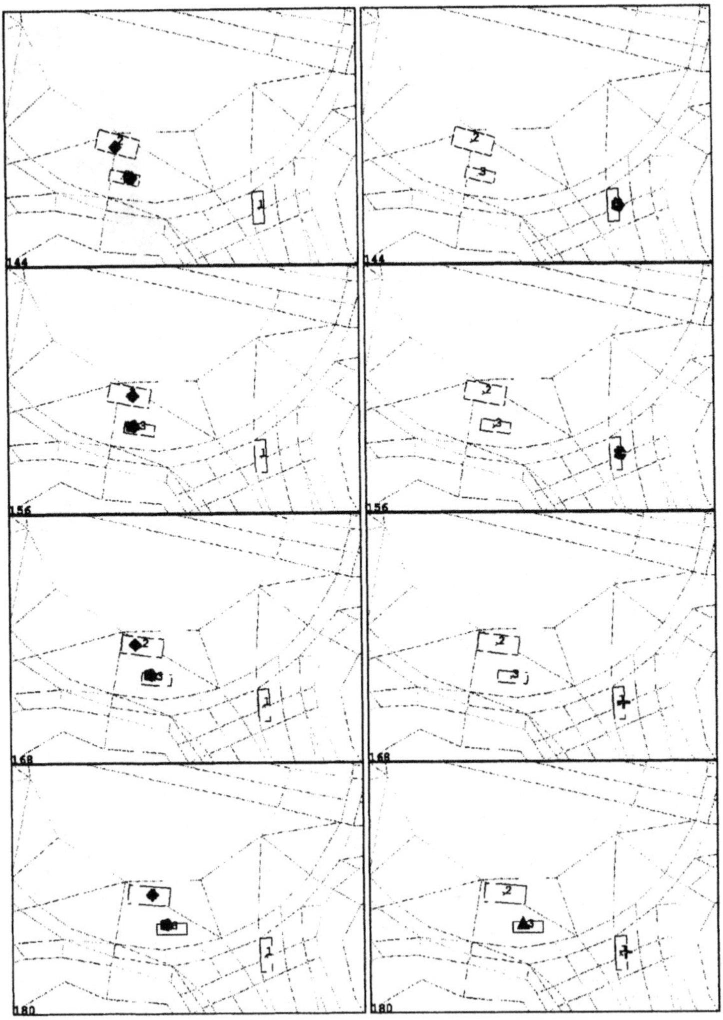

Fig. 3. Part 2 (overtaking policy left, giveway policy right).

generate the contents of the kernel activation planes. Frames 204–258 display the activation plane. Frame 228 shows the removal of ***head-marker*** following a successful intersection.

4.4 General features

The traditional separation made in cognitive science between input and central systems provides a description of the two tightly coupled components in HIVIS-WATCHER. The input system obtains object aspects, while the central system

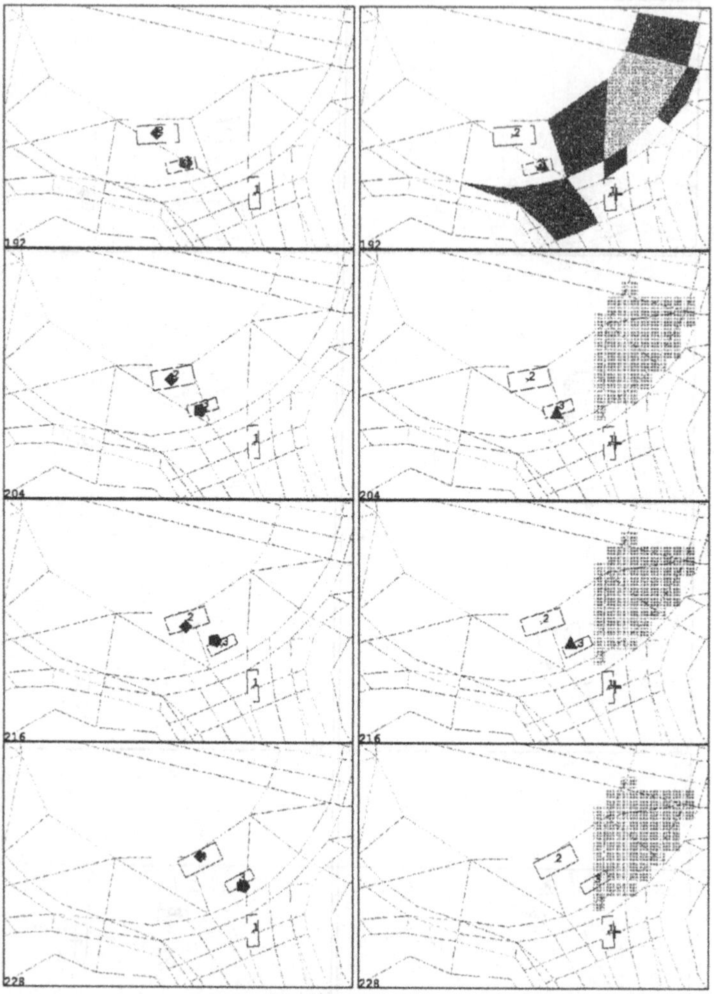

Fig. 4. Part 3 (overtaking policy left, giveway policy right).

controls which objects should be attended so their aspects can fulfill a given surveillance task. The separation of preattentive and attentive processing, and the use of a task directed central mechanism here provides what is needed for the official-observer to watch out for selected behaviours. HIVIS-WATCHER thus provides timely surveillance information about what is happening in the scene.

5 Conclusion

The main benefit of HIVIS-WATCHER over HIVIS-MONITOR is that its task orientedness reduces runtime representation, reasoning and complexity. In HIVIS-WATCHER: (1) the deictic representation has simplified the computational model of behaviour; (2) the situated approach has taken into account both the evolving context of the dynamic scene objects and also the task-oriented observer's context; (3) the use of selective attention provides a more viable form of real-time processing.

Other key points of this paper concern: (1) the distinction between script-based and more situated approaches; (2) the separation and integration of global and local reasoning in the context of a single official-observer, together with the illustration of how both play complementary roles in developing different levels of understanding; (3) the propagation of reasoning in the "here-and-now" through to the control mechanism in order to reflect the reactive quality of dynamic object behaviour.

Current work is addressing two important issues. The first concerns ways to control perceptual processing so that the task-level knowledge will influence when model-matching is performed. The second concerns learning the behavioural information, removing the hand coded element in choosing preattentive and attentive cues in HIVIS-WATCHER. Although this research has been illustrated by using data from a road-traffic surveillance, the intention is that the general framework should be applicable to other domains.

References

1. Philip E. Agre and David Chapman. Pengi: An implementation of a theory of activity. In *Sixth AAAI Conference*, pages 268–272. AAAI Press, 1987.
2. Dana H. Ballard. Animate vision. *Artificial Intelligence*, 48:57–86, 1991.
3. Hilary Buxton and others. VIEWS: Visual Inspection and Evaluation of Wide-area Scenes. *IJCAI-91 Videotape Program*, Morgan Kaufmann, 1991.
4. Hilary Buxton and Shaogang Gong. Visual Surveillance in a Dynamic and Uncertain World. *Artificial Intelligence*, 78:371–405, 1995.
5. David Chapman. *Vision, Instruction and Action*. The MIT Press, 1991.
6. Anthony N. Clark. Pattern recognition of noisy sequences of behavioural events using functional combinators. *The Computer Journal*, 37(5):385–398, 1994.
7. David R. Corrall, Anthony N. Clark, and A. Graham Hill. Airside ground movements surveillance. In *NATO AGARD Symposium on Machine Intelligence in Air Traffic Management*, pages 29:1–29:13, 1993.
8. David R. Corrall and A. Graham Hill. Visual surveillance. *GEC Review*, 8(1):15–27, 1992.
9. Li Du, Geoffery D. Sullivan, and Keith B. Baker. Quantitative analysis of the viewpoint consistency constraint in model-based vision. In *Fourth ICCV*, pages 632–639. IEEE Press, 1993.
10. Margaret M. Fleck. Representing space for practical reasoning. *Image and Vision Computing*, 6(2):75–86, 1986.

11. Margaret M. Fleck. *Boundaries and Topological Algorithms*. PhD thesis, MIT AI Lab., 1988. AI-TR 1065.
12. R.A. Frost. Constructing programs as executable attribute grammars. *The Computer Journal*, 35(4):376–387, 1992.
13. James J. Gibson. *The Ecological Approach to Visual Perception*. Houghton Mifflin Company, 1979.
14. Shaogang G. Gong and Hilary Buxton. On the expectations of moving objects. In *Tenth ECAI Conference*, pages 781–784, 1992.
15. Ian Horswill. Visual routines and visual search: a real-time implementation and an automata-theoretic analysis. In *Fourteenth IJCAI Conference*, pages 56–62. Morgan Kaufmann, 1995.
16. Richard J. Howarth. *Spatial Representation, Reasoning and Control for a Surveillance System*. PhD thesis, QMW, University of London, 1994.
17. Richard J. Howarth. Interpreting a dynamic and uncertain world: high-level vision. *Artificial Intelligence Review*, 9(1):37–63, 1995.
18. Richard J. Howarth and Hilary Buxton. An analogical representation of space and time. *Image and Vision Computing*, 10(7):467–478, 1992.
19. Richard J. Howarth and Hilary Buxton. Selective attention in dynamic vision. In *Thirteenth IJCAI Conference*, pages 1579–1584. Morgan Kaufmann, 1993.
20. Edwin Hutchins. Understanding micronesian navigation. In G. Dedre and A.L. Stevens *Mental Models*, pages 191–225. Lawrence Erlbaum Associates, 1983.
21. Simon King, Sophie Motet, Jérôme Thoméré, and François Arlabosse. A visual surveillance system for incident detection. In *AAAI workshop on AI in Intelligent Vehicle Highway Systems*, pages 30–36. AAAI Press, 1994.
22. Stephen M. Kosslyn, Rex A. Flynn, Jonathan B. Amsterdam, and Gretchen Wang. Components of high-level vision: a cognitive neuroscience analysis and accounts of neurological syndromes. *Cognition*, 34:203–277, 1990.
23. Hans-Hellmut Nagel. From image sequences towards conceptual descriptions. *Image and Vision Computing*, 6(2):59–74, May 1988.
24. Bernd Neumann. Natural language descriptions of time-varying scenes. In David L. Waltz, editor, *Semantic Structures: Advances in Natural Language Processing*, pages 167–206. Lawrence Erlbaum Associates, 1989.
25. Nils J. Nilsson. Teleo-reactive programs for agent control. *Journal of Artificial Intelligence Research*, 1:139–158, 1994.
26. Raymond D. Rimey and Christopher M. Brown. Control of selective perception using Bayes nets and dicision theory. *International Journal of Computer Vision*, 12(2/3):173–207, April 1994.
27. Roger C. Schank and Robert P. Abelson. *Scripts, Plans, Goals and Understanding*. Lawrence Erlbaum Associates, 1977.
28. Andrew F. Toal and Hilary Buxton. Spatio-temporal reasoning within a traffic surveillance system. In G. Sandini, editor, *Computer Vision -ECCV'92*, pages 884–892. Springer-Verlag, 1992.
29. John K. Tsotsos. Toward a computational model of attention. In T. Papathomas, C. Chubb, A. Gorea, and E. Kowler, editors, *Early Vision and Beyond*, pages 207–218. The MIT Press, 1995.
30. Shimon Ullman. Visual routines. In Steven Pinker, editor, *Visual Cognition*, pages 97–159. The MIT Press, 1985.
31. Anthony D. Worrall, Geoffery D. Sullivan, and Keith B. Baker. Advances in model-based traffic vision. In *British Machine Vision Conference 1993*, pages 559–568. BMVA Press, 1993.

A Robust Active Contour Model for Natural Scene Contour Extraction with Automatic Thresholding

Kian Peng, Ngoi[1] and Jiancheng, Jia[2]

[1] Defence Science Organisation, 20 Science Park Drive, Singapore 0511
[2] Nanyang Technological University, Nanyang Avenue, Singapore 2263

Abstract. An active contour model is proposed for contour extraction of objects in complex natural scenes. Our model is formulated in an analytical framework that consists of a colour contrast metric, an illumination parameter and a blurring scale estimated using hermite polynomials. A distinct advantage of this framework is that it allows for automatic selection of thresholds under conditions of uneven illumination and image blur. The active contour is also initialised by a single point of maximum colour contrast between the object and background. The model has been applied to synthetic images and natural scenes and shown to perform well.

1 Introduction

Shape analysis is very important in computer vision and has been widely used for recognition and matching purposes in many applications. Important information about the shape of the object can be provided by its contour. Active contour models have been successfully applied to the problem of contour extraction and shape analysis since its introduction by Kass et. al. [4] as snakes. These active contour models [1,11] are attracted to image features such as lines and edges, whereas internal forces impose a smoothness constraint. However, existing models are not well adapted for object segmentation in a complex image scene as they are easily distracted by uneven illumination, image blur, texture and noise. Several noteworthy variants were therefore proposed to allow active contour models to handle a complex scene. One class of active contour models [14] uses *a priori* knowledge about the expected shape of the features to guide the contour extraction process. When no knowledge of the expected object geometry is available, Lai et. al. [6] proposed a global contour model based on a regenerative shape matrix obtained by shape training. Region-based strategies to guide the active contours to the object boundaries were also proposed by Ronfard [10]. To distinguish the target object in a scene comprising of multiple objects, a local Gaussian law based on textural characteristics of objects [2] was introduced.

However, there are still several unresolved issues in applying active contour models to a complex natural scene. One issue is that current active contour models have not considered the effects of blurring. For example, Dubuisson et. al. [3] assumed that moving objects in an outdoor scene are not blurred. The reason is that blurring corrupt salient features like edges, lines and image contrast which is used to guide the active contour to the object's boundary. Blurring also makes the location the exact edge's position difficult. Unfortunately image blur cannot be prevented in many applications. In a scene consisting of multiple objects, the limited depth of field of the camera's lens will cause some objects to be out of focus. Objects can also be blurred by motion when lighting conditions do not permit a fast shutter speed.

The performance of current active contour models also depends on the initial contour position and the proper selection of internal parameters and thresholds [13]. The common experience among users is that if the initial position and thresholds are not chosen properly, the active contour may be trapped in a local minima resulting in a sub-optimal solution. The selection of internal parameters and thresholds are done heuristically through several times of trial and error. Thus, for a complex natural scene, the active contour has to be initialised near the object's boundary and this requires human interaction and knowledge of the object's shape. To overcome the problem of initial position, Lai et. al. [6] had proposed the use of Hough Transform to initialise the active contour. However, computations and processing time are increased. Vieren [12] used a "border snake" to initialise and enclosed the object as it entered the scene by its periphery.

In this work, an active contour model robust to uneven illumination and image blur has been proposed to extract object contours in complex natural scenes. Our model is formulated in an analytical framework consisting of a colour contrast metric, a blurring parameter estimated using hermite polynomials and an illumination parameter. The distinct advantage of our model over current models is its robustness to an arbitrary complex natural scene achieved by the automatic threshold selection. Our active contour model is initialised by a single point in the object of interest with no *a priori* knowledge of the shape. However, the point of initialisation must lie in the region with maximum colour contrast between the object and background. It is argued that locating a point with maximum contrast is computationally more efficient than initialising an entire contour near the object's boundary.

2 Model Formulation

In this section, the proposed automatic threshold framework will be presented. Details of the other aspects of the model can be found in [8].

2.1 Colour Contrast

The logarithmic colour intensity has been chosen to determine the minimum threshold value. The purpose is to provide a high gain at the low and central portion of the camera's response and decreasing gain at the positive extremes. This is consistent with our human eye which is sensitive to low and normal lighting but saturates when the lighting is bright. The colour contrast between a point of colour intensity (r,g,b) and a reference point with colour intensity (r_o,g_o,b_o) is defined as

$$C = \ln^2[(1+r)/(1+r_o)] + \ln^2[(1+g)/(1+g_o)] + \ln^2[(1+b)/(1+b_o)] \quad (1)$$

2.2 Automatic Threshold Selection

The threshold of the active contour has to be adaptively determined because of uncertainties introduced by noise, uneven illumination and image blur. Setting the correct threshold will ensure that the active contour remain at the desired edges of the object. For real images, we characterise edges by a Gaussian model with a scale σ_b. This scale parameter is related to our human experience of blurness. A high value of σ_b will result in a very blurred edge while a sharp edge has a small value. The Gaussian edge model used is

$$I(x,y) = L_e + \frac{C}{2}\,\mathrm{erf}\,[(x\cos\theta + y\sin\theta - d)/\sigma_b] \quad (2)$$

where

L_e	: *the mean value of the edge's intensity*
C	: *the colour contrast of the edge in (1)*
θ	: *the orientation of the edge*
d	: *perpendicular distance from the centre of window function*
erf	: *error function*

The Gaussian blur which is one of the most commonly encountered description of blur has been assumed. It has been shown that the blur due to camera defocuses can well be approximated by a 2-D Gaussian kernel [9]. The case of Gaussian blurring is also easy to tackle mathematically. In this respect, the Hermite polynomials are chosen as a basis since it is orthogonal to the Gaussian window [7].

An input image *f(x,y)* can be decomposed into the sum of several images using a window function *w(x,y)* given as

$$f(x,y) = \frac{1}{h(x,y)} \sum_{(p,q)\in S} f(x,y) \bullet w(x-p, y-q) \quad (3)$$

where *h(x,y)* is the periodic weighting function

$$h(x,y) = \sum_{(p,q)\in S} w(x-p, y-q) \quad (4)$$

for all sampling position *(p,q)* in sampling lattice *S*. The windowed image is then approximated by a polynomial with orthonormal the basis functions φ given as

$$w(x-p, y-q)[f(x,y) - \Sigma\Sigma f_{m,n-m}(p,q)\varphi_{m,n-m}(x-p, y-q)] = 0 \quad (5)$$

If the window function is chosen to be Gaussian with a window spread of σ, the coefficients of the polynomial can be evaluated as

$$f_{m,n-m}(p,q) = \int_{-\infty}^{+\infty}\int_{-\infty}^{+\infty} f(x,y)\,a_{m,n-m}(x-p, y-q)\,dxdy \quad (6)$$

where

$$a_{m,n-m}(x,y) = \frac{1}{\sqrt{2^n m! n!}} \frac{1}{\pi\sigma^2} H_m(\tfrac{x}{\sigma}) H_{n-m}(\tfrac{y}{\sigma}) \exp(\tfrac{-(x^2+y^2)}{\sigma^2})$$

$H_n(x)$ is the Hermite polynomial of degree n in x.

It has been shown [5,7] that the analysis function $a_{m,n-m}$ of order n is the nth order derivative of the Gaussian i.e.

$$a_n(x) = \frac{1}{\sqrt{2^n n!}} d^n/d(\tfrac{x}{\sigma})^n \left[\frac{1}{\sigma\sqrt{\pi}} \exp(-x^2/\sigma^2)\right] \tag{7}$$

From (6) and (7), the hermite coefficients of a blurred edge with a scale of σ_b up to the third order can be written as

$$\begin{aligned}
f_0 &= L_e + (C/2)\mathrm{erf}[(d/\sigma)/\sqrt{1+(\sigma_b/\sigma)^2}\,] \\
f_1 &= [1/\sqrt{1+(\sigma_b/\sigma)^2}\,](C/\sqrt{2\pi}\,)\exp(-(d/\sigma)^2/(1+(\sigma_b/\sigma)^2)) \\
f_2 &= [1/\sqrt{1+(\sigma_b/\sigma)^2}\,]^2(C/\sqrt{2\pi}\,)[(d/\sigma)/\sqrt{1+(\sigma_b/\sigma)^2}\,]\exp(-(d/\sigma)^2/(1+(\sigma_b/\sigma)^2)) \\
f_3 &= [1/\sqrt{1+(\sigma_b/\sigma)^2}\,]^3(C/\sqrt{2\pi}\,)(\tfrac{1}{\sqrt{6}})[(2(d/\sigma)/1+(\sigma_b/\sigma)^2)-1]\exp(-(d/\sigma)^2/(1+(\sigma_b/\sigma)^2))
\end{aligned} \tag{8}$$

We derived the edge parameters from the hermite coefficients as

$$\sigma_b/\sigma = [(2e_2^2/e_1^2 \pm \sqrt{6}\,e_3/e_1) - 1]^{1/2} \tag{9}$$

$$d/\sigma = e_2/e_1(2e_2^2/e_1^2 \pm \sqrt{6}\,e_3/e_1)^{-1} \tag{10}$$

$$C = e_1\sqrt{2\pi}\,(2e_2^2/e_1^2 \pm \sqrt{6}\,e_3/e_1)^{-1/2}\exp(e_2^2/e_1^2(2e_2^2/e_1^2 \pm \sqrt{6}\,e_3/e_1)^{-1}) \tag{11}$$

where the nth order local energy is $e_n^2 = \Sigma f_n^2$

From (9), there are two possible solutions. The first solution known as the near solution

$$\sigma_b/\sigma = [(2e_2^2/e_1^2 + \sqrt{6}\,e_3/e_1) - 1]^{1/2} \tag{12}$$

is valid when the centre of the window function is near the edge i.e.

$$(d/\sigma)^2/(1+(\sigma_b/\sigma)^2) \le 1/2 \tag{13}$$

and the far solution

$$\sigma_b/\sigma = [(2e_2^2/e_1^2 - \sqrt{6}\,e_3/e_1) - 1]^{1/2} \tag{14}$$

is valid for all other cases.

The threshold value for our Gaussian edge model occurs when the rate of change of gradient is maximum. If the window is sufficiently close to the edge point such that (12) is satisfied, an analytical expression to select the threshold automatically can be written as

$$\begin{aligned} \text{thld} &= |C| \exp(-d/\sigma_b)^2 \\ &= \sqrt{2\pi}\,\sqrt{1+(\sigma_b/\sigma)^2}\,\exp[-1/(2(\sigma_b^2/\sigma^2+1))]\,e_1 \end{aligned} \tag{15}$$

The parameter e_1 in (15) is the energy of the first order hermite polynomials and is obtained by convoluting the Gaussian mask with the image. Physically, e_1 can be interpreted as an illumination measure. In other words, a strong illumination increases the value of e_1 while a weak illumination reduces e_1 correspondingly. The value of e_1 therefore acts as a scaling agent of the threshold when illumination changes. For a constant illumination, the threshold of the active contour also increases with image blur. Setting the threshold under constant illumination is known as scale thresholding. It is observed in (15) that when the ratio of the image blur and the window function (σ_b/σ) is greater than 1, the threshold (*thld*) increases almost linearly with blurring.

3 Model Evaluation

In this section, the automatic threshold framework proposed in the previous section will be evaluated using synthetically generated images. The performance between our model and current active contour models using fixed thresholds is also compared.

3.1 Effects of Uncorrelated Errors on Image Blur estimates

The uncorrelated errors in the measured energies of e_1, e_2 and e_3 occur mainly due to image noise, quantisation of filter or polynomial coefficients . For small uncorrelated errors in energies, it is possible to express the error in the estimate of image blur as

$$\begin{aligned} \Delta\sigma_b &= (\partial\sigma_b/\partial e_1)\Delta e_1 + (\partial\sigma_b/\partial e_2)\Delta e_2 + (\partial\sigma_b/\partial e_3)\Delta e_3 \\ &= [\sigma^2/2\sigma_b][2(e_2^2/e_1^2) + \sqrt{6}\,e_3/e_1]^{-2}[(4e_2^2/e_1^2 + \sqrt{6}\,e_3/e_1^2)\Delta e_1 - (4e_2/e_1^2)\Delta e_2 - (\sqrt{6}\,e_3/e_1)\Delta e_3] \end{aligned} \tag{16}$$

Figure 1 shows the error in the estimation of the edge's scale for an error of 5% in e_1, e_2 *and* e_3 plotted by the analytical expression in (16). From Figure 1, the error is minimum when the window spread is approximately equal to the image blur ($\sigma_b/\sigma = 1$). After that, the error increases rapidly as the ratio of σ_b/σ becomes smaller. This is due to the large spread of the window function resulting in information redundancy. Estimation accuracy is poor because the edge's information forms only a small portion of the overall data. However if the spread of the window function is smaller than the image blur ($\sigma_b/\sigma > 1$) estimation accuracy is still more reliable as compared to ($\sigma_b/\sigma < 1$). For a small spread of the window function ($\sigma_b/\sigma > 1$), all the information obtained still belongs to the edge. This shows that incomplete information has greater reliability than redundant information. From Figure 1, we set the following bound to ensure an accuracy of 86%

$$0.5 \leq (\sigma_b/\sigma) \leq 2.0 \tag{17}$$

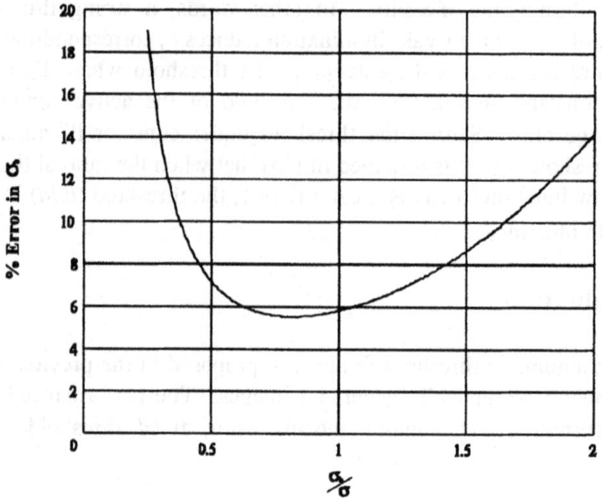

Fig. 1. : Error in the estimation of the image blur for $C = 5, d = 0.25, \sigma = 8$

3.2 Minimum Threshold

It can be seen from Figure 1 that when $\sigma_b/\sigma < 0.5$, the estimate of the image blur and the threshold of the active contour is unreliable. This happens when the object's boundary is not within the region of interest of the active contour. In this instance, a minimum threshold value ($thld_{min}$) is set to allow the active contour to move towards the object's boundary.

The contrast of a colour (r,g,b) from a reference colour (r_o,g_o,b_o) for a maximum colour deviation of $\{\Delta r, \Delta g, \Delta b\}$ and assuming that r,g,b [illegible] 1, can be expressed as

$$thld_{min} = \ln^2(1 + \Delta r/r_o) + \ln^2(1 + \Delta g/g_o) + \ln^2(1 + \Delta b/b_o) \tag{18}$$

If $\max\{|\Delta r/r_o|, |\Delta g/g_o|, |\Delta b/b_o|\} < 1$, expanding the above equation gives

$$thld_{min} \approx (\Delta r/r_o)^2 - (\Delta r/r_o)^3 + 0.25(\Delta r/r_o)^4 + (\Delta g/g_o)^2 - (\Delta g/g_o)^3 + 0.25(\Delta g/g_o)^4 + (\Delta b/b_o)^2 - (\Delta b/b_o)^3 + 0.25(\Delta b/b_o)^4 \quad (19)$$

If $|\Delta r/r_o|, |\Delta g/g_o|, |\Delta b/b_o|$ is sufficiently small, the minimum threshold ($thld_{min}$) required for the proper working of the active contour can be obtained from (19) as

$$thld_{min} = (\Delta r/r_o)^2 + (\Delta g/g_o)^2 + (\Delta b/b_o)^2 \quad (20)$$

The above equation serves as a guide to determine the minimum threshold value of the active contour in our model in terms of the maximum colour intensity fluctuations allowable within the object. When the ratio of σ_b/σ is less than 0.5, the minimum threshold criterion in (20) is used. Scale thresholding in (15) is used for the rest of the image edge's scale $(0.5 \le \sigma_b/\sigma \le 2.0)$.

3.3 Performance Evaluation on Image Blur

Simulations of synthetic images had been performed to evaluate the performance of the automatic thresholding framework on image blur. Firstly, a reference threshold must be obtained. This reference threshold is the value the active contour has to be set to remain at the object's boundary as the object blurs. Using the reference threshold, the performance of three thresholding methods (i.e. fixed thresholding, scale thresholding and the proposed automatic thresholding) under blurring is evaluated. The performances of current active contour models are similar to fixed thresholding because their thresholds do not change according to image blur.

The reference threshold is obtained using the following procedure. A step edge with a colour contrast of 4.84 is generated at a predetermined location. The image is subsequently blurred by a series of Gaussian kernels with a known scale σ_b. Each time the step edge is blurred by a Gaussian kernel, the contrast at the predetermined location is measured. For example, if the step edge at location 400 in the x-axis have an initial contrast of zero, a Gaussian blurring scale of 5 will give a contrast of approximately 0.5 at location 400 (see Figure 2). Setting the active contour threshold below 0.5 yields an edge location below 400 and a threshold higher than 0.5 results in an edge location greater than 400. Therefore, the active contour will contract if the threshold is less than 0.5 and expands outward if the threshold is greater than 0.5.

The performance of automatic thresholding, scale thresholding and fixed thresholding under image blurring is also shown on Figure 2. A window spread σ of 12 is used for the experiment. Scale thresholding is less reliable for small image blur $(0 < \sigma_b < 6)$ but gives better results for image blur from 6 to 24 in comparison to the reference threshold. The most accurate estimate occurs at an image blur of approximately 12 $(\sigma_b/\sigma = 1)$. This is consistent with the theoretical results in Figure 1. However, the performance of a using a fixed threshold for an active contour is better at smaller image blur and becomes progressively unreliable at higher image scale. The propose automatic thresholding framework combines the advantages of scale thresholding and fixed thresholding. In the framework, fixed thresholding is used when the ratio of σ_b/σ is between 0.0 and 0.5 while scale thresholding is used when the ratio is between 0.5 and 2.0.

Figure 3 gives a comparison of the performance under blurring between the our model which uses automatic thresholding and current models with fixed thresholds under image blur. A window function with spread of 12 and fixed threshold of 0.5 is again used. It can be seen from Figure 3 that fixed thresholding gives an error of 6 pixels for an image scale of 24. This is compared to an error of only 2 pixels for our automatic thresholding scheme. Therefore our proposed model is more robust to the effects of blurring than current active contour models.

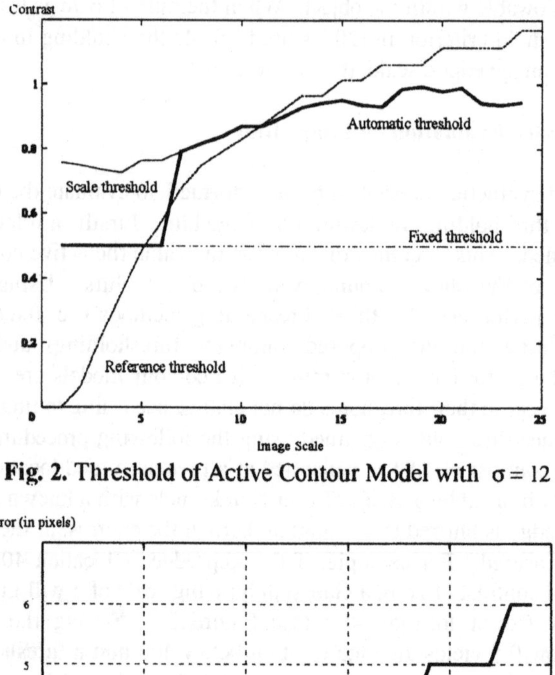

Fig. 2. Threshold of Active Contour Model with $\sigma = 12$

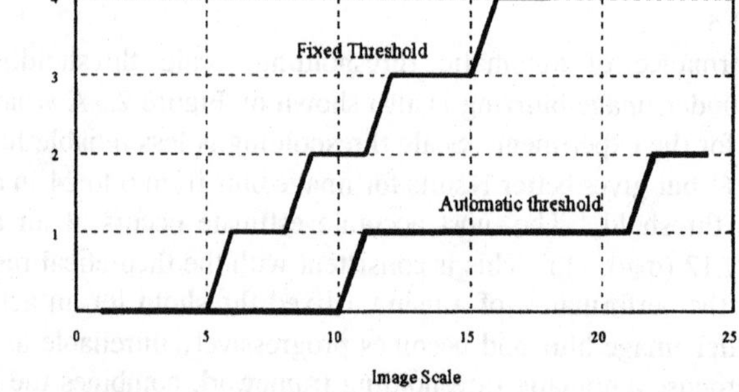

Fig. 3. Performance of automatic threshold against a fixed threshold

4. Experimental Results

Our active contour model has been applied to natural scenes. Figure 4 shows a complex image of a reptile on a relatively uniform background. The active contour is initialised at the body of the reptile as indicated by a cross. The high contrast between the reptile (brown hues) and the blue sky allowed the reptile's profile to be extracted accurately. Figure 5 shows a monochromatic image of three rocks with large intensity variations. The active contours are again initialised at the centre of the rocks and the intensity difference between the rocks and their background allowed reliable contours to be extracted. Figure 6 shows the contour of a vehicle extracted from a complex background with illumination and colour fluctuations.

For the contour extraction of a blurred vehicle in Figure 7, some rocks on the river bank have very similar colour and intensity as the vehicle. Our active contour cannot differentiate between the object and the background at these points without any *a priori* knowledge of the car's shape. Therefore, two volcanoes [4] denoted by crosses have been manually placed to push back the contours. It is observed that a reliable contour is still extracted despite the effects of image blur.

Figure 8 shows the performance of our model for a complex object (cottage) in a cluttered background. Noticed the different textures of the roof, windows, door and the background trees. Furthermore, shadows also reduce the colour contrast between some parts of the roof and the trees. The overall performance of our active contour model is still reliable despite some distortions at the roof and chimney. The active contour is initialised near the door of the cottage. It is noted that the point of initialisation is important for this image. For example, if the initialisation is at the roof of the cottage, the active contour will not be able to extract the lower half of the cottage. The reason is that the blue doors and windows have a greater colour contrast as compared to the background trees with respect to the colour of the roof. Therefore, under these circumstances, the active contour must be initialised at the point in the cottage where the colour contrast with the background is maximum.

5. Conclusion

In this paper, we present an active contour model for the contour extraction of objects in a natural image scene. Only local region computations are involved and no pre-processing, feature detection or knowledge of the object's geometry is required. Our model is formulated in an analytical framework that allows thresholds to be selected automatically. This makes our model more robust than current models which have heuristic thresholds. In the contour extraction of a natural scene, our active contour model is initialised by a single point with maximum colour contrast between the object and the background. Locating the point of maximum contrast is computationally more efficient than initialising the active contour near the object boundary as done by current models. Experimental results on synthetic and real images showed that the model is robust to blurred images as well as complex natural outdoor scenes.

Fig. 4. Contour extracted of reptile

Fig. 5. Contour extracted of monochromatic rocks image

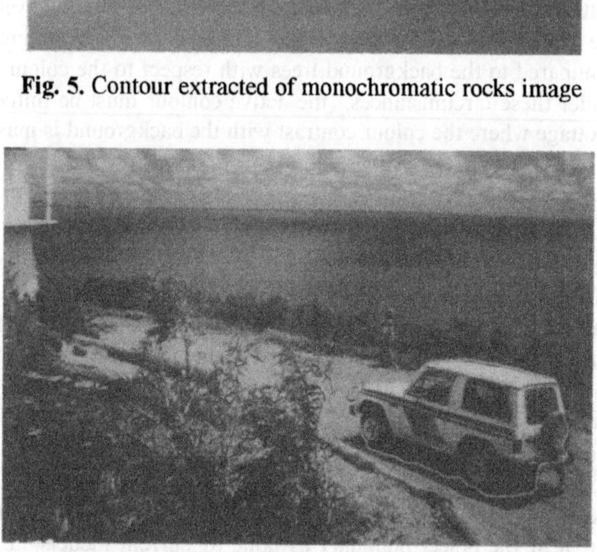

Fig. 6. Contour extracted of a vehicle

Fig. 7. Contour extracted of a blurred vehicle

Fig. 8. Contour extracted of a cottage

References

1. Cohen, L. D., Cohen, I.: Deformable Models for 3D Medical Images Using Finite Element and Balloons. Proc. IEEE Conf. on Computer Vision and Pattern Recognition (1992) 592-598
2. Delagnes, P., Benois, J., Barba, D.: Active contours approach to object tracking in image sequences with complex background. Pattern Recognition Letters **16** (1995) 171-178
3. Dubuisson, M.P., Jain, A. K.: Contour Extraction of Moving Objects in Complex Outdoor Scenes. Int. Journal of Computer Vision **14** (1995) 83-105
4. Kass, M., Witkin, A., Terzopoulos, D.: Snakes : Active Contour Models. First Int. Conf. on Computer Vision (1987) 259-269

5. Kayargadde, V., Martens, J. B.: Estimation of Edge Parameters and Image Blur Using Polynomial Transforms. CVGIP:Graphical Models and Image Processing. **56** (1994) 442-461
6. Lai, K. F., Chin,R. T.: Deformable Contours: Modelling and Extraction. Int. Conf. on Computer Vision and Pattern Recognition (1994) 601-608
7. Martens, J. B.: The Hermite transform-theory. IEEE Trans. Accoust. Speech Signal Process **38** (1990) 1595-1606
8. Ngoi, K. P., Jia, J. C..: An Analytical Active Contour Model with Automatic Threshold Selection. Second Asian Conf. on Comp. Vision (1995) III72 - III76
9. Pentland, A.: A New sense for depth of field. IEEE Trans. Pattern Anal. Mach. Intelligence **9** (1987) 523-531
10. Ronfard, R.: Region-Based Strategies for Active Contour Models. Int. Journal of Computer Vision **13** (1994) 229-251
11. Terzopoulos, D., Szeliski, R.: Tracking with Kalman Snakes. Active Vision, MIT Press (1992) 3- 20
12. Vieren, C., Cabestaing, F., Postaire, J. G.: Catching moving objects with snakes for motion tracking. Pattern Recognition Letters **16** (1995) 679-685
13. Xu, G., Segawa, E., Tsuji, S.: A Robust Active Contour Model with Insensitive Parameters. Proc. of 4th Int. Conf. on Computer Vision (1993) 562-566
14. Yuille, A. L., Hallinan, P. W.,Cohen, D. S.: Feature Extraction from Faces Using Deformable Templates. Int. Journal of Computer Vision **8** (1992) 99-111

A Maximum-Likelihood Approach to Visual Event Classification

Jeffrey Mark Siskind* and Quaid Morris

Department of Computer Science, University of Toronto, Toronto Ontario M5S 1A4 CANADA

Abstract. This paper presents a novel framework, based on maximum likelihood, for training models to recognise simple spatial-motion events, such as those described by the verbs *pick up*, *put down*, *push*, *pull*, *drop*, and *throw*, and classifying novel observations into previously trained classes. The model that we employ does not presuppose prior recognition or tracking of 3D object pose, shape, or identity. We describe our general framework for using maximum-likelihood techniques for visual event classification, the details of the generative model that we use to characterise observations as instances of event types, and the implemented computational techniques used to support training and classification for this generative model. We conclude by illustrating the operation of our implementation on a small example.

1 Introduction

People can describe what they see. Not only can they describe the objects that they see, they can also describe the events in which those objects participate. So if you were to see a person pick up a pen, you could describe that event by saying *The person picked up the pen*. In doing so you classify the two participant objects as a person and a pen respectively. You also classify the observed event as a picking-up event. Almost all recognition work in machine vision has focussed on object classification. In contrast, this paper is an attempt to address the problem of event classification.

While we are not the first to address this problem, our work differs from prior approaches (Badler, 1975; Nagel, 1977; Tsuji, Morizono, & Kuroda, 1977; Okada, 1979; O'Rourke & Badler, 1980; Rashid, 1980; Tsotsos, Mylopoulos, Covvey, & Zucker, 1980; Abe, Soga, & Tsuji, 1981; Marburger, Neumann, & Novak, 1981; Waltz, 1981; Marr & Vaina, 1982; Neumann & Novak, 1983; Thibadeau, 1986; Yamamoto, Ohya, & Ishii, 1992; Pinhanez & Bobick, 1995) in a number of important ways. First, we apply our methods to camera input, in contrast to synthetic input. Second, we group naturally occurring events into classes that correspond to pre-theoretic notions described by simple spatial-motion verbs like *pick up*, *put down*, *push*, *pull*, *drop*, *throw*, and so forth. While people may be able to perceive many other kinds of differences between two motion sequences,

* Current address: Department of Electrical Engineering, Technion, Haifa 32000, ISRAEL

we are only interested in detecting those differences that can be described using ordinary verbs. Finally, we use an approach based on maximum likelihood. We determine the parameters of a general model empirically from training data instead of formulating detailed logical and geometric descriptions of event classes by hand.

The vision community has done much prior work on processing image sequences, particularly sequences involving motion. Examples of such work include optical flow, 2D and 3D object tracking, and shape from motion, to name a few. While our work bears superficial resemblance to such work, we wish to stress that we are concerned with a fundamentally different problem that is orthogonal to the ones addressed by such prior work. For example, we are interested in event classification and not 3D tracking. For us, techniques like 3D tracking are relevant only in so far as they might facilitate—or be facilitated by—event classification. In fact, the techniques that we describe in this paper do not perform detailed shape recovery, object classification, or 3D pose estimation.

Siskind (1992, 1995) proposed a technique for classifying events by recovering changing support, contact, and attachment relations between participant objects using a kinematic simulator driven from the output of 3D tracking. A tacit assumption behind this prior work was that object recognition is a prerequisite for event recognition. While attempting to implement the aforementioned techniques, we conducted a simple experiment that calls into question the validity of this assumption.

We took several movies of simple spatial-motion events in ordinary desk-top environments. These events included picking-up, putting-down, pushing, and pulling boxes, dropping erasers, and various collisions between objects. These movies were filmed using an off-the-shell SunVideo system recording image sequences with a resolution of 320×240 at 30 frames-per-second compressed using Sun's CellB format. We then applied an edge detector and line finder to each of the images in each movie and animated the resulting output. One original movie and the edge-detected images that correspond to that movie are shown in figure 1.

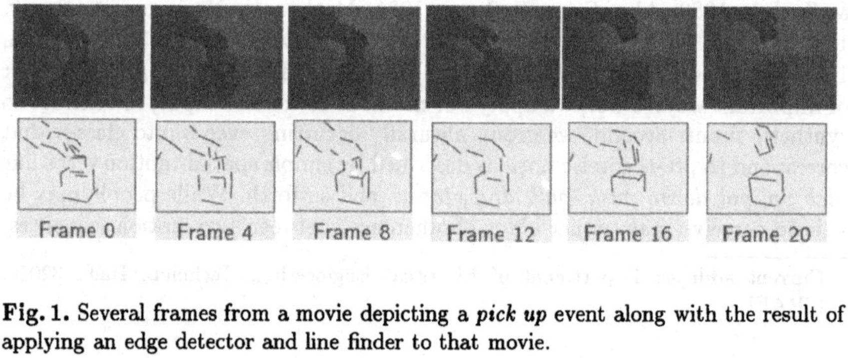

Fig. 1. Several frames from a movie depicting a ***pick up*** event along with the result of applying an edge detector and line finder to that movie.

The results of this experiment are striking. Edge detection and line finding collectively reduce the amount of information in the movie from 76,800 bytes per frame to roughly 25 line segments, or 100 bytes, per frame. This corresponds to a factor-of-768 lossy data compression. When watching such edge-detected output, either as isolated frames or as animated line movies, humans cannot reliably recognise the objects that appear in the images. Yet despite such lossy data compression, humans can still reliably recognise the depicted events when viewing an animation of the line drawings.

In retrospect, these results are not surprising. They are very much in-line with the point-light studies of animate-body motion performed by Johansson (1973) and others. They do, however, call into question the assumption of prior work, namely that event recognition presupposes object recognition. The information required to classify objects and recover their 3D pose over time is simply not available in the animated line drawings that we constructed. Yet even without such information, event recognition is a very robust process.

This leads us to conjecture, instead, that human event perception does not presuppose object recognition. We conjecture that event recognition is performed by a visual pathway that is separate from object recognition. Furthermore, we conjecture that this pathway requires far lower information bandwidth than object recognition. If this is true, then it may be the case that event recognition is an *easier* problem than object recognition and more amenable, in the short term, to synthetic engineered implementations. The rest of this paper offers precisely that: one possible framework for building an event-recognition engine.

2 The Framework

Linguistic evidence indicates that humans characterise events in terms of characteristic changes in the properties of, and relations between, objects that participate in those events. For example, a *pick up* event typically consists of a sequence of two subevents. During the first subevent, the hand of the *agent* moves toward the *patient*,[2] the object being picked up, while the patient rests on the *source* object. The agent then comes into contact with, and grasps, the patient. During the second subevent, the agent moves together with the patient away from the source, while supporting the patient. Similarly, a *throw* event typically consists of a sequence of two subevents. During the first subevent, the patient moves with the agent while the agent grasps, supports, and applies force to the patient. This subevent ends when the agent releases the patient. During the second subevent, the patient continues in unsupported motion after leaving the patient's hand in a trajectory that results, in part, from the force applied by the agent during the first subevent. The types and properties of the participant objects have little importance in defining these event types. Any agent can throw any patient or pick up any patient from any source. Objects simply fill *roles* of an event type. While some event types are characterised by the types

[2] The term 'patient' is being used here to denote the object affected by an action.

and potentially changing properties of participant objects, events described by simple spatial-motion verbs are largely characterised by the changing spatial and force-dynamic relations between objects. This paper focuses solely on such event types. Our long-term goal is to characterise and recognise simple spatial-motion events by recovering changing force-dynamic relations in addition to changing spatial relations. Mann, Jepson, and Siskind (1996) present some work along these lines. This paper, however, describes techniques for event recognition that are based solely on modelling the characteristic motion profiles, changing over time, of objects that participate in different simple spatial-motion events.

We partition the event recognition task into two independent subtasks. The lower-level task performs 2D object tracking, taking image sequences as input and producing as output a stream of 2D position, orientation, shape, and size values for each participant object in the observed event. This 2D pose information takes the form of a set of parameters for ellipses that abstractly characterise the position, orientation, shape, and size of the participant objects. This lower-level tracking is done without any constraint from event models. The upper-level task takes as input the 2D pose stream produced by lower-level processing, without any image information, and classifies such a pose stream as an instance of a given event type. We use a maximum-likelihood approach to perform the upper-level task. In this approach we use a supervised learning strategy to train a model for each event class from a set of example events from that class. We then classify novel event observations as the class whose model best fits the new observation.

3 Tracking

Our tracker uses a mixture of colour-based and motion-based techniques. Colour-based techniques are used to track objects with uniform distinctive colour, such as the blocks, even though such objects might not be in motion, or might be in motion for only part of the event. Motion-based techniques are used to track moving objects, such as the hand of the agent, even though the colour of such objects might not be uniform or distinctive and thus would not be detected by the colour-based techniques.

Our tracker operates on a frame-by-frame basis, tracking coloured objects and moving objects independently. To track coloured objects it first determines a set of 'coloured pixels' in each frame. Pixels are considered to be coloured if their saturation and value are above specified thresholds. Figure 2(b) shows the coloured pixels derived from the input image in figure 2(a). Such pixels are then classified into bins using a histogram clusterer based on hue.

After finding coloured regions in each frame, our tracker then finds moving regions. To do so it determines a set of 'moving pixels' for each frame by thresholding the absolute value of the difference between the grey scale values of corresponding pixels in adjacent frames. Figure 2(d) shows the set of moving pixels recovered by this technique for the image in figure 2(a).

Each hue cluster might, however, be spread over noncontiguous regions of the image. There might also be multiple moving objects, so the set of moving pixels

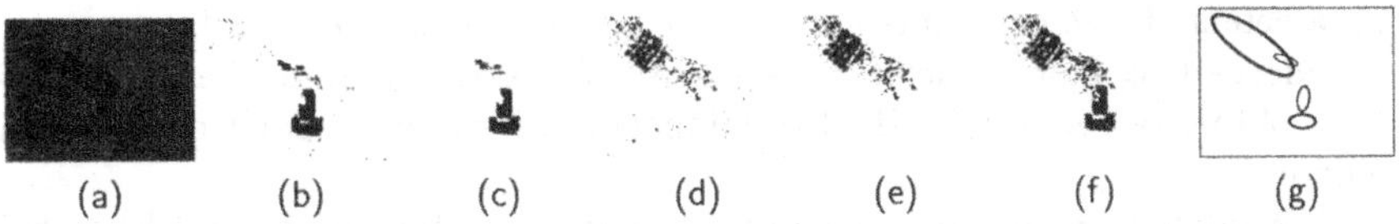

Fig. 2. The processing stages of our tracker. (a) shows an input image. (b) shows the coloured pixels. (c) shows the output of the region grower on (b). (d) shows the moving pixels. (e) shows the output of the region grower on (d). (f) shows the combination of (c) and (e). (g) shows the ellipses that are fit to the regions from (f).

might also be spread over noncontiguous regions. We apply a proximity clustering technique to divide each hue cluster and the set of moving pixels into contiguous subclusters. We employ a simple region-growing algorithm that groups pixels into equivalence classes. Two pixels are placed in the same equivalence class when the Euclidean distance between the $\langle x, y \rangle$ coordinates of those two pixels is less than a specified threshold. We then discard equivalence classes that have fewer than a specified number of pixels. This eliminates small spurious regions, such as those that appear in figures 2(b) and 2(d). Figure 2(c) shows the result of applying our region-grower to the hue clusters in the image in figure 2(b). Figure 2(e) shows the result of applying our region-grower to the image in figure 2(d) while figure 2(f) shows the combined colour-based and motion-based output of our tracker. At this point, a movie is represented as a set of regions for each frame, where each region is a set of pixels.

We now abstract each region in each frame as an ellipse. To do so, we compute the mean and covariance matrix of the two-dimensional $\langle x, y \rangle$ coordinate values of the pixels in a given region. We take the ellipse for that region to be centered at the mean and to follow a contour one standard deviation out from the mean. Thus the orientation of the major axis of the ellipse is along the primary eigenvector of the covariance matrix and the lengths of the major and minor axes of the ellipse are given by the eigenvalues of the covariance matrix. Figure 2(g) shows the ellipses generated for the regions in figure 2(f). All subsequent processing ignores the underlying image pixel data and uses only the derived ellipse parameters.

The operation of fitting ellipses to image data is done independently for each frame in the movie. Such independent processing suffers from two limitations. First, it does not recover the intra-frame correspondence between ellipses. We require this *internal correspondence* in order to track object position over time. Second, different frames can contain different numbers of ellipses. There may be situations, as is the case in figure 2(g), where the tracker produces spurious ellipses that do not correspond to objects that participate in events in the movie. There may also be situations where the tracker fails to produce an ellipse to represent an object that does participate in an event. Such drop outs can happen for a variety of reasons, such as inappropriate settings for the various threshold parameters. These drop outs and spurious ellipses make the intra-frame correspondence task more difficult. Subsequent processing can overcome the limitations of the tracker by robustly determining intra-frame correspondence between ellipses.

We employ a simple technique to determine intra-frame correspondence between ellipses in a single movie. First, we group ellipses believed to track the same object between frames into ellipse chains. We use a weighted five-dimensional Euclidean distance metric on ellipse parameters to determine object continuity between adjacent frames. The chains are contiguous sequences that track the motion of a single ellipse for a subrange of frames in the movie. Due to noise, chains might not span the entire movie. Thus we subsequently relax the frame-adjacency requirement and attempt to attach the ends of the chains together to produce contiguous sequences of ellipses that span the entire movie. Relaxation of the frame-adjacency requirement reduces the problem of ellipse drop outs, while elimination of short chains reduces the problem of spurious ellipses. The final result of this technique is a set of ellipse sequences that track all of the participant objects in the event through the entire movie.

The result of applying our tracker to several complete movies is shown along with the original movies in figure 3. Only a small subset of key frames for each movie is shown. These movies depict *pick up*, *put down*, *push*, *pull*, *drop*, and *throw* events respectively. In these movies, the intra-frame ellipse correspondence is indicated by line thickness. Notice that it is fairly easy for a human to recognise the depicted event solely from the ellipse data. Our event recogniser attempts to mimic this ability.

4 Event Recognition

The output of our tracker consists of a stream of five parameters for each ellipse in each frame of each movie. Since movies typically have from two to five objects, this constitutes roughly 10 to 25 floating point numbers per frame. From this data stream we compute a larger feature vector. This feature vector contains both absolute, and relative, ellipse positions, orientations, velocities, and accelerations. More specifically, our feature vector includes the following features for each frame:

- Absolute features
  1. the magnitude of the velocity vector of the centre of each ellipse,
  2. the orientation of the velocity vector of the centre of each ellipse,
  3. the angular velocity of each ellipse,
  4. the first derivative of the area of each ellipse,
  5. the first derivative of the eccentricity of each ellipse,
  6. the first derivatives of each of the above five features,
- Relative features
  1. the distance between the centres of every pair of ellipses,
  2. the orientation of the vector between the centres of every pair of ellipses,

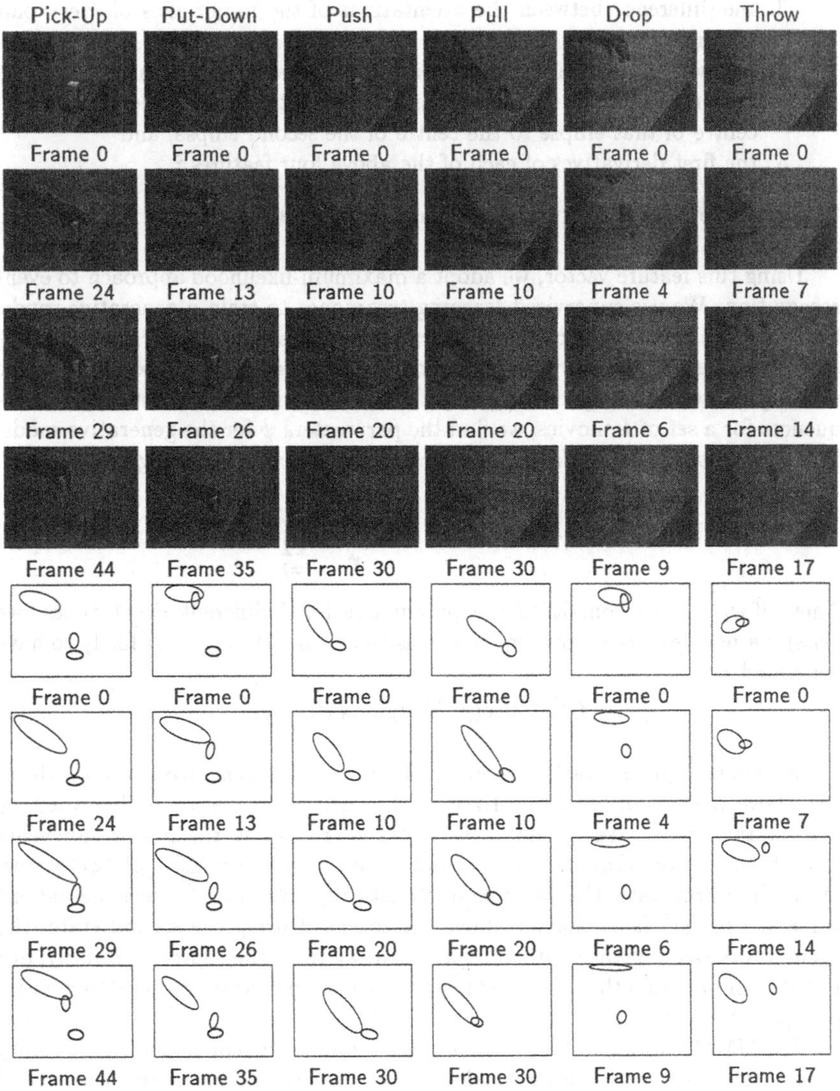

Fig. 3. Sample frames from several movies depicting six different events along with the result of applying our tracker to those movies.

3. the difference between the orientations of the major axes of every pair of ellipses,
4. for every pair of ellipses, the difference between the orientation of the major axis of the first ellipse and the orientation of a vector from the centre of that ellipse to the centre of the second ellipse, and
5. the first derivatives of each of the above four features.[3]

In the above, all derivatives are calculated as finite differences between two adjacent frames.

Using this feature vector, we adopt a maximum-likelihood approach to event recognition. We use supervised learning techniques to train a generative model for each class of events, from a set of training examples for each class, and subsequently use the derived generative models to classify new observations into existing classes. More specifically, if $p_1, \ldots, p_L$ constitute the feature vector sequences for a set of L movies, we find the parameters ψ for the generative model that maximise the joint likelihood of generating all of the training sequences.

$$\textsc{Train}(\{p_1, \ldots, p_L\}) \triangleq \operatorname*{argmax}_{\psi} \prod_{l=1}^{L} P(p_l|\psi) \tag{1}$$

Then, if $\psi_1, \ldots, \psi_J$ constitute the parameters for J different event types, we classify a new feature vector sequence p as the class j that is most likely to have generated p.

$$\textsc{Classify}(p) \triangleq \operatorname*{argmax}_{j} P(p|\psi_j) \tag{2}$$

We adopt Hidden Markov Models (HMMs) as the generative model within the above maximum-likelihood framework. Each event class j is described by a u_j-state model.[4] We intend these states to represent the major subevents of each event type. For instance, a *pick up* event might consist of two states. During the first state, the agent characteristically will move toward the patient while the patient is stationary above the source. During the second state, the agent will move together with the patient away from the source. Similarly, our HMMs can partition the other event types into distinct states representing easily interpretable subevents.

The HMMs in our implementation assume independent normal probability distributions over each feature in the feature vector. The mean and variance of each feature in each state of each model are adjusted in order to maximise the likelihood of that model generating the training data. The assignment of high and low variances to different features is tantamount to learning which features

[3] The results that we report in this paper are based on a simpler feature vector that contains only the first two of the aforementioned absolute and relative features along with their first derivatives. While we obtain good results with this smaller feature set, when classifying a small set of different event types, we believe that the larger feature set will be necessary to classify a larger set of event types.

[4] Currently, the number of states used to represent each event type is specified manually as a parameter to the training process.

are relevant to each subevent of each event type, and which are not, since the likelihood is more sensitive to changes in the values of those features having low variance than it is to those features having high variance. For example, the variance of the first derivative of the distance between the agent and the patient during the first state of a *pick up* event will be low, indicating that that feature is relevant, while the variance of the orientation of the velocity vector of the agent will be high, since the angle of approach during a *pick up* can vary significantly, indicating that that feature is not relevant. In contrast, the variance of that feature, namely the orientation of the velocity vector of an object, will be low during a *drop* event, since objects typically fall straight downward.

We train the parameters of the HMMs with the Baum-Welch reestimation procedure (Baum, Petrie, Soules, & Weiss, 1970) and use the Viterbi procedure (Viterbi, 1967) for classification. During training, we restrict the state-transition matrix to be upper triangular, thus disallowing non-self cycles in the state-transition graph. Our experience shows that such non-ergodic models generalise much better to new observations. This is not a severe restriction, since none of the event types that we have considered require repetitive subevents.

The training and classification procedures are somewhat more complex than the standard Baum-Welch and Viterbi procedures for a number of reasons. First, our tracker does not provide object-identity information. While our tracker does track the position of objects through time in the movie with ellipse sequences, it does not identify the objects that it is tracking. Second, our tracker can produce an ellipse sequence for an object that does not participate in the event or an object that does not really exist, for example a shadow. This is problematic because we need to know which tracked objects in the movie are participating in the event and what role those objects play within the event. Each event type specifies a fixed number of participant objects and each object plays a well-defined role in a given event type. Roles must be assigned to tracked objects both during training, to group together ellipse sequences from different movies that play the same role, and during classification, to assign ellipse sequences to existing roles in a model. We refer to this grouping problem as the *external correspondence* problem.

We determine the external correspondence among members of the training set by examining a set of candidate correspondences. Each candidate correspondence contains a subset of the ellipse sequences for each movie in the training set. Each subset is ordered to match corresponding ellipse sequences across the training movies. Conceptually, an HMM can be trained on each candidate correspondence, and the best correspondence can be chosen to be that which leads to a model that has the highest likelihood of generating the event instances in the training set.

Since evaluating all possible candidate correspondences would be intractable, we use a variant of the greedy algorithm to choose which candidate correspondences to evaluate. We choose one particular movie from the training set to be the canonical event. The movie chosen to represent the canonical event must contain the same number of ellipse sequences as there are participant objects in the given event type. An arbitrary order is chosen for the set of ellipse sequences in the movie representing the canonical event. Ordered subsets of the set of ellipse sequences in every other movie are then chosen to establish a candidate correspondence with the ordered set of ellipse sequences in the canonical event. This choice is done incrementally. First, a candidate correspondence is established between the canonical event movie and one other movie chosen from the training set. This correspondence is chosen by training an HMM on every possible correspondence between the canonical event and the other movie and choosing the HMM that has the highest likelihood of generating these two event instances. We then explore successively larger candidate correspondences. At each stage we maintain a set of r candidate correspondences for n movies and choose as the candidate correspondences for $n + 1$ movies the r best ways of augmenting a previous candidate with a new movie. At the last iteration we choose the best HMM among the r candidates correspondences constructed for the entire training set.

During classification, ellipse-sequence role assignments are determined in the following manner. We examine all possible permutations of all subsets of the set of ellipse sequences derived from a new observation. We then compute the likelihood that that sequence was generated by each of the models that specifies the same number of participant objects as the number of ellipse sequences in that candidate. The new observation is classified as being generated by the model that assigns the highest likelihood to some permutation of some subset of the set of ellipse sequences derived from that observation.

5 Experiments

To test our techniques, we filmed 72 movies using a SunVideo system to record image sequences with a resolution of 320 × 240 at 30 frames-per-second compressed in JPEG format. These 72 movies comprised twelve instances for each of the six event types *pick up*, *put down*, *push*, *pull*, *drop*, and *throw*. All of these movies were filmed in a relatively uncluttered desk-top environment, though some movies contain extraneous objects. The camera position changed somewhat from movie to movie. Four coloured foam blocks (one red, one blue, one green, and one yellow) were used as props to enact the six different event types. We clipped these movies by hand so that the beginning and end of the event coincided with the beginning and end of the movie. The *pick up* and *put down* movies were uniformly clipped to be 50 frames long, the *push* and *pull* movies were uniformly clipped to be 35 frames long, while the *drop* and *throw* movies were uniformly clipped to be 20 frames long. We then processed each of the 72 movies with our tracker. Of these 72 movies, 36 were randomly selected as training movies, six movies for each of the six event types. We constructed six five-state non-ergodic Hidden Markov Models, one for each of the six event classes. We then classified all 72 movies, i.e. both the original training data as well as the data not used for training, against all six event classes. Our model correctly classified all 36 of the training movies and correctly classified 35 out of the 36 test movies. One *drop* movie was misclassified as a 'throwing' event. Our classifier, however, selected 'drop' as the second-best choice for this movie. This misclassification appears to result from poor tracking.[5]

6 Discussion

We are cautiously optimistic with our results. We plan to test our techniques with a larger set of event types using a wider range of participant objects filmed in different length movies in a wider range of environments. We also plan to evaluate the robustness of these techniques in the face of event occurrences that exhibit larger variance in motion profile. Furthermore, while our event recogniser can work with output from any tracker, the class of events that we can classify is to some extent dependent on the properties of our current tracker. We have built a variety of trackers to drive our recogniser and are currently tuning a more sophisticated tracker that we believe will allow us to apply our event recogniser to a wider class of events and environments. A longer-term objective is to eliminate the need for manually specifying the beginning and end of each event by automatically performing temporal segmentation among events that potentially overlap in space and time.

Our work has surprising similarities with earlier work on event recognition reported in Siskind (1992, 1995). That work suggests that event classes can be described with temporal-logic formulas, such as

$$\textsc{PickUp}(x, y) \triangleq \exists w \left(\begin{array}{c} \textsc{Part}(w, x) \wedge \\ \left[\left(\exists z \begin{pmatrix} \textsc{Disjoint}(z, w) \wedge \\ \textsc{Supported}(y) \wedge \\ \textsc{Supports}(z, y) \wedge \\ \textsc{Contacts}(z, y) \end{pmatrix} \right) ; \begin{pmatrix} \textsc{Translating}(w) \wedge \\ \textsc{Contacts}(w, y) \wedge \\ \textsc{Attached}(w, y) \wedge \\ \textsc{Supports}(x, y) \wedge \\ \textsc{Translating}(y) \end{pmatrix} \right] \end{array} \right)$$

over a set of primitives that describe spatial relations, motion, and force-dynamic relations such as support, contact, and attachment. On the surface, the approach taken in that work might appear to be fundamentally different from the approach taken here. There are deep similarities, however. The features used as input to our HMMs are analogous to conceptual-structure representations proposed by Jackendoff (1990). For example, the orientation of the velocity of the centre of each ellipse can be viewed as a quantified representation of the conceptual structures $\mathrm{GO}(x, \mathrm{UP})$ and $\mathrm{GO}(x, \mathrm{DOWN})$. Similar analogies can be drawn for all of the remaining features.

Temporal-logic formulas can be viewed as regular expressions over symbols that represent the occurrence of primitive events described by these Jackendovean representations. Such symbols can be viewed as thresholded feature values, while the temporal-logic formulas can be viewed as finite-state recognisers over these thresholded values. Our HMMs can be viewed as probabilistic finite-state recognisers over the same features. Thus, while Siskind (1992, 1995) only presents techniques for using event descriptions to recognise events, assuming that event descriptions are produced by hand, the work presented here

[5] All of our training and test data, as well and source code for our implementation is available from `http://www.cs.toronto.edu/~qobi`.

extends this approach to learn event descriptions in a supervised fashion. The work presented here, however, uses only spatial motion features and not force-dynamic ones. We pursued this approach because of our initial concern that it would be difficult to robustly recover force-dynamic relations from camera input. Mann et al. (1996), however, present some encouraging results that show that recovery of force-dynamics relations from camera input is feasible. In the future we hope to integrate the force-dynamics recovery techniques described in that paper with the event-recognition techniques described here.

7 Conclusion

Event perception is a fundamental cognitive faculty. Any solution to the ultimate goal of artificial intelligence, namely endowing a computer with human-level intelligence, must embody a mechanism for perceiving events. Event recognition plays a central role in human cognition, perhaps even a more central role than object recognition. Consider, for instance, the fact the sentence structure of natural languages is built around verbs rather than nouns. Furthermore, event recognition appears to be easier than object recognition, and to rely on less information. It may even be the case that event recognition is an evolutionarily more primitive faculty.

Earlier work adopted the tacit assumption that object recognition precedes event recognition. It is interesting to consider the reverse assumption. If event recognition is easier and evolutionarily more primitive than object recognition, perhaps object recognition can be facilitated by information provided by prior event recognition. For instance, it might be difficult to reliably segment, recognise, and track a hammer on the basis of an object model, geometric or otherwise. Yet, it might be easy to recognise the characteristic motion of a hammering event even without detecting a hammer. Having hypothesised a hammering event, and produced a course abstract object pose sequence in the process of forming such a hypothesis, it might be possible to use that knowledge to guide the search for a hammer to confirm that hypothesis. Prior event recognition could prune the space of object models to consider and also provide course grouping and tracking data, using a novel source of top-down constraint.

We have applied a particular strategy to event classification, namely maximum likelihood. To the best of our knowledge this approach has not been previously applied to this task. Maximum-likelihood methods, however, have proven extremely successful in machine analysis of speech. These methods provide several important methodological advantages. First, they provide graded levels of performance rather than monolithic success or failure. This is crucial when working on hard problems for which there is no unequivocally successful method in sight. Second, they provide two well-defined methods for improving performance: additional training data and more accurate generative models. The course of speech recognition research over the past few decades can be seen as long intervals of shallow but steady performance growth punctuated by a small number of discontinuous paradigm changes. Third, systems that maintain graded decisions internally, delaying all categorical decisions as long as possible, are more robust than those that operate as a pipe of categorical decision processes. Systems in the later category are brittle, since a minor inaccuracy can affect a categorical decision early in the pipe and that decision can irrevocably affect the ultimate outcome. Visual event perception shares much in common with speech analysis in that both deal with time-varying signals. It is not surprising that techniques that have been applied to speech could also be applied to visual event perception. The success of maximum-likelihood methods should not be taken to imply that they are the only feasible approach to performing event classification. Rather, we intend the results of this paper to be taken simply as encouragement that the problem of event classification can be solved. Given this realisation, we encourage other researchers to explore alternate techniques for solving this problem.

Acknowledgments

We wish to thank Peter Dayan and Sven Dickinson for invaluable assistance on this project and Richard Mann for offering his comments on an earlier draft of this paper. Any remaining errors are, of course, the sole responsibility of the authors. This work was supported by a grant from the Natural Sciences and Engineering Research Council of Canada.

References

Abe, N., Soga, I., & Tsuji, S. (1981). A Plot Understanding System on Reference to Both Image and Language. In *Proceedings of the Seventh International Joint Conference on Artificial Intelligence*, pp. 77–84, Vancouver, BC.

Badler, N. I. (1975). Temporal Scene Analysis: Conceptual Descriptions of Object Movements. Tech. rep. 80, University of Toronto Department of Computer Science.

Baum, L. E., Petrie, T., Soules, G., & Weiss, N. (1970). A Maximization Technique Occuring in the Statistical Analysis of Probabilistic Functions of Markov Chains. *The Annals of Mathematical Statistics*, *41*(1), 164–171.

Jackendoff, R. (1990). *Semantic Structures*. Cambridge, MA: The MIT Press.

Johansson, G. (1973). Visual Perception of Biological Motion and a Model for its Analysis. *Perception and Psychophysics*, *14*(2), 201–211.

Mann, R., Jepson, A., & Siskind, J. M. (1996). The Computational Perception of Scene Dynamics. In *Proceedings of the Fourth European Conference on Computer Vision*, Cambridge, UK: Springer-Verlag.

Marburger, H., Neumann, B., & Novak, H. (1981). Natural Language Dialogue About Moving Objects in an Automatically Analyzed Traffic Scene. In *Proceedings of the Seventh International Joint Conference on Artificial Intelligence*, pp. 49–51, Vancouver, BC.

Marr, D., & Vaina, L. (1982). Representation and Recognition of the Movements of Shapes. *Proc. R. Soc. Lond. B*, *214*, 501–524.

Nagel, H. (1977). Analysing Sequences of TV-Frames. In *Proceedings of the Fifth International Joint Conference on Artificial Intelligence*, p. 626, Cambridge, MA.

Neumann, B., & Novak, H. (1983). Event Models for Recognition and Natural Language Description of Events in Real-World Image Sequences. In *Proceedings of the Eighth International Joint Conference on Artificial Intelligence*, pp. 724–726, Karlsruhe.

Okada, N. (1979). SUPP: Understanding Moving Picture Patterns Based on Linguistic Knowledge. In *Proceedings of the Sixth International Joint Conference on Artificial Intelligence*, pp. 690–692, Tokyo.

O'Rourke, J., & Badler, N. I. (1980). Model-Based Image Analysis of Human Motion Using Constraint Propagation. *IEEE Transactions on Pattern Analysis and Machine Intelligence*, *2*(6), 522–536.

Pinhanez, C., & Bobick, A. (1995). Scripts in Machine Understanding of Image Sequences. In *AAAI Fall Symposium Series on Computational Models for Integrating Language and Vision.*

Rashid, R. F. (1980). Towards a System for the Interpretation of Moving Light Displays. *IEEE Transactions on Pattern Analysis and Machine Intelligence*, *2*(6), 574-581.

Siskind, J. M. (1992). *Naive Physics, Event Perception, Lexical Semantics, and Language Acquisition*. Ph.D. thesis, Massachusetts Institute of Technology, Cambridge, MA.

Siskind, J. M. (1995). Grounding Language in Perception. *Artificial Intelligence Review*, *8*, 371–391.

Thibadeau, R. (1986). Artificial Perception of Actions. *Cognitive Science*, *10*(2), 117–149.

Tsotsos, J. K., Mylopoulos, J., Covvey, H. D., & Zucker, S. W. (1980). A Framework for Visual Motion Understanding. *IEEE Transactions on Pattern Analysis and Machine Intelligence*, *2*(6), 563–573.

Tsuji, S., Morizono, A., & Kuroda, S. (1977). Understanding a Simple Cartoon Film by a Computer Vision System. In *Proceedings of the Fifth International Joint Conference on Artificial Intelligence*, pp. 609–610, Cambridge, MA.

Viterbi, A. J. (1967). Error Bounds for Convolutional Codes and an Asymtotically Optimum Decoding Algorithm. *IEEE Transactions on Information Theory*, *13*, 260–267.

Waltz, D. L. (1981). Toward A Detailed Model of Processing for Language Describing the Physical World. In *Proceedings of the Seventh International Joint Conference on Artificial Intelligence*, pp. 1–6, Vancouver, BC.

Yamamoto, J., Ohya, J., & Ishii, K. (1992). Recognizing Human Action in Time-Sequential Images Using Hidden Markov Model. In *Proceedings of the 1992 IEEE Conference on Computer Vision and Pattern Recognition*, pp. 379–385. IEEE Press.

Applications and Recognition

Measures for Silhouettes Resemblance and Representative Silhouettes of Curved Objects

Yoram Gdalyahu and Daphna Weinshall

Institute of Computer Science, The Hebrew University, 91904 Jerusalem, Israel

Abstract. We claim that the task of object recognition necessitates a measure for image likelihood, that is: a measure for the probability that a given image is obtained from a familiar (pre-investigated) object. Moreover, in a system where objects are represented by 2D images, the best performance is achieved if those images are selected according to a maximum likelihood principle. This is equivalent to maximum stability of the image, or minimal change under a viewpoint perturbation. All of those qualities involve a quantitative comparison of similarity between images. We propose different metric functions which can be imposed on the image space of curved three dimensional objects. We use these metrics to detect the representative views (most stable and most likely views) of three test models. We find the same representative views under all the investigated metrics, suggesting that local maxima of stability and likelihood are metric independent. Our method of image comparison is based solely on the appearance of the occluding contour, hence our method is suitable for object recognition from silhouettes.

1 Introduction

Three dimensional objects can be projected to an infinite number of images due to the infinite number of possible viewing directions. In the absence of any particular symmetry, all those images may be different from each other, but some of them can be thought of as better representing the three dimensional object. This leads to the intuitive notions of "canonical", "characteristic" or "generic" views, which frequently appear in the literature. Recently, an objective definition has been given for those notions [14, 13]. According to it, a representative view has the properties of maximum likelihood (it is similar, as an image, to a large number of other images), and maximum stability (it doesn't vary a lot when the viewing direction is perturbed). We have already shown that a local maximum of the likelihood function is obtained at the same points (views) for which a local maximum of the stability function is obtained [13]. Therefore, the same single view is (locally) both the most likely and the most stable one, and it is naturally proposed as a representative view.

Explicit identification of such views requires an explicit quantitative measure of similarity between images. Such a measure converts the image space into a metric space, where the distance between every two images is proportional to their dissimilarity. Finding an appropriate metric is a research subject of its own [6], and for complicated images it is still an open question how to measure their resemblance. It becomes easier to measure image similarity if the family of the objects is constrained. For example, a discrete object which is made of a cloud of points. This case and other simple cases are reviewed in section 2. Our present work is devoted to the study of the case of curved objects (not necessarily convex) which lack intrinsic feature points. We discuss and compare some possible metric functions to judge the similarity between images of such objects. We apply our proposed metrics to three different test models, perform a numerical study of their views stability, and compute their representative views.

Representative views play an important role in viewer-centered approaches to 3D shape representation, where three dimensional information is not represented explicitly. In these approaches a novel view of the object is recognized by comparing it to stored views, and storing "good" views is essential for successful performance. If an angle is to be recognized, Ben-Arie [2] and Burns et al. [5] already identified the more common views of it (see section 2). A related measure of view genericity, or view likelihood, also underlies the Bayesian decision strategy suggested by Freeman [7] to interpret correctly ambiguous scenes. Freeman showed how to *use* measures of view "genericity" in image understanding, but did not propose a general way to *compute* such measures directly from images and models.

The rest of the paper is organized as follow: In section 2 we summarize our definitions of stability and likelihood, for a given image metric. We also present a few early results obtained with these definitions. In section 3 we propose metric functions which are suitable for the case of curved objects. The application of these metrics to identify representative views of some 3D models is presented in section 4. In section 5 we draw our conclusions.

2 Stability and likelihood of views

2.1 Definitions

Our framework consists of a weak perspective projection scheme, hence the image space of a given object is two dimensional, spanned by the two angles ϑ and φ, respectively specifying the elevation and azimuth of the camera axis with respect to the center of the object. The image space can be therefore mapped to a unit sphere, where each point on that sphere is called a view, and each such point represents a set of images related to each other by a 2D similarity transformation (2D rotation, translation and scale). We call this sphere "the viewing sphere".

The parameterization of the viewing sphere demands some preselection of a pole, relative to which the angles of elevation and azimuth are measured. Since the definitions of stability and likelihood that we are to present involve the comparison between neighboring views, it is useful to measure the elevation and azimuth angles of one view with respect to the other. Let $V_{\vartheta,\varphi}$ be the view whose elevation ϑ and azimuth φ are measured with respect to the z axis pole, and let $V_{\delta,\psi}^{\vartheta,\varphi}$ be the view whose elevation δ and azimuth ψ are measured with respect to the previous view.

By definition, the geodesic distance on the viewing sphere between $V_{\vartheta,\varphi}$ and $V_{\delta,\psi}^{\vartheta,\varphi}$ is δ. However, the similarity between these two views, as 2D images, is not necessarily proportional to the geodesic distance, and we denote their image distance by $d(\vartheta,\varphi,\delta,\psi)$. This image distance can change abruptly with δ, usually if features appear or disappear. In regions of continuous behavior the image distance can be expressed as $d(\vartheta,\varphi,\delta,\psi) = D(\vartheta,\varphi,\psi)\delta + O(\delta^2)$, where $D(\vartheta,\varphi,\psi)$ is the derivative of the distance function with respect to δ in the direction ψ, and $D(\vartheta,\varphi,\psi) \geq 0$

The variability of a view reflects the amount of image change caused by a perturbation of the viewing direction. More precisely, we define the ε-variability of a view $V_{\vartheta,\varphi}$ as the average change in the images which belong to an ε-neighborhood of $V_{\vartheta,\varphi}$:

$$R_\varepsilon(\vartheta,\varphi) = < d(\vartheta,\varphi,\delta,\psi) >_{0\leq\delta\leq\varepsilon,\ 0\leq\psi\leq 2\pi} = \frac{\int_0^{2\pi}\int_0^{\varepsilon} d(\vartheta,\varphi,\delta,\psi) sin(\delta) d\delta d\psi}{\int_0^{2\pi}\int_0^{\varepsilon} sin(\delta) d\delta d\psi}$$

The smaller the view variability is - the grater is its stability. In [14] the ε-stability $s_\varepsilon(\vartheta,\varphi)$ of a view was defined simply as $1/R_\varepsilon$. Here we are primarily interested in the comparison between different metric functions, so we prefer another definition which is scale invariant:

$$s_\varepsilon(\vartheta,\varphi) = \frac{R_\varepsilon^{max} - R_\varepsilon(\vartheta,\varphi)}{R_\varepsilon^{max} - R_\varepsilon^{min}}$$

The notion of stability captures a local property of a view, namely its level of invariance under a small change of the viewing position. On the other hand, the notion of likelihood is defined with respect to the whole image space, and it measures how common a view is. We start by simply "counting" the number of views which are at least ε-similar to a given view, according to some measure d. This yields the definition of an ε-area:

$$A_\varepsilon(\vartheta,\varphi) = \int_0^{2\pi}\int_0^{\frac{\pi}{2}} \chi_{\{d(\vartheta,\varphi,\delta,\psi)\leq\varepsilon\}} sin(\delta) d\delta d\psi$$

where χ_b is 1 when b is true, and 0 otherwise.

The total surface area of the viewing sphere is 4π, hence $A_\varepsilon/4\pi$ can be considered as the conditional probability $f(\varepsilon|\vartheta,\varphi)$ of an image error ε, given a reference view $V_{\vartheta,\varphi}$. Using Bayes rule we get the inverse relation:

$$f(\vartheta,\varphi|\varepsilon) = \frac{f(\vartheta,\varphi) f(\varepsilon|\vartheta,\varphi)}{f(\varepsilon)}$$

where $f(\vartheta,\varphi)$ is the prior distribution of camera positions. If the viewer position is not constrained, then $f(\vartheta,\varphi)$ is of course uniform. The part of $f(\vartheta,\varphi|\varepsilon)$ which depends on the viewing direction is defined as the view ε-likelihood $l_\varepsilon(\vartheta,\varphi)$:

$$l_\varepsilon(\vartheta,\varphi) = f(\vartheta,\varphi)f(\varepsilon|\vartheta,\varphi) \simeq f(\vartheta,\varphi)\int_0^{2\pi}\int_0^{\frac{\pi}{2}} \chi_{\{d(\vartheta,\varphi,\delta,\psi)\leq\varepsilon\}} sin(\delta) d\delta d\psi$$

We note that it is also possible to define both the stability and the likelihood in the limit of $\varepsilon \to 0$, as we do in [13]. This means that stability is measured in the limit of a zero perturbation of camera position, and likelihood is measured in the limit of zero image error.

2.2 General results and some special cases

It can be shown [13] that if the prior distribution of viewpoints of a given object is uniform, namely all camera positions are equally likely, then the most stable and the most likely views are the same view. In other words, the local maxima of the stability function and of the likelihood function are obtained for the same views.

In previous works we applied our definitions to simple objects, for which there is a natural metric function [13, 14]. The first case is that of objects represented by feature points in 3D space. A natural measure for the distance between two images is obtained by measuring residual distances between corresponding points, after the two images have been normalized and aligned. We used two different alignment schemes, with either an affine or a similarity image transformation, and obtained two metric functions (which are positive, symmetric and obey the triangular rule). We showed that under both of them, the most stable and likely view depends only on the three principal moments [1], and not on the specific location of all the features. Moreover, this view is unique and it is the "flattest view", obtained when the three dimensional object has its minimal spread along the viewing direction [2].

A second case which was studied is that of a planar angle that can be viewed from an arbitrary direction in space. This case was studied empirically by Ben-Arie [2] and Burns et. al [5], who found the "picking effect": the most probable value of an angle to be picked from a random viewing direction is the value of the angle itself. We imposed a natural metric on the image space of such an angle, namely the difference between the two projected angles, and we were able to find the same result analytically.

A third metric function is based on extraction of extreme curvature points along the occluding contour of an object. We discuss this metric in section 3.2. Here we note, that for the special case of an ellipsoid, we found analytically that under this metric the views which are local maxima of stability and likelihood are located along the three main axes of the ellipsoid.

[1] We defined the three principal moments of a discrete cloud of feature points as the three eigenvalues of their autocorrelation matrix.

[2] More precisely, if we assume that the z axis is the viewing direction, then the flattest view is the view for which the autocorrelation matrix is diagonal, with the three principle moments ordered in decreasing order.

3 Metrics to compare silhouettes

In this work the 3D objects are assumed to be topologically simple (without holes), which means that any image can be divided into two connected and non overlapping areas: the interior and the exterior regions. The border between those two areas is called the occluding contour, and the 3D curve which lies on the surface and is projected to the occluding contour is called the rim.

The shape of the occluding contour depends only on the object's shape and the viewing direction, in contrast to internal contours that may be sensitive to illumination. Therefore, the analysis of image geometrical variability here is solely based on the shape variations of the occluding contour. It goes without saying that all our objects are recognized according to their silhouettes.

Usually one wouldn't like to distinguish between images which differ by only an irrelevant image transformation. Here we assume a weak perspective projection, hence it is natural to choose the group of similarity transformations in the image plane (translation, rotation and scaling) as the group which divides the images into equivalence classes. Hence, All the images which belong to the same class (equivalent images) are related to each other by a 2D similarity transformation. Those images are represented by the same point on the viewing sphere. Moreover, when two images are not equivalent, the distance between them is determined by alignment, which means that the two closest images in the two respective equivalent classes determine the distance for all the members in those classes. Evidently, the distance between equivalent images is zero.

In the following sections we introduce the metric functions which we use in order to measure distances between views (closed curves). Two different strategies are taken: either a whole curve comparison (section 3.1), or comparison via interest points (section 3.2). The second mechanism permits a huge reduction of information due to the usage of only a zero measure set of points instead of a continuous line, but an undesirable result is the unification of equivalence classes into larger ones: two curves whose sets of interest points are related by a similarity transformation have a zero distance, and it is not necessary that this transformation maps the whole one curve onto the other. On the other hand, methods based on feature extraction more easily deal with occlusions and can be integrated with database indexing methods.

3.1 Whole contour comparison

Our first proposed method to compare two images (two planar closed curves), is simply a comparison of their tangent angle transforms. This transform is a representation of the curve by its tangent angle versus its normalized arc length [12, 1]. To construct such a representation, we start from an arbitrary point on the curve and follow it in an arbitrary direction, measuring in each point the angle of the curve tangent with respect to some arbitrary reference direction. While traveling along the curve we record this angle as a function of the accumulated arc length. We obtain a one dimensional graph which can always be transformed into a "canonical" form, which starts at the point [0,0] and ends at [1,1]. To achieve this form we may need to: (1) normalize the x-axis to one (equivalent to an image scaling operation), (2) reflect the function around its middle (equivalent to selection of travel direction along the curve), (3) shift the y-axis (equivalent to a rotation operation in the image plane), and (4) rescale the y-axis by $1/2\pi$ (which is no more than a cosmetic manipulation). We note that for the first operation to be valid we assume that the object is not occluded (see section 5). Figure 1 shows an example of an occluding contour extracted from an image, and its canonical form.

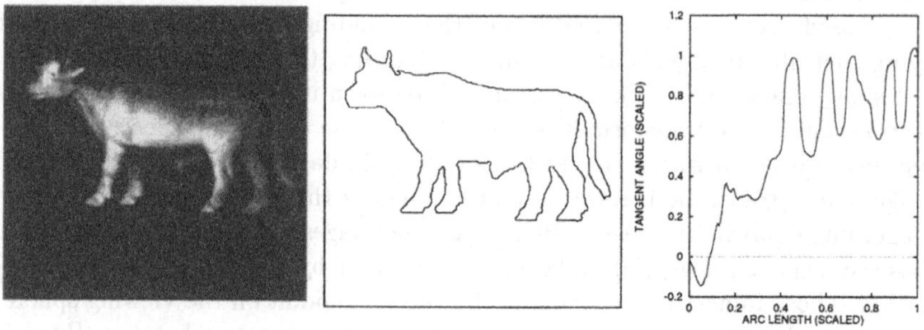

Fig. 1. Example of an occluding contour and a canonical form

A canonical form of a curve is not unique, since it depends on the arbitrary chosen starting point. However, The family of all such forms, obtained for all possible selections of the starting point, is invariant under a similarity transformation, and by sliding the origin along one such form, the family of all canonical forms is obtained. Hence for given two curves, a proper alignment algorithm performs a one dimensional search along one of them, to find the optimal starting point. The power of this alignment algorithm is that it combines the search for the proper rotation and translation into a one dimensional search of the starting point. It requires, however, having an algorithm to travel along curves.

The distance between canonical forms, which is to be minimized by the alignment procedure, is defined as follow: For a given desired resolution, the two functions are represented by two ordered sets of equally spaced points P_1 and P_2. Their distance is $min(d_{12}, d_{21})$, where d_{12} is the sum of distances of P_1 points from the spline formed by P_2, and similarly for d_{21}. The two sets are assumed to be of equal size.

3.2 Contour comparison using features

In the following we describe some methods of feature extraction that we use to locate interest points on a given smooth curve. The evaluation of distance between two curves is then composed of the following stages: feature extraction, alignment of feature points by a similarity transformation, and measuring the average residual distance between corresponding points. Note, however, the difference from the case of an object equipped with intrinsic features: here, two corresponding interest points in two different images *do not* correspond to the same single 3D point on the surface of the object.

Extreme curvature points Points of maximum (and perhaps also minimum) curvature along the curve are probably the most intuitive features that one can think of. However, extreme curvature points are known to be unstable [8] and hard to extract, due to their sensitivity to resolution and to the presence of noise. In general, a reliable extraction requires some smoothing mechanism, which enhances the extraction robustness and reduces its locality. Nevertheless extreme curvature points are the most local feature that we use (i.e, their position is determined by a small fraction of the curve around the point, which is advantageous in a case of occlusion).

By definition, curvature computation involves the second derivatives of the curve with respect to arc length. For digital curves, we find it more robust to use the following method of maximum curvature extraction: let p denotes one of the curve points, and let $\mathbf{u}_n$ and $\mathbf{v}_n$ denote the vector sums of the n vectors pointing from p to its n successive neighbors to the left and to the right. Then the angle between $\mathbf{u}_n$ and $\mathbf{v}_n$ is minimal when the curvature is maximal. This method avoids any computations of arc length or derivatives with respect to it. The vector summation gives more weight to distant points, for which errors due to the finite resolution are less important.

The polygonal approximation which is obtained by the extracted points may be refined by adding more interest points. At each refinement step, the curve point which is most distant from the polygonal approximation is added, until the desired accuracy is obtained. Figure 2 shows the interest points extracted by this method, together with the results of other extraction methods that we describe below.

Fig. 2. Feature points which are extracted by maximum curvature plus refinement (left), by transformed self intersections (middle), and by unit signature (right).

Transformed self intersections The occluding contour, which in our view here is the only image content, is a simple closed curve which does not intersect itself. But it was recently shown [9] how such a curve can be transformed into another one, which does intersect itself. In this method the tangent angle transform (see above) is multiplied by a constant factor and a new curve, which in general does intersect itself, is reconstructed (figure 3). The exceptional conditions for which self intersections can't be made, or their number becomes infinite, are discussed in [9] and are ignored here.

The self intersection points can be mapped back to the original image. Note that each intersection is both a starting point and an ending point of a loop on the transformed curve, hence it is mapped back to *two* points on the original curve. Clearly, only the part of the curve which is enclosed between these two points has any influence on the intersection existence. Hence locality is controlled by the multiplication factor. More local features are obtained for larger multiplication factors, at the cost of losing robustness against noise.

Note that for the extraction of self intersections one only needs the first derivative of the curve (the tangent angle), while for the extraction of extreme curvature points the third derivative is needed.

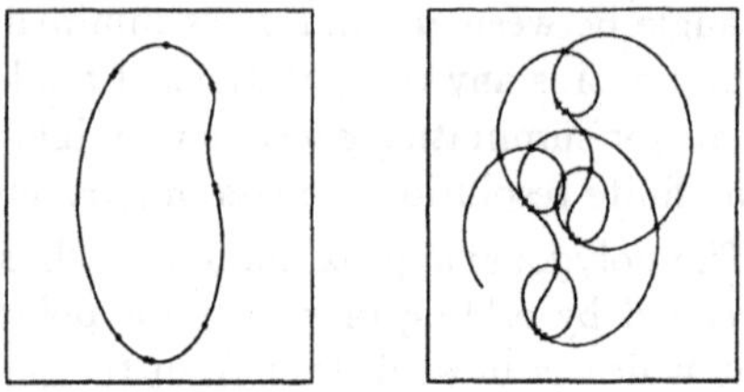

Fig. 3. An original curve (left) and the transformed curve (right) obtained with a multiplication factor of 3. To avoid losing intersections due to end effects, the transformed curve is obtained by tracing the original curve twice.

Points of unit signature The term "invariant signature" was assigned in [4] to functions that can be computed from a given curve, and have an invariance property under a certain transformation group. For example, had we been interested merely in rigid transformation, we could adopt the curvature function as a signature of the curve, and extract features by a certain curvature value. However, since the curvature function is not invariant with respect to scale, we could rely above only on its extreme points, but not on the function values themselves.

More sophisticated signature functions, that remain invariant under similarity, affine and projective transformations, can be found in the literature (see [3] and references therein). Some of them suffer from the involvement of high order derivatives, which is clearly an undesired property when using digitized curves. Our similarity invariant signature function is computed as follow: we place a right angle on a curve point s, such that the angle is symmetric with respect to the normal at s, and it is directed toward the interior region (figure 4). The right angle intersects with the curve at points $s1$ and $s2$, and the signature at s is defined as the ratio of lengths $||s1 - s||/||s2 - s||$. We extract points for which this ratio has the value of one.

Fig. 4. Left: a right angle positioned on a given curve to measure the signature at that point. Right: the signature function obtained for the given curve.

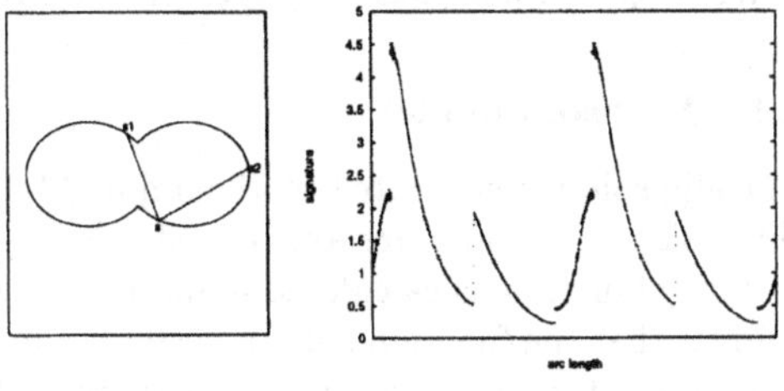

The features locality can be controlled by the angle sharpness. Robustness against noise is achieved due to the fact that only the first derivative of the curve is needed, and there is no measurement of arc length.

4 Representative views of smooth objects

Imposing a metric on the image space of a three dimensional object is a crucial preliminary stage in the process of identifying representative views. In this work we examine the relation between representative views and the metric being used. We calculate numerically the views stability function for three different test objects, applying for each object the different metric functions which are proposed above. The investigated objects are: an ellipsoid (which obeys reflection symmetries), a sandglass model (which is axial symmetric), and a model of an eggplant (which lacks any kind of symmetry). All the models are represented by implicit functions and they are shown is figure 5.

Fig. 5. The three test models: an ellipsoid, a sandglass and an eggplant.

According to our numerical simulations with different test objects and different metric functions, we conjecture that *the representative views do not depend on the metric being used* [3]. We find that the positions of the local maxima of the stability function are metric independent, hence the "best" viewing directions are an intrinsic property of the object.

We also find, as should be expected, an inverse relation between robustness and locality: using extreme curvature points is most sensitive to noise, while the other proposed methods have greater noise resistance.

4.1 Ellipsoid model

The ellipsoid is represented by the implicit function $x^2+2y^2+5z^2=1$, hence the three axes x, y and z are axes of reflection symmetry. Figure 6 shows the views stability function as calculated numerically using extreme curvature points. It is evident that the function has six local maxima in the entire image space (the two maxima at the poles appear in the graph as two lines). The six local maxima are in fact three pairs of points, since the stability function must obey the same reflection symmetries as the ellipsoid itself. Furthermore, each one of those pairs of viewing points is located on one of the ellipsoid axes. It turns out that the most stable view, which corresponds to the highest local maximum, is obtained when the ellipsoid is viewed along its minor axis (z). The most stable view has minimal depth and maximal image area. These results are in agreement with the analytical study reviewed in section 2.2.

Fig. 6. Stability of ellipsoid views (also plotted on the viewing sphere). Note that the gray levels are not ordered from bright to dark, to enhance the contrast in the figure.

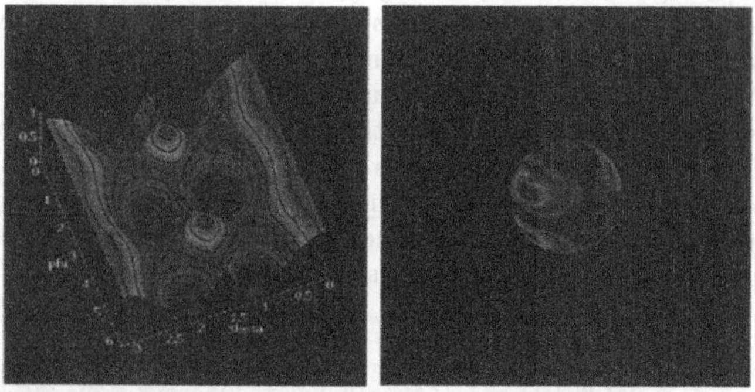

The relation between representative views and maximal image area was pointed out in the past on psychophysical ground. However, assigning to views a "genericity level" which is proportional to their apparent area, necessarily leads to the identification of the least representative view as the one with the minimal area. In the ellipsoid case, the minimal area is obtained when the viewing point is located on the main axis line (x), in sharp contrast to its being locally stable and likely.

[3] We exclude here metrics which are not related to the image content, like the discrete metric that assign a zero distance between an image to itself and a unit distance between any non identical images. For such metrics our conjecture certainly cannot hold.

Using the other metrics to estimate views stability, we get the same results regarding the position of local maximum points. The representative views are therefore invariant under the four examined metrics. The question of which metric to use should be addressed according to the application (section 5).

4.2 Sandglass model

The second object that we consider is represented by the implicit function $(x^2 + y^2 + 0.3z^2 - .09) * (x^2 + y^2 + 0.3z^2) - 0.3z^2 = 0$, and we refer to it as a model of a sandglass (see figure 5). In this representation, the z axis is an axis of rotation symmetry, so views which have the same altitude angle ϑ are all equivalent and are equally stable. It is therefore sufficient to investigate the views stability with respect to this angle alone, as is shown in figure 7 (transformed self intersection points are used). Again, using any one of our proposed metrics

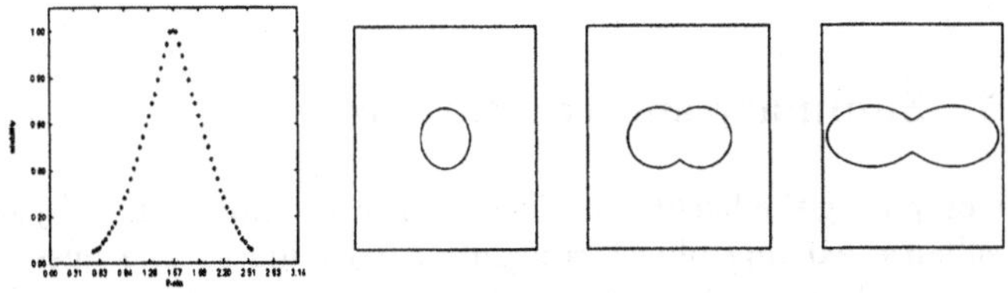

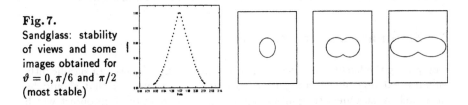

Fig. 7. Sandglass: stability of views and some images obtained for $\vartheta = 0, \pi/6$ and $\pi/2$ (most stable)

produced the same result, that the most representative view of the sandglass model is obtained (like before) when the principle axis of the object lies in the image plane (figure 7).

We note that with perfectly circular curves, all the proposed methods for feature extraction must fail, since the curvature and signature functions become uniform, and the number of transformed self intersections approaches infinity. The stability value for views which are close to the z axis could therefore be computed only with the method of a whole curve comparison, since for those viewing directions the object appears as a circle.

4.3 Eggplant model

The last test object is a model of an eggplant, which is represented by a fourth degree polynomial whose 35 coefficients were fitted experimentally to a real eggplant [10]. The eggplant model lacks any kind of symmetry, except the trivial reflection symmetry of silhouettes which is obtained for any two antipode viewing directions.

The stability map and the corresponding representative and non representative views are shown in figure 8. As can be seen, our results seem to agree with human intuition. Similar results are obtained with the different metric functions, but we note that with the metric based on extreme curvature extraction it was almost impossible to get reliable results with the given amount of noise in the

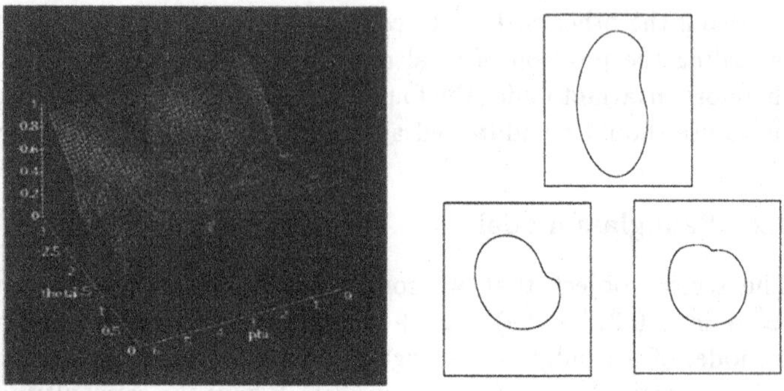

Fig. 8. Stability of the eggplant views. The upper view on the right is found as most stable (hence representative) while the other two views are in local minima of stability.

images. (The origin of noise in our case is the limit put on the accuracy of image generation from the implicit function model. Due to the complexity of the model and the large number of images needed to be generated during the simulation, we chose a rather coarse image resolution, which had introduced noise).

5 Summary and conclusions

Computing the likelihood of views of three dimensional objects can significantly enhance existing object recognition techniques. The question that should be answered by an object recognition algorithm is the identity of an object which appears in an image, but this question is frequently ambiguous, and should be answered statistically. The framework of maximal likelihood decision provides the appropriate statistical tool, and leads to the following obvious rule: among all the possible interpretations of a (noisy) image, select the one which is the most likely.

In this work we demonstrated how to use the quality of view stability to identify characteristic views of general curved objects, that can be recognized according to their silhouettes. Very simple objects could be dealt with analytically, but for more general cases we used numerical methods. We estimated the views stability function for three qualitatively different test models, and obtained their representative views according to the points of local maximum of stability.

Likelihood and stability cannot be evaluated without a quantitative measure of similarity between views. We based our similarity measure on the shape variation of the occluding contour, which depends only on the object's shape and the viewing direction, and does not depend on illumination. Four different metric functions were proposed, either using a whole curve comparison or using feature points. We found the same representative views under all metrics, hence we conjectured that *the representative views do not depend on the metric being used* (as long as the metric is "reasonable"), hence representative views are an intrinsic property of the object.

Since the selection of a metric does not seem to alter the most likely views, the important criteria for a metric selection become its locality and noise resistance. These two properties were shown to be inversely related, as one would expect. Locality is important as it best resists occlusions, where the given contour is not complete. In this work we didn't consider cases of occlusion, hence we could normalize the perimeter of a given curve to a unit length (section 3.1). This operation is crucial only for the metric 3.1 (a whole contour comparison). The other metric functions can be applied also to non normalized curves, and therefore can deal with occlusions.

Another benefit of comparison methods based on feature extraction is the possibility to combine them with database indexing methods [11]. The extracted features can serve not only for distance evaluation against stored arrays of feature points in the memory, but their invariant coordinates can also be used to generate pointers into the database, to enhance the search.

References

1. E. M. Arkin, L. P. Chew, D. P. Huttenlocher, K. Kedem, and J. S. B. Mitchell. An efficiently computable metric for comparing polygonal shapes. *T-PAMI*, pages 209–216, 1991.
2. J. Ben-Arie. The probabilistic peaking effect of viewed angles and distances with application to 3-d object recognition. *T-PAMI*, pages 760–774, 1990.
3. A. M. Bruckstein, R. J. Holt, A. N. Netravali, and T. J. Richardson. Invariant signature for planar shape recognition under partial occlusion. *CVGIP: Image Understanding*, pages 49–65, 1993.
4. A. M. Bruckstein, N. Katzir, M. Lindenbaum, and M. Porat. Similarity invariant recognition of partially occluded planar curves and shapes. *IJCV*, pages 271–285, 1992.
5. J. B. Burns, R. S. Weiss, and E. M. Riseman. View variation of point set and line segment features. *T-PAMI*, pages 51–68, 1993.
6. S. Duvdevani-Bar and S. Edelman. On similarity to prototypes in 3d object representation. Technical Report CS-TR 95-11, Weizmann Institute, July 1995.
7. W. T. Freeman. Exploiting the generic view assumption to estimate scene parameters. In *proc. of ICCV*, pages 347–356, 1993.
8. D. P. Huttenlocher and S. Ullman. Object recognition using alignment. In *proc. of ICCV*, pages 102–111, 1987.
9. N. Katzir, M. Lindenbaum, and M. Porat. Planar curve segmentation for recognition of partially occluded shapes. In *proc. of ICPR*, 1990.
10. D. Keren, D. Cooper, and J. Subrahmonia. Describing complicated objects by implicit polynomials. *T-PAMI*, pages 38–53, 1994.
11. Y. Lamdan, J. T. Schwartz, and H. J. Wolfson. Affine invariant model based object recognition. *T-RA*, pages 578–589, 1990.
12. P. J. van Otterloo. *A Contour-Oriented Approach to Shape Analysis.* Prentice Hall Intrernational, 1991.
13. D. Weinshall and M. Werman. Disambiguation techniques for recognition in large databases and for under-constrained reconstruction. In *proc. of Intr. Symp. on Comp. Vision*, pages 425–430, Coral Gables, USA, 1995.
14. D. Weinshall, M. Werman, and N. Tishby. Stability and likelihood of views of three dimensional objects. *Computing Suppl.*, 11:237–256, 1996.

Real-Time Lip Tracking for Audio-Visual Speech Recognition Applications

Robert Kaucic, Barney Dalton, and Andrew Blake

Department of Engineering Science, University of Oxford, Oxford OX1 3PJ, UK

Abstract. Developments in dynamic contour tracking permit sparse representation of the outlines of moving contours. Given the increasing computing power of general-purpose workstations it is now possible to track human faces and parts of faces in real-time without special hardware. This paper describes a real-time lip tracker that uses a Kalman filter based dynamic contour to track the outline of the lips. Two alternative lip trackers, one that tracks lips from a profile view and the other from a frontal view, were developed to extract visual speech recognition features from the lip contour. In both cases, visual features have been incorporated into an acoustic automatic speech recogniser. Tests on small isolated-word vocabularies using a dynamic time warping based audio-visual recogniser demonstrate that real-time, contour-based lip tracking can be used to supplement acoustic-only speech recognisers enabling robust recognition of speech in the presence of acoustic noise.

1 Introduction

Since verbal communication is the easiest and most natural method of conveying information, the possibility of communicating with computers through spoken language presents an opportunity to change profoundly the way humans interact with machines. Voice interactive systems could relieve users of the burden of entering commands via keyboards and mice and prove indispensable in situations where the operator's hands are occupied such as when driving a car or operating machinery. Much research has focused on the development of spoken language systems and rapid advances in the field of automatic speech recognition (ASR) have been made in recent years [7, 23]. Although progress has been impressive, researchers have yet to overcome the inherent limitations of purely acoustic-based systems, particularly their susceptibility to environmental noise. Such systems readily degrade when exposed to non-stationary or unpredictable noise as might be encountered in a typical office environment with ringing telephones, background radio music, and disruptive conversations. Acoustic solutions typically employ noise compensation methods during preprocessing or recognition to reduce the effect of the noise. The preprocessing approaches often use spectral subtraction or adaptive filtering techniques to remove the additive noise from the signal [14]. Hidden Markov Model (HMM) decomposition, where separate models are used for the clean speech and noise, is a common method used to provide compensation during recognition [21, 11]. While these approaches have proven to be effective, they ignore a basic tenet, that is, the multi-modal nature of human communication. Here we attempt to exploit this by using visual information in the form of the outline of the lips to improve upon acoustic speech recognition performance.

The majority of automatic lipreading research to date has focused primarily on establishing that visual information can be used to supplement acoustic speech recognition on small isolated-word vocabularies. The extraction of features in real-time has been largely ignored by lipreading researchers—deferring that complication to the future. Real-time feature extraction is obviously required for practical audio-visual language understanding systems. However, to track in real-time, it is often necessary to reduce the dimensionality of the image data through parameterisation which could result in the loss of important recognition information. This work demonstrates that, despite this loss of information, visual features obtained from tracking the lips in real-time can supplement automatic acoustic speech recognisers.

Two lip trackers, one that tracks lips from a profile view and the other from a frontal view, have been developed. Both are capable of locating, tracking, and compactly representing the lip outline in real-time at full video field rate (50/60 Hz). The 'profile lip tracker' follows the outline of the upper and lower lips and needs no cosmetic assistance. Tracking from the frontal view is more difficult as the lips are set against flesh-tones with consequently weak contrast. Therefore, when the frontal view was used, the speaker wore lipstick to enhance the contrast around the lips. The tracker framework is identical for tracking assisted and unassisted lips and thus for the frontal view, assisted lips were used to demonstrate the feasibility of using real-time dynamic contour-based lip trackers in audio-visual speech recognition applications. Preliminary work has begun on tracking natural lips from the frontal view and a tracking sequence using this tracker is presented as well, although, to date, no recognition experiments have been conducted using it.

Visual features extracted from the lip trackers are incorporated into a dynamic time warping (DTW) based isolated-word recogniser. Recognition performance is evaluated using acoustic only, visual only, and audio-visual information with and without added artificial acoustic noise. The experiments demonstrate that visual information obtained from tracking the lip contour from either view can improve upon acoustic speech recognition, especially in speech degraded by acoustic noise.

2 Lipreading

It is well known that human speech perception is enhanced by seeing the speaker's face and lips—even in normal hearing adults [9, 17]. Several researchers [9] have demonstrated that the primary visible articulators (teeth, tongue, and lips) provide useful information with regard to the place of articulation and Summerfield [20] concluded that such information conveyed knowledge of the mid- to high-frequency part of the speech spectrum—a region readily masked by background noise.

Motivated by this complementary contribution of visual information, researchers have recently developed audio-visual speech recognisers which have proven to be robust to acoustic noise [15, 22, 19, 5, 6]. These systems can be classified by the visual features they extract into three categories—pixel-based systems, lip-velocity systems, and lip-outline/measurement systems. The pixel-based systems [15, 22, 5] maximise retention of information about the visible articulators by using directly or indirectly the grey-level pixel data around the mouth region. Unfortunately, these systems tend to be highly susceptible to changes in lighting, viewing angle, and speaker head movements. They also usually employ computationally expensive processing algorithms to locate the mouth and/or extract relevant recognition features. While these systems serve as excellent research platforms, the extensive processing required limits their use in real-time or near real-time applications. The lip-velocity systems [12] assume that it is the *motion* of the lips that contains the relevant recognition information especially with respect to determining syllable or word boundaries and thus extract the velocities of different portions of the lips. A similar limitation exists for this approach where computationally expensive procedures like optical flow analysis and morphological operations are used to extract the lip velocities which prevents their use in near real-time applications. The lip outline/measurement systems [10, 19] extract geometrical features from the lip outline or oral cavity. Typical features include the height, width, area, and spreading (width/height) of the mouth. These systems are able to extract visual features in real-time, although they avoid many of the complications of tracking in real-world images by tracking strategically placed reflective dots on the face.

The recognition systems presented here fall into this last category, however, real-time feature extraction is achieved without the need for markers by parameterising the lip outline and learning the dynamics of moving lips.

3 Lip tracker

The lip trackers resulted from the tailoring of Blake et al.'s [2, 3] general purpose dynamic contour tracker to the specific task of tracking lips. The 2D outline of the lips is parameterised by quadratic B-splines which permits sparse representation of the image data. Motion of the lips is represented by the x and y coordinates of B-Spline control points, $(\mathbf{X}(t), \mathbf{Y}(t))$, varying over time. Stability of the lip tracker is obtained by constraining the lip movements to deformations of a lip template, $(\overline{\mathbf{X}}, \overline{\mathbf{Y}})$. Lip motion is modelled as a second order process driven by noise with dynamics that imitate typical lip motions found in speech. These dynamics are learned using a Maximum Likelihood Estimation (MLE) algorithm [4] from representative sequences of connected speech. Temporal continuity is provided by a Kalman filter which blends predicted lip position with measurement/observation features taken from the image. To enable real-time tracking, the search for image features is confined to one dimensional lines along normals to the lip curve. The profile lip tracker uses high contrast edges for image features while the frontal lip tracker uses a combination of edges and intensity valleys.

4 Tracking

The profile view is favourable for tracking because the mouth appears sharply silhouetted against the background whereas, in a frontal view, the lips are set against flesh-tones with consequently weak contrast—a problem for visual trackers. However there is, of course, a potential loss of information in profile viewing in that the tongue and teeth are no longer visible. There may also be a loss of shape information in the lip contour itself, since its width is no longer directly observable in profile, and our experiments suggest that lip width is significant for audio-visual speech analysis. Figure 1 shows that the tracker can follow the lips

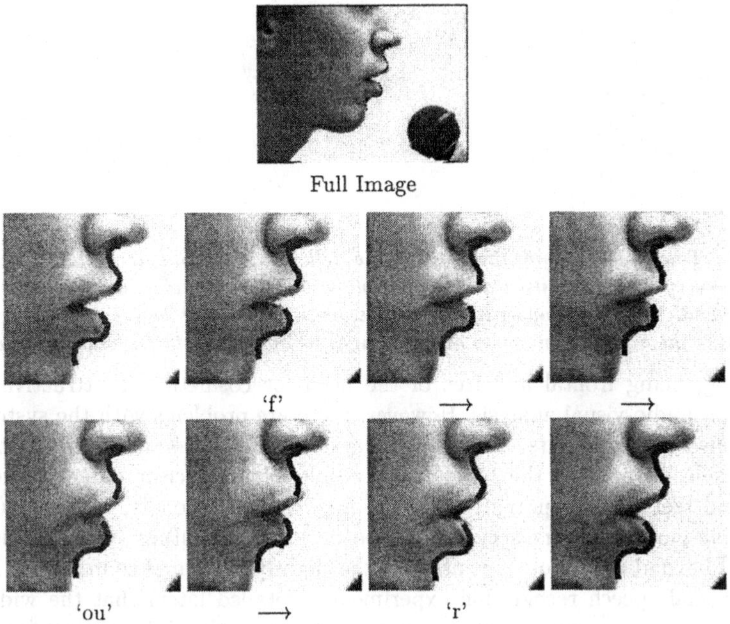

Fig. 1.: *Tracking the word "four". Snapshots taken approximately every 40 ms. The tracker accurately follows the lower lip during the f-tuck (curling of the lip to form the 'fa' sound) in tracked frames 3 and 4 and continues tracking through the lowering of the jaw necessary for the 'our' sound.*

even during subtle lip movements such as the f-tuck in the word 'four'. Similar tracking results were obtained using a frontal view when lipstick was worn to enhance the contrast around the lips [8].

Natural (un-aided) lips can also be tracked from the frontal view using the dynamic contour framework; however, instead of using edges for image features, the intensity valley between the lips is used to locate the corners of the mouth and upper lip. This valley has been shown to be robust to variations in lighting, viewpoint, identity and expression [13] and proves to be a reliable feature for lip tracking. Figure 2 shows a tracked sequence of the word ‘five’ using valley features for the upper lip and edge features for the lower lip. This mode of

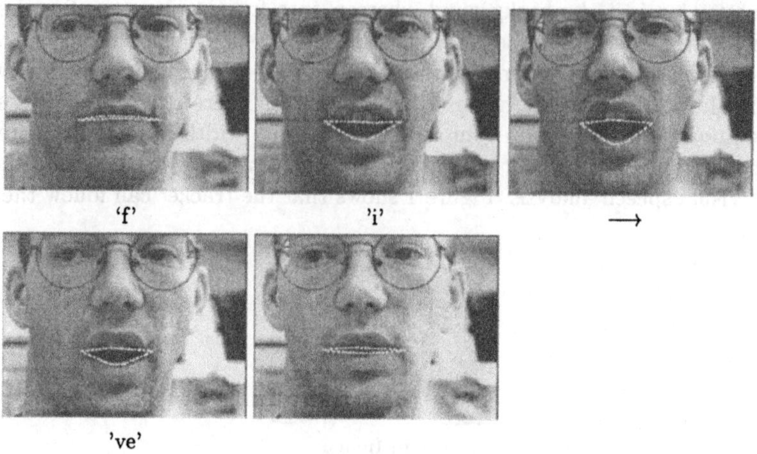

Fig. 2.: *Tracking the word "five" using the valley tracker. Snapshots taken approximately every 60 ms. Whilst tracking is stable and the outline closely approximates the inner mouth region, the upper lip contour becomes confused by the presence of the teeth mistaking them for the inner lip and continues to track them throughout the sequence.*

tracking, being frontal and free of the need for cosmetics, is attractive as a basis for audio-visual analysis. However, there are problems with the system as developed thus far. First, the upper tracked contour has an affinity both for the inside lip and for the teeth when visible, whereas clear differentiation of lips and teeth is a requirement for the application. Secondly, it is difficult to pinpoint mouth corners accurately—the dark visual feature (valley) tends to extend beyond the mouth, resulting in the slightly elongated contour. We know from visual speech recognition experiments (detailed later) that the width of the mouth (oral cavity) contains important recognition information for word discrimination tasks, so further work is needed before this tracker is entirely adequate for speech recognition applications.

Incidental head movements do not affect tracking performance as long as the lip tracker remains locked, however, rapid or large head movements may cause the tracker to lose lock and become unstable. Additionally, since the position of the head naturally influences the position of the lips, head movements may corrupt the recognition data. To compensate for this we are investigating the coupling of a head tracker to the lip tracker [18].

5 Feature Extraction

An essential part of any recognition system is the extraction of features that reliably represent the objects in the data set. The features must compactly represent the data in a suitable form for recognition. For acoustic speech signals the features are typically the result of spectral analysis on the waveform [16]. Thus the acoustic pre-processing consisted of the extraction of 8 "mel-scale" filter-bank coefficients from overlapping 32ms windows and 20ms frames. The 20ms frame interval was chosen to coincide with the 50 Hz video rate to facilitate integration of the two modalities without additional sub-sampling or linear interpolation.

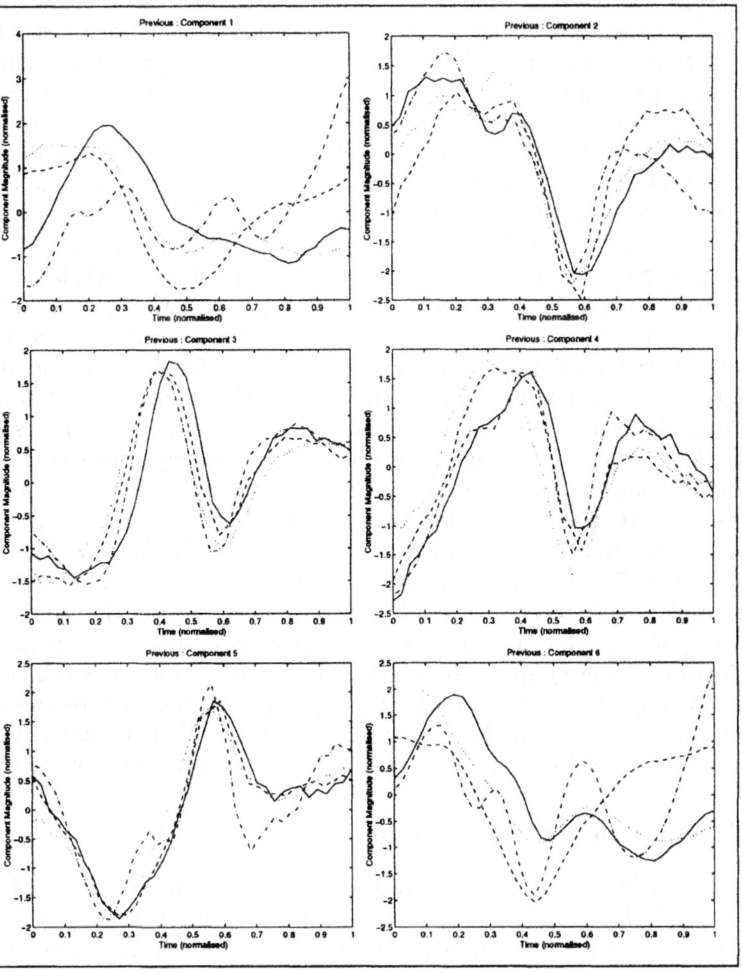

Fig. 3.: *Affine components 1 through 6 for four repetitions of the word 'previous'. Each component has been thresholded, set to zero mean and unity variance and linear time normalised. Significant shape correlation exists in components 2 (Y translation), 3 (X scale), 4 (Y scale) and 5 (Y shear) across all four repetitions suggesting that they may contain useful recognition information.*

Several different visual processing methods were examined in order to gain insight into which lip movements/deformations would be most beneficial for speech recognition. All visual features resulted from the projection of the lip outline, represented as a sequence of control points, onto a sub-space spanned by a reduced basis. The first basis chosen was the affine basis. Our experience in tracking had shown that the affine basis was insufficient to model all deformations of the lips, but given a successfully tracked lip sequence, we felt that projection onto the affine basis would still provide useful recognition information. Others [10, 19] have reported success using similar features obtained through tracking dots on the face. Additional visual feature representations were obtained through principal components analysis (PCA) where it was found that 99% of the deformations of the outer lip contour were accounted for by the first 6 principal components [8]. Furthermore, since we believed that global horizontal displacement of the lip centroid was not necessary for speech production and only a bi-product of spurious head movements (global vertical displacement is present as a result of the asymmetrical movement of the upper and lower lips), a third recognition basis was created by subtracting horizontal displacement (X translation) from each set of control points and then performing PCA on the remaining data. Similar to the original principal components analysis, it was found that 99% of the remaining lip motions were accounted for by just 6 components.

When choosing features to be used in recognition experiments it is important that the features chosen be repeatable across multiple repetitions of the same token (word), yet be sufficiently different between repetitions of dissimilar words. This was of special concern as Bregler at al. [5] had concluded that the outline of the lip was not sufficiently distinctive to give reliable recognition performance. However, several of the features in the affine basis do in fact satisfy these criteria. This can be seen in figure 3 where traces of the six affine features for multiple repetitions of the word 'previous' are shown. In the figure we see that components 2 (vertical translation), 3 (horizontal scale), 4 (vertical scale) and 5 (vertical shear) are consistent across all four repetitions which suggests that they may contain useful recognition information. Similarly, components 1 (horizontal translation) and 6 (horizontal shear) show little consistency which was expected as neither appears to play a role in the production of speech.

6 Recognition Experiments

Both the profile and the frontal lip tracker (with lipstick) were used to explore the extent to which lip contour information could aid speech recognition. Separate isolated-word, audio-visual recognition experiments were conducted using visual features extracted from each of the trackers. Raw visual and audio data were gathered simultaneously and in real-time (50 Hz) on a Sun IPX workstation with Datacell S2200 framestore. The visual data consisted of the mouth outline represented as (x, y) control points (10 for the side view and 13 for the frontal view) and the audio data 8-bit μ-law sampled at 8 KHz.

Recognition experiments were conducted using audio-only, visual only, and combined audio-visual DTW recognition. Composite feature vectors were created by concatenating the acoustic and vision features, although it was possible to vary the relative weighting between the two modalities during recognition. Each feature was normalised to zero mean and unity variance over the entire frame sequence. The 20 repetitions of each word were partitioned into three sets. Two repetitions were used as exemplar patterns for matching, seven were used as a training set, and eleven as a test set.

6.1 Recognition using the profile view

Although the main experiments were done using the frontal view, it seemed important to run at least a pilot experiment using the side-view, given that tracking in profile is robust even without cosmetic aids. This was done to demonstrate that real-time (50Hz), unaided visual tracking for audio-visual speech analysis is indeed a possibility, albeit currently on a modest scale. A 10-word digit database was used.

Significant improvements in error rate were realised by incorporating the visual data-stream. Figure 4 shows the error rates for experiments conducted with

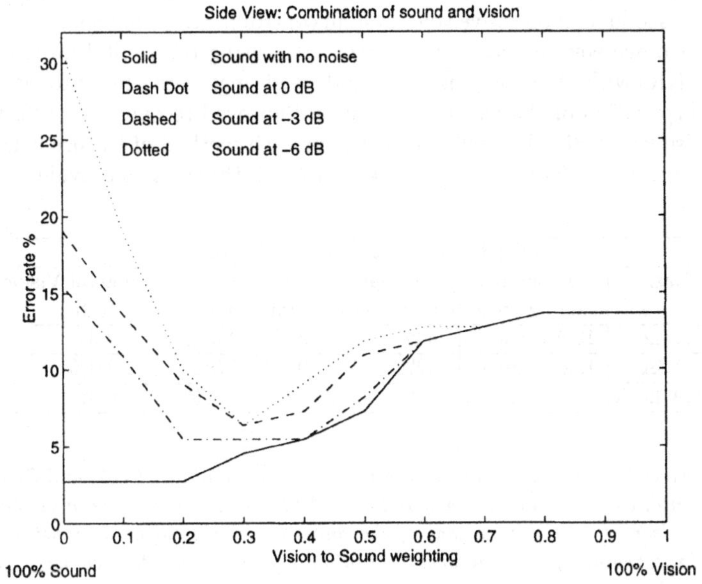

Fig. 4.: *Side View: Error rate variation on the test set as sound to vision weighting is varied. The incorporation of visual features extracted from the lip profile improves recognition performance at all noise levels. With a clean audio signal, vision is only marginally beneficial. However, as the audio signal becomes noisy, the contribution of vision is noticeably improved with a reduction in error rate from 15% to 5.5% for the 0 dB signal and from 19% to 6.5% for -3 dB. With the audio quality further degraded to -6 dB, the error rate drops from 31% to 6.5%.*

audio signals at various SNRs with different vision to sound weightings. Several points are evident from this graph. The first is that the audio-only recogniser performs better than the visual-only recogniser at high signal to noise ratios. This merely reflects the higher information content in audio data with respect to speech recognition in typical noise-free dialogue. Secondly, incorporation of the vision information improved performance at all noise levels—with the largest improvements occurring at the lowest signal to noise ratios—a key finding of this research. It is this increase in recognition performance due to the incorporation of visual information—a term we refer to as the *incremental vision rate*—that is a true measure of the added benefit of lip reading. As an example, one sees that at a SNR of -3 dB, an incremental vision rate (error rate reduction) of 12.5% (19% to 6.5%) is achieved for optimally combined vision/speech compared with speech alone.

6.2 Recognition using the frontal view

In the frontal view, vision data was represented in each of the three previously discussed bases—affine, pca, and pca minus X translation. Experiments were conducted on a 40-word database consisting of numbers and commands that might be used in an interactive voice system controlling a car phone, fax machine, or similar office equipment. Plots of error rates using the frontal lip tracker at various SNRs with different vision to sound weightings were similar to figure 4 in that incorporation of the vision information improved recognition performance at all noise levels [8]. The best error rates for each method of feature extraction are shown in figure 5 on sound at -3 dB SNR. All three bases provide a similar

Best error rates for each basis							
Basis	Acoustic		Visual		Combined		Incremental Vision
	training	test	training	test	training	test	Rate
affine	13.9%	16.6%	44%	52%	8.2%	9.3%	7.3%
PCA	13.9%	16.6%	42%	51%	9.6%	9.3%	7.3%
PCA no X	13.9%	16.6%	41%	49%	9.6%	9.8%	6.8%

Fig. 5.: *Frontal View: Best recognition error rates for the affine, PCA, and PCA without horizontal translation bases on sound at -3 dB SNR. The error rate of the acoustic-only recogniser is nearly twice that of the audio-visual recogniser demonstrating the benefit of incorporating visual information into the acoustic speech recogniser. All three bases provide a similar increase in recognition performance. This is encouraging as the geometrically derived affine basis presents an opportunity for speaker-independent recognition while the PCA bases are particular to a given speaker.*

increase in recognition performance. These results demonstrate that there is useful recognition information contained in the lip outline contrary to Bregler et al. [5] who claims that the outline of the lip is too coarse for accurate recognition. Furthermore, the comparable performance of the affine basis with respect to the derived bases suggests the possibility of developing a speaker independent recognition system with the visual features represented as affine transformations of the lip template.

6.3 Evaluating visual shape components

Having determined the utility of lip shape information, the recognition performance of individual motion components was measured in order to determine which contribute most to recognition performance. It was hoped that a coherent picture would result yielding the lip movements most beneficial for speech recognition. Figure 6 shows the recognition performance achieved using vision

Best error rates using only a single vision component					
Basis component	Vision only training	Vision only test	Combined training	Combined test	Incremental Vision Rate
Full affine	44%	52%	8.2%	9.3%	7.3%
X Trans	93%	93%	14%	17%	0.0%
Y Trans	76%	81%	13%	14%	3.0%
X Scale	59%	63%	9%	12%	5.0%
Y Scale	75%	79%	14%	16%	0.7%
Y Shear	77%	86%	14%	13%	3.2%
X Shear	86%	90%	14%	17%	0.0%
Full PCA	42%	51%	9.6%	9.3%	7.3%
1 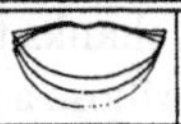	70%	74%	11%	12%	4.6%
2 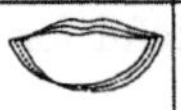	91%	91%	14%	17%	0.0%
3 	82%	88%	14%	17%	0.0%
4 	70%	75%	9%	11%	5.2%
5 	73%	82%	12%	12%	4.6%
6 	89%	94%	14%	14%	0.0%

Fig. 6.: *Results of recognition performance using only one vision component from each of the bases. Recognition using sound alone at -3 dB was 14% for the training set and 17% for the test set. Full affine and Full PCA refer to overall recognition performance using all six components of each basis. The lip deformations represented by PCA components 1,4,5 and affine components Y Trans, X Scale, and Y Shear contribute the most to recognition performance implying that the recognition information of the lip outline can be expressed with just a few shape parameters.*

components from each of the bases singly. Error rates are shown for the components used individually and in concert with the acoustic features. The tests were conducted on speech at a SNR of -3 dB. These results suggest that most of the recognition information is contained in only a few (2-4) shape parameters.

7 Conclusions and Future Work

Despite doubts expressed by other researchers [5], it has been shown that dynamic contours can be used to track lip outlines, with sufficient accuracy to be useful in visual and audio-visual speech recognition. Moreover, tracking can be performed at real-time video rates (50 Hz). Recognition experiments conducted on a 40-word database demonstrated that isolated words could be accurately recognised in speech severely degraded by artificial noise. Experiments reported here used Dynamic Time Warping as the recognition algorithm; however, given the state of the art in speech analysis [16], it is natural to try Hidden Markov Model recognition. Such experiments are in progress and initial indications are that vision similarly makes a significant contribution to lowering error-rates in accordance with results from others [1, 6].

It is known that human lip-readers rely on information about the presence/absence of the teeth and the tongue inside the lip contour [20]. For this reason it is likely that the best recognition results will ultimately be obtained from frontal views with this additional information extracted. Towards this end, we are developing a real-time un-assisted frontal lip tracker capable of extracting the lip contour as well as determining the presence/absence of the teeth and tongue; furthermore, we are investigating how the coupling of a head tracker to the lip tracker can be used to compensate for global head movements during tracking and recognition [18].

Acknowledgements: The authors wish to express special thanks to Dave Reynard, Michael Isard, Simon Rowe, and Andrew Wildenberg whose assistance and elegant software made this research possible. We are grateful for the financial support of the US Air Force and the EPSRC.

References

1. A. Adjoudani and C. Benoit. On the integration of auditory and visual parameters in an HMM-based ASR. In *Proceedings NATO ASI Conference on Speechreading by Man and Machine: Models, Systems and Applications.* NATO Scientific Affairs Division, Sep 1995.
2. A. Blake, R. Curwen, and A. Zisserman. A framework for spatio-temporal control in the tracking of visual contours. *Int. Journal of Computer Vision*, 11(2):127–145, 1993.
3. A. Blake and M.A. Isard. 3D position, attitude and shape input using video tracking of hands and lips. In *Proc. Siggraph*, pp. 185–192. ACM, 1994.
4. A. Blake, M.A. Isard, and D. Reynard. Learning to track the visual motion of contours. *Artificial Intelligence*, 78:101–134, 1995.
5. C. Bregler and Y. Konig. Eigenlips for robust speech recognition. In *Proc. Int. Conf. on Acoust., Speech, Signal Processing*, pp. 669–672, Adelaide, 1994.
6. C. Bregler and S.M. Omohundro. Nonlinear manifold learning for visual speech recognition. In *Proc. 5th Int. Conf. on Computer Vision*, pp. 494–499, Boston, Jun 1995.
7. R. Cole, L. Hirschmann, L. Atlas, et al. The challenge of spoken language systems: Research directions for the nineties. *IEEE Trans. on Speech and Audio Processing*, 3(1):1–20, 1995.

8. B. Dalton, R. Kaucic, and A. Blake. Automatic Speechreading using dynamic contours. In *Proceedings NATO ASI Conference on Speechreading by Man and Machine: Models, Systems and Applications*. NATO Scientific Affairs Division, Sep 1995.
9. B. Dodd and R. Campbell. *Hearing By Eye : The Psychology of Lip Reading*. Erlbaum, 1987.
10. E. K. Finn and A. A. Montgomery. Automatic optically based recognition of speech. *Pattern Recognition Letters*, 8(3):159–164, 1988.
11. M.J.F. Gales and S. Young. An improved approach to the Hidden Markov Model decomposition of speech and noise. In *Proc. Int. Conf. on Acoust., Speech, Signal Processing*, pp. 233–239, San Franciso, Mar 1992.
12. M.W. Mak and W.G. Allen. Lip-motion analysis for speech segmentation in noise. *Speech Communication*, 14(3):279–296, 1994.
13. Y. Moses, D. Reynard, and A. Blake. Determining facial expressions in real-time. In *Proc. 5th Int. Conf. on Computer Vision*, pp. 296–301, Boston, Jun 1995.
14. J.P. Openshaw and J.S. Mason. A review of robust techniques for the analysis of degraded speech. In *Proc. IEEE Region 10 Conf. on Comp., Control, and Power Engr.*, pp. 329–332, 1993.
15. E.D. Petajan, N.M. Brooke, B.J. Bischofy, and D.A. Bodoff. An improved automatic lipreading system to enhance speech recognition. In E. Soloway, D. Frye, and S.B. Sheppard, editors, *Proc. Human Factors in Computing Systems*, pp. 19–25. ACM, 1988.
16. L. Rabiner and J. Bing-Hwang. *Fundamentals of speech recognition*. Prentice-Hall, 1993.
17. D. Reisberg, J. McLean, and A. Goldfield. Easy to hear but hard to understand: A lip-reading advantage with intact auditory stimuli. In B. Dodd and R Campbell, editors, *Hearing By Eye : The Psychology of Lip Reading*, pp. 97–113. Erlbaum, 1987.
18. D. Reynard, A. Wildenberg, A. Blake, and J. Marchant. Learning dynamics of complex motions from image sequences. In *Proc. 4th European Conf. on Computer Vision*, Cambridge, England, Apr 1996.
19. D.G. Stork, G. Wolff, and E. Levine. Neural network lipreading system for improved speech recognition. In *Proceedings International Joint Conference on Neural Networks*, volume 2, pp. 289–295, 1992.
20. Q. Summerfield, A. MacLeod, M. McGrath, and M. Brooke. Lips, teeth and the benefits of lipreading. In A.W. Young and H.D. Ellis, editors, *Handbook of Research on Face Processing*, pp. 223–233. Elsevier Science Publishers, 1989.
21. A.P. Varga and R.K. Moore. Hidden Markov Model decomposition of speech and noise. In *Proc. Int. Conf. on Acoust., Speech, Signal Processing*, pp. 845–848, 1990.
22. B.P. Yuhas, M.H. Goldstein, T.J. Sejnowski, and R.E. Jenkins. Neural network models of sensory integration for improved vowel recognition. *Proceedings of the IEEE*, 78(10):1658–1668, 1990.
23. V. Zue, J. Glass, D. Goodine, L. Hirschman, H. Leung, M. Phillips, J. Polifroni, and S. Seneff. From speech recognition to spoken language understanding: The development of the MIT SUMMIT and VOYAGER systems. In R.P. Lippman, J.E. Moody, and D.S. Touretzky, editors, *Advances in Neural Information Processing 3*, pp. 255–261. Morgan Kaufman, 1991.

Matching Object Models to Segments from an Optical Flow Field

Henner Kollnig[1] and Hans-Hellmut Nagel[1,2]

[1]Institut für Algorithmen und Kognitive Systeme
Fakultät für Informatik der Universität Karlsruhe (TH)
Postfach 6980, D-76128 Karlsruhe, Germany

[2]Fraunhofer-Institut für Informations- und Datenverarbeitung (IITB),
Fraunhoferstr. 1, D-76131 Karlsruhe, Germany
Telephone +49 (721) 6091-210 (Fax -413), E-Mail hhn@iitb.fhg.de

Abstract. The temporal changes of gray value structures recorded in an image sequence contain significantly more information about the recorded scene than the gray value structures of a single image. By incorporating optical flow estimates into the measurement function, our 3D pose estimation process exploits *interframe* information from an image sequence in addition to *intraframe* aspects used in previously investigated approaches. This increases the robustness of our vehicle tracking system and facilitates the correct tracking of vehicles even if their images are located in low contrast image areas. Moreover, partially occluded vehicles can be tracked *without* modeling the occlusion explicitly. The influence of interframe and intraframe image sequence data on pose estimation and vehicle tracking is discussed systematically based on various experiments with real outdoor scenes.

1 Introduction

Many computer vision approaches to pose refinement match model features only to stationary data features, for example edge elements or edge segments which are extracted from a *single* image frame. Applying such an approach to evaluate an image *sequence*, the temporal aspects of image sequence data appear to be insufficiently exploited.

In order to avoid matching moving objects to stationary image features that exhibit coincidentally the same gray value structure as the object image under scrutiny we do no longer match polyhedral vehicle models only to stationary image gradients which is described in more detail in [Kollnig & Nagel 95]. Rather than restricting the update step only to these *intraframe* data, we extend the update step to evaluate *optical flow* vectors as *interframe* image contributions. Optical flow estimates the apparent shift of gray value structures. In this contribution, the estimated optical flow is matched to the *motion field* (also denoted as *displacement rate*), i.e. the image plane velocity of projected scene points.

The domain of discourse of our investigations is illustrated by an image sequence recording gas station traffic (see Figure 1). The tracking of vehicle C is a

Fig. 1. 350th frame of an image sequence showing gas station traffic. Vehicle `A` has stopped in front of the petrol pump area, a dark vehicle `B` has stopped on the rear lane behind the petrol pump area. An additional vehicle `C` is just passing the vehicle `B` in order to pull up to the leftmost petrol pump. This significantly occluded moving vehicle `C` will be discussed in more detail.

challenge since it is significantly occluded by other vehicles as well as by stationary scene components. In addition, its image is located in a low contrast image environment. The image features which provide significant cues for the vehicle image cannot, therefore, be found only in the *spatial* gray value gradients. As Figure 2 illustrates, important information is contained in the optical flow vectors estimated from *spatio-temporal* gray value gradients. This information enables the image sequence analysis system to track partially occluded vehicles *without* explicitly modeling the occlusion.

2 Related Publications

A review of relevant literature can be found in [Sullivan 92; Koller et al. 93; Sullivan *et al.* 95; Cédras & Shah 95], for optical flow estimation see [Barron et al. 94; Otte & Nagel 94; Otte & Nagel 95]. We can, therefore, confine ourselves to recent publications.

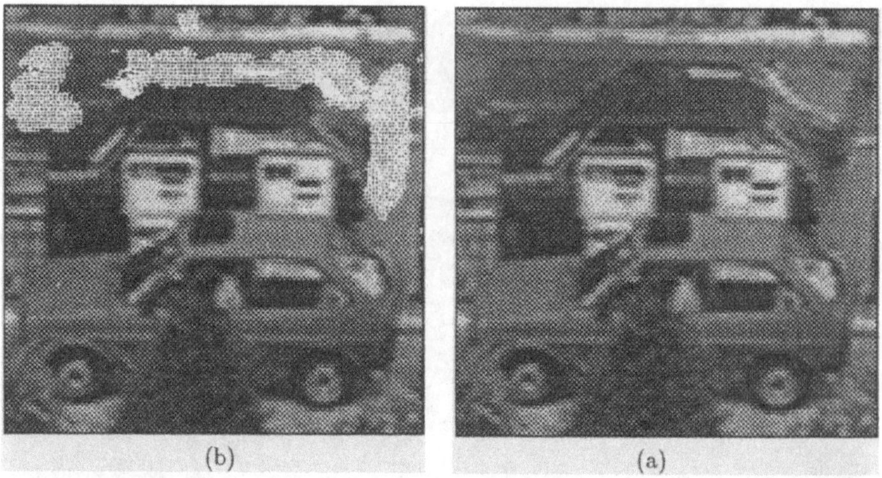

Fig. 2. An enlarged section of Figure 1. The significantly occluded vehicle `C` is moving from right to left. The estimated optical flow vectors shown in (b) are better cues for the vehicle image than the gray values of the current image frame (a) themselves. The optical flow vectors overlap most of the visible parts of the image of vehicle `C`.

[Otte & Nagel 94] recorded a three-dimensional polyhedral scene with a moving camera on a robot and compared the results of various optical flow estimation approaches with ground truth about image motion, but did not exploit the measured differences in order to update a pose estimate.

[Schirra *et al.* 87] as well as [Gong & Buxton 93] track vehicles by clustering and chaining displacement vectors. After determining optical flow and its discontinuities using non-linear diffusion, [Proesmans *et al.* 94] estimate 2D vehicle motion parameters. In distinction to our 3D pose estimation and tracking, these three groups confine themselves to techniques in the 2D image domain.

Supposing that an initial value for the object pose is available, for instance chosen interactively, [Worrall *et al.* 94] proposed a pose refinement of active models using forces in 3D without an extraction of line segments. [Tan *et al.* 94] localize vehicles without feature extraction, too, based on a 1D correlation technique. Both use a histogram voting and peak searching process in order to determine the vehicle pose whereas we compute the *Jacobian of the measurement function* to update the 3D pose with a Maximum-A-Posteriori (MAP) estimation process. Moreover, in distinction to the approach of [Worrall *et al.* 94; Tan *et al.* 94], we obtain initial pose estimates automatically by segmenting an optical flow field. The framework of our approach can be seen as analogous to the work of [Lowe 87], [Koller et al. 93], and [Kollnig *et al.* 94]. However, none of the cited approaches exploits optical flow estimates in order to update a 3D pose estimate.

3 Computing the Motion Field

The five-dimensional state vector

$$\boldsymbol{x}(t) = \Big(p_x(t), \quad p_y(t), \quad \phi(t), \quad v(t), \quad \omega(t)\Big)^T \tag{1}$$

to be estimated by our pose estimation algorithm comprises the position $\boldsymbol{p} = (p_x, p_y, 0)^T$ and orientation ϕ of the vehicle model relative to a reference (world) coordinate system in the road plane as well as the magnitudes of the translational velocity v and angular velocity ω.

In our implementation, the trajectory of the vehicle model reference point $\boldsymbol{p}(t)$ is assumed to be (locally) described by a simple circular motion model with constant magnitudes of the translational and angular velocities:

$$\boldsymbol{p}(t) = \boldsymbol{C} + \frac{v}{\omega}\begin{pmatrix} \sin\phi(t) \\ -\cos\phi(t) \\ 0 \end{pmatrix} \quad , \tag{2}$$

where $\boldsymbol{C}$ denotes the center of the circular trajectory and $\rho = v/\omega$ its radius. This model contains a straightforward movement as special case (infinite radius of the circle).

Let $\boldsymbol{x}_m$ denote the coordinates of a point in the (vehicle) model coordinate system, $\boldsymbol{x}_w$ its position vector in the world coordinate system and $\boldsymbol{x}_c$ its coordinates in the camera coordinate system, respectively.

The trajectory of a scene point on the vehicle surface with the model coordinates $\boldsymbol{x}_m$ is given by the following equation:

$$\boldsymbol{x}_w = \boldsymbol{x}_w(\boldsymbol{x}(t), \boldsymbol{x}_m) = \boldsymbol{p} + R_{wm}(\phi)\boldsymbol{x}_m \quad ,$$

where $R_{wm}(\phi)$ denotes a orthonormal 3×3 matrix describing the rotation of the model coordinate system with respect to the world coordinate system. The 3D motion $\dot{\boldsymbol{x}}_w$ of a scene point with model coordinates $\boldsymbol{x}_m$ is given by

$$\dot{\boldsymbol{x}}_w = \dot{\boldsymbol{x}}_w(\boldsymbol{x}(t), \boldsymbol{x}_m) = v\begin{pmatrix} \cos\phi \\ \sin\phi \\ 0 \end{pmatrix} + \frac{\partial R_{wm}(\phi)}{\partial \phi}\omega \boldsymbol{x}_m \quad . \tag{3}$$

The 2D image coordinates $\boldsymbol{\xi}$ of a 3D point with the world coordinates $\boldsymbol{x}_w$ are obtained by the following chain of operations:

$$\boldsymbol{x}_c = R_{kw}\boldsymbol{x}_w + \boldsymbol{t}_{kw} \quad , \tag{4}$$

$$\boldsymbol{\xi} = \begin{pmatrix} \xi \\ \eta \end{pmatrix} = \begin{pmatrix} f_x \frac{x_c}{z_c} + x_0 \\ f_y \frac{y_c}{z_c} + y_0 \end{pmatrix} . \tag{5}$$

The rotation matrix R_{kw} as well as the translation vector $\boldsymbol{t}_{kw}$ contain *external* camera parameters, whereas the focal length $(f_x, f_y)^T$ as well as the principal point $(x_0, y_0)^T$ represent *internal* camera parameters. External and internal camera parameters are to be estimated a priori by a calibration step.

The derivation of operations given in equations 4 and 5 with respect to time expresses the dependence of the 2D motion field vector $\dot{\boldsymbol{\xi}}$ at the pixel location $\boldsymbol{\xi}$ on the 3D motion $\dot{\boldsymbol{x}}_w$ of the corresponding scene point with the model coordinates $\boldsymbol{x}_m$:

$$\dot{\boldsymbol{\xi}} = \dot{\boldsymbol{\xi}}(\boldsymbol{x}(t), \boldsymbol{\xi}) = \frac{\partial \boldsymbol{\xi}}{\partial \boldsymbol{x}_c} \frac{\partial \boldsymbol{x}_c}{\partial \boldsymbol{x}_w} \dot{\boldsymbol{x}}_w(\boldsymbol{x}(t), \boldsymbol{x}_m) \quad . \tag{6}$$

Exploiting equations 4 and 5, we obtain:

$$\frac{\partial \boldsymbol{x}_c}{\partial \boldsymbol{x}_w} = R_{kw} \quad , \tag{7}$$

$$\frac{\partial \boldsymbol{\xi}}{\partial \boldsymbol{x}_c} = \begin{pmatrix} f_x \frac{1}{z_c} & 0 & -f_x \frac{x_c}{z_c^2} \\ 0 & f_y \frac{1}{z_c} & -f_y \frac{y_c}{z_c^2} \end{pmatrix} . \tag{8}$$

In order to estimate the motion field vector $\dot{\boldsymbol{\xi}}(\boldsymbol{x}(t), \boldsymbol{\xi})$ at each pixel location $\boldsymbol{\xi}$ at time point t, we determine the model coordinates $\boldsymbol{x}_m$ of the corresponding 3D scene point by means of a ray tracing algorithm and by exploiting a priori knowledge about the scene and the current state vector $\boldsymbol{x}(t)$. Substituting equations 3, 7, and 8 into equation 6 yields the motion field vector $\dot{\boldsymbol{\xi}}$ at the pixel location $\boldsymbol{\xi}$.

Analogous considerations hold for the estimation of a motion field vector $\dot{\boldsymbol{\xi}}(\boldsymbol{x}(t), \boldsymbol{\xi})$ at a pixel location $\boldsymbol{\xi}$ which represents the image position of a shadow point cast by a vehicle, except that we have to perform *two* ray tracing steps: one to get the corresponding scene coordinates of the shadow point and a second one to get the vehicle point which is projected onto the shadow point.

4 Update Step

Let $\boldsymbol{x}_k = \boldsymbol{x}(t_k)$ denote the state vector (see equation 1) to be estimated at halfframe time point t_k. We adopt the usual dynamic system notation (see, e.g., [Gelb 74]) denoting by $(\hat{\boldsymbol{x}}_k^-, P_k^-)$ and $(\hat{\boldsymbol{x}}_k^+, P_k^+)$, respectively, the estimated state vectors and their covariances before and after an update which incorporates the measurement at halfframe time point t_k.

Let an initial guess $\hat{\boldsymbol{x}}_0^-$ about state vector $\boldsymbol{x}_0$ be provided by a data driven motion segmentation step as described by, e.g., [Bouthemy & François 93; Gong & Buxton 93; Proesmans *et al.* 94; Kollnig *et al.* 94] or by a predicted estimate $\hat{\boldsymbol{x}}_k^-$ exploiting the state vector $\hat{\boldsymbol{x}}_{k-1}^+$ at the previous halfframe time point t_{k-1} and a state transition function with respect to a vehicle motion model. A view sketch is then generated, i.e. a set of model edge segments, by projecting edges of a 3D polyhedral vehicle model from the scene into the image plane and by removing invisible edge segments by a hidden-line algorithm. At each pixel location $\boldsymbol{\xi}$ of the frame at time point t_k, let $h_{||\nabla g||}(\boldsymbol{x}_k, \boldsymbol{\xi})$ denote the synthetic gradient norm as described by [Kollnig & Nagel 95] and let $(h_u(\boldsymbol{x}_k, \boldsymbol{\xi}), h_v(\boldsymbol{x}_k, \boldsymbol{\xi}))^T = \dot{\boldsymbol{\xi}}(\boldsymbol{x}_k, \boldsymbol{\xi})$ denote the motion field according to equation 6. The measurement function

$$\boldsymbol{h}(\boldsymbol{x}_k, \boldsymbol{\xi}) = \begin{pmatrix} h_{||\nabla g||}(\boldsymbol{x}_k, \boldsymbol{\xi}) \\ h_u(\boldsymbol{x}_k, \boldsymbol{\xi}) \\ h_v(\boldsymbol{x}_k, \boldsymbol{\xi}) \end{pmatrix} \tag{9}$$

is nonlinear in the state vector $\boldsymbol{x}_k$ since the perspective projection is nonlinear.

At each pixel location $\boldsymbol{\xi}$ of the halfframe at time point t_k, we estimate

$$\boldsymbol{z}_k(\boldsymbol{\xi}) = \begin{pmatrix} ||\nabla g_k(\boldsymbol{\xi})||_2 \\ u(\boldsymbol{\xi}) \\ v(\boldsymbol{\xi}) \end{pmatrix} \tag{10}$$

the Euclidean norm $||\nabla g_k||_2$ of the gray value gradient as well as the optical flow $(u, v)^T$.

We assume that the measurement $\boldsymbol{z}_k(\boldsymbol{\xi})$ at the current halfframe time point t_k at the pixel location $\boldsymbol{\xi}$ is equal to the measurement function $\boldsymbol{h}(\boldsymbol{x}_k, \boldsymbol{\xi})$ plus white Gaussian measurement noise $\boldsymbol{v}_k$ with covariance R_k:

$$\boldsymbol{z}_k(\boldsymbol{\xi}) = \boldsymbol{h}(\boldsymbol{x}_k, \boldsymbol{\xi}) + \boldsymbol{v}_k. \tag{11}$$

Assuming the state vector $\boldsymbol{x}_k$ is normally distributed around the estimate $\hat{\boldsymbol{x}}_k^-$ with covariance P_k^-, a MAP estimation can be stated as the minimization of the following objective function:

$$\frac{1}{2n} \sum_{\boldsymbol{\xi}} \left(\boldsymbol{z}_k(\boldsymbol{\xi}) - \boldsymbol{h}(\boldsymbol{x}_k, \boldsymbol{\xi})\right)^T R_k^{-1} \left(\boldsymbol{z}_k(\boldsymbol{\xi}) - \boldsymbol{h}(\boldsymbol{x}_k, \boldsymbol{\xi})\right)$$
$$+ \frac{1}{2} \left(\boldsymbol{x}_k - \hat{\boldsymbol{x}}_k^-\right)^T P_k^{-\,-1} \left(\boldsymbol{x}_k - \hat{\boldsymbol{x}}_k^-\right) \to \min_{\boldsymbol{x}_k} \tag{12}$$

resulting in an update step of an *iterated extended Kalman Filter* (IEKF) [Bar-Shalom & Fortmann 88; Gelb 74]. n denotes the size of the image region to be summed over. In our actual implementation, each term of the sum is weighted by the confidence (the estimated singular value) of the corresponding optical flow vector.

5 Results

5.1 Downtown Intersection Sequence

We discuss the results of our experiments with an image sequence illustrated by its first frame in Figure 3.

Fig. 3. An enlarged section of the 3rd frame of an image sequence recording downtown intersection traffic. The location and shape of detected moving image regions are described by an enclosing rectangle with one side parallel to the direction of motion. The lines inside the rectangle form an arrow in the direction of motion. By means of an off-line calibration process, these vehicle hypotheses are backprojected from the 2D image plane into the 3-D world. The resulting pose estimates are used to initialize a Kalman-Filter tracking process. The vehicles are referred to by numbers indicated in this frame.

In order to systematically analyze the influence of the optical flow measurements on the pose update and tracking, we first exploit only the optical flow, neglecting a match of image gradients, i.e. ignoring the first component $h_{||\nabla g||}(\boldsymbol{x}_k, \boldsymbol{\xi})$ of the measurement function given in eq. 9. A potential deficiency of the approach based purely on the new measurements is thus not covered by the robustness of the previously used approach [Kollnig & Nagel 95].

Although we could track a remarkable number of vehicles exploiting only optical flow estimates as measurements (see, too, Section 5.3), the object #12 could not be correctly tracked in this mode.

In order to assess the state–estimations, we compare the root of the estimated state covariance diagonal elements (standard deviations) for object #12 for three different tracking methods (see Figure 4): using only image gradients in the measurement function (first row in Figure 4), using only optical flow (second row), and finally combining these two approaches (last row). The three diagrams in the *left* column show the temporal development of the estimated standard deviations for the (estimated) *position* of the object in the road plane. The middle diagram shows, that the estimated position is not very accurate if we exploit only optical flow measurements: the estimated standard deviations remain in the vicinity of their initial values. This is taken as an indication that we cannot rely only on the optical flow measurements. The estimated motion field depends on the vehicle position, but the vehicle images are too small in the image sequence under scrutiny for the motion field to vary significantly along a vehicle image. Combining the two kinds of measurements, the position estimation becomes accurate even without modeling occlusion, see Figure 4 (bottom, left). The *orientation* can be better estimated by exploiting optical flow than image gradients, see Figure 4 (*right* column). In the latter case (first row), the estimated orientation oscillates in the initial phase with a significant amplitude. In both cases (position and orientation), the combination of image gradients and optical flow yields the lowest values for the estimated standard deviations and thus yields the most reliable state estimations. Similar considerations hold for the estimation of the remaining state components (speed and angular velocity).

Figure 5 demonstrates the evaluation of the entire sequence with superimposed results.

5.2 Gas Station Image Sequence

In a second experiment, we have tested our new approach by means of an image sequence recorded at a gas station (see Figure 1). Due to space limitations, we focus on the tracking of vehicle `C` which turns out to be the most challenging case: it is partially occluded by stationary scene components and by the vehicles `A` and `B`. Moreover, vehicle `C` is moving in a low contrast image area under the gas station roof. The tracking is shown in detail in Figure 6. The left column shows the result of the updated pose estimation, while the right column depicts the estimated optical flow field. The resulting trajectories are shown in Figure 7.

5.3 Testing Benchmark Image Sequences

In order to assess our approach, we tried to track vehicles only on the basis of optical flow matching in image sequences in which we are able to track all moving vehicles with our former approach (i. e. only using the image gradient).

Based only on optical flow estimates, we could correctly track 11 out of 12 moving objects in the Durlacher Tor sequence and 10 out of 12 moving objects in the Ettlinger Tor sequence. In the latter sequence, we could even track the partially occluded vehicle moving in front of a bus, while we got problems with our former approach.

Moreover we noticed, that by exploiting only optical flow vectors, a rough initial estimation for the *orientation* of the bus can be corrected in 3 iteration steps, significantly less steps than those which are necessary in the approach exploiting only image gradients [Kollnig & Nagel 95]. However, by exploiting only optical flow, the *position* component in the driving direction is not correctly updated.

6 Conclusion

As far as we know, we present the first approach in which exploiting *interframe* image data leads to an improvement of tracking objects in the 3D scene domain

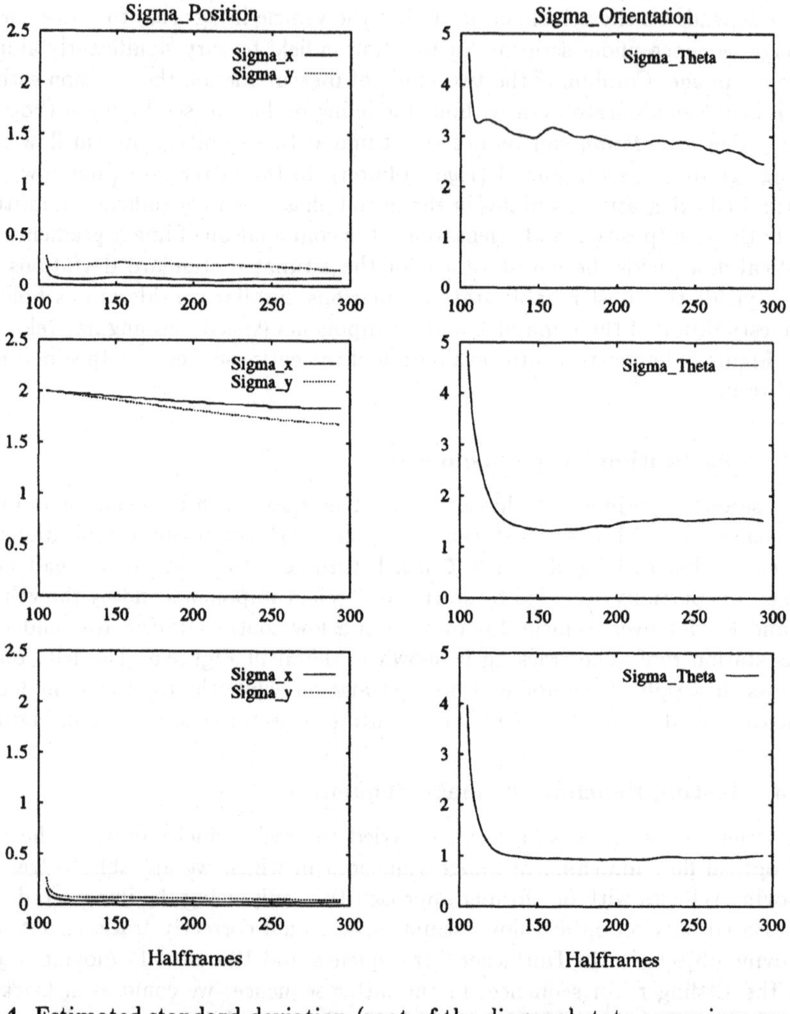

Fig. 4. Estimated standard deviation (root of the diagonal state covaraince matrix) for position (left column) and orientation (right column) of vehicle #12. In the first row, only image gradients are used in the measurement function. Here we had to explicitly model the occlusion in order to track the object properly. In the second row only optical flow is used *without* explicitly modelling the occlusion. In the last row, image gradients and optical flow are combined in the measurement function. In the case of the single use of optical flow, the estimated *position* is not very accurate: the estimated standard deviations remain in the vicinity of their initial values (second row, left), while the single use of image gradients results in a more reliable estimate (top left diagram). On the contrary, the *orientation* can be estimated better by means of optical flow than with image gradients. In both cases (position and orientation) the combined use of image gradients and optical flow reduces the estimated standard deviations in distinction to the isolated use of either measurement (last row).

Fig. 5. The results of uninterrupted tracking of the objects recorded in the downtown intersection sequence until halfframe time point 250th.

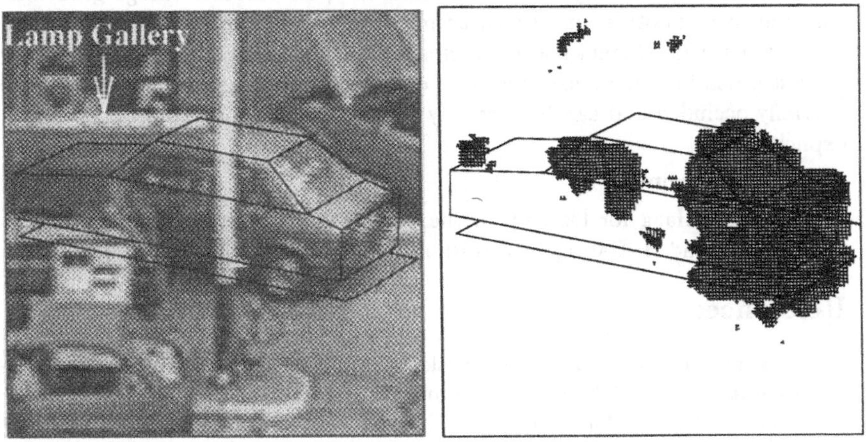

Fig. 6. Evaluation of the image of vehicle C of the gas station sequence. It is occluded by other cars and by stationary scene components, for instance a lamp gallery. The left image shows the updated pose estimate, while the right image depicts the estimated flow vectors for halfframe #700. The occluding scene components can be recognized in the optical flow field although they are not modeled explicitly.

Fig. 7. Computed vehicle trajectories for vehicles A, B, and C.

compared to using *intraframe* image data only, for instance image gradients. The influence of both kinds of measurements on the pose estimation is discussed systematically in different real-world image sequences. The robustness of the new approach has been demonstrated on a considerable range of vehicles which are partially occluded and can be correctly tracked *without* modeling the occlusion explicitly.

Acknowledgment

We thank M. Haag for his help in the preparation of the final version of this contribution and H. Damm for recording the gas station image sequence.

References

[Bar-Shalom & Fortmann 88] Y. Bar-Shalom, T. E. Fortmann, *Tracking and Data Association*, Academic Press, Inc., Boston/MA, Orlando/FL, and others, 1988

[Barron et al. 94] J.L. Barron, D.J. Fleet, and S.S. Beauchemin, Performance of Optical Flow Techniques, *International Journal of Computer Vision* **12** (1994) 43-77

[Bouthemy & François 93] P. Bouthemy, E. François, Motion Segmentation and Qualitative Dynamic Scene Analysis from an Image Sequence, *International Journal of Computer Vision* **10** (1993) 157–182.

[Cédras & Shah 95] C. Cédras, M. Shah, Motion-Based Recognition: A Survey, *Image and Vision Computing* **13**:2 (1995) 129–155.

[Gelb 74] A. Gelb (ed.), *Applied Optimal Estimation*, MIT Press, Cambridge/MA, 1974.

[Gong & Buxton 93] S. Gong, H. Buxton, From Contextual Knowledge to Computational Constraints, in *Proc. British Machine Vision Conference*, Guildford/UK, Sept. 21-23, 1993, pp. 229–238.

[Koller et al. 93] D. Koller, K. Daniilidis, H.-H. Nagel, Model-Based Object Tracking in Monocular Image Sequences of Road Traffic Scenes, *International Journal of Computer Vision* **10** (1993) 257–281.

[Kollnig *et al.* 94] H. Kollnig, H.-H. Nagel, and M. Otte, Association of Motion Verbs with Vehicle Movements Extracted from Dense Optical Flow Fields, in J.-O. Eklundh (ed.), *Proc. Third European Conference on Computer Vision ECCV '94*, Vol. II, Stockholm, Sweden, May 2-6, 1994, Lecture Notes in Computer Science **801**, Springer-Verlag, Berlin, Heidelberg, New York/NY, and others, 1994, pp. 338–347.

[Kollnig & Nagel 95] H. Kollnig and H.–H. Nagel, 3D Pose Estimation by Fitting Image Gradients Directly to Polyhedral Models, *Proc. Fifth International Conference on Computer Vision ICCV '95*, Cambridge/MA, 20–23 June 1995, pp. 569–574

[Lowe 87] D.G. Lowe, Three-Dimensional Object Recognition from Single Two-Dimensional Images, *Artificial Intelligence* **31** (1987) 355–395.

[Otte & Nagel 94] M. Otte, H.–H. Nagel, Optical Flow Estimation: Advances and Comparisons, *Proc. Third European Conference on Computer Vision ECCV '94*, Vol. I, Stockholm / Sweden, 2–6 May 1994, J.-O. Eklundh (ed.), Lecture Notes in Computer Science **800**, Springer-Verlag Berlin Heidelberg New York/NY 1994, pp. 51–60.

[Otte & Nagel 95] M. Otte, H.–H. Nagel, Estimation of Optical Flow Based on Higher–Order Spatiotemporal Derivatives in Interlaced and Non–Interlaced Image Sequences, *Artificial Intelligence* **78** (1995) 5–43

[Proesmans *et al.* 94] M. Proesmans, L. Van Gool, E. Pauwels, A. Oosterlinck, Determination of Optical Flow and Its Discontinuities Using Non-Linear Diffusion, in J.-O. Eklundh (Ed.), *Proc. Third European Conference on Computer Vision ECCV '94*, Vol. II, Stockholm, Sweden, May 2-6, 1994, Lecture Notes in Computer Science **801**, Springer-Verlag, Berlin, Heidelberg, New York/NY and others, 1994, pp. 295–304.

[Schirra *et al.* 87] J.R.J. Schirra, G. Bosch, C.K. Sung, G. Zimmermann, From Image Sequences to Natural Language: A First Step towards Automatic Perception and Description of Motion, *Applied Artificial Intelligence* **1** (1987) 287–307.

[Sullivan 92] G.D. Sullivan, Visual Interpretation of Known Objects in Constrained Scenes, *Philosophical Transactions Royal Society London* (B) **337** (1992) 361–370.

[Sullivan *et al.* 95] G.D. Sullivan, A.D. Worrall, and J.M. Ferryman, Visual Object Recognition Using Deformable Models of Vehicles, in *Proc. Workshop on Context-Based Vision*, 19 June 1995, Cambridge/MA, pp. 75–86.

[Tan *et al.* 94] T.N. Tan, G.D. Sullivan, K.D. Baker, Fast Vehicle Localization and Recognition without Line Extraction and Matching, in *Proc. British Machine Vision Conference*, York/UK, Sept. 13-16, 1994, pp. 85–94.

[Worrall *et al.* 94] A.D. Worrall, G.D. Sullivan, and K.D. Baker, Pose Refinement of Active Models Using Forces in 3D, in J.-O. Eklundh (ed.), *Proc. Third European Conference on Computer Vision (ECCV '94)*, Vol. I, Stockholm, Sweden, May 2-6, 1994, Lecture Notes in Computer Science **800**, Springer-Verlag, Berlin, Heidelberg, New York/NY, and others, 1994, pp. 341–350.

Quantification of Articular Cartilage from MR Images Using Active Shape Models

Stuart Solloway[1], Chris J. Taylor[1], Charles E. Hutchinson[1], John C. Waterton[2]

[1] Departments of Medical Biophysics and Radiology,
University of Manchester. Manchester. M13 9PT, UK
email: ss,ctaylor,ch@sv1.smb.man.ac.uk
[2] Department of Vascular, Inflammatory, Musculoskeletal Research,
Zeneca Pharmaceuticals, Alderley Park, Macclesfield, Cheshire SK10 4TG.

Abstract. Osteoarthritis is the most common cause of disability in the developed world. One the most important features of the disease is the progressive thinning and eventual loss of articular cartilage which can be visualised using Magnetic Resonance (MR) imaging. A major goal of research in osteoarthritis is the discovery and development of drugs which preserve the articular cartilage. To guide this research, accurate and automatic methods of quantifying the articular cartilage are needed. All previous attempts to do this have used manual or semi-automated data-driven segmentation strategies. These approaches are labour-intensive and lack the required accuracy. We describe a model-driven approach to segmentation of the articular cartilage using Active Shape Models (ASMs) and show how measurements of mean thickness of the cartilage can be obtained. We have applied the technique to 2D slices taken from T_1-weighted 3D MR images of the human knee. We give results of systematic experiments designed to determine the accuracy and reproducibility of the automated system. In summary, the method has been shown to be sufficiently robust and accurate for use in drugs trials.

1 Introduction

This paper describes a novel application of Active Shape Models (ASMs) to the problem of segmentation and quantification of the articular cartilage of the femoral condyles from T_1-weighted 3D magnetic resonance (MR) images.

Articular cartilage is found in all synovial joints, where it serves two main purposes: it protects the bones through its ability to withstand compressive loading and facilitates joint mobility, allowing smooth interaction of the articulating surfaces. When cartilage is lost in arthritic disease, particularly osteo-arthritis, joint movement becomes extremely difficult, if not impossible. Pharmaceutical companies are interested in developing drugs that either slow down or halt the loss of cartilage that accompanies such disease. Large-scale clinical trials, costing several millions of pounds, need to be performed to determine the efficacy or otherwise of these drugs. To assess the outcome, accurate and consistent measurements of cartilage volume or thickness are needed. Sufficiently automated, accurate and consistent measurement techniques do not currently exist.

2 Background

Previous attempts to quantify articular cartilage have adopted either manual delineation [1–5] or a data-driven automated strategy [4, 6] to segment the cartilage from surrounding tissue. Manual and semi-automatic methods have produced more reproducible results than automatic ones. However, these methods are more time-consuming and are difficult to perform on a large-scale. The most reproducible volume measurements have been made by Peterfy *et al.* [4] using a Saturation Transfer Subtraction (STS) imaging sequence. The time needed to generate these images is considerably longer than that required to obtain standard T_1-weighted images. The ideal technique for quantifying the articular cartilage would be one that made accurate and automatic quantitative measurements from standard image sequences. We show how this can be achieved for 2D slices from 3D T_1-weighted images, using a model-based technique.

3 Segmenting the Cartilage

Figure 1 shows an annotated 2D slice through a T_1-weighted 3D image of the knee. The cartilage, particularly the femoral articular cartilage which envelops the base of the femur (the femoral condyles), is difficult to segment. This is due mainly to its shape: it is typically only a few pixels thick at its widest point and is often difficult to distinguish from neighbouring structures, particularly in the region where it meets the anterior fat pad.

Previous studies have used manual or data-driven segmentation techniques. Manually marking the cartilage boundary in many slices is a pains-taking process that is prone to subjective errors. Data-driven segmentation produces errors where the cartilage boundary is not sufficiently distinct. We have attempted to overcome these problems by using a *model-based segmentation* technique which guarantees an anatomically plausible result. In particular, we have used flexible models - templates that deform to fit the image evidence. Given the inherent shape variability of anatomical objects, this approach is well-suited to segmentation of medical imagery.

Many kinds of flexible template have been described in the literature, for example [7–14]. The major drawback with many of these is that they lack specificity: they can produce illegal examples of the desired objects because they lack global shape constraints. To apply such constraints, the model needs to incorporate *a priori* knowledge of the objects being segmented. We have used Active Shape Models (ASMs) [15] - flexible models that incorporate such constraints.

4 Active Shape Models

ASMs use explicit models of the shape variation and grey-level appearance of a given class of objects. Knowledge of the objects is gained by building the models from training sets of images that contain examples of the objects.

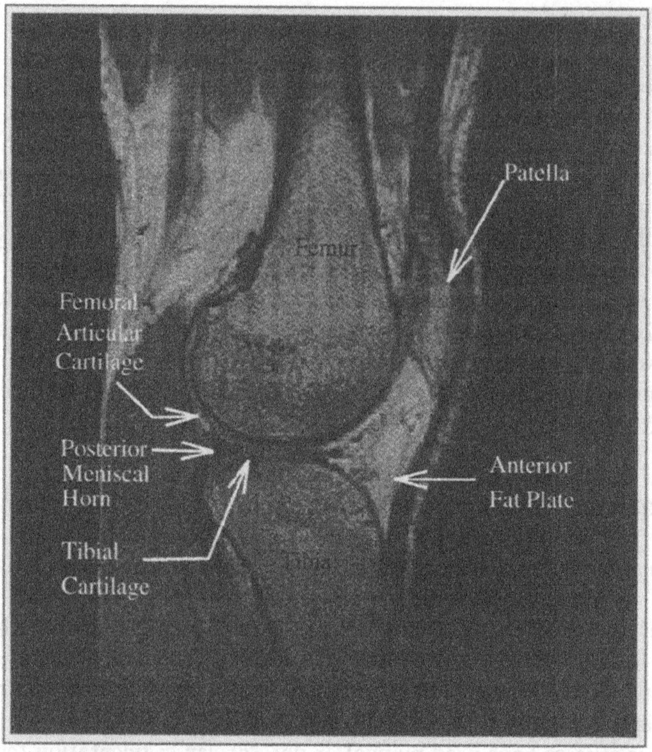

Fig. 1. Sagittal slice taken from a T_1-weighted 3D MR image of the knee.

Models of the variability of object shape are built by representing the objects as sets of labelled *landmark points*. By performing a statistical analysis of the variation in location of the landmark points over the training set, a *Point Distribution Model* (PDM) can be built. A PDM represents the class of objects for which it has been trained using a mean shape and a set of linearly independent *modes* describing the main ways in which the training shapes varied. New shape examples, $\boldsymbol{x}$, are created from linear superpositions of the mean shape, $\bar{\boldsymbol{x}}$, and the modes of variation,

$$\boldsymbol{x} = \bar{\boldsymbol{x}} + \boldsymbol{P}\boldsymbol{b} \tag{1}$$

where $\boldsymbol{P}$ is the matrix describing a subset of the modes of variation and $\boldsymbol{b}$ is a vector of weights controlling the influence of each mode.

The grey-level appearance of the objects across the training set is modelled by examining the pixel values in small patches around the landmark points. Similarly to the shape model, the grey-level landscape around each point can be represented as a superposition of the mean grey-level landscape and a set of linear modes of variation. Using the normalised derivative of the image intensity for each pixel gives the model invariance to shifts in image intensity and scaling. The details of both shape and grey-level modelling have been published previously [15].

Having built models of the shape and grey-level variation of the objects of interest, an ASM can be created and used to search for new examples of the objects in images. The method of fitting the model to the data is iterative and starts by projecting an approximation to the correct solution into the given image. Grey-level information from around each model landmark point is used to suggest the direction in which it should move in order to better fit the image evidence. Evidence from all the landmark points is combined to calculate an overall deformation of the shape in order to produce a better model fit. ASMs are constrained to deform *only* in ways in which the objects have been observed to deform in the training set - global shape constraints are applied. A multi-resolution version of this search process has been described [16] which can improve its accuracy. ASMs have been applied successfully to a range of image analysis problems in medicine [17–20] and industry [21].

5 Method

5.1 Building the Models

For our experiments, 2D slices taken from 14 T_1-weighted 3D MR images of the knee were used to create an ASM. The purpose of the experiments was to demonstrate the ability of the ASM to locate precisely the femoral articular cartilage. For this reason, only subjects with no diagnosed serious knee problems were included in the model. All images were produced from adult male patients in the age range 20 to 30.

A potential drawback of making 2D measurements on slices taken from 3D acquisitions is that the measurements are critically dependent upon the angle of the scan plane. Measures were taken during acquisition to ensure consistency in the choice of scan plane orientation. To build a 2D model of the articular cartilage, single sagittal slices were extracted from similar locations in the 3D volume for each subject. Slices were chosen from the centre of the lateral femoral condyle, as viewed in the sagittal plane. When the choice of slice was not immediately obvious, two candidate slices were chosen for inclusion in the model. This was done to avoid errors due to slice selection when the model was used in image search. This led to a total of 16 slices being included in the model.

Training images were marked up by an experienced radiologist who placed landmark points around the boundary of the femoral cartilage. Points were placed at each end of the cartilage and joined by two splines [22], each containing 20 points. These points were manually adjusted until the splines delineated the inner and outer surfaces of the cartilage. The points were then equally spaced along the splines (see figure 2).

In addition to building a model consisting of points around the femoral cartilage, a second model, which also incorporated points placed around the endosteal surface of the femoral condyles, was built; an example is shown in figure 3. The

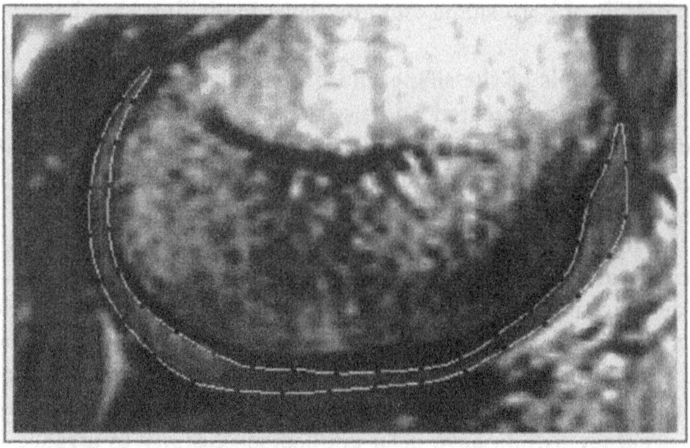

Fig. 2. A T_1-weighted MR sagittal slice taken from a 3D image of the knee with landmark points overlaid. The model consists of landmark points placed on the femoral cartilage boundary only.

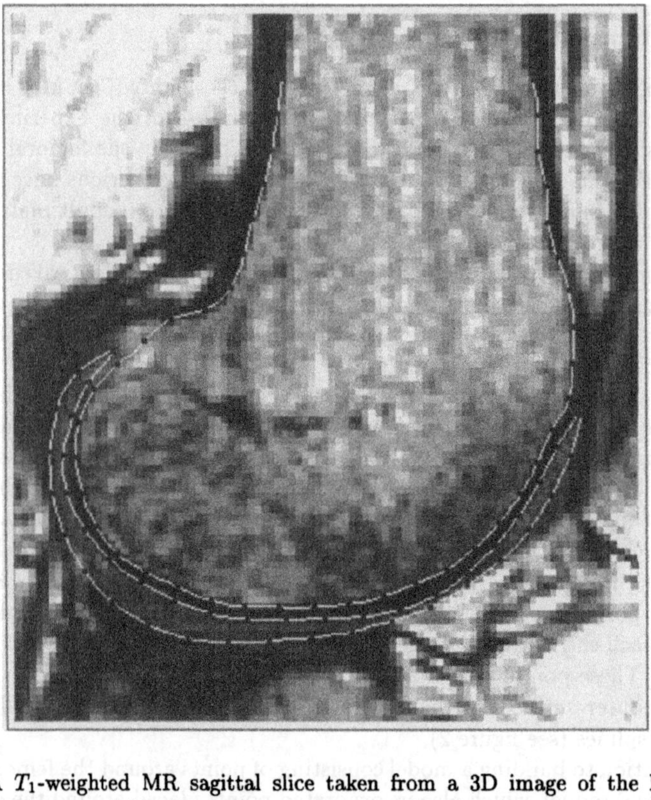

Fig. 3. A T_1-weighted MR sagittal slice taken from a 3D image of the knee with landmark points overlaid. The model consists of landmark points placed on the femoral cartilage boundary and the endosteal surface of the femoral condyle.

edges of the femoral condyles are relatively strong and can be used as cues to help locate the femoral cartilage when the starting point for image search is poor.

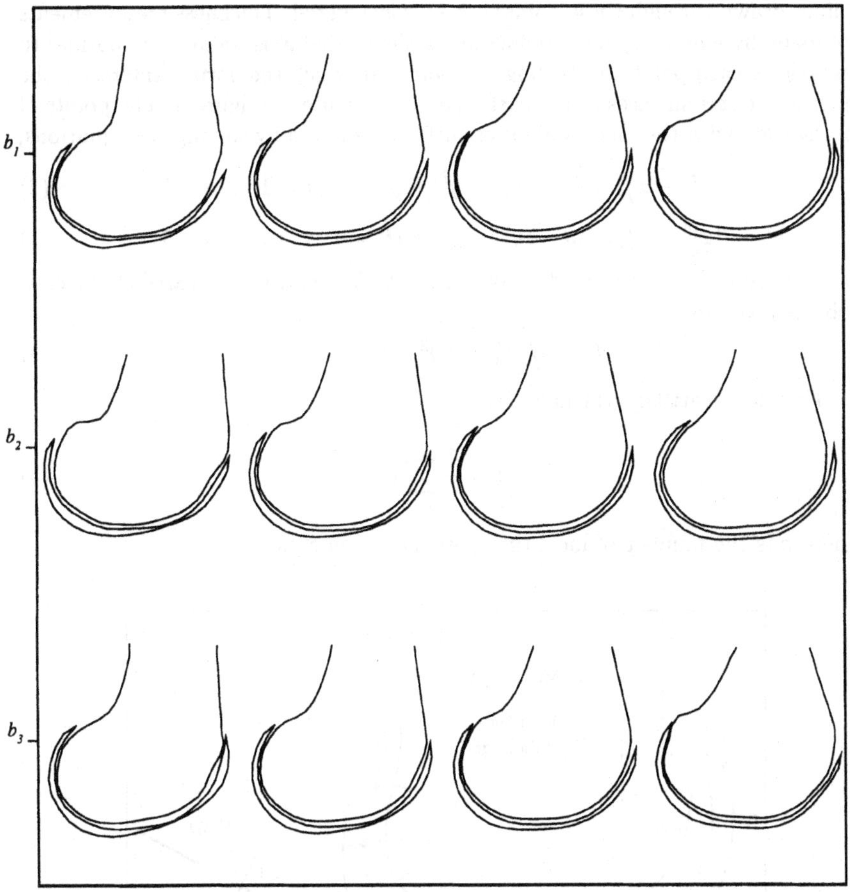

Fig. 4. The first three modes of variation, b_1, b_2, b_3, of a 2D PDM comprising the femoral cartilage and the femoral condyles.

Figure 4 shows the most significant modes of shape variation for the model comprising the femoral cartilage and the femoral condyles. The 12 shapes displayed correspond to 4 different values for each of the weights, b_1, b_2 and b_3 used in the linear combination of the modes of variation in equation 1. Results obtained for image search using the model are presented in section 6.

5.2 Making Thickness Measurements

Once image search has been completed, the segmented outline of the cartilage, as in the marked up versions in the training set, consists of a set of points which define the inner and outer surfaces of the cartilage. The medial axis of the splines drawn through these points can be constructed. Thickness measurements are made by equally spacing points along the medial axis, taking the normal to the axis at each point and finding the points at which the normal intersects the inner and outer surfaces of the cartilage, as illustrated in figure 5. The points of intersection with the inner and outer surface are found by solving the equations,

$$(x_j^m - x_j^i)\cos\theta + (y_j^m - y_j^i)\sin\theta = 0 \quad j = 1, \ldots n-1 \tag{2}$$

$$(x_j^m - x_j^o)\cos\theta + (y_j^m - y_j^o)\sin\theta = 0 \quad j = 1, \ldots n-1 \tag{3}$$

for each point, j, on the medial axis. The thickness, t, of the cartilage at each point is given by,

$$t_j = \sqrt{(x_j^i - x_j^o)^2 + (x_j^i - x_j^o)^2} \tag{4}$$

The mean cartilage thickness, $\bar{t}$, is,

$$\bar{t} = \frac{1}{n}\sum_{j=1}^{n} t_j \tag{5}$$

where n is the number of local thickness measurements.

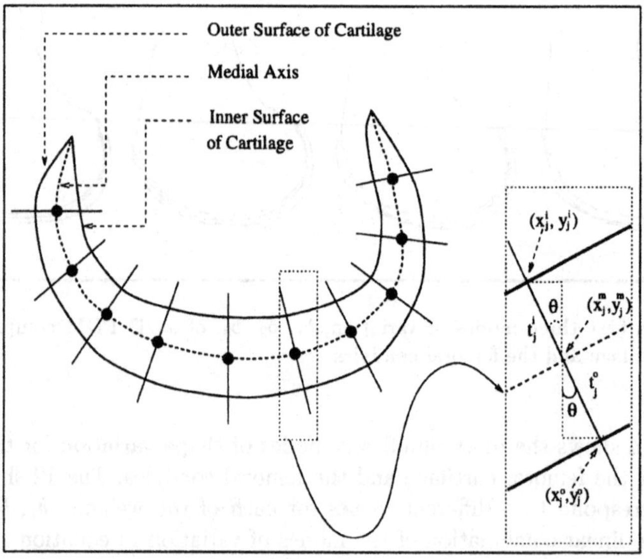

Fig. 5. The method for making thickness measurements of the femoral cartilage.

6 Results

Experiments were performed to test the accuracy and precision of the segmentation performed using the ASM. Accuracy was determined by comparing the cartilage boundaries found during image search with those marked by the radiologist during training. This is a surrogate method - ideally, histological comparisons are needed to assess accuracy. Measurement accuracy was assessed similarly by comparing the mean thickness obtained from the automatically fitted model with that derived from the boundaries delineated by the radiologist. Measurement precision was determined from thickness measurements made from repeated scans of several healthy volunteers.

6.1 Segmentation Accuracy

In order to assess the accuracy of the ASM search, a set of leave-one-out experiments were performed, providing a pessimistically biased estimate of the errors. Tests were performed by systematically excluding each image from the training set and using the excluded image to test the performance of the model built without it. To perform the tests, the model was initialised to the mean shape and perturbed a known amount from the correct position, (x_0, y_0), orientation, θ_0, and scale, s_0. Exhaustive combinations of these test conditions were used with each image. A multi-resolution search process was used, with the coarsest level containing pixels with linear dimensions 32 times larger than the originals, and the finest being 4 times smaller than the originals. Grey-level models of 5×5 pixels were built for each landmark point at each level. Accuracy was assessed by determining how well the resulting model found the cartilage in the excluded test image using the mean point-to-line error, $\bar{l}$, illustrated in figure 6.

A range of initial perturbations of the model were used: translations of $x_0 \pm 30$, $x_0 \pm 10$, $y_0 \pm 30$, $y_0 \pm 10$ pixels, scaling of $0.5s_0$, s_0 and $1.5s_0$ and rotations of $\theta_0 \pm 45^o$, $\theta \pm 15^o$. When exhaustive combinations of these initial perturbations were used, (leading to 192 tests on each image), we found that the majority of experiments resulted in small point-to-line errors (good fits), but that the models failed to find the femoral cartilage in a small proportion of experiments.

To characterise the performance of the system more thoroughly, we assumed that the distribution of $\bar{l}$ was a mixture of Gaussian distributions. The EM algorithm [23],was used to calculate the number and parameters of these underlying distributions. For the model based on the cartilage alone, experiments that produced good fits were accurately modelled by a Gaussian having mean, $\mu_1 = 0.52$mm and standard deviation, $\sigma_1 = 0.14$mm. 83% of all experiments produced fits less than $\mu_1 + 3\sigma_1$. Most of the errors occurred for starting points far from the correct position. The distribution of good fits for the model combining the femoral cartilage and the femoral condyles, was described by a Gaussian having mean, $\mu_2 = 0.59$mm and standard deviation, $\sigma_2 = 0.14$mm. 93% of all experiments gave fits of less than $\mu_2 + 3\sigma_2$.

The results show that the model comprising the femoral cartilage *and* the femoral condyles is more accurate over a wide range of starting positions, but

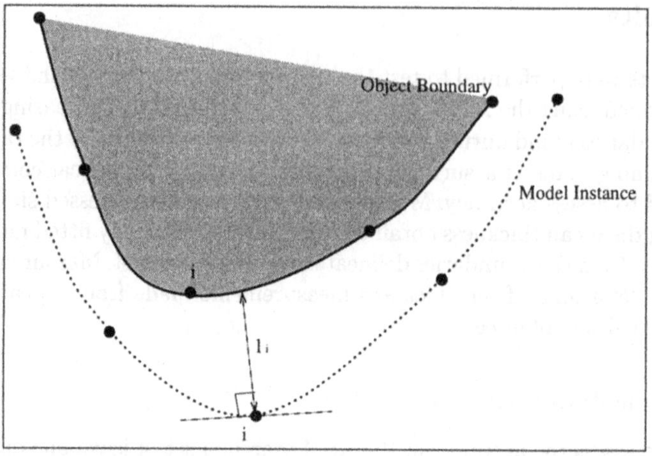

Fig. 6. Illustration of the point-to-line distance, l_i, for the landmark point, i, on the model instance. The mean point-to-line distance, $\bar{l} = \frac{1}{n}\Sigma_{i=1}^{i=n} l_i$, where n= number of landmark points.

that the model comprising the femoral cartilage only can produce a better fit, given a reasonable starting point. This suggests that an effective strategy would be to begin the search with the more complicated model, changing to the simpler model when a reasonable approximation to the result had been achieved.

6.2 Measurement Accuracy

A second set of leave-one-out experiments using smaller initial perturbations to initialise the ASM (exhaustive combinations of $x_0 \pm 10$ pixels, $y_0 \pm 10$ pixels, $0.9s_0$, $\theta_0 \pm 15^o$, leading to 8 experiments on each image) were performed. In this case, all experiments led to good fits, with a mean point-to-line error of 0.49±0.01mm. For each experiment, the mean cartilage thickness was calculated. The group mean of the calculations for each image was compared with the cartilage thickness calculated from the set of points placed by the radiologist during training. The mean difference between the two calculations over the 16 training images was 7.841% for the model comprising the femoral cartilage only.

6.3 Measurement Precision

To test the precision of automated measurements made using the ASM, five healthy, adult male volunteers were chosen. The age range of the volunteers was 22 to 30. None of the volunteers had any diagnosed knee problems. The left knee of each volunteer was imaged twice using a 3D imaging sequence at a single session. After the first image was produced, the subject was taken out of the scanner. They were then re-placed in the scanner and a second image was produced. The images from the volunteers were *not* included in the ASM training set.

One of the subjects (volunteer 1) was imaged on two occasions, producing a total of four images; the imaging sessions were separated by two weeks, but the imaging was performed at the same time of day.

In order to begin searching, the ASM must be given an initial estimate. In every case, this was done by manually placing the mean shape in roughly the correct position on the image. The model comprising the femoral cartilage only was used.

For each of the chosen 2D slices, a mean thickness, $\bar{t}$, was calculated from 20 thickness measurements along the medial axis of the cartilage. To calculate the coefficient of variation of the mean thickness measurements for each volunteer, the group mean thickness, $\bar{T}$, was calculated:

$$\bar{T} = \frac{1}{N}\sum_{j=1}^{N} \bar{t}_j \qquad (6)$$

where N is the number of images produced for each patient, and $\bar{t}_j$ is the mean thickness for j^{th} image collected. The results for the five volunteers are shown in table 1.

Table 1. The mean femoral cartilage thickness measurements, $\bar{T}$, for five healthy adult male volunteers. The bottom row of the table shows the mean coefficient of variation (CoV) for the six measurements.

Volunteer	$\bar{T} \pm \Delta T$(mm)	(CoV %)
1a	2.355±0.013	(0.552%)
1b	2.161±0.360	(16.66%)
2	2.202±0.078	(3.542%)
3	2.035±0.087	(4.275%)
4	2.449±0.009	(0.368%)
5	2.116±0.046	(2.174%)
Mean CoV		(4.595%)

The results obtained for volunteer 1 on the second occasion (1b) were less reproducible than those obtained in the other cases. This appeared to be due to poor adherence to the imaging protocol, leading to an inappropriate slice orientation in the 3D images. This viewing-angle dependence is a weakness of the 2D technique. If the poor result for volunteer 1b was discounted, the mean coefficient of variation for the 5 subjects was 2.18%.

7 Conclusions

We have described a technique for measuring cartilage thickness from 2D slices taken from 3D T_1-weighted images of the human knee. The results demonstrate the effectiveness of the ASM technique for making reproducible measurements of femoral cartilage thickness. These results compare favourably with those previously reported. The advantage of the ASM approach over data-driven techniques and less specific model-driven approaches lies in its ability to generate only *legal* interpretations of the modelled objects. In regions where evidence for the cartilage boundary is weak, stronger evidence from other regions acts as a constraint on the overall interpretation. A second important advantage is that the segmentation is fully automatic once the model has been built. The current system is sufficiently reliable to be useful in drug trials.

The results for volunteer 1b highlighted the shortcomings of the 2D approach. To remove alignment dependencies, models need to be built in 3D. The effectiveness of 3D ASMs has previously been demonstrated [19]. A 3D model of the femoral cartilage would allow volume measurements to be made, giving a more complete picture of the degree of cartilage loss. Future work will address the problem of building 3D models of the cartilage.

References

1. L. Pilch, C. Stewart, D. Gordon, R. Inman, K. Parsons, I. Pataki, J. Stevens, Assessment of Cartilage Volume in the Femorotibial Joint with Magnetic Resonance Imaging and 3D Computer Construction. *Journal of Rheumatology* **21**, 2307–2319 (1994).
2. J. C. Waterton, V. Rajanayagam, B. D. Ross, D. Brown, A. Whittemore, D. Johnstone, Magnetic Resonance Methods for Measurement of Disease Progression in Rheumatoid Arthritis. *Magnetic Resonance Imaging* **11**, 1033–1038 (1993).
3. F. Eckstein, H. Sittek, S. Milz, M. Reiser, The Morphology of Articular Cartilage Assessed by Magnetic Resonance Imaging: Reproducibility and Anatomical Correction. *Surgical and Radiologic Anatomy* **16**, 429–438 (1994).
4. C. G. Peterfy, C. F. van Dijke, D. L. Janzen, C. C. Glüer, R. Namba, S. Majumdar, P. Lang, H. K. Genant, Quantification of Articular Cartilage in the Knee with Pulsed Saturation Transfer Subtraction and Fat-suppressed MR Imaging: Optimization and Validation. *Radiology* **192**, 485–491 (1994).
5. K. Jonsson, K. Buckwalter, M. Helvie, L. Nicklason, W. Martel, Precision of Hyaline Cartilage Thickness Measurements. *Acta Radiologica* **33**, 234–239 (1992).
6. J. Hodler, D. Trudell, M. N. Pathria, D. Resnick, Width of the Articular Cartilage of the Hip: Quantification by Using Fat-Suppression Spin-Echo MR Imaging in Cadavers. *American Journal of Radiology* **159**, 351–355 (1992).
7. M. Kass, A. Witkin, D. Terzopoulos, Snakes: Active Contour Models. *International Journal of Computer Vision* **1**, 133–144 (1987).
8. L. Staib, J. Duncan, Parametrically Deformable Contour Models. In *IEEE Computer Society Conference on Computer Vision and Pattern Recognition*, 427–430, San Diego (1989).
9. R. Bajcsy, S. Kovačič, Multiresolution Elastic Matching. *Computer Vision, Graphics and Image Processing* **46**, 1–21 (1989).
10. P. Karaolani, G. Sullivan, K. Baker, M. Baines, A Finite Element Method for Deformable Models. In *Proceedings of the Fifth Alvey Vision Conference*, 73–78, Reading, England. (1989).

11. D. Terzopoulos, D. Metaxas, Dynamic 3D Models with Local and Global Deformations: Deformable Superquadrics. *IEEE Transactions on Pattern Analysis and Machine Intelligence* **13**, 703–714 (1991).
12. F. L. Bookstein, Principal warps: Thin-plate splines and the decomposition of deformations. *IEEE Transactions on Pattern Analysis and Machine Intelligence* **11**, 567–585 (1989).
13. C. Nastar, N. Ayache, Fast segmentation, tracking and analysis of deformable objects. In 4th *International Conference on Computer Vision*, 275–279, Berlin (May 1993).
14. S. Sclaroff, A. Pentland, A modal framework for correspondence and description. In 4th *International Conference on Computer Vision*, 308–313, Berlin (May 1993).
15. T. F. Cootes, C. J. Taylor, D. H. Cooper, J. Graham, Active Shape Models - Their Training and Application. *Computer Vision and Image Understanding* **61**, 38–59 (1995).
16. T. F. Cootes, C. Taylor, A. Lanitis, Active Shape Models: Evaluation of a Multi-Resolution Method for Improving Image Search. In 5th *British Machine Vision Conference* (Ed. E. Hancock), 327-336, York, England (1994), BMVA Press.
17. T. F. Cootes, A. Hill, C. J. Taylor, J. Haslam, The Use of Active Shape Models for Locating Structures in Medical Images. *Image and Vision Computing* **12**, 276–285 (1994).
18. A. Hill, C. J. Taylor, Model-Based Image Interpretation using Genetic Algorithms. In 2nd *British Machine Vision Conference* (Ed. P. Mowforth), 265–274, Glasgow, Scotland (1991), Springer-Verlag.
19. A. Hill, A. Thornham, C. J. Taylor, Model-Based Interpretation of 3D Medical Images. In 4th *British Machine Vision Conference* (Ed. J. Illingworth), 339–348, Guildford, England (1993), BMVA Press.
20. A. Hill, T. F. Cootes, C. J. Taylor, K. Lindley, Medical Image Interpretation: A Generic Approach using Deformable Templates. *Journal of Medical Informatics* **19**, 47–59 (1994).
21. J. J. Hunter, User Programmable Visual Inspection. In 5th *British Machine Vision Conference*, 661–670, York, England (1994), BMVA Press.
22. W. H. Press, S. A. Teukolsky, W. T. Vetterling, B. P. Flannery, *Numerical Recipes in C: The Art of Scientific Computing*. Cambridge University Press, 2nd edn. (1992).
23. G. McLachlan, K. Basford, *Mixture Models: Inference and Applications to Clustering*, vol. 84 of *Statistics: Textbooks and Monographs*. Dekker (1988), pp. 37-70.

Calibration/ Focus/Optics

Direct Methods for Self-Calibration of a Moving Stereo Head

M.J. Brooks[1,3] L. de Agapito[2] D.Q. Huynh[1] L. Baumela[3]

[1] Centre for Sensor Signal and Information Processing, Signal Processing Research Institute, Technology Park, Adelaide, SA 5095, Australia
[2] Instituto de Automática Industrial, CSIC, La Poveda, Madrid 28500, Spain
[3] Departamento de Inteligencia Artificial, Facultad de Informática, Universidad Politécnica de Madrid, 28660 Boadilla del Monte, Spain
Email correspondence: mjb@cs.adelaide.edu.au

Abstract. We consider the self-calibration problem in the special context of a stereo head, where the two cameras are arranged on a lateral rig with coplanar optical axes, each camera being free to vary its angle of vergence. Under various constraints, we derive explicit forms for the epipolar equation, and show that a static stereo head constitutes a degenerate camera configuration for carrying out self-calibration in the sense of Hartley [4]. The situation is retrieved by consideration of a special kind of motion of the stereo head in which the baseline remains confined to a plane. New closed-form solutions for self-calibration are thereby obtained, inspired by an earlier discrete motion analysis of Zhang et al. [11]. Key factors in our approach are the development of explicit, analytical forms of the fundamental matrix, and the use of the vergence angles in the parameterisation of the problem.

Keywords: Self-calibration, stereo head, degeneracy, epipolar equation, fundamental matrix, ego-motion.

1 Introduction

The simultaneous recovery of intrinsic and extrinsic parameter values from both static and dynamic imagery has in recent times attracted considerable attention (see [2], [3], [4], [7], [8], [9], [10], [11], [12], [13]). In this paper, we consider this self-calibration problem in the special context of a stereo head, perhaps the most commonly adopted binocular camera configuration in robotics. First, however, we recall the epipolar geometry which underpins the analysis.

We adopt a notation similar, but not identical, to that of Faugeras et al. [3]; see the Appendix for a summary of the differences. Let $\mathbf{m}$ and $\mathbf{m}'$ denote corresponding points, in homogeneous coordinates, in the left and right images, respectively. We may express the epipolar equation as

$$\mathbf{m}^T \mathbf{F}\, \mathbf{m}' = 0, \tag{1}$$

where $\mathbf{F}$ is the fundamental matrix [3, 6], defined as

$$\mathbf{F} = \mathbf{A}^T \mathbf{T} \mathbf{R} \mathbf{A}'. \tag{2}$$

Here, $\mathbf{R}$ embodies the pure rotation that renders the left image parallel with the right image, $\mathbf{T}$ is a skew-symmetric matrix formed from the baseline vector connecting the left and right optical centres, and $\mathbf{A}$ and $\mathbf{A}'$ are the intrinsic parameter matrices of the left and right cameras. (Note again the use in equations (1) and (2) of non-standard, but convenient, definitions of the various matrices. See the Appendix.)

As is well known, no more than 7 imaging parameters may be recovered from the epipolar equation, due to the special properties of the fundamental matrix [3]. In particular, Hartley [4] and Pan et al. [8] have shown that, under favourable conditions, two focal lengths and 5 relative orientation parameters may be recovered by a process of self-calibration.

In the next section, we consider the special situation in which two cameras form a stereo head assembly. We show that, in the absence of motion, this commonly adopted configuration is degenerate in that self-calibration may no longer be carried out in the sense of Hartley [4]. Direct methods of self-calibration are then explored in the context of a moving stereo head. Note that an extended version of this work appears in [1].

2 Stereo head assembly

Consider the special case of a stereo head in which a pair of cameras is mounted on a lateral rig. The cameras are free to vary their angles of vergence. The y-axes of the two images are parallel, and are orthogonal to the baseline vector, as depicted in Figure 1. The optical axes and the baseline are therefore coplanar. The matrices $\mathbf{R}$, $\mathbf{T}$ and $\mathbf{A}$ now take the forms

$$\mathbf{R} = \begin{pmatrix} \cos\beta & 0 & -\sin\beta \\ 0 & 1 & 0 \\ \sin\beta & 0 & \cos\beta \end{pmatrix}, \quad \mathbf{T} = \begin{pmatrix} 0 & -t_z & 0 \\ t_z & 0 & -t_x \\ 0 & t_x & 0 \end{pmatrix}, \quad \mathbf{A} = \begin{pmatrix} 1 & 0 & -u_0 \\ 0 & 1 & -v_0 \\ 0 & 0 & -f \end{pmatrix}.$$

Here, β is the angular rotation about the y-axis that renders the left image parallel with the right image; $\mathbf{T}$ is formed out of the baseline vector $\mathbf{t} = (t_x, 0, t_z)^T$; and, for each camera, the focal length and the principal point are the only unknown intrinsic parameters, denoted by f and (u_0, v_0), with the image coordinate system axes assumed to be orthogonal and similarly scaled.

In view of (2), the fundamental matrix is now given by

$$\mathbf{F} = \begin{pmatrix} 0 & -t_z & t_z v_0' \\ \phi & 0 & -u_0'\phi - f'\sigma \\ -v_0\phi & u_0 t_z - f t_x & u_0' v_0 \phi + v_0'(f t_x - u_0 t_z) + f' v_0 \sigma \end{pmatrix}, \tag{3}$$

where $\phi = t_z \cos\beta - t_x \sin\beta$, and $\sigma = -t_x \cos\beta - t_z \sin\beta$.

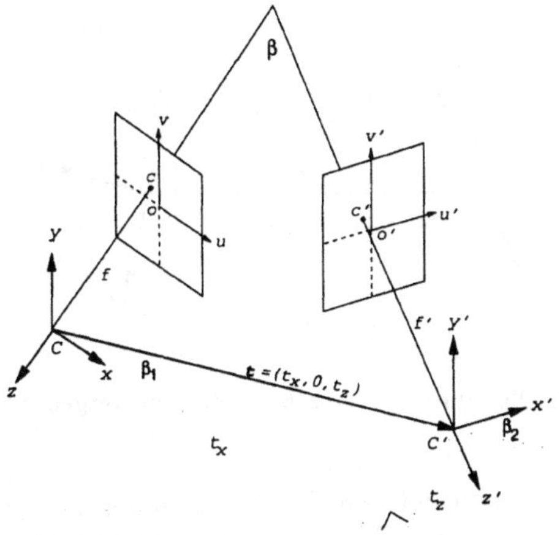

Fig. 1. Stereo head configuration.

As is well known, absolute dimensions of depth cannot be determined solely from knowledge of corresponding points and the associated fundamental matrix. Accordingly, without loss of generality, we set the baseline length to unity, and note that the direction of the baseline vector is now effectively described by 1 parameter. There are therefore 8 unknowns encoded within **F**, these being $\beta, u_0, v_0, f, u_0', v_0', f'$ and either t_x or t_z.

2.1 Vergence-angle parameterisation

The form of **F** is simplified if an adjustment is made to the parameterisation by incorporating the left and right vergence angles β_1 and β_2, where

$$t_x = \cos\beta_1, \qquad t_z = \sin\beta_1, \qquad \beta = \beta_1 + \beta_2. \tag{4}$$

Here, β_1 and β_2 specify the extent to which the left and right optical axes point inwards from the direction 'straight-ahead'. (Note, therefore, that the left and right vergence angles are measured in an opposite sense. See Figure 2.) Relative orientation in this situation is now determined by the pair β_1, β_2, instead of β, t_x. Equation (3) is then expressed as

$$\mathbf{F} = \begin{pmatrix} 0 & -\sin\beta_1 & v_0'\sin\beta_1 \\ -\sin\beta_2 & 0 & (u_0'\sin\beta_2 + f'\cos\beta_2) \\ v_0\sin\beta_2 & (u_0\sin\beta_1 - f\cos\beta_1) & \begin{matrix} -v_0(u_0'\sin\beta_2 + f'\cos\beta_2) \\ -v_0'(u_0\sin\beta_1 - f\cos\beta_1) \end{matrix} \end{pmatrix}, \tag{5}$$

in which the 8 unknowns are $\beta_1, \beta_2, u_0, v_0, f, u_0', v_0', f'$.

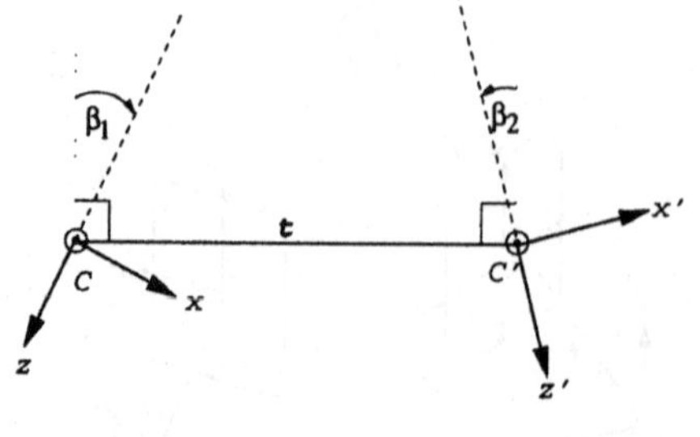

Fig. 2. Plan view of stereo head showing vergence angles.

In the event that sufficiently-many corresponding points can be located in the two images, it may be possible to obtain a numerical estimate, $\mathbf{F}_{est}$, of the matrix $\mathbf{F}$. Let

$$\mathbf{F}_{est} = \begin{pmatrix} \delta_1 & \delta_2 & \delta_3 \\ \delta_4 & \delta_5 & \delta_6 \\ \delta_7 & \delta_8 & \delta_9 \end{pmatrix}. \tag{6}$$

Noting that $\mathbf{F}_{est}$ may only be determined up to a scale factor, we may form the equation

$$\mathbf{F} = \lambda \mathbf{F}_{est}. \tag{7}$$

Here, the unknown λ aligns the scales of the two matrices. We can now obtain 7 equations by linking respective elements of the matrices. However, these equations are not all independent, since $F_{33} = F_{23}F_{31}/F_{21} + F_{32}F_{13}/F_{12}$. Thus we may obtain up to 6 independent equations, one of which will be utilised in eliminating λ. We therefore observe that, of the 8 unknowns encoded within $\mathbf{F}$, at most 5 may be determined provided the remaining 3 are known. Further constraints are therefore needed if we are to solve for the various parameters.

3 Self-calibration of a static stereo head

We now consider how (7) may be solved under additional assumptions. In doing so, we shall gain insight into the feasibility of self-calibration of a static rig. Our aim is to develop closed-form expressions for the various parameters.

Case 1: u_0, v_0, u'_0, v'_0 known

If the locations of the principal points are known, we may without loss of generality offset the coordinates of image points. This simple form of image rectification then permits the setting of $u_0 = v_0 = u'_0 = v'_0 = 0$ in (7), giving

$$\mathbf{F} = \begin{pmatrix} 0 & -\sin\beta_1 & 0 \\ -\sin\beta_2 & 0 & f'\cos\beta_2 \\ 0 & -f\cos\beta_1 & 0 \end{pmatrix} = \lambda \begin{pmatrix} 0 & \delta_2 & 0 \\ \delta_4 & 0 & \delta_6 \\ 0 & \delta_8 & 0 \end{pmatrix}. \tag{8}$$

This case is relevant either to a pair of static stereo cameras having independent, unknown focal lengths, or to a single mobile camera in which the focal length may be varied. We now have 5 unknown parameters, including λ, but are able to generate only 4 independent equations. Thus, whereas in general we may obtain via self-calibration 2 focal lengths and 5 relative orientation parameters, we are unable to fix any of the unknown parameters in this special situation. The camera configuration is therefore degenerate for Hartley self-calibration.

Case 2: u_0, v_0, u_0', v_0' known; $f = f'$

Here we assume that the left and right focal lengths are equal, and seek only the 3 unknown parameters f, β_1, β_2. Equation (7) now reduces to

$$\mathbf{F} = \begin{pmatrix} 0 & -\sin\beta_1 & 0 \\ -\sin\beta_2 & 0 & f\cos\beta_2 \\ 0 & -f\cos\beta_1 & 0 \end{pmatrix} = \lambda \begin{pmatrix} 0 & \delta_2 & 0 \\ \delta_4 & 0 & \delta_6 \\ 0 & \delta_8 & 0 \end{pmatrix}, \tag{9}$$

yielding 4 independent equations. All 3 imaging parameters can now be determined, viz:

$$f = \sqrt{(\delta_8^2 - \delta_6^2)/(\delta_4^2 - \delta_2^2)}, \quad \tan\beta_1 = f\,\delta_2/\delta_8, \quad \tan\beta_2 = -f\,\delta_4/\delta_6. \tag{10}$$

Note here that f is computed in the same units as the coordinates of the corresponding points used to estimate the fundamental matrix.

Case 3: u_0, v_0, u_0', v_0' known; $\beta_1 = \beta_2 = \beta/2$

We now seek to determine β, f and f', given the very special situation in which the vergence angles are equal, with the principal axes of the cameras and the baseline forming an isosceles triangle. Our equation is now

$$\mathbf{F} = \begin{pmatrix} 0 & -\sin\beta/2 & 0 \\ -\sin\beta/2 & 0 & f'\cos\beta/2 \\ 0 & -f\cos\beta/2 & 0 \end{pmatrix} = \lambda \begin{pmatrix} 0 & \delta_2 & 0 \\ \delta_4 & 0 & \delta_6 \\ 0 & \delta_8 & 0 \end{pmatrix}. \tag{11}$$

Noting that $F_{12} = F_{21}$, we see that none of the unknown parameters may be determined without more information being provided. Remarkably, if the focal lengths are known to be equal, it remains impossible to recover any of the parameters. Note, however, that the ratio of the focal lengths may be determined.

Case 4: $(u_0, v_0, f) = (u_0', v_0', f')$

Here we assume left and right cameras have identical focal length and principal point locations. This also corresponds to a mobile camera moving horizontally. The 5 parameters $\beta_1, \beta_2, u_0, v_0, f$ are now free, (7) reducing to:

$$\begin{pmatrix} 0 & -\sin\beta_1 & v_0\sin\beta_1 \\ -\sin\beta_2 & 0 & u_0\sin\beta_2 + f\cos\beta_2 \\ v_0\sin\beta_2 & \begin{matrix} u_0\sin\beta_1 \\ -f\cos\beta_1 \end{matrix} & \begin{matrix} -u_0v_0(\sin\beta_1 + \sin\beta_2) \\ +v_0 f(\cos\beta_1 - \cos\beta_2) \end{matrix} \end{pmatrix} = \lambda \begin{pmatrix} 0 & \delta_2 & \delta_3 \\ \delta_4 & 0 & \delta_6 \\ \delta_7 & \delta_8 & \delta_9 \end{pmatrix}. \tag{12}$$

Note that $F_{31} = F_{21}F_{13}/F_{12}$ and $F_{33} = (F_{23} + F_{32})F_{13}/F_{12}$, and so only 5 independent equations may be generated. Given the need to eliminate λ, at most 4 of the 5 parameters may be determined, provided the remaining parameter is known.

Further discussion of self-calibration and the effect of either the tilting or rotation about the optical axis of one camera is to be found in [1].

4 Self-calibration of a horizontally moving stereo head

Having seen that a static stereo head, with coplanar optical axes, is a degenerate configuration for self-calibration, we now assess the consequences of moving the head. Specifically, we permit:

- motion of the head such that the optical axes of the cameras are confined to a plane. This therefore captures the situation in which an upright robot head may translate or rotate in the horizontal plane.
- independent vergence angles of the head that may vary with the motion.
- each camera to have an unknown but fixed focal length.

The following analysis adopts a technique of Zhang et al. [12] in which various fundamental matrices are utilised.

4.1 Formulating the fundamental matrices

Let the rig move from an initial position to a final position. Let the left-right pair of images in the initial position be termed I_1 and I_2, and let the left-right images in the final position be termed I_3 and I_4 (see Figure 3). The left camera is thus responsible for the successive images I_1 and I_3. Assume that the determining of corresponding points has led to estimates for the fundamental matrices linking the following image pairs: (I_1, I_2), (I_3, I_4), (I_1, I_3), (I_2, I_4). Let the associated analytical fundamental matrices be termed $\mathbf{F}^{12}$, $\mathbf{F}^{34}$, $\mathbf{F}^{13}$, $\mathbf{F}^{24}$. We shall not here make use of $\mathbf{F}^{14}$ and $\mathbf{F}^{23}$. As before, we aim to solve for the parameters embedded within these matrices by exploiting the fact that the analytical and the estimated forms of the fundamental matrix are directly proportional. Note that, in this regard, the approach pursued in Zhang et al. [12] is quite different in that a least-squares approach is used to solve a more general problem in which motion is not confined to the plane (although, unlike here, the relative orientation of the head is assumed fixed).

Recalling (8), the initial position of the rig gives rise to the fundamental matrix, $\mathbf{F}^{12}$, given by

$$\mathbf{F}^{12} = \begin{pmatrix} 0 & -\sin\beta_1^{12} & 0 \\ -\sin\beta_2^{12} & 0 & f'\cos\beta_2^{12} \\ 0 & -f\cos\beta_1^{12} & 0 \end{pmatrix}. \tag{13}$$

Assuming that the focal lengths of the respective cameras remain fixed, and that the vergence angles are free to shift, we obtain the following fundamental matrix,

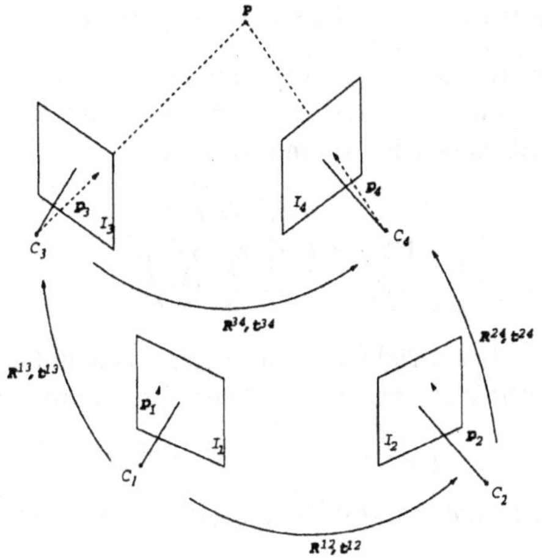

Fig. 3. Motion of the stereo head.

$\mathbf{F}^{34}$, capturing the epipolar relationship between the left and right images of the rig in its final position:

$$\mathbf{F}^{34} = \begin{pmatrix} 0 & -\sin\beta_1^{34} & 0 \\ -\sin\beta_2^{34} & 0 & f'\cos\beta_2^{34} \\ 0 & -f\cos\beta_1^{34} & 0 \end{pmatrix}. \tag{14}$$

Similarly, the fundamental matrix relating the image pair (I_1, I_3) is given by

$$\mathbf{F}^{13} = \begin{pmatrix} 0 & -\sin\beta_1^{13} & 0 \\ -\sin\beta_2^{13} & 0 & f\cos\beta_2^{13} \\ 0 & -f\cos\beta_1^{13} & 0 \end{pmatrix}, \tag{15}$$

with the fundamental matrix for the image pair (I_2, I_4) being

$$\mathbf{F}^{24} = \begin{pmatrix} 0 & -\sin\beta_1^{24} & 0 \\ -\sin\beta_2^{24} & 0 & f'\cos\beta_2^{24} \\ 0 & -f'\cos\beta_1^{24} & 0 \end{pmatrix}. \tag{16}$$

Here we note that the focal lengths of the respective cameras remain unchanged in the movement of the rig from its initial to final position. We observe that, under the above parameterisation, image I_1 undergoes a rotation of $(\beta_1^{13} + \beta_2^{13})$, relative to its own local coordinate system, in becoming oriented in parallel with image I_3.

4.2 Solving the fundamental matrix equations

It is now necessary to further enhance our notation so as to be able to deal simultaneously with various fundamental matrices. Let the numerical estimate, $\mathbf{F}^{ij}_{est}$, of fundamental matrix $\mathbf{F}^{ij}$ be represented as

$$\mathbf{F}^{ij}_{est} = \begin{pmatrix} \delta_1^{ij} & \delta_2^{ij} & \delta_3^{ij} \\ \delta_4^{ij} & \delta_5^{ij} & \delta_6^{ij} \\ \delta_7^{ij} & \delta_8^{ij} & \delta_9^{ij} \end{pmatrix}, \tag{17}$$

and let $\mu_k^{ij} = (\delta_k^{ij})^2$. Let a right bar and superscript indicate the fundamental matrix from which the elements derive. Thus, for example, note that

$$\mu_k + \mu_l|^{ij} = (\delta_k^{ij})^2 + (\delta_l^{ij})^2. \tag{18}$$

In view of the earlier analysis, resulting in (10), we may immediately infer that

$$f = \sqrt{\frac{\mu_8 - \mu_6}{\mu_4 - \mu_2}}\Bigg|^{13}, \qquad f' = \sqrt{\frac{\mu_8 - \mu_6}{\mu_4 - \mu_2}}\Bigg|^{24}$$

$$\tan\beta_1^{13} = f\frac{\delta_2}{\delta_8}\Bigg|^{13}, \tan\beta_2^{13} = -f\frac{\delta_4}{\delta_6}\Bigg|^{13}, \tan\beta_1^{24} = f'\frac{\delta_2}{\delta_8}\Bigg|^{24}, \tan\beta_2^{24} = -f'\frac{\delta_4}{\delta_6}\Bigg|^{24}$$

It then follows that

$$\tan\beta_1^{12} = f\frac{\delta_2}{\delta_8}\Bigg|^{12}, \tan\beta_2^{12} = -f'\frac{\delta_4}{\delta_6}\Bigg|^{12}, \tan\beta_1^{34} = f\frac{\delta_2}{\delta_8}\Bigg|^{34}, \tan\beta_2^{34} = -f'\frac{\delta_4}{\delta_6}\Bigg|^{34}$$

We therefore have closed-form solutions for the 2 focal lengths, and the rotations between images. Implicit in the above are the directions of the various translations between perspective centres. Note that we have so far not made any assumption about the rigidity or otherwise of the rig.

4.3 Solving the baseline constraint equation

We have yet to completely determine the relative orientation of all image pairs as we have still to compute the relative magnitudes of the baselines. (As noted earlier, it is not possible to compute absolute scale of the baselines only from corresponding points.) These relative magnitudes will complete the description of the motion of the head.

Let the magnitude of the head's baseline vector in the initial position be unity. The baseline vector, $\mathbf{t}^{12}$, may therefore be written as

$$\mathbf{t}^{12} = (\cos\beta_1^{12}, 0, \sin\beta_1^{12})^T. \tag{19}$$

Letting L^{ij} denote the length of the baseline vector $\mathbf{t}^{ij}$, we may immediately write down the remaining baseline vectors as

$$\mathbf{t}^{13} = L^{13}(\cos\beta_1^{13}, 0, \sin\beta_1^{13})^T \tag{20}$$
$$\mathbf{t}^{24} = L^{24}(\cos\beta_1^{24}, 0, \sin\beta_1^{24})^T \tag{21}$$
$$\mathbf{t}^{34} = L^{34}(\cos\beta_1^{34}, 0, \sin\beta_1^{34})^T. \tag{22}$$

Our task is now to determine the lengths L^{13}, L^{24}, L^{34}. Returning to Figure 3, we observe (after [12]) the baseline constraint equation

$$\mathbf{R}^{12}\,\mathbf{t}^{24} = \mathbf{t}^{13} - \mathbf{t}^{12} + \mathbf{R}^{13}\mathbf{t}^{34}. \tag{23}$$

Expanding this, we have

$$L^{24}\begin{pmatrix}\cos(\beta^{12}+\beta_1^{24})\\ 0\\ \sin(\beta^{12}+\beta_1^{24})\end{pmatrix} = \begin{pmatrix} L^{13}\cos\beta_1^{13} - \cos\beta_1^{12} + L^{34}\cos\beta_1^{34}\cos\beta^{13} \\ -L^{34}\sin\beta_1^{34}\sin\beta^{13} \\ 0 \\ L^{13}\sin\beta_1^{13} - \sin\beta_1^{12} + L^{34}\cos\beta_1^{34}\sin\beta^{13} \\ +L^{34}\sin\beta_1^{34}\cos\beta^{13}\end{pmatrix}.$$

Here β^{13} rotates I_1 parallel to I_3 and is such that $\beta^{13} = \beta_1^{13} + \beta_2^{13}$. Recall that $\beta^{12} = \beta_1^{12} + \beta_2^{12}$. Clearly, the 3 unknown lengths may not be determined from the above equation. But, on the assumption that the baseline length of the rig remains constant, so that L^{34} also has unit length, we may readily infer that

$$L^{13} = \left(\sin(\omega - \beta^{13}) - \sin\phi\right)/\sin\tau, \tag{24}$$

where $\omega = \beta^{12} + \beta_1^{24} - \beta_1^{34}$, $\phi = \beta_2^{12} + \beta_1^{24}$ and $\tau = \beta_1^{13} - \beta^{12} - \beta_1^{24}$. The formula for L^{24} then follows directly from the baseline constraint equation. We have therefore described the motion of the rig. Further discussion is given in [1].

5 Adding camera tilt to the moving stereo head

Our analysis here is a generalisation of that considered in the previous section in that the head may now tilt up or down, by a rotation about the baseline. We note that the baseline remains confined to a plane, and that the optical axes of the two cameras are at all times coplanar, but are not confined to the same plane in consecutive head positions. Critically, in the analysis presented here, either the initial or final position of the head should have zero tilt.

We now consider how a rotation of θ about the baseline maps the left image to a new position. A rotation of θ about the baseline is equivalent, in the left image's coordinate system, to three composite rotations: a rotation of β_1^{34} about the y-axis, followed by a rotation of θ about the x-axis, and then a rotation of $-\beta_1^{34}$ about the y-axis. In addition to this tilting, the previous rotation in the plane may still take place. The fundamental matrix may therefore be expressed as $\mathbf{F}^{13} = \mathbf{A}^T\,\mathbf{T}\,\mathbf{R}\mathbf{A}$, where

$$\mathbf{T} = \begin{pmatrix} 0 & -\sin\beta_1^{13} & 0\\ \sin\beta_1^{13} & 0 & -\cos\beta_1^{13}\\ 0 & \cos\beta_1^{13} & 0\end{pmatrix}, \qquad \mathbf{A} = \begin{pmatrix}1 & 0 & 0\\ 0 & 1 & 0\\ 0 & 0 & -f\end{pmatrix}, \tag{25}$$

and $\mathbf{R} = R_y(\beta_1^{13} + \beta_2^{13})R_y(\beta_1^{34})R_x(\theta)R_y(-\beta_1^{34})$. Here, we adopt the convention that $R_m(\psi)$ signifies a rotation of ψ about the m-axis.

A rather complex fundamental matrix results (given in [1]) from which the following equations may now be derived:

$$\tan\beta_1^{13} = f\frac{\delta_2}{\delta_8}\bigg|^{13}, \qquad \tan\beta_1^{34} = f\frac{\delta_1}{\delta_3}\bigg|^{13}, \qquad \tan\theta = -\frac{\delta_7}{\delta_8\sin\beta_1^{34}}\bigg|^{13}$$

$$f^2 = \frac{-\delta_3\delta_6}{\delta_2\delta_5+\delta_1\delta_4}\bigg|^{13}, \qquad \sec^2(\beta_1^{34}+\beta_2^{13}) = \sin^2\theta\left(1+\frac{\mu_4+\mu_6/f^2}{\mu_5}\right)\bigg|^{13}$$

We therefore have closed-form solutions for the 5 unknowns f, θ, β_1^{34}, β_1^{13}, β_2^{13}.

Consideration of the fundamental matrix $\mathbf{F}^{24}$ yields symmetric formulae for the right camera vergence angles. The analysis is completed when we note that the previous formulae for baseline lengths are precisely applicable here, since the moving baseline has remained confined to a plane.

6 Experimental Results

We now describe synthetic tests carried out on the method of self-calibration. A cloud of 35 points was randomly generated within a cubic volume of side 2400mm lying approximately 600mm in front of the stereo head. These points were then projected onto each of the 4 image planes arising in the two positions of the stereo head. The location of each image point was then perturbed in a random direction by a distance governed by a Gaussian distribution with zero mean and standard deviation, σ, expressed in pixel units. Such a distribution results in an expected value for the perturbation distance of approximately 0.8 σ. As a matter of interest, in the many tests carried out here, the highest perturbation distance was found to be 3.7 σ.

Left and right focal lengths were set at 6mm and 8mm, with a fixed baseline length of 300mm. Vergence angles were 15 deg and 17 deg in the initial position, and 18 deg and 22 deg in the final position. The motion of the head was such that the upward tilt was 10 deg, rotation of the baseline in the plane was 12 deg, with the length of the translation vector mapping the left camera from initial to final position being approximately the same as the baseline length of the head. Image sizes were 1000 × 1000 pixels.

Experiments were conducted with σ varying from 0.0 to 1.2 in steps of 0.1. For each value of σ, self-calibration was run 20 times (each time operating on a different set of images) and the root-mean-square (rms) error of each parameter was computed. Figure 4 gives a brief summary of how self-calibration is affected by increasing noise, in the case considered. Errors (rms) in lengths and tilt rotation are given as percentages of the true values, while errors (rms) in the vergence angles are expressed in degrees. We can see from the figures that errors in the estimates of the various parameters vary approximately linearly with the extent of the introduced noise, over the range considered. The lengths of the translation vectors L^{13} and L^{24} are the parameters most affected by noise, with relative errors of up to 14% occurring with noise $\sigma = 1.2$. At this high noise level, the rms error of the right camera's focal length is 5.2%, the error in the

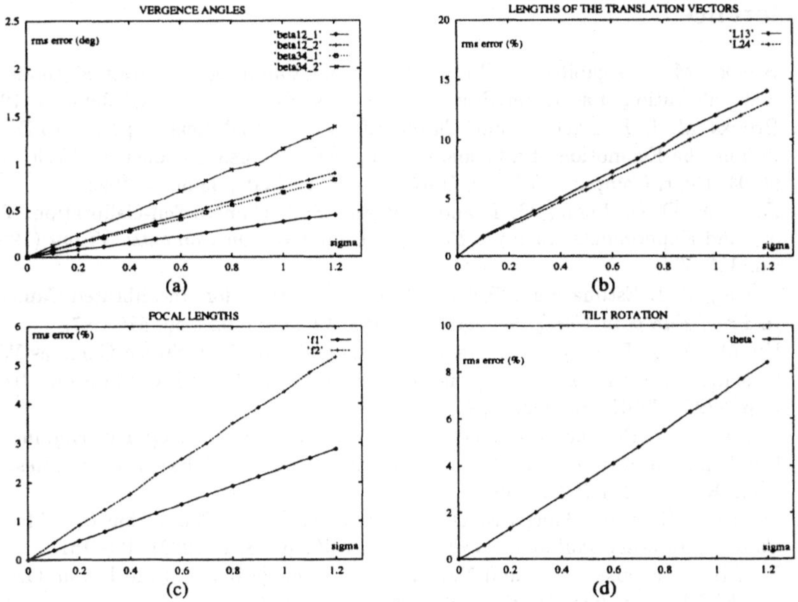

Fig. 4. Results of the experiments.

estimated tilt rotation is 8.5%, and the maximum rms error of the vergence angles is 1.4 deg. Comparable results were obtained for similar head movements.

Given that points can routinely be located by automated techniques with an accuracy of better than $\sigma = 0.5$, the above results suggest that this approach holds promise. Note that no special efforts have been made to optimise the estimates obtained via this process of self-calibration. Thus, for example, no effort has been expended to generate more accurate estimates of fundamental matrices by reducing the deleterious impact of poor localisation of points, nor has any optimisation of the estimates been attempted as a post-process.

Acknowledgements

The authors are grateful for comments of Zhengyou Zhang, Heping Pan and Charlie Wiles. Michael Schneider provided helpful advice. Mike Brooks thanks Prof. D. Maravall for his encouragement and provision of an excellent research environment in Madrid. This work was in part supported by Grant SAB95-0264, issued by the Dirección General de Investigación Científica y Técnica of Spain.

References

1. Brooks, M. J., Agapito, L., Huynh, D. Q. and Baumela, L. Direct Methods for Self-Calibration of a Moving Stereo Head. Tech. Rep. 1/96, CSSIP, January, 1996.
2. Brooks, M. J., Baumela, L. and Chojnacki, W. An Analytical Approach to Determining the Egomotion of a Camera having Free Intrinsic Parameters. Tech. Rep. 96-04, Dept. Computer Science, University of Adelaide, January, 1996.
3. Faugeras, O. D., Luong, Q. T. and Maybank, S. J. Camera Self-Calibration: Theory and Experiments. In *Proc. European Conference on Computer Vision* (1992), pp. 321–334.
4. Hartley, R. I. Estimation of Relative Camera Positions for Uncalibrated Cameras. In *Proc. European Conference on Computer Vision* (1992), pp. 579–587.
5. Huynh, D. Q., Brooks, M. J., Agapito, L. de and Pan. H-P. Stereo Cameras With Coplanar Optical Axes: a Degenerate Configuration for Self-Calibration. Tech. Rep. 2/96, CSSIP, January, 1996.
6. Luong, Q.-T., Deriche, R., Faugeras, O. and Papadopoulo, T. On Determining the Fundamental Matrix: Analysis of Different Methods and Experimental Results. Tech. Rep. 1894, INRIA, 1993.
7. Maybank, S. J. and Faugeras, O. D. A Theory of Self-Calibration of a Moving Camera. *International Journal of Computer Vision*, 8, 2 (1992), 123–151.
8. Pan, H.-P., Brooks, M. J. and Newsam, G. N. Image Resituation: Initial Theory. In *SPIE Videometrics* (1995), vol. 2598, pp. 162–173.
9. Viéville, T., Luong, Q. and Faugeras, O. Motion of Points and Lines in the Uncalibrated Case. *International Journal of Computer Vision* 17, 1 (1994).
10. Viéville, T. and Faugeras, O. Motion Analysis with a Camera with Unknown, and Possibly Varying Intrinsic Parameters. In *ICCV'95* (Cambridge, MA, June 1995), IEEE, pp. 750–756.
11. Zhang, Z., Deriche, R., Faugeras, O. and Luong, Q. T. A Robust Technique for Matching Two Uncalibrated Images Through the Recovery of the Unknown Epipolar Geometry. Tech. Rep. 2273, INRIA, 1994.
12. Zhang, Z., Luong, Q. T. and Faugeras, O. D. Motion of an Uncalibrated Stereo rig: Self-Calibration and Metric Reconstruction. Tech. Rep. 2079, INRIA, 1993.
13. Zisserman, A., Beardsley, P. A. and Reid, I. Metric Calibration of a Stereo Rig. Proc. IEEE Workshop on Representation of Visual Scenes, Boston, June, 1995.

A Notation semantics

Our notation differs from the standard notation of Faugeras et al. [3] (henceforth termed the Faugeras notation). Symbols $\mathbf{F}$, $\mathbf{T}$, $\mathbf{R}$ and $\mathbf{A}$ denote in this work the fundamental, translation, rotation and intrinsic-parameter matrices, respectively. Let the corresponding matrices in Faugeras notation be denoted F, T, R and A. Herein, the epipolar equation has the form $\mathbf{m}^T\,\mathbf{F}\,\mathbf{m}' = 0$, where $\mathbf{F} = \mathbf{A}^T\,\mathbf{T}\,\mathbf{R}\,\mathbf{A}'$. This contrasts with Faugeras notation, where $\mathbf{m}'^T\,F\,\mathbf{m} = 0$, and $F = A'^{-T}\,T\,R\,A^{-1}$. The full list of notational relationships is now given:

$$\mathbf{F} = \sqrt{\det(A)\,\det(A')}\,F^T, \quad \mathbf{A} = -\sqrt{\det(A)}\,A^{-1}, \quad \mathbf{R} = R^T, \quad \mathbf{T} = -R^T\,T\,R.$$

See [1] for further discussion.

Dense Reconstruction by Zooming

C Delherm, JM Lavest, M Dhome, JT Lapresté

Université Blaise-Pascal de Clermont-Ferrand,
LAboratoire des Sciences et Matériaux pour l'Electronique, et d'Automatique
URA 1793 of the CNRS, 63177 Aubière Cedex, France

Abstract. Reconstruction by zooming is not an unachievable task. As it has been previously demonstrated, axial stereovision technics allows to infer 3D information, but involves very small triangulation angles. Accurate calibration, data matching and reconstruction have to be performed to obtain satisfactory modelling results. In this paper, a new approach is proposed to realize dense reconstruction using a static camera equipped with a zoom lens.
The proposed algorithm described in the following sections is divided in three major steps:

- First of all, the matching problem is solved using a correlation algorithm that explicitely takes into account the zooming effect through the images set. An intensity-based multiscale algorithm is applied to the feature points in the first image, to obtain unique point correspondences in all the other images.
- Then, using pixels matched by the previous method, an iterative process is proposed to obtain a sub-pixel matching.
- Finally, the 3D surface is reconstructed using image point correspondances. The modelling algorithm does not require any explicit calibration model and the computations involved are straightforward. This approach uses several images of accurate regular grids placed on a micrometric table, as a calibration process [1]. Complete experiments on real data are provided and show that it is possible to compute 3D dense information from a zooming image set.

Key-words : Correlation, Dense Reconstruction, Axial Stereovision, Implicit Calibration.

1 Introduction

Zoom-lens are commonly used to capture precise details of a global scene but more recently [2], [3] have shown that is also possible to infer 3D information from a zoom-lens mounted on a static camera.

In this paper, using high quality calibration grids, the modelling problem is transferred into a metric space defined between two grids. In [1], experimentations on real data show that this new calibration approach leads to a much more accurate reconstruction than in previous experimentations. The accuracy achieved is better than 1 mm in z-coordinate for a planar grid located 1 meter in front of the camera. Even data close to the optical axis lead to relevant reconstruction. However, in experiments described in [1], points matching observed through the whole images set have been carried out manually, by a sub-pixel cross detection.

In order to improve this step, this article proposes a matching algorithm based on a surfacic correlation method that takes into account the zooming effect.

Then, in order to valid this approach, dense reconstructions on real objects are presented and some statistical analysis performed to measure the modelling reliability.

2 Calibration Algorithm Using Two Reference Grids

Let us suppose we have a distorted grid image (Figure 1b); distortion phenomena can be produced by optics, electronics or sampling problems. A classical non linear pin-hole calibration tries to insert these phenomena into the mathematical model of projection but this kind of model can never be perfectly accurate. If it is possible, as shown in Figure 1a, to compute a transformation from the distorted grid to the regular one, then distortion effects can be corrected. In this way, any new image taken with the camera can be mapped in the reference grid, which is assumed to be as perfect as possible.

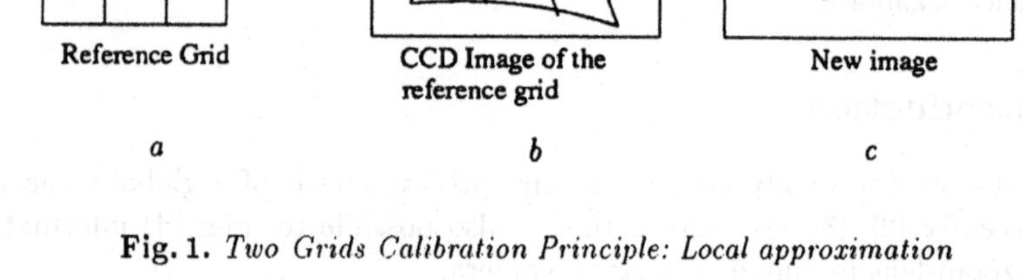

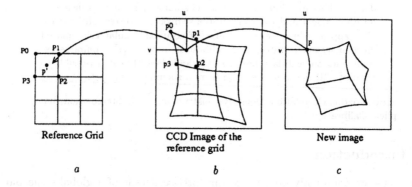

Fig. 1. *Two Grids Calibration Principle: Local approximation*

Let us suppose now that two parallel grids are used, , for which relative space positions can be accurately determined. A micrometric table is used during calibration and an optical control can ensure these assumptions. These two simple images correspond to the basis of the calibration process and define a metric space. The grid closest to the camera defines the origin of this metric space, x and y coordinates correspond respectively to the row and column of the grid, z coordinate is defined along the grid axis of translation.

If a new object is digitized with the camera , it is possible to compute, for each image data, its corresponding point in the first and the second grid. By this way, we have estimated the 3D optical ray of the considered image data in the metric space defined by the two reference grids, without knowing any camera parameter. Of course, it is possible to use more than two grids. The accuracy of the computed 3D optical ray will be improved; nevertheless it will require more computation time and storage capacity.

This calibration process called 'Two Grids Calibration' is not new; [4] introduced it few years ago and [5], [6] have proposed several ways to solve this problem. Two distinct principles can be found:

- the first one tries to compute a global transformation from the distorted image content to the reference point located on each reference grid,
- the second one estimates a local approximation that maps an image point to its corresponding point on the reference grids.

From a computational point of view, global approximation is economical in execution time and storage capacity but errors are averaged over all points. Obviously, local interpolation provides more accurate results than global interpolation. In the local model, the closest calibration points (at least three) are searched around each point detected in the new image, and a local function is used to perform an interpolation over each local region.

Differences between results presented in literature depend on:

- local functions used to establish the relationship between the distorted image and the reference frame,
- but also, on the way the reference points are detected on the CCD matrix.

As indicated above, one of the main difficulties is to compute accurately each point position (p_0, p_1, p_2, p_3 in Figure 1b) in the distorted grid image. [7] has developped an original algorithm that fits directly into the grey level image a mathematical model of the photonic point response. Similar works have been realized by [8], [9] et [10]. In this way, the point location is obtained with high accuracy (more than 1/100 pixel) [7].

Let us consider a point $p(u, v)$ (Figure 1c) in a new distorted image. Knowing its coordinates (u, v), it is possible to find in the stored CCD grid image, the 4 points (p_0, p_1, p_2, p_3 Figure 1b) around it. [11] has proposed several interpolation models and proved that a bi-cubic function leads to better results in the case of high distortion phenomena (X-Ray image), but for a standard camera, linear approximation leads to the same order of accuracy.

In the following experiments, a local function using the anharmonic ratio will be used to interpolate the coordinates of a new point p relatively to the local base (p_0, p_1, p_2, p_3). In order to avoid instability problem, calculus are realized in an homogeneous coordinates system.

3 Points Matching by Correlation

Correlation technics are commonly used to establish the correspondences between edge points in two images from a standard camera device. In case of images taken with a zoom-lens (i.e different magnification), correlation scores need to be modified to take into account the scale effect throught the image set. Furthermore, the triangulation angles involved in axial stereovision require an accurate matching between features observed in the images set. The matching algorithm, proposed in this article, is done in two steps:

- The first one is called "basic correlation"; The correlation is applied from the less zoomed image (wide-angle field) to the most zoomed one (narrow-angle field). Matching is performed with an accuracy at least equals to one pixel.
- The second one is called "sub-pixel correlation". Using previous matched data, an iterative process is applied. Images are considered in the opposite way, it means from the most zoomed to the less zoomed image. Scale effect is explicitely introduced in the correlation parameters to obtain a sub-pixel accuracy.

3.1 Basic Correlation

The aim of this first algorithm is to follow a feature point (not necessarily an edge point) along all an images set taken with a zoom-lens. The correlation is performed from the image taken with the shortest focal-length (wide-angle field) to the longest one. Matching is realized step by step between two successive images I^k and I^{k+1}. The point $p_i^k(u_i^k, v_i^k)$ belonging to the image I^k is associated to the point $p_i^{k+1}(u_i^{k+1}, v_i^{k+1})$ of I^{k+1} that maximizes the correlation coefficient.

In order to look for point p_i^{k+1} in a sufficiently large area in the image I^{k+1} (without spending too many time), a pyramidal research process [12] has been implemented. It works with different images extracted from (I^k, I^{k+1}) but at a lower resolution.

3.2 Sub-Pixel Correlation

Results from the previous correlation algorithm, which is working with two consecutive images $((I_1, I_2), (I_2, I_3), ..(I_l, I_m))$, can diverge along the images set. In order to avoid this phenomena, the sub-pixel matching will be done according to the same reference image along all the sequence. Let us define $\{p_i^1, .., p_i^k, .., p_i^m\}$ a set of matched point p_i, viewed from image I^1 to image I^m and computed by the first correlation algorithm. The last point p_i^m and its grey-level neighbourhood will be used as the reference point during all the sub-pixel process. This assumption seems to be logical, because most information is contained in the image I^m (taken with the longest focal length). Of course, to find accurately two matched points between images I^m and I^l ($l \in [1 \ldots m-1]$), zooming effect has to be taken into account. Therefore, the image I^m will be under-sampled and smoothed automatically, and the algorithm will compute the best location of the correlation window in I^l that fits the data. Parameters are estimated using an iterative process; the following criterion is minimised:

$$C(I^m, I^l, \Delta, f, du, dv) = \sum_{i=-n}^{+n} \sum_{j=-n}^{+n} \left(I^m(u_i^m + \Delta.i,\ v_i^m + \Delta.j,\ f) - I^l(u_i^l + du + i,\ v_i^l + dv + j,\ 1)\right)^2 \quad (1)$$

where parameters (Δ, f, du, dv) are real values (we need sub-pixel accuracy) and take into account respectively the undersampling Δ (i.e. the zooming effect), the smoothing f (due to undersampling), and the subpixel location du, dv of the correlation window in the second image.

In equation [1], $I^k(u, v, \rho)$ is a function that defines the grey-level value according to real point coordinates (u, v) in the image I^k. As image I^k can be undersampled, a smoothing effect is realized on a window $(\rho \times \rho)$ centrered in (u, v). It gives a ponderation to the grey level information (of each pixel) contained in the image I^k, by a coefficient proportionnal to the pixel surface (greater than 1 due to the zooming phenomena) actually included in the considered window.

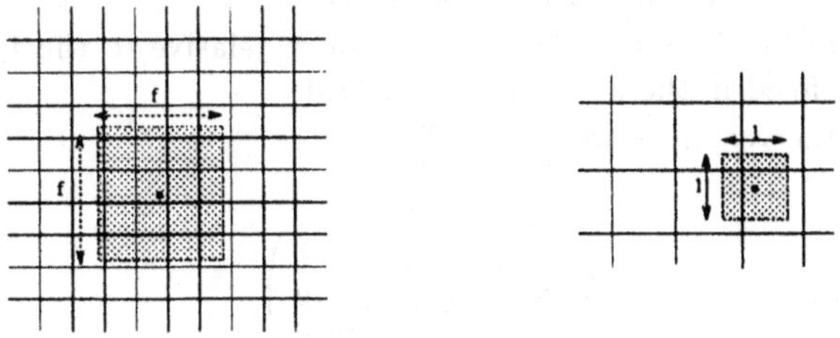

Fig. 2. *Calculus area associated with* $I^m(u, v, f)$ *Calculus for* $I^l(u, v, 1)$

Solution of equation [1] is found by searching the value of coefficients (Δ, f, du, dv) that minimises criterion C. We have to solve a non linear optimisation problem ; as it is not posssible to obtain an analytical expression of the derivatives, the first derivatives of the criterion C are numerically estimated.

After the completion of the sub-pixel algorithm, data matched along the images set, are available. Each point coordinates are estimated with a sub-pixel accuracy. Using this information, the reconstruction process will provide the 3D coordinates of all the feature points, computed in the metric space of calibration grids.

4 Modelling Process

Using considerations developped in section 2, the modelling algorithm does not involve any difficulties. For each focal-length position used during the zooming

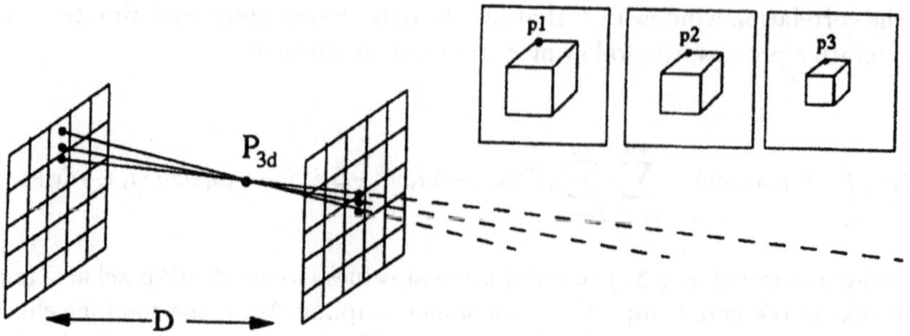

Fig. 3. *Modelling Process.*

sequence, two calibration grids are previously digitized and each calibration point accurately detected and memorized. This step represents the calibration process.

Given a same physical point p^i, matched in all the new images set (Figure 3), (i represents the number of different focal-length used during the zooming sequence), it is possible to estimate the 3D optical rays relative to each image and expressed in the metric space of the grids.

Let (P_1^i, P_2^i) be the interpolated points of p^i relative to the i_{th} image and respectively defined in the first and second grid.

Let $\boldsymbol{V_i}$ be the vector between (P_1^i, P_2^i):

$$\boldsymbol{V_i} = \begin{pmatrix} P_2^i.x - P_1^i.x \\ P_2^i.y - P_1^i.y \\ P_2^i.z - P_1^i.z \end{pmatrix} \tag{2}$$

Let $\mathcal{P}_H^i$ be the quasi horizontal plane and $\mathcal{P}_V^i$ be the quasi vertical plane that contain $\boldsymbol{V_i}$.

The plane equations are given as follows:

$$\begin{cases} A_1^i x + B_1^i y + C_1^i z + D_1^i = 0 \\ A_2^i x + B_2^i y + C_2^i z + D_2^i = 0 \end{cases} \tag{3}$$

Obvioulsy, the three-dimensional coordinates of the observed point $P_{3d}(x, y, z)$ correspond to the intersection of all the optical rays (Figure 3) and are computed by solving the overdeterminated linear system :

$$A_j^i x + B_j^i y + C_j^i z = -D_j^i \quad j = 1 \ldots 2,\ i = 1 \ldots n \tag{4}$$

5 Experiments

5.1 Experimental conditions

All experimental results presented in this section are taken from real images. We have used a *Angénieux* zoom lens T14 × 9;

The reconstruction process is realized in three steps: first of all calibration, then matching by correlation, finally 3D reconstruction.

- *Calibration*
  Given 16 focal-lengths (between 25 à 90 *mm* according to the zoom length parameters), two accurate patterns are digitized, to create the grid metric space. Aperture and focus remain constant during images acquisition. The size of a unit square of the grid is equal to 6.2×6.2 *mm*. The distance between the two grids used during the calibration is set to 15 *cm*. The first one is located at approximatively 90 *cm* from the camera. Each grid is preprocessed and each calibration point (row-column intersection) located, matched and stored in memory.
- *Matching and reconstruction using a new set of images*
  After the calibration step, the new object to be reconstructed is located in front of the camera at approximatively 1 meter. This distance corresponds to the metric space defined between the two grids. Of course, it is not necessary to put the object accurately between the two grids. As it is described in [13], it is possible to achieve good reconstruction results even if the object is not exactely located in the metric space defined by the grids. However, if the object is too far from this space, the reconstruction accuracy is decreasing. Sixteen images are digitized using the same focal length as those used during the calibration step. A mechanical device makes possible to accurately recover the different focal length position on the zoom.
  On the first images of the new set, a given number of points to be reconstructed are manually selected. They cover all the image surface. As they are randomely selected, this step could easily be automated.
  The correlation algorithm is performed for all the points. After convergence, it gives the coordinates of all paired points along the image set. As it can be noticed, some points located near the image border, quickly disappear due to the magnification effect and are only matched in a few images. In order to achieve accurate reconstruction results, only points that appear in at least five images will be taken into account for reconstruction purpose. It means that the 3D coordinates of a reconstructed point correspond to the best intersection, using a least square approximation, of at least five optical rays defined in the grids metric space.

In the following sections, two experimentations of differents reconstructed objects will be presented : a 3D cube (size 10 × 10 × 10 *cm*) and a pebble (whose the size is approximately 10 × 8 × 5 *cm*). 3D coordinates are obtained for each point taken into account in the first image. As it is really difficult to recognize the object shape from a set of independant 3D points, the reconstuction results will be presented on different ways according to the object, in order to evaluate the modelling accuracy.

5.2 Cube reconstruction

Fig. 4. *Three views of the cube image set*

The cube is placed in front of the camera in order to observe three faces (Figure 4). To perform statistical studies, points belonging to each face are independently selected (approximatively one hundred points per face). By doing so, it is possible to estimate the reconstructed points face after face. Angle and planarity between them are estimated using a least square approximation.

Fig. 5. *Cube Reconstruction (dense map of 3D points)*

Figure 5 shows the dense reconstructed map obtained with data collected on the three faces.

- Orthogonality : To verify the modelling accuracy, the angles between each plane are estimated (Table 1):
  Reconstructed planes are not perfectely perpendicular. Some points are badly reconstructed and introduce major errors in the statistical study. As the inital selection of point to be reconstructed is randomely done, some of them are located in area where grey level information is poor; these points matchings are inaccurate and lead to bad reconstruction results.

Table 1. *Angles in degrees between the cube faces*

	Right-Left	*Right-Top*	*Left-Top*
$nbMatch > 4$			
$threshold = \sigma_i$	92.46	91.94	88.07
$threshold = 3\sigma_i$	92.38	92.50	87.37
$nbMatch > 9$			
$threshold = \sigma_i$	89.14	93.50	83.43
$threshold = 3\sigma_i$	89.35	93.37	84.55
$nbMatch > 15$			
$threshold = \sigma_i$	87.39	97.95	64.25
$threshold = 3\sigma_i$	87.60	97.13	62.63

In Table 1, parameter ($nbMatch$) gives the minimum number of matched data to reconstruct a given point; for example ($nbMatch > 9$) means that a point will be reconstructed if and only if it appears at least in 9 images. The threshold value σ_i is used to supress some bad reconstructed points before computing angles between planes.
It can be noticed that angles are close to 90 *degrees*. Mean error increases when statistical studies are realized with ($nbMatch > 15$). But in this case, only points visible during all the images set (it means points close to the optical axis) are taken into account. In this case the triangulation angles are the smallest.

- Planarity : As for previous example, a least square estimation allows to establish the planarity error of each faces. Results are presented in (Tables 2, 3) and expressed in millimeters.
  - (nbMatch) gives the minimum number of images where a point should be visible to be reconstructed.
  - (nbPtsRec) gives the number of points taken into account according to (nbMatch) and the threshold values.
  - (threshold) allows to suppress points too far from the mean plane.

In Figure 6 different points of view of the reconstrcuted object are presented. To have an easier representation of dense data, 3D reconstructed points are presented using couple of faces. As it can be noticed, results are quite good, the real shape is almost estimated even for data close to the image center. A complete study of the triangulation angles influence is realized in [1].

5.3 Pebble reconstruction

In previous experiments, the object to be reconstructed was composed by a set of plane surfaces. Statistical studies could easily be performed to estimate the

Table 2. *Cube Reconstruction : Top Face*

	$nbMatch > 4$	$nbMatch > 9$	$nbMatch > 15$
$nbPts\ Rec$	194	194	67
$\sigma_i(mm)$	4.16	4.16	5.79
$threshold = 3\sigma_i$			
$nbPts\ Rec$	191	191	65
$\sigma(mm)$	2.97	2.97	3.61
$threshold = 2\sigma_i$			
$nbPts\ Rec$	184	184	62
$\sigma(mm)$	2.44	2.44	2.35
$threshold = \sigma_i$			
$nbPts\ Rec$	164	164	59
$\sigma(mm)$	1.98	1.98	1.78

Table 3. *Cube Reconstruction : Left Face / Right Face*

	$nbMatch > 4$	$nbMatch > 9$	$nbMatch > 15$	$nbMatch > 4$	$nbMatch > 9$	$nbMatch > 15$
$nbPts\ Rec$	187	168	32	181	164	31
$\sigma_i(mm)$	7.61	4.18	2.04	2.04	1.99	1.03
$threshold = 3\sigma_i$						
$nbPts\ Rec$	182	164	32	180	162	31
$\sigma(mm)$	4.51	2.84	2.04	1.97	1.86	1.03
$threshold = 2\sigma_i$						
$nbPts\ Rec$	180	158	31	173	156	28
$\sigma(mm)$	3.93	2.52	1.95	1.69	1.61	0.65
$threshold = \sigma_i$						
$nbPts\ Rec$	155	137	22	127	114	22
$\sigma(mm)$	2.59	1.99	1.03	1.16	1.01	0.49

reconstruction quality. In the following experiment, a pebble reconstruction is performed. We present the modelling results obtained with a curved surface, using a set of pebble images.

Figures 8 shows three different views of the reconstructed object. In spite of the object texture, reconstruction errors still remain, due to a lack of accuracy during the matching algorithms. However, the global object shape is well estimated.

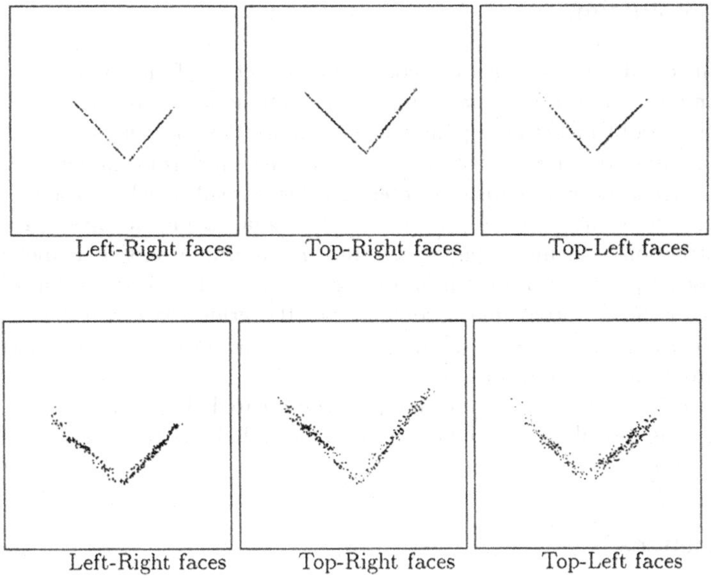

Fig. 6. *Cube Reconstruction*

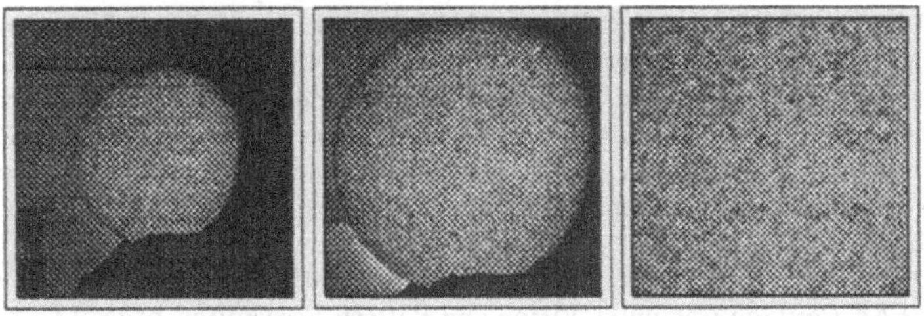

Fig. 7. *Three images of the pebble set*

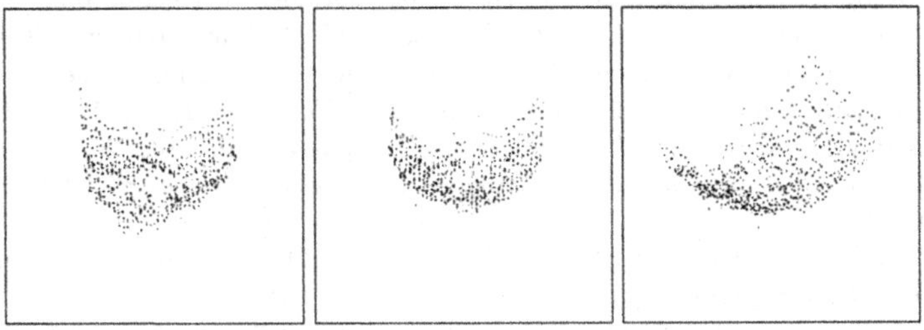

Fig. 8. *Pebble Reconstruction (dense map of 3D points)*

6 Conclusion

This article describes a dense reconstruction method from images taken with a zoom-lens. Calibration, matching and reconstruction steps are detailed. The matching problem between points viewed along the zooming set is performed with an iterative correlation algorithm. Zooming effect is explicitely introduced in the correlation parameters in order to achieve a sub-pixel accuracy.

Such an accuracy is essential due to the small angles involved in zooming reconstruction. The modelling algorithm is realized in a particular metric space composed by two accurate calibration grids. A local approximation allows to transform the distorted image content into the grids space which is supposed distortion free. By doing so, all distortion phenomena (optical and electronics) are automatically taken into account.

Several experiments on real images are presented. Promising results are obtained ; the global object shapes are well estimated, even for data close to the image center.

References

1. JM Lavest, C Delherm, B Peuchot, and N Daucher. Implicit Reconstruction by Zooming. *To appear in Computer Vision, Graphics and Image Processing*, 1995.
2. JM Lavest, G Rives, and M Dhome. 3D Reconstruction by Zooming. *IEEE Transactions on Robotics and Automation*, 9(2):196–208, April 1993.
3. JM Lavest, G Rives, and M Dhome. Modeling an Object of Revolution by Zooming. *IEEE Trans. on Robotics and Automation*, 11(2):267–271, April 1995.
4. HA Martins, JR Birk, and RB Kelley. Camera Models Based on Data from Two Calibration Planes. *Computer Graphics and Image Processing*, 17:173–180, 1981.
5. KD Gremban, CH Thorpe, and T Kanade. Geometric Camera Calibration using Systems of Linear Equations. *Proc. of IEEE Robotics and Automation*, pages 562–567, 1988.
6. GQ Wei and SD Ma. Two Plane Camera Calibration: a Unified Model. *in Proc. of IEEE Conf. on Computer Vision and Pattern Recognition*, pages 133–138, June 1991.
7. B Peuchot. Utilisation de détecteurs sub-pixels dans la modélisation d'une caméra. *9ème congrès AFCET RFIA*, pages 691–695, Paris, January 1994.
8. P Brand, R Mohr, and P Bobet. Distorsions optiques : correction dans un modèle projectif. *9ème congrès AFCET RFIA*, pages 87–98, Paris, January 1994.
9. R Deriche and G Giraudon. A Computational Approach for Corner and Vertex Detection. *International Journal of Computer Vision*, 10(2):101–124, 1993.
10. HA Beyer. Accurate Calibration of CCD Cameras. *in Proc. of Conference on Computer Vision and Pattern Recognition, Urbana Champaign, USA*, pages 96–101, 1992.
11. B Peuchot. Camera Virtual Equivalent Model, 0.01 Pixel Detector. *Computerized Medical Imaging and Graphics*, 17(4-5):289–294, 1993.
12. CS Zhao. Reconstruction de surfaces tridimensionnelles en vision par ordinateur. *Thèse de doctorat, Institut National Polytechnique de Grenoble, France*, 1993.
13. C Delherm, JM Lavest, B Peuchot, and N Daucher. Reconstruction implicite par zoom. *To appear in Traitement du signal*, 1995.

Telecentric Optics for Computational Vision

Masahiro Watanabe[1] **and Shree K. Nayar**[2]

[1] Production Engineering Research Lab., Hitachi Ltd.,
292 Yoshida-cho, Totsuka, Yokohama 244, Japan
[2] Department of Computer Science,
Columbia University, New York, NY 10027, USA

Abstract. An optical approach to constant-magnification imaging is described. Magnification variations due to changes in focus setting pose problems for pertinent vision techniques, such as, depth from defocus. It is shown that magnification of a conventional lens can be made invariant to defocus by simply adding an aperture at an analytically derived location. The resulting optical configuration is called "telecentric." It is shown that most commercially available lenses can be turned into telecentric ones. The procedure for calculating the position of the additional aperture is outlined. The photometric and geometric properties of telecentric lenses are discussed in detail. Several experiments have been conducted to demonstrate the effectiveness of telecentricity.

1 Introduction

The problem of magnification variation due to change in focus setting has significance in machine vision. The classical approach to solving this problem has been to view it as one of *camera calibration* [5, 17]. Willson and Shafer [17] conducted a careful analysis of the interaction between focus and magnification. They proposed a joint calibration approach that measures the relation between zooming and focusing for a given lens. Through this calibration model, it becomes possible to hold magnification constant while focusing; the calibration results are used to apply zoom adjustments so as to correct magnification changes caused by focusing. Though this approach provides a general scheme to tackle the problem, it has its drawbacks. One requires an expensive computer-controlled zoom lens and, of course, the extensive calibration procedure mentioned above, even if one only needs to vary the focus setting and not any other parameter. Further, the necessity to change two physical parameters (focus and zoom) simultaneously tends to increase errors caused by backlashes in the lens mechanism and variations in lens distortion.

An alternative approach to the magnification problem is a computational one, commonly referred to as *image warping*. Darrell and Wohn [3] proposed the use of warping to correct image shifts due to magnification changes caused by focusing. This method is simple and effective for some applications, but can prove computationally intensive for real-time ones. Furthermore, since warping is based on spatial interpolation and resampling techniques, it generally introduces undesirable effects such as smoothing and aliasing. These can be harmful for applications that rely on precise spatial-frequency analysis, such as, depth from focus/defocus.

Depth from focus/defocus methods provide a powerful means of getting a range map of a scene from two or more images taken from the same viewpoint but with different optical settings. Depth from focus uses a sequence of images taken by incrementing the focus setting in small steps. For each pixel, the focus setting that maximizes image contrast is determined. This in turn can be used to compute the depth of the corresponding scene point [3, 7, 9, 10, 11, 17]. Magnification variations due to defocus, however, cause additional image variations in the form of translations and scalings. Estimation of image contrast in the presence of these effects could result in erroneous depth estimates.

In contrast to depth from focus, depth from defocus uses only two images with different optical settings [2, 4, 13, 14, 15, 18]. As it attempts to compute depth from this minimal number of images, it requires all aspects of the image formation process to be precisely modeled. Since the two images taken correspond to different levels of defocus, they are expected to differ in magnification. Given that two images are all we have in the case of defocus analysis, the magnification variations prove particularly harmful during depth computation. As a result, most investigators of depth from defocus have been forced to vary aperture size [2, 4, 14, 15, 18] rather than the focus setting (for example, the distance between the sensor and the lens) to induce defocus variations between the two images. Aperture size changes have the advantage of not introducing magnification variations. However, this approach has two major drawbacks when compared with changing the focus setting. (a) One of the two apertures used must be significantly smaller than the other. This decreases the brightness of the corresponding image and consequently degrades its signal-to-noise ratio. (b) The sensitivity of aperture induced defocus to depth is lower and less uniform than in the case of focus setting changes. Therefore, depth from defocus based on focus change is clearly desirable if there exists a simple and inexpensive way of making magnification invariant to defocusing.

In this paper, an approach to constant-magnification imaging is described. Rather than resorting to the traditional approaches of detailed calibration or computational warping, we suggest a simple but effective optical solution to the problem. The magnification problem is eliminated in its entirety by the introduction of an optical configuration, referred to as *telecentric optics*. Though telecentricity has been known for long in optics [1, 8], it has not been exploited in the realm of computational vision. With this optics, magnification remains constant despite focus changes. The attractive feature of this solution is that commercially available lenses (used extensively in machine vision) are easily transformed to telecentric ones by adding an extra aperture. We analytically derive the positions of aperture placement for a variety of off-the-shelf lenses. Further, extensive experimentation is conducted to verify the invariance of magnification to defocus in four telecentric lenses that were constructed from commonly used commercial lenses. We have successfully incorporated a telecentric lens into a real-time active range sensor that is based on depth from defocus [13]. Recently, we demonstrated the application of telecentric optics to passive depth from defocus [16]. The performance of the telecentric lens was found to be far superior to that of a conventional one.

2 Telecentricity

2.1 Conventional Lens Model

To begin with, we discuss the lens model that is widely used in computer vision. Figure 1 shows the commonly used image formation model, where the main assumptions are that the lens is thin and the aperture position coincides with the lens. All light rays that are radiated by scene point P and pass the aperture A are refracted by the lens to converge at point Q on the image plane. The relationship between the object distance d, focal length of the lens f, and the image distance d_i is given by the Gaussian lens law:

$$\frac{1}{d} + \frac{1}{d_i} = \frac{1}{f}. \tag{1}$$

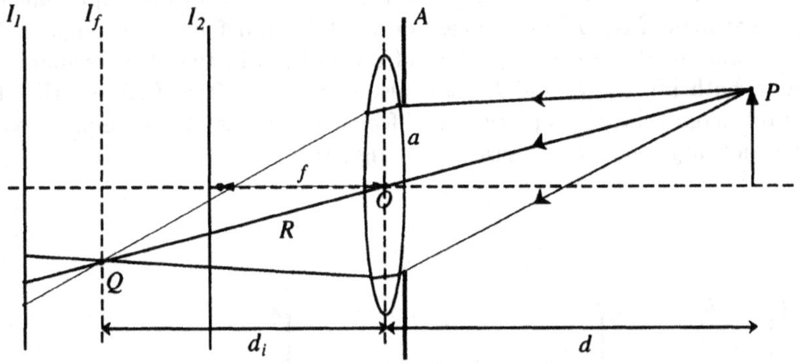

Fig. 1. Image formation using a conventional thin lens model.

Each point P on the object plane is projected onto a single point Q on the image plane, causing a clear or *focused* image I_f to be formed. When the focus setting is changed by displacing the sensor plane, for example, to I_1 or I_2 from I_f, the energy received from P by the lens is distributed over a circular patch on the sensor plane[3]. Although this causes a blurred image, the effective image location of point P can be taken to be the center of the circle. This center lies on

[3] Here, focus change is modeled as a translation of the sensor plane. This model is also valid for the case of lens translation which is used in most non-zoom lens systems, where the distance d between lens and object is typically much larger than the focal length f.

the ray R which is radiated from point P and passes through the center O of the lens. This ray is called the ***principal ray*** [1, 8] in optics. Since its intersection with the sensor plane varies with the position of the sensor plane, image magnification varies with defocus.

2.2 Telecentric Optics

Keeping in mind the image formation model shown in Figure 1, we proceed to discuss the constant magnification configuration, called telecentric optics. Figure 2 illustrates the principle underlying telecentric projection. The only modification made with respect to the conventional lens model of Figure 1 is the use of the external aperture A'. The aperture is placed at the ***front-focal plane***, i.e. a focal length in front of the ***principal point*** O of the lens. This simple addition solves the problem of magnification variation with the distance of the sensor plane from the lens. Straightforward geometrical analysis reveals that the ray of light R' from any scene point that passes through the center O' of aperture A', i.e. the ***principal ray***, emerges parallel to the optical axis on the image side of the lens [8]. Furthermore, this parallel ray is the axis of a cone that includes all light rays radiated by the scene point, passed through by A', and intercepted by the lens. As a result, despite defocus blurring, the effective image coordinates of point P on the sensor plane stay constant irrespective of the displacement of the sensor plane from I_f. In the case of depth from defocus, the magnification of any point in the scene, regardless of its position in the scene, remains the same in both images, I_1 and I_2. It is also easy to see from Figure 2 that this constant-magnification property is unaffected by the aperture radius a' as far as it is not large enough to cause severe vignetting.

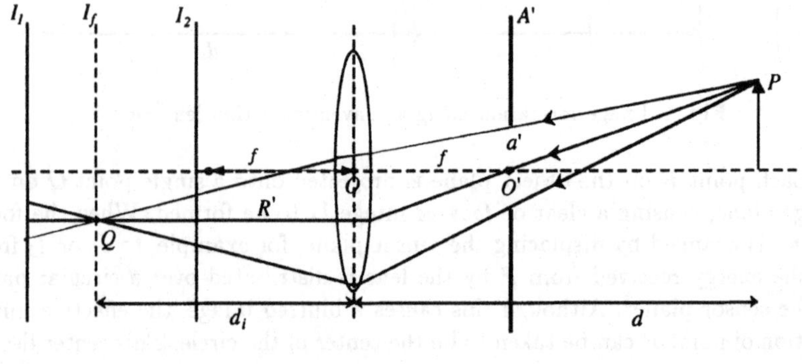

Fig. 2. Telecentric optics achieved by adding an aperture to a conventional lens. This simple modification causes image magnification to be invariant to the position of the sensor plane, i.e. the focus setting.

2.3 Properties of Telecentric Optics

While the nominal and effective *F-numbers* for the conventional lens model in Figure 1 are $f/2a$ and $d_i/2a$, respectively, they are both equal to $f/2a'$ in the telecentric case. The reason for this invariance of the F-number for telecentric optics is the following. The effective F-number is defined as the ratio of the height to the diameter of the base of the light cone that emerges out of the lens. The nominal F-number is equal to the effective F-number when the lens is focused at infinity. The above-mentioned light cone is bordered by the family of *marginal rays* which pass through the aperture A' touching its circumference. Consider these rays as they emerge from the lens on the image side. If the scene point P is displaced, the marginal rays on the image side only shift in a parallel fashion, keeping the angle subtended by the apex of the cone constant. This is due to the aperture being located at the front focal plane (see Figure 2): Light rays which pass through the same point in the front focal plane emerge parallel from the lens. As a result, the effective F-number does not change when the distance of the focused point from the lens changes. Further, when the lens is focused at infinity, the effective F-number is $f/2a'$ because the marginal rays pass through the aperture A' parallel to one another and strike the lens with the same diameter a' as the aperture, and converge on the image side of the lens at the focal length f.

This above fact results in another remarkable property of the telecentric lens: The image brightness stays constant in a telecentric lens, while, in a conventional lens, brightness decreases as the effective focal length d_i increases. This becomes clear by examining the image irradiance equation [6]:

$$E = L \frac{\pi}{4} \frac{\cos^4 \theta}{{F_e}^2}, \tag{2}$$

where, E is the irradiance of the sensor plane, L is the radiance of the surface in the direction of the lens and F_e is the effective F-number. In the case of a conventional lens, $F_e = d_i/2a$, while $F_e = f/2a'$ for a telecentric lens. θ is the angle between the optical axis and the principal ray which originates from object point P and passes through the center of the aperture.

In summary, telecentric optics provides us a way of taking multiple images of a scene at different focus settings while keeping magnification constant between the images. In practice, this can be accomplished by using a beam splitter (or for more than two images, a sequence of beam splitters) behind the telecentric lens [12]. In a real-time application, all images can be digitized simultaneously and processed in parallel as in [13].

3 Aperture Placement

Although the discussion on telecentricity in section 2 was based on the thin lens model, the results holds true for compound lenses. The thin lens is just a special case when the two *principal planes* [1, 6] coincides. Figure 1 is easily modified for the compound lens case, by replacing the thin lens with the two principal planes, U and U', of the compound lens, as shown in Figure 3 (a). The Gaussian lens law (1) remains valid when the distances are measured as follows; the scene point distance d from plane U and the image distance d_i from plane U'.

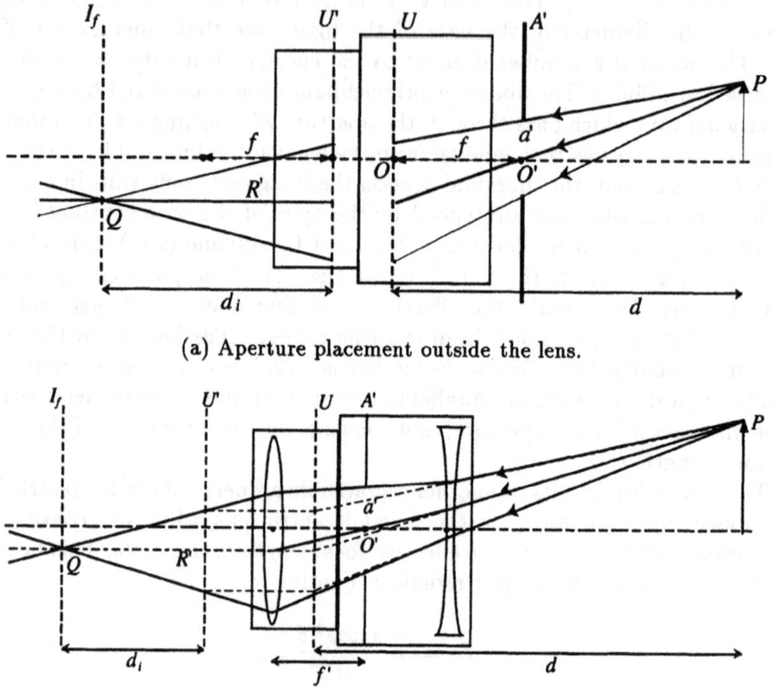

(a) Aperture placement outside the lens.

(b) Aperture placement inside the lens.

Fig. 3. Telecentric aperture placement in the case of compound lenses.

Given an off-the-shelf lens, the additional telecentric aperture can be easily appended to the casing of the lens as far as the front focal plane is outside the lens (see Figure 3(a)). This is the case with most lenses including telephoto ones. However, for wide angle lenses with focal lengths shorter than the *back focal length* (distance from lens mount plane to sensor plane), the front focal plane is likely to be inside the lens (see Figure 3(b)). In such cases, one can still make the lens telecentric by placing an aperture inside the lens. The procedure is as follows. First, consider a set of parallel rays entering from the image side of the lens and find the point in the lens system where the parallel rays converge. In Figure 3(b), this would correspond to the point O'. If this point does not lie inside any of the physical lenses that comprise the compound lens, one can open the lens and place the telecentric aperture at the plane passing through the point and normal to the optical axis of the lens. Fujinon's CF12.5A f/1.4 is an example such a lens, which we converted to telecentric by placing an aperture inside and used to develop a real-time focus range sensor [13].

In practice, the exact location where the additional aperture needs to be placed can be determined in the following way. In some cases, the lens manufacturer provides information regarding the front focal position, which is customarily denoted as F in the schematic diagram. For instance, Nikon provides this data for their old line of SLR lenses, two of which, Micro-Nikkor 55mm f/2.8 and Nikkor 85mm f/2, are used in our experiments reported in section 4.1. If this information is not available from the manufacturer, it can be determined by the following procedure. Hold the lens between a screen (say, a white piece of paper) and a far and bright source (such as the sun or a distant lamp). In this setup the lens is held in the direction opposite to normal use; light enters from the back of the lens and the image is formed on the screen in front of the lens. The screen is shifted around to find the position that provides the clearest image. This position is the front focal plane where the telecentric aperture should be placed.

The lens is then mounted on an image sensor with the telecentric aperture attached to it, and the actual magnification variation due to defocus is used as feedback to refine the aperture position so as to drive magnification variation to zero. The method we used to measure magnification change is detailed in section 4.1. The above refinement process is recommended even when the front focal plane position is available from the manufacturer. This is because the precise position of the front focal plane may vary slightly between lenses of the same model. As we will see, after conversion to telecentric, magnification variations produced by a lens can be reduced to well below 0.1%.

While using the telecentric lens, the original aperture of the lens should be opened fully and the diameter of the telecentric aperture should be chosen so as to minimize vignetting. The degree of vignetting can be gauged by reducing the original aperture by one step from the fully open position and measuring image brightness in the corners of the image. If vignetting is significant, even this small reduction in the original aperture size will change image brightness. Table 1 summarizes all the information needed to convert five popular lenses into telecentric ones. We have in fact converted all five lenses and the telecentric versions of three of them are shown in Figure 4. In each case, except Fujinon's CF12.5A, the telecentric aperture resides in the aluminum casing appended to the front of lens. In the case of Fujinon's CF12.5A, the telecentric aperture resides inside the lens body.

There is yet another way to convert a conventional lens into a telecentric one. This is by placing an additional convex lens between the original (compound) lens and the image sensor to make the principal ray parallel to the optical axis. Though this is better in the sense that one can use the full aperture range without worrying about vignetting, it changes the focal length and the position of the image plane. In addition, the convex lens may have to reside deep inside the camera.

It is worth mentioning that there are some commercially available lenses that are called telecentric. But these lenses are telecentric on the object side, i.e.

Table 1. Aperture placement for five off-the-shelf lenses. Outside aperture position is measured along the optical axis from the surface of outermost lens. Inside aperture position is measured from the stray light stop aperture toward the scene direction.

Lens	Focal length	F-number	Aperture position	Max. aperture†
Fujinon CF12.5A	12.5mm	1.4	Inside : 4mm	F/8
Cosmicar B1214D-2	25mm	1.4	Outside: 4.7mm	F/8
Nikon AF Nikkor	35mm	2	Outside: 3.3mm	F/8.5
Nikon Micro-Nikkor	55mm	2.8	Outside: 46mm (38.5)*	F/13
Nikon Nikkor	85mm	2	Outside: 67mm (57.8)*	F/6.8

* Number in parentheses is the maker supplied front focal position.
† Max. aperture is the maximum aperture (minimum F-number) which does not cause vignetting (converted into F-number).

(a)

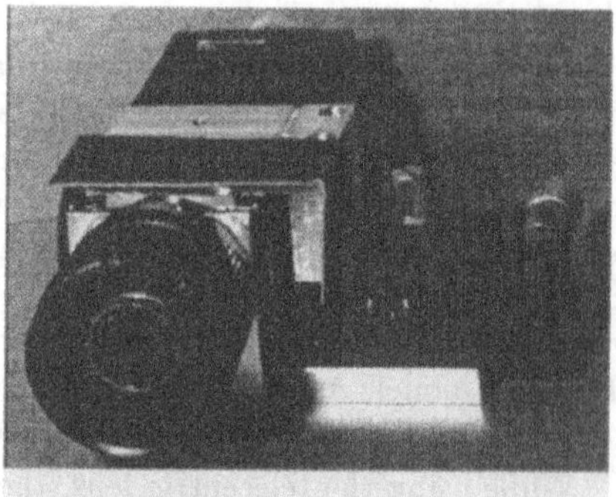

(b)

Fig. 4. Popular lenses converted into telecentric ones: (a) The Nikon Nikkor f=85mm SLR (left) and the Nikon AF Nikkor f=35mm SLR (right), and (b) The Fujinon CF12.5A f=25mm lens on a micrometer stage. In each case, the telecentric aperture resides in the aluminum casing appended to the front of the lens. The micrometer stage in (b) allows constant-magnification focus variation by displacing the image sensor from the lens.

principal rays come into the lenses parallel to the optical axis, which is opposite to what has been discussed here. These lens are used in profile projectors where magnification changes caused by the variation of object distance from the lens is a serious problem. In effect, these lenses realize precise orthographic projection and not constant-magnification focusing.

It turns out that zoom lenses for 3-CCD color cameras are made telecentric on the image side to avoid color shading caused by RGB color separation. One can tell this by looking at the exit pupil position in the specification sheet provided by the manufacturer. If the exit pupil position is at ∞, the lens is image-side telecentric [1, 8]. An example is Fujinon's H12×10.5A. But this does not mean that this zoom lens has magnification that is invariant to focus change, because zoom lenses usually change focus by complex movements of some of the lens components. To achieve constant magnification in such cases, a special mechanism must be added to shift the relative position between the zoom lens and CCD sensor.

4 Experiment

4.1 Magnification

To verify the constant-magnification capability of telecentric lenses, we have taken a series of images by varying focus and measured the magnification. To detect magnification change between images, the following method was used.

1. **FFT-phase-based local shift detection:** We compute the Fourier Transform of corresponding small areas in the two images and find the ratio of the two resulting spectra. Then a plane is fitted to the phases of the divided spectra. The gradient of the estimated plane is nothing but the image shift vector. As the two image areas should contain the same scene areas to get sub-pixel accuracy, the area used for FFT computation is refined iteratively by using the computed image shift. The image window used to compute local shifts has 64×64 pixels.
2. **Object pattern:** A crisp and dense scene pattern is designed to ensure the phase estimates are accurate. The pattern we have used is shown in Figure 5 (b). The period of the pattern must be larger than the FFT computation area to avoid phase ambiguities.
3. **Needle diagram of local shifts:** A sparse needle map (5×4 needles) of the local shift vectors are chosen for display over the 640×480 pixel image.
4. **Magnification change and translation:** A shift vector $(\Delta x, \Delta y)$ at image coordinate (x, y), where x and y are measured from the center point of the image, is modeled as a similarity transformation:

$$\Delta x = a_x + m\,x\,, \quad \Delta y = a_y + m\,y\,. \tag{3}$$

   Least-mean-square fitting is used to estimate the parameters m and (a_x, a_y). This way we separate out the global magnification change m from the global translation (a_x, a_y). The translation factor (a_x, a_y) is introduced in the above model for two reasons; (a) the optical center of the image is assumed to be unknown, and (b) the optical center itself can shift between images due to any possible misalignment or wobble in the focus adjustment mechanism. The residual error of the above fit reveals the local shift detection accuracy and the validity of the above transformation model.

Figures 5 and 6 show the magnification change for the f=25mm lens without and with the telecentric aperture, respectively. In each of these two figures, (a) is the image focused at infinity, (b) at 787mm from the lens, and (c) at 330mm from the lens. Figures (d) and (e) are needle diagrams that show the local shifts of image (b) relative to image (a) and that of image (c) relative to image (a), respectively. The needles are magnified by a factor of 5 for clarity of display.

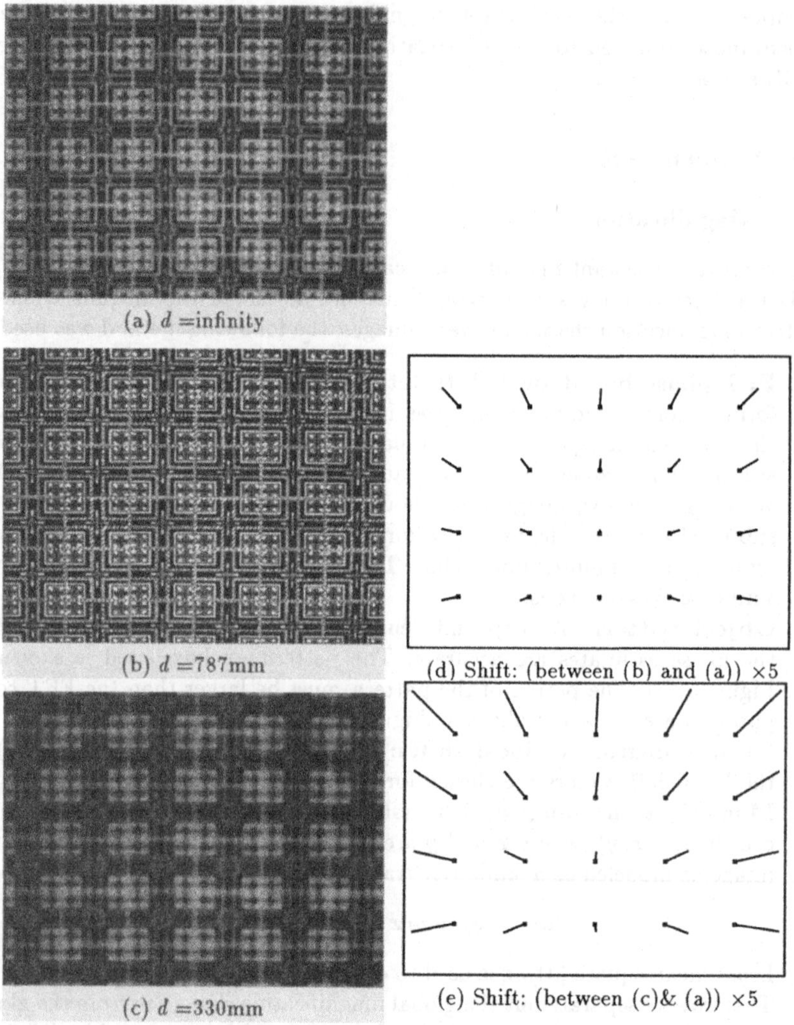

(a) d =infinity

(b) d =787mm

(c) d =330mm

(d) Shift: (between (b) and (a)) ×5

(e) Shift: (between (c)& (a)) ×5

Fig. 5. Magnification variations in non-telecentric optics due to focus change.

In the same manner, images were taken and processed for the 35mm, 55mm and 85mm lenses and the results are summarized in Table 2. As is clear from the table, magnification variations in the telecentric versions of the lenses are remarkably small. Pixel shifts due to large focus setting changes are as small as 0.5 pixels. With careful tuning of the aperture position, this number can be further reduced to 0.1 pixels.

The above experiments relate to the geometric properties of both conventional and telecentric lenses. As explained in section 2.3, image brightness varies with focus change in the case of the conventional lens while it remains con-

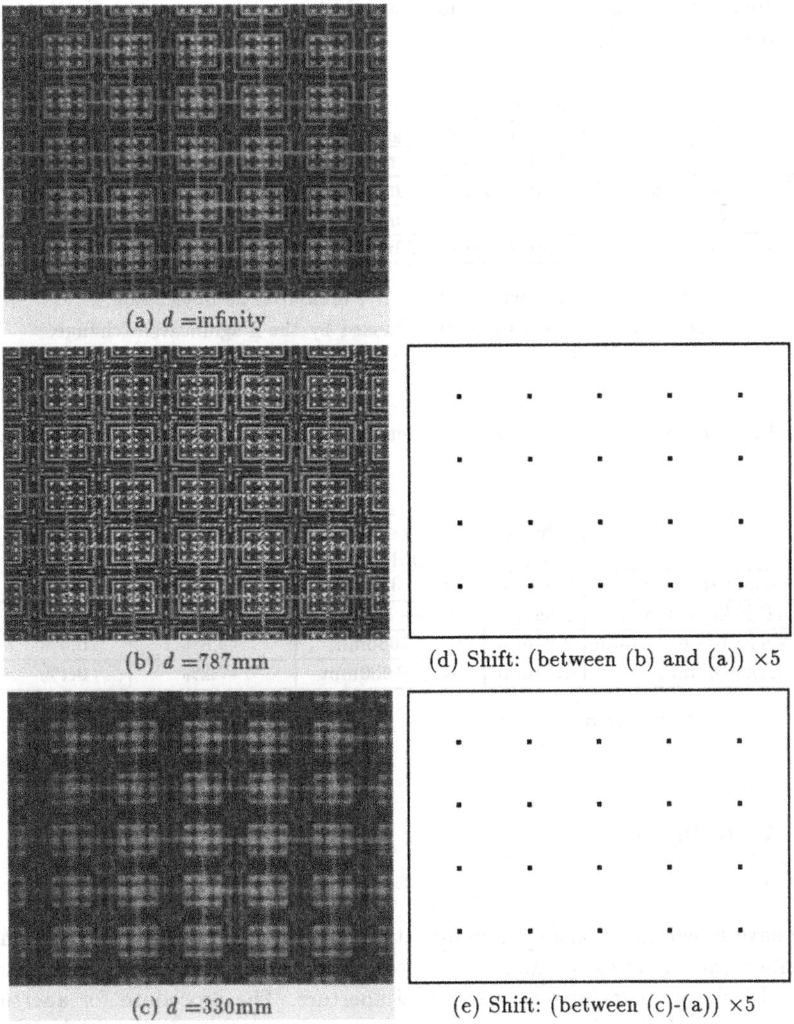

(a) d =infinity

(b) d =787mm

(d) Shift: (between (b) and (a)) ×5

(c) d =330mm

(e) Shift: (between (c)-(a)) ×5

Fig. 6. Magnification variations in telecentric optics due to focus change. The needle maps can be compared with those in Figure 5 to see that magnification remains unaffected by focus change.

stant in the telecentric case due to the invariance of the effective F-number (see equation 2). Therefore, in a conventional lens, the brightness values for different focus settings need to be normalized prior to further processing. This process is avoided in the telecentric case. Table 3 summarizes our experiments on the photometric properties of conventional and telecentric lenses.

Table 2. Magnification variations for four widely used lenses and their telecentric versions.

Lens	Shift of CCD*	Change in focused distance†	Non-telecentric Mag.(max. shift‡)	Telecentric Mag.(max. shift‡)
Cosmicar 25mm	2.05mm	∞ – 330mm	5.9% (18.9pix)	-0.03% (0.1pix)
AF Nikkor 35mm	2.08mm	4022– 550mm	4.5% (14.3pix)	-0.15% (0.5pix)
Micro-Nikkor 55mm	2.94mm	∞ –1085mm	5.8% (18.6pix)	-0.07% (0.2pix)
Nikkor 85mm	2.85mm	1514–1000mm	4.7% (14.9pix)	0.01% (0.03pix)

* Measured on image side. † Measured on object side.
‡ max. shift is the maximum shift caused by the magnification change.

Table 3. Brightness variation due to focus change for four widely used lenses and their telecentric versions.

Lens	Shift of CCD*	Change in focused distance†	Brightness variation	
			Non-telecentric	Telecentric
Cosmicar 25mm	2.05mm	∞ – 330mm	-7.6%	0.0%
AF Nikkor 35mm	2.08mm	4022– 550mm	-5.1%	0.7%
Micro-Nikkor 55mm	2.94mm	∞ –1085mm	-9.5%	0.6%
Nikkor 85mm	2.85mm	1514–1000mm	-7.2%	0.1%

* Measured on image side. † Measured on object side.

5 Conclusion

We have shown that constant-magnification imaging can be achieved using what is called *telecentric optics*. Most commercial lenses can be turned into telecentric ones by simply attaching an additional aperture. The procedure for aperture placement and the photometric and geometric properties of telecentric lenses were discussed in detail. We have demonstrated that magnification changes can be reduced to as low as 0.03%, i.e. maximum magnification induced shift of 0.1 pixel, by the proposed aperture placing method. The only drawback of this optics is that the available F-number that does not cause vignetting is larger (aperture must be smaller) than in the lens prior to conversion. This needs to be compensated by using either brighter illumination or a more sensitive image sensor. However, it should be noted that a lens designed from scratch can be made telecentric without substantial reduction in brightness. We have applied telecentric optics to the depth from defocus problem [13, 16]. The computed depth maps were found to be far superior to ones computed using a conventional lens.

Acknowledgements

We thank one of the anonymous reviewers for detailed comments on the paper and Simon Baker for his help in preparing the final manuscript. This research was conducted at the Center for Research in Intelligent Systems, Department of Computer Science, Columbia University. It was supported in part by the Production Engineering Research Laboratory, Hitachi, and in part by the David and Lucile Packard Fellowship.

References

1. M. Born and E. Wolf. *Principles of Optics*. London:Permagon, 1965.
2. V. M. Bove, Jr. Entropy-based depth from focus. *Journal of Optical Society of America A*, 10:561–566, April 1993.
3. T. Darrell and K. Wohn. Pyramid based depth from focus. *Proc. of IEEE Conf. on Computer Vision and Pattern Recognition*, pages 504–509, June 1988.
4. J. Ens and P. Lawrence. A matrix based method for determining depth from focus. *Proc. of IEEE Conf. on Computer Vision and Pattern Recognition*, pages 600–609, June 1991.
5. B. K. P. Horn. Focusing. Technical Report Memo 160, AI Lab., Massachusetts Institute of Technology, Cambridge, MA, USA, 1968.
6. B. K. P. Horn. *Robot Vision*. The MIT Press, 1986.
7. R. A. Jarvis. A perspective on range finding techniques for computer vision. *IEEE Trans. on Pattern Analysis and Machine Intelligence*, 5(2):122–139, March 1983.
8. R. Kingslake. *Optical System Design*. Academic Press, 1983.
9. A. Krishnan and N. Ahuja. Range estimation from focus using a non-frontal imaging camera. *Proc. of AAAI Conf.*, pages 830–835, July 1993.
10. E. Krotkov. Focusing. *International Journal of Computer Vision*, 1:223–237, 1987.
11. S. K. Nayar and Y. Nakagawa. Shape from focus. *IEEE Trans. on Pattern Analysis and Machine Intelligence*, 16(8):824–831, August 1994.
12. S. K. Nayar and M. Watanabe. Passive bifocal vision sensor. Technical Report (in preparation), Dept. of Computer Science, Columbia University, New York, NY, USA, October 1995.
13. S. K. Nayar, M. Watanabe, and M. Noguchi. Real-time focus range sensor. *Proc. of Intl. Conf. on Computer Vision*, pages 995–1001, June 1995.
14. A. Pentland. A new sense for depth of field. *IEEE Trans. on Pattern Analysis and Machine Intelligence*, 9(4):523–531, July 1987.
15. G. Surya and M. Subbarao. Depth from defocus by changing camera aperture: A spatial domain approach. *Proc. of IEEE Conf. on Computer Vision and Pattern Recognition*, pages 61–67, June 1993.
16. M. Watanabe and S. K. Nayar. Telecentric optics for constant-magnification imaging. Technical Report CUCS-026-95, Dept. of Computer Science, Columbia University, New York, NY, USA, September 1995.
17. R. G. Willson and S. A. Shafer. Modeling and calibration of automated zoom lenses. Technical Report CMU-RI-TR-94-03, The Robotics Institute, Carnegie Mellon University, Pittsburgh, PA, USA, January 1994.
18. Y. Xiong and S. A. Shafer. Moment and hypergeometric filters for high precision computation of focus, stereo and optical flow. Technical Report CMU-RI-TR-94-28, The Robotics Institute, Carnegie Mellon University, Pittsburg, PA, USA, September 1994.

Tracking (3)

Recognition, Pose and Tracking of Modelled Polyhedral Objects by Multi-Ocular Vision

P. Braud J.-T. Lapresté M. Dhome

LASMEA, URA 1793 of the CNRS, Université Blaise Pascal
63177 Aubière Cedex, FRANCE
dhome@le-eva.univ-bpclermont.fr

Abstract. We developped a fully automated algorithmic chain for the tracking of polyhedral objects with no manual intervention. It uses a multi-cameras calibrated system and the 3D model of the observed object.
The initial phase of the tracking is done according to an automatic location process using graph theoretical methods. The originality of the approach resides mainly in the fact that compound structures (triple junction and planar faces with four vertices) are used to construct the graphs describing scene and model. The association graph construction and the search of maximal cliques are greatly simplified in this way. The final solution is selected among the maximal cliques by a prediction-verification scheme.
During the tracking process, it is noticeable that our model based approach does not use triangulation although the basis of the multi-ocular system is available. The knowledge of calibration parameters (extrinsic as well as intrinsic) of the cameras enables to express the various equations related to each images shot in one common reference system. The aim of this paper is to prove that model based methods are not bound to monocular schemes but can be used in various multi-ocular situations in which they can improve the overall robustness.

KeyWords : **3D model - multi-cameras - graph theory - prediction-verification - localisation without triangulation - tracking**

1 Introduction

Model based recognition technics has received a growing attention from many artificial vision researchers; one can consult [CD86] for a detailled state of the art. Various recognition systems exist that can be classified by the kind of primitives used (2D or 3D) to describe the model and extracted from the images. The more frequently used methods are based on the partionning of the parameters space [Ols94], on the traversal of a tree of matches between model and images primitives [Gri89], or on the direct search of compatibles transforms between model and images [HU90].

These various approaches share the property of using an objects models database. Our aim is quite different, as it consists in the computation of the pose of a polyhedral object which is known to pertain to the scene.

The recognition proposed algorithm follows S. Pollard, J. Porrill, J. Mayhew and J. Frisby [PPMF87] and is based on a research of constraints between 3D primitives extracted from the images and of the model. The originality of the approach resides mainly in the fact that compound structures (triple junction and planar faces with four vertices) are used to construct the graphs describing scene and model. The association graph construction and the search of maximal cliques are greatly simplified in this way.

Actually, the paper intends to deal with the more general problem of temporal tracking of polyhedral objects along multi-ocular images sequences. The automatic initial pose estimate of the 3D object allows to begin the tracking process with no manual intervention.

We, thus propose a new approach to localization and tracking in a multi-cameras system: this approach does not involve triangulation. Our aim is to prove that model based methods are not bound to monocular schemes but can be used in various multi-ocular situations in which they can improve the overall robustness.

To solve the general problem of inverse perspective, various searchers have used the model based paradigm in monocular vision [Kan81] [RBPD81] [FB81] [HCLL89] [Bar85] [SK86] [Hor87] [DRLR89] [Rei91] [Low85].

To compensate the loss of information due to the perspective projection, it is necessary to provide a priori knowledges about what is seen. If a model is available it is then necessary to match primitives arising from the images treatments to parts of the model and write down the projection equations relating these primitives. The methods can be classified using the kind of primitives involved (points, segments, elliptical contours, limbs, ...) and the projection chosen to modelize the image formation process (orthographic or perspective projection for example).

When more than one camera are at disposal, generally, a 3D reconstruction of the observed object is made by triangulation. During a tracking process, the reconstructed model is filtered along the images sequence [Aya89]. Few articles, in such cicumstances, assume the model knowledge. M. Anderson and P. Nordlund [AN93] improve incrementally the reconstructed model and match it with the known 3D model to increase the pose accuracy. Our method is similar, but we show that the step of triangulation is not necessary.

This work is based on a previous paper where it was proven that the localization and tracking of a rigid body by monocular vision can be performed using images segments and model ridges as primitives [DDLR93]. The pose estimate was done using a Lowe-like algorithm ([Low85]), coupled with a constant velocity Kalman filter. We had shown the robustness of the process even if the actual velocity was far from the constant model wired in the Kalman filter.

However such a method has limits: object occultations along the sequence can lead to divergence of the tracking. [Gen92] has shown that more than one camera can help if the object is not simultaneously lost by all cameras... but at each iteration he only makes use of one of the images to minimize computation times. We have shown (which is intuitively obvious) in [BDLD94] that using simultaneously several cameras improves pose estimation as well as robustness.

The paper organization is the following: section 2 describe the recognition algorithm. The multi-ocular localization algorithm is the subject of section 3. The tracking is treated in section 4. Finally we present experimental results in section 5 and then conclusions.

2 Recognition

N. Daucher, M. Dhome, J.-T. Lapresté et G. Rives [DDLR93] have developped a method for localization and tracking of polyhedral objects by monocular vision using a surfacic 3D model. As in every tracking system, it is necessary to know the initial pose of the object. Up to now, this knowledge was obtained from a manual match of image segments and model ridges. From a multi-cameras system and also using a surfacic 3D model, we have build a recognition algorithm that can be summarized by the 6 following steps:

- polygonal segmentation of the images (to be used further);
- corner detection in the images from the N cameras at subpixel precision;
- stereo matching by cameras pairs of the detected corners,
- 3D reconstruction of the matched corners using the cameras parameters. At this step we have 3D points and 2D connection between these points obtained from images segments (from the polygonal segmentation);
- 3D matching between these reconstructed corners and the model corners;
- pose computation using the matchings deduced between the images and the model.

We shall not discuss here of the polygonal segmentation which can now be considered as a classic.

2.1 Corners extraction

We detect corners in two phases: coarse and loose then fine and severe.

1. the method of C. Harris et M. Stephens [HS88] has been implemented because it offers a good trade off between accuracy and simplicity. Corners are extracted at pixel accuracy and many false ones are present.
2. the model based method of R. Deriche et T. Blaszka [DB93] is then used to refine or suppress the previously detected corners. After this step, in each image we possess a list of angular points and triple junctions.

2.2 Stereo matching

We thus produce matches between images pairs with the algorithm from Z. Zhang, R. Deriche, O. Faugeras et Q.-T. Luong [ZDFL94]. They have developped a robust matching scheme based on the computation of the epipolar geometry. In fact, in our case we know the epipolar geometry and it is only the final part of their method we will borrow.

2.3 3D reconstruction

If for instance one has a three cameras system, the outputs of the previous step must be treated to combine the pair of image results. It is necessary to group matches sharing one primitive. To disambiguate, we use the trinocular epipolar constraint.

The reconstruction will be obtained by a least square minimization of the distances between all the lines that have been grouped together as defining the same spatial primitive.

Then, an algorithm relying on previous polygonal segmentations is able to etablish "physical" links between the reconstructed points.

2.4 3D matching

The recognition problem consists to establish the list of the matches between vertices of the model and corners of the reconstructed scene. The major drawback of such an approach is the combinatorial explosion. Thus, to limit the complexity, we have chosen to define new elaborated structures both in the scene and the model: triple junctions and four edges planar faces.

From the graph of the model and the graph of the scene constructed from the elaborated structures, we establish the association graph $\mathcal{A}$.

At last, when $\mathcal{A}$ is constructed it remains to find the maximal cliques. Their exists many algorithms [Fau93] [Sko88] which solve this problem, and they are all based on the extension of an initial clique.

2.5 Localization

We are ready know to choose in the cliques list the best one. It is not necessarily the largest one because false matches can still occur. To choose the best solution we use a prediction-verification scheme.

From the matches a multi-ocular localization is performed computing the model pose compatible with the matches and the associated covariance matrix. The verification algorithm tests the exactness of the pose: an automatic matching scheme between image segments and projection of model ridges controled by the Mahalanobis distance is used. Finally we retain the pose which lead to the maximal number of such matches.

3 Localization

3.1 General formulation

The multi-ocular localization algorithm is an extension of a monocular one described in a paper by N. Daucher, M. Dhome, J.-T. Lapresté et G. Rives [DDLR93].

The problem we intend to address here consists to find the extrinsic parameters giving the pose of the viewed object in the acquisition reference system related to the first camera (arbitrarily chosen) and the uncertainties associated to these parameters.

3.2 Criterium definition

We are looking for the 6-dimensional vector $\mathcal{A}_1 = (\alpha_1, \beta_1, \gamma_1, t_{x_1}, t_{y_1}, t_{z_1})^t$ defining the rotation R_1 (by the three Euler angles) and the translation T_1 that minimize the sum of the squared distances of the vertices of the model to the interpretation planes of the images segments.

3.3 Solving the equations

We simply use a classical Newton-Raphson iterative scheme. Starting from an initial guess, to the previous equations we substitute at step k the set of linear equations given by their first order developpement at the current solution. Then we solve the linear system writing down the normal equations.

4 Tracking

4.1 The tracking algorithm

From a multi-cameras system and a 3D object model we have developped a tracking algorithm that can be split in three parts. At each shot k of the images acquisition sequence, we perform succesively:

- An automatic matching between the various images segments and the model ridges. The process is made independantly on each image of the shot looking for informations directly in the grey level data. The search is guided by the prediction of the model pose and its covariance matrix from the previous step.
- The new estimation of the model pose in the global reference system determined from the matchings using the algorithm described in the previous section.
- The prediction of the model pose at the next step of the tracking and the associated covariance matrix obtained by Kalman filtering.

4.2 Automatic matching in grey-level images

At each shot we have the previous pose of the object as well as the predicted new pose. The automatic grey-level match consists in finding independantly for each camera the correspondances between model ridges and image segments.

We have seen previously that a matching could be made using a global polygonal segmentation of the image: this was done in the initialisation phase where no pose estimates were at hand. As the possess the previous pose and the prediction of the new one, we can predict:

- the nature of the grey level transition leading to a segment detection (from the previous image),
- the approximate position in the new image of a comparable signal.

The new location of the segment in the new image will be obtained by correlation with the corresponding signal in the previous image. To be accurate let us say the matching is performed in three phases (independantly on each image):

- We compute the perspective projection of each visible model ridges at computed at step k and predicted for step $k+1$. We also determine with the help of the covariance matrix of the pose parameters, the size of a square window centered on the extremities predicted on the images from step $k+1$.
- In each of these windows we look for the point homologous to the one which is the projection at step k of the same model vertex. This is done by a pyramidal correlation algorithm.
- Between each pair of extremities supposed to belong to the projection of the same model ridge we look for contour points to verify the presence of a ridge projection. The points are extracted by convolution with an ideal model of transition (step or line) choosen from the previous image. Finally the points alignment is tested by a RANSAC paradigm[1] [FB81].

5 Experimentations

5.1 Recognition

The recognition algorithm has been tested on a real images trinocular sequence of a cassette. Figure 1 present images acquired by the three cameras at the beginning of the trinocular sequence. Figure 2 present the final result of the recognition algorithm obtained from this three images. In this sequence, the triangulation basis is weak, so the 3D reconstruction is less accurate. Nevertheless the algorithm converges to the right solution.

In [BDL96], we have shown results of the recognition algorithm obtained from trinocular sequences of others objects.

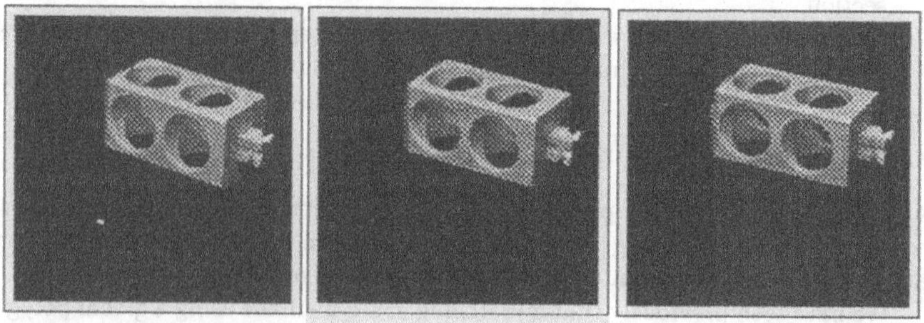

Fig. 1. Left, center and right images: grey level images of the cassette given by cameras #1, #2 and #3 respectively.

[1] RANdom SAmple Consensus

Fig. 2. Final solution obtained by the recognition algorithm on the cassette: superimposition of the grey level image from camera #1 and of the model projection.

5.2 Tracking

During the contract ESTEC #8019/88/NL/PP with the European Spatial Agency, SAGEM provided a trinocular sequence of real images and the CAO model of an "Orbital Replacement Unit" (ORU). The tracking algorithm has been tested on this trinocular sequence. Process has been initiated with the result of the recognition algorithm. The results can be seen in figures 3

In [BDLD94], we have shown by comparing experiments on trinocular and monocular sequences that:

- the multi-ocular approach was more robust,
- The pose computation was more accurate in the multi-cameras case.

In fact, we have shown that the monocular approach was robust if the Kalman filtering had properly modelled the object kinematics. If the object is occulted along some images and the movement has not been properly modelled yet, when the visual information is restored, the pose prediction can be wrong and imply a divergence of the tracking.

Of course the multi-cameras approach is more robust and does not break down in the case of an occultation occuring on one of the cameras. This property could be used in real application to request from some of the cameras a transitory task as (re)calibration.

Speaking of accuracy, the more sensible parameter in the monocular case is obviously z: the depth. Comparing the results, we have found an increased accuracy in trinocular localization.

6 Conclusion

We have developped an algorithm of temporal tracking of polyhedral objects with no human intervention (see [Bra96] for more details). The major lack in existing methods is the initialisation step : we have adressed it.

The initial phase of the tracking is done according to an automatic location process based on a multi-cameras system and using graph theoretical methods.

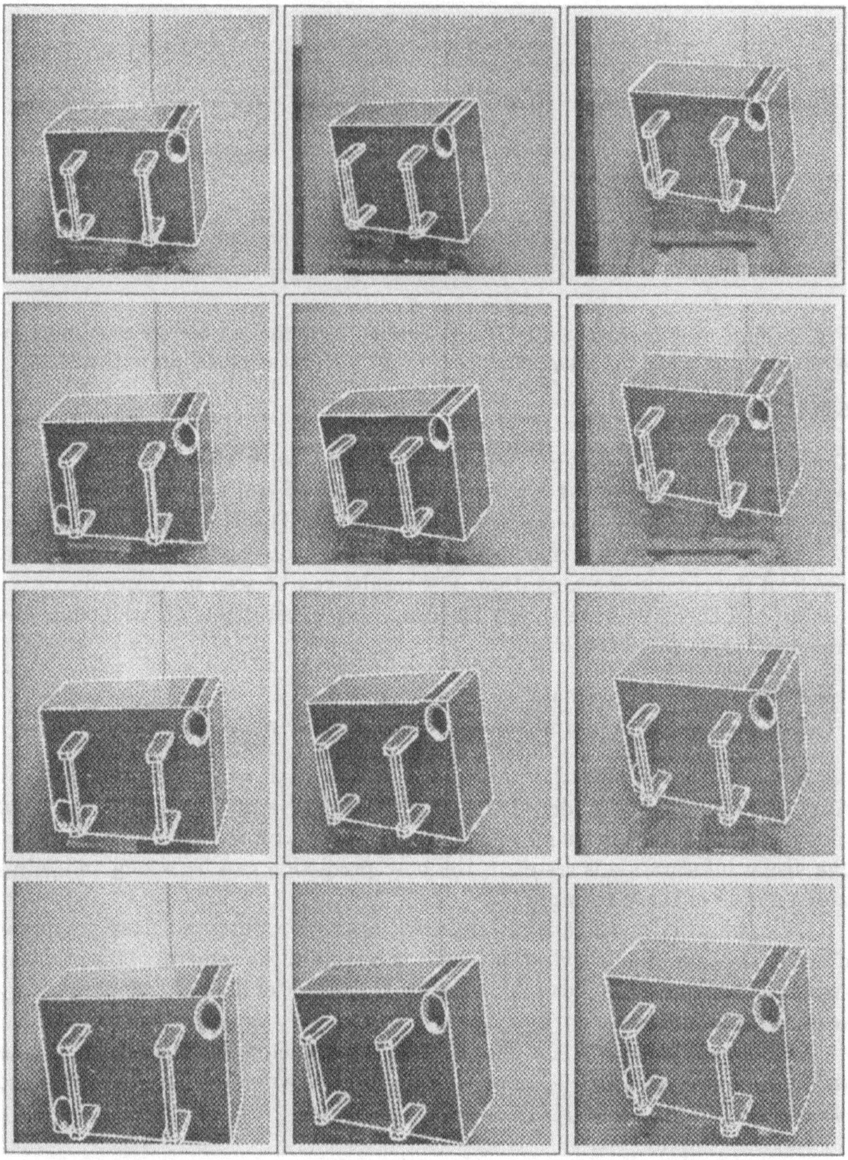

Fig. 3. trinocular tracking of the ORU. Left, center and right are cameras #1, #2 and #3 respectively; from top to bottom : shots 0, 13, 26 and 39 superimposition of the grey level image and the perspective projection of the models ridges.

The originality of the approach resides mainly in the fact that compound structures (triple junction and planar faces with four vertices) are used to construct the graphs describing scene and model. The steps of construction of the association graph and the search of maximal cliques have been greatly simplified in this way.

The algorithm has also proven its robustness facing a poor triangulation basis and thus a less accurate reconstruction.

Moreover we have shown for the tracking process that model based methods that are generally viewed as purely monocular ones can be extended to multioculars by merging data without using the triangulation basis of the system (after the initialisation phase).

Up to now, few articles have treated of the model based approach in a multicamera environment. It appears interesting to have an alternative solution to the stereographic reconstruction. This method allows to disconnect a camera or to assign it to a different task as long as at least one camera goes on providing tracking informations.

References

[AN93] M. Anderson and P. Nordlund. A model based system for localization and tracking. In *Workshop on Computer Vision for Space Applications*, pages 251–260, September 1993.

[Aya89] N. Ayache. *Vision stéréoscopique et perception multisensorielle*. InterEditions, Science Informatique, 1989.

[Bar85] S.T. Barnard. Choosing a basis for perceptual space. *Computer Vision Graphics and Image Processing*, 29(1):87–99, 1985.

[BDL96] P. Braud, M. Dhome, and J.-T. Lapresté. Reconnaissance d'objets polyédriques par vision multi-oculaire. In 10ème *Congrès Reconnaissance des Formes et Intelligence Artificielle*, Rennes, France, January 1996.

[BDLD94] P. Braud, M. Dhome, J.T. Lapresté, and N. Daucher. Modelled object pose estimation and tracking by a multi-cameras system. In *Int. Conf. on Computer Vision and Pattern Recognition*, pages 976–979, Seattle, Washington, June 1994.

[Bra96] P. Braud. *Reconnaissance, localisation et suivi d'objets polyédriques modélisés par vision multi-oculaire*. PhD thesis, Université Blaise Pascal de Clermont-Ferrand, France, January 1996.

[CD86] R.T. Chin and C.R. Dyer. Model-based recognition in robot vision. *ACM Computing Surveys*, 18(1):67–108, March 1986.

[DB93] R. Deriche and T. Blaszka. Recovering and characterizing image features using an efficient model based approach. In *Int. Conf. on Computer Vision and Pattern Recognition*, pages 530–535, New-York, June 1993.

[DDLR93] N. Daucher, M. Dhome, J.T. Lapresté, and G. Rives. Modelled object pose estimation and tracking by monocular vision. In *British Machine Vision Conference*, volume 1, pages 249–258, October 1993.

[DRLR89] M. Dhome, M. Richetin, J.T. Lapresté, and G. Rives. Determination of the attitude of 3d objets from a single perspective image. *IEEE Trans. on Pattern Analysis and Machine Intelligence*, 11(12):1265–1278, December 1989.

[Fau93] O.D. Faugeras. *Three Dimensional Computer Vision : A Geometric View-Point*, chapter 12. MIT Press, Boston, 1993.

[FB81] M.A. Fischler and R.C. Bolles. Random sample consensus: A paradigm for model fitting with applications to image analysis and automated cartography. *Com. of the ACM*, 24(6):381–395, June 1981.

[Gen92] D.B. Gennery. Tracking of known three-dimensional objects. *International Journal of Computer Vision*, 7(3):243–270, April 1992.

[Gri89] W.E.L. Grimson. On the recognition of parameterized 2d objects. *International Journal of Computer Vision*, 2(4):353–372, 1989.

[HCLL89] R. Horaud, B. Conio, O. Leboulleux, and B. Lacoll. An analytic solution for the perspective 4 points problem. In *Computer Vision Graphics and Image Processing*, 1989.

[Hor87] R. Horaud. New methods for matching 3-d objects with single perspective view. *IEEE Trans. on Pattern Analysis and Machine Intelligence*, 9(3):401–412, May 1987.

[HS88] C.G. Harris and M. Stephens. A combined corner and edge detector. In *Fourth Alvey Vision Conference*, pages 189–192, Manchester, United Kingdom, August 1988.

[HU90] D.P. Huttenlocher and S. Ullman. Recognizing solid objects by alignment with an image. *International Journal of Computer Vision*, 5(2):195–212, 1990.

[Kan81] T. Kanade. Recovery of the three dimensional shape of an object from a single view. *Artificial Intelligence, Special Volume on Computer Vision*, 17(1-3), August 1981.

[Low85] D.G. Lowe. *Perceptual Organization and Visual Recognition*, chapter 7. Kluwer Academic Publishers, Boston, 1985.

[Ols94] C.F. Olson. Time and space efficient pose clustering. In *Int. Conf. on Computer Vision and Pattern Recognition*, pages 251–258, Seattle, Washington, June 1994.

[PPMF87] S.B. Pollard, J. Porrill, J. Mayhew, and J. Frisby. Matching geometrical descriptions in three space. *Image and Vision Computing*, 5(2):73–78, 1987.

[RBPD81] P. Rives, P. Bouthemy, B. Prasada, and E. Dubois. Recovering the orientation and position of a rigid body in space from a single view. Technical report, INRS-Télécommunications, 3, place du commerce, Ile-des-soeurs, Verdun, H3E 1H6, Quebec, Canada, 1981.

[Rei91] P. Reis. *Vision Monoculaire pour la Navigation d'un Robot Mobile dans un Univers Partiellement Modélisé*. PhD thesis, Université Blaise Pascal de Clermont-Ferrand, March 1991.

[SK86] T. Shakunaga and H. Kaneko. Perspective angle transform and its application to 3-d configuration recovery. In *Int. Conf. on Computer Vision and Pattern Recognition*, pages 594–601, Miami Beach, Florida, June 1986.

[Sko88] T. Skordas. *Mise en correspondance et reconstruction stéréo utilisant une description structurelle des images*. PhD thesis, Institut National Polytechnique de Grenoble, October 1988.

[ZDFL94] Z. Zhang, R. Deriche, O. Faugeras, and Q.-T. Luong. A robust technique for matching two uncalibrated images through the recovery of the unknown epipolar geometry. Technical Report 2273, INRIA Sophia-Antipolis, France, May 1994.

Locating Objects of Varying Shape Using Statistical Feature Detectors

T.F. Cootes and C.J.Taylor

Department of Medical Biophysics, University of Manchester, Oxford Road, Manchester. M13 9PT, UK
email: bim,ctaylor@sv1.smb.man.ac.uk

Absract. Most deformable models use a local optimisation scheme to locate their targets in images, and require a 'good enough' starting point. This paper describes an approach for generating such starting points automatically given no prior knowledge of the pose of the target(s) in the image. It relies upon choosing a suitable set of features, candidates for which can be found in the image. Hypotheses are formed from sets of candidates, and their plausibility tested using the statistics of their relative positions and orientations. The most plausible are used as the initial position of an Active Shape Model, which can then accurately locate the target object. The approach is demonstrated for two different image interpretation problems.

1 Introduction

Image search using deformable models has been shown to be an effective approach for interpreting images of objects whose shape can vary [1,2,3]. Usually the object of interest is located by some form of local optimisation so a 'good enough' starting approximation is required. Such starting points are either supplied by user interaction or obtained in some application–specific manner. We wish to develop a system which can automate the generation of such starting points for a general class of models. This paper proposes a framework for generating hypotheses for all plausible instances of variably shaped objects in a scene. The approach is to determine a set of key features, use statistical feature detectors to locate all examples of these in a scene and then to generate a ranked list of all plausible combinations of features. We systematically consider all possible sets of features, ranking or eliminating each by considering the statistics of the relative positions and orientation of feature points using statistical shape models [4]. Missing features are dealt with, allowing robustness to occlusion. The best feature sets are then used to instantiate a deformable model known as an Active Shape Model [3,4] which can be run to locate the full structure. This leads to a generally applicable method, which can locate multiple instances of a target in a given scene.

In the following we will describe the approach in more detail. We will consider related approaches and describe the statistical feature detectors we use. We will then cover the statistical shape models and show how they can be used to determine how plausible a set of features is. Finally we will show the method working on real data and give results of systematic experiments.

2 Background

Many people have studied the problem of locating rigid objects in images. A review is given in Grimson [5]. In most approaches an image is preprocessed to locate features such as edges or corners, and the best matches of these to a model are located by a suitably pruned tree search. In general the simpler the features, the more possible matches there are with the model and the more expensive the resulting combinatorial explosion. The Local Feature Focus method of Bolles and Cain [6] attempts to deal with this problem both by finding sets of more distinctive features and by choosing sub–sets of model features which are sufficient to identify and position the model. Work by Ashbrook *et al* [7] on Geometric Histograms is an example of an attempt to reduce the combinatorial explosion by making the feature models more detailed, and thus to generate fewer responses for each detector (usually at the expense of greater cost of locating the features).

Where the objects of interest can vary in shape, deformable models such as the 'snakes' of Kass *et al* [1], the finite element models of Pentland and Sclaroff [2] or Active Shape Models [3] have proved useful. However such models are usually used in a local optimisation schemes, requiring a suitable initial position to be provided. Most of the methods proposed for locating rigid objects rely on tight constraints between the positions and orientations of features, which are violated when the objects can deform. Loosening the constraints in simple ways often leads to a combinatorial explosion [5]. More complex models of the constraints are required.

Yow and Cipolla [8] describe a system for locating faces which uses a gaussian derivative filter to locate candidates for the lines of the eyes, nostrils and mouth, then uses a belief network to model the face shape and select a set of features most likely to be those of the face.

Burl *et al* [9] combine feature detectors with statistical shape models to generate a set of plausible configurations for facial features. Their (orientation independent) feature detectors match template responses with the output of multi–scale gaussian derivative filters. They use distributions of shape statistics given by Dryden and Mardia [10] to test configurations of outputs from detectors which locate possible locations of the eyes and nostrils. They build up hypothesis sets by first considering all pairs of features. From each pair they determine the regions in which they would expect candidates for other features to lie, and consider all sets of points which lie in these regions. They allow for missing features in their derivations, to give robustness to feature detector failure.

Our general approach is similar to that of Burl *et al*. However, we use more complex statistical feature detectors in order to minimise the number of false positive responses. Our detectors are orientation dependent, making them more discriminating. Although they must be run at multiple angles, they return both the position and orientation of the found features. We use a simpler method of calculating the shape statistics to test sets of points, and include the feature orientation in our tests. We systematically calculate all plausible sets of features using a depth first tree search, pruning sub–trees as soon as they become implausible. Our aim is to rapidly determine sets good enough to initialise an Active Shape Model to locate the object of interest accurately.

3 Overview Of Approach

The approach we use is as follows. We assume that we have sets of training images, in which points are labelled on the the objects of interest. In advance we determine a sub–set of points which can be used as features to detect the objects of interest, by considering the performance of feature detectors trained at every point. We train feature detectors for the chosen points, and build statistical models of both the shape of the whole model and that of the sub–set of feature points.

Given a new image, we proceed as follows:

- We find all responses for the feature detectors
- We systematically search through all these responses to determine all plausible sets of features. How plausible a set is is determined by the relative positions of the candidate features and their relative orientations
- We fit the full shape model to the best sub–set, and use this as the starting point for an Active Shape Model, which can then accurately locate the object of interest.

4 Statistical Feature Detectors

In order to locate features in new images we use statistical models of the grey levels in regions around the features, and use a coarse-to-fine search strategy to find all plausible instances in a new image [14].

4.1 Statistical Feature Models

We wish to locate accurately examples of a given feature in a new image. To deal with variations in appearance of features we use statistical models derived from a set of training examples. These are simply rectangular regions of image containing instances of the feature of interest. For each n_x x n_y patch we sample the image at pixel intervals to obtain an $n = n_x n_y$ element vector, **g**.

A statistical representation of the grey-levels is built from a set of s example patches, $\mathbf{g}_i$ ($i = 1..s$). A Principle Component Analysis is applied to obtain the mean, $\bar{\mathbf{g}}$, and t principle modes of variation represented by the n x t matrix of eigenvectors, $\mathbf{Q}$. The value of t is chosen so that the model represents a suitable proportion of the variation in the training set (eg 95%) [4].

Our statistical model of the data is

$$\mathbf{g} = \bar{\mathbf{g}} + \mathbf{Qc} + \mathbf{r}_g \tag{1}$$

where the elements of **c** are zero mean gaussian with variance λ_i, the elements of **r** are zero mean gaussian with variance v_j and the columns of $\mathbf{Q}$ are mutually orthogonal. This is the form of a Factor Model [12], with **c** as the common factors and $\mathbf{r}_g$ as the errors.

Given a new example, **g**, we wish to test how well it fits to the model. We define two quality of fit measures as follows,

$$f_1 = M_t + \frac{R^2}{V_r} \tag{2}$$

$$f_2 = M_t + \sum_{j=1}^{j=n} \frac{r_j^2}{v_j} \tag{3}$$

where

$$M_t = \sum_{i=1}^{i=t} \frac{c_i^2}{\lambda_i} \qquad \mathbf{c} = \mathbf{Q}^T(\mathbf{g} - \bar{\mathbf{g}}) \tag{4}$$

$$R^2 = \mathbf{r}_g^T \mathbf{r}_g = (\mathbf{g} - \bar{\mathbf{g}})^T(\mathbf{g} - \bar{\mathbf{g}}) - \mathbf{c}^T\mathbf{c} \qquad \mathbf{r}_g = \mathbf{g} - (\bar{\mathbf{g}} + \mathbf{Qc}) \tag{5}$$

In [14] we show that the distribution of these fit values is a scaled chi-squared distribution of degree k, $p(f) = (n/k)\mathrm{X}^2(kf/n, k)$ where k is the number of degrees of freedom of the pixel intensities. k can be estimated from a verification set of examples; $k = 2(n/\sigma)^2$, where σ is the standard deviation of the distribution of fit values across the set.

The time to calculate f_1 is about half that for f_2, but gives slightly less predictable distributions and poorer discrimination between true positive and false positive responses. Knowledge of $p(f)$ allows us to set thresholds which will produce predictable numbers of false negatives (missed true features). More details of the statistical feature models are given in [14].

4.2 Properties of the Statistical Feature Detectors

The quality of fit of the feature models to an image is sensitive to position, orientation and scale. By systematically displacing the models from their true positions on the training set we can quantify this sensitivity. This information allows us to calculate the number of different scales and orientations at which the detector should be run in order to cover a given range of object orientation and scales.

4.3 Searching for Features

Given a new image we wish to locate all plausible instances of a given feature. We train feature models at several levels of a gaussian pyramid [13] and determine their sensitivity to angle and scale. We then use a coarse–to–fine search strategy as follows;

- Test every pixel of a coarse resolution image with the matching feature model at a set of angles and scales. Determine the peaks in response which pass the statistical threshold.
- Refine the position and angle estimates of each response on finer resolution images. The accuracy in angle required is determined by the (pre-computed) sensitivity of the model.

Those which pass the threshold at the finest resolution are candidate features. We can choose suitable resolutions at which to perform the search by performing tests on the training or verification set. The total work required to run the detector starting and ending at different resolution levels can be calculated. We choose those ranges of levels which can accurately relocate the true features, require least work and generate the fewest false positive responses.

4.4 Choice of Features

We wish to be able to select automatically a set of features suitable for locating the object of interest. The suitability of a given feature depends on a number of factors including the success rate of the feature detector, the number of false positives it tends to generate and the time it takes to run the detector over a given region.
For the experiments described below we have used training sets labelled with many points on and around structures of interest, which we use for building statistical shape models [3]. In order to obtain a set of suitable features we build models for every shape model point and test each one. By running each over the training set we can estimate the success rate, the number of false positives the detector generates and the computational cost for that detector. Those detectors which have the highest success rate (hopefully 100% on the training set) are ranked by the number of false positives.
For instance, Figure 1 shows the 40 best features for detectors trained on 122 landmark points on 40 face examples. Figure 2 shows the 8 best features for detectors trained on points on the outline of the brain stem in a 2D slice of a 3D MR image of the brain.

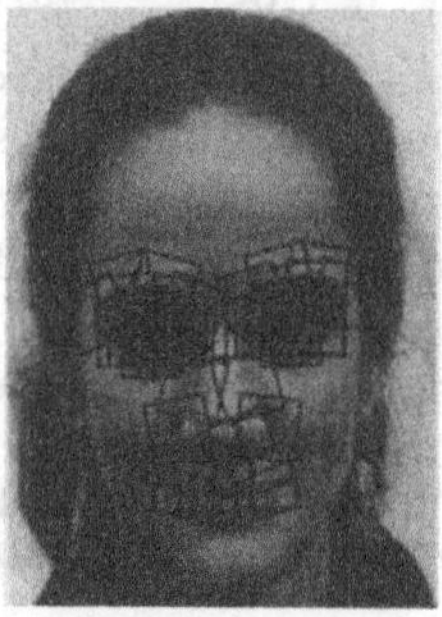

Fig. 1. The 40 best features for locating a face. (Tested 122 points)

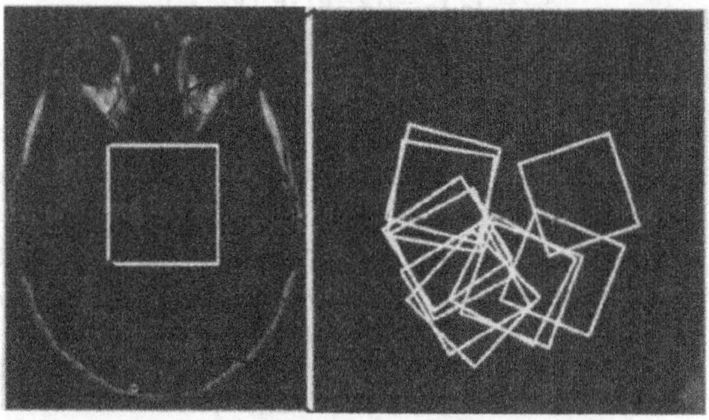

Fig. 2. The 10 best features for locating the brain stem in an MR image. (Tested 60 points).

We are experimenting with algorithms for automatically deciding on sets of features, but at present the sets are chosen by the user, guided by these rankings and the desire to have the features reasonably spread out over the target object.

5 Use of Statistical Shape Models to Test Hypotheses

When presented with a new image we apply the selected feature detectors over target regions to generate a number of candidates for each feature. By choosing one candidate for each feature we can generate a hypothesis for sets of features belonging to the same object. We will use statistical shape models to determine the plausible sets by considering the relative positions of the features and the orientations at which each was detected.

5.1 Statistical Shape Models

We have previously described methods for building statistical shape models. Given a training set of shapes, each representing n labelled points, we can find the mean configuration (shape) and the way the points tend to vary from the mean [3,4]. The approach is to align each example set into a common reference frame, represent the points as a vector of ordinates in this frame and apply a PCA to the data. We can use the same formulation as for the grey-level models above,

$$\mathbf{x} = \bar{\mathbf{x}} + \mathbf{Pb} + \mathbf{r} \tag{6}$$

where $\mathbf{x} = (x_1 \ldots x_n\, y_1 \ldots y_n)^T$, $\mathbf{P}$ is a $2n$ x t matrix of eigenvectors and $\mathbf{r}$ is a set of residuals whose variance is determined by miss–one–out experiments. In this case t is the number of *shape* parameters required to explain say 95% of the shape variation in the training set.
Again, the quality of fit measure for a new shape is given by

$$f_{shape} = \sum_{i=1}^{t} \frac{b_i^2}{\lambda_i} + \sum_{j=1}^{j=2n} \frac{r_j^2}{v_j} \qquad \mathbf{b} = \mathbf{P}^T(\mathbf{x} - \bar{\mathbf{x}}) \qquad \mathbf{r} = \mathbf{x} - (\bar{\mathbf{x}} + \mathbf{Pb}) \tag{7}$$

Which should be distributed approximately as chi–squared of degree $2n-4$.
In the case of missing points, we can reformulate this test using weights (1.0 for point present, 0.0 for point missing);

$$f_{shape} = \sum_{i=1}^{t} \frac{b_i^2}{\lambda_i} + \sum_{j=1}^{j=2n} w_i \frac{r_j^2}{v_j} \tag{8}$$

where in this case $\mathbf{b}$ is obtained as the solution to the linear equation

$$(\mathbf{P}^T\mathbf{W})(\mathbf{x} - \bar{\mathbf{x}}) = (\mathbf{P}^T\mathbf{WP})\mathbf{b} \tag{9}$$

($\mathbf{W}$ is a diagonal weight matrix).
This measure will be distributed as chi–squared of degree $2n_v - 4$ where n_v is the number of points present.

5.2 Models of the Feature Sets

Our features represent a sub–set of the points making up the full shape model for the object of interest. For each such sub–set we can generate statistical models of the configurations of the feature positions as described above. For instance for the face model we choose features at four of the 122 points of the full model and build statistical models both of the whole set and of the four points. Each shape model has its own co–ordinate frame (usually centred on the centre of gravity of the points and with some suitably normalised scale and orientation [3]).
To test the validity of a set of image points forming a shape, $\mathbf{X}$, we must calculate the shape parameters $\mathbf{b}$ and the pose Q (mapping from model frame to image) which minimise the distance of the transformed model points, $\mathbf{X}'$, to the target points

$$\mathbf{X} \approx \mathbf{X}' = Q(\bar{\mathbf{x}} + \mathbf{Pb}) \tag{10}$$

(Q is a 2D Euclidean transformation with four parameters, t_x, t_y , s and Θ.)

This is a straightforward minimisation problem [4,3]. Having solved for Q and **b** we can project the points into the model frame using Q^{-1} and calculate the residual terms and hence the quality of fit, f_{shape} . We can test the plausibility of the shape probabalistic limits both to the overall quality of fit f_{shape} and, if desired, to the individual shape parameters b_i . The latter have zero mean and a variance of λ_i , the eigenvalues obtained from the PCA.

By considering the training set we can calculate the average mapping between the co-ordinate frame for the full model and that for a sub-set of points. This allows us to propagate any known constraints on the pose of the whole object model to test the pose of the sub-set of points representing the current feature set. In addition, we can learn the expected orientation and scale of each feature relative to the scale and orientation of the set as a whole, allowing further discrimination tests.

If we assume that the configuration of the sets is independent of the errors in feature orientation and scale we can estimate the probability density for a configuration as follows;

$$p = p(shape)\prod_{i=1}^{i=n_f}[p(\theta_i)p(s_i)] \tag{11}$$

where the probabilities for shape, angle and scale terms are determined from the estimated distributions. If we assume normal distributions for the measured orientations and scales, then

$$ln(p) = const + \frac{f_{shape}}{2} + \sum_{i=1}^{i=n_f}[\frac{(a_i - \bar{a}_i)}{2\sigma_{ai}^2} + \frac{(s_i - \bar{s}_i)^2}{2\sigma_{si}^2}] \tag{12}$$

This allows us to sort any plausible hypotheses by their estimated probability.

5.3 Systematic Hypothesis Generation and Testing

If we have n_f feature detectors, and detector i produces m_i candidates, then there are $\prod m_i$ possible sets. If we allow for the detectors missing true features, then there are $\prod(m_i + 1)$ possible sets (allowing a wildcard feature match). Selecting plausible sets of features given this potential combinatorial explosion has received much attention [5,9].

We have used a relatively simple scheme amounting to a depth first tree search. The feature candidates are sorted by quality of fit and, if missing features are to be allowed, a wildcard is added. We then recursively construct sets, starting with a candidate from the first detector and adding each candidate from the second detector in turn. The pose and shape of each pair is tested. If a pair passes the tests, each candidate from the third detector is added and the three points tested. Those sets which pass are extended with candidates from the fourth detector and so on. In this manner all possible plausible sets can be generated fairly efficiently. (This approach has the advantage that it can be implemented in a recursive algorithm in a small number of lines of code). We record all the sets which have at least three valid features and pass the statistical tests.

Burl *et al* [9] calculate a probability for each set of candidates which takes into account missing features. We feel that it is difficult to correctly assign probabilities for missing features and instead simply sort our hypotheses first by the number of features present, and secondly by their probability. This avoids comparing sets with different numbers of features directly. In practice it is those which have the fewest missing which tend to be the correct responses.

5.4 Verification of Plausible Feature Sets

Given a plausible set of features, we find the least-squares fit of the full object shape model to these points. This can be achieved by solving a weighted version of (10), with zero weights for all but the points corresponding to the found features. This gives the starting point for an Active Shape Model. We can run the ASM to convergence, using a multi-resolution search scheme to locate all the points [11]. The ASM has grey-level models of what it expects in the region around every one of its points. By considering the quality of fit ot these models to the image after convergence we can determine whether a good example of the object of interest has been found. Where there are several equally plausible sets of features, the ASM can be run for each and the one with the best final fit accepted. To detect multiple instances of the model the best examples which do not overlap in pose space should be accepted.

6 Results of Experiments

6.1 Performance of Hypothesis Tester

We have performed experiments to study how well the hypothesis testing scheme works. We generated synthetic feature detector responses by taking one true set of 5 feature positions and adding varying numbers of random responses to each candidate list. We then ran the hypothesis tester, recording the number of tests required to find all plausible feature sets. We tried this both allowing and not allowing missing features, and either using or not using information on the orientation of the features. Each experiment was performed 10 times. Figure 3 summarises the results for the number of tests required. Allowing missing features increases the number of tests significantly, but including angle information reduces the number by almost as much. Our current implementation tests about 500 sets each second. Of course, ordering the features so that those with the fewest responses are tested first can significantly reduce the work required. The number of plausible sets resulting varied between 1 (the correct one) and 3 (with >20 false responses) for each case except for the case of missing features with no angle information, which produced many spurious plausible sets when the number of false responses was large.

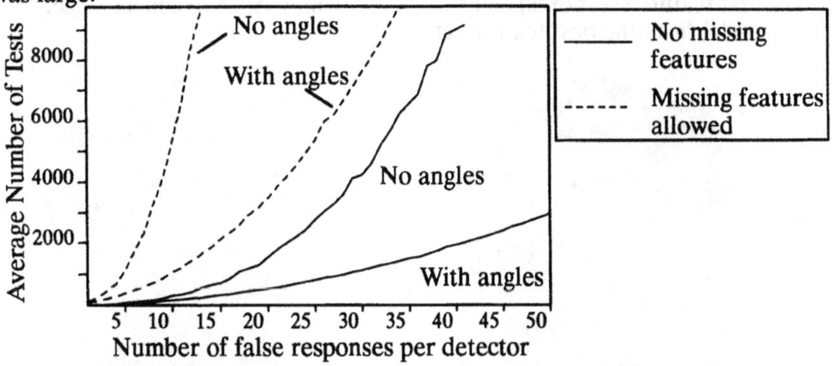

Fig. 3. Average number of tests required to find all plausible sets of candidates, given varying numbers of false responses for each feature detector.

6.2 Locating Facial Features

We have used the system to locate features as a way of initialising a face shape model which can be refined with ASM search. The full model is trained on 122 points marked in each of 40 images (a subset of those used by Lanitis *et al* [15]). Four features based around the eyes and nose were chosen from the set of most distinctive points shown above (Figure 1). Figure 4 shows an new face image, the positions of the detected features, the best set of such features and the position of the full 122 point shape model determined from this set. Figure 5 shows the points after running an ASM to convergence from this starting position. We assumed that we knew the approximate scale, but that the orientation and position were unknown. Each feature detector was run over the whole image and allowed any orientation. It took about 10 seconds on a Sun Sparc20 to run each feature detector over the 512^2 images, then one to two seconds to consider all plausible sets of features. In a real face detection system the orientation is likely to be better constrained, but the features would have to be allowed to vary in scale. Figures 6 and 7 show results for a different image, in which one of the feature detectors has failed to locate a satisfactory candidate. The quality of the full model fit to the found features is worse, but still quite adequate for the ASM to converge to a good solution. This demonstrates the robustness to missing features (and thus to occlusions).

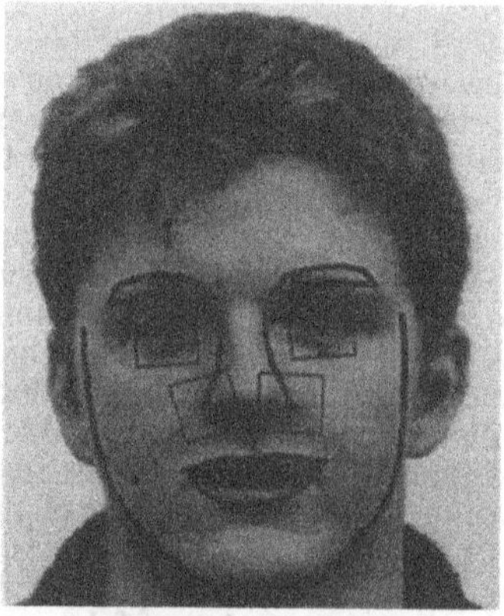

Fig. 4. Candidate features (crosses), best feature set (boxes) and full model fit to the best feature set.

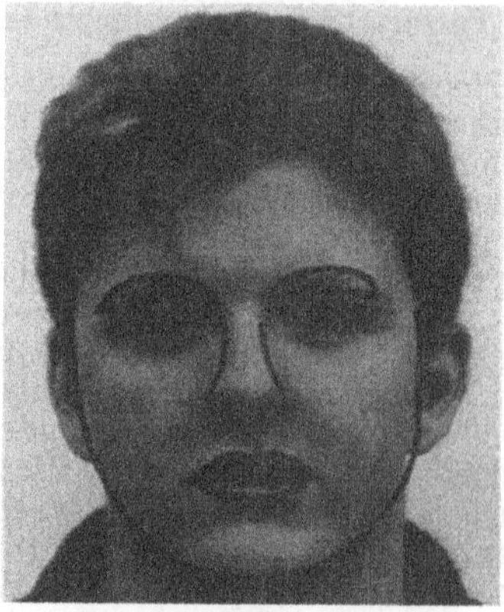

Fig. 5. Full model after running ASM to convergence.

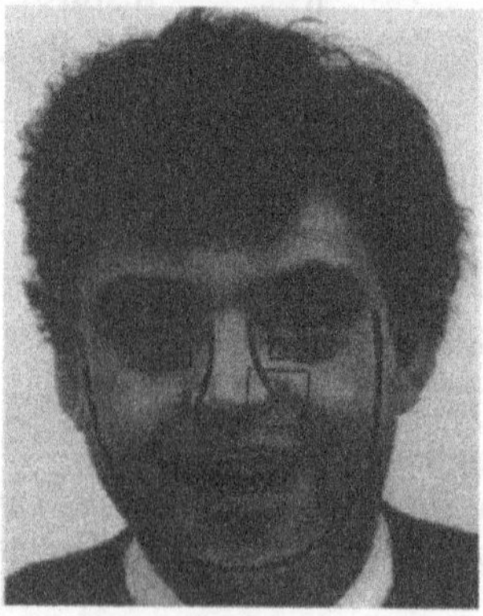

Fig. 6. Candidate features (crosses), best feature set (boxes) (only 3 of 4 found) and full model fit to the best feature set.

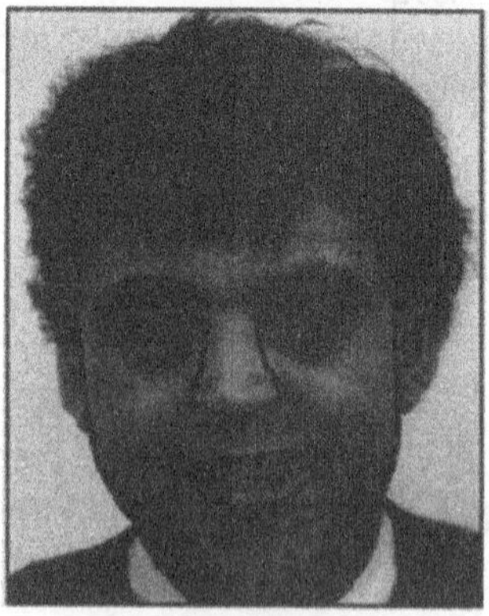

Fig. 7. Full model after running ASM to convergence.

To test the performance more systematically we ran the system on a test set of 40 different images, which had been marked up with the target point positions by hand. On 5 of the images it failed to find any plausible sets, due to multiple feature detector failure. On one image false positives conspired to give a plausible (but wrong) result. On the other 35 the best set gave a good fit. The mean distance between the (known) target points and the full set of points estimated from the best features was 8.7 pixels. The mean error after running the ASM was 6.5 pixels. On average 3.5 of the 4 features were used, and each feature detector found 12 candidates in the image. By more careful choice of size of feature detectors, and by using more detectors (giving more robustness) and a larger training set we expect to improve the results significantly.

6.3 Locating the Brain Stem in MR Slices

We used the same approach to locate the outline of the brain stem in 2D MR slices. The full outline is represented by 60 points, and models trained from those points marked in 15 images. Five features were chosen from the set of the most distinctive points shown in Figure 2. Figure 8 shows candidates for these 5 in a new image, the best set of features and the full model fitted to these features. Figure 9 shows the points after running an ASM to convergence from this starting position. Again we assumed the scale was fixed but that the position and orientation of the brain stem were unknown. It took about 5 seconds on a Sun Sparc20 to run each feature detector over the 256^2 images, then a fraction of a second to consider all plausible sets of features. Of course in practice the position and orientation of the structure is fairly well constrained, so the system would run far quicker. In cases in which some of the feature detectors fail to locate satisfactory candidates the quality of the full model fit to the found features is worse, but again still quite adequate for the ASM to converge to a good solution.

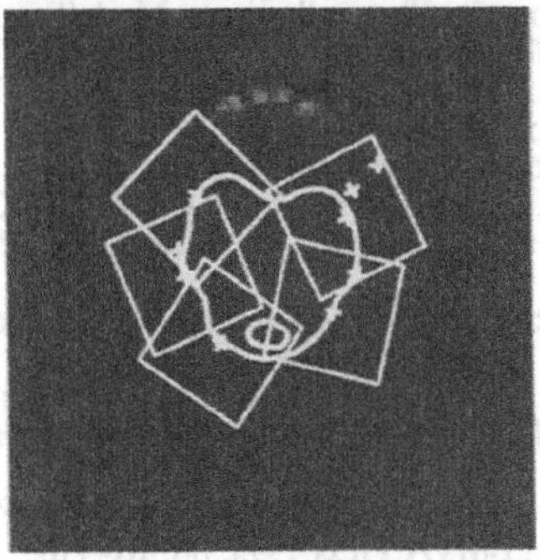

Fig. 8. Candidate features (crosses), best feature set (boxes) and full model fit to the best feature set.

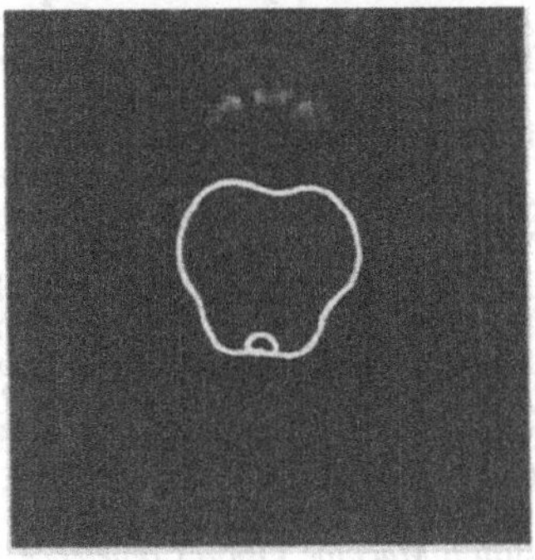

Fig. 9. Full model after running the ASM to convergence.

Again we ran systematic tests over a set of 15 new images. The system failed to find any plausible feature sets on 2 of them, but gave good results on the remaining 13. The mean difference between target points and the full set of points estimated from the best features was 1.5 pixels, falling to 1.1 pixels after the ASM was run to convergence. On average 3.8 of the 5 features were used, and each feature detector found 3 candidates in the image. We expect to improve on these results by using a larger training set and more careful choice of feature model size.

7 Discussion

Although we have used one particular form of statistical feature detector, any approach which located the features of interest could be used.

The calculation of plausible feature sets can take a long time if a large number of candidates are to be tested. We are currently interested in generating all plausible hypotheses. However, since the candidates are sorted by quality of fit, we usually find the best overall set quite early in the search. We intend to investigate early verification strategies, terminating the combinatorial search when we find a 'good enough' solution to explain the data.

8 Conclusions

We have shown how statistical feature detectors can be used to find good starting positions for deformable models, given no prior information on the position or orientation of the object of interest in the image. We used statistical models of the relative positions and orientations of the detected features to determine the plausible sets of features and to limit the possible combinatorial explosion by pruning bad sets as soon as possible. The plausible feature sets can be used to instantiate a statistical shape model representing the whole of the object of interest, which can then be refined using an Active Shape Model. This approach can locate multiple instances of the objects of interest in an image, and can be applied to a wide variety of image interpretation problems.

Acknowledgements

T.F.Cootes is funded by an EPSRC Advanced Fellowship Award.
The face images were gathered and annotated by A. Lanitis. The brain stem images were gathered by C. Hutchinson and his team. They were annotated by A.Hill.

References

1 M. Kass, A. Witkin and D. Terzopoulos , Snakes: Active Contour Models, in *Proc. First International Conference on Computer Vision*, pp 259-268 IEEE Comp. Society Press, 1987.
2 A. Pentland and S. Sclaroff, Closed-Form Solutions for Physically Based Modelling and Recognition, *IEEE Trans. on Pattern Analysis and Machine Intelligence*. **13**, 1991, 715-729.
3. T.F.Cootes, C.J.Taylor, D.H.Cooper and J.Graham, Active Shape Models - Their Training and Application. *CVIU* Vol. 61, No. 1, 1995. pp.38-59.
4. T.F.Cootes, A.Hill, C.J.Taylor, J.Haslam, The Use of Active Shape Models for Locating Structures in Medical Images. *Image and Vision Computing* Vol.12, No.6 1994, 355-366.
5. W.E.L. Grimson. Object Recognition By Computer : The Role of Geometric Constraints. Pub. MIT Press, Cambridge MA, 1990.
6. R.C. Bolles and R.A. Cain, Recognising and locating partially visible objects : The Local Feature Focus Method. *Int. J. Robotics Res.* **1**(3) 1982, pp. 57-82.
7. A.P.Ashbrook, N.A. Thacker, P.I. Rockett and C.I.Brown. Robust Recognition of Scaled Shapes using Pairwise Geometric Histograms. *in Proc. British Machine Vision Conference*, (Ed. D.Pycock) BMVA Press 1995, pp.503-512.
8. K.C.Yow, R.Cipolla, Towards an Automatic Human Face Localisation System. *in Proc. British Machine Vision Conference*, (Ed. D.Pycock) BMVA Press 1995, pp.701-710.
9. M.C.Burl, T.K.Leung, P.Perona. Face Localization via Shape Statistics. Proc. Int. Workshop on Automatic Face- and Gesture-Recognition (ed. M.Bichsel), 1995. pp. 154-159
10. I.L.Dryden and K.V.Mardia, General Shape Distributions in a Plane. *Adv. Appl. Prob.* 23, 1991. pp.259-276.
11. T.F.Cootes , C.J.Taylor, A.Lanitis, Active Shape Models : Evaluation of a Multi-Resolution Method for Improving Image Search, *in Proc. British Machine Vision Conference*, (Ed. E.Hancock) BMVA Press 1994, pp.327-338.
12. R.A.Johnson, D.W. Wichern, Applied Multivariate Statistical Analysis. *Prentice-Hall* 1982.
13. P.J.Burt, The Pyramid as a Structure for Efficient Computation. in *Multi-Resolution Image Processing and Analysis*. Ed. Rosenfield, pub. Springer-Verlag. 1984. pp. 6 - 37.
14. T.F.Cootes , G.J.Page, C.B.Jackson, C.J.Taylor, Statistical Grey-Level Models for Object Location and Identification, *in Proc. British Machine Vision Conference*, (Ed. D.Pycock) BMVA Press 1995, pp.533-542. (Also submitted to IVC)
15. A.Lanitis, C.J.Taylor and T.F.Cootes, A Unified Approach to Coding and Interpretting Face Images. Proc. 5th ICCV, IEEE Comp. Soc. Press, 1995, pp. 368-373.

Generation of Semantic Regions from Image Sequences

Jonathan H. Fernyhough, Anthony G. Cohn and David C. Hogg

Division of Artificial Intelligence, School of Computer Studies
University of Leeds, Leeds, LS2 9JT
{jfern,agc,dch}@scs.leeds.ac.uk

Abstract. The simultaneous interpretation of object behaviour from real world image sequences is a highly desirable goal in machine vision. Although this is rather a sophisticated task, one method for reducing the complexity in stylized domains is to provide a context specific spatial model of that domain. Such a model of space is particularly useful when considering spatial event detection where the location of an object could indicate the behaviour of that object within the domain. To date, this approach has suffered the drawback of having to generate the spatial representation by hand for each new domain. A method is described, complete with experimental results, for automatically generating a region based context specific model of space for *strongly* stylized domains from the movement of objects within that domain.
Keywords: spatial representation, scene understanding.

1 Introduction

Event recognition provides a significant challenge for high-level vision systems and explains the impetus behind the work described in this paper. Nagel (1988) outlines several previous applications that connect a vision system to a natural language system to provide retrospective descriptions of analysed image sequences. Typically the vision system is used to provide a 'geometric scene description' (GSD) containing a complete description of the spatial structure within the domain (i.e. the area in view of the camera) and the spatial coordinates of the objects in the scene at each instance of time. A generic event model (Neumann & Novak 1983), characterizing a spatio-temporal representation for that event, can be matched against the GSD in order to recognize instances of that event which can then be expressed in natural language.

More recent work demonstrates a simultaneous analysis of image sequences to provide the incremental recognition of events within a football game (Retz-Schmidt 1988, André, Herzog & Rist 1988). This enables the system to provide a running commentary of the actions within the domain including perceived intentions. A model of the world representing the static background of the scene is supplied manually so that the system can recognize situated events, for example realizing the difference between passing the ball and attempting to score a goal.

Although not necessary for *all* event recognition tasks, a spatial model providing a context specific representation of the domain is certainly beneficial. In strongly stylized domains, such as road traffic environments where vehicles' movements are governed by strict constraints, a spatial model containing semantic information would allow the interpretation of object behaviour from the sequenced position of objects within the domain, for example areas where vehicles turn or where pedestrians cross the road. Fig. 1 shows an example to illustrate how a context specific region based model of space can be used to facilitate the recognition of a vehicle waiting to turn right. The region occupied by the vehicle in fig. 1*b* is an area of behavioural significance representing the location where vehicles must await oncoming traffic before turning right.

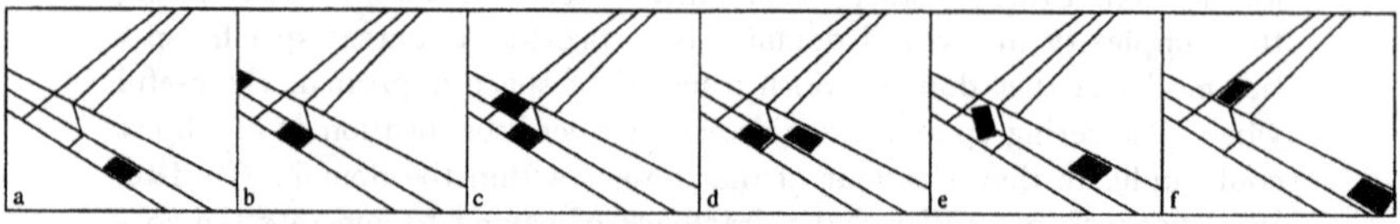

Fig. 1. A simplified spatial model of a road junction showing a sequence of object locations. A vehicle approaches a junction (*a*), reaches it (*b*) and then awaits oncoming traffic (*c & d*) before turning right into the new road (*e & f*).

Howarth & Buxton (1992) introduced such a spatial model for spatial event detection in the domain of traffic surveillance. This representation of space is a *hierarchical* structure based on *regions,* where a region is a spatial primitive defined as a (closed) two-dimensional area of space with the spatial extent of a region controlled by the continuity of some property.

There are two kinds of regions:

- *Leaf regions* are the finest granularity of region. They are areas of space that tile the entire scene and do not overlap. Leaf regions are used to structure space and are completely defined by how *composite regions* overlap.
- Concatenations of adjacent leaf regions form *composite regions* expressing areas sharing the same significance, for example region types (i.e. roads and footpaths) and regions with similar behavioural significance (i.e. give-way zones.) It is possible for different composite regions to share leaf regions (i.e. they may overlap) providing the hierarchical structure to the spatial layout. In terms of the domain context, a composite region represents the area described by the movement of objects within the domain (i.e. a *path.*)

Howarth (1994) produced such representations of space manually for each new domain: a time consuming and painstaking process. This paper demonstrates a method to generate such a spatial structure automatically for strongly stylized domains through the monitoring of object movement over extended periods.

Li-Qun, Young & Hogg (1992) describe a related method of constructing a model of a road junction from the trajectories of moving vehicles. However, this deals only with straight road lanes and is unable to handle the fine granularity of region required for a detailed behavioural analysis – such as regions where a vehicle turns left. The method described here is not limited in this way.

Johnson & Hogg (1995) demonstrates a related approach in which the distribution of (partial) trajectories in a scene is modelled automatically by observing long image sequences.

2 Outline of the Method

The system accepts live video images from a static camera to produce shape descriptions corresponding to moving objects within the scene. This dynamic scene data is then analysed, in real-time, to build a database of paths used by the objects, before being further processed to generate the regions required for the spatial model. A diagram outlining this system is shown in fig. 2.

There are three main stages:

- A *tracking* process obtains shape descriptions of moving objects (§3).
- *Path generation* builds a model corresponding to the course taken by moving objects and subsequently updates the database of paths (§4).
- *Region generation* accesses the database of paths so that leaf and composite regions can be constructed for the spatial model within the domain (§5).

Latter sections will provide implementation details and results for the test domains as well as a discussion of future applications.

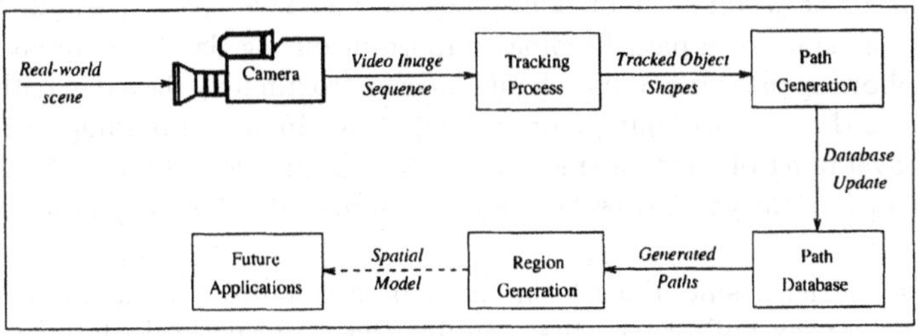

Fig. 2. Overview of the system

3 Tracking

The first step in automatically generating the spatial representation is the analysis of *dynamic scene* data. Visual information is provided through live video images from a static camera. The current test domains include an elevated view of a busy junction containing both pedestrians and vehicles (fig. 5*a*) as well as a predominantly pedestrian scene (fig. 5*b*). A list of objects is provided on a frame by frame basis using the tracking process described in Baumberg & Hogg (1994*b*). A combination of background subtraction, blurring and thresholding is used to obtain object silhouettes for each frame. The outline of each silhouette is then described by a number of uniformly spaced control points for a closed cubic B-spline and assigned a label by considering object size and proximity in the previous frame. Although this method does not handle occlusion and is not particularly robust, it provides sufficient information for our purposes and it proves significantly faster than the active shape model described in Baumberg & Hogg (1994*a*).

4 Path Generation

A *path* is defined as the course that an object takes through the domain. More specifically, a path is represented by all pixels covered by the object along its course through the domain. To enable real-time processing from the tracking output and to reduce storage requirements, a list of active paths is maintained from frame to frame. With each new frame, the latest location of each object is combined with its respective existing active path.

Object location can be taken just from the outline of the object provided from the tracking process. However, these outlines are subject to various forms of noise. In particular, light reflections can alter the object silhouette dramatically (fig. 3*a*.) When combined, such object locations may produce a jagged path (fig. 3*b*).

Under ideal conditions, an object moving along a straight line trajectory will produce a convex path (except possibly at the ends) and although an object with a curved trajectory will obviously not have a convex path it will be 'locally convex'. The state of a path becomes important during database update – two objects following the same course should have approximately the same path which may not be the case without preprocessing them. Image smoothing techniques (such as averaging or median smoothing) enhance the condition of the path by filling in some of the gaps. However they are, in real-time terms, computationally expensive.

Instead of using smoothing techniques, path condition is enhanced by generating the convex hull of the object outline (fig. 3*c*.) Such calculations are *not* computationally expensive – the convex hull of any polygon can be found in linear time, $O(n)$ (see Melkman 1987). Although Baumberg & Hogg's (1994*b*) tracking program supplies a cubic B-spline representation of the object outlines, it is relatively easy to convert them to a polygonal representation (Sonka, Hlavac & Boyle 1993, Chapter 6.2.5, pp. 212–214).

The convex hulls combine to give a significantly smoother path (fig. 3*d*,) that is more likely to be correctly matched during database update.

Once an active path becomes complete it is merged into the database of existing paths. There are two possibilities when merging a new path into the database :

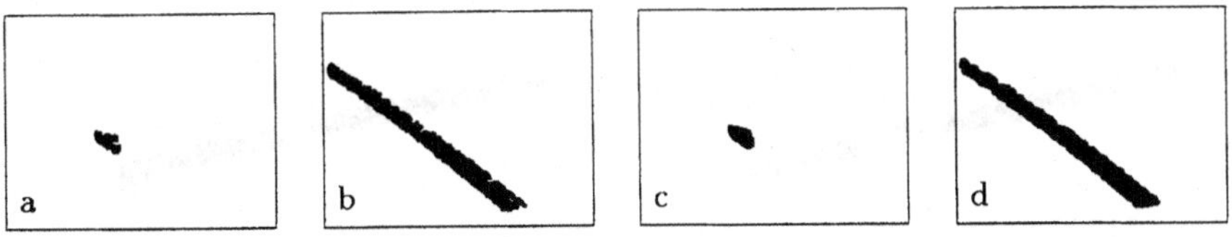

Fig. 3. (a) Object outline, (b) Path generated using object outline, (c) Convex hull of object outline, (d) Path generated using convex hull of object outline.

- an equivalent path already exists and should be updated to accommodate the new path.
- no equivalent path is found and the new path should be allocated a unique identity.

Equivalence is based on the percentage overlap between the new path and the paths contained within the database. *Path overlap* occurs when the constituent pixels of two paths coincide. Two paths are considered to be equivalent if a specified proportion of their paths overlap. When the specified percentage overlap is too low it is possible that two different paths will be found equivalent – for example, two adjacent road lanes may be matched and seen as just one wide lane. Alternatively, if the overlap is too high there may be no equivalences identified within a satisfactory time scale. Experimental results within the test domains have shown that a tolerable compromise appears to be an overlap of 80% – this allows a sufficient duration for the training period without undesirable equivalences being identified. Of course, this value is scene specific and will be discussed more in $ 6.

When updating the database, a new path could be merged with an existing database path using a function analogous to the binary *or* operation – the value of each pixel representing a database path would indicate if any equivalent path has occupied that pixel. However, the update function used is analgous to arithmentic *addition* – allowing the value of each pixel for a database path to indicate the number of equivalent paths sharing that pixel.

At the end of the training period, each path held in the database will contain frequency distribution information for that path, fig. 4*a*. This representation has two benefits :

- "noise" can easily be identified from low distribution areas.
- it is possible to extract the most "common" path by thresholding the distribution, fig. 4*b*.

5 Region Generation

At *any* time during the training period it is possible to generate regions for the spatial model. Effectively this halts the database generation process (although it may be resumed) and uses that information to build the regions. A new region

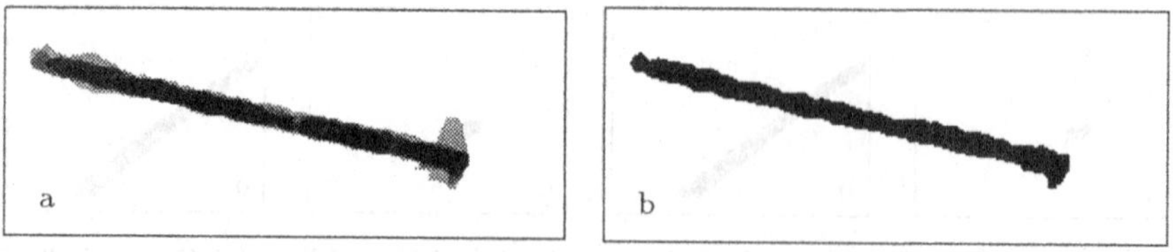

Fig. 4. (a) Path displaying a grey scale representation of the frequency distribution (b) Path obtained from most "common" usage

model can be created during the path generation stage each time a path becomes complete and is merged into the database. However, it is unclear how useful this continuous region generation may be as the model is constantly changing and the most recent state is unlikely to have any connection to the previous state.

When regions are generated only as required, path verification may also be accomplished. Each database path is tested against all other paths in the database to verify that no path equivalences have been created through the database update process – the merging of equivalent paths may alter the original shape enough that a previously unmatched path may now be found equivalent. Should any 'new' equivalences be discovered they are merged together as before.

Although this step is not entirely necessary, it has the advantage that a previously "weak" path may be strengthened by a 'new' equivalence. Without this operation, such paths will be strengthened with extra training – essentially, this step allows a shorter training period and as such provides an advantage over continous region generation.

To reduce "noise", any path with a uniformly low frequency distribution is discarded. Although low frequency distribution may represent infrequent object movement rather than "noise", it is also possible that abnormal or unusual behaviour is being displayed. In some applications this information may be useful; however, the method described in this paper relies on behavioural evidence and it is safe to reject these paths as they are not statistically frequent enough.

The remaining paths are then processed to obtain a binary representation of the 'best' or most 'common' route used – this depends on the database path update function being 'addition' rather than 'or' (see previous section). Thresholding is used to provide a binaray representation where the threshold is selected from the *cumulative* frequency histogram of each database path and the percentage overlap value employed in the test for path equivalence. An 80% overlap value is required to merge a path into the database and indicates the percentage of pixels shared by equivalent paths. This is reflected in the cumulative frequency histogram where the 'common' path forms the highest 80% of the histogram. So, the frequency value found at 20% of the histogram provides the value for the threshold operation.

These binary path representations express the composite regions for the spatial model – they describe each area of similar behavioural significance from objects following the same course through the domain. From §1, the leaf regions can be completely defined by how the binary path representations overlap. Each binary path is allocated a unique identification before being added to the region map. Overlapping segments form separate leaf regions and are reassigned a new unique identification. When all the paths have been processed each *leaf region* will have been identified and labelled.

Occasionally, adjacent paths may share small areas of common ground – perhaps from shadows or the occasional large vehicle. This can generate very small regions that are not actually useful and the last step in leaf region generation is to remove such small regions by merging them with an adjacent region – selected by considering the smoothness of the merge. Smoothness is checked by considering the boundary of the small region and the proportion shared with the adjacent leaf regions. The adjacent region sharing the highest proportion of the small region's boundary is selected for the merge, e.g. if the small region has a border length of seven pixels and shares five with region A and only two with region B, the combination with region B would form a 'spike' whereas region A may have a 'local concavity' filled and subsequently be smoother. Fig. 5 displays the leaf regions obtained for the test domains.

To complete the spatial model, it is necessary to discover the union of leaf regions which make up each composite region (based on the binary representations of the database paths.) A complication in this process results from the previous merge of small "useless" regions which may now be part of a larger leaf region that should not be a member of the composite region for the path under consideration. Each composite region should contain only those leaf regions that are completely overlapped by the path it represents. A selection of composite regions is displayed in fig. 5 along with the identified leaf regions.

6 Experimental Results

The tracking program processes about 5 frames/second on a regular UNIX platform. The video image sequence used for the traffic junction is about 10 minutes in length and averages 5 or 6 objects each frame. In comparison, the pedestrian scene is roughly double the length with at most 3 objects in any frame and often with periods of no object movement.

At the end of the training period the traffic junction has entered 200 paths into the database which reduces to 70 after checking for equivalences. Of these paths, 28 prove frequent enough to be used in region generation so giving 28 composite regions and initially over 400 leaf regions. The removal of small regions reduces this number to around 150. After only 2 minutes, many of the significant routes have already been identified with 16 paths strong enough to be considered composite regions and generating a total of 87 leaf regions. For the pedestrian scene about 120 leaf regions are generated from 23 recognized paths.

These results rely on three threshold paramaters we were unable to eliminate from the system. Thresholds remain necessary for the overlap value in the path equivalence test, the actual threshold operation and the size of leaf regions that are to be merged into an adjacent region. As previously indicated, the overlap

Fig. 5. (a) Road junction and (b) pedestrian scene displaying identified leaf regions along with a selection of composite regions.

value for path equivalence and the threshold operation are linked – one being the dual of the other. Experimental results indicated that an overlap value of 80% was suitable for both test domains. It is possible that the percentage overlap value is related to the camera angle for the scene. As the angle is reduced, objects in adjacent lanes will naturally overlap more. This means that when attempting the path equivalence test a higher overlap percentage value will be needed to distinguish equivalent paths from those that are actually adjacent lanes. The value used to determine small regions is passed on from the tracking program – here the minimum tracked object size is 10 pixels otherwise problems can arise. Ten pixels is less than 0.02 percent of the total image area size which does not seem too excessive.

7 Further Work / Work in Progress

The process as described is real-time as far as the training period is concerned and is able to generate the regions at any time during the training sequence. However, once generated the spatial model becomes a static entity which may cause problems in a changing world. For instance, if the model is used for traffic surveillance and road works subsequently alter traffic flow, the spatial model becomes inaccurate. In such situations it is desirable to have an adaptive model of space that is able to learn during use. Such an adaptive model may also prove useful for robot navigation with a non-static camera where the domain is constantly changing and in need of updating. It should be possible to enhance the method described here to provide an adaptive model of space.

The driving force behind the development of this technique to automatically generate semantic regions for a scene was the desire to provide a spatial model to assist event recognition procedures. Dynamic scene analysis has traditionally been quantitative and typically generates large amounts of temporally evolving data. Qualitative reasoning methods (Zimmermann & Freksa 1993, Cui, Cohn & Randell 1992) would be able to provide a more manageable way of handling this data if a formal framework for the given situation exists. The spatial model described here, being topologically based, is able to provide such a qualitative formal framework. This allows generic event models to be provided using qualitative logic descriptions. Typically such generic event models are provided as part of the *a priori*. However, it is our intention to provide a method to determine these event models automatically from a statistical analysis of object location, movement and the relationships to other objects.

This representation of space could provide control for a tracking process by reducing the search space for moving objects – the spatial representation contains the paths followed by objects. The spatial model could also identify the potential location of new objects in the scene, again reducing the search space.

Other possible areas where such a spatial layout could be used are stereo image matching and fusing of multiple overlapping views. The topology of the spatial model is largely invariant to small changes in the viewing angle and provides sets of corresponding regions.

8 Conclusion

By using an existing tracking program that produces (2D) shape descriptions for tracked objects from a real image sequence, we have demonstrated an effective method for the real-time generation of a context specific model of a (2D) area of space. The domain is required to be strictly stylized for this method to be suitable; for example in the traffic surveillance domain there is typically a constrained set of possibilities for the movement of vehicles. This may not be the case for less stylized domains like the movement of fish in a pond. However, the extent of such stylized domains is sufficiently frequent for the method to be widely applicable.

The spatial model can be considered to be "data-centered" due to its construction from real image data. This means that an alternative tracking application could be used that provides object outlines projected onto the ground plane rather than the image plane to produce a spatial model representing a ground plane projection of the viewed scene which could prove useful[1]. Howarth & Buxton (1992) use a ground plane projection of the image plane to "better facilitate reasoning about vehicle interactions, positions and shape." Similarly, by using a tracking process that provides 3D shape descriptions the method would require relatively few changes to provide a complete 3D spatial model.

Previous contextually relevant spatial models have been generated by hand and as a consequence the domain is subject to human interpretation and occasionally misconception so the generated spatial model may not be entirely accurate. Our method relies only on observed behavioural evidence to describe the spatial model. As long as a sufficiently broad representation of object behaviour occurs throughout the training period the derived spatial model should be accurate without being prone to any misconceptions.

[1] A ground plane projection could also be obtained by back projection of the derived spatial model.

Statistical analysis allows the most used routes to be extracted from the database. This means that the length of the training period depends on the volume of object movement as well as representative object behaviour – for a quiet scene, a much longer image sequence will be necessary than with a busy scene. As long as the image sequence is of a sufficient length and demonstrates typical behaviour it is possible to obtain a reasonable representation of a (2D) area of space that is contextually relevant to the viewed scene.

References

André, E., Herzog, G. & Rist, T. (1988), On the simultaneous interpretation of real world image sequences and their natural language description: The system soccer, *in* 'Proc. ECAI-88', Munich, pp. 449–454.

Baumberg, A. M. & Hogg, D. C. (1994*a*), An efficient method for contour tracking using active shape models, *in* 'IEEE Workshop on Motion of Non-rigid and Articulated Objects', I.C.S. Press, pp. 194–199.

Baumberg, A. M. & Hogg, D. C. (1994*b*), Learning flexible models from image sequences, *in* 'European Conference on Computer Vision', Vol. 1, pp. 299–308.

Cui, Z., Cohn, A. & Randell, D. (1992), Qualitative simulation based on a logical formalism of space and time, *in* 'Proceedings of AAAI-92', AAAI Press, Menlo Park, California, pp. 679–684.

Howarth, R. J. (1994), Spatial Representation and Control for a Surveillance System, PhD thesis, Queen Mary and Westfield College, The University of London.

Howarth, R. J. & Buxton, H. (1992), 'An analogical representation of space and time', *Image and Vision Computing* **10**(7), 467–478.

Johnson, N. & Hogg, D. (1995), Learning the distribution of object trajectories for event recognition, *in* D. Pycock, ed., 'Proceedings of the 6th British Machine Vision Conference', Vol. 2, BMVA, University of Birmingham, Birmingham, pp. 583–592.

Li-Qun, X., Young, D. & Hogg, D. (1992), Building a model of a road junction using moving vehicle information, *in* D. Hogg, ed., 'Proceedings of the British Machine Vision Conference', Springer-Verlag, London, pp. 443–452.

Melkman, A. V. (1987), 'On-line construction of the convex hull of a simple polyline', *Information Processing Letters* **25**(1), 11–12.

Nagel, H. H. (1988), 'From image sequences towards conceptual descriptions', *Image and Vision Computing* **6**(2), 59–74.

Neumann, B. & Novak, H.-J. (1983), Event models for recognitions and natural language description of events in real-world image sequences, *in* 'Proceedings of the Eigth IJCAI Conference', pp. 724–726.

Retz-Schmidt, G. (1988), A replai of soccer: Recognizing intentions in the domain of soccer games, *in* 'Proc. ECAI-88', Pitman, Munich, pp. 455–457.

Sonka, M., Hlavac, V. & Boyle, R. (1993), *Image Processing, Analysis and Machine Vision*, Chapman & Hall.

Zimmermann, K. & Freksa, C. (1993), Enhancing spatial reasoning by the concept of motion, *in* A. Sloman, D. Hogg, G. Humphreys, A. Ramsay & D. Partridge, eds, 'Prospects for Artificial Intelligence', IOS Press, pp. 140–147.

Tracking of Occluded Vehicles in Traffic Scenes

Thomas Frank[1], Michael Haag[1], Henner Kollnig[1], and Hans-Hellmut Nagel[1,2]

[1]Institut für Algorithmen und Kognitive Systeme
Fakultät für Informatik der Universität Karlsruhe (TH)
Postfach 6980, D-76128 Karlsruhe, Germany

[2]Fraunhofer-Institut für Informations- und Datenverarbeitung (IITB),
Fraunhoferstr. 1, D-76131 Karlsruhe, Germany
Telephone +49 (721) 6091-210 (Fax -413), E-Mail hhn@iitb.fhg.de

Abstract. Vehicles on downtown roads can be occluded by other vehicles or by stationary scene components such as traffic lights or road signs. After having recorded such a scene by a video camera, we noticed that the occlusion may disturb the detection and tracking of vehicles by previous versions of our computer vision approach. In this contribution we demonstrate how our image sequence analysis system can be improved by an explicit model–based recognition of 3D *occlusion* situations. Results obtained from real world image sequences recording gas station traffic as well as inner-city intersection traffic are presented.

1 Introduction

Occlusion causes truncated image features and, therefore, may mislead the object recognition process. In the sensitivity analysis of [Du *et al.* 93] occlusion is, therefore, treated as a source of error. [Du *et al.* 93] compare truncated image features with features which are contaminated by image noise or clutter. But unlike for the two latter effects, the analysis of the image sensing process is not adequate in order to cope with occlusion, nor is a pure 2D picture domain analysis. We model occlusion, therefore, in the 3D scene domain.

Our investigations are illustrated by an image sequence recording gas station traffic where intrinsically many occlusions occur, and a second one showing the traffic at a much frequented inner–city intersection.

2 Related Work

Related publications by other groups have been surveyed in [Sullivan 92; Koller *et al.* 93; Sullivan *et al.* 95; Cédras & Shah 95]. Inner-city traffic scenes are evaluated by [Koller *et al.* 93; Meyer & Bouthemy 94; Sullivan *et al.* 95]. Rather than repeating the bulk of this material, we concentrate on a small set of selected publications. In our previously published system (see [Koller *et al.* 93]), initial model hypotheses are generated using information extracted from optical flow fields in order to initialize a Kalman Filter tracking process. By projecting a hypothesized 3D polyhedral vehicle model into the image plane, 2D model edge segments are obtained which are matched to straight-line edge segments, so called data segments, extracted from the image. In [Koller *et al.* 94] the camera is mounted above the road on a bridge, looking down along the driving direction of the vehicles. Exploiting this special camera pose, [Koller *et al.* 94] can decide about partial occlusion of vehicles by comparing the vertical image coordinates of the candidate regions for moving vehicles.

[Meyer & Bouthemy 94] present a purely picture domain approach, without requiring three-dimensional models for the vehicles or for their motion. By comparing a predicted image region with a measured image region, [Meyer & Bouthemy 94] detect occlusions. In contrast to this geometric approach, [Toal & Buxton 92] used spatio-temporal reasoning to analyze occlusion behavior. In this latter approach, temporarily occluded vehicles are correctly relabeled after re-emerging from behind the occluding object rather than being treated as completely independent vehicles. So far it has been assumed by these authors, however, that vehicles had already been tracked and classified prior to the onset of occlusion.

[Sullivan 92] documents an approach which tracks one vehicle that is occluded by a lamp post — which is part of his scene model — as well as by another car which is tracked simultaneously. In contrast to his approach, we do not match edge segments but match synthetic model gradients to gray value gradients. The investigations of Baker, Sullivan, and coworkers are discussed in more detail in [Sullivan *et al.* 95] who report recent improvements of an important sub-step in the Reading approach, namely refinement of vehicle pose hypotheses.

[Dubuisson & Jain 95] present a 2D approach, where vehicles are detected by means of combining image subtraction and color segmentation techniques. In distinction to our approach, their 2D vehicle model requires images showing a side view of the vehicles.

3 Scene Model

We focus our investigations for the moment on the evaluation of the gas station scene depicted in Figure 1 where many occlusions by static scene objects occur. It appears not to be easy to track vehicle images properly, if significant parts of those images are occluded by stationary scene components. In this case, the matching process, which tries to minimize the difference between synthetic model gradients and the actual gray value gradient of the image at the estimated position of the vehicle, will associate at least parts of the model gradient field of the occluded vehicle to the gradient of the image of the occluding stationary scene object. Figure 3 (a) shows an enlarged section containing a vehicle which moves from right to left on the front lane of the gas station. Due to the occlusion by the advertising post and the bush, the superimposed vehicle model falls back compared to the actual position of the vehicle image. In this case, the synthetic gradient of the vehicle model has been associated with the gray value gradient of the stationary post.

In order to take potential occlusions caused by *static* scene components into account, we first had to extend a previous 2D road model to a 3D scene model of the depicted gas station.

The static objects which can cause occlusion of moving cars are represented by generic 3D surface models (see Figure 1). We thus need only one prototype model for each class of scene component, e.g., petrol pumps or bushes. The actual scene object is then represented by an instantiation of this generic model. Each model can be described by its corners (specified by length parameters relative to a fixed object coordinate system) and its surfaces (specified by the corners). The object coordinates have to be transformed to the common world coordinate system. With exception of the gas station roof which is slanted, the remaining 3D objects need only to be translated. The roof model is rotated and translated.

All mentioned 3D scene objects including parts of the gas station building are projected into the image, see Figure 1.

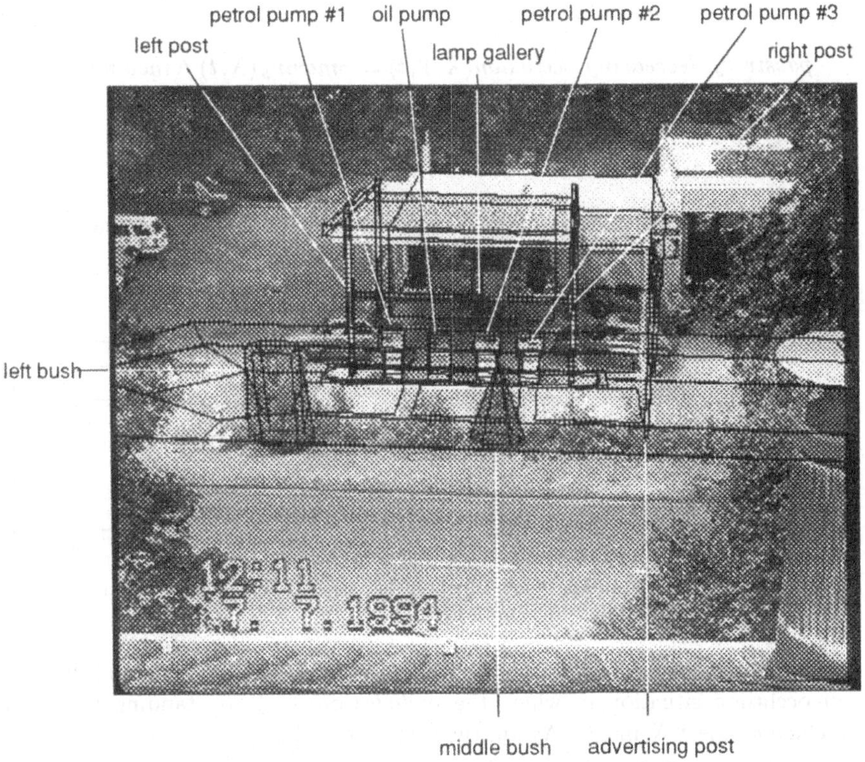

Fig. 1.: Frame #3605 of the gas station image with superimposed 3D scene model as well as road model.

By means of a pixel–oriented raytracer procedure, the distance of each modeled surface point from the projection center is computed, yielding a depth map of the observed scene. With this information, we are able to take the occluded parts of the car into consideration during our tracking process.

4 Occlusion Characterization

If we were able to *predict* the temporal development of a considered occlusion between two objects, we could exploit this knowledge in order to stabilize the tracking process. Therefore, we began to classify possibly occurring occlusions. This yielded occlusion situations which we arranged in a transition diagram in order to obtain all potential transitions between the occlusion situations due to the movements of the concerned objects. In order not to exceed the scope of this contribution, we demonstrate the approach with only a single occlusion predicate, namely *passively_decreasing_occlusion(X, Y,t)* which expresses the fact that at timepoint t the moving object Y *is occluded* by the stationary object X, so that the occlusion decreases with respect to time (see Figure 2):

$$passively\_decreasing\_occlusion(X,Y,t) = \neg moving(X,t) \land moving(Y,t) \land occlusion(X,Y,t)$$

The predicate *moving(X,t)* implies a non-zero velocity of Object X and *occlusion(X, Y,t)* holds if object X occludes object Y.

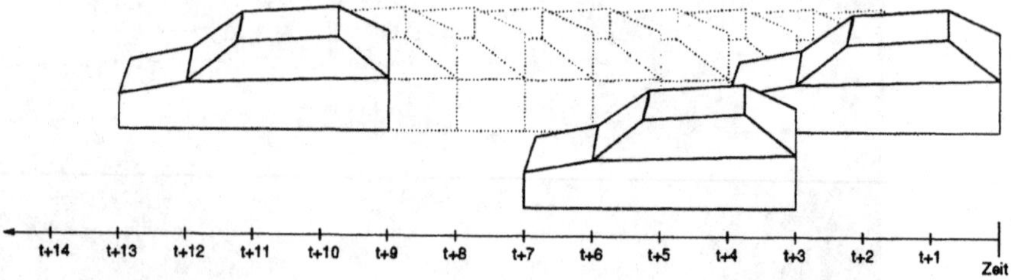

Fig. 2.: Object X is standing while object Y is passing. The occlusion decreases.

Similarly, the predicate actively_decreasing_occlusion(X,Y,t) would describe an occlusion situation in which the *occluded* object Y is standing while the *occluding* object X moves. Again, the occlusion will decrease.

Altogether, we found seven different occlusion predicates (composed out of four primitives) and 15 transition predicates which control the transitions between these occlusion situations. The (primitive and composed) predicates are modelled by means of *fuzzy sets* in order to abstract from quantitative details like *velocity*.

5 Results

5.1 Gas Station Image Sequence

Our first experiments focus on the gas station scene (see Figure 1). Without modeling the occurring occlusions, we had difficulties in tracking such vehicle images properly. For instance, the former tracking process lost the image of the vehicle shown in Figure 3 (a) due to the occlusion by the advertising post. After we introduced an explicit 3D model of the static scene components and excluded the occluded parts of the vehicle image from the matching process, we were able to track this vehicle without any problems (see Figure 3 (b)).

(a) (b)

Fig. 3.: (a) Enlarged section showing a vehicle driving on the front lane of the gas station. The vehicle has been occluded by the advertising post on the right hand side and, therefore, could not be tracked properly by the tracking process. (b) Same vehicle as in (a), but now with the superimposed trajectory obtained by our automatic tracking process which took modeled 3D scene components, like the advertising post, into account.

A vehicle which is even more severely occluded is shown in Figure 4 (a). This car drives on the back lane of the gas station from the right to the left. On its way to the petrol pumps, it is occluded by the advertising post, by the right post carrying the roof, by two petrol pumps, by the oil pump, and by the vehicle driving on the front lane. The superimposed trajectory and the vehicle model in Figure 4 (b) show that the tracking process succeeds once we take the occluding scene components into account.

5.2 Downtown Image Sequence

In a second experiment, our approach is illustrated by a test image sequence of a much frequented multi-lane inner-city street intersection (see Figure 5).

Fig. 4.: (a) Enlarged section of frame #314 of the gas station image sequence. Due to the occlusion, the tracking process associates parts of the vehicle model gradients to the stationary post and looses the moving vehicle image. (b) Enlarged section of frame #463 with superimposed trajectory for the vehicle driving on the back lane. Although the vehicle is heavily occluded, it can be tracked properly after the 3D models of the relevant scene components have been taken into account.

Fig. 5.: First frame of an image sequence recorded at an inner-city intersection.

The vehicle image contained in the marked rectangle in Figure 5 covers only 35×15 pixels. Even a human observer has problems to classify it as a saloon or a fast-back. In our experiments it turned out that the vehicle could not be correctly tracked by optimizing the size and shape of the vehicle model (Figure 6 (a)). After considering the occlusion by the street post in the pose estimation process, we were able to track this vehicle, see Figure 6 (b).

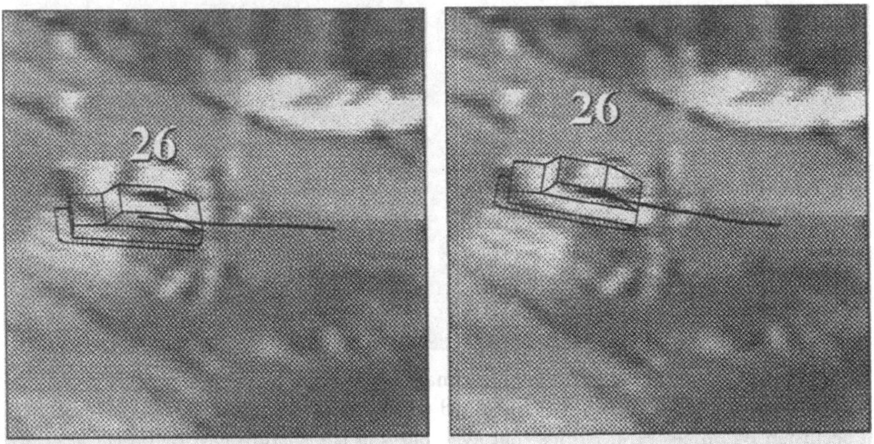

(a) (b)

Fig. 6.: Enlarged section of the upper right quadrant of the image sequence depicted by its first frame in Figure 5 showing the tracking results of object #26 at halfframe #195. (a) The traffic light post which occludes the vehicle is *not* considered in the Kalman-Filter update step: the tracking fails. (b) As soon as the occlusion is modeled the vehicle is correctly tracked.

Figure 8 (a) shows the results of the tracking process applied to the bright transporter (object #12) which is driving side by side with the truck. Since the transporter is permanently occluded by the big truck while it is turning left, the tracking fails. In this case, the occlusion is caused by a *dynamic* scene component, namely the moving truck. Therefore, we could not determine an a–priori scene–model as in the case of the gas station scene. In this case, we first computed automatically the trajectory of the big truck where no occlusion problems arise. After this, we knew the estimated position and orientation of the truck for each time instant. We thus were able to determine the occluded parts of the image of the bright transporter by applying our raytracing algorithm to each frame and by considering the current position and orientation of the truck model at the corresponding point of time. During the tracking of the bright transporter, those parts of the vehicle image were excluded from the Kalman filter update step which were currently occluded by the projected truck model. Figure 7 shows in which parts of the vehicle image the model gradient has been suppressed due to the occlusion.

Moreover, one can see that we do not only exploit the surfaces of each car within the matching process, but in addition the information contained in its shadow. The case of an occluded shadow is treated similarly to the occlusion of the vehicle image itself. Figure 8 (b) shows the results of a successful tracking of the occluded bright transporter and the truck.

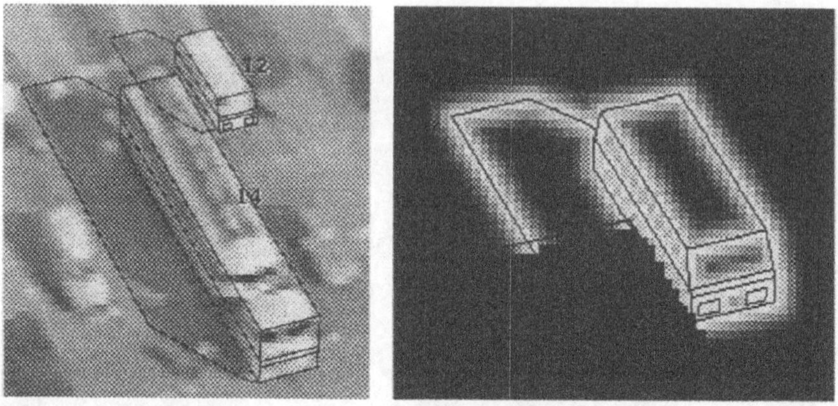

Fig. 7.: The model gradient norm of the image of the minivan (object #12) is constrained to such image points which correspond to the image of object #12 considering the occlusion by object #14. Notice that even parts of the shadow cast by object #12 are suppressed in the tracking process, due to the occlusion by object #14, although the unoccluded parts of the shadow cast by object #12 contribute to the update process.

We currently extend this approach to a *parallel* tracking of all object candidates in the scene in order to treat even bilateral occlusions where first one object occludes the other and subsequently a converse occlusion relation may occur.

5.3 Occlusion Characterization

Due to space limitations, we concentrate on the occlusion characterization between the advertising post and object #12 (see Figure 3). In order to examine the validity of the predicate *passively_decreasing_occlusion*(*advertising_post*, *object* #12, t) defined in section 4, we have to consider the estimated speed of object #12 as a function of the (half–) frame number as shown in Figure 9. The predicate $\neg$*moving*(*advertising_post*, t) obviously holds for all t. The vehicle is occluded by the advertising post between halfframes #5 and #49. Due to the low degree of validity for *occluded*(*advertising_post*, *object* #12, t), the degree of validity of all predicates which characterize the occlusion between the post and the vehicle will be zero in subsequent halfframes (#50 through #175). The resulting fuzzy membership function, after evaluating the considered predicate, is shown in Figure 9.

Fig. 8.: (a) Tracking of object #12 of the downtown image sequence fails if the occlusion of the minivan by the truck remains unconsidered. (b) Halfframe #293 with overlaid trajectory obtained by modelling the occlusion.

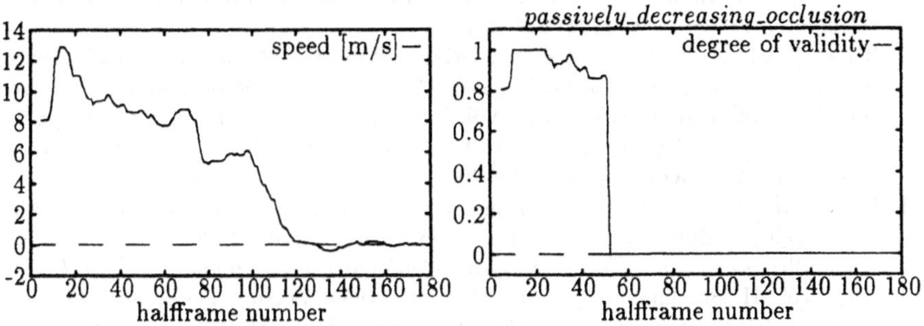

Fig. 9.: Estimated speed of object #12 in the gas station scene during the halfframes #5 through #172 (left) and the resulting characterization of the occlusion between the advertising post and the vehicle (right).

6 Conclusion

In this contribution, we explicitly modeled occlusions occuring between objects in real traffic-scenes. We showed in several experiments that the performance of the vehicle tracking process could be improved significantly by taking occlusions between static and dynamic scene components into account. We could extend this approach even to the case of moving occluding objects and to the shadow cast by a partially occluded object.

However, there still exist cases in which the occlusion modeling is not sufficient for proper vehicle tracking, e.g., when there are significant disturbances in the underlying image data or if there are influences not yet modeled, like image features in the background of the scene, for example strong road markings.

It is still an open question how we can obtain 3D models of the scenes *automa-*

tically, i.e. how we can eliminate the need for modeling the 3D scene components interactively.

Acknowledgment

We thank H. Damm for providing the gas station image sequence and for the 2D road model of the gas station.

References

[Cédras & Shah 95] C. Cédras, M. Shah, Motion-Based Recognition: A Survey, *Image and Vision Computing* **13**:2 (1995) 129–155.

[Du *et al.* 93] L. Du, G.D. Sullivan, K.D. Baker, Quantitative Analysis of the Viewpoint Consistency Constraint in Model-Based Vision, in *Proc. Fourth International Conference on Computer Vision (ICCV '93)*, Berlin, Germany, 11–14 May 1993, pp. 632–639.

[Dubuisson & Jain 95] M.-P. Dubuisson, A.K. Jain, Contour Extraction of Moving Objects in Complex Outdoor Scenes, *International Journal of Computer Vision* **14**:1 (1995) 83–105.

[Koller *et al.* 93] D. Koller, K. Daniilidis, H.-H. Nagel, Model-Based Object Tracking in Monocular Image Sequences of Road Traffic Scenes, *International Journal of Computer Vision* **10**:3 (1993) 257–281.

[Koller *et al.* 94] D. Koller, J. Weber, J. Malik, Robust Multiple Car Tracking with Occlusion Reasoning, in J.-O. Eklundh (Ed.), *Proc. Third European Conference on Computer Vision (ECCV '94)*, Vol. I, Stockholm, Sweden, May 2-6, 1994, Lecture Notes in Computer Science **800**, Springer-Verlag, Berlin, Heidelberg, New York/NY and others, 1994, pp. 189–196.

[Kollnig *et al.* 94] H. Kollnig, H.-H. Nagel, and M. Otte, Association of Motion Verbs with Vehicle Movements Extracted from Dense Optical Flow Fields, in J.-O. Eklundh (ed.), *Proc. Third European Conference on Computer Vision ECCV '94*, Vol. II, Stockholm, Sweden, May 2-6, 1994, Lecture Notes in Computer Science **801**, Springer-Verlag, Berlin, Heidelberg, New York/NY, and others, 1994, pp. 338–347.

[Kollnig & Nagel 95] H. Kollnig, H.-H. Nagel, 3D Pose Estimation by Fitting Image Gradients Directly to Polyhedral Models, in *Proc. Fifth International Conference on Computer Vision (ICCV '95)*, Cambridge/MA, June 20-23, 1995, pp. 569–574

[Meyer & Bouthemy 94] F. Meyer, P. Bouthemy, Region-Based Tracking Using Affine Motion Models in Long Image Sequences, *CVGIP: Image Understanding* **60**:2 (1994) 119–140.

[Sullivan 92] G.D. Sullivan, Visual Interpretation of Known Objects in Constrained Scenes, *Philosophical Transactions Royal Society London* (B) **337** (1992) 361–370.

[Sullivan *et al.* 95] G.D. Sullivan, A.D. Worrall, and J.M. Ferryman, Visual Object Recognition Using Deformable Models of Vehicles, in *Proc. Workshop on Context-Based Vision*, 19 June 1995, Cambridge/MA, pp. 75–86.

[Toal & Buxton 92] A. F. Toal, H. Buxton, Spatio-temporal Reasoning within a Traffic Surveillance System, in G. Sandini (Ed.), *Proc. Second European Conference on Computer Vision (ECCV '92)*, S. Margherita Ligure, Italy, May 18-23, 1992, Lecture Notes in Computer Science **588**, Springer-Verlag, Berlin, Heidelberg, New York/NY and others, 1992, pp. 884–892.

Imposing Hard Constraints on Soft Snakes

P. Fua[1]* and C. Brechbühler[2]

[1] SRI International, 333 Ravenswood Avenue, Menlo Park, CA 94025 fua@ai.sri.com
[2] ETH-Zürich, Gloriastr. 35, CH-8092 Zürich, Switzerland, brech@vision.ee.ethz.ch

Abstract. An approach is presented for imposing generic hard constraints on deformable models at a low computational cost, while preserving the good convergence properties of snake-like models. We believe this capability to be essential not only for the accurate modeling of individual objects that obey known geometric and semantic constraints but also for the consistent modeling of sets of objects.

Many of the approaches to this problem that have appeared in the vision literature rely on adding penalty terms to the objective functions. They rapidly become untractable when the number of constraints increases. Applied mathematicians have developed powerful constrained optimization algorithms that, in theory, can address this problem. However, these algorithms typically do not take advantage of the specific properties of snakes. We have therefore designed a new algorithm that is tailored to accommodate the particular brand of deformable models used in the Image Understanding community.

We demonstrate the validity of our approach first in two dimensions using synthetic images and then in three dimensions using real aerial images to simultaneously model terrain, roads, and ridgelines under consistency constraints.

1 Introduction

We propose an approach to imposing generic hard constraints on "snake-like" deformable models [9] while both preserving the good convergence properties of snakes and avoiding having to solve large and ill-conditioned linear systems.

The ability to apply such constraints is essential for the accurate modeling of complex objects that obey known geometric and semantic constraints. Furthermore, when dealing with multiple objects, it is crucial that the models be both accurate and consistent with each other. For example, individual components of a building can be modeled independently, but to ensure realism, one must guarantee that they touch each other in an architecturally feasible way. Similarly when modeling a cartographic site from aerial imagery, one must ensure that the roads lie on the terrain—and not above or below it—and that rivers flow downhill.

* This work was supported in part by contracts from the Advanced Research Projects Agency.

A traditional way to enforce such constraints is to add a penalty term to the model's energy function for each constraint. While this may be effective for simple constraints this approach rapidly becomes intractable as the number of constraints grows for two reasons. First, it is well known that minimizing an objective function that includes such penalty terms constitutes an ill-behaved optimization problem with poor convergence properties [4, 8]: the optimizer is likely to minimize the constraint terms while ignoring the remaining terms of the objective function. Second, if one tries to enforce several constraints of different natures, the penalty terms are unlikely to be commensurate and one has to face the difficult problem of adequately weighing the various constraints.

Using standard constrained optimization techniques is one way of solving these two problems. However, while there are many such techniques, most involve solving large linear systems of equations and few are tailored to preserving the convergence properties of the snake-like approaches that have proved so successful for feature delineation and surface modeling. Exceptions are the approach proposed by Metaxas and Terzopoulos [10] to enforce holonomic constraints by modeling the second order dynamics of the system and the technique proposed by Amini *et al.* [1] using dynamic programming.

In this work we propose a new approach to enforcing hard-constraints on deformable models without undue computational burden while retaining their desirable convergence properties. Given a deformable model, the state vector that defines its shape, an objective function to be minimized and a set of constraints to be satisfied, each iteration of the optimization performs two steps:

- Orthogonally project the current state toward the constraint surface, that is the set of all states that satisfy the constraints.
- Minimize the objective function in a direction that belongs to the subspace that is tangent to the constraint surface.

This can be achieved by solving small linear systems. This algorithm is closely related to the two-phase algorithm proposed by Rosen [11] and is an extension of a technique developed in [2]. The corresponding procedure is straightforward and easy to implement. Furthermore, this approach remains in the spirit of most deformable model approaches: they can also be seen as performing two steps, one attempting to fit the data and the other to enforce global constraints [3].

We view our contribution as the design of a very simple and effective constrained-optimization technique that allows the imposition of hard constraints on deformable models at a very low computational cost.

We first present the generic constrained optimization algorithm that forms the basis of our approach. We then specialize it to handle snake-like optimization. Finally, we demonstrate its ability to enforce geometric constraints upon individual snakes and consistency constraints upon multiple snakes.

2 Constrained Optimization

Formally, the constrained optimization problem can be described as follows. Given a function f of n variables $S = \{s_1, s_2, .., s_n\}$, we want to minimize it under a set of m constraints $C(S) = \{c_1, c_2, .., c_m\} = 0$. That is,

$$\text{minimize } f(S) \text{ subject to } C(S) = 0 \tag{1}$$

While there are many powerful methods for nonlinear constrained minimization [8], we know of none that are particularly well adapted to snake-like optimization: they do not take advantage of the locality of interactions that is characteristic of snakes. We have therefore developed a robust two-step approach [2] that is closely related to gradient projection methods first proposed by Rosen [11] and can be extended to snake optimization.

2.1 Constrained Optimization in Orthogonal Subspaces

Solving a constrained optimization problem involves satisfying the constraints and minimizing the objective function. For our application, it has proved effective to decouple the two and decompose each iteration into two steps:

1. Enforce the constraints by projecting the current state onto the constraint surface. This involves solving a system of nonlinear equations by linearizing them and taking Newton steps.
2. Minimize the objective function by projecting the gradient of the objective function onto the tangent subspace to the constraint surface and searching in the direction of the projection, so that the resulting state does not stray too far away from the constraint surface.

Figure 1 depicts this procedure. Let C and S be the constraint and state vectors of Equation 1 and A be the $n \times m$ Jacobian matrix of the constraints. The two steps are implemented as follows:

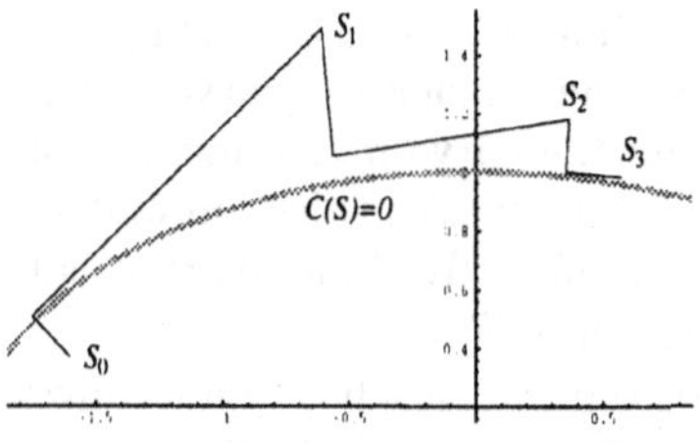

Fig. 1. Constrained optimization. Minimizing $(x - 0.5)^2 + (y - 0.2)^2$ under the constraint that $(x/2)^2 + y^2 = 1$. The set of all states that satisfy the constraint $C(S) = 0$, i.e. the constraint surface, is shown as a thick gray line. Each iteration consists of two steps: orthognal projection onto the constraint surface followed by a line search in a direction tangent to the surface. Because we perform only one Newton step at each iteration, the constraint is fully enforced only after a few iterations

1. To project S, we compute dS such that $C(S + dS) \approx C(S) + A^t dS = 0$ and increment S by dS. The shortest possible dS is found by writing dS as AdV and solving the equation $A^t A dV = -C(S)$.

2. To compute the optimization direction, we first solve the linear system $A^T(S)A(S)\lambda = A^T(S)\nabla f$ and take the direction to be $\nabla f - A\lambda$. This amounts to estimating Lagrange multipliers, that is, the coefficients that can be used to describe ∇f as closely as possible as a linear combination of constraint normals.

These two steps operate in two locally orthogonal subspaces, in the column space of A and in its orthogonal complement, the null space of A^T. Note that $A^T(S)A(S)$ is an $m \times m$ matrix and is therefore small when there are more variables than constraints, which is always the case in our application.

3 Snake Optimization

We first introduce our notations and briefly review traditional "snake-like" optimization [9]. We then show how it can be augmented to accommodate our constrained-optimization algorithm to impose hard constraints on single and multiple snakes.

3.1 Unconstrained Snake Optimization

In our work, we take 2–D features to be outlines that can be recovered from a single 2–D image while we treat 3–D features as objects whose properties are computed by projecting them into several 2–D images. We model 2–D and 3–D linear features as polygonal curves and 3–D surfaces as hexagonally connected triangulations.

We will refer to S, the vector of all x, y, and z coordinates of the 2–D or 3–D vertices that define the deformable model's shape as the model's *state vector*.

We recover a model's shape by minimizing an objective function $\mathcal{E}(S)$ that embodies the image-based information. For 2–D linear features, $\mathcal{E}(S)$ is the average value of the edge gradient along the curve. For 3–D linear features, $\mathcal{E}(S)$ is computed by projecting the curve into a number of images, computing the average edge-gradient value for each projection and summing these values [5]. For 3–D surfaces, we use an objective function that is the sum of a stereo term and a shape-from-shading term. [7].

In all these cases, $\mathcal{E}(S)$ typically is a highly nonconvex function, and therefore difficult to optimize. However, it can effectively be minimized [9] by

- introducing a quadratic regularization term $\mathcal{E}_D = 1/2S^tK_SS$ where K_S is a sparse stiffness matrix,
- defining the total energy $\mathcal{E}_T = \mathcal{E}_D(S) + \mathcal{E}(S) = 1/2S^tK_SS + \mathcal{E}(S)$,
- embedding the curve in a viscous medium and iteratively solving the dynamics equation $\frac{\partial \mathcal{E}_T}{\partial S} + \alpha \frac{dS}{dt} = 0$, where α is the viscosity of the medium.

Because $\mathcal{E}_D$ is quadratic, the dynamics equation can be rewritten as

$$K_SS_t + \left.\frac{\partial \mathcal{E}}{\partial S}\right|_{S_{t-1}} + \alpha(S_t - S_{t-1}) = 0 \Rightarrow (K_S + \alpha I)S_t = \alpha S_{t-1} - \left.\frac{\partial \mathcal{E}}{\partial S}\right|_{S_{t-1}} . \quad (2)$$

In practice, α is computed automatically at the start of the optimization procedure so that a prespecified average vertex motion amplitude is achieved [6]. The optimization proceeds as long as the total energy decreases. When it increases, the algorithm backtracks and increases α, thereby decreasing the step size. In the remainder of the paper, we will refer to the vector $dS_t = S_t - S_{t-1}$ as the "snake step" taken at iteration t.

In effect, this optimization method performs implicit Euler steps with respect to the regularization term [9] and is therefore more effective at propagating smoothness constraints across the surface than an explicit method such as conjugate gradient. It is this property that our constrained-optimization algorithm strives to preserve.

3.2 Constraining the Optimization

Given a set of m hard constraints $C(S) = \{c_1, c_2, .., c_m\}$ that the snake must satisfy, we could trivially extend the technique of Section 2 by taking the objective function f to be the total energy $\mathcal{E}_T$. However, this would be equivalent to optimizing an unconstrained snake using gradient descent as opposed to performing the implicit Euler steps that so effectively propagate smoothness constraints.

In practice, propagating the smoothness constraints is key to forcing convergence toward desirable answers. When a portion of the snake deforms to satisfy a hard constraint, enforcing regularity guarantees that the remainder of the snake also deforms to preserve it and that unwanted discontinuities are not generated.

Therefore, for the purpose of optimizing constrained snakes, we decompose the second step of the optimization procedure of Section 2 into two steps. We first solve the unconstrained Dynamics Equation (Equation 2) as we do for unconstrained snakes. We then calculate the component of the snake step vector—the difference between the snake's current state and its previous one—that is perpendicular to the constraint surface and subtract it from the state vector. The first step regularizes, while the second prevents the snake from moving too far away from the constraint surface.

As in the case of unconstrained snakes, α, the viscosity term of Equation 2, is computed automatically at the start of the optimization and progressively increased as needed to ensure a monotonic decrease of the snake's energy and ultimate convergence of the algorithm.

An iteration of the optimization procedure therefore involves the following three steps:

1. Take a Newton step to project S_{t-1}, the current state vector, onto the constraint surface.

   $$S_{t-1} \leftarrow S_{t-1} + AdV \text{ where } A^T AdV = -C(S_{t-1})$$

   If the snake's total energy has increased, back up and increase viscosity.
2. Take a normal snake step by solving

   $$(K_S + \alpha I)S_t = \alpha S_{t-1} - \left.\frac{\partial \mathcal{E}}{\partial S}\right|_{S_{t-1}}$$

3. Ensure that dS, the snake step from S_{t-1} to S_t, is in the subspace tangent to the constraint surface.

$$S_t \leftarrow S_t - A\lambda \text{ where } A^t A\lambda = A^T(S_t - S_{t-1}) \ ,$$

so that the snake step dS becomes $dS = (S_t - A\lambda) - S_{t-1} \Rightarrow A^T dS = 0$.

3.3 Multiple Snakes

Our technique can be further generalized to the simultaneous optimization of several snakes under a set of constraints that bind them. We concatenate the state vectors of the snakes into a composite state vector S and compute for each snake the viscosity coefficient that would yield steps of the appropriate magnitude if each snake was optimized individually. The optimization steps become:

1. Project S onto the constraint surface as before and compute energy of each individual snake. For all snakes whose energy has increased, revert to the previous position and increase the viscosity.
2. Take a normal snake step for each snake individually.
3. Project the global step into the subspace tangent to the constraint surface.

Because the snake steps are taken individually we never have to solve the potentially very large linear system involving all the state variables of the composite snake but only the smaller individual linear systems. Furthermore, to control the snake's convergence via the progressive viscosity increase, we do not need to sum the individual energy terms. This is especially important when simultaneously optimizing objects of a different nature, such as a surface and a linear feature, whose energies are unlikely to be commensurate so that the sum of these energies would be essentially meaningless.

In effect, the optimization technique proposed here is a decomposition method and such methods are known to work well [8] when their individual components, the individual snake optimizations, are well behaved, which is the case here.

4 Results

We demonstrate the ability of our technique to impose geometric constraints on 2–D and 3–D deformable models using real imagery.

4.1 2–D Features

To illustrate the convergence properties of our algorithm, we introduce two simple sets of constraints that can be imposed on 2–D snakes. The most obvious one forces the snake to go through a specific point (a_0, b_0). It can be written as the two constraints

$$x_i - a_0 = y_i - b_0 = 0 \ , \tag{3}$$

where i is the index of the snake vertex that is closest to (a_0, b_0) at the beginning of an iteration. In practice, the constraint always remains "attached" to the vertex that was closest initially and we refer to this constraint as an "attractor constraint." A slightly more sophisticated set of constraints achieves a similar purpose while allowing the point at which the snake is attached to slide. It is designed to force the snake to be tangent to a segment $((a_0, b_0), (a_1, b_1))$, and we will refer to it as a "tangent constraint." It can also be written as a set of two constraints

$$\begin{vmatrix} x_i & a_0 & a_1 \\ y_i & b_0 & b_1 \\ 1 & 1 & 1 \end{vmatrix} = \begin{vmatrix} x_{i+1} - x_{i-1} & a_1 - a_0 \\ y_{i+1} - y_{i-1} & b_1 - b_0 \end{vmatrix} = 0 \tag{4}$$

where i in the index of the snake vertex that is both closest to the line segment and between the endpoints at the beginning of an iteration. The first constraint ensures that (x_i, y_i), (a_0, b_0), and (a_1, b_1) are collinear. The second ensures that the finite-difference estimate of the tangent vector is parallel to the segment's direction. The vertex at which the constraint is attached can slide along the segment and can slide off its edges so that a different vertex may become attached.

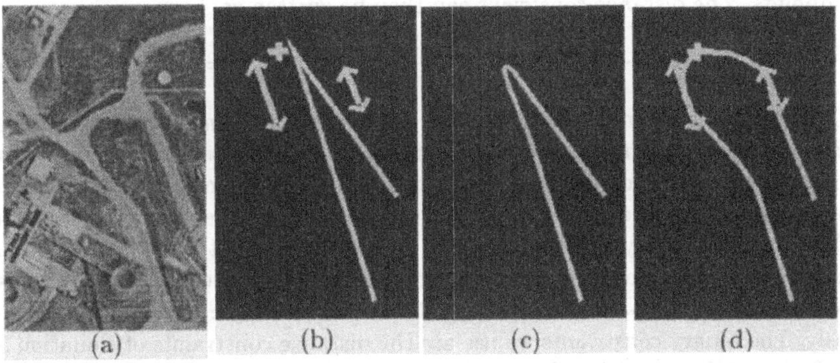

Fig. 2. Modeling the edge of a road. (a) Aerial image of a set of roads. (b) A very rough approximation of one of the road's edges and a set of constraints. The two-sided arrows represent tangent constraints (Equation 4) while the crosshair depicts an attractor constraint (Equation 3). (c) The result of unconstrained snake optimization. (d) The result of constrained snake optimization using the constraints depicted by (b).

Figure 2(b) depicts the very rough outline of the edge of a road. The outline is too far from the actual contour for a conventional snake to converge toward the edge. However, using two of the tangent constraints of Equation 4 and one of the attractor constraints of Equation 3, we can force convergence toward the desired edge.

To demonstrate the behavior of the multiple snake optimization, we also introduce a "distance" constraint between two snakes. Given a vector of length d, such as the ones depicted by arrows in Figure 3(a) and two snakes, let (x_i^1, y_i^1, z_i^1) and (x_j^2, y_j^2, z_j^2) be the vertices of each snake that are closest to the vector's

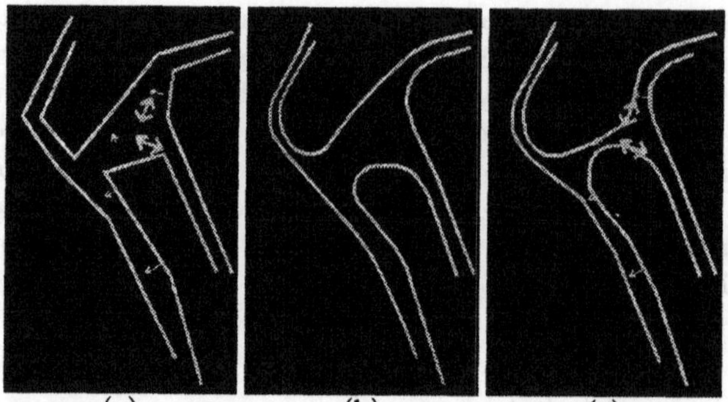

Fig. 3. Modeling a set of road edges. (a) A set of three contours roughly approximating the edges of the main roads and a set of constraints. As before, the two-sided arrows represent "tangent constraints" (Equation 4) that apply to individual contours, while the thinner one-sided arrows represent distance constraints (Equation 5) that bind pairs of contours. (b) The result of unconstrained snake optimization. (c) The result of constrained snake optimization.

endpoints. The distance constraint can then be written as

$$(x_i^1 - x_j^2)^2 + (y_i^1 - y_j^2)^2 + (z_i^1 - z_j^2)^2 - d^2 = 0 \ . \tag{5}$$

Using the same images as before, we can model the main road edges starting with the three rough approximations shown in Figure 3(a). Here again, these initial contours are too far away from the desired answer for unconstrained optimization to succeed. To enforce convergence toward the desired answer, in addition to the unary constraints—that is, constraints that apply to individual snakes—of the previous example, we can introduce binary constraints—that is, constraints that tie pairs of snakes—and optimize the three contours simultaneously. The binary constraints we use are the distance constraints of Equation 5.

In both of these examples, we were able to mix and match constraints of different types as needed to achieve the desired result without having to worry about weighting them adequately. Furthermore the algorithm exhibits good convergence properties even though the constraints are not linear but quadratic.

4.2 3–D Features

We now turn to the simultaneous optimization of 3–D surfaces and 3–D features. More specifically, we address the issue of optimizing the models of 3–D linear features such as roads and ridgelines and the terrain on which they lie under the constraint that they be consistent with one another. In Figures 4 and 5 we present two such cases where recovering the terrain and the roads independently of one another leads to inconsistencies.

Because we represent the terrain as a triangulated mesh and the features as 3–D polygonal approximations, consistency can be enforced as follows. For each

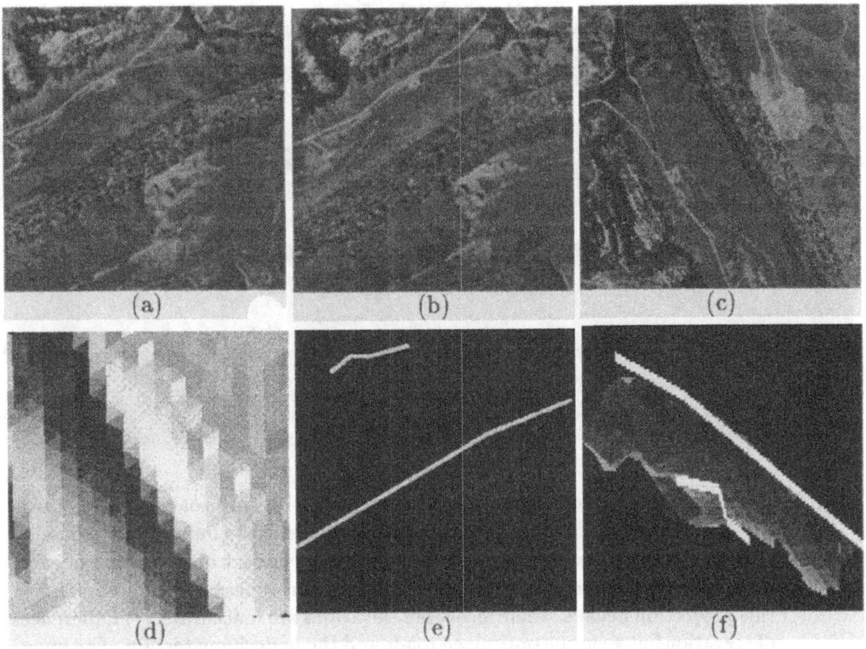

Fig. 4. Rugged terrain with sharp ridge lines. (a,b,c) Three images of a mountainous site. (d) Shaded view of an initial terrain estimate. (e) Rough polygonal approximation of the ridgelines overlaid on image (a). (f) The terrain and ridgeline estimates viewed from the side (the scale in z has been exaggerated).

edge $((x_1, y_1, z_1), (x_2, y_2, z_2))$ of the terrain mesh and each segment $((x_3, y_3, z_3), (x_4, y_4, z_4))$ of a linear feature that intersect when projected in the (x, y) plane, the four endpoints must be coplanar so that the segments also intersect in 3–D space. This can expressed as

$$\begin{vmatrix} x_1 & x_2 & x_3 & x_4 \\ y_1 & y_2 & y_3 & y_4 \\ z_1 & z_2 & z_3 & z_4 \\ 1 & 1 & 1 & 1 \end{vmatrix} = 0 \ , \qquad (6)$$

which yields a set of constraints that we refer to as consistency constraints.

In both examples shown here, we follow a standard coarse-to-fine strategy. We start with a rough estimate of both terrain and features and reduced versions of the images. We then progressively increase the resolution of the images being used and refine the discretization of our deformable models. In Figures 6 and 7, we show that the optimization under the constraints of Equation 6 avoids the discrepancies that result from independent optimization of each feature.

In the example of Figure 6, the "ridge-snake" attempts to maximize the average edge gradient along its projections in all three images. In the case of Figures 5 and 7 the roads are lighter than the surrounding terrain. At low resolution, they

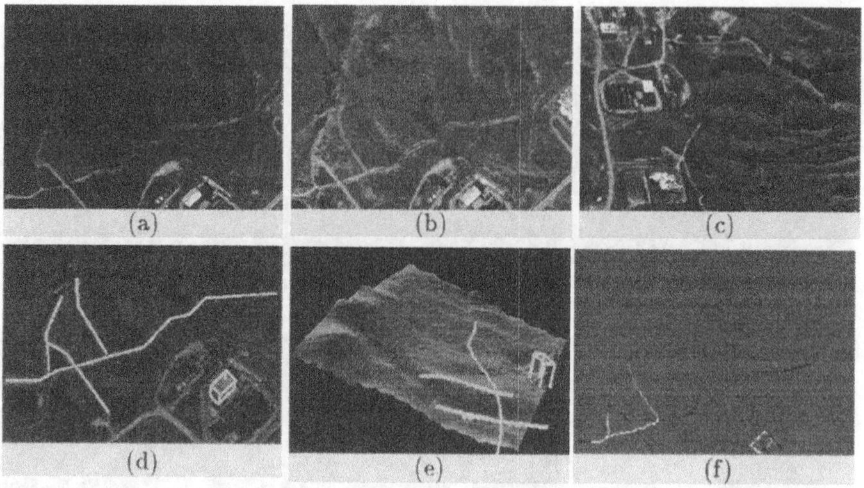

Fig. 5. Building a site model. (a,b,c) Three images of a site with roads and buildings. (d) A rough sketch of the road network and of one of the buildings. (e) Shaded view of the terrain with overlaid roads after independent optimization of each. Note that the two roads in the lower right corner appear to be superposed in this projection because their recovered elevations are inaccurate. (f) Differences of elevation between the optimized roads and the underlying terrain. The image is stretched so that black and white represent errors of minus and plus 5 meters, respectively.

can effectively be modeled as white lines, and the corresponding snakes attempt to maximize image intensity along their projections. We also introduce a building and use its base to further constrain the terrain. Figures 7(a,b) depict the result of the simultaneous optimization of the terrain and low-resolution roads. By supplying an average width for the roads, we can turn the lines into ribbons and reoptimize terrain and features under the same consistency constraints as before, yielding the result of Figure 7(c).

These two examples illustrate the ability of our approach to model different kinds of features in a common reference framework and to produce consistent composite models.

5 Conclusion

We have presented a constrained optimization method that allows us to enforce hard constraints on deformable models at a low computational cost, while preserving the convergence properties of snake-like approaches. We have shown that it can effectively constrain the behavior of linear 2–D and 3–D snakes as well as that of surface models. Furthermore, we have been able to use our technique to simultaneously optimize several models while enforcing consistency constraints between them.

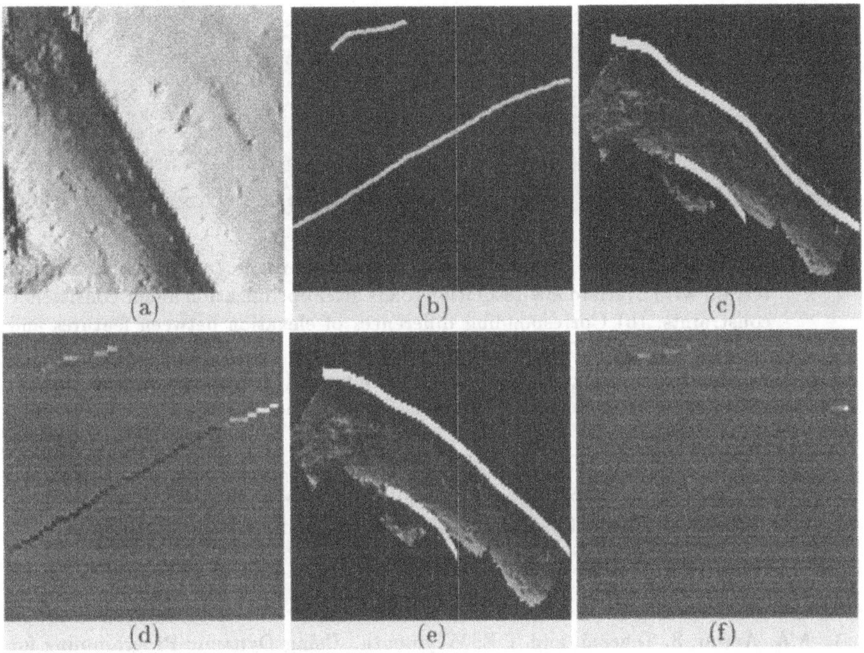

Fig. 6. Recovering the 3–D geometry of both terrain and ridges. (a) Shaded view of the terrain after refinement. (b) Refined ridgeline after 3–D optimization. (c) Side view of the ridgeline and terrain after independent optimization of each one. Note that the shape of the ridgeline does not exactly match that of the terrain. (d) Differences of elevation between the recovered ridge-line and the underlying terrain. The image is stretched so that black and white represent errors of minus and plus 80 feet, respectively. (e) Side view after optimization under consistency constraints. (f) Corresponding difference of elevation image stretched in the same fashion as (d).

We believe that these last capabilities will prove indispensable to automating the generation of complex object databases from imagery, such as the ones required for realistic simulations or intelligence analysis. In such databases, the models must not only be as accurate—that is, true to the data—as possible but also consistent with each other. Otherwise, the simulation will exhibit "glitches" and the image analyst will have difficulty interpreting the models. Because our approach can handle nonlinear constraints, in future work we will use it to implement more sophisticated constraints than the simple geometric constraints presented here. When modeling natural objects, we intend to take physical laws into account. For example, rivers flow downhill and at the bottom of valleys; this should be used when modeling both the river and the surrounding terrain. In addition, when modeling man-made objects, we intend to take advantage of knowledge about construction practices such as the fact that roads do not have arbitrary slopes.

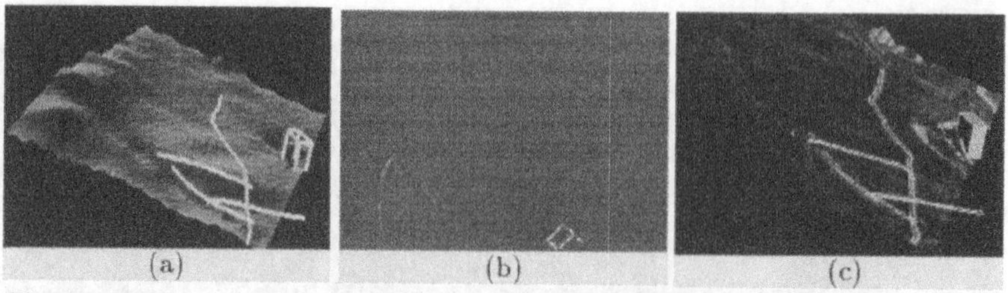

Fig. 7. Recovering the 3–D geometry of both terrain and roads. (a) Shaded view of the terrain with overlaid low-resolution roads after optimization under consistency constraints. (b) Corresponding differences of elevation between features and underlying terrain. The image is stretched as the one of Figure 5(f). Only the building's roof is significantly above the terrain. (c) Synthetic view of the site with the roads modeled as ribbons overlaid on the image.

Eventually, we hope that the technique presented in this paper will form the basis for a suite of tools for modeling complex scenes accurately while ensuring that the model components satisfy geometric and semantic constraints and are consistent with each other.

References

1. A.A. Amini, S. Tehrani, and T.E. Weymouth. Using Dynamic Programming for Minimizing the Energy of Active Contours in the Presence of Hard Constraints. In *International Conference on Computer Vision*, pages 95–99, 1988.
2. C. Brechbühler, G. Gerig, and O. Kübler. Parametrization of Closed Surfaces for 3-D Shape Description. *Computer Vision, Graphics, and Image Processing: Image Understanding*, 61(2):154–170, March 1995.
3. L. Cohen. Auxiliary Variables for Deformable Models. In *International Conference on Computer Vision*, pages 975–980, Cambridge, MA, June 1995.
4. R. Fletcher. *Practical Methods of Optimization.* John Wiley & Sons, Chichester, New York, Brisbane, Toronto, Singapore, 2nd edition, 1987. A Wiley-Interscience Publication.
5. P. Fua. Parametric Models are Versatile: The Case of Model Based Optimization. In *ISPRS WG III/2 Joint Workshop*, Stockholm, Sweden, September 1995.
6. P. Fua and Y. G. Leclerc. Model Driven Edge Detection. *Machine Vision and Applications*, 3:45–56, 1990.
7. P. Fua and Y. G. Leclerc. Object-Centered Surface Reconstruction: Combining Multi-Image Stereo and Shading. *International Journal of Computer Vision*, 16:35–56, September 1995.
8. P.E. Gill, W. Murray, and M.H. Wright. *Practical Optimization.* Academic Press, London a.o., 1981.
9. M. Kass, A. Witkin, and D. Terzopoulos. Snakes: Active Contour Models. *International Journal of Computer Vision*, 1(4):321–331, 1988.
10. D. Metaxas and D. Terzopoulos. Shape and Norigid Motion Estimation through Physics-Based Synthesis. *IEEE Transactions on Pattern Analysis and Machine Intelligence*, 15(6):580–591, 1991.
11. Rosen. Gradient projection method for nonlinear programming. *SIAM Journal of Applied Mathematics*, 8:181–217, 1961.

X Vision: Combining Image Warping and Geometric Constraints for Fast Visual Tracking

Gregory D. Hager and Kentaro Toyama

Department of Computer Science
Yale University, P.O. Box 208285
New Haven, CT, 06520

Abstract. In this article, we describe X Vision, a modular, portable framework for visual tracking. X Vision is designed to be a programming environment for real-time vision which provides high performance on standard workstations outfitted with a simple digitizer. X Vision consists of a small set of image-level tracking primitives and a framework for combining tracking primitives to form complex tracking systems. Efficiency and robustness are achieved by propagating geometric and temporal constraints to the feature detection level, where image warping and specialized image processing are combined to perform feature detection quickly and robustly. We illustrate how useful, robust tracking systems can be constructed by simple combinations of a few basic primitives with the appropriate task-specific constraints.

1 Introduction

Real-time visual feedback is an essential tool for implementing systems that interact dynamically with the world, but recovering visual motion quickly and robustly is difficult. Full-frame vision techniques such as those based on optical flow or region segmentation tend to emphasize iterative processing which requires slow offline computation or expensive, inflexible, specialized hardware. A viable alternative is feature tracking, which concentrates on spatially localized areas of the image. Feature tracking minimizes bandwidth requirements between the host and the framegrabber and reduces the amount of input data to a level where it can be efficiently dealt with by off-the-shelf hardware. Such systems are cost-effective, and since most of the tracking algorithms reside in software, extremely flexible.

Feature tracking has already found wide applicability in the vision and robotics literature. Subfields which utilize feature tracking include structure-from-motion research, robotic hand-eye applications, and model-based tracking. Although tracking is necessary for such research, however, it is generally not a focus of the work and is often solved in an *ad hoc* fashion. Many so-called "tracking systems" are in fact only applied to pre-stored video sequences and do not operate in real time [8].

In response to this perceived need, we have constructed a fast, portable, reconfigurable tracking system called "X Vision." Experience from several teaching and research applications suggests that this system reduces the startup time for new vision applications, makes real-time vision accessible to "non-experts," and demonstrates that interesting research utilizing real-time vision can be performed with minimal hardware. The remainder of the article is organized into three sections: Section 2 describes the essential features of X Vision, Section 3 supplies evidence that our system facilitates real-time vision-based research, and Section 4 discusses future research directions.

2 Tracking System Design and Implementation

X Vision embodies the principle that "vision is inverse graphics." The system is organized around a small set of image-level tracking primitives referred to as *basic features.* Each of these features is described in terms of a small set of parameters, a *state vector*, which specifies the feature's position and appearance. *Composite features* track their own, possibly more complex, state vectors by computing functions of their constituent features' states. Conversely, given the state vector of a complex feature, constraints are imposed on the state of its constituent features and the process recurses until image-level primitives are reached. The image-level primitives search for features in the neighborhood of their expected locations which produces a new state vector, and the cycle repeats.

Ease of use has also been a goal in X Vision's design. To this end, X Vision incorporates a data abstraction mechanism that dissociates the information carried in the feature state from the tracking mechanism used to acquire it. Details on this mechanism can be found in [4].

2.1 Image-Level Feature Tracking

At the image level, tracking a feature means that the algorithm's "region of interest" maintains a fixed, pre-defined relationship (*e.g.* , containment) to the feature. In X Vision, a region of interest is referred to as a *window.* The key to efficient window-based image processing is the use of geometric and temporal constraints to supply a natural change of coordinates on images. This change of coordinates places the feature of interest in a canonical configuration in which imaging processing is simple and fast. *Image warping* can be optimized to perform this change of coordinates quickly, and to apply it exactly once for each pixel.

The low-level features currently available in X Vision include solid or broken contrast edges detected using several variations on standard edge-detection, general grey-scale patterns tracked using SSD methods [1, 9], and a variety of color and motion-based primitives used for initial detection of objects and subsequent match disambiguation. The remainder of this section briefly describes these features and supplies timing figures for an SGI Indy workstation equipped with a 175Mhz R4400 SC processor and an SGI VINO digitizing system.

Edges X Vision provides a tracking mechanism for linear edge segments of arbitrary length. The state of an edge segment is characterized by its position and orientation in framebuffer coordinates as well as its filter response. Given prior state information, image warping is used to acquire a window which, if the prior estimate is correct, leads to an edge which is vertical within the warped window. Convolving each row of the window with a scalar derivative-based kernel will then produce an aligned series of response peaks. These responses can be superimposed by summing down the columns of the window. Finding the maximum value of this 1-D response function localizes the edge. Performance can be improved by commuting the order of the convolution and summation. In an $n \times m$ window, edge localization with a convolution mask of width k can be performed with just $m \times (n + k)$ additions and mk multiplications.

The detection scheme described above requires orientation information to function correctly. If this information cannot be supplied from higher-level geometric constraints, it is estimated at the image level. An expanded window at the predicted orientation is acquired, and the summation step is performed along the columns and along two diagonal paths corresponding to two bracketing orientation offsets. This approximates the effect of performing the convolution at three different orientations. Quadratic interpolation of the three curves is used to estimate the orientation of the underlying edge. In practice, the subsequent estimate of edge location will be biased, so edge location is computed as the weighted average of the edge location of all three peaks. Additional sub-pixel accuracy can be gained by finding the zero-crossing of the second derivative at the computed orientation.

Tracking robustness can be increased by making edge segments as long as possible. Long segments are less likely to become completely occluded, and changes in the background tend to affect a smaller proportion of the segment with a commensurately lower impact on the filter response. Speed is maintained by subsampling the window in the direction of the edge segment. Likewise, the maximum edge motion between images can be increased by subsampling in the horizontal direction. In this case, the accuracy of edge localization drops and the possibility of an ambiguous match increases.

Ultimately, very little can be done to handle distraction, occlusion, and aperture problems for local, single edges. For this reason, simple edge segments are almost never used alone or in the absence of higher level geometric constraints. Some results on different edge matching schemes can be found in [10].

Figure 1 shows timings for simple edge tracking that were obtained during test runs.

Line	Length Sampling		
Length, Width	Full	1/2	1/4
20, 20	0.44	0.31	0.26
40, 20	0.73	0.43	0.29
40, 40	1.33	0 .69	0.49

Fig. 1. Timing in milliseconds for one iteration of tracking an edge segment.

Region-Based Tracking In region-based tracking, we consider matching a pre-defined "reference" window to a region of the image. The reference is either a region taken from the scene itself, or a pre-supplied "target" template. If the region of interest is roughly planar and its projection relatively small compared to the image as a whole, the geometric distortions of a region can be modeled using an affine transformation. The state vector for our region tracker includes these six geometric parameters and a value which indicates how well the reference image and the current image region match.

The process proceeds using the usual SSD optimization methods [7, 8]. Let $I(\mathbf{x}, t)$ denote the value of the pixel at location $\mathbf{x} = (x, y)^T$ at time t in an image sequence. At time t, a surface patch projects to an image region with a spatial extent represented as a set of image locations, $\mathcal{W}$. At some later point, $t+\tau$, the surface path projects to an affine transformation of the original region. If illumination remains constant, the geometric relationship between the projections can be recovered by minimizing:

$$O(\mathbf{A}, \mathbf{d}) = \sum_{\mathbf{x} \in \mathcal{W}} (I(\mathbf{A}\mathbf{x} + \mathbf{d}, t + \tau) - I(\mathbf{x}, t))^2 w(\mathbf{x}), \quad \tau > 0, \tag{1}$$

where $\mathbf{d} = (u, v)^T$, $\mathbf{x} = (x, y)^T$, $\mathbf{A}$ is an arbitrary positive definite 2×2 matrix and $w(\cdot)$ is an arbitrary weighting function. In order to avoid bias due to changes in brightness and contrast we normalize images to have zero first moment and unit second moment. With these modifications, solving (1) for rigid motions (translation and rotation) is equivalent to maximizing normalized correlation.

Suppose that a solution at time $t + \tau$, $(\mathbf{A}_\tau, \mathbf{d}_\tau)$, is known and we define the warped image $J_\tau(\mathbf{x}, t) = I(\mathbf{A}_\tau \mathbf{x} + \mathbf{d}_\tau, t + \tau)$. It should be the case that $J_\tau(\mathbf{x}, t) = I(\mathbf{x}, t)$ and subsequent deviations can then be expressed as small variations on the reference image. We can then write $\delta\mathbf{A} = \text{offdiag}(-\alpha, \alpha) + \text{diag}(\mathbf{s}_x, \mathbf{s}_y) + \text{offdiag}(\gamma, 0)$ where α is a differential rotation, $\mathbf{s}$ is a vector of differential scale changes, and γ is a differential shear. Applying this observation to (1) and linearizing I at the point $(\mathbf{x}, t)$ yields

$$O(\mathbf{d}, \alpha, \mathbf{s}, \gamma) = \sum_{\mathbf{x} \in \mathcal{W}} (\ (I_x(\mathbf{x}, t), I_y(\mathbf{x}, t)) \cdot (\delta\mathbf{A}\mathbf{x} + \mathbf{d}) \ + \ I_t(\mathbf{x}, t)\delta t)^2 w(\mathbf{x}), \tag{2}$$

where $(I_x, I_y)^T$ are the image spatial derivatives, and $I_t(\mathbf{x}, t) = J_\tau(\mathbf{x}, t + \delta t) - I(\mathbf{x}, t)$.

The parameter groups are then solved for individually. First, we ignore $\delta\mathbf{A}$ and solve (2) for $\mathbf{d}$ by taking derivatives and setting up a linear system in the unknowns. Defining

$$\mathbf{g}_x(\mathbf{x}) = I_x(\mathbf{x}, t)\sqrt{w(\mathbf{x})} \quad \mathbf{g}_y(\mathbf{x}) = I_y(\mathbf{x}, t)\sqrt{w(\mathbf{x})} \quad \mathbf{h}_0(\mathbf{x}) = J_t(\mathbf{x}, t)\sqrt{w(\mathbf{x})},$$

the linear system for computing translation is

$$\begin{bmatrix} \mathbf{g}_x \cdot \mathbf{g}_x & \mathbf{g}_x \cdot \mathbf{g}_y \\ \mathbf{g}_x \cdot \mathbf{g}_y & \mathbf{g}_y \cdot \mathbf{g}_y \end{bmatrix} \mathbf{d} = \begin{bmatrix} \mathbf{h}_0 \cdot \mathbf{g}_x \\ \mathbf{h}_0 \cdot \mathbf{g}_y \end{bmatrix}. \tag{3}$$

Since the spatial derivatives are only computed using the original reference image, $\mathbf{g}_x$ and $\mathbf{g}_y$ and the inverse of the matrix on the left hand side are constant over time and can be computed offline.

Once $\mathbf{d}$ is known, the image differences are adjusted according to the vector equation

$$\mathbf{h}_1 = \mathbf{h}_0 - \mathbf{g}_x u - \mathbf{g}_y v. \quad (4)$$

If the image distortion arises from pure translation and no noise is present, then we expect that $\mathbf{h}_1 = 0$ after this step. Any remaining residual can be attributed to linearization error, noise, or other geometric distortions. The solutions for the remaining parameters operate on $\mathbf{h}_1$ in a completely analogous fashion. Details can be found in [4]. This incremental solution method is not likely to be as accurate as solving for all parameters simultaneously, but it does provide a method for solving underdetermined systems quickly with a controlled bias.

In order to guarantee tracking of motions larger than a fraction of a pixel, these calculations must be carried out at varying levels of resolution. Resolution reduction effectively consumes a vector multiply and k image additions, where k is the reduction factor. The tracking algorithm changes the resolution adaptively based on image motion. If the interframe motion exceeds $0.25k$, the resolution for the subsequent step is halved. If the interframe motion is less than $0.1k$, the resolution is doubled. This leads to a fast algorithm for tracking fast motions and a slower but more accurate algorithm for tracking slower motions.

Size	40 × 40		60 × 60		80 × 80		100 × 100	
Reduction	4	2	4	2	4	2	4	2
Trial A	0.8	1.7	1.7	3.9	3.2	7.3	5.0	12.7
Trial B	3.2	7.2	7.3	13.0	13.0	23.6	20.3	37.0
Trial C	1.5	5.6	3.2	9.5	6.1	17.7	9.4	28.3
Trial D	2.9	4.1	6.4	9.2	11.1	16.9	17.5	27.9
Trial E	2.8	3.8	6.2	8.6	10.8	15.8	17.5	26.0
Trial F	3.7	8.4	8.1	15.6	14.4	28.5	22.5	43.1

Fig. 2. The time in milliseconds consumed by one cycle of tracking for various instantiations of the SSD tracker. Trial A computed translation with aligned, unscaled images. Trial B computed translation with unaligned, scaled images. Trial C computed rotation with rotated images. Trial D computed scale with scaled images. Trial E computed shear with sheared images. Trial F computed all affine parameters under affine distortions.

To get a sense of the time consumed by these operations, we consider several test cases as shown in Figure 2. The first two rows of this table show the timings for various size and resolution images when computing only translation. The trials differ in that the second required full linear warping to acquire the window.

The subsequent rows present the time to compute translation and rotation while performing rotational warping, the time to compute scale while performing scale warping, the time to compute shear while performing shear warping, and the time to compute all affine parameters while performing linear warping. With the exception of two cases under 100×100 images at half resolution, all updates require less than one frame time (33.33 msec) to compute.

2.2 Networks of Features

Recall that composite features compute their state from other basic and composite features. We consider two types of feature composition. In the first case information flow is purely bottom-up. Features are composed solely in order to compute information from their state vectors without altering their tracking behavior. For example, given two point features it may be desirable to present them as the line feature passing through both. A feature (henceforth *feature* refers to both basic and composite features) can participate in any number of such constructions.

In the second case, the point of performing feature composition is to exploit higher level geometric constraints in tracking as well as to compute a new state vector. In this case, information flows both upward and downward. An example is given in §3, where a rectangle's corner positions are used to compute the orientation of the edge segments which compose each corner.

More generally, we define a ***feature network*** to be a set of nodes connected by two types of directed arcs referred to as ***up-links*** and ***down-links***. Nodes represent features, up-links represent information dependency to a composite feature from its constituent features, and down-links represent constraints on features from above. For example, a corner is a graph with three nodes. There are both up-links and down-links between the corner node and the edge nodes since the position of the corner depends upon the states of the constituent edges, and the positions of the edges, in turn depends on the location of the corner.

A complete tracking cycle then consists of: 1) traversing the down-links from nodes which are down-link sources applying state constraints until basic features are reached; 2) applying low-level detection in every basic feature; and 3) traversing the up-links of the graph computing the state of composite features. State prediction can be added to this cycle by including it in the downward propagation. Thus, a feature tracking system is completely characterized by the topology of the network, the identity of the basic features, and a state computation and constraint function for each composite feature.

Composite features that have been implemented within this scheme range from simple edge intersections as described above, to Kalman snakes, to three-dimensional model-based tracking using pose estimation [6] as well as a variety of more specialized object trackers, some of which are described in Section 3.

3 Applications

In this section, we present several applications of X Vision which illustrate how the tools it provides—particularly image warping, image subsampling, constraint propagation and typing—can be used to build fast and effective tracking systems. Additional results can be found in [2, 3, 5, 6, 10].

3.1 Pure Tracking

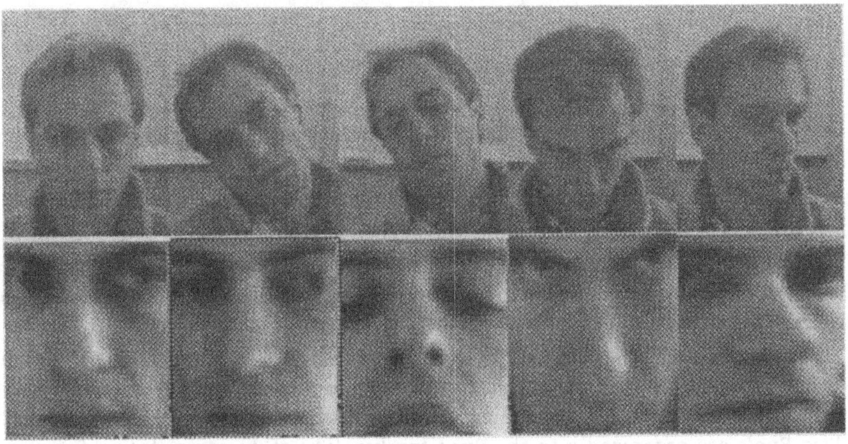

Fig. 3. Above several images of the a face and below the corresponding warped images used by the tracking system.

Face Tracking A frontal view of a human face is sufficiently planar to be tracked as a single SSD region undergoing affine deformations. Figure 3 shows several image pairs illustrating poses of a face and the warped image resulting from tracking. Despite the fact that the face is nonplanar, resulting for example in a stretching of the nose as the face is turned, the tracking is quite effective. However, it is somewhat slow (about 40 milliseconds per iterations), it can be confused if the face undergoes nonrigid distortions which cannot be easily captured using SSD methods, and it is sensitive to lighting variations and shadowing across the face.

We have also constructed a face tracker that relies on using SSD trackers at the regions of high contrast—the eyes and mouth—which provides much higher performance as well as the ability to "recognize" changes in the underlying features. The tracker is organized as shown in Figure 4. We first create a `MultiSSD` composite tracker which performs an SSD computation for multiple reference images. The state vector of the `MultiSSD` tracker is state of the tracker which

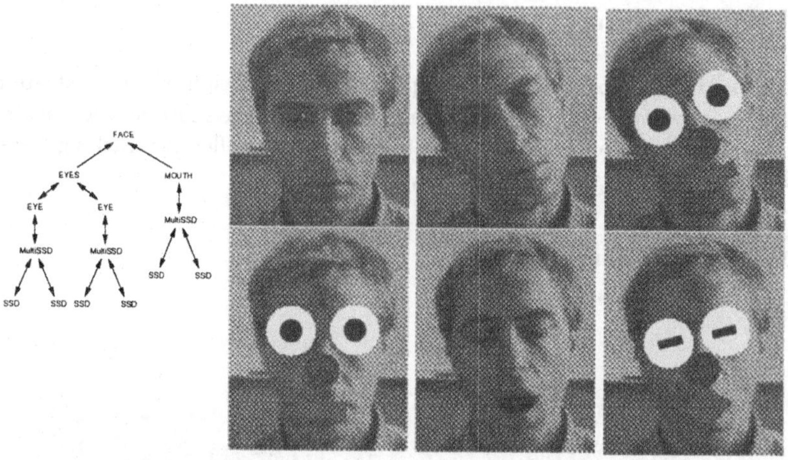

Fig. 4. The tracking network used for face tracking and the resulting output of the "clown face" tracker.

has lowest (best) match value and the identity of this tracker. The constraint function propagates this state to all subsidiary features with the result that the "losing" SSD trackers follow the "winner." In effect, the `MultiSSD` feature is a tracker which includes an n-ary switch.

From `MultiSSD` we derive an `Eye` tracker which modifies the display function of `MultiSSD` to show an open or closed eye based on the state of the binary switch. We also derive `Mouth` which similarly displays an open or closed mouth. Two `Eye`'s are organized into the `Eyes` tracker. The state computation function computes the location of the center of the line joining the eyes and its orientation. The constraint function propagates the orientation to the low-level trackers, obviating the need to compute orientation from image-level information. Thus, the low-level `Eye` trackers only solve for translation. The `Mouth` tracker computes both translation and orientation. The `Face` tracker is then a combination of `Eyes` and `Mouth`. It imposes no constraints on them, but its state computation function computes the center of the line joining the `Eyes` tracker and the `Mouth`, and its display function paints a nose there.

The tracker is initialized by clicking on the eyes and mouth and memorizing their appearance when they are closed and open. When run, the net effect is a graphical display of a "clown face" that mimics the antics of the underlying human face—the mouth and eyes follow those of the operator and open and close as the operator's do as shown in Figure 4. This system requires less than 10 milliseconds per iteration disregarding graphics.

Disk Tracking One important application for any tracking system is model-based tracking of objects for applications such as hand-eye coordination or virtual reality. While a generic model-based tracker for three-dimensional objects for this system can be constructed [6], X Vision makes it possible to gain additional speed and robustness by customizing the tracking loop using object-specific geometric information.

In several of our hand-eye experiments we use small floppy disks as test objects. The most straightforward rectangle tracker is simply a composite tracker which tracks four corners, which in turn are composite trackers which track two edges each. Four corners are then simply tracked one after the other to track the whole rectangle with no additional object information.

Increased speed and robustness can be attained by subsampling of the windows along the contour and by computing edge orientations based on corner positions. The latter is an example of a top-down geometric constraint.

As shown in Figure 5, there is no loss of precision in determining the location of of the corners with reasonable sampling rates, since the image is sampled less frequently only in a direction parallel to the underlying lines. We also see a 10% to 20% speedup by not computing line orientations at the image level and a nearly linear speedup with image subsampling level.

line length (pixels)	sampling rate	tracking speed (msec/cycle)		σ of position (pixels)	
		A	B	A	B
24	1	9.3	7.7	0.09	0.01
12	2	5.5	4.5	0.10	0.00
8	3	3.7	3.3	0.07	0.03
6	4	3.0	2.7	0.05	0.04
2	12	1.9	1.6	0.05	0.04
1	24	1.5	1.3	0.05	0.07

Fig. 5. Speed and accuracy of tracking rectangles with various spatial sampling rates. The figures in column A are for a tracker based on four corners computing independent orientation. The figures in column B are for a tracker which passes orientation down from the top-level composite feature.

3.2 An Embedded Application

As an illustration of the use of X Vision embedded within a larger system, we briefly describe some results of using X Vision within a hand-eye coordination we have recently developed [3, 5]. The system relies on image-level feedback from two cameras to control the relative pose between an object held in a robot end-effector and a static object in the environment.

The typing capabilities of X Vision make it possible to abstract the function of the hand-eye coordination primitives from their visual inputs. The hand-eye system implements a set of *primitive skills* which are vision-based regulators for attaining a particular geometric constraint between the pose of a robot-held object and a target object. For example, two primitive skills are point-to-point positioning and point-to-line positioning. Each of these skills must acquire inputs from a stereo pairs of trackers for point features or line features as appropriate. Hence, they are written in terms of the point-type features and the line-type features as defined by the X Vision typing system. In this way the same feedback methods can be used with a variety of application-specific tracking configurations without change. For example, touching two disks corner to corner and positioning a robot in a point on a plane defined by four points both operate with the same point-to-point controller (Figure 6). The only difference is in the tracking, which in these cases tracks two point-type features and five point-type features (one for the robot and four for the plane), respectively, where each point-feature may be composites of the necessary basic features.

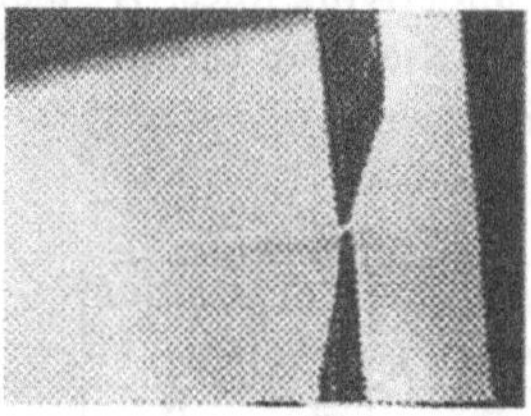

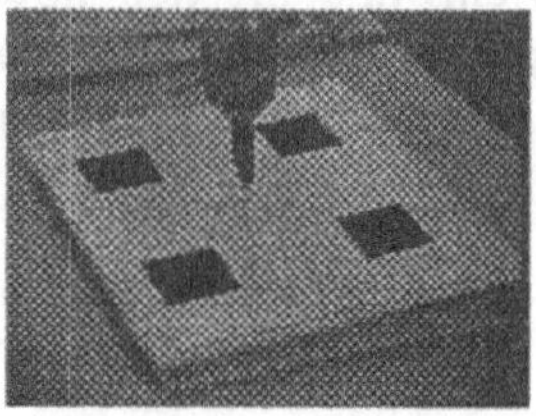

Fig. 6. The results of performing point-to-point positioning to observable features (left) and to a setpoint defined in the plane (right).

Because the accuracy of primitive skills depends only on the accuracy of feature location in the image [3], the physical accuracy of hand-eye experiments can be used to directly determine the accuracy of our feature localization algorithms. At a camera baseline of approximately 30cm at distances of 80 to 100cm, repeated experiments show that positioning accuracy is typically within a millimeter of position (see Figure 6 (left) for two 2.5mm thick diskettes). Simple calculations show that this positioning accuracy implies a corner localization accuracy of ±0.15 pixels.

Alignment tasks are performed similarly by use of line-type features and line-to-line controllers. By combining positioning and alignment skills, we have been able to fit screwdrivers on screws and insert floppy disks into disk drives.

4 Conclusions

We have presented the X Vision system for fast visual tracking. The main features of this system are the strong use of image warping, highly optimized low-level tracking methods, and a simple notion of feature combination. In addition, modularity, portability, and flexibility have made it an ideal framework for vision and robotics research, both by in-house researchers and others.

We believe that this paradigm will have a large impact on real-time vision applications. We are currently continuing to advance the state of the art by considering how to build tracking methods that are faster and more robust [10]. We are also extending the capabilities of the system toward a complete vision programming environment. Information on the current version is available at `http://www.cs.yale.edu/HTML/YALE/CS/AI/VisionRobotics/YaleAI.html`.

Acknowledgments This research was supported by ARPA grant N00014-93-1-1235, Army DURIP grant DAAH04-95-1-0058, by National Science Foundation grant IRI-9420982, and by funds provided by Yale University.

References

[1] P. Anandan. A computational framework and an algorithm for the measurement of structure from motion. *Int'l Journal of Computer Vision*, 2:283–310, 1989.
[2] G. D. Hager. Real-time feature tracking and projective invariance as a basis for hand-eye coordination. In *Proc. IEEE Conf. Comp. Vision and Patt. Recog.*, pages 533–539. IEEE Computer Society Press, 1994.
[3] G. D. Hager. A modular system for robust hand-eye coordination. DCS RR-1074, Yale University, New Haven, CT, June 1995.
[4] G. D. Hager. The "X-vision" system: A general purpose substrate for real-time vision applications. DCS RR-1078, Yale University, New Haven, CT, December 1995. Submitted to Image Understanding.
[5] G. D. Hager, W-C. Chang, and A. S. Morse. Robot hand-eye coordination based on stereo vision. *IEEE Control Systems Magazine*, 15(1):30–39, February 1995.
[6] C.-P. Lu. *Online Pose Estimation and Model Matching.* PhD thesis, Yale University, 1995.
[7] B. D. Lucas and T. Kanade. An iterative image registration technique with an application to stereo vision. In *Proc. Int. Joint Conf. Artificial Intelligence*, pages 674–679, 1981.
[8] J. Shi and C. Tomasi. Good features to track. In *Computer Vision and Patt. Recog.*, pages 593–600. IEEE Computer Society Press, 1994.
[9] C. Tomasi and T. Kanade. Shape and motion from image streams: a factorization method, full report on the orthographic case. CMU-CS 92-104, CMU, 1992.
[10] K. Toyama and G. D. Hager. Distraction-proof tracking: Keeping one's eye on the ball. In *IEEE Int. Workshop on Intelligent Robots and Systems*, volume I, pages 354–359. IEEE Computer Society Press, 1995. Also Available as Yale CS-RR-1059.

Human Body Tracking by Monocular Vision

F.Lerasle, G.Rives, M.Dhome, A.Yassine

Université Blaise-Pascal de Clermont-Ferrand,
LAboratoire des Sciences et Matériaux pour l'Electronique, et d'Automatique
URA 1793 of the CNRS, 63177 Aubière Cedex, France

Abstract. This article describes a tracking method of 3D articulated complex objects (for example, the human body), from a monocular sequence of perspective images. These objects and their associated articulations must be modelled. The principle of the method is based on the interpretation of image features as the 3D perspective projections points of the object model and an iterative Levenberg-Marquardt process to compute the model pose in accordance with the analysed image.
This attitude is filtered (Kalman filter) to predict the model pose relative to the following image of the sequence. The image features are extracted locally according to the computed prediction.
Tracking experiments, illustrated in this article by a cycling sequence, have been conducted to prove the validity of the approach.

Key-words : monocular vision, articulated polyhedric model, matchings, localization, tracking.

1 Introduction

In the last ten years, many researchers have tried to locate human body limbs by video or cine-film techniques. Their methods can be classified in function of the type of markers which are used. We can mention the marker free and marker based methods.

With regard to marker based methods, digitized data analysis can be done from anatomical landmarks by the placement of markers on the specific joints. The body can then be modelled using a multi link chain model (Winter [1]). After matchings between the primitives of the model and the digitized data, joint centers are reconstructed (Yeadon [2]). Automatic tracking system (Elite 1989) are also widely used. This system combines the real-time analysis of video-images, the signals acquisition relative to the muscular activity and external forces. The markers are passive (reflector markers) or active (infrared blankers, electro-luminescent Light Emitting Diodes). The active methods make the recognition and the tracking of the markers easier because each LED emission can be analysed separately. Yet, the physical constraints are greater.

Using these kind of markers are problematic. The non-rigidity wrapping during movement causes a relative body/markers displacement and induces uncertainty in the results. Moreover, wearing this kind of marker is quite easy for the ankles and wrists, but is difficult for complex articulations like shoulders, knees, hips. Moreover, adding passive or active markers induces some psychological effects on the subject such as rigidity in movements. To obtain accurate measurement, it is better to reduce the constraints on the subject as much as possible.

With regard to marker free methods, a well known technique in image processing is based on model matching using Distance Transformation (DT). It is described for instance in [3]. The method consists in making a DT image in which the value of each pixel is the distance to the nearest point in the object. The optimal position and orientation of the model can be found by minimizing some criteria function of these pixel values. Persson [4] uses this method with a simple 2D model of a leg prothesis.

Like Persson, Geurtz [5] developed a method based to a 2D representation of the body. The body limbs of the model are restricted to elliptical curves describing the segment contours. The image feature, used in the movement estimation, is only constituted by the contour data. His method is interesting but sensitive to noise. Consequently, results obtained on real images are somewhat inaccurate.

To solve the ambiguity (movement-depth), some researchers have proposed volumic models based on a priori knowledge of the human body. Rohr [6] introduces a model based approach for the recognition of pedestrians. He represents the human body by a 3D model consisting of cylinders, whereas for modelling the movement of walking he uses data from medical motion studies. The estimation of model parameters in consecutive images is done by applying a kalman filter. Wang [7] gives models of the different body limbs with simple geometrical primitives (such as cylinders, planar surfaces...) connected together by links which simulate articulations. The different images of the sequence are divided into regions by motion segmentation. From these detected regions and an affine model of the a priori movement, Wang deduces the 2D movement in each image of the set. With the knowledge of the volumic model, he estimates the 3D parameters of the model pose.

The modelisation error, due to representation with simple geometrical primitives, can have a negative effect on the interpretation result. A more precise modelisation should improve the results of the analysis.

In Robotic field, very few papers address the current problem which corresponds to the estimation of the spatial attitude of an articulated object from a single perspective image. Mulligan [8] presents a technique to locate the boom, the stick and the bucket of an excavator.

Kakadiaris [9] presents a novel approach to segmentation shape and motion estimation of articulated objects. Initially, he assumes the object consists of a single part. As the object attains new postures, he decides based on certain criteria if and when to replace the initial model with two new models. This approach is applied iteratively until all the object's moving parts are identified. Yet, two observed object's moving parts are supposed to be linked by only one inner degree of freedom.

2 Aim of the method

The research deals with the automatic analysis of 3D human movement based on a vision system. Like most of the developed methods, we need some prerequisites which are the knowledges of the observed object, free matchings between 2D image primitives and 3D model elements, and assumption about the projection of the real world on the image (perspective in our case). Our analysis will be based on the articulated volumic model and a localization process which computes the attitude of the 3D object model such that the selected model elements are projected on the matched 2D image primitives. A similar approach can be found in [10] where Lowe proposes a technic to fit a parametrized 3D model to a perspective image. Nevertheless, our method differs from Lowe 's one in the minimized criterion: Lowe uses a 2D criterion calculated in the image plane. Our criterion is 3D one which permits to greatly simplify the computations involved.

To give more precision about the model, the human body model will be deduced from Resonnance Magnetic Imaging (RMI) measurements and will be manipulate like a set of rigid parts which are articulated to each other. A cycling sequence was taken to illustrate this research.

3 Method

3.1 Model description

The modelisation of the human body is similar to the results of work carried out in the laboratory on articulated objects (Yassine [11]). We describe them briefly.

In our system, the articulated object comprises several Computer Assisted Design models (CAD) connected by articulations which describe the possible relative displacements between parts. Each CAD model corresponds to the polyhedrical approximation of the real part of the object. It is composed by a set of 3D vertices, a set of links between vertices to define the ridges and a set of links between ridges to build the different faces.

Each articulation is characterized by a set of degrees of freedom. Each inner degree of freedom is defined by its type (rotation and translation) and its 3D axis.

To animate the model, we have defined operators which allow us to place, in the observer frame, a 3D model vertice which initially defined in the model frame (see Yassine [11]).

3.2 Matchings 2D-3D

Before the localization process, we must extract some image features and match them with the associated geometrical primitives of the articulated model. Dhome [12] computes the pose of a simple polyhedrical object from matchings between all the visible model ridges and segments extracted from grey images. In our application, where the CAD model associated with each part (shank, thigh...) has a repetitive structure (comparable to a skew netted surface), such an approach is not applied. Only ridges, which are limbs, will be matched. A limb is a ridge common both to a visible surface and an invisible surface, after projection of the model in the image plane. Obviously, the non-rigidity of the bodily wrapper, during the movement, causes an incoherence comparatively to the static model (and so comparatively to the detected limbs). Yet, for smooth movement (like pedaling), deformations are quite insignificant.

To bring more constraints, some random points are chosen on the model surface and are matched with their features detected in the image. This tracking of specific points is comparable to a classic technique of marker tracking, but the advantages are obvious : the number and the location of these characteristic points are not pre-defined compared to markers. These points can be ignored or replaced by other random points during the tracking process.

The primitives extracted from the image are also straight segments or points.

Matchings 2D-3D from points: Given the model pose relative to the image I^1, some 2D points p_i^1 are selected in this first image and their 3D coordinates are deduced by reverse perspective projection. The equivalents p_i^n ($i = 1..p$) of the points p_i^1 are searched in the consecutive images (noted I^n) of the sequence by analysis of the images grey levels.

A priori, it is necessary to search the points p_i^n in a sufficiently large zone of the image I^n to include the displacement existing between the homologous points of the two images. To reduce the combinative, we take the predicted position estimated for the image I^n into account. This prediction will be computed by a Kalman filter ([13], [14]). The points p_i^n in the image I^n will be researched in proximity of the predict projected point (noted $p_i^n pred$) of the model. Each point p_i^n is obtained after maximizing correlation scores on grey levels windows included respectively in images 1 and n (Lerasle [15]). In fact, the predict pose makes it possible to restrict the research domain for the correlation and thus to reduce the processing cost.

To sum up the method, some 3D points of the model are associated to textured figures (centred at points p_i^1) of the initial grey level image. These points are localized in the following images of the sequence after correlation on the grey levels. The localization of these 2D points, in the different images of the sequence, will be better if the texture included in the images is rich. For this reason, during the film, the observed subject is wearing a pair of tights with non repetitive texture.

Moreover, to manage the possible vanishing of these 3D points and thus the textured designs associated with the initial image, this list of 3D model points will be modified during tracking. From one image to another, from the computed attitude, some points are removed from the list and replaced by others chosen randomly in their proximities to respect the initial spatial spread. This modification is managed like a stack Last In First Out so that after n localizations corresponding to the n first images of the sequence, all the points in the initial list have been removed. It is in fact, a markers system (or rather textured designs markers) sliding.

Matchings 2D-3D from limbs: At step n of the tracking (image I^n), for the attitude according to the image I^{n-1}, a Z-buffer or depth-buffer method (Catmull [16]) allows to extract the model ridges which correspond with limbs.

The idea of wearing a textured dark pair of tights moving in front of a white background can also be used. Projections of the points extremities (noted $A_i^{n-1}B_i^{n-1}$) of the limbs L_i are removed on to the white/dark transitions of the image I^{n-1}. Then, the correspondents $A_i^nB_i^n$ (in the image I^n) of the points $A_i^{n-1}B_i^{n-1}$ will be deduced after correlation on the respective grey levels. The segment $A_i^nB_i^n$ will be matched with the limb L_i and so with the 3D associated ridge of the model.

3.3 Localization process

We describe the problem of the localization of an articulated object, give briefly the mathematical equations and the algorithm used to solve them. The validity of this algorithm was proved by Yassine [11] on articulated polyhedrics objects like an operator arm. The pose estimation of an articulated object from a monocular perspective image depends on $10 + q$ parameters. The first four ones are the intrinsic camera parameters. The six following ones are the extrinsic parameters (noted $\alpha, \beta, \gamma, u, v, w$). These parameters correspond to the three rotations and the three translations around and along the observer frame axis which permit to locate a rigid object. In the present process, these parameters determine the pose of the reference part of the viewed articulated object. The q following ones (noted $a_1, ..., a_q$) represent the inner degrees of freedom.

The problem we intend to address can be described as follows. We suppose known : the perspective projection as a model of image formation, the intrinsic parameters of the acquisition system, the CAD model of the viewed object and a sufficient set of matchings between image features and model primitives.

Then, we must obtain the $6 + q$ parameters which define the object location minimizing the sum of the distance between the matched model primitives and the interpretation planes(planes through the optical center of the camera and the considered segments). Thus, for the model primitives like ridges, only the extremities of these ridges are considered. For each matched ridge, we minimize the sum of the distance between the two extremities of the model ridges and the correspondent interpretation plane. For the matchings on points, the matched points in the image ($p1$ in figure 1) will be replaced by two segments which will be perpendicular in the image. Then, we will minimize the distance between the 3D point ($P1$ in the figure 1) of the model and these two perpendicular interpretation planes.

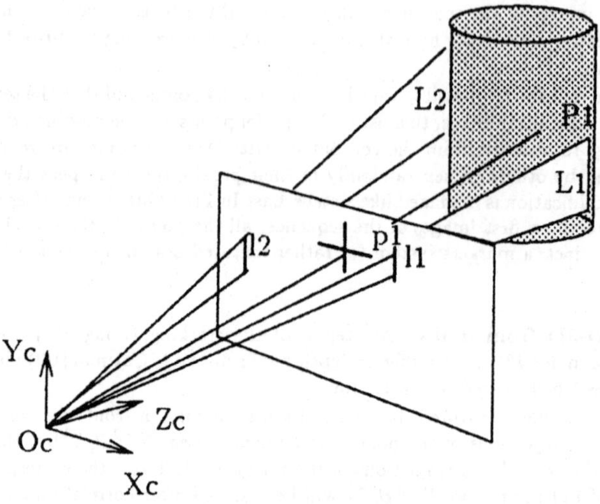

Fig. 1. example of interpretation planes

Distances to interpretation planes: In the following equations, the index c refers to the observer frame, m refers to the model frame and the variable A refers to the position vector with $A = (\alpha, \beta, \gamma, u, v, w, a_1, .., a_q)$. Let p image segments l_i be matched respectively with p model points P_i^m. It was assumed that all the vectors and points are expressed in the camera coordinate system. To compute the vector A, we merely express that the transformed point P_i^c by the transformation represented by A of the point P_i^m must lie in the interpretation plane Π_i (normal $\vec{N_i}$) of the corresponding image segment. This can be written by a product scalar function. A method to linearize this function consists in approximing this function near the value $(A)_k$, A at step k, by a first order Taylor development :

$$F(A, P_i^m) = (\vec{N_i}.\vec{OP_i^c}) \approx F(A_k, P_i^m) + \frac{\partial F(A_k, P_i^m)}{\partial A}(A - A_k)$$

where $i = 1..n$ is the index of the matched point (n the number of matched points).

Resolution of the system: If A is the solution, $F(A, P_i^m) = 0$ and thus the set of 3D matched points allows to construct a system with n linear equations : $[J]_k^T.(E_k) = [J]_k^T.[J]_k.(\Delta A)_k$ with

$$(\Delta A)_k = \begin{pmatrix} \alpha_k - \alpha \\ . \\ . \\ a_{1k} - a_1 \\ . \\ . \\ a_{qk} - a_q \end{pmatrix}, (E)_k = \begin{pmatrix} F(A_k, P_i^1) \\ F(A_k, P_i^2) \\ . \\ . \\ F(A_k, P_i^n) \end{pmatrix}, (J)_k = \begin{pmatrix} \frac{\partial F(A_k, P_1^m)}{\partial \alpha} & \dots & \frac{\partial F(A_k, P_1^m)}{\partial a_q} \\ \frac{\partial F(A_k, P_2^m)}{\partial \alpha} & \dots & \frac{\partial F(A_k, P_2^m)}{\partial a_q} \\ . & \dots & . \\ . & \dots & . \\ \frac{\partial F(A_k, P_n^m)}{\partial \alpha} & \dots & \frac{\partial F(A_k, P_n^m)}{\partial a_q} \end{pmatrix}$$

This system will be solved by an iterative Levenberg-Marquardt approach [17]. By this way, the global criterion to minimize is :

$$Error = \sum_{i=1}^{p} (\vec{N_i}.\overrightarrow{OP_i^c})^2$$

At each iteration of the iterative process, we obtain a correction vector to apply to the position vector $(A)_k$: $(A)_{k+1} = (A)_k + (\Delta A)_k$

The iterative process is repeated until a stable attitude is reached meaning $Error < \epsilon$. In any case, the matrix of partial derivatives $(J)_k$ must be computed and these calculations will be set out in detail in [15]. The computation of the covariance associated with the model attitude have been detailed in [15] too.

4 Experiments and results on tracking

The matchings between 3D model primitives in the attitude linked to an image of the sequence and 2D primitives in the following image are automatic. At the step n of the tracking (image I^n), the matchings process takes the attitude computed for the image I^{n-1} into account. So, the attitude linked to the first image of the sequence and the set of markers linked to this image must be known in advance. In fact, initialization of the tracking process requires two steps :

1. manually, the operator superimposed the model to the first image of the sequence,
2. for this attitude, the operator manually selects some visible points from the model. A good spatial spread of the set of these points increases the constraints caused by these points.

For example, the following figure represents the first image of the cycling sequence and the markers which have been choosen.

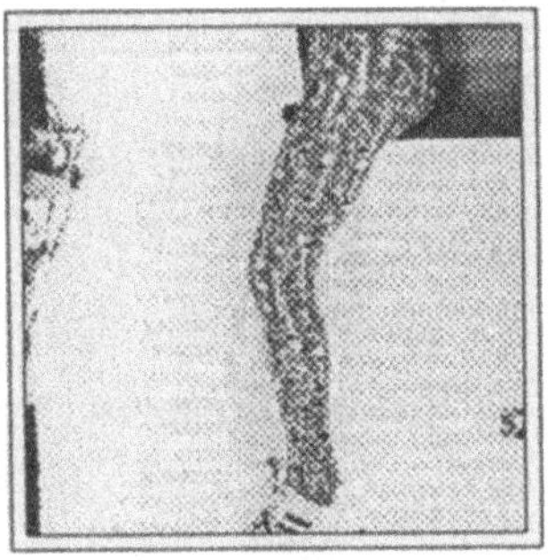

Fig. 2. initial grey level image with selected 3D points projection

The tracking process have been validated on real images sequences. We have choosen to present the results obtained on a cycling sequence. First, we have built the shank and thigh CAD models of the cyclist. These models have been deduced from images provided by an RMI scan. This scan consisted in 34 cross-sections of the legs. Low level treatement (smoothing and contours detection) makes it possible to extract peripheral contours of each cross-section. The contour points coordinates x and y of each cross-section, associated with its height z, enables us to determine the 3D vertices of the model (see figure 3).

Moreover, we had to compute the articulation model of the knee which is quite complex. An articulation model with three rotations, corresponding to the flexion-extension (axis Oy), to the internal rotation (axis Oz) and the valrus-valgus rotation (axis Ox) has been choosen.

The figure 4 represents the projection of the attitudes computed during the sequence from a point of view located in the cycling plan. The six following figures (end of article) represent the model projection superimposition on different images of the sequence. Obviously, the model surfaces which are not really compatible with the grey-level image target are distort surfaces like the back of the leg or calf.

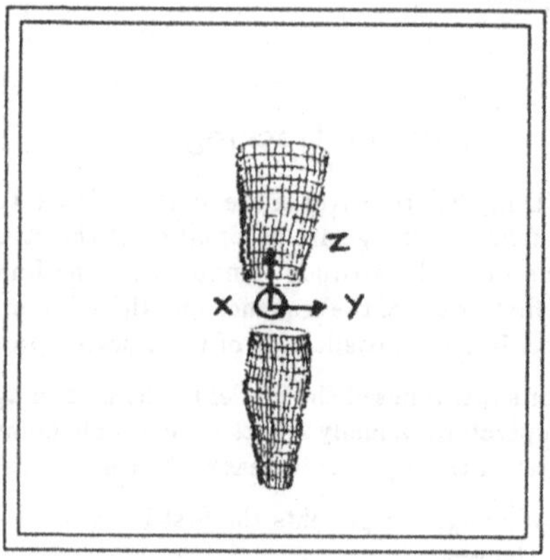

Fig. 3. leg model used for the pedaling sequence

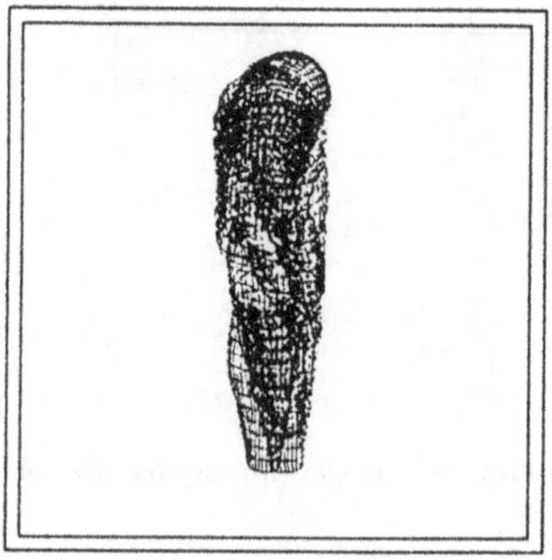

Fig. 4. model tracking from a point of view situed in front of the bicycle

5 Conclusion

This paper presents a new method for estimating, in the viewed coordinate system, the spatial attitude of an articulated object from a single perspective view.

The approach is based :

- on the a priori knowledge of the visualised object meaning the knowledge of its model,
- on the interpretation of some image points as the perspective projection of 3D model points,
- on an iterative search of the model attitude consistent with these projections,
- on a calculation of the covariance matrix associated with these projections.

The presented method is quite different from the markers method because we don't use real but fictitious markers (through a textured pair of tights). Thus, the proposed method is more flexible because the number of these markers and their emplacements are not a priori fixed. Moreover, wearing a pair of tights causes no psychological effects, no physical constraints and no added techniques (comparatively to active markers method).

The inaccuracies of the method are in process initialization and especially in the approximate estimation of the model attitude compatible with the first image. This first pose has an effect on the quality of the localizations obtained during the tracking. Moreover, the defined articulated static model is not always consistent with the image target because the body wrapping is not always constant during the movement.

Our next purpose will be to improve both the initialization and the modelisation, and to analyze occulting movement, for example the occulting of one leg by one another. The implemented Kalman formalism should help us to cope with this kind of problem.

To our knowledge, such marker free method has never been applied to human movement and could be useful to further study biomechanical and energetics of muscular activity aspects.

References

1. A.D. Winter. A new definition of mechanical work done in human movement. *J. Aplli. Physiol.*, 46:79–83, 1979.
2. M.R. Yeadon. A method for obtaining three-dimensionnal data on ski jumping using pan and tilt camera. *International Journal of Sport Biomechanics*, 5:238–247, 1989.
3. G. Borgefors. On hierarchial edge matching in digital images using distance transformations. *Internal Report of the Royal Inst. of Technology, Stockholm*, 1986.
4. T. Persson and H. Lanshammar. Estimation of an object's position and orientation using model matching. In *Proc. of the sixth I.S.B Congress*, 1995.
5. A. Geurtz. *Model Based Shape Estimation.* PhD thesis, Ecole Polytechnique de Lausanne, 1993.
6. K. Rohr. Towards model based recognition of human movements in image sequences. *Image Understanding*, 59(1):94–115, January 1994.
7. J. Wang. *Analyse et Suivi de Mouvements 3D Articulés : Application à l'Etude du Mouvement Humain.* PhD thesis, IFSIC, Université Rennes I, 1992.
8. I.J. Mulligan, A.K. Mackworth, and P.D. Lawrence. A model based vision system for manipulator position sensing. In *Proc. of Workshop on Interpretation of 3D scenes, Austin, Texas*, pages 186–193, 1990.
9. I.Kakadiaris, D.Metaxas, and R.Bajcsy. Active part-decomposition, shape and motion estimation of articulated objects : a physics-based approach. In *Computer Vision and Pattern Recognition*, 1994.
10. D.G. Lowe. Fitting parameterized three-dimensional models to images. *PAMI*, 13(5):441–450, May 1991.
11. A. Yassine. *De la Localisation et du Suivi par Vision Monoculaire d'Objets Polyédriques Articulés Modélisés.* PhD thesis, Université Blaise Pascal de Clermont-Ferrand, 1995.
12. M. Dhome, M. Richetin, J.T. Lapresté, and G. Rives. Determination of the attitude of 3d objects from a single perspective image. *I.E.E.E Trans. on P.A.M.I*, 11(12):1265–1278, December 1989.

13. N. Daucher, M. Dhome, J.T. Lapresté, and G.Rives. Modelled object pose estimation and tracking by monocular vision. In *British Machine Vision Conference*, volume 1, pages 249–258, 1993.
14. N. Ayache. *Vision Stéréoscopique et Perception Multisensorielle.* Inter Editions, 1987.
15. F. Lerasle, G. Rives, M. Dhome, and A. Yassine. Suivi du corps humain par vision monoculaire. In *10th Conf. on Reconnaissance des Formes et Intelligence Artificielle*, January 1996.
16. E. Catmull. *A Subdivision Algorithm for Computer Display of Curved Surfaces.* PhD thesis, University of Utah, 1974.
17. D.W. Marquardt. *Journal of the Society for Industrial and Applied Mathematics*, 11:431–441, 1963.

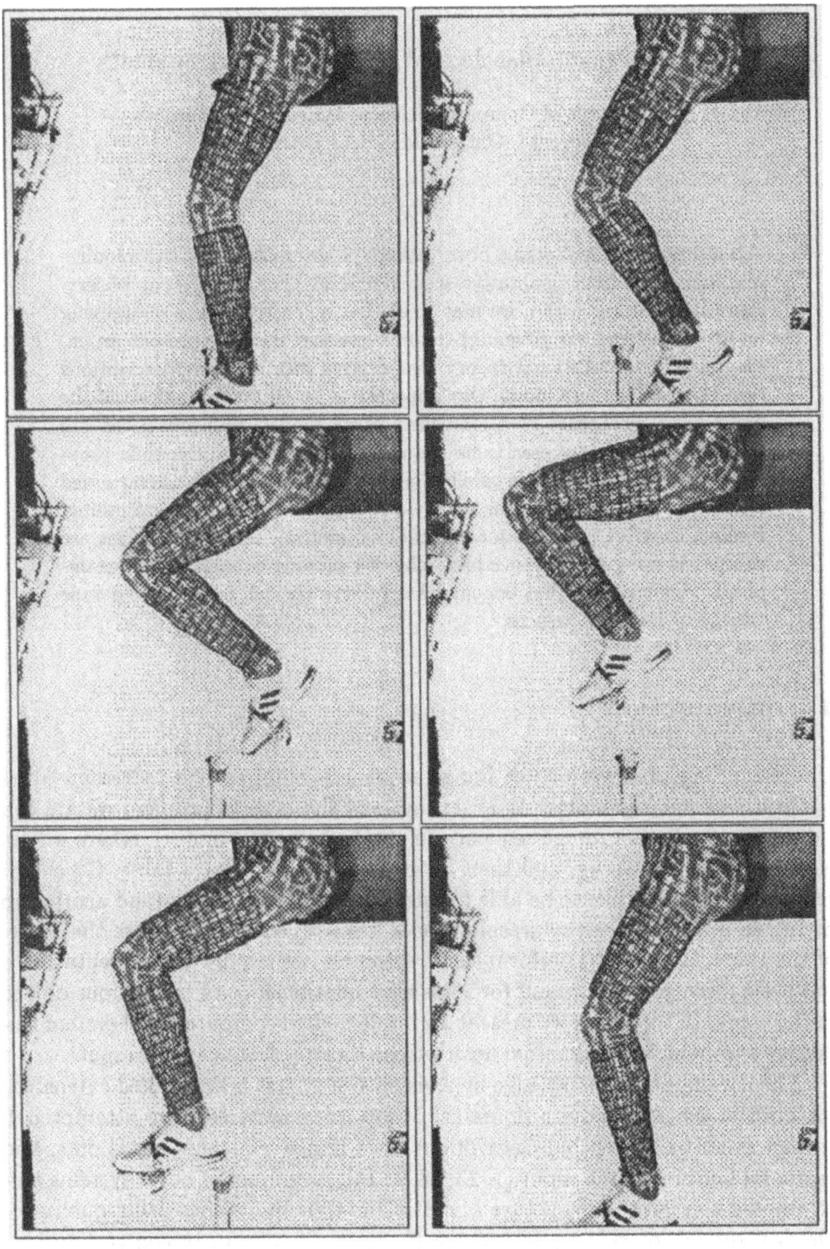

Fig. 5. model projection superimposed to the images 1,10,20,30,40,50 of the sequence

Computational Perception of Scene Dynamics

Richard Mann, Allan Jepson*, and Jeffrey Mark Siskind**

Department of Computer Science, University of Toronto
Toronto, Ontario M5S 1A4 CANADA

Abstract. Understanding observations of interacting objects requires one to reason about qualitative scene dynamics. For example, on observing a hand lifting a can, we may infer that an 'active' hand is applying an upwards force (by grasping) to lift a 'passive' can. We present an implemented computational theory that derives such dynamic descriptions directly from camera input. Our approach is based on an analysis of the Newtonian mechanics of a simplified scene model. Interpretations are expressed in terms of assertions about the kinematic and dynamic properties of the scene. The feasibility of interpretations can be determined relative to Newtonian mechanics by a reduction to linear programming. Finally, to select plausible interpretations, multiple feasible solutions are compared using a preference hierarchy. We provide computational examples to demonstrate that our model is sufficiently rich to describe a wide variety of image sequences.

1 Introduction

Understanding observations of image sequences requires one to reason about qualitative scene dynamics. As an example of the type of problem we are considering, refer to the image sequence in the top row of Figure 1, where a hand is reaching for, grasping, and then lifting a coke can off of a table. Given this sequence, we would like to be able to infer that an 'active' hand (and arm) is applying an upward force (by grasping) on a 'passive' coke can to raise the can off of the table. In order to perform such reasoning, we require a representation of the basic force generation and force transfer relationships of the various objects in the scene. In this work we present an implemented computational system that derives symbolic force-dynamic descriptions directly from camera input.

The use of domain knowledge by a vision system has been studied extensively for both static and motion domains. Many prior systems have attempted to extract event or conceptual descriptions from image sequences based on spatio-temporal features of the input [1, 23, 18, 4, 14]. A number of other systems have attempted to represent structure in static and dynamic scenes using qualitative physical models or rule-based systems [6, 8, 13, 20, 22, 5]. In contrast to both

* Also at Canadian Institute for Advanced Research.

** Current address: Department of Electrical Engineering, Technion, Haifa 32000, ISRAEL of these approaches, our system uses an explicit physically-based representation based on Newtonian physics.

A number of other systems have used physically-based representations. In particular, Ikeuchi and Suehiro [10] and Siskind [21] propose representions of events based on changing kinematic relations in time-varying scenes. Also, closer to our approach, Blum *et. al.* [3] propose a representation of forces in static scenes. Our system extends these approaches to consider both kinematic and dynamic properties in time-varying scenes containing rigid objects.

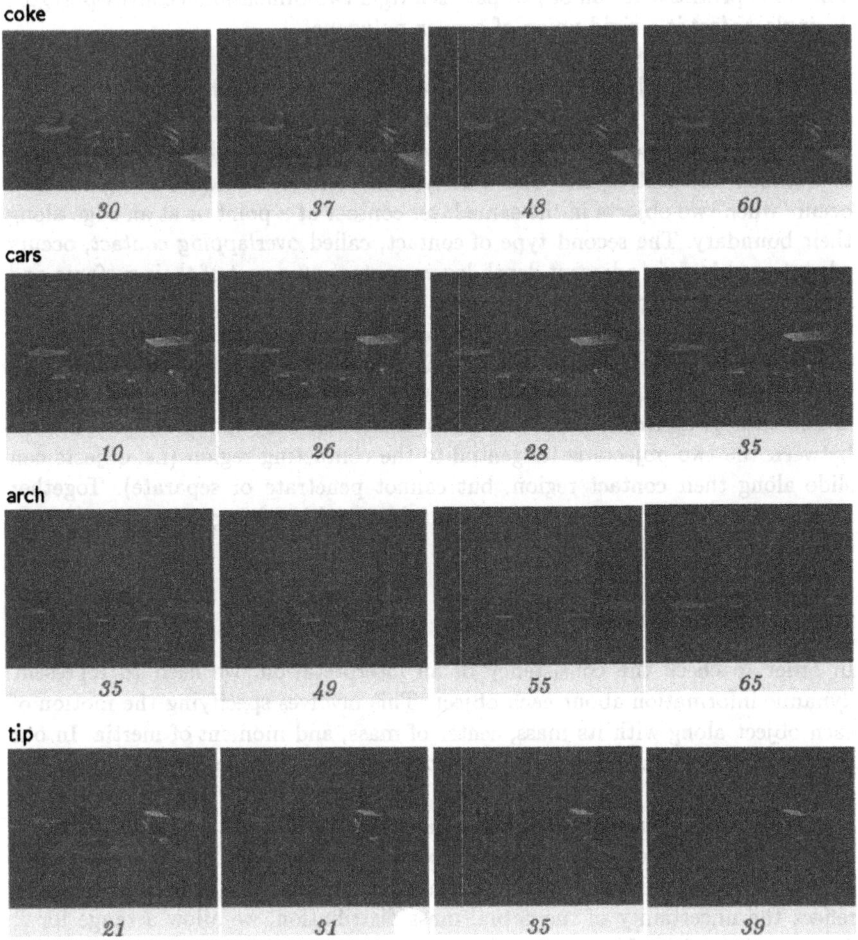

Fig. 1. The example sequences: coke, cars, arch, and tip. The frame numbers are given below each image.

2 Ontology

In this section we describe the form of the system's representation for its domain. This representation must be suitable for specifying the geometry of the scene interpretation and the type of forces that can be generated on the various objects. Moreover, in order to avoid unphysical interpretations, there must be a notion of consistency for particular scene models. We describe the representation of the geometry, the types of forces, and the notion of consistency in the next sections.

2.1 Kinematic Model

The basic primitive for an object part is a rigid two-dimensional convex polygon. A single *object* is a rigid union of convex polygons.

To represent the spatial relationship between objects in the scene we use a *layered* scene model. In our layered model there is no depth ordering. Instead, we represent only whether two objects are in the same layer, in adjacent layers, or in layers separated in depth. Objects can contact either within the same layer or between adjacent layers. The first type of contact, called *abutting contact*, occurs when two objects in the same layer contact at a point or at an edge along their boundary. The second type of contact, called *overlapping contact*, occurs when two objects in adjacent depth layers contact over part of their surfaces and the region of overlap has non-zero area.

In order for a given assignment of contacts to be *admissible* two types of constraints must be satisfied. First, each pair of objects considered to be contacting must actually intersect (but possibly just on their boundary). Second, in the case of abutment, the contact is admissible only if the relative motion between the two objects is tangential to the contacting region (i.e. objects can slide along their contact region, but cannot penetrate or separate). Together these constraints provide a weak kinematic model involving only pairwise constraints between objects.

2.2 Dynamic Model

In order to check the consistency of an interpretation, we need to represent dynamic information about each object. This involves specifying the motion of each object along with its mass, center of mass, and moment of inertia. In our system the 2D velocities, angular velocities, and accelerations of the objects are all provided by the image observations. An object's total mass is taken to be a positive, but otherwise unknown, parameter. We take each object's center of mass to be at the object's geometric center. For the case of two-dimensional motion considered in this paper the inertial tensor I is a scalar. In order to reflect the uncertainty of the actual mass distribution, we allow a range for I. An upper bound for I is provided by considering an extreme case where all of the mass is placed at the furthest point from the center. A lower bound is provided by considering an alternate case where all of the mass is distributed uniformly inside a disk inscribed in the object.

An object is subject to gravitational and inertial forces, and to forces and torques resulting from contact with other objects. The dynamics of the object under these forces is obtained from the physics-based model described in §3.

Finally, particular objects may be designated as *ground*. We typically use this for the table top. Forces need not be balanced for objects designated as ground.

It is convenient to define a *configuration* to be the set of scene properties that are necessarily present, given the image data and any restrictions inherent in the ontology. For example, in the current system, the positions, velocities, and accelerations of the objects are provided by the image observations, and the positions of the centers of mass are fixed, by our ontology, to be at the object centroids.

2.3 Assertions

In order to supply the information missing from a configuration, we consider *assertions* taken from a limited set of possibilities. These assertions correspond to our hypothesis about the various contact relations and optional types of force generation and force transfer relationships between objects.

Currently, our implementation uses the following *kinematic assertions* which describe the contact relationships between objects:

- CONTACT(o_1, o_2, c) — objects o_1 and o_2 contact in the scene with the region of contact c;
- ATTACH(o_1, o_2, p) — objects o_1 and o_2 are attached at some set p of points in the contact region.

The intuitive meaning is that attachment points are functionally equivalent to rivets, fastening the objects together. Attached objects can be pulled, pushed, and sheared without coming apart while, without the attachment, the contacting objects may separate or slide on each other depending on the applied forces and on the coefficient of friction.

In addition we consider the following *dynamic assertions* which determine the types of optional forces which might be generated:

- BODYMOTOR(o) — object o has a 'body motor' that can generate an arbitrary force and torque on itself;
- LINEARMOTOR(o_1, o_2, c) — a linear motor exists between the abutting objects o_1 and o_2. This motor can generate an arbitrary tangential shear force across the motor region c. This region must be contained within the contact region between the objects;
- ANGULARMOTOR(o_1, o_2, p) — an angular motor exists at a single point p that can generate an arbitrary torque about that point. The point p must be within the contact region between the objects.

The intuitive meaning of a BODYMOTOR is that the the object can generate an arbitrary force and torque on itself, as if it had several thrusters. LINEARMOTORS are used to generate a shear force across an abutment (providing an abstraction for the tread on a bulldozer). ANGULARMOTORS are used to generate torques at joints.

We apply the following admissibility constraints to sets of assertions. First the contact conditions described in §2.1 must be satisfied for each assertion of contact. Second, linear motors are admissible only at point-to-edge and edge-to-edge abutments but not at point-to-point abutments or overlapping contacts. Finally, angular motors are admissible only at a single point within the contact region between two objects and the objects must be attached at this point.

We define an interpretation $i = (C, A)$ to consist of the configuration C, as dictated by the image data, along with a complete set of assertions A. (A set of assertions is complete when every admissible assertion has been specified as being true or false.) In the next section we will show how to test the feasibility of various interpretations.

3 Feasible Interpretations

Given an interpretation $i = (C, A)$ we can use a theory of dynamics to determine if the interpretation has a feasible force balance. In particular, we show how the test for consistency within the physical theory can be expressed as a set of algebraic constraints that, when provided with an admissible interpretation, can be tested with linear programming. This test is valid for both two and three dimensional scene models.

For rigid bodies under continuous motion, the dynamics are described by the Newton-Euler equations of motion [9] which relate the total applied force and torque to the observed accelerations of the objects. Given a scene with convex polygonal object parts, we can represent the forces between contacting parts by a set of forces acting on the vertices of the convex hull of their contact region [7, 2]. Under this simplification, the equations of motion for each object can be written as a set of equality constraints which relate the forces and torques at each contact point to the object masses and accelerations.

The transfer of forces between contacting objects depends on whether the objects are in resting contact, sliding contact, or are attached. Attached objects have no constraints on their contact forces. However, contacts which are not asserted to be ATTACHed are restricted to have a positive component of normal force. In addition, contact points that are not part of a LINEARMOTOR have tangential forces according to the *Coulombic* model of friction. In particular, the magnitude of the tangential force is bounded by some multiple of the magnitude of the normal force. Both sliding and resting friction are modeled.

An interpretation is dynamically feasible if these motion equations can be satisfied subject to the contact conditions and the bounds on the mass and inertia described in §2.2. Since we can approximate these constraints by a set of linear equations and inequalities, dynamic feasibility can be tested using linear programming (see [17] for details).

4 Preferences

Given a fairly rich ontology, it is common for there to be multiple feasible interpretations for a given scene configuration. For example, for the lifting phase of the coke sequence in Figure 1 there are five feasible interpretations, as shown in Figure 2. Indeed, for any scene configuration there is always at least one trivial interpretation in which every object has a body motor, and thus multiple interpretations can be expected.

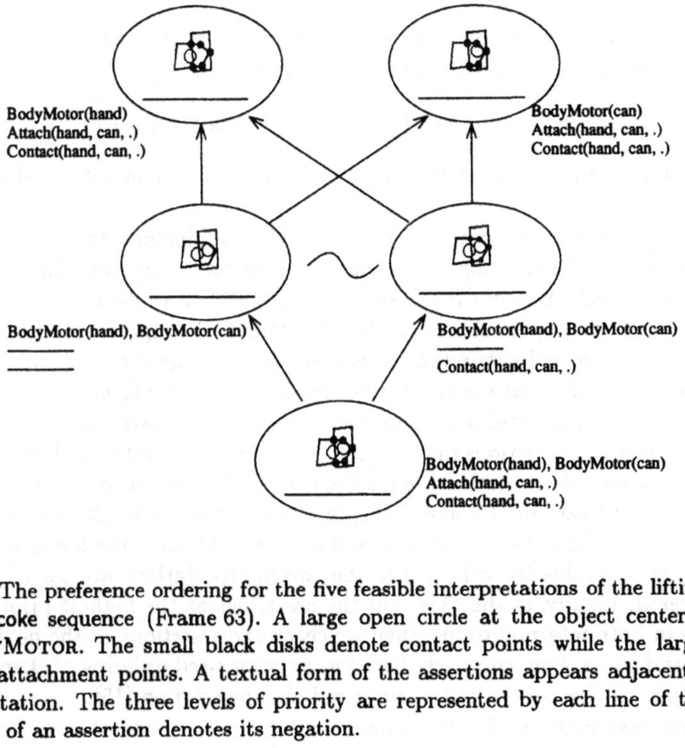

Fig. 2. The preference ordering for the five feasible interpretations of the lifting phase in the coke sequence (Frame 63). A large open circle at the object center denotes a BODYMOTOR. The small black disks denote contact points while the larger disks denote attachment points. A textual form of the assertions appears adjacent to each interpretation. The three levels of priority are represented by each line of text. The absence of an assertion denotes its negation.

Rather than searching for all interpretations, we seek interpretations that require, in some specified sense, the *weakest* properties of the various objects. We use model preference relations, as discussed by Richards, Jepson, and Feldman [19], to express a suitable ordering on the various interpretations. The basic idea is to compare pairs of interpretations using a prioritised set of elementary preference relations.

Our current ontology includes the following elementary preferences for the *absence* of any motor:

- $P_{bodymotor}(o) : \neg\textsc{BodyMotor}(o) \succ \textsc{BodyMotor}(o)$;

- $P_{linearmotor}(c)$: $\neg$LINEARMOTOR$(o_1, o_2, c) \succ$ LINEARMOTOR(o_1, o_2, c);
- $P_{angularmotor}(c)$: $\neg$ANGULARMOTOR$(o_1, o_2, p) \succ$ ANGULARMOTOR(o_1, o_2, p).

Here $\neg$ denotes the negation of the predicate that follows. These elementary preference relations all encode the specification that it is preferable not to resort to the use of a motor, all else being equal. The absence of a motor is considered to be a weaker assumption about an object's properties. These elementary preference relations appear at the highest priority.

At the next level of priority we have

- $P_{attach}(o_1, o_2, p)$: $\neg$ATTACH$(o_1, o_2, p) \succ$ ATTACH(o_1, o_2, p),

so the absence of an attachment assertion is also preferred. Finally, at the lowest level of priority, we have the indifference relation

- $P_{contact}(o_1, o_2, c)$: $\neg$CONTACT$(o_1, o_2, c) \sim$ CONTACT(o_1, o_2, c),

so the system is indifferent to the presence or absence of contact, all else being equal.

All of the above preferences, except for the indifference to contact, have the form of a preference for the negation of an assertion over the assertion itself. It is convenient to use the absence of an assertion to denote its negation. When the elementary preferences can be written in this simple form, the induced preference relation on interpretations is given by prioritised subset ordering on the sets of assertions made in the various feasible interpretations. As illustrated in Figure 2, we can determine the preference order for any two interpretations by first comparing the assertions made at the highest priority. If the highest priority assertions in one interpretation are a subset of the highest priority assertions in a second interpretation, the first interpretation is preferred. Otherwise, if the two sets of assertions at this priority are not ordered by the subset relation, that is neither set contains the other, then the two interpretations are considered to be unordered. Finally, in the case that the assertions at the highest priority are the same in both interpretations, then we check the assertions at the next lower priority, and so on. This approach, based upon prioritised ordering of elementary preference relations, is similar to prioritised circumscription [15].

To find maximally-preferred models, we search the space of possible interpretations. We perform a breadth-first search, starting with the empty set of assertions, incrementally adding new assertions to this set. Each branch of the search terminates upon finding a minimal set of assertions required for feasible force balancing. Note that because we are indifferent to contacts, we explore every set of admissible contact assertions at each stage of the search. While in theory this search could require the testing of every possible interpretation, in practice it often examines only a fraction of the possible interpretations since the search terminates upon finding minimal models.

Moreover, when the assertions are stratified by a set of priorities we can achieve significant computational savings by performing the search over each priority level separately. For example, under our preference ordering, we can search for minimal sets of motors using only interpretations that contain all admissible attachments. It is critical to note that this algorithm is only correct because of the special structure of the assertions and the domain. The critical property is that if there is a feasible interpretation $i = (C, A)$, and if A' is the set obtained by adding all of the admissible attachments to A, then the interpretation $i = (C, A')$ is also feasible. This property justifies the algorithm above where we set all of the lower priority assertions to the most permissive settings during each stage of the minimization. In general we refer to this property as *monotonicity* [16].

5 Examples

We have applied our system to several image sequences taken from a desktop environment (see Figure 1). The sequences were taken from a video camera attached to a SunVideo imaging system. MPEG image sequences were acquired at a rate of thirty frames per second and a resolution of 320×240 pixels. The 24-bit colour image sequences were converted to 8-bit grey-scale images used by the tracker.

As described in §2.1, we model the scene as a set of two-dimensional convex polygons. To obtain estimates for the object motions we use a view-based tracking algorithm similar to the optical flow and stereo disparity algorithms described in [12, 11]. The input to the tracker consists of the image sequence, a set of object template images (including a polygonal outline for each object), and an estimate for the object positions in the first frame of the sequence. In addition, we provide an estimate for the position of the table top which is designated as a *ground* object in our ontology. The tracking algorithm then estimates the position and orientation of these initial templates throughout the image sequence by successively matching the templates to each frame. The position of the object polygons is obtained by mapping the original outlines according to these estimated positions. Finally, the velocity and acceleration of the polygons are obtained using a robust interpolation algorithm on these position results.

In the current system we consider interpretations for each frame in isolation. Given estimates for the shapes and motions of the objects in each frame, we determine possible contact relations assuming a layered model as described in §2.1. For each possible contact set[1] we determine the admissible attachment and motor assertions described in §2.3. Finally, a breadth-first search is performed to find the preferred interpretations for each frame.

Figure 3 shows the preferred interpretations found for selected frames from each sequence. (Note that the selected frames do not necessarily match those shown in Figure 1.) For each sequence we show frames ordered from left to

[1] In the current system we consider only a single maximal contact set in which every admissible contact is added to the assertion set. Since there are no depth constraints in our layered model, this single contact hypothesis will not disallow any of the remaining assertions.

right.[2] While the preferred interpretations are often unique, at times there are multiple interpretations, particularly when objects interact. We highlight frames with multiple preferred interpretations by grey shading.

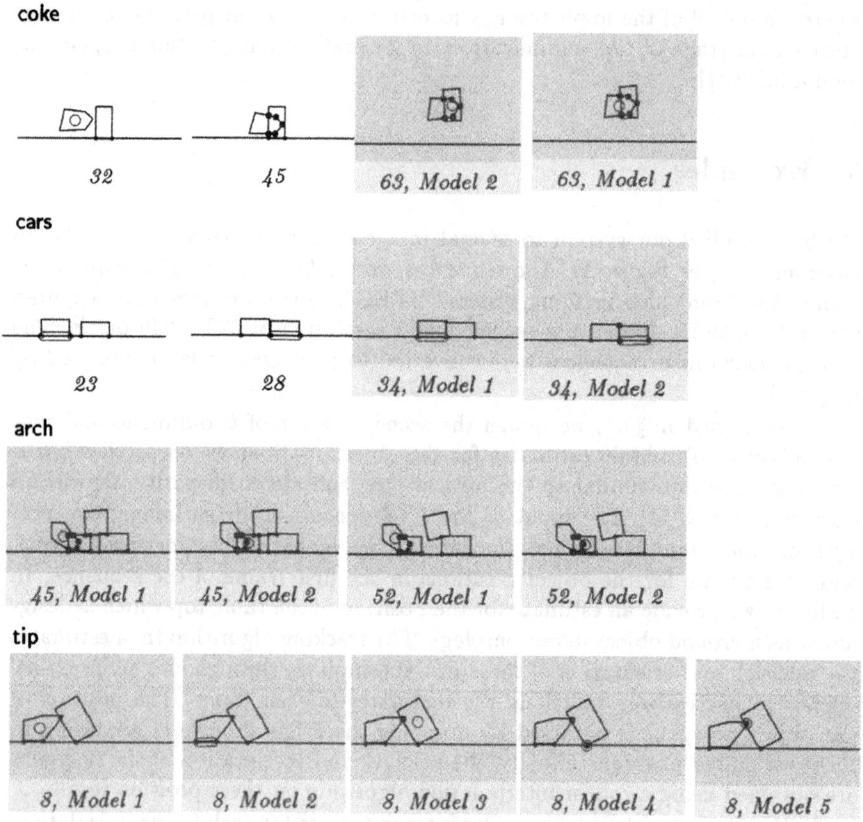

Fig. 3. Some preferred models for: coke, cars, arch, and tip. Frames with a non-unique maximally-preferred interpretation are shown with a grey background. We use the following symbols. For contacts, small disks denote contact points at rest while small circles denote sliding contacts. Larger disks denote attachment points. Motors are denoted by either a circle at the center of the object (BODYMOTOR), a circle around a contact point (ANGULARMOTOR), or by a closed curve around a contact region (LINEARMOTOR).

[2] For the cars and arch sequences there is an ambiguity in the type of motors used, with body motors being interchangeable with linear motors (except on the hand). For clarity we show only linear motors in the cars sequence, and only body motors in the arch sequence.

Our machine interpretations are surprisingly intuitive. For example, the difference between models 1 and 2 in frame 63 of the **coke** sequence can be interpreted as the hand 'lifting' the can versus the can 'lifting' the hand. Similarly, the difference between models 1 and 2 in frame 34 of the **cars** sequence can be interpreted as the rear car 'pushing' the front car versus the front car 'pulling' the rear car. (Note that the system correctly hypothesises an attachment between the front and rear cars in the 'pulling' interpretation, but does not do so in the 'pushing' interpretation.) The third row of Figure 3 shows the interpretations for the **arch** sequence in which a hand removes the left block from an arch causing the top block to tip over. The system correctly infers that the top block is *supported* in frame 45, and *tipping* in frame 52, but is not able to determine whether the hand is 'pulling' the left block or whether the left block is 'carrying' the hand. Finally, the last row of Figure 3 shows the results for the **tip** sequence where a hand raises a box onto its corner and allows it to tip over. There are five interpretations corresponding to various assertions of an active hand, active box, and various types of linear and angular motors.

While encouraging, our current implementation exhibits a number of anomalies. These anomalies generally fall into three classes. The first problem is that because we consider single frames in isolation, in many cases the system cannot find unique interpretations. In particular, since the system does not have any prior information about the objects in the scene, it cannot rule out interpretations such as an active coke can lifting the passive hand in the **coke** sequence or an active block pulling a passive hand in the **tip** sequence. In addition, because of our preference for minimal sets of assertions, certain degenerate interpretations may occur. An example of this is shown in frame 45 of the **coke** sequence, where the hand is interpreted as a passive object (which is attached to the coke can). Since the system does not have any prior information about object properties and since it considers single frames in isolation, all of these interpretations are reasonable.

A second problem concerns the detection of collisions and changing contact relations between objects. In particular, when objects collide, the estimates for relative velocity and acceleration at their contact points may differ, resulting in the contact relation being deemed inadmissible. An example of this is shown in frame 28 of the **cars** sequence where the contact between the colliding cars is missed. Note that the acceleration of the cars should be equal (since they remain in contact after the collision), but the interpolator has smoothed over this discontinuity and given unreliable estimates of the acceleration.

Finally, a third problem occurs because we do not use a complete kinematic model, as mentioned in §2.1. An example of this problem is shown in the **tip** sequence in Figure 3. While all of the interpretations have a feasible force balance, the last three are not consistent with rigid-body motion since it is not *kinematically* feasible for the hand to be both attached to the box and in sliding contact with the table. Since our system considers only pairwise constraints between contacting objects, it does not check for global kinematic consistency. Further tests could be implemented to rule out these interpretations.

6 Conclusion

We have presented an implemented computational theory that can derive force-dynamic representations directly from camera input. Our system embodies a rich ontology that includes both kinematic and dynamic properties of the observed objects. In addition, the system provides a representation of uncertainty along with a theory of preferences between multiple interpretations.

While encouraging, this work could be extended in several ways. First, in order to work in a general environment, 3D representations are required. While our current system is able to represent 3D scenes provided it has suitable input, further work will be required to determine what type of 3D representation is suitable and how accurate the shape and motion information will have to be. Second, in order to deal with collisions and changing contact relations, a theory of impulses (transfer of momentum) will be required. Third, as indicated by the tip example, a more complete kinematic model is needed. Finally, in order to represent the structure of time-varying scenes, we require a representation of object properties and a method to integrate such information over multiple frames. We believe our current system provides the building blocks for such a representation, but additional work will be required to show how our ontology can be built into a more complex system.

Acknowledgments

The authors would like to thank Whitman Richards, Michael Black and Chakra Chennubhotla for helpful comments on this work.

References

1. Norman I. Badler. Temporal scene analysis: Conceptual descriptions of object movements. Technical Report 80, University of Toronto Department of Computer Science, February 1975.
2. David Baraff. Interactive simulation of solid rigid bodies. *IEEE Computer Graphics and Applications*, 15(3):63–75, May 1995.
3. M. Blum, A. K. Griffith, and B. Neumann. A stability test for configurations of blocks. A. I. Memo 188, M. I. T. Artificial Intelligence Laboratory, February 1970.
4. Gary C. Borchardt. Event calculus. In *Proceedings of the Ninth International Joint Conference on Artificial Intelligence*, pages 524–527, Los Angeles, CA, August 1985.
5. Matthew Brand, Lawrence Birnbaum, and Paul Cooper. Sensible scenes: Visual understanding of complex scenes through causal analysis. In *Proceedings of the Eleventh National Conference on Artifical Intelligence*, pages 588–593, July 1993.
6. Scott Elliott Fahlman. A planning system for robot construction tasks. *Artificial Intelligence*, 5(1):1–49, 1974.
7. Roy Featherstone. *Robot Dynamics Algorithms*. Kluwer, Boston, 1987.
8. Brian V. Funt. Problem-solving with diagrammatic representations. *Artificial Intelligence*, 13(3):201–230, May 1980.
9. Herbert Goldstein. *Classical Mechanics*. Addison-Wesley, second edition, 1980.

10. Katsushi Ikeuchi and T. Suehiro. Towards an assembly plan from observation, part i: Task recognition with polyhedral objects. *IEEE Transactions on Robotics and Automation*, 10(3):368–385, 1994.
11. Michael Jenkin and Allan D. Jepson. Detecting floor anomalies. In *Proceedings of the British Machine Vision Conference*, pages 731–740, York, UK, 1994.
12. Allan D. Jepson and Michael J. Black. Mixture models for optical flow. In *Proceedings of the IEEE Conference of Computer Vision and Pattern Recognition*, pages 760–761, 1993.
13. Leo Joskowicz and Elisha P. Sacks. Computational kinematics. *Artificial Intelligence*, 51(1–3):381–416, October 1991.
14. Yasuo Kuniyoshi and Hirochika Inoue. Qualitative recognition of ongoing human action sequences. In *IJCAI93*, pages 1600–1609, August 1993.
15. Vladimir Lifschitz. Computing circumscription. In *IJCAI85*, pages 121–127, 1985.
16. Richard Mann. PhD thesis, Department of Computer Science, University of Toronto. To appear.
17. Richard Mann, Allan Jepson, and Jeffrey Mark Siskind. The compuational perception of scene dynamics. 1996. Submitted.
18. Bernd Neumann and Hans-Joachim Novak. Event models for recognition and natural language description of events in real-world image sequences. In *Proceedings of the Eighth International Joint Conference on Artificial Intelligence*, pages 724–726, August 1983.
19. Whitman Richards, Allan D. Jepson, and Jacob Feldman. Priors, preferences and categorical percepts. In Whitman Richards and David Knill, editors, *Perception as Bayesian Inference*. Cambridge University Press. To appear.
20. Jeffrey Mark Siskind. *Naive Physics, Event Perception, Lexical Semantics, and Language Acquisition*. PhD thesis, Massachusetts Institute of Technology, Cambridge, MA, January 1992.
21. Jeffrey Mark Siskind. Axiomatic support for event perception. In Paul McKevitt, editor, *Proceedings of the AAAI-94 Workshop on the Integration of Natural Language and Vision Processing*, pages 153–160, Seattle, WA, August 1994.
22. Jeffrey Mark Siskind. Grounding language in perception. *Artificial Intelligence Review*, 8:371–391, 1995.
23. John K. Tsotsos, John Mylopoulos, H. Dominic Covvey, and Steven W. Zucker. A framework for visual motion understanding. *IEEE Transactions on Pattern Analysis and Machine Intelligence*, 2(6):563–573, November 1980.

A Filter for Visual Tracking Based on a Stochastic Model for Driver Behaviour

Stephen J. Maybank, Anthony D. Worrall and Geoffrey D. Sullivan

Dept. of Computer Science, University of Reading, RG6 6AY, UK.
Email: (S.J.Maybank, A.D.Worrall, G.D.Sullivan)@reading.ac.uk.

Abstract. A driver controls a car by turning the steering wheel or by pressing on the accelerator or the brake. These actions are modelled by Gaussian processes, leading to a stochastic model for the motion of the car. The stochastic model is the basis of a new filter for tracking and predicting the motion of the car, using measurements obtained by fitting a rigid 3D model to a monocular sequence of video images. Experiments show that the filter easily outperforms traditional filters.

1 Introduction

The abilities to track and to predict car motion are important in any vision system for road traffic monitoring. Accurate tracking is required to simplify the measurement process. Tracking and prediction rely on filters which contain a model for the vehicle motion. Traditional filters such as the extended Kalman filter (EKF) [4] or the $\alpha - \beta$ filter [1] have very poor models for the car motion. For this reason they perform badly whenever the car carries out a complicated manœuvre, for example a three point turn. This paper describes a new filter which contains a simple but accurate model for car motion. Experiments show that the filter easily outperforms the $\alpha - \beta$ filter.

The model is based on the behaviour of a driver who controls the motion of a car by turning the steering wheel, by pressing the accelerator or the brake, and by changing to a reverse gear, as appropriate. The velocity of the car depends on the steering angle θ and the signed magnitude v of the velocity. The position and orientation of the car are obtained by integrating the velocity over time.

The new filter uses a Gaussian approximation to the probability density function for the state of the car, conditional on the measurements. This density is propagated forward in time using the model for the motion of the car. Measurements of the position and orientation of the car are obtained by fitting the projection of a rigid 3D wire frame to individual video images taken by a single fixed camera; the measurements are incorporated into the approximating density using the standard Kalman filter update.

The treatment developed here is fundamentally different from that of other stochastic filters such as the Kalman filter, in which the state variables are assumed to have a linear interaction. In the new filter the covariances of the state variables are updated with errors of order only $O(t^3)$, where t is the time step. For this reason, we call the filter the Covariance-Updating (CU) filter.

In comparison with the $\alpha - \beta$ filter, the CU filter has a greatly improved response to changes of state which are not explicitly included in the model, for example changes involving higher order terms, which are only modelled implicitly as process noise. Thus, the CU filter is more robust when the vehicle switches from a constant forward acceleration to a turning or a stopping motion.

2 The Filter

The CU filter is based on a motion model which enforces the links between the driver behaviour and the car motion, and which also enforces the simple geometric constraints applicable to car motion.

2.1 Driver Behaviour and the Motion of the Car

It is assumed that the car driver can vary the steering angle θ, and the signed magnitude v of the velocity of the car. Changes in θ are achieved by turning the steering wheel and changes in v are achieved by pressing on the accelerator or the brake, or by using the reverse gear. In the absence of any further prior knowledge, the changes over time in θ, v are modelled by Gaussian stochastic processes Θ, V in $\mathbb{R}$ defined by

$$(\Theta_t, V_t)^\top = (\Theta_0, V_0)^\top + \int_0^t (\dot{\Theta}_s, \dot{V}_s)^\top \, ds \qquad (0 \leq t) \tag{1}$$

The random variables Θ_0, V_0 in (1) are Gaussian. The rate of change of steering angle, $\dot{\Theta}$, is an Ornstein-Uhlenbeck process [2] which satisfies the Itô stochastic differential equation

$$d\dot{\Theta}_t = -\alpha \dot{\Theta}_t \, dt + \sigma \, dC_t \qquad (0 \leq t) \tag{2}$$

where $\alpha > 0$, $\sigma > 0$ are constants, and C is a Brownian motion in $\mathbb{R}$ such that $C_0 = 0$. The random variable $\dot{\Theta}_0$ is Gaussian and independent of C. The process $\dot{V}$ in (1) is given by

$$\dot{V}_t = \dot{V}_0 + qB_t \qquad (0 \leq t) \tag{3}$$

where $q > 0$ is constant and B is a Brownian motion in $\mathbb{R}$ independent of C and $B_0 = 0$. The random variable $\dot{V}_0$ is Gaussian and independent of B, C.

The parameters σ, α, q have the following effects: if σ, q are large, then the variances of $\dot{\Theta}_t$, $\dot{V}_t$ tend to be large. The filter becomes more responsive to recent measurements and rapidly discards past measurements. If α is large then the variance of $\dot{\Theta}_t$ tends to be small, and the estimates of Θ_t, $\dot{\Theta}_t$ become less responsive to all the measurements. At the same time the trajectory of the car, as generated by (5), shows a reduced curvature.

It remains to link the driver behaviour to the motion of the car. It is assumed that the car moves in the ground plane, and that the motion is completely described by a time dependent state vector $(x, y, \theta, v, \dot{\theta}, \dot{v})^\top$, where $(x, y)^\top$ is the position of the car in the ground plane, θ is the orientation of the car, v is the signed magnitude of the velocity of the car, and $\dot{\theta}$, $\dot{v}$ are the time derivatives of θ, v respectively.

The model for the time evolution of the state vector is a stochastic process

$$M = (X, Y, \Theta, V, \dot{\Theta}, \dot{V})^\top \qquad (4)$$

in $\mathbb{R}^6$, where $(X, Y)^\top$ models the position of the vehicle on the ground plane, and Θ, V, $\dot{\Theta}$, $\dot{V}$ are as defined by (1), (2) and (3). The process $(X, Y)^\top$ is given in terms of $(\Theta, V)^\top$ by

$$(X_t, Y_t)^\top = (X_0, Y_0)^\top + \int_0^t V_s(\cos(\Theta_s), \sin(\Theta_s))^\top \, ds \qquad (0 \le t) \qquad (5)$$

Equation (5) enforces a fundamental geometric constraint: relative to the car body, the direction of the velocity is fixed up to sign (cars cannot move sideways). In this case the direction of the velocity $(\dot{X}_t, \dot{Y}_t)^\top$ fixes the steering angle Θ_t through the equations $\dot{X}_t = V_t \cos(\Theta_t)$, $\dot{Y}_t = V_t \sin(\Theta_t)$ for $0 \le t$.

In practice, v and $\dot{\theta}$ are correlated for v small: it is difficult to steer rapidly a slowly moving car. The performance of the CU filter was improved by making α a function of the expected tangential velocity $\hat{v}$, $\alpha = 0.01 + 2.0 \exp(-\hat{v}^2/2)$. At high $\hat{v}$, α is small, to ensure that the filter can track a car successfully while it is turning. At low values of $\hat{v}$, α is large, thus dragging $\dot{\theta}$ towards zero. Very large values of α are avoided, to prevent the filter locking on to an erroneous estimate of θ.

2.2 Propagation of the Density and the Measurement Update

The task of the filter is to estimate the probability density function for the state M_t of the car at a time t, given the measurements obtained at times prior to and including t. In this application the measurements are of x, y, θ. It is assumed that the measurement errors are Gaussian with zero expected value and a known covariance matrix.

The strategy in the filter is to approximate the true conditional density with a Gaussian density. The approximation is reliable provided the pose of the car is well localised by the measurements. The filter propagates the approximating density forward in time, incorporating each measurement as it arises. It is convenient to discuss propagation from time zero, to reduce the notation. Let p_0 be the Gaussian approximation to the density of M_0. Then it is required to find a Gaussian approximation p_t to the density of M_t, $t \ge 0$, assuming that no measurements are taken in the interval $[0, t)$. If t is small, then the expectation and covariance of p_t are calculated correct to $O(t^3)$ using the formulae given in the next section. If t is large, then $[0, t)$ is divided into small subintervals $0 = t_1 < \ldots < t_n = t$, and the p_{t_i} are calculated in turn until p_t is reached.

A model of the vehicle is instantiated into the image in the pose predicted by the current state of the filter, as a wire-frame with hidden lines removed. Evidence for directed edges in the image that fall near to the visible "wires" is pooled to provide a scalar value associated with the pose (see [6] and [8] for further details). This defines a function on the space of vehicle poses. A search is carried out to find a local maximum of the function under small perturbations of the pose from the initial pose, using the simplex search technique [5]. The recovered location $(x, y)^\top$ and orientation θ of the vehicle form the measurement vector z.

Let a measurement z_t be obtained at time $t \geq 0$. The measurement update is the standard one used in the Kalman filter [3]. Let the estimated density p_t prior to the measurement have expectation m_t and covariance C_t. Let $w \mapsto Hw$, $w \in \mathbb{R}^6$ be the measurement function, and let R be the covariance matrix for the measurement errors, which are assumed to be Gaussian with expectation zero. The 3×6 matrix H is defined by $H_{ii} = 1$, $1 \leq i \leq 3$ and $H_{ij} = 0, i \neq j$. The expected value n_t and the covariance matrix D_t of the estimated density of M_t conditional on z_t are fixed by

$$(w - n_t)^\top D_t^{-1}(w - n_t) \equiv (w - m_t)^\top C_t^{-1}(w - m_t) + (z_t - Hw)^\top R^{-1}(z_t - Hw) + c_t$$

where c_t is independent of w.

3 Propagation of Expectations and Covariances

Let p_0 be the Gaussian approximation to the density of M_0. A Gaussian approximation p_t to the density of M_t is obtained under the assumption that $t > 0$ is small. The expectation and covariance of p_t are correct to $O(t^3)$, provided the density of M_0 is exactly p_0.

3.1 Expectation and Covariance of $(\Theta_t, V_t, \dot{\Theta}_t, \dot{V}_t)^\top$

The expected values and the covariances of Θ_t, V_t, $\dot{\Theta}_t$, $\dot{V}_t$ are easy to evaluate, in part because $\dot{\Theta} - \dot{\Theta}_0$ and $\dot{V} - \dot{V}_0$ are independent Gaussian processes. It follows from (2) and (3) that

$$\begin{aligned}
E(\dot{\Theta}_t, \dot{V}_t) &= E(\exp(-\alpha t)\dot{\Theta}_0, \dot{V}_0) \\
\mathrm{Cov}(\dot{\Theta}_t, \dot{\Theta}_t) &= \exp(-2\alpha t)\mathrm{Cov}(\dot{\Theta}_0, \dot{\Theta}_0) + \alpha^{-1}\sigma^2 \exp(-\alpha t)\sinh(\alpha t) \\
\mathrm{Cov}(\dot{V}_t, \dot{V}_t) &= \mathrm{Cov}(\dot{V}_0, \dot{V}_0) + q^2 t \\
\mathrm{Cov}(\dot{\Theta}_t, \dot{V}_t) &= \exp(-\alpha t)\mathrm{Cov}(\dot{\Theta}_0, \dot{V}_0)
\end{aligned} \qquad (6)$$

Let $r = t - \alpha t^2/2$. It follows from (1) and (6) that

$$\begin{aligned}
E(\Theta_t) &= E(\Theta_0) + E(\dot{\Theta}_0)r + O(t^3) \\
E(V_t) &= E(V_0) + E(\dot{V}_0)t \\
\mathrm{Cov}(\Theta_t, \Theta_t) &= \mathrm{Cov}(\Theta_0, \Theta_0) + 2\mathrm{Cov}(\Theta_0, \dot{\Theta}_0)r + \mathrm{Cov}(\dot{\Theta}_0, \dot{\Theta}_0)t^2 + O(t^3) \\
\mathrm{Cov}(V_t, V_t) &= \mathrm{Cov}(V_0, V_0) + 2\mathrm{Cov}(V_0, \dot{V}_0)t + \mathrm{Cov}(\dot{V}_0, \dot{V}_0)t^2 + O(t^3) \\
\mathrm{Cov}(\Theta_t, V_t) &= \mathrm{Cov}(\Theta_0, V_0) + \mathrm{Cov}(\dot{\Theta}_0, V_0)r + \mathrm{Cov}(\Theta_0, \dot{V}_0)t + \mathrm{Cov}(\dot{\Theta}_0, \dot{V}_0)t^2
\end{aligned}$$

The cross covariances between $(\Theta_t, V_t)^\top$ and $(\dot{\Theta}_t, \dot{V}_t)^\top$ are

$$\begin{aligned}
\mathrm{Cov}(\dot{\Theta}_t, \Theta_t) &= \exp(-\alpha t)(\mathrm{Cov}(\dot{\Theta}_0, \Theta_0) + \mathrm{Cov}(\dot{\Theta}_0, \dot{\Theta}_0)r + \sigma^2 t^2/2) \\
\mathrm{Cov}(\dot{V}_t, \Theta_t) &= \mathrm{Cov}(\dot{V}_0, \Theta_0) + \mathrm{Cov}(\dot{V}_0, \dot{\Theta}_0)r \\
\mathrm{Cov}(\dot{\Theta}_t, V_t) &= \exp(-\alpha t)(\mathrm{Cov}(\dot{\Theta}_0, V_0) + \mathrm{Cov}(\dot{\Theta}_0, \dot{V}_0)t) \\
\mathrm{Cov}(\dot{V}_t, V_t) &= \mathrm{Cov}(\dot{V}_0, V_0) + \mathrm{Cov}(\dot{V}_0, \dot{V}_0)t + 2^{-1}q^2t^2
\end{aligned}$$

3.2 Expectation of $(X_t, Y_t)^\top$

The covariances and expectations involving X_t, Y_t are more difficult to evaluate than those involving only Θ_t, V_t, $\dot{\Theta}_t$, $\dot{V}_t$. The calculations are reduced by combining X and Y to make a complex valued stochastic process, $Z = X + iY$, where $i^2 = -1$. It follows from (5) that to an error $O(t^3)$,

$$Z_t = Z_0 + \exp(i\Theta_0)\left(V_0 t + 2^{-1}\dot{V}_0 t^2\right) + 2^{-1} i \exp(i\Theta_0) V_0 \dot{\Theta}_0 t^2 \qquad (0 \le t) \qquad (7)$$

Let $c = E(\exp(i\Theta_0))$. It follows from (7) that

$$\begin{aligned}
\frac{E(Z_t)}{c} = \frac{E(Z_0)}{c} &+ t[i\mathrm{Cov}(\Theta_0, V_0) + E(V_0)] + \frac{t^2}{2}[i\mathrm{Cov}(\Theta_0, \dot{V}_0) + E(\dot{V}_0)] \\
&+ \frac{it^2}{2}[\mathrm{Cov}(V_0, \dot{\Theta}_0) - \mathrm{Cov}(\Theta_0, V_0)\mathrm{Cov}(\Theta_0, \dot{\Theta}_0) + iE(V_0)\mathrm{Cov}(\Theta_0, \dot{\Theta}_0)] \\
&- \frac{t^2}{2}[E(\dot{\Theta}_0)\mathrm{Cov}(\Theta_0, V_0) - iE(V_0)E(\dot{\Theta}_0)] + O(t^3) \qquad (8)
\end{aligned}$$

3.3 Covariances Involving X_t or Y_t

To simplify the notation, let $W \mapsto g(W)$ be the function defined on Gaussian random variables by

$$g(W) = \mathrm{Cov}(Z_t, W) \qquad (0 \le t) \qquad (9)$$

It follows from (7) and (9) that

$$\begin{aligned}
\frac{g(W)}{c} = \frac{\mathrm{Cov}(Z_0, W)}{c} &+ t[\mathrm{Cov}(V_0, W) - \mathrm{Cov}(\Theta_0, V_0)\mathrm{Cov}(\Theta_0, W)] \\
&+ itE(V_0)\mathrm{Cov}(\Theta_0, W) + 2t^2[\mathrm{Cov}(\dot{V}_0, W) - \mathrm{Cov}(\Theta_0, \dot{V}_0)\mathrm{Cov}(\Theta_0, W)] \\
&+ 2t^2[iE(\dot{V}_0)\mathrm{Cov}(\Theta_0, W) - \mathrm{Cov}(\dot{\Theta}_0, W)\mathrm{Cov}(\Theta_0, V_0)] \\
&- 2t^2[\mathrm{Cov}(V_0, \dot{\Theta}_0)\mathrm{Cov}(\Theta_0, W) + \mathrm{Cov}(W, V_0)\mathrm{Cov}(\Theta_0, \dot{\Theta}_0)] \\
&+ 2t^2[\mathrm{Cov}(\Theta_0, W)\mathrm{Cov}(\Theta_0, \dot{\Theta}_0)\mathrm{Cov}(\Theta_0, V_0) + iE(V_0)\mathrm{Cov}(\dot{\Theta}_0, W)] \\
&- 2it^2[E(V_0)\mathrm{Cov}(\Theta_0, \dot{\Theta}_0)\mathrm{Cov}(\Theta_0, W) + E(\dot{\Theta}_0)\mathrm{Cov}(\Theta_0, V_0)\mathrm{Cov}(\Theta_0, W)] \\
&+ 2t^2[iE(\dot{\Theta}_0)\mathrm{Cov}(V_0, W) - E(V_0)E(\dot{\Theta}_0)\mathrm{Cov}(\Theta_0, W)] + O(t^3)
\end{aligned}$$

The cross covariances of Z_t with Θ_t, V_t, $\dot{\Theta}_t$, $\dot{V}_t$ are (to $O(t^3)$)

$$\begin{aligned}
\mathrm{Cov}(Z_t, \Theta_t) &= g(\Theta_0) + g(\dot{\Theta}_0)t \qquad & \mathrm{Cov}(Z_t, \dot{\Theta}_t) &= g(\dot{\Theta}_0) \\
\mathrm{Cov}(Z_t, V_t) &= g(V_0) + g(\dot{V}_0)t \qquad & \mathrm{Cov}(Z_t, \dot{V}_t) &= g(\dot{V}_0) \qquad (10)
\end{aligned}$$

It remains to find the covariance of $(X_t, Y_t)^\top$. To simplify the notation, let the functions $W \mapsto f_1(W)$ and $W \mapsto f_2(W)$, $\Theta \mapsto h(\Theta)$ of the Gaussian random variables W, Θ be defined by

$$\begin{aligned}
f_1(W) &= \mathrm{Cov}(Z_0, \exp(i\Theta_0)W) \\
&\equiv c[\mathrm{Cov}(W, Z_0) - \mathrm{Cov}(W, \Theta_0)\mathrm{Cov}(Z_0, \Theta_0) + iE(W)\mathrm{Cov}(Z_0, \Theta_0)] \\
f_2(W) &= \mathrm{Cov}(Z_0, \exp(-i\Theta_0)W) \\
&\equiv \overline{c}[\mathrm{Cov}(W, Z_0) - \mathrm{Cov}(W, \Theta_0)\mathrm{Cov}(Z_0, \Theta_0) - iE(W)\mathrm{Cov}(Z_0, \Theta_0)] \\
h(\Theta) &= \mathrm{Cov}(Z_0, \exp(i\Theta)V_0\dot{\Theta}_0) \qquad (11)
\end{aligned}$$

Let $c_\theta = E(\exp(i\Theta))$. It follows from the last equation of (11) that

$$\begin{aligned}
c_\theta^{-1}h(\Theta) = {} & i[\mathrm{Cov}(Z_0, V_0)\mathrm{Cov}(\Theta, \dot{\Theta}_0) + \mathrm{Cov}(Z_0, \dot{\Theta}_0)\mathrm{Cov}(\Theta, V_0)] \\
& + i[\mathrm{Cov}(Z_0, \Theta)\mathrm{Cov}(V_0, \dot{\Theta}_0) - \mathrm{Cov}(Z_0, \Theta)\mathrm{Cov}(\Theta, \dot{\Theta}_0)\mathrm{Cov}(\Theta, V_0)] \\
& + E(V_0)[\mathrm{Cov}(Z_0, \dot{\Theta}_0) - \mathrm{Cov}(Z_0, \Theta)\mathrm{Cov}(\Theta, \dot{\Theta}_0)] \\
& + E(\dot{\Theta}_0)[\mathrm{Cov}(Z_0, V_0) - \mathrm{Cov}(Z_0, \Theta)\mathrm{Cov}(\Theta, V_0)] \\
& + iE(V_0)E(\dot{\Theta}_0)\mathrm{Cov}(Z_0, \Theta) \qquad (12)
\end{aligned}$$

Let A, B be the random variables defined by

$$A = \exp(i\Theta_0)V_0 \qquad B = \exp(i\Theta_0)(\dot{V}_0 + iV_0\dot{\Theta}_0) \qquad (13)$$

It follows from (7) and (13) that

$$Z_t = Z_0 + At + Bt^2/2 + O(t^3) \qquad (0 \le t) \qquad (14)$$

thus (to order $O(t^3)$)

$$\begin{aligned}
\mathrm{Cov}(Z_t, Z_t) &= \mathrm{Cov}(Z_0, Z_0) + 2t\mathrm{Cov}(Z_0, A) + t^2[\mathrm{Cov}(Z_0, B) + \mathrm{Cov}(A, A)] \\
&= \mathrm{Cov}(Z_0, Z_0) + 2tf_1(V_0) + t^2[f_1(\dot{V}_0) + ih(\Theta_0) + \mathrm{Cov}(A, A)] \\
\mathrm{Cov}(\overline{Z}_t, Z_t) &= \mathrm{Cov}(\overline{Z}_0, Z_0) + t(\mathrm{Cov}(\overline{Z}_0, A) + \mathrm{Cov}(\overline{A}, Z_0)) + t^2\mathrm{Cov}(\overline{A}, A) \\
&\quad + 2^{-1}t^2[\mathrm{Cov}(\overline{Z}_0, B) + \mathrm{Cov}(\overline{B}, Z_0)] \\
&= \mathrm{Cov}(\overline{Z}_0, Z_0) + 2t\mathrm{Re}(f_2(V_0)) + t^2\mathrm{Re}(f_2(\dot{V}_0) - ih(-\Theta_0)) \\
&\quad + t^2\mathrm{Cov}(\overline{A}, A) \qquad (15)
\end{aligned}$$

A short calculation yields

$$\begin{aligned}
\mathrm{Cov}(A, A) &= c^3\overline{c}[\mathrm{Cov}(V_0, V_0) - 4\mathrm{Cov}(\Theta_0, V_0)^2 + 4iE(V_0)\mathrm{Cov}(\Theta_0, V_0) + E(V_0)^2] \\
&\quad + c^2[\mathrm{Cov}(\Theta_0, V_0)^2 - 2iE(V_0)\mathrm{Cov}(\Theta_0, V_0) - E(V_0)^2] \\
\mathrm{Cov}(\overline{A}, A) &= \mathrm{Cov}(V_0, V_0) + E(V_0)^2 - c\overline{c}(\mathrm{Cov}(\Theta_0, V_0)^2 + E(V_0)^2) \qquad (16)
\end{aligned}$$

The covariances $\mathrm{Cov}(Z_t, Z_t)$, $\mathrm{Cov}(\overline{Z}_t, Z_t)$ can be evaluated to an accuracy of $O(t^3)$ using (11), (12), (15) and (16). The covariance of $(X_t, Y_t)^\top$ is given by

$$\begin{aligned}
\mathrm{Cov}(X_t, X_t) &= 2^{-1}[\mathrm{Cov}(\overline{Z}_t, Z_t)) + \mathrm{Re}(\mathrm{Cov}(Z_t, Z_t))] \\
\mathrm{Cov}(X_t, Y_t) &= 2^{-1}\mathrm{Im}(\mathrm{Cov}(Z_t, Z_t)) \\
\mathrm{Cov}(Y_t, Y_t) &= 2^{-1}[\mathrm{Cov}(\overline{Z}_t, Z_t) - \mathrm{Re}(\mathrm{Cov}(Z_t, Z_t))]
\end{aligned}$$

3.4 Comparison with the Extended Kalman Filter

The extended Kalman filter (EKF) is a popular generalisation of the Kalman filter to nonlinear systems [4]. In spite of its widespread use, the EKF often gives an astonishingly bad approximation to the true density of the system state conditional on the measurements. The EKF differs radically from the CU filter, for example the latter is accurate to within an error of $O(t^3)$. In the EKF the estimate of $E(M_t)$ is independent of $\mathrm{Cov}(M_0, M_0)$. However, it is clear from (5) that X_t, Y_t depend on products of functions of Θ_s, V_s, $0 \leq s \leq t$. It follows that an accurate approximation to $E(M_t)$ includes terms depending on the covariances of Θ_s, V_s, $0 \leq s \leq t$. In fact, the covariances are needed even if the approximation is correct only to $O(t^2)$.

In more detail, let M_0 have a Gaussian density, and let $Z_t = X_t + iY_t$. It follows from (7) that the estimate $\hat{Z}_t$ of $E(Z_t)$ produced by the EKF is

$$\hat{Z}_t = E(Z_0) + \exp(iE(\Theta_0))E(V_0)t + O(t^2)$$

As an approximation to $E(Z_t)$, $\hat{Z}_t$ is incorrect to first order in t, because

$$E(Z_t) = E(Z_0) + E(\exp(i\Theta_0))(i\mathrm{Cov}(\Theta_0, V_0) + E(\dot{V}_0))t + O(t^2)$$

4 Vehicle Tracking

The CU filter for driving behaviour has been implemented in Mathematica [7], and a C-code procedure was obtained, using the `Splice` command in Mathematica. The C code was incorporated, after trivial changes of syntax, into the pop-11 code used to control the model-based tracking system. The performance of the CU filter was then assessed in comparison with a simple $\alpha - \beta$ filter, by applying both filters to a video sequence of a car performing a three-point turn in a cluttered scene (see Fig. 1). In the past this sequence has proved unusually difficult to track, largely because of the complexity of the manœuvre. In the absence of an accurate dynamical model, the measurement process is easily distracted by the background clutter.

Figure 1 (left) shows the trace on the ground-plane of the estimated trajectory, using the CU filter (grey line) updated by measurements (dots) taken at 25 Hz. The three outlying measurements near frame 120 were gated automatically using the Mahanalobis distance with a threshold of 4 standard deviations. Examples of the fit of the model to the image are also given in Fig. 1. The performance seems to be very good; the filter accurately followed the measurements, with significant smoothing and minimal overshoot at the cusps of the track. These filter states are taken as the "ground-truth" in subsequent experiments.

Figure 2 (top) shows the effects of using the CU filter to track the vehicle with increasing time gaps between the measurements for each run, using measurement intervals of 1, 2, 4 and 8 frames (at 25Hz). The propagation of the density was at 25Hz as described in Sect. 2.2. In each run the differences between the filter prediction and the ground truth are shown for the state variables x, y and

θ. It can be seen that the predictions usually stay very close to the ground-truth (ordinate = 0.0 in each graph). Some errors are seen up to frame 180, as the car turns sharply and comes to a halt, as well as between frames 300 and 400, where the car was viewed directly rear-on, and the measurement was more than usually unstable (see Fig. 1). The prediction errors became worse as the intervals between measurements increased. None the less, the predictions were always sufficiently accurate for the measurement process to recover a good solution, and the tracking succeeded.

Figure 2 (bottom) illustrates the performance of an $\alpha - \beta$ tracker, used to filter $(x, y, \theta, v, \dot{\theta})^{\mathrm{T}}$ with $\alpha = 0.5$ and $\beta = 0.1$. Performance is much poorer. Somewhat paradoxically, the best performance was obtained with measurements every second frame (dotted); the car was tracked successfully, though with noticeably more instability than with the CU filter. With measurements taken every frame (solid line), tracking failed towards the end (note the consistent errors in x and y beyond frame 380). With measurements every 4th frame (grey), tracking failed spectacularly; near frame 100 the recovered pose spun through 90°, and thereafter the car was tracked with the model at the wrong orientation.

4.1 Discussion

This paper has demonstrated a new method for building a dynamic filter to track vehicles in road scenes. The filter explicitly models driver behaviour by stochastic processes and it includes a realistic updating of the state covariances. When applied to pose data obtained from a model based vision system, initial results indicate that the covariance updating (CU) filter has significantly better performance than a simple linear filter such as the $\alpha - \beta$ filter.

The CU filter is fundamentally different from the usual stochastic filters such as the extended Kalman filter. The (assumed) dynamic model for the system is represented more accurately, because unlike the Kalman treatment, the evolution of the state variables is not assumed to be linear. The covariances which describe the nonlinear interactions between state variables are handled correctly up to and including terms of order $O(t^2)$, where t is a small time step. This greatly improves the performance of the filter when the evolution of the system is strongly nonlinear or when there is a short run of unreliable measurements.

References

1. Bar-Shalom, Y. and Fortmann, T.E.: Tracking and Data Association. Mathematics in Science and Engineering Series **179** (1988). Academic Press: San Diego, CA.
2. Kloeden, P.E., Platen, E.: Numerical Solution of Stochastic Differential Equations. Applications of Mathematics Series: stochastic modelling and applied probability **23** (1992). Springer-Verlag: Berlin.
3. Maybeck, P.S.: Stochastic Models, Estimation, and Control, **1**. Mathematics in Science and Engineering Series **141-1** (1979). Academic Press: San Diego, CA.
4. Maybeck, P.S.: Stochastic Models, Estimation, and Control **3**. Mathematics in Science and Engineering Series **141-3** (1982). Academic Press: San Diego, CA.

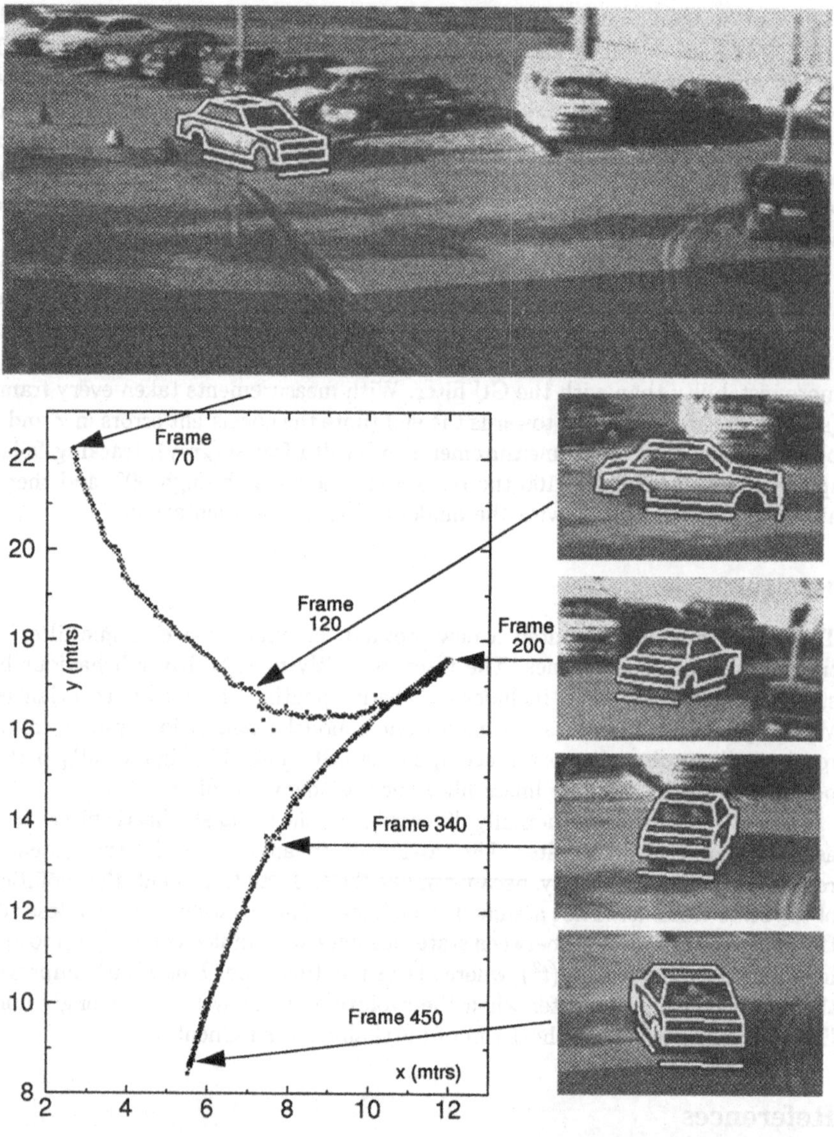

Figure 1 Top & right: Fragments of frames showing the car in ground-truth poses. Left: Trace of (x,y) ground-truth (grey line) and measured (dots) positions on the ground during 3-point turn manœuvre.

5. Press, W.H.: Numerical Recipes. (1986) CUP: Cambridge, UK.
6. Sullivan, G.D.: Visual interpretation of known objects in constrained scenes. Phil. Trans. R. Soc. London, Series B **337** (1992) 361-370.

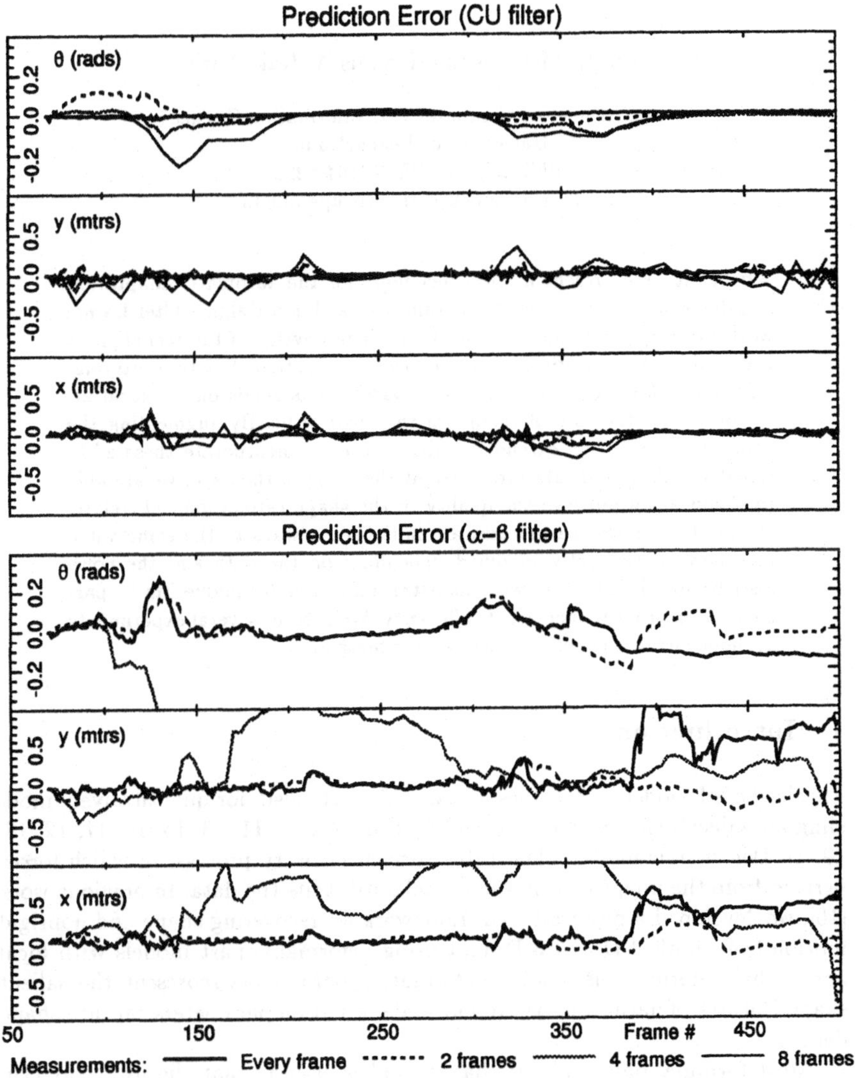

Figure 2: Prediction errors for pose parameters θ, x and y (w.r.t the ground-truth) as a function of frame number, using different measurement intervals.
Top: The CU filter, for intervals of 1, 2, 4 and 8 frames (@ 25Hz).
Bottom: The α–β filter, for intervals of 1, 2 and 4 frames (@ 25Hz).

7. Wolfram, S.: Mathematica, a system for doing mathematics by computer (1991) 2nd edition. Addison Wesley: Redwood City, CA.
8. Worrall, A.D., Ferryman, J.M., Sullivan, G.D. and Baker, K.D.: Pose and structure estimation using active models. Editor Pycock, D.: British Machine Vision Conference 1995 **1** (1995) 137-146. BMVA.

Elastically Adaptive Deformable Models

Dimitri Metaxas and Ioannis A. Kakadiaris

Department of Computer and Information Science
University of Pennsylvania
Philadelphia, PA 19104-6228
{dnm, ioannisk}@grip.cis.upenn.edu

Abstract. We present a novel technique for the automatic adaptation of a deformable model's elastic parameters within a Kalman filter framework for shape estimation applications. The novelty of the technique is that the model's elastic parameters are not constant, but time varying. The model for the elastic parameter variation depends on the local error of fit and the rate of change of the error of fit. By augmenting the state equations of an extended Kalman filter to incorporate these additional variables and take into account the noise in the data, we are able to significantly improve the quality of the shape estimation. Therefore, the model's elastic parameters are initialized always to the same value and they subsequently modified depending on the data and the noise distribution. In addition, we demonstrate how this technique can be parallelized in order to increase its efficiency. We present several experiments to demonstrate the effectiveness of our method.

1 Introduction

Physics-based modeling provides a powerful mechanism for quantitatively modeling an object's shape, structure and motion [3, 5, 9, 11, 13, 15, 16, 17, 12, 18, 19, 1]. Deformable models offer a data-driven recovery process, in which forces derived from the image deform the model until it fits the data. In previous work [16], we developed a physics-based framework for recovering shape and nonrigid motion from both 2-D and 3-D data using deformable part models with local and global deformations. Global deformation parameters represent the salient shape features of natural parts, and local deformation parameters capture shape details.

Most formulations of physics-based models assume that the user correctly initializes the model's parameters which determine the goodness of fit of the model to the given data. Recently, there have been several attempts to overcome the above problem. Blake et al. [1] developed a new technique based on ideas from adaptive control theory and maximum likelihood estimation, to learn the model dynamics for tracking in real time 2D curve motions similar to those in the training set. Samadani [19] provides rules to estimate and adjust the parameters of a snake to avoid instabilities and improve the accuracy of shape estimation. Larsen [13] proposes guidelines to determine the bounds of the optimal elastic parameters.

In this paper, we propose a new formal methodology to automatically determine a deformable model's elastic parameter values [16] which affect the accuracy of the shape estimation. The technique is useful in applications where accurate shape estimation is desired and the shape characteristics of the data may vary significantly in space and/or time. Such applications include accurate contour estimation from biomedical data, and extraction of 3D shape from range data for reverse engineering. In all these applications, minimal human intervention in terms of defining the model's elastic parameters and accuracy is desired.

Our technique is based on the use of a model for the modification of the model's elastic parameters and the stiffness matrix. The characteristic of this model is that each of the elastic parameters are modified based on the local error of fit and the local rate of change of the error of fit. If the model's initial elastic parameters are sufficient to fit the given data within a user specified tolerance, then their change during the fitting process will be minimal. Otherwise, they will gradually change based on the above criteria. In particular, the elastic parameters decrease when the model has not fit the data, and increase when the model is close to the data. The increase of the elastic parameters in the latter case, has the effect of anchoring the model to the portion of the data it has fit and also improves the continuity of the solution.

Kalman filtering has also been used by many researchers in computer vision to account for noise in the data [1, 2, 5, 14, 15, 18]. Based on our previous experience with incorporating a dynamic deformable model to an extended Kalman filter framework, we develop a *modified* extended Kalman filter. This filter allows the simultaneous modification of the model's degrees of freedom and its elastic parameters. In particular, we extend the state vector of the dynamical system which corresponds to our deformable model, to include the model's elastic parameters. This modification is based on the theory of dynamic system parameter identification [8].

We present a series of experiments with two and three dimensional data to demonstrate the effectiveness of our technique in accurate shape estimation, where the elastic parameters are always initialized to the same value.

2 Deformable Models

Geometrically, deformable models are defined on a domain Ω. The positions of points on the model relative to an inertial frame of reference Φ in space are given by a vector-valued, time varying function. We set up a noninertial, model-centered reference frame ϕ and express the position function as

$$\mathbf{x} = \mathbf{c} + \mathbf{R}\mathbf{p} = \boldsymbol{\xi}(\mathbf{q}), \tag{1}$$

where $\mathbf{c}(t)$ is the origin of ϕ at the center of the model and the rotation matrix $\mathbf{R}(t)$ gives the orientation of ϕ relative to Φ. We further express $\mathbf{p}$ as the sum of a reference shape $\mathbf{s}(\mathrm{u}, t)$ and a displacement $\mathbf{d}(\mathrm{u}, t)$. Finally, we incorporate into the vector $\mathbf{q}$ the degrees of freedom of our model which consist of the parameters necessary to define the translation, rotation, global and local deformations of the model [16].

Our goal when fitting the model to visual data is to recover the vector of degrees of freedom $\mathbf{q}$. The velocity of a point on the model is computed based on $\dot{\mathbf{x}} = \mathbf{L}\dot{\mathbf{q}}$, where $\mathbf{L}$ is the model's Jacobian matrix. This is achieved based on forces that the data exert to the surface of the model [16]. We make our model dynamic in $\mathbf{q}$ by introducing mass, damping, and a deformation strain energy. Based on Lagrangian dynamics, the simplified equations of motion [16] take the general form

$$\mathbf{D}\dot{\mathbf{q}} + \mathbf{K}\mathbf{q} = \mathbf{f}_q, \tag{2}$$

where $\mathbf{D}$, and $\mathbf{K}$ are the damping, and stiffness matrices, respectively. The generalized forces $\mathbf{f}_q$ are computed from $\mathbf{f}_q(\mathrm{u}, t) = \int \mathbf{L}^\top \mathbf{f}$, where $\mathbf{f}$ are the external three-dimensional forces that are exerted on the model. The stiffness matrix $\mathbf{K}$ determines the elastic properties of the model and is calculated from a deformation energy which is a linear combination of a membrane and a thin-plate. [16].

3 Elastically Adaptive Deformable Models

According to the theory of elasticity, the relationship between the stresses ($\boldsymbol{\sigma}$) and strains ($\boldsymbol{\epsilon}$) of an elastic material is expressed as $\boldsymbol{\sigma} = dW/d\boldsymbol{\epsilon} = \mathbf{C}\boldsymbol{\epsilon}$ for a linear material, and as $d\boldsymbol{\sigma} = \mathbf{C}(\boldsymbol{\epsilon})d\boldsymbol{\epsilon}$ for a non-linear material. Furthermore, by assuming a small stress-strain displacement[1] and the use of finite elements, we can further write $\boldsymbol{\epsilon} = \mathbf{D}\mathbf{d} = \mathbf{D}\mathbf{S}\mathbf{q}_d$, where $\mathbf{d}$ is the material displacement, $\mathbf{S}$ is the finite element shape matrix, and $\mathbf{q}_d$ are the FEM nodal displacements. Based on the above definitions and the theory of elasticity we can express the (linear or nonlinear) elastic deformation energy W as

$$W = \int \boldsymbol{\epsilon}^\top \boldsymbol{\sigma}\, du = \mathbf{q}_d^\top \left[\int (\mathbf{DS})^\top \mathbf{C}(\boldsymbol{\epsilon})(\mathbf{DS})\, du \right] \mathbf{q}_d, \tag{3}$$

where the stiffness matrix $\mathbf{K}$ is defined as

$$\mathbf{K} = \int (\mathbf{DS})^\top \mathbf{C}(\boldsymbol{\epsilon})(\mathbf{DS})\, du. \tag{4}$$

In the past we have used a combined membrane and a thin-plate energy deformation energy which can be written in the general form:

$$W = \frac{1}{2} \int (a_1 e_{11}^2 + a_2 e_{22}^2 + a_3 e_{33}^2 + a_4 e_{12}^2 + a_5 e_{23}^2 + a_6 e_{13}^2)\, du, \tag{5}$$

where e_{ij} are the components of the strain vector $\boldsymbol{\epsilon}$. In almost all implementations of finite elements to computer vision, the elastic parameters a_i are assumed constant across the deformable model and during model fitting, and are initialized manually at the beginning of the shape estimation process. This approach

[1] The theory easily generalizes to large stress-strains[20]

requires that the user may have to experiment for a long time before the correct initial values are identified. Second, since these parameters are assumed constant across the model, accurate shape estimation may never be achieved in case of complex data. Clearly a technique for automatically adjusting a deformable model's elastic parameters in a local fashion is necessary.

The first contribution of this paper is the development of a new method for automatically modifying the elastic parameters a_i for each of the model's finite elements. The model for the modification of each of the model's elastic parameters is based on ideas from the theory of PD (Proportional-Derivative) control. In particular, we modify during the fitting process the model's elastic parameters based on the local error of fit and the rate of change of the error of fit. In all of our experiments, we always start with the same initial value $a_i = a_0 = 0.005$ for all the model's elastic parameters. In our implementation, each finite element $j, j = 1...k$, has its own elastic parameters a_{i_j}. We then fit the deformable model to the given data based on (2). For each finite element j, we define the error of fit $||\mathbf{e}_j||$ to be the average of the error of fit of its nodes $l, l = 1...n$. In particular

$$||\mathbf{e}_j|| = \frac{\sum_{l=1}^{n} ||\mathbf{d}p_{l_j} - \mathbf{x}_{l_j}(\mathbf{q})||}{n}, \tag{6}$$

where $\mathbf{x}_{l_j}(\mathbf{q})$ is the position of node i and $\mathbf{d}p_{l_j}$ is the average position of the datapoints assigned to this node, based on algorithms defined in [16].

During the fitting process, the values of each of the elastic parameters a_{i_j} for each finite element j are modified based on

$$a_{i_j} = (a_0 - a_{min})\ e^{-sgn(\mathbf{e}_j.\dot{\mathbf{e}}_j)(||\mathbf{e}_j||+||\dot{\mathbf{e}}_j||)} + a_{min}, \tag{7}$$

where a_{min} is the minimum value for all the elastic parameters which for a membrane and/or thin-plate deformable model is 10^{-5}, and sgn is the *sign* function. Notice that $sgn(\mathbf{e}_j.\dot{\mathbf{e}}_j)$ is positive or zero when the model is converging towards the data and negative otherwise. The whole process of fitting and elastic parameter adaptation terminates when the error of fit for each finite element is below a tolerance e_{tol} specified by the user.

This way of modifying the elastic parameters has the following desired properties. In (7), the change in each of the a_{i_j}'s is always w.r.t. to their initial value a_0. Initially, since the error of fit is large, while the rate of change of the error of fit is small or zero, the value of the a_{i_j}'s decrease exponentially to quickly improve the fitting. In the intermediate steps of the fitting, the values of the a_{i_j}'s stabilize and are not modified significantly, since the sum of the error of fit and the rate of change of the error of fit does not change significantly. When the model is very close to the desired data, the sum of the error of fit and the rate of change of the error of fit decrease (the forces assigned to each node are now small) which result in an increase of the a_{i_j}'s towards a_0. This results in a model that achieves a smooth solution to the data where necessary, and also better "holds" the model to the desired data.

In summary, the behavior of this model is such that initially all its elastic parameters are equal. During the intermediate steps of the fitting the parameters exponentially decrease to improve the shape estimation, while when the model has almost fit the data they start to exponentially increase again towards their initial value. Therefore, the elastic parameters oscillate mostly between a_0 and a_{min}. Due to the introduction of $sgn(\mathbf{e}_j.\dot{\mathbf{e}}_j)$ in (7), the model's elastic parameters are automatically increased beyond a_0, if the model has fit the data and tries to deviate from them. Therefore, the model resists deviation from the data once it has fit them. This is an additional desired property in cases where the model has partially fit the data. It will allow the portion of the model that has not fit the data to become more elastic and fit the data, while the portion that has fit them will not be modified or become more stiff in case there is any deviation from the data.

4 Dynamic Shape Estimation

The above model for the modification of the model's elastic parameters does not take into account the noise in the data. In the past we have shown how the dynamics of a deformable model can be incorporated into an extended Kalman filtering framework to formally account for noise in the data [15, 16]. However, an extension to this formulation is necessary in order to incorporate the the model for the modification of the model's elastic parameters into a Kalman filler. The reason is that the elastic parameters are not degrees of freedom, since they do not appear in $\mathbf{q}$, but are the unknown parameters that determine the value of the stiffness matrix $\mathbf{K}$. Therefore, the problem we have is that of parameter identification in a dynamic system. In [7, 8], the theory of parameter identification in dynamic systems is presented. Based on this approach, with the addition that we have a model for the modification of the deformable model's elastic parameters[2] we augment the state vector of our system to include the model's elastic parameters. Therefore, the new state vector is of the form

$$\mathbf{u} = \begin{bmatrix} \mathbf{a} \\ \mathbf{q} \end{bmatrix}, \tag{8}$$

where $\mathbf{a}$ is the vector of the model's elasticity parameters with components a_{i_j}.

Based on the above modification, we make the following definitions necessary to define the equations of a modified extended Kalman filter. Let the observation vector $\mathbf{z}(t)$ denote time-varying input data. We can relate $\mathbf{z}(t)$ to the model's state vector $\mathbf{u}(t)$ through the nonlinear observation equation

$$\mathbf{z} = \mathbf{h}(\mathbf{u}) + \mathbf{v}, \tag{9}$$

where $\mathbf{v}(t)$ represents uncorrelated measurement errors as a zero mean white noise process with known covariance $\mathbf{V}(t)$, i.e., $\mathbf{v}(t) \sim \mathbf{N}(0, \mathbf{V}(t))$. If $\mathbf{z}$ consists

[2] According to that theory the time derivative of the elastic parameters should have been zero, as opposed to the one we use.

of observations of time varying positions of model points at material coordinates u_k on the model's surface, the components of $\mathbf{h}$ are computed using (1) evaluated at u_k[3]. In case of computing an image potential, then what is being measured at every node of the model is the difference $\mathbf{z} - \mathbf{h}(\mathbf{u})$, which is what is needed for an extended Kalman filter formulation. In addition let's also assume that $\mathbf{w}(t)$ represents uncorrelated modeling errors as a zero mean white noise process with known covariance i.e., $\mathbf{w}(t) \sim \mathbf{N}(\mathbf{0}, \mathbf{Q}(t))$.

Based on the above definitions and (2), the *modified* extended Kalman filter equations for our dynamic system take the form

$$\dot{\mathbf{u}} = \mathbf{f}(\mathbf{u}) + \mathbf{w}, \ \ \mathbf{z} = \mathbf{h}(\mathbf{u}) + \mathbf{v}, \tag{10}$$

where

$$\mathbf{f}(\mathbf{u}) = \begin{bmatrix} \dot{\mathbf{a}} \\ -\mathbf{D}^{-1}\mathbf{K}\mathbf{q} \end{bmatrix}, \dot{\mathbf{a}} = (\dot{a}_{1_1}, ..., \dot{a}_{i_j}, ..., \dot{a}_{6_k})^{\mathsf{T}}, \tag{11}$$

$$\dot{a}_{i_j} = -(a_0 - a_{min}) \ e^{-sgn(\mathbf{e}_j . \dot{\mathbf{e}}_j)(||\mathbf{e}_j|| + ||\dot{\mathbf{e}}_j||)} \ sgn(\mathbf{e}_j . \dot{\mathbf{e}}_j) \ (||\dot{\mathbf{e}}_j|| + ||\ddot{\mathbf{e}}_j||). \tag{12}$$

Notice that due to the modification in the state vector we now have a fully nonlinear extended Kalman filter as opposed to our previous formulations [16]. However, the filter converges to the right solution since we impose a correct behavior to the model for the adaptation of the model's elastic parameters and the model dynamics are appropriate for our applications.

The state estimation equation for uncorrelated system and measurement noises (i.e., $E[\mathbf{w}(t)\mathbf{v}^{\mathsf{T}}(\tau)] = 0$) is

$$\dot{\hat{\mathbf{u}}} = \mathbf{f}(\hat{\mathbf{u}}) + \mathbf{P}\mathbf{H}^{\mathsf{T}}\mathbf{V}^{-1}(\mathbf{z} - \mathbf{h}(\hat{\mathbf{u}})), \tag{13}$$

where $\mathbf{H}$ is computed from

$$\mathbf{H} = \left. \frac{\partial \mathbf{h}(\mathbf{u})}{\partial \mathbf{u}} \right|_{\mathbf{u}=\hat{\mathbf{u}}}. \tag{14}$$

The expression $\mathbf{G}(t) = \mathbf{P}\mathbf{H}^{\mathsf{T}}\mathbf{V}^{-1}$ is known as the Kalman gain matrix. The symmetric error covariance matrix $\mathbf{P}(t)$ is the solution of the matrix Riccati equation

$$\dot{\mathbf{P}} = \mathbf{F}\mathbf{P} + \mathbf{P}\mathbf{F}^{\mathsf{T}} + \mathbf{Q} - \mathbf{P}\mathbf{H}^{\mathsf{T}}\mathbf{V}^{-1}\mathbf{H}\mathbf{P}, \tag{15}$$

where

$$\mathbf{F}(\mathbf{u}) = \frac{\partial \mathbf{f}(\mathbf{u})}{\partial \mathbf{u}}. \tag{16}$$

The improvement that is offered from the Kalman filter formulation is that we can formally introduce data noise statistics into the model.

[3] Note that the definition of function $\boldsymbol{\xi}$ in (1) does not depend on $\mathbf{a}$.

5 Implementation

Since the model's equations of motion of the model are numerically well-conditioned, we partition, for computational efficiency, the full Kalman filter formulated above into two separate filters. One that includes the translation, rotation and global deformations, and another that includes only the local deformations. While the computation to the solution of the Riccati equation for the first filter is fast since the associated degrees of freedom are few, this is not the case for the second filter whose state vector includes the model's elastic parameters and the local degrees of freedom.

To solve the matrix Riccati equations at interactive rates in the latter case, we take advantage of the decomposition of the model's surface into finite elements. We use a similar approach to the one we use for the computation of the stiffness matrix $\mathbf{K}$. Based on the covariance of each component of $\mathbf{u}$ that corresponds to the variable of the second Kalman filter at a given step, we compute the contribution of each element to (15), by computing the right hand side of this equation for each element. The right hand side of (15) written for each finite element results in matrices of very small dimensions compared to the size of the respective matrices for the whole system. This per element computation can be done in parallel and once we loop through all the elements and we place the result for each element to the appropriate location of $\mathbf{P}$ (in an identical way to the computation of $\mathbf{K}$), we have computed the right hand side of (15) for the second filter. This significantly improves the speed of the calculations and is justified since we have partitioned the model's surface into finite elements.

While the observation covariance matrix $\mathbf{V}$ can be computed based on the noise in the data, the system covariance matrix $\mathbf{Q}$ serves the purpose of expressing our "belief" in the appropriateness of the dynamic model we use for our applications. This covariance matrix serves as a way of tuning the filter and achieve fast convergence [7]. In our applications however, the dynamic system we use is appropriate and we did not have to tune the filter by experimenting with various values for the elements of $\mathbf{Q}$. In particular we used a value of $\mathbf{Q} = 0.01\mathbf{I}$ which assumes a small system noise.

6 Results

Based on the above implementation, all our experiments run at interactive rates on a Silicon Graphics R4400 Indigo workstation. Furthermore, we always started from a unit covariance matrix $\mathbf{P}$. Even though that was the initial value, the subsequent structure of $\mathbf{P}$ is not diagonal and has a form similar to $\mathbf{K}$. Notice that any other reasonable initial condition will work if our dynamic model is appropriate for our applications. For the global deformations we used a superellipsoid, the elastic parameters were always initialized to $a_0 = 0.005$, and we used an adaptive Euler integration method for increased stability. In addition, the noise in our data is small and we defined $\mathbf{V}$ as $\mathbf{V} = 0.1\mathbf{I}$.

In the first experiment, we applied our technique to the semi-automated identification of the myocardial borders from breath-hold MRI. The data were obtained from the Department of Radiology at the University of Pennsylvania. The dataset included sixteen slice locations, from the Left Ventricle apex to the level of the aortic valve. In order to determine the location of the borders with higher accuracy we magnify each image four times, and then we convolve it with an 8x8 Gaussian mask. An initial superellipsoid model of the model was placed manually at the vicinity of the border of the first slice. In this study, we concentrated at the identification of the LV endocardial contour for locations 4 to 11 in which the contour is visible. During the fitting process, the results from fitting one slice, were used as the initial model for the next slice, as if we had an evolving curve over time. Therefore the user only initializes the model in the first slice. In the second experiment our goal is to fit the shape of the

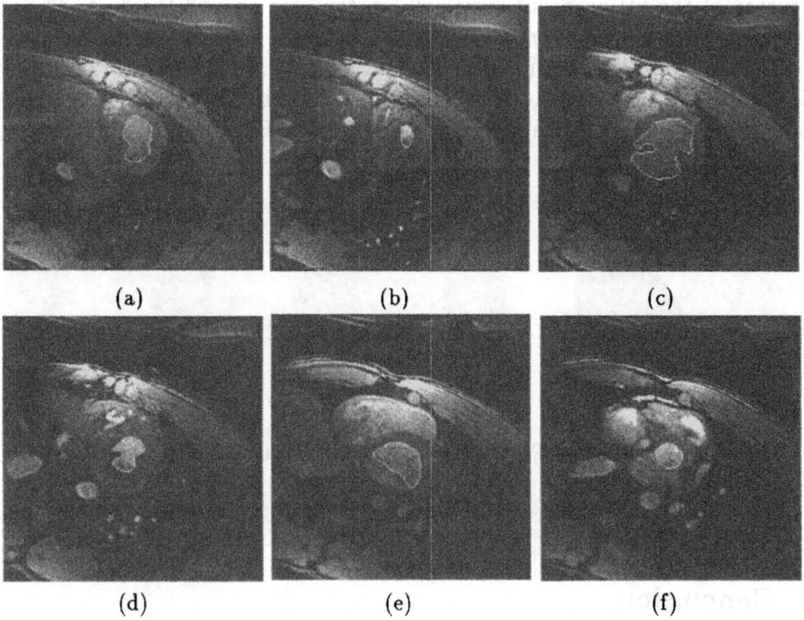

Fig. 1. Semi-automated Identification of the myocardial borders. (a), (c), (e) Locations 4, 7, 11 - End diastole; (b), (d), (f) Locations 4, 7, 11 - End systole.

outline of a head. Figure 2(a) shows the original image. The model recovered using a superellipsoid with only global deformations can be seen overlaid to the original data in Figure 2(b), while Figure 2(c) shows the contours of the data and the model. In Figure 2(d) the model has reached an equilibrium state and the error does not change over time. By varying the elastic parameters according to our technique we obtain the result in Figure 2(e) that can be compared to the original image in Figure 2(f).

In the third experiment 3D data from a human head were fitted using a deformable superquadric. Figure 3(a) shows the 1269 3D datapoints of a Viewpoint model of a human head. Figure 3(b) shows an intermediate result in fitting the

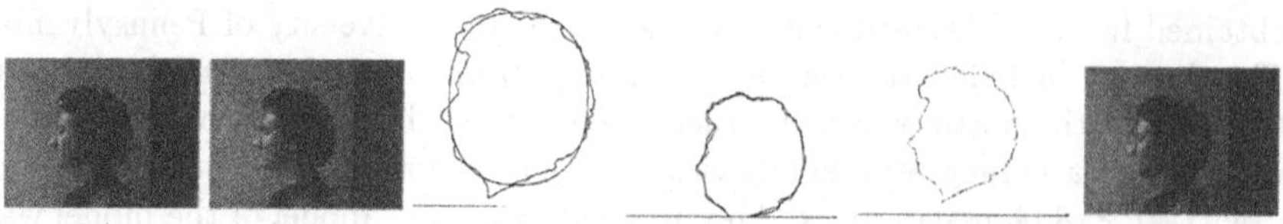

(a) (b) (c) (d) (e) (f)

Fig. 2. Fitting 2D head contours.

data using global deformations only. Figures 3(c-d) show two views of the result obtained using the initial elastic parameters ($a_i = 0.005$), and Figures 3(e-f) show the same views of the final fitting. One can notice the difference in the accuracy of the fitting result, especially in the area around the eyes. The reason, despite the automatic modification of the model's elastic parameters is that the node distribution was insufficient. Our above technique has to be coupled with our earlier developed technique [17] for automatically adapting the model's nodes in order to further improve the fitting results. But this is beyond the scope of this paper.

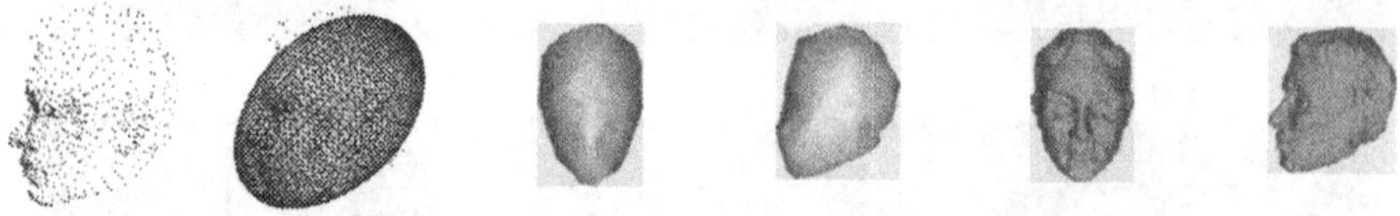

(a) (b) (c) (d) (e) (f)

Fig. 3. Fitting 3D head data.

7 Conclusion

We have presented a new method for the automatic adjustment of the elastic parameters of a deformable model. This method coupled with our method [17] for automatically adapting a model's nodes to better fit a given dataset, is very promising towards automating the process of object shape estimation based on deformable models.

Acknowledgments

This work has been supported in part by the following grants: ARO Grant DAAL03-89-C-0031PRI, ARPA Grants N00014-92-J-1647 and DAAH-049510067, NSF Grants CISE/CDA-88-22719, MIP94-20397, IRI93-09917 and MIP94-20393.

References

1. A. Blake and M. Isard, "3D position, attitude and shape input using video tracking of hands and lips", Proc. Siggraph'94, pp. 185-192, 1994.
2. T. Broida and R. Chellappa, "Estimation of object motion parameters from noisy images", IEEE Transactions on Pattern Analysis and Machine Intelligence, 8(1):90–98, January 1986.
3. L. D. Cohen. "On Active Contour Models and Balloons". CVGIP: IU, 53(2), 1991.
4. C. W. Chen and T. S. Huang, "Epicardial Motion and Deformation Estimation from Coronary Artery Bifurcation Points", IEEE Proc. Third International Conference on Computer Vision (ICCV'90), pp. 456–459, Osaka, Japan, Dec. 1990.
5. E. D. Dickmanns and Volker Graefe, "Dynamic Monocular Machine Vision", Machine Vision and Applications, 1:223–240, 1988.
6. J. S. Duncan and R. L. Owen and P. Anandan, "Measurement of Nonrigid Motion Using Contour Shape Descriptors", IEEE Computer Vision and Pattern Recognition Conference (CVPR'91), pp. 318–324, Hawaii, 1991.
7. A. Gelb. "Applied Optimal Estimation", MIT Press, Cambridge, MA, 1974.
8. M. S. Grewal and A. P. Andrews. Kalman Filtering: Theory and Applications. Prentice Hall, 1993.
9. D. B. Goldgof and H. Lee and T. S. Huang, "Motion Analysis of Nonrigid Structures", IEEE Computer Vision and Pattern Recognition Conference (CVPR'88), pp. 375–380, 1988.
10. T. S. Huang, "Modeling, Analysis and Visualization of Nonrigid Object Motion", IEEE 10th International Conference on Pattern Recognition, Atlantic City, NJ, pp. 361–364, 1990.
11. M. Kass and A. Witkin and D. Terzopoulos, "Snakes: Active Contour Models", International Journal of Computer Vision, 1(4), pp. 321–331, 1988.
12. I. A. Kakadiaris and D. Metaxas, 3D Human Body Model acquisition from Multiple views, Proc. ICCV'95, pp. 618-623, June 20 -23, Boston, MA, 1995.
13. O. V. Larsen, P. Radeva, and E. Marti, Guidelines for choosing optimal parameters of elasticity for snakes, 6th International Conference on Computer Analysis of Images and Patterns, pp. 106-113, Praga, 1995.
14. L. Matthies, T. Kanade, and R. Szeliski, "Kalman Filter-based Algorithms for Estimating Depth from Image Sequences", International Journal of Computer Vision, 3:209–236, 1989.
15. D. Metaxas and D. Terzopoulos. Recursive Estimation of Nonrigid Shape and Motion. *Proc. IEEE Motion Workshop, Princeton, NJ*, pp. 306–311, October 1991.
16. D. Metaxas and D. Terzopoulos. Shape and Nonrigid Motion Estimation Through Physics-Based Synthesis. *IEEE Trans. Pattern Analysis and Machine Intelligence*, June, 1993.
17. D. Metaxas and E. Koh. "Flexible Multibody Dynamics and Adaptive Finite Element Techniques for Model Synthesis and Estimation" Computer Methods in Applied Mechanics and Engineering, in press.
18. A. Pentland and B. Horowitz, "Recovery of Non-rigid Motion and Structure", IEEE Trans. Pattern Analysis and Machine Intelligence, 13(7):730–742, 1991.
19. R. Samadani. Adaptive Snakes: Control of damping and material parameters. SPIE Geometric Methods in Computer Vision, Vol 1570, 1991.
20. O. Zienkiewicz. *The Finite Element Method.* McGraw-Hill, 1977.

Statistical Feature Modelling for Active Contours

Simon Rowe and Andrew Blake

Dept. of Engineering Science,
Oxford University,
Parks Road, Oxford OX1 3PJ UK

Abstract. *A method is proposed of robust feature-detection for visual tracking. Frequently strong background clutter competes with foreground features and may succeed in pulling a tracker off target. This effect may be avoided by modelling the appearance of the foreground object (the target). The model consists of probability density functions of intensity along curve normals—a form of statistical template. The model can then be located by the use of a dynamic programming algorithm—even in the presence of substantial image distortions. Practical tests with contour tracking show marked improvement over simple feature detection techniques.*

1 Introduction

This paper details recent work aimed at making curve matchers and trackers more robust. A major problem in achieving robust curve tracking is the distracting effect of background objects—clutter. Strong features in the background compete for the attention of the tracked curve and may eventually succeed in pulling it away from the foreground object. This effect is clearly visible in figure 1 Immunity to distraction can be enhanced both by modelling of the foreground and of the background. A foreground model may include a shape template, object dynamics [3, 10] and intensity profiles for certain object features [12, 4]. A background model may use simple differencing with a gray level reference image, or a more detailed statistical model [9]. However background modelling does not help with a moving background (other than a rotating camera[9]).

In this paper we propose a method to reduce the effect of clutter by using a model of the gray level intensity of the target. This model is built along *lines* in an image, typically normals to an estimated curve position.We concentrate on the problem of finding a *feature* along these lines in the image. If this feature is an intensity profile, the problem is one of template matching. One technique which has been very successfully applied to this problem is correlation [2], unfortunately this suffers from some problems related to none-linear changes in the image of the target. This paper proposes a more flexible way to find the best match for the template on the image line.

The gray level profile of an image line overlying a moving target will vary for many reasons, such as shadows, sub-sampling of its texture and rotation of the target perpendicular to the image plane. Since these effects interact it will be

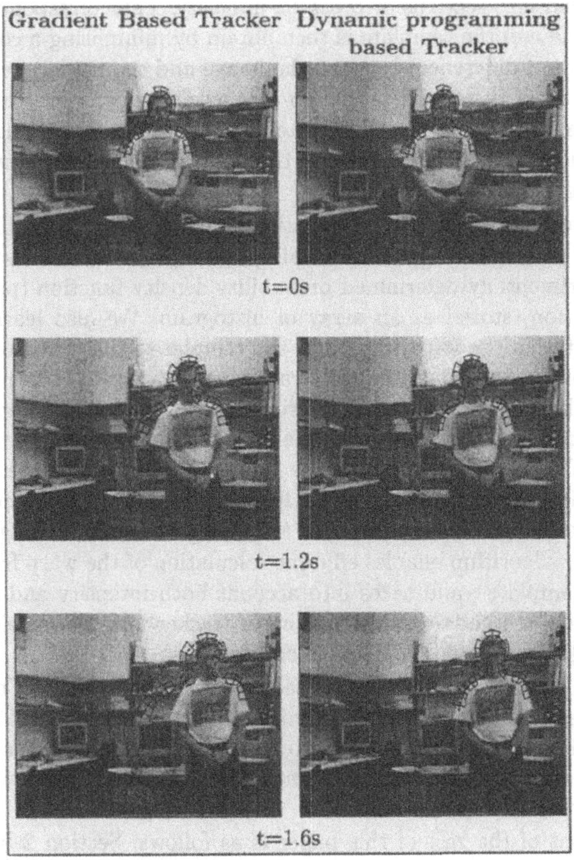

Fig. 1. A dynamic programming based tracker outperforms a gradient based one. The left column shows a tracking sequence obtained using a gradient based tracker - The tracker is distracted by the strong edge and fails to track the target. The right column shows the same sequence, but using the dynamic programming based feature search. The tracker continues to track the target past the strong distracting clutter.

almost impossible to predict the profile exactly. It may however be possible to build a statistical model incorporating the unknown elements of this variation, and use this model to locate the target—this is the approach used in this paper.

Woods, Taylor *et al* [11] use a Gaussian model for the intensity of pixels along an image line as their template. They first divide the image line into sections of constant intensity and then form statistics as to how well each section fits the constant intensity model. This intensity template is based on an analysis of hand picked regions in typical images. In the matching phase each image line is first divided into sections of *constant* intensity. The correspondence between the image line and the template is then obtain by minimising a cost function (a weighted sum of differences between the image and the model) over all possible interpretations of the model (although how exactly this is done is unspecified). The model is shown to work well in matching image lines containing large regions of constant intensity (as are typical in their application), but its applicability to more general models is doubtful.

In our work we similarly use a statistical model for our template. However rather than making any (possibly invalid) assumptions about distributions, we use an experimentally determined probability density function (pdf) for the intensity function, stored as an array of histogram. We also learn the pdf for stretches between the search line and the template. The pdf's are learnt from *real* data obtained *automatically* by tracking the target on a live video stream, rather than being hand crafted from proto-typical images, or given *a priori*. Image search lines are then matched to the template using a dynamic programming algorithm. The use of a pdf enables the algorithm to be more or less strict about how tightly intensities along the image line need to match the template (depending on where along the template they are trying to match). Using the dynamic programming algorithm enables efficient calculation of the warp from the image line to the template—and takes into account both intensity and stretching effects. The new methods are tested using a tracker based on snakes deforming over time [8], represented by B-spline curves.

Dynamic Programming has been used previously with Active Contours but in a very different context. Whereas Amini et al. [1] applied dynamic programming to the problem of matching curves to models under certain constraints, here we use dynamic programming to identify features by matching templates to intensity data along certain lines.

The layout of the rest of this paper is as follows. Section 2 introduces the notation we use to describe the feature search algorithm. Section 3 then actually explains the algorithm. Sections 4 and 5 show how the probability density functions used by the algorithm may actually be determined in practice. Section6 gives a brief explanation of the active contour tracker used to obtain the results given throughout this paper. Finally section 7 draws some conclusions on the method.

2 Feature search using a statistical foreground model

Before introducing the template matching algorithm, we first introduce notation. We then describe the algorithm and the acquisition of the gray level statistical template. The image line (or search line) under consideration at time t is denoted I_t, and the intensity a distance r along it as $I_t(r)$. It is defined it from $\pm\rho_t$ (this length being a function of time). The template is a pdf of intensity. It exists at a discrete set of positions (spaced $\bar{\delta}$ apart) along a given normal to the active contour. We define the template to be a density function, $\bar{\phi}$, so that the probability that intensity γ was generated by $\bar{\phi}$ will be denoted:

$$\bar{\phi}(\bar{r},\gamma), \qquad -\bar{\rho} \leq \bar{r} \leq \bar{\rho} \tag{1}$$

where $\bar{r}$ is distance along the template and this is illustrated in figure 2. Details of how $\bar{\phi}$ may be estimated are given later in Section 4 .

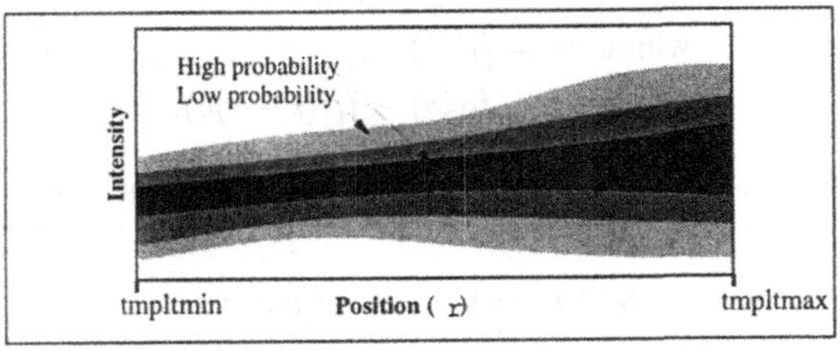

Fig. 2. The statistical template, $\bar{\phi}$. The template is a pdf for the intensity function along a prototypical search line. The darker colours in the figure indicate higher probability intensities. It is the task of the feature search under discussion here to find the warping which produces the closest fit of a search line to the template, given some constraints on the warps which are actually allowed.

The problem facing us when searching for a feature is to match a search line (I_t) to the template ($\bar{\phi}$).The search line is deformed relative to the template; to *undo* these deformations we need to find the warp which takes I_t and matches it, as far as possible, to $\bar{\phi}$. This is done by considering all possible warps of the search line and picking the one which matches the template with the highest probability. The warp function $f_t(r)$ is defined by $\bar{r} = f_t(r)$ and the warped intensity function is $\bar{I}_t$ given by:

$$\bar{I}_t(f_t(r)) = I_t(r) \tag{2}$$

where $f_t(\bar{r})$ is the mapping, assumed invertible, from the search line to the template. In general the ends of the search line will not map to the ends of the template. If fact only a sub-section of the search line may actually correspond to part of the template, with one or both ends of the search line lying off the end of the template. The beginning and end of the section of search line which maps onto the template will be denoted as ρ^{min} and ρ^{max} respectively. The positions on the template corresponding to the ends of the mapped section will be denoted at $\bar{\rho}^{min}$ and $\bar{\rho}^{max}$.

Since the I_t corresponds to the intensity pattern on physical objects viewed with a camera, there are constraints on the possible deformations which it can have undergone in the imaging process. These constraints can be expressed as the pdf, $\bar{\alpha}(\bar{r}, d)$, that a length δ (the discretisation interval), on the search line will map to length d of the template. This pdf may vary with distance along the template. Mathematically, $\bar{\alpha}(\bar{r},d)$ is given by:

$$\bar{\alpha}(\bar{r},d) = \text{prob}\left((f_t(r) - f_t(r-\delta)) = d \mid \bar{r} = f_t(r)\right) \tag{3}$$

We assume $\bar{\alpha}(\bar{r},d)$ and $\bar{\alpha}(\bar{r}+\bar{\delta},d')$ are independent of each other, then the probability of a given mapping of the entire search line is given (discretely) by:

$$\begin{aligned} \text{prob}(f_t) &= \prod_{(-\rho_t/\delta)\leq i\leq(\rho_t/\delta)} \bar{\alpha}(\bar{r},d) \\ \text{where} \quad \bar{r} &= f_t(i\delta) \\ d &= f_t(i\delta) - f_t((i-1)\delta). \end{aligned} \tag{4}$$

An example of $\bar{\alpha}(r,d)$, representing the case where the search line is a translated copy of the template, is:

$$\begin{aligned} \bar{\alpha}(a,b) &= 1, \qquad \text{iff } b = \delta \\ &= 0, \qquad \text{otherwise} \end{aligned} \tag{5}$$

Section 5 explains how $\bar{\alpha}(r,d)$ may be estimated in practice.

To find the ***best*** warp of the search line to the template we use the dynamic programming algorithm.

3 The dynamic programming algorithm

Dynamic programming works in a recursive fashion [5]. Considering each point r along the search line and $\bar{r}$ along the template in turn, the algorithm finds the probability that r corresponds to $\bar{r}$ by considering the set of possible partial mappings over $(-\rho_t, r-\delta)$. We denote the ***highest probability*** of r matching $\bar{r}$ by $P_t(\bar{r},r)$.

We obtain $P_t(\bar{r},r)$ by applying the following dynamic programming algorithm:

$$\begin{aligned} P_t(\bar{r},-\rho_t) &= \bar{\phi}(\bar{r},I_t(-\bar{\rho})), \qquad -\bar{\rho} \leq \bar{r} \leq \bar{\rho} \\ P_t(\bar{r},r) &= \max_{-\bar{\rho}\leq\bar{k}\leq\bar{\rho}} \bar{\phi}(\bar{r},I_t(r)).\bar{\alpha}(\bar{r},\bar{r}-\bar{k}).P_t(\bar{k},r-\delta) \quad \begin{matrix} -\bar{\rho} < \bar{r} \leq \bar{\rho} \\ -\rho_t < r \leq \rho_t \end{matrix} \\ \mathcal{W}_t(\bar{r},r) &= \arg \max_{-\bar{\rho}\leq\bar{k}\leq\bar{\rho}} (\bar{\alpha}(\bar{r},\bar{r}-\bar{k})P_t(\bar{k},\bar{r}-\bar{\delta})) \end{aligned} \tag{6}$$

where the "policy function" $\mathcal{W}$ is defined for later use in recovering the most probable warp.

Before we can recover the warping, we need to find the most probable position on the template to which the end of the template mapped ($\bar{\rho}^{max}$). Since the search line may actually overlap the end of the template, we also need to find the position on the search line which maps to $\bar{\rho}^{max}$: we denoted this ρ^{max}. We find these positions by examining the probabilities $P_t(\bar{r},r)$ of the warps which contain either the end of the search line or the end of the template (i.e. either

$r = \rho_t$ or $\bar{r} = \bar{\rho}$), and thus finding the warp with the highest probability. $\bar{\rho}^{max}$ and ρ^{max} are obtained using the equation:

$$(\bar{\rho}^{max}, \rho^{max}) = \arg\max_{(a,b)} \{ \; P_t(a,b) | \, ((a = \bar{\rho}) \wedge (-\rho_t \leq b \leq \rho_t)) \vee ((-\bar{\rho} \leq a \leq \bar{\rho}) \wedge (b = \rho_t)) \}$$

Once $\bar{\rho}^{max}$ and ρ^{max} have been found, $\tilde{I}_t(r)$ can be found by backtracking using the path implicitly stored in $\mathcal{W}_t(a,b)$:

$$\begin{aligned} \tilde{I}_t(r) &= I_t(f_t^{-1}(r)) \\ \text{where} \quad f_t^{-1}(r) &= \mathcal{W}_t(r+\delta, f_t^{-1}(r+\delta)) \\ f_t^{-1}(\bar{\rho}^{max}) &= \rho^{max} \end{aligned} \tag{7}$$

In order to use this algorithm estimates are needed for template intensity distribution $\bar{\phi}(r,\gamma)$ and warp probabilities $\bar{\alpha}(r,d)$. The following sections explain how these estimates may be obtained.

4 Estimating Intensity PDF's

We would like to estimate the match probabilities $\bar{\phi}(r,\gamma)$ from data obtained while tracking the target in a representative environment. This training data will include variations along the image line similar to those which we are likely to see once tracking using the intensity model. The training data can be collected by a bootstrap tracker (a tracker able to follow the target as it undergoes some limited motion[1]) tracking the target from time $t = 1$ to $t = T$. The training data will contain variations in I_t due to tracking error and scale changes in the object. This means that in order to obtain an reliable estimate for $\bar{\phi}(r,\gamma)$, each item of training data needs to be pre-warped, or transformed to remove these errors (see Figure 3). We perform this transformation by using the dynamic programming algorithm and a rough estimate $\hat{\phi}(r,\gamma)$ for $\bar{\phi}(r,\gamma)$. One method[2] for obtaining the estimate $\hat{\phi}(r,\gamma)$ is to align the training data based on the feature positions ν_t obtained using the bootstrap tracker. The estimated intensity pdf, $\hat{\phi}(r,\gamma)$, can then calculated using simple statistics.

Using $\hat{\phi}(r,\gamma)$ to estimate $\bar{\phi}(r,\gamma)$ We define $\hat{I}_t$ to be the warp of I_t using the dynamic programming algorithm and the estimates $\hat{\phi}(t,\hat{r})$ and $\hat{\alpha}(r,d)$ (see section 5). $\hat{I}_t(\hat{r})$ is the intensity at distance $\hat{r}$ along this warped search line and exists for $\hat{\rho}^{min} \leq \hat{r} \leq \hat{\rho}^{max}$. We can now obtain an estimate for $\bar{\phi}(r,\gamma)$ by

[1] Such a tracker may use image differencing, or a strong model of the targets motion to determine the target's position.

[2] This method assumes that only small changes in the scale of the target occurred during $1 \leq t \leq T$

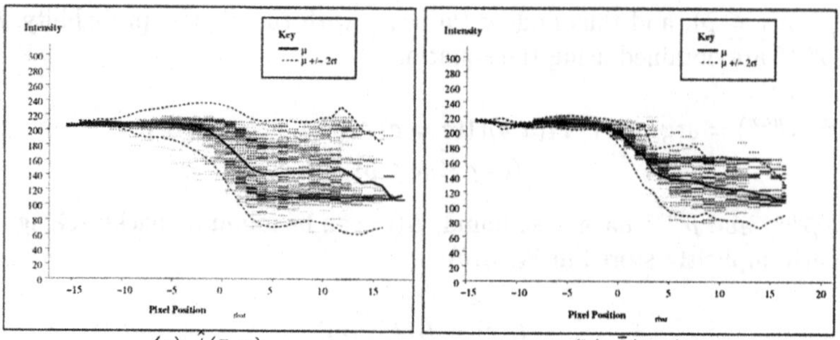

(a) $\hat{\phi}(\bar{r},\gamma)$ (b) $\bar{\phi}(\bar{r},\gamma)$

Fig. 3. The effect of pre-warping on the distribution of the template. Dark regions on the graphs show regions of high probability. Overlaid on the templates are the mean (μ) and $\pm$ 2 standard deviations (σ). Graph (a) shows the estimate $\hat{\phi}(\bar{r},\gamma)$ obtained using the raw data. (b) shows the corresponding $\bar{\phi}(\bar{r},\gamma)$ obtained by pre-warping the training data using the estimate $\hat{\phi}(\bar{r},\gamma)$. Note how the template in (b) is much sharper and tighter than the estimate (a)—this is because the search lines in the *pre-warped* training dataset are better aligned with one another.

analysing the results of warping the training data to the rough template:

$$\bar{\phi}(r,\gamma) = \frac{1}{N_d(r)} \sum_{t=1}^{T} h\left(\hat{I}_t(r),\gamma,r\right)$$

$$\text{where} \quad h(\lambda,\gamma,r) = \begin{cases} 1, & \text{iff } (\lambda=\gamma) \wedge (\hat{\rho}^{min} \le (r-\nu_t) \le \hat{\rho}^{max}) \\ 0, & \text{otherwise} \end{cases}$$

$$N_d(r) = \sum_{t=0}^{T} g(t,r)$$

$$g(t,r) = \begin{cases} 1, & \text{iff } \hat{\rho}^{min} \le r-\nu_t \le \hat{\rho}^{max} \\ 0, & \text{otherwise} \end{cases}$$

Figure 3 shows the effect that the pre-warping has on the estimate of $\bar{\phi}(\bar{r},\gamma)$. Note that it is perfectly feasible to use the histogram based approach to storing pdf in this case. A 50 pixel long template will only need 32k of memory to store $\bar{\phi}(\bar{r},\gamma)$ in a histogram form (to an accuracy of $\frac{1}{256}$. If a similar approach was used to store the pdf of intensity in a 768 × 576 background model, it would require 108MB of memory!

5 Estimating stretching PDF's

We also need an estimate for $\bar{\alpha}(r,d)$. As with estimating $\bar{\phi}(r,\gamma)$ this can be done by analysing data collected with a bootstrap tracker from time $t=1$ to $t=T$. The function $f_t(u)$ proves to be particularly useful in estimating $\bar{\alpha}(r,d)$ – we can simply examine what length each δ section of search line mapped to on the template and build a pdf from this. This pdf may vary along the length of the template.

$$\bar{\alpha}(\bar{r}, d) = \frac{1}{N_D} \sum_{t=1}^{T} \sum_{r=-\rho_t+\delta}^{\rho_t} h\left(f_t(r), f_t(r) - f_t(r-\delta)\right)$$

$$\text{where} \quad h(a,b) = \begin{cases} 1, & \text{iff } (a = \bar{r}) \wedge (b = d) \\ 0, & \text{otherwise} \end{cases} \tag{8}$$

$$N_D = \sum_{t=1}^{T} \sum_{r=-\rho_t}^{\rho_t} g(r)$$

$$g(a) = \begin{cases} 1, & \text{iff } a = \bar{r} \\ 0, & \text{otherwise} \end{cases}$$

Of course, in order to obtain $f_t^{-1}(u)$ used in the above equation we need to have already estimated $\bar{\alpha}(r, d)$. We get around this problem by supplying an initial estimate, $\hat{\alpha}(r, d)$, of $\bar{\alpha}(r, d)$—a Gaussian centered around $d = \delta$ with standard deviation 2δ.

Unfortunately, $\bar{\alpha}(\bar{r}, d)$ will not be constant over time, but will vary depending on the scale of the target being tracked–the stretches necessary to fit the target when it is small in the image will be different from those when it is large. Fortunately as the occluding contour of the target is being tracked, an estimate of the scale of the object is available. This means that $\bar{\alpha}(\bar{r}, d)$ can be adjusted depending on the expected scale of the object. Initial experiments suggest that simply scaling the d-axis of $\bar{\alpha}(\bar{r}, d)$ by the relative scale of the target is sufficient.

6 Tracking

In order to test the algorithm we implemented an active contour tracker [3] and applied the feature search by dynamic programming. The tracker used in the experiments reported here is an estimator for a N span B-spline image curve of order 3 (quadratic) [6]. A given curve is represented as $\mathbf{x}_{s,t} = (x_{s,t}, y_{s,t})$ with length parameter s (the t subscript signifies that the curve's position and shape may alter over time). The coordinates $(x_{s,t}, y_{s,t})$ are given by the equation:

$$\begin{aligned} x_{s,t} &= B_s \mathbf{X}_t \\ y_{s,t} &= B_s \mathbf{Y}_t \end{aligned} \qquad 0 \le s \le N \tag{9}$$

where $\mathbf{X}_t$ and $\mathbf{Y}_t$ are stacked vectors of the X and Y coordinates of the B-spline's control points respectively. B_s is the standard B-spline basis function matrix [6].

The target is assumed to be described in the same form as the curve, with control points $\mathbf{X}_t$ and $\mathbf{Y}_t$ varying over time. The active contour tracker generates estimates of these over time, $\hat{\mathbf{X}}_t, \hat{\mathbf{Y}}_t$, based on measurements of the actual position of the curve. We also define a shape template, $(\bar{\mathbf{X}}, \bar{\mathbf{Y}})$, which is the *average* shape of the target, and helps to stabilise the tracker[12]. The state of the tracker (position and velocity) at time t is denoted $\mathcal{X}_t$. We assume that this state evolves based on a 2nd order dynamical model.

In order to estimate the state of the curve, measurements are made of its position at each time-step. These measurements are made by casting rays along normals $\hat{\mathbf{n}}_{s,t}$ to the estimated curve, and, simultaneously at certain points s along the curve measuring the relative position, $\nu_{s,t}$ of a feature[3] along the ray so that:

$$\nu_{s,t} = \left[\mathbf{r}_{s,t} - \hat{\mathbf{r}}_{s,t}\right].\hat{\mathbf{n}}_{s,t} + v_{s,t} \tag{10}$$

where $v_{s,t}$ is a scalar noise variable, assumed Gaussian, that is taken as constant, both spatially and temporally. The measurement is defined along the normal as displacement tangentially is unobservable, the well-known aperture problem.

The tracking process is performed using a Kalman Filter based around the motion and measurement model. One important feature is that each state estimate is accompanied by an estimate of covariance. This enables a validation, or range gate $\rho_{s,t}$ to be defined [7] which allows measurements to be included only if they lie within a certain distance ($\rho_{s,t}$) of the predicted position. This distance $\rho_{s,t}$ changes over time as the tracker is more or less certain about its prediction. The range-gate mechanism of the Kalman filter means that the feature will only be considered if

$$-\rho_{s,t} \leq \nu_{s,t} \leq \rho_{s,t} \tag{11}$$

This means that we only need consider the segment of $\mathbf{n}_{s,t}$ which lies within $\rho_{s,t}$ of the curve. These segments of the normals form are the *search lines* of the active contour.

The grey-level intensity of the pixel at a position r along $\mathbf{n}_{s,t}$ will be denoted as $I_{s,t}(r)$, and the set of intensities along the whole search line as $I_{s,t}$. Note that $I_{s,t}(r)$ is the intensity of the pixel at location $\mathbf{x}_{s,t} + r\mathbf{n}_{s,t}$.

When using a gradient based feature search the tracker is able to track at 50Hz using an 8 span template on a Sun IPX. Unfortunately when tracking with the dynamic programming based feature search, this drops to 0.2Hz. This is because the complexity of the dynamic programming algorithm is $\bigcirc(\bar{\rho}^2 \rho_t)$. Ways of improving the speed of the algorithm are being pursued, but in the mean time application is restricted to recorded video sequences.

Figure 1 shows a tracking problem. A target moves in front of a cluttered background. A gradient based tracker is distracted by the clutter and loses the target. When the foreground is incorporated the tracker is able to reliably detect the boundary of the target, and able to stay locked to it.

---

[3] This feature should represent the actual underlying position of the curve on the target. Traditionally high contrast edges have been used as "the feature", this paper proposes a different model based on the gray level profile of the target.

7 Conclusions

An algorithm has been proposed to improve the feature detection for active contours so that they can track robustly in a wider range of applications. The use of a statistical template and the dynamic programming algorithm enables the feature search to locate the target, even in the presence of heavy clutter. Results have been shown for hard tracking sequences which clearly demonstrate the possible improvements. Future work will address ways of reducing the computational burden of the algorithm to enable real time tracking. It will also investigate the effect which normalising the probabilities $P_t(\bar{r}, r)$ by the length of mapped search line has on the warps chosen by the algorithm.

References

1. A. Amini, S. Tehrani, and T. Weymouth. Using dynamic programming for minimizing the energy of active contours in the presence of hard constraints. In *Proc. 2nd Int. Conf. on Computer Vision*, 95–99, 1988.
2. B. Bascle and R. Deriche. Region tracking through image sequences. In *Proc. 5th Int. Conf. on Computer Vision*, 302–307, Cambridge, MA. IEEE, IEEE Computer Society, June 1995.
3. A. Blake, R. Curwen, and A. Zisserman. Affine-invarient contour tracking with automatic control of spatiotemporal scale. In *Proc. 4th Int. Conf. on Computer Vision*, 66–95, Berlin. IEEE, IEEE Computer Society, May 1993.
4. T. Cootes, C. Taylor, A. Lanitis, D. Cooper, and J. Graham. Buiding and using flexible models incorporating grey-level information. In *Proc. 4th Int. Conf. on Computer Vision*, 242–246, Berlin. IEEE, IEEE Computer Society, May 1993.
5. T. H. Cormen, C. E. Leiserson, and R. L. Rivest. *Introduction to Algorithms*. McGraw-Hill, 1992.
6. J. Foley, A. van Dam, S. Feiner, and J. Hughes. *Computer Graphics, principles and practice*. Addison Wesley, 2nd edition, 1990.
7. A. Gelb, editor. *Applied Optimal Estimation*. MIT Press, Cambridge, MA, 1974.
8. M. Kass, A. Witkin, and D. Terzopoulos. Snakes:active contour models. In *Proc. 1st Int. Conf. on Computer Vision*, 259–268, 1987.
9. S. Rowe and A. Blake. Statistical background models for tracking with a virtual camera. In D.Pycock, editor, *British Machine Vision Conference 1995*, volume 2, 423–432. BMVA, Sept 1995.
10. D. Terzopoulos and R. Szeliski. Tracking with Kalman snakes. In A. Blake and A. Yuille, editors, *Active Vision*, 3–20. The MIT Press, 1993.
11. P. Woods, C. Taylor, D. Cooper, and R. Dixon. The use of geometric and grey-level models for industrial inspection. *Pattern Recognition Letters*, 5(1):11–17, Jan 1987.
12. A. Yuille and P. Hallinan. Deformable templates. In A. Blake and A. Yuille, editors, *Active Vision*, 20–38. The MIT Press, 1993.

Applications

Motion Deblurring and Super-resolution from an Image Sequence

B. Bascle, A. Blake, A. Zisserman

Department of Engineering Science, University of Oxford, Oxford OX1 3PJ, England

Abstract. In many applications, like surveillance, image sequences are of poor quality. Motion blur in particular introduces significant image degradation. An interesting challenge is to merge these many images into one high-quality, estimated still. We propose a method to achieve this. Firstly, an object of interest is tracked through the sequence using region based matching. Secondly, degradation of images is modelled in terms of pixel sampling, defocus blur and motion blur. Motion blur direction and magnitude are estimated from tracked displacements. Finally, a high-resolution deblurred image is reconstructed. The approach is illustrated with video sequences of moving people and blurred script.

1 Introduction

Real images of a moving object can each be regarded as a degraded representation of the ideal image that would have been captured at a certain instant by an ideal camera. These degradations include: (i) optical blur (ii) image sampling by the CCD array (iii) motion blur. Often changing the cameras to improve the quality of the images is not an option, so post-processing is needed to restore the images. The aim of this paper is to recover from a sequence of images a higher resolution deblurred image that is as close as possible to an ideal image, by removing the blur due to the real image formation process. Firstly, the object is tracked using area-based deformable regions. Secondly, given an initial estimate of the ideal image, the physical image formation process is simulated. Finally the ideal image is estimated recursively by minimising the difference between the real images and the simulated ones. The originality of our approach is in: (i) avoiding explicitly calculating an inverse filter of the blurring process, which is ill-conditioned for a single image (ii) addressing the problem of removing motion blur for non-purely translational motions and using an image sequence rather than a single image, contrary to previous approaches (iii) studying the problem of removing a combination of motion blur and optical blur, which have only been studied separately before.

The paper is organised as follows: firstly a general image formation model is presented. Secondly, we review related approaches in the literature. Thirdly, the approach proposed by this paper is described. Lastly, experimental results on real images are shown.

2 Model of the image formation process

This section introduces the model of image formation used in this paper. It will be used to recover a high-resolution unblurred image from an image sequence.

Let I^{ideal} be the "ideal" image that a perfect pinhole camera would produce at time $t = 0$. This is the image that we'll try to estimate as closely as possible. Three types of distortions occurring during the image formation process and leading from the ideal image to the observed image sequence are considered: optical blur, motion blur and spatial sampling (see fig. 1). They can be described as follows:

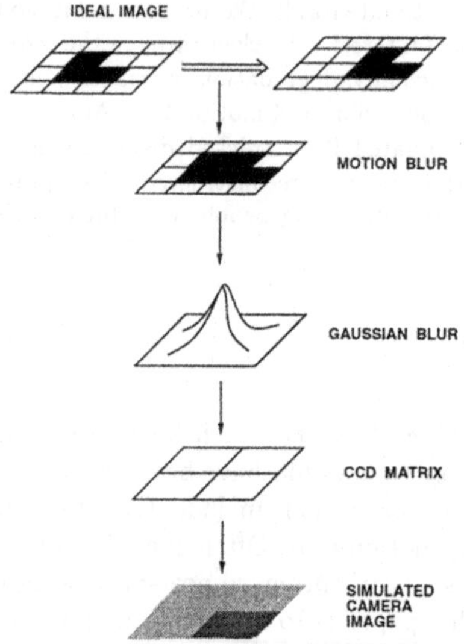

Fig. 1. Model of the image formation process used for deblurring. The ideal image has arbitrarily high resolution. First, since the object is moving, it is integrated over time by the camera, so that the resulting image is degraded by motion blur. This blur cannot be treated as purely translational and is modelled as affine in our approach. Second, given that the camera may be defocussed, and other image formation process defects, spatial blur is introduced. It is modelled to a first approximation as Gaussian. Third the CCD matrix limits the resolution of the image and performs averaging over pixel blocks of the ideal image. This model allows us to simulate the real images.

MOTION A 3D motion of the object induces a 2D motion in the image, so that the ideal image I_t^{ideal} that would be observed at time t is a 2D transform of the ideal image I^{ideal} at $t = 0$. In the general case, this image transform requires a 3D model [3]. In more restricted cases, this transform can be written as a pure 2D transform. Here we will consider a 2D affine motion model, though the approach can be easily applied to other pure 2D transforms such as homographic. Let $M(b)$ be the motion model with parameters b chosen to describe the image transform corresponding to the object 3D motion. Then:

$$I_t^{\text{ideal}}(x,y) = I_0^{\text{ideal}}(u_t,v_t) \quad \text{with:} \quad \begin{pmatrix} x \\ y \end{pmatrix} = M(b)_0^t \begin{bmatrix} u_t \\ v_t \end{bmatrix}$$

MOTION BLUR The transformed versions I_t^{ideal} of the ideal image I^{ideal} are integrated by the camera during its integration time T. If the object is moving significantly during this time, the resulting image I_t^{Mot} is smeared by motion blur. This effect is particularly noticeable when the camera is not equipped with an electronic shutter, as often in surveillance applications. The motion-blurred image I_t^{Mot} can be written as follows:

$$I_t^{Mot}(x,y) = \int_{\tau=t-T}^{\tau=t} I_{t-T}^{\text{ideal}}(u_\tau,v_\tau)\,d\tau \quad \text{with} \quad \begin{pmatrix} x \\ y \end{pmatrix} = M(b)_{t-T}^{\tau} \begin{bmatrix} u_\tau \\ v_\tau \end{bmatrix} \tag{1}$$

where $M(b)_{t-T}^{\tau}$ is the estimated motion of the object between time $t-T$ and t - with parameter values b - using the chosen motion model.

OPTICAL BLUR In addition to motion blur, the physics of image formation by a camera create other blurring effects, that we will refer to as "optical blur", and that affect as much static objects as moving ones. It groups several effects: (i) the blur introduced by the optics and thick lenses of the camera. Radial distortions in the periphery of the image are not considered here (ii) the blur induced by the camera being out-of-focus. To a first approximation, the "optical blur" will be modelled by 2D Gaussian blur.

SPATIAL SAMPLING The image is then sampled by the camera CCD array. The characteristics of this array determine the image resolution and the amount of averaging occurring when the signal is integrated by each cell of the CCD array. This sampling effect is modelled by averaging of the image over blocks of pixels. The CCD cells' point spread function is supposed to be already modelled by Gaussian blur (see previous paragraph).

Given the motion-blurred image I_t^{Mot}, the image S_t resulting after optical blur and spatial sampling can be written as:

$$S_t = \uparrow \left[S_{CCD} * G_y(x,y) * G_x(x,y) * I_t^{Mot}(x,y)\right]$$

where G_x and G_y are 1D Gaussian kernels along respectively the x and y directions. These two kernels apply 2D Gaussian blur to the image (in a separable way). S_{CCD} represents the averaging over pixel blocks introduced by the CCD matrix. This image is further sampled by the operator $\uparrow$ so that only 1 pixel out of the block of averaged pixels is retained. S_t simulates the low-resolution and blurred image that should be observed by the camera at time t.

3 Review of related approaches

The aim of this paper is to retrieve the "ideal" image of an object, of which real cameras give only a degraded version, using a sequence of images. The previous section describes our model of the image formation process, including both optical blur and motion blur. This section presents related approaches in the literature. To our knowledge, the removal of the combined effects of optical and motion blur has never been done before. However, these effects have been studied separately.

Inverse filtering of optical blur and super-resolution: The approaches aiming at removing the effects of "optical blur" and subsampling are called "super-resolution methods". They do not consider motion blur. Tsai and Huang [6] solved the problem in the frequency domain, disregarding blurring and assuming inter-frame translations. It is difficult to generalise to non-translational motions. Gross [5] merged the low resolution images by interpolation and obtained the high-resolution image by deblurring this merged image. He also assumed translational motion. Peleg and Keren [9] simulate the imaging process, and optimise the high-resolution image so as to minimise the difference between the observed and simulated low-resolution images. However the minimisation method is relatively simple: each pixel is examined in turn, and its value is incremented by 1, kept constant, or decreased by 1, so that the global criterion is decreased. Irani and Peleg [7, 8] minimise the same difference between observed and simulated images, using a back-projection method similar to that used in Computer Aided Tomography. However, the use of a back-projection operator limits the method to blurring processes for which such an operator can be calculated or approximated. Zevin and Werman consider directly a 3D model of the world with perspective cameras. Berthod *et al* [2] improved this method by considering reflectance models and their method is partly based on height from shading.

Inverse filtering of motion blur: The inverse filtering of motion blur has also been addressed in the literature, separately from the super-resolution problems. It has mainly been studied for translational motions and for single images. The approaches can be decomposed into two main groups, depending whether filtering is done in the spatial domain (Sondhi's filter [10]) or frequency space (inverse Wiener filtering [4]). Both these types of approaches are difficult to generalise to more complicated motions than purely translational.

4 Motion deblurring and Super-resolution from an image sequence

In this section, we describe our work in detail. This extends the work of [8] who showed for the case of optical blur that, though restoring degraded images is an ill-conditioned problem, the use of a sequence of images to accumulate information about the object can help to partly overcome this indeterminacy. Here we also consider distortions introduced by motion blur. Motion blur is known to be a particularly ill-conditioned blur and has previously been studied for purely translational motions and single images only (see [4]).

Object tracking approach: First the object is tracked through the sequence of images using an approach combining area-based and contour-based deformable models [1]. The tracking approach can be described as follows: (i) First the region is tracked by a deformable region based on texture correlation and constrained by the use of an affine motion model. The use of texture correlation ensures the robustness of tracking for textured images, and is also more reliable than deformable contours for blurred images (ii) Then the region contour is refined by a deformable contour. Thus the detection of the region edges is more precise. It also helps to correct tracking errors made by the deformable region in the case of occlusions and specularities. This refinement of the region contour is very useful if the image texture is poor. But it must be turned off in case of major occlusions or if too much blur renders the detection of edges unreliable.

This tracking approach gives a reliable and sub-pixel segmentation of a moving object. Such precision is necessary to recover a higher resolution image from a sequence of images. It also gives an estimation of the region 2D apparent motion. This motion estimation is used to register the images of the sequence back to the first frame, and to determine motion blur. Let $M(b)_{t-T}^{T} = (A, B)$ be the affine motion measured between the two images taken at time $t - T$ and T. It is supposed that to a first approximation the motion $M(b)_{t-T}^{\tau}$ varies linearly between the two images. Then, according to eq. 1, the image blurred by the motion (A, B) can be written as:

$$I_t^{Mot}(x, y) = \int_{\tau=t-T}^{\tau=t} I_{t-T}^{\text{ideal}}(u_\tau, v_\tau)\, d\tau$$

with

$$\begin{pmatrix} x \\ y \end{pmatrix} = A(\tau) \begin{pmatrix} u_\tau \\ v_\tau \end{pmatrix} + B(\tau)$$

and

$$A(\tau) = (A - I)\ \tau + I \quad , \quad B(\tau) = B\ \tau$$

Note that, once the region 2D motion is estimated, the transform between an image and its motion-blurred version is linear (but not spatially invariant). Thus it can be written in matrix form as: $I^{Mot} = \mathcal{B}\ I^{\text{ideal}}$.

Deblurring approach: Once the moving object is tracked and its motion estimated, the information gathered from the multiple images can be merged to enhance the images of the object, as shown by Irani and Peleg [8]. To this end, we use the image formation process described earlier. A high-resolution deblurred image of the object is iteratively estimated by minimising the difference between the real observed images and the corresponding images that are predicted by applying the modelled image formation process to the current estimation of the high-resolution image (see fig. 2). Our optimisation approach differs from [7] in that it doesn't use an explicit inverse filter - which can be difficult to approximate - of the image degradations in the feedback loop.

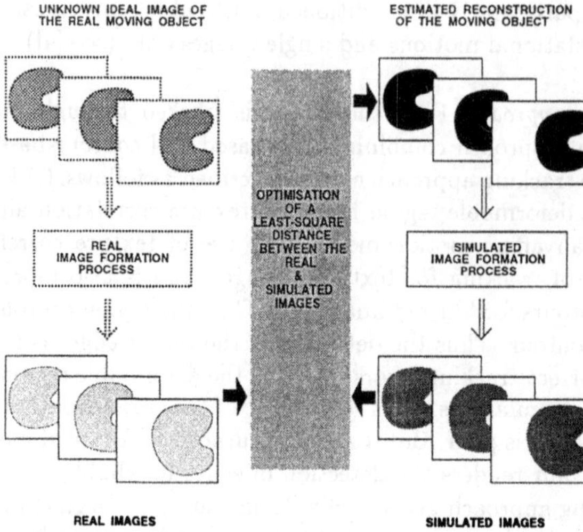

Fig. 2. Deblurring and super-resolution from image sequences. Given an initial estimate of the ideal image of the object, and having estimated the object's 2D motion by tracking, a prediction of the observed image sequence can be constructed by simulation of the image formation process. The optimisation of the reconstructed ideal image is then performed by minimising the difference between this predicted and the actual image sequence. This is done by minimisation of the corresponding least-square criterion by conjugate gradient descent. A regularisation term is also added, which ensures that the reconstructed image is not dominated by periodic noise, as can happen if the deblurring of motion blur is not done carefully. The advantage of this minimisation approach is that it does not require the construction of inverse filters for the different image formation and degradation processes.

Cost function The criterion to optimise measures the discrepancy between the real observed images R_i and the images S_i simulated by applying our image formation model (described in section 2 and eq. 4) to the estimated "ideal" image I:

$$
\begin{aligned}
E = & \textstyle\sum_{i=0}^{N} \sum_{(x,y)\in R_i} [R_i(x,y) - S_i(x,y)]^2 \\
& + \lambda . \textstyle\sum_{(x,y)\in I} [2 * I(x,y) - I(x+1,y) - I(x-1,y)]^2 \\
& + \lambda . \textstyle\sum_{(x,y)\in I} [2 * I(x,y) - I(x,y+1) - I(x,y-1)]^2 \\
& \text{where} \quad S_i = \uparrow [S_{CCD} * G_y(x,y) * G_x(x,y) * \mathcal{B}_i . I = \mathcal{D}_i I]
\end{aligned}
$$

where N is the number of images in the sequence. The second term adds second-order smoothness constraints on the reconstructed image I. These constraints improve the robustness of the method to noise. It also improves the conditioning and stability of the minimisation, which can be ill-conditioned, especially because of motion blur.

Optimisation Minimisation is performed by a multidimensional conjugate gradient method. Rewriting E, its gradient can be written as:

$$E = \sum_{i=0}^{N} [R_i - \mathcal{D}_i I]^2 + \lambda . IHI \quad \text{and} \quad \nabla = \sum_{i=0}^{N} 2[(\mathcal{D}_i^T . \mathcal{D}_i - \mathcal{D}_i R_i) . I] + 2\lambda H . I$$

where H is a matrix containing the smoothness constraints.

5 Experimental results

This section illustrates our approach to achieve super-resolution and deblurring of a moving object using a sequence of images, considering both optical blur and motion blur. An affine motion model is considered. Double resolution of the reconstructed image with respect to the real images is achieved. The deblurring algorithm has been applied only to the region of interest in the image, and the results are shown as blow-ups of these regions.

The first example is related to surveillance applications. A person is running in front of the camera, inducing motion blur in the image (see fig. 3). The person's face is tracked through four images and its 2D affine motion is estimated (see fig. 3). These multiple regions and motion estimations are then used to perform the inverse filtering of the motion blur on the face and to achieve double resolution for it. The results (see fig. 4) show a significant improvement of the level of details visible on the face (observe the ear and cheeks) and greatly reduce the smearing effect induced by motion blur. It can be noted that, when motion blur occurs, averaging the image sequence after registration (which is a standard method to increase resolution) gives an even more smeared image.

The second example (see fig. 5) shows the removal of blur from script images. The original images (a,b, detail A) are low-resolution and motion blurred. Our method (B) produces a deblurred and double resolution image of the label "pasta", using the ten images of the sequence. The label is visibly improved.

6 Conclusion

This paper has presented a new method for motion deblurring, focus deblurring and super-resolution from image sequences. Previous research has addressed copiously the problem of recovery from these individual degradations, but here it has proved crucial to address the three degradations above simultaneously. In particular, motion blur plays a key role in image degradation. An affine motion blur is considered, contrary to standard studies limited to purely translational motions. The motion blur is not assumed to be known; indeed it is actually derived from the estimated object motion. The methodology has been illustrated by experimental results. In the future, it is proposed to extend modelling to 3D motion, necessary for larger baselines with non-planar objects. Another area of future research concerns more ambitious applications of the reconstruction method. One that is particularly appealing is to reconstruct not merely a single frame but an entire sequence, by reintegrating the estimated "ideal" image over the recovered affine motion field. This would allow, for instance, the regeneration of slow-motion "action replay" sequences undegraded by excessive motion blur.

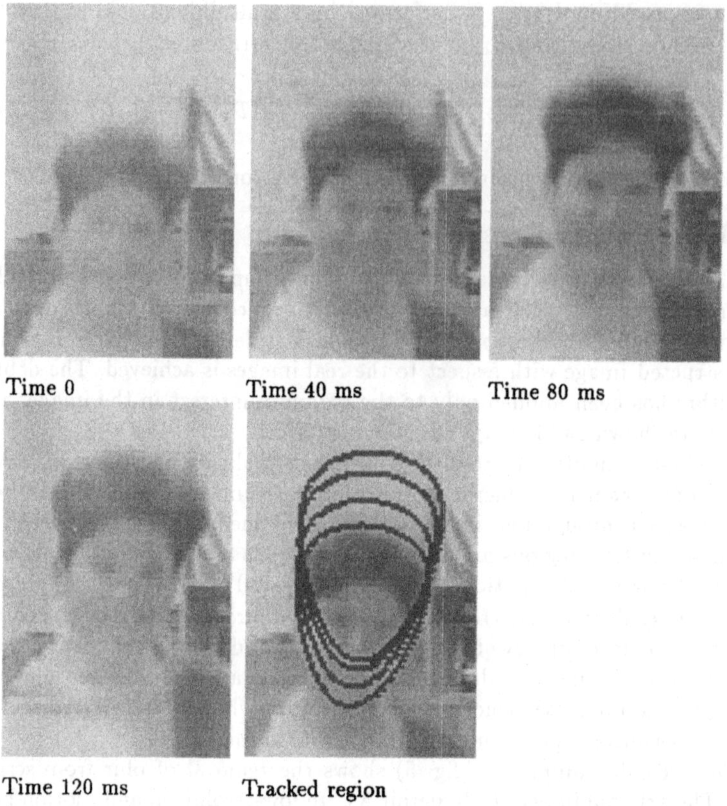

Fig. 3. Deblurring motion blur and super-resolution from an image sequence. Initial low-resolution image sequence, taken at video rate. Only even fields are retained. Because the camera is not equipped with an electronic shutter, motion blur in the image is significant. Here the face is tracked by an area-based deformable region with an affine motion model. As can be seen from the trajectories, here the motion is really affine and not purely translational.

References

1. B. Bascle and R. Deriche. Region tracking through image sequences. *Proceedings of the 5th International Conference on Computer Vision (ICCV'95), Boston, USA*, June 1995.
2. M. Berthod, H. Shekarforoush, M. Werman, and J. Zerubia. Reconstruction of High Resolution 3D Visual Information. Rapport de Recherche 2142, INRIA, Dec 1993.
3. Faugeras, O. *Three-dimensional computer vision: a geometric viewpoint.* MIT Press, 1993.
4. R.C. Gonzalez and R.E. Woods. *Digital Image Processing.* Addison Wesley, 1993.

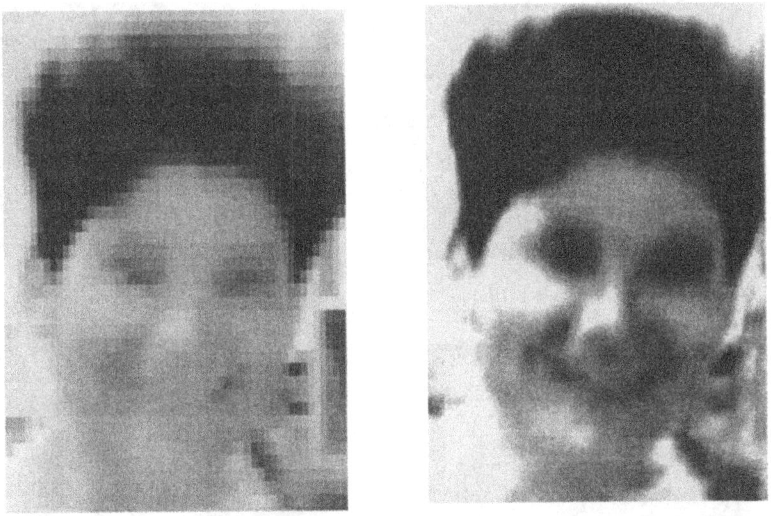

a. 4th image   b. Deblurred image

Fig. 4. Deblurring motion blur and super-resolution from an image sequence. The original images (a) are low-resolution and motion blurred. Our method (b) produces a deblurred and super-resolution (here double) image of the face, using the four images of the sequence. Details such as the nose and ear are retrieved.

5. D. Gross. Super-resolution from Sub-pixel Shifted Pictures. Master's thesis, Tel-Aviv University, Oct 1986.
6. T. Huang and R. Tsai. *Advances in Computer Vision and Image Processing*, volume 1, chapter Multiframe Image Restoration and Registration, pages 317–339. JAI Press Inc, 1984.
7. M. Irani and S. Peleg. Improving Resolution by Image Registration. *CVGIP:GMIP*, 53(3):231–239, May 1991.
8. M. Irani and S. Peleg. Motion Analysis for Image Enhancement: Resolution, Occlusion, and Transparency. *Journal of Visual Communication and Image Representation*, 4(4):324–335, December 1993.
9. D. Keren, S. Peleg, and R. Brada. Image Sequence Enhancement Using Sub-pixel Displacement. In *Proc. of International Conference on Computer Vision and Pattern Recognition (CVPR'88)*, pages 742–746, Ann Arbor, Michigan, June 1988.
10. M.M. Sondhi. Image Restoration: The Removal of Spatially Invariant Degradations. *Proc. IEEE*, 60(7):842–853, 1972.

a. Image 1

b. Image 10

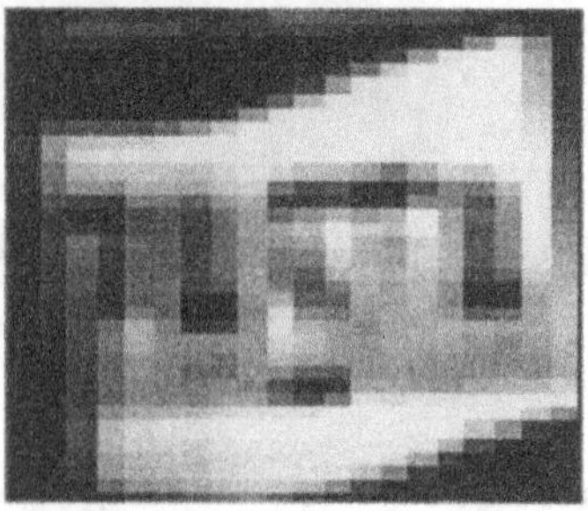

A. Detail of Image 1

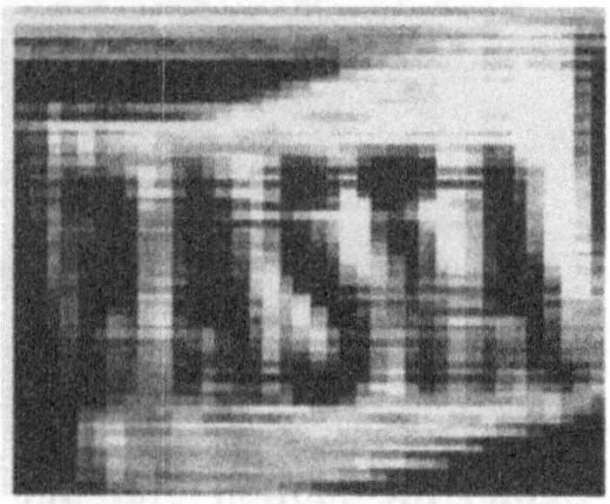

B. Deblurred detail

Fig. 5. Deblurring motion blur and super-resolution from an image sequence. The original images (a,b,A) are very low-resolution and motion blurred, so that it is difficult to extract significant edges from them (I). Our method (B) produces a deblurred and double resolution image of the label "pasta", using the 10 images of the sequence. The label is now more easily readable than before

Uncalibrated Visual Tasks via Linear Interaction

Carlo Colombo[1] and James L. Crowley[2]

[1] Dipartimento di Elettronica per l'Automazione, Università di Brescia, Via Branze 38, I-25123 Brescia, Italy***
[2] LIFIA-IMAG, Institut National Polytéchnique de Grenoble, 46 Avenue Félix Viallet, F-38031 Grenoble Cedex, France

Abstract. We propose an approach for the design and control of both reflexive and purposive visual tasks with an uncalibrated camera. The approach is based on the bi-dimensional appearance of the objects in the environment, and explicitly takes into account independent object motions. The introduction of a linear model of camera-object interaction dramatically simplifies visual analysis and control by reducing the size of the visual representation. We discuss the implementation of three tasks of increasing complexity, based on active contour analysis and polynomial planning of image contour transformations. Real-time experiments with a robot wrist-mounted camera demonstrate that the approach is conveniently usable for visual navigation, active exploration and perception, and man-robot interaction.

1 Introduction

Active vision systems, often borrowing from biological systems, combine selective sensing strategies and motor control techniques to optimize the execution of complex tasks. The simplest visual tasks can be regarded as reactive transformations from perception to action, where motor actions are reflexes to incoming visual data [6]. In addition, active tasks involve the purposive planning of visuo-motor strategies, and require an a priori knowledge of the visual environment [8, 3]. The problem of the integration of multiple tasks is of key importance for the design of active vision systems and more general robotic systems as well [2]. Some recent implementations of interacting and cooperating tasks – e.g. using saccadic shifts to recover from pursuit errors [12], or executing reactive saccades before an active recognition "scanpath" [5] – are explicitly inspired from the human visual system. A general framework for the integration of reactive visual processes was presented recently, in which the problem of the hierarchical organization of control processes was addressed [7]. Much work has been done, in the last few years, on the design of architectures for active camera control (visual servoing). A modern approach to visual servoing is to close the visual loop at the image level instead than in the tri-dimensional (3D) work-space, so as to reduce the system sensitivity to uncertainties in camera calibration, kinematic modeling, etc. [10].

*** Formerly at the ARTS Lab of Scuola Superiore Sant'Anna, Pisa, Italy.

In this paper, we present an approach to the design and control of active and reactive visual tasks with an uncalibrated camera. The approach, which is based on the bi-dimensional (2D) visual appearance of rigid objects in the work-space, allows independent object motions and features a task layering mechanism, has evolved from an earlier framework called Affine Visual Servoing (AVS) [4]. One of the distinguishing features of AVS is the combination of differential control and an affine model of camera-object interaction which, once that the ambiguities intrinsic to the linearization are solved, dramatically simplifies both object representation and visual servoing. We discuss a system implementation with a manipulator-mounted camera which uses active contours as image primitives and includes three different tasks: fixation, motion imitation, and relative positioning. In the latter case, we show how to generate camera displacements from polynomial planning of image contours. The techniques described here have natural applications in landmark-based visual navigation, active exploration and perception, and man-robot interaction.

2 Overview and Control of Visual Tasks

Given a model of camera-object interaction, a *visual representation* $\{\mathbf{p}, \mathbf{d}\}$ can be defined where, at each time t:

- $\mathbf{p}(t)$ is an m-dimensional parameterization of visual appearance;
- $\mathbf{d}(t)$ is a set of n differential parameters describing 2D changes of image appearance caused by the 3D relative velocity twist of camera and object.

Any visual task can be described as a desired evolution $\widetilde{\mathbf{p}}(t)$ of object appearance. From a differential viewpoint, the task is specified as a trajectory $\widetilde{\mathbf{d}}(t)$. This is nonzero only in the case of active tasks, while reactive tasks do not require planning.

At run-time, the current representation is estimated as $\{\widehat{\mathbf{p}}, \widehat{\mathbf{d}}\}$ via visual analysis and an image tracking process which we refer to as *passive tracking*, as it takes place also when the camera is fixed.

The $n \times 6$ *interaction matrix* $\mathcal{L}$ encodes the differential transformation from relative twist $\Delta\mathbf{V}$ to appearance changes:

$$\mathbf{d} = \mathcal{L}\,\Delta\mathbf{V} = \mathcal{L}\,(\mathbf{V}_C - \mathbf{V}_O) \ , \qquad (1)$$

where $\mathbf{V}_C = [\mathbf{T}_C^T\ \Omega_C^T]^T$ is the camera velocity twist and $\mathbf{V}_O = [\mathbf{T}_O^T\ \Omega_O^T]^T$ is the object velocity twist.

A differential strategy is adopted for task control:

$$\widetilde{\mathbf{V}}_C = \widehat{\mathbf{V}}_O + \mathcal{L}^+(\widetilde{\mathbf{d}} + k\,\mathbf{e}(\widetilde{\mathbf{p}}, \widehat{\mathbf{p}})) \ , \qquad (2)$$

where $\widetilde{\mathbf{V}}_C$ is the desired camera motion, $\widehat{\mathbf{V}}_O$ is an estimate of object motion, $\mathbf{e}(\widetilde{\mathbf{p}}, \widehat{\mathbf{p}})$ is an n-dimensional error resulting from the comparison of the desired and estimated appearances, $k \in [0, 1]$ is the feedback gain, and $\mathcal{L}^+$ denotes the $6 \times n$ pseudo-inverse of $\mathcal{L}$. The anticipation $\widehat{\mathbf{V}}_O$ is obtained as $\widehat{\mathbf{V}}_O = \widehat{\mathbf{V}}_C - \mathcal{L}^+\widehat{\mathbf{d}}$, where the camera motion $\widehat{\mathbf{V}}_C$ is estimated directly from joint data and robot kinematics. Position feedback, if k is properly tuned, compensates for various modeling and estimation inaccuracies (robot kinematics, interaction model, camera parameters, finite differences approximation, initial conditions, etc.). A regulation-to-zero (no planning) scheme for the control of a full-perspective camera based on the interaction matrix concept is introduced in [10], where $\mathbf{d} = \dot{\mathbf{p}}$, i.e. $n = m$. As the size of $\mathcal{L}$ ($\propto n$) is directly related to the number of visual features used to represent an object ($\propto m$) – mainly points and lines –, the use of this scheme is limited to rather simple object shapes. Below we show how, thanks to a careful modeling of the interaction, it is possible to decouple control complexity from shape complexity – n independent of m – and easily augment regulation with a suitable planning strategy based on contour features, with significant improvements over the basic scheme both in terms of loop time and stability.

As several tasks may be executed independently in parallel, there is a danger of tasks issuing conflicting commands to hardware and computing resources. Such conflicts can be resolved by organizing the tasks into a hierarchy based on the processing time (or bandwidth) of the transformations and, in ultimate analysis, on the feedback gain of each task. With such techniques, slower tasks, working in more abstract reference spaces, provide the reference signal to lower level tasks.

3 Models and Measurements

3.1 Interaction Model

The differential interaction between camera and object can be expressed, at a generic image point $\mathbf{x} = [x\ y]^T$, in terms of the 2×6 matrix $\mathcal{V}(x, y)$ s.t.

$$\mathbf{v}(x, y) = \mathcal{V}(x, y)\,\Delta\mathbf{V}\ , \tag{3}$$

relating image velocity $\mathbf{v} = \dot{\mathbf{x}}$ (motion field) to 3D relative velocity. Under full perspective and unit focal length, the motion field matrix evaluates as

$$\mathcal{V}(x, y) = \begin{bmatrix} -1/z & 0 & x/z & xy & -(1+x^2) & y \\ 0 & -1/z & y/z & (1+y^2) & -xy & -x \end{bmatrix}\ , \tag{4}$$

with $z = z(x, y)$ s.t. $z(X/Z, Y/Z) = Z$ bringing into play the depth of the visible surface $Z = Z(X, Y)$.

Under para-perspective projection (a linearization of perspective [11]), the visible surface is approximated by a plane $Z(X, Y) = pX + qY + c$ passing through the object's centroid $[X_B\ Y_B\ Z_B]^T$ (*object plane*), with

$$c/z = 1 - px - qy\ . \tag{5}$$

Besides, for any object point para-projected in $\mathbf{x}$, it holds $(\mathbf{x} - \mathbf{x}_B)^T(\mathbf{x} - \mathbf{x}_B) \simeq 0$, where $\mathbf{x}_B$ is the centroid's image. Thus we can neglect quadratic and higher order terms in the Taylor's development of $\mathbf{v}(x, y)$ around (x_B, y_B), and obtain a linear motion field:

$$\mathbf{v}(x, y) = \mathbf{v}_B + \mathcal{M}_B [x - x_B \; y - y_B]^T . \tag{6}$$

In ultimate analysis, the dynamic evolution of any object image patch has six degrees of freedom (DOF), namely the two components of $\mathbf{v}^B$ (rigid translation of the whole patch), and the motion parallax

$$[\mathrm{m}_{11} \; \mathrm{m}_{12} \; \mathrm{m}_{21} \; \mathrm{m}_{22}]^T = \mathbf{w}_B \leftrightarrow \mathcal{M}_B = \begin{bmatrix} \mathrm{m}_{11} & \mathrm{m}_{12} \\ \mathrm{m}_{21} & \mathrm{m}_{22} \end{bmatrix} , \tag{7}$$

which accounts for affine image shape transformations.

The linearization of eq. (6) allows a compact representation of dynamic interaction (n small and independent of m), not achievable with a full-perspective model. Indeed, by combining eqs. (3) through (7), we can easily construct the three interaction matrices $\mathcal{V}_B$ $(n = 2)$, $\mathcal{W}_B$ $(n = 4)$ and $\mathcal{U}_B$ $(n = 6)$ s.t.

$$\mathbf{v}_B = \mathcal{V}_B \Delta\mathbf{V} , \quad \mathbf{w}_B = \mathcal{W}_B \Delta\mathbf{V} , \quad \mathbf{u}_B = [\mathbf{v}_B^T \; \mathbf{w}_B^T]^T = \mathcal{U}_B \Delta\mathbf{V} , \tag{8}$$

and use them for designing visual tasks (see Sect. 4). Notice that the matrix $\mathcal{U}_B$ establishes a one-one correspondence between the six object DOF in the image and those in the work-space.

3.2 Passive Tracking and Feedback

To estimate at each time the current visual representation of the object (visual appearance and differential parameters) we use quadratic B-spline active contours [1]. These use a Kalman filter to robustly track affine deformations of a template contour, and allow to compactly represent object shape – small values of m for a fixed shape complexity – in terms of their M control points $\mathbf{x}_i$ and optimize image processing computations.

The six parameters of the affine transformation $\widehat{\mathbf{u}}_B$ between two successive contour estimates, $\{\widehat{\mathbf{x}}_i(t)\}$ and $\{\widehat{\mathbf{x}}_i(t+1)\}$, are obtained via least squares. The feedback error (centroid, shape) is evaluated analogously, as the least squares matching of the desired visual appearance, $\{\widetilde{\mathbf{x}}_i\}$, against $\{\widehat{\mathbf{x}}_i\}$, the estimated appearance.

To enhance the quality of all visual measurements (visual representation, object motion), simple mobile-mean filters are used [4].

3.3 Initializing and Updating the Interaction Matrix

The interaction matrix embeds, in the object plane coefficients p, q and c, information on 3D relative camera-object pose and translation (extrinsic camera parameters). Fig. 1 (left) shows an object-centered frame $\{X_{obj}, Y_{obj}, Z_{obj}\}$ fixed on the centroid $[X_B \; Y_B \; Z_B]^T$, s.t. $Z_{obj} = 0$ denotes the object plane. Relative pose is uniquely determined by the three angles $\sigma \in [0, \pi/2]$ (slant), $\tau \in [-\pi, \pi]$ (tilt) and $\varphi \in [-\pi, \pi]$, to which plane parameters are related as follows:

$$p = -\tan\sigma \cos\tau \; , \quad q = -\tan\sigma \sin\tau \; , \quad c = Z_B \cdot (1 - p x_B - q y_B) \; . \tag{9}$$

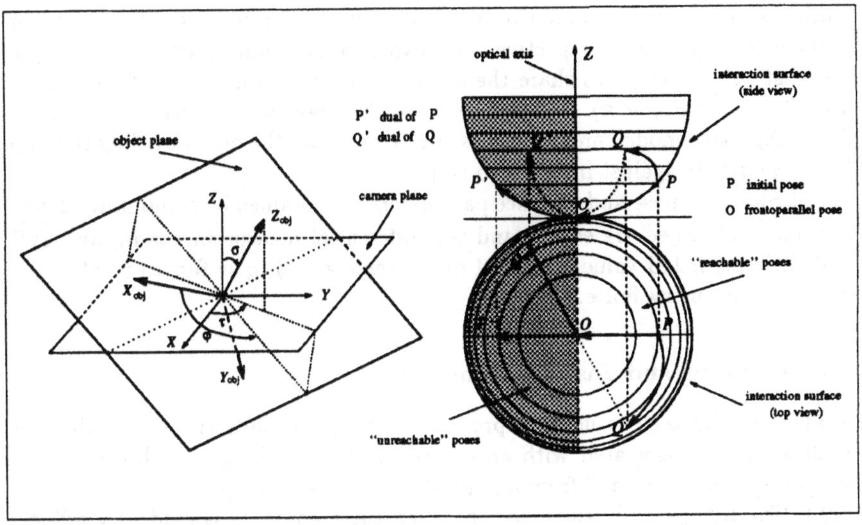

Fig. 1. *Left*: Definition of pose parameters. The camera frame has been translated for convenience in the object frame's origin. *Right*: Interaction surface and pose ambiguity for weak perspective (see Subsect. 3.4).

It can be shown [9] that the loop closure in the image rather than in the workspace greatly enhances stability w.r.t. conventional servoing approaches, and convergence is ensured even for bad initial estimates of either intrinsic and extrinsic camera parameters (uncalibrated camera). Still, having the interaction matrix even roughly estimated at start-up and updated at run-time improves the speed of convergence of the control scheme. A raw initial estimate of object pose and centroid depth is obtained using the simple weak perspective camera model $Z(X,Y) \simeq Z_B$ in the place of para-perspective, which yields $[x - x_B \; y - y_B]^T = \mathcal{T}^{wp} [X_{obj} \; Y_{obj}]^T$, with

$$\mathcal{T}^{wp} = \frac{1}{Z_B} \begin{bmatrix} \cos\tau & -\sin\tau \\ \sin\tau & \cos\tau \end{bmatrix} \begin{bmatrix} \cos\sigma & 0 \\ 0 & 1 \end{bmatrix} \begin{bmatrix} \cos\varphi & \sin\varphi \\ -\sin\varphi & \cos\varphi \end{bmatrix} . \quad (10)$$

An estimate of the weak perspective matrix, $\widehat{\mathcal{T}}^{wp} = \begin{bmatrix} t_{11} & t_{12} \\ t_{21} & t_{22} \end{bmatrix}$, is easily obtained from the least squares comparison of the current appearance of the object and an a priori model, e.g. the frontoparallel view of the object at unit distance and scale. Once that this is known, we can estimate both pose and scale by solving the following nonlinear system:

$$\begin{cases} t_{11} + t_{22} = 1/Z_B \cos(\tau - \varphi)(\cos\sigma + 1) \\ t_{21} - t_{12} = 1/Z_B \sin(\tau - \varphi)(\cos\sigma + 1) \\ t_{11} - t_{22} = 1/Z_B \cos(\tau + \varphi)(\cos\sigma - 1) \\ t_{21} + t_{12} = 1/Z_B \sin(\tau + \varphi)(\cos\sigma - 1) \end{cases} . \quad (11)$$

Notice that – a fact common to all perspective linearizations – a pose ambiguity exists for weak perspective. I.e., two dual solutions exist to eq. (11), as two distinct object poses share the same visual appearance: $\mathcal{T}^{wp}(Z_B, \tau, \sigma, \varphi) = \mathcal{T}^{wp}(Z_B, \tau + \pi, \sigma, \varphi + \pi)$. To disambiguate the pose, we can refer back to the full-perspective model, and choose as the "true" pose the one providing the best least squares fit against image data [11].

At run-time, pose and distance parameters are obtained by combining current estimates, obtained via eq. (1) and the differential measurements $\widehat{\mathbf{u}}_B$ and $\widehat{\Delta\mathbf{T}}$, and their predicted value, obtained by expressing $\dot{p}$, $\dot{q}$, $\dot{c}$ as functions of p, q, c and $\Delta\mathbf{T}$, and using finite differences [4].

3.4 Planning and Task Disambiguation

A planning strategy is used to produce a viewpoint shift (pose, translation). Such a shift is associated with an according smooth change of object's visual appearance of duration T from an initial contour $\{\mathbf{x}_i^o\}$ to a final desired contour $\{\mathbf{x}_i^T\}$. The mapping "in the large" between these contours is evidently affine – $\mathbf{x}_i^T = \mathbf{x}_B^T + \mathcal{A}_T(\mathbf{x}_i^o - \mathbf{x}_B^o)$ – as the result of a sequence of affine transformations "in the small." The reference contour evolution is planned according to a trajectory for each of the control points, which is polynomial (degree $h \geq 1$) in time and linear in the image space:

$$\widetilde{\mathbf{x}}_i(t) = [\mathbf{a}(t) + \mathbf{x}_B^o] + \mathcal{A}(t)\,(\mathbf{x}_i^o - \mathbf{x}_B^o) \ , \tag{12}$$

with $\mathbf{a}(t) = \sum_{l=0}^{h} \mathbf{c}_l t^l$ a 2-vector and $\mathcal{A}(t) = \sum_{l=0}^{h} C_l t^l$ a 2×2 matrix, $\mathbf{c}_l$ and C_l being constants to be determined based on boundary conditions. The conditions $\mathbf{a}(0) = \mathbf{0}$, $\mathcal{A}(0) = \mathcal{I}$, $\mathbf{a}(T) = \mathbf{x}_i^T - \mathbf{x}_i^o$, and $\mathcal{A}(T) = \mathcal{A}_T$ ensure that the contour evolution starts with the initial contour and terminates with the desired contour. From the solution

$$\mathbf{a}(t) = \xi(t)(\mathbf{x}_B^T - \mathbf{x}_B^o) \ , \quad \mathcal{A}(t) = \xi(t)\mathcal{A}_T + [1 - \xi(t)]\mathcal{I} \ , \tag{13}$$

where $\xi(t) \in [0,1]$, the desired differential reference is computed as $\widetilde{\mathbf{v}}_B(t) = \dot{\mathbf{a}}(t)$ and $\widetilde{\mathbf{w}}_B(t) \leftarrow \widetilde{\mathcal{M}}_B(t) = \dot{\mathcal{A}}(t)\mathcal{A}^{-1}(t)$. Additional constraints on the derivatives of $\mathbf{a}(t)$ and $\mathcal{A}(t)$ – with beneficial effects on contour tracking, visual analysis and camera velocities and accelerations at the expense of a slower convergence – can be imposed at the trajectory endpoints with $h \geq 3$. Smooth trajectories are obtained with cubic ($h = 3$) or quintic ($h = 5$) polynomials by imposing zero endpoint velocity and acceleration:

$$\xi(t) = \begin{cases} \chi^2(t)[3 - 2\chi(t)] & \text{if } h = 3 \ , \\ \chi^3(t)[6\chi^2(t) - 15\chi(t) + 10] & \text{if } h = 5 \ , \end{cases} \tag{14}$$

with $\chi(t) = (t/T) \in [0,1]$ the normalized task time.

The smooth pose shift produced by the planning strategy can be represented as a curvilinear path on the *interaction surface* – the semi-sphere of all possible relative orientations between camera and object plane (Fig. 1, right). As it is, planning produces always, of the two dual poses $\mathbf{Q}$ and $\mathbf{Q}'$ sharing the same goal appearance under weak perspective, the one which is closest to the initial pose moving along a geodesic path on the interaction surface ($\mathbf{Q}$). To reach the farthest pose ($\mathbf{Q}'$) instead, we split the path $\mathbf{P} \mapsto \mathbf{Q}'$ in two parts, $\mathbf{P} \mapsto \mathbf{O}$ and $\mathbf{O} \mapsto \mathbf{Q}'$, and pass through a suitably scaled frontoparallel view of the object $\mathbf{O}$.

4 Implementation and Results

Three Visual Tasks: Definition and Composition. Tab. 1 introduces, in order of complexity, the three tasks implemented using the interaction matrices defined in eq. (8). Indices of task computational complexity are loop time, degree of object representation and initial conditions required.

TASK SYNOPSIS	Fixation Tracking	Reactive Tracking	Active Positioning
Initial conditions	$\{\mathbf{x}_i^0\}$	$\{\mathbf{x}_i^0\}$	$\{\mathbf{x}_i^0\}, \{\mathbf{x}_i^T\}$
Visual Representation	$\{\mathbf{x}_B, \mathbf{v}_B\}$	$\{\{\mathbf{x}_i\}, \mathbf{u}_B\}$	$\{\{\mathbf{x}_i\}, \mathbf{u}_B\}$
Task Description	$\widetilde{\mathbf{x}}_B = \mathbf{0}$	$\{\widetilde{\mathbf{x}}_i(t)\} = \{\mathbf{x}_i^0\}$	$\{\widetilde{\mathbf{x}}_i(T)\} = \{\mathbf{x}_i^T\}$
Interaction Matrix	$\mathcal{V}_B$	$\mathcal{U}_B$	$\mathcal{U}_B$
Task Type	reactive	reactive	active

Table 1. Task Synopsis.

During *fixation tracking*, a reactive task important in both artificial and biological vision systems [6], the camera is constrained to fix always a specific point of the object. Thanks to the linear interaction model, the centroid of the object's visible surface, chosen here as fixation point, is tracked by forcing the imaged object's centroid to be zero. The goal of *reactive tracking* is to imitate the motion of the object in the visual environment. An estimate of object motion can be also derived directly from joint data. Such a task can be useful to human-robot interfacing (mimicking human gestures, person following, etc.). Differently from fixation, the image point with constant zero speed is not, in general, the image origin, and the direction of gaze does not normally coincide with the direction of attention. Such an attentive shift is possible also in humans, but only if voluntary. The *active positioning* task consists in purposively changing the relative spatial configuration (pose, distance) of the camera with respect to a fixed or moving object. This task can be essential for the optimal execution of more complex perceptive and explorative tasks, for instance vision-based robot navigation. The linear transformation required to plan the task (see Subsect. 3.4) is obtained from the least squares comparison of the initial and goal object appearances. Shifts of visual appearance can be related to corresponding attentional shifts from a region to another of the image. The reactive tracking task can thus be regarded as a particular case of active positioning, where the goal appearance always coincides with the initial one.

As mentioned earlier, the tasks can be composed based on the value of their feedback gains (the higher the gain, the faster the task). Thus, fixation can be composed with positioning to yield the task of positioning w.r.t. a fixated object. After completion, the composite task degenerates into a reactive tracking task, which attempts to preserve the relative position and orientation between the camera and the fixated object.

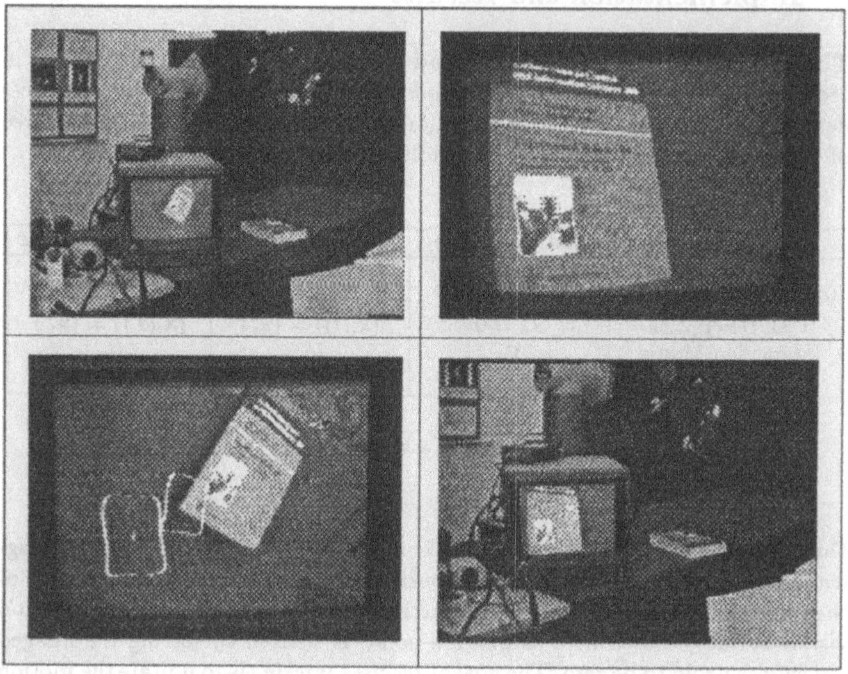

Fig. 2. A positioning experiment (see text). The monitor upon the table displays the current scene as viewed by the camera.

Setup and Parameter Setting. The system is implemented on an eye-in-hand robotic setup featuring a PUMA 560 manipulator equipped with a wrist-mounted off-the-shelf camera. Frame grabbing and control routines run on a 80486/66 MHz PC using an Imaging Technology VISIONplus-AT CFG board. The PUMA controller runs VAL II programs and communicates with the PC via the ALTER real-time protocol using an RS232 serial interface. New velocity setpoints are generated by the PC with a sampling rate $T_2 = N T_1$, where $T_1 = 28$ ms is the sampling rate of the ALTER protocol and the integer N depends on the overall loop time. Using $M = 16$ control points for B-spline contours, loop time is about 100 ms, hence $N = 4$.

Camera optics data-sheets provide a raw value for focal length and pixel dimensions; the remaining intrinsic parameters of the camera are ignored. Smoothing filters and feedback gain, all tuned experimentally, are set to $k_{pos} = 0.1$ (position) and $k_{vel} = 0.01$ (velocity), and $k = 0.1$, respectively. Cubic planning ($h = 3$) is used, which offers a good compromise between smoothness and contour inertia.

The System at Work. Fig. 2 summarizes the execution of an active positioning task, the most complex task among the three, with respect to a planar object – a book upon a table. The top left part of the figure shows the initial relative configuration, and the top right the goal image appearance. At the bottom left the planned trajectory between the initial and goal contours is sketched. At the bottom right of the figure, the obtained final configuration is shown which, as a typical performance, is reached with an error of within a few degrees (pose) and millimeters (translation).

To assess the stability characteristics of control, and tune up the feedback gain so as to obtain a slightly underdamped behavior, active positioning is run in *output regulation mode*. This mode is characterized by the absence of planning ($\widetilde{\mathbf{p}} = \mathbf{0}$ *and* $\widetilde{\mathbf{d}} = \mathbf{0}$), while the system brings itself, thanks to the feedback, from the initial to the goal configuration. The *servoing mode* ($\widetilde{\mathbf{d}} = \mathbf{0}$) is used instead to assess the tracking performance of the control scheme, as the system is forced to compensate the feedback error $\mathbf{e}(\widetilde{\mathbf{p}}, \widehat{\mathbf{p}})$. Fig. 3 (left) shows the camera velocity as obtained with the servoing mode. Cubic-based planning and stable tracking contribute to obtain graceful relative speed and acceleration profiles.

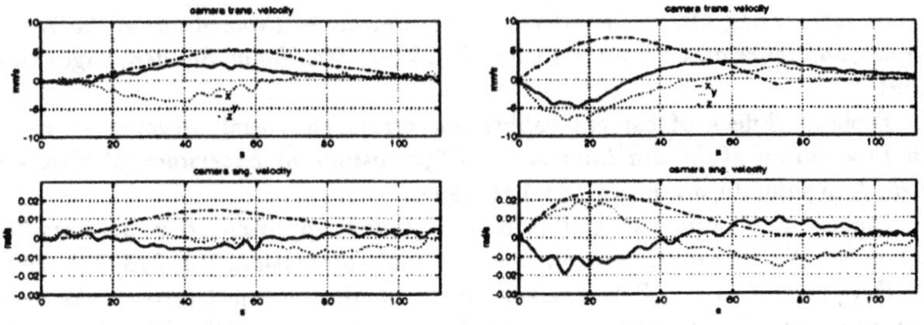

Fig. 3. Camera velocities. *Left*: Without feedforward (servoing mode). *Right*: With feedforward.

Introducing the feedforward term $\widetilde{\mathbf{d}}$ in the control scheme significantly alleviates the job of the feedback and reduces the tracking lag; yet, system performance gets more sensitive to 3D interaction data – in Fig. 3 (right), feedforward was dropped out after about 75 s. Still, both control schemes (with and without feedforward) exhibit a nice behavior, also considering that the tests were run without estimating and updating on-line the interaction matrix.

Acknowledgements

Part of this work was done during a stay of C. Colombo at LIFIA-IMAG, Grenoble, as a visiting scientist supported by a fellowship from the EC HCM network SMART. The authors warmly thank Dr. B. Allotta of ARTS Lab, Pisa for his useful comments and help in the experiments.

References

1. A. Blake, R. Curwen, and A. Zisserman. A framework for spatiotemporal control in the tracking of visual contours. *International Journal of Computer Vision*, 11(2):127–145, 1993.
2. R. Brooks. A robust layered control system for a mobile robot. *IEEE Journal of Robotics and Automation*, pages 14–23, 1986.
3. R. Cipolla and N.J. Hollinghurst. Visual robot guidance from uncalibrated stereo. In C.M. Brown and D. Terzopoulos, editors, *Realtime Computer Vision*. CUP, 1994.
4. C. Colombo, B. Allotta, and P. Dario. Affine Visual Servoing: A framework for relative positioning with a robot. In *Proc. IEEE International Conference on Robotics and Automation ICRA'95, Nagoya, Japan*, 1995.
5. C. Colombo, M. Rucci, and P. Dario. Attentive behavior in an anthropomorphic robot vision system. *Robotics and Autonomous Systems*, 12(3–4):121–131, 1994.
6. J.L. Crowley, J.M. Bedrune, M. Bekker, and M. Schneider. Integration and control of reactive visual processes. In *Proceedings of the 3rd European Conference on Computer Vision ECCV'94, Stockholm, Sweden*, pages II:47–58, 1994.
7. J.L. Crowley and H.I. Christensen. *Vision as Process*. Springer Verlag Basic Research Series, 1994.
8. S.J. Dickinson, H.I. Christensen, J. Tsotsos, and G. Olofsson. Active object recognition integrating attention and viewpoint control. In *Proceedings of the 3rd European Conference on Computer Vision ECCV'94, Stockholm, Sweden*, pages II:3–14, 1994.
9. B. Espiau. Effect of camera calibration errors on visual servoing in robotics. In *Proceedings of the 3rd International Symposium on Experimental Robotics ISER'93, Kyoto, Japan*, pages 182–192, 1993.
10. B. Espiau, F. Chaumette, and P. Rives. A new approach to visual servoing in robotics. *IEEE Transactions on Robotics and Automation*, 8(3):313–326, 1992.
11. R. Horaud, S. Christy, F. Dornaika, and B. Lamiroy. Object pose: Links between paraperspective and perspective. In *Proceedings of the 5th IEEE International Conference on Computer Vision ICCV'95, Cambridge, Massachusetts*, pages 426–433, 1995.
12. I.D. Reid and D.W. Murray. Tracking foveated corner clusters using affine structure. In *Proceedings of the 4th IEEE International Conference on Computer Vision ICCV'93, Berlin, Germany*, pages 76–83, 1993.

Finding Naked People

Margaret M. Fleck[1], David A. Forsyth[2], and Chris Bregler[2]

[1] Department of Computer Science, University of Iowa, Iowa City, IA 52242
[2] Computer Science Division, U.C. Berkeley, Berkeley, CA 94720

Abstract. This paper demonstrates a content-based retrieval strategy that can tell whether there are naked people present in an image. No manual intervention is required. The approach combines color and texture properties to obtain an effective mask for skin regions. The skin mask is shown to be effective for a wide range of shades and colors of skin. These skin regions are then fed to a specialized grouper, which attempts to group a human figure using geometric constraints on human structure. This approach introduces a new view of object recognition, where an object model is an organized collection of grouping hints obtained from a combination of constraints on geometric properties such as the structure of individual parts, and the relationships between parts, and constraints on color and texture. The system is demonstrated to have 60% precision and 52% recall on a test set of 138 uncontrolled images of naked people, mostly obtained from the internet, and 1401 assorted control images, drawn from a wide collection of sources. **Keywords:** Content-based Retrieval, Object Recognition, Computer Vision, Erotica/Pornography, Internet, Color

1 Introduction

The recent explosion in internet usage and multi-media computing has created a substantial demand for algorithms that perform *content-based retrieval*—selecting images from a large database based on what they depict. Identifying images depicting naked or scantily-dressed people is a natural content-based retrieval problem. These images frequently lack textual labels adequate to identify their content but can be effectively detected using simple visual cues (color, texture, simple shape features), of the type that the human visual system is known to use for fast (preattentive) triage [19]. There is little previous work on finding people in static images, though [9] shows that a stick-figure group can yield pose in 3D up to limited ambiguities; the work on motion sequences is well summarised in [4].

Several systems have recently been developed for retrieving images from large databases. The best-known such system is QBIC [15], which allows an operator to specify various properties of a desired image. The system then displays a selection of potential matches to those criteria, sorted by a score of the appropriateness of the match. Searches employ an underlying abstraction of an image as a collection of colored, textured regions, which were manually segmented in advance, a significant disadvantage. Photobook [17] largely shares QBIC's model of an image as a collage of flat, homogenous frontally presented regions, but incorporates more sophisticated representations of texture and a degree of automatic segmentation. A version of Photobook ([17], p. 10) incorporates a simple notion of objects, using plane matching by an energy minimisation strategy. However, the approach does not adequately address the range of variation in object shape and appears to require manually segmented images for all but trivial cases. Appearance based matching is also used in [8], which describes a system that forms a wavelet based decomposition of an image and matches based on the coarse-scale appearance. Similarly, Chabot [16] uses a combination of visual appearance and text-based cues to retrieve images, but depends strongly on text cues to identify objects. However, appearance is not a satisfactory notion of content, as it is only loosely correlated with object identity.

Current object recognition systems represent models either as a collection of geometric measurements or as a collection of images of an object. This information is then compared with image information to obtain a match. Most current systems that use geometric models use invariants of an imaging transformation to index models in a model library, thereby producing a selection of recognition hypotheses. These hypotheses are combined as appropriate, and the result is back-projected into the image, and verified by inspecting relationships between the back-projected outline and image edges. An extensive bibliography of this approach appears in [12].

Systems that recognize an object by matching a view to a collection of images of an object proceed in one of two ways. In the first approach, correspondence between image points and points on the model of some object is assumed known and an estimate of the appearance in the image of that object is constructed from correspondences. The hypothesis that the object is present is then verified using the estimate of appearance [20]. An alternative approach computes a feature vector from a compressed version of the image and uses a minimum distance classifier to match this feature vector to feature vectors computed from images of objects in a range of positions under various lighting conditions [13]. Neither class of system copes well with models that have large numbers of internal degrees of freedom, nor do they incorporate appropriate theories of parts. Current part-based recognition systems are strongly oriented to recovering cross-sectional information, and do not treat the case where there are many parts with few or no individual distinguishing features[22].

Typical images of naked people found on the internet: have uncontrolled backgrounds; may depict multiple figures; often contain partial figures; are static images; and have been taken from a wide variety of camera angles, e.g. the figure may be oriented sideways or may viewed from above.

2 A new approach

Our system for detecting naked people illustrates a general approach to object recognition. The algorithm:

- first locates images containing large areas of skin-colored region;

- then, within these areas, finds elongated regions and groups them into possible human limbs and connected groups of limbs, using specialised groupers which incorporate substantial amounts of information about object structure.

Images containing sufficiently large skin-colored groups of possible limbs are reported as potentially containing naked people.

2.1 Finding Skin

The appearance of skin is tightly constrained. The color of a human's skin is created by a combination of blood (red) and melanin (yellow, brown) [18]. Therefore, human skin has a restricted range of hues. Skin is somewhat saturated, but not deeply saturated. Because more deeply colored skin is created by adding melanin, the range of possible hues shifts toward yellow as saturation increases. Finally, skin has little texture; extremely hairy subjects are rare. Ignoring regions with high-amplitude variation in intensity values allows the skin filter to eliminate more control images.

The skin filter starts by subtracting the zero-response of the camera system, estimated as the smallest value in any of the three colour planes omitting locations within 10 pixels of the image edges, to avoid potentially significant desaturation. The input R, G, and B values are transformed into a log-opponent representation (cf e.g. [6]). If we let L represent the log transformation, the three log-opponent values are $I = L(G)$, $R_g = L(R) - L(G)$, and $B_y = L(B) - \frac{L(G)+L(R)}{2}$. The green channel is used to represent intensity because the red and blue channels from some cameras have poor spatial resolution.

Next, smoothed texture and color planes are extracted. The R_g and B_y arrays are smoothed with a median filter. To compute texture amplitude, the intensity image is smoothed with a median filter, and the result subtracted from the original image. The absolute values of these differences are run through a second median filter. [3]

The texture amplitude and the smoothed R_g and B_y values are then passed to a tightly-tuned skin filter. It marks as probably skin all pixels whose texture amplitude is small, and whose hue and saturation values are appropriate. The range of hues considered to be appropriate changes with the saturation, as described above. This is very important for good performance. When the same range of hues is used for all saturations, significantly more non-skin regions are accepted.

Because skin reflectance has a substantial specular component, some skin areas are desaturated or even white. Under some illuminants, these areas appear as blueish or greenish off-white. These areas will not pass the tightly-tuned skin filter, creating holes (sometimes large) in skin regions, which may confuse geometrical analysis. Therefore, output of the initial skin filter is refined to include adjacent regions with almost appropriate properties.

Specifically, the region marked as skin is enlarged to include pixels many of whose neighbors passed the initial filter. If the marked regions cover at least 30%

[3] All operations use a fast multi-ring approximation to the median filter [5].

of the image area, the image will be referred for geometric processing. Finally, to trim extraneous pixels, the algorithm unmarks any pixels which do not pass a more lenient version of the skin filter, which imposes no constraints on texture amplitude and uses less exacting constraints on hue and saturation.

3 Grouping People

The human figure can be viewed as an assembly of nearly cylindrical parts, where both the individual geometry of the parts and the relationships between parts are constrained by the geometry of the skeleton. These constraints on the 3D parts induce grouping constraints on the corresponding 2D image regions. These induced constraints provide an appropriate and effective model for recognizing human figures.

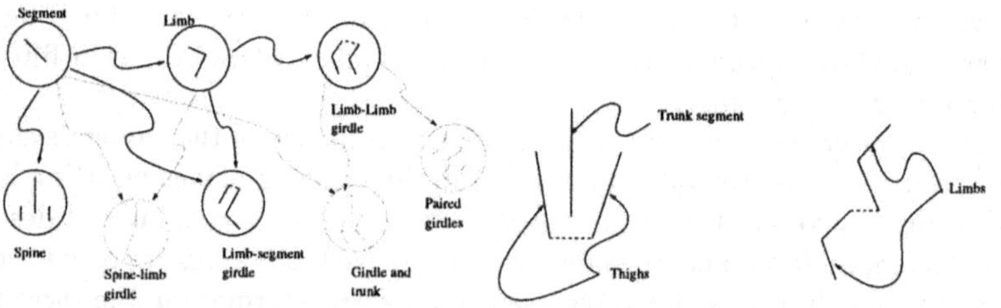

Fig. 1. Left: grouping rules (arrows) specify how to assemble simple groups (e.g. body segments) into complex groups (e.g. limb-segment girdles). These rules incorporate constraints on the relative positions of 2D features, induced by constraints on 3D body parts. Dashed lines indicate grouping rules that are not yet implemented. Middle: the grouper rejects this assembly of thighs and a spine (the dashed line represents the pelvis) because the thighs would occlude the trunk if a human were in this posture, making the trunk's symmetry impossible to detect. Right: this hip girdle will also be rejected. Limitations on hip joints prevent human legs from assuming positions which could project to such a configuration.

The current system models a human as a set of rules describing how to assemble possible girdles and spine-thigh groups (Figure 1). The input to the geometric grouping algorithm is a set of images, in which the skin filter has marked areas identified as human skin. Sheffield's version of Canny's [3] edge detector, with relatively high smoothing and contrast thresholds, is applied to these skin areas to obtain a set of connected edge curves. Pairs of edge points with a near-parallel local symmetry [1] are found by a straightforward algorithm. Sets of points forming regions with roughly straight axes ("ribbons" [2]) are found using an algorithm based on the Hough transformation.

Grouping proceeds by first identifying potential segment outlines, where a segment outline is a ribbon with a straight axis and relatively small variation in average width. Ribbons that may form parts of the same segment are merged, and suitable pairs of segments are joined to form limbs. An affine imaging model is satisfactory here, so the upper bound on the aspect ratio of 3D limb segments induces an upper bound on the aspect ratio of 2D image segments corresponding to limbs. Similarly, we can derive constraints on the relative widths of the 2D segments.

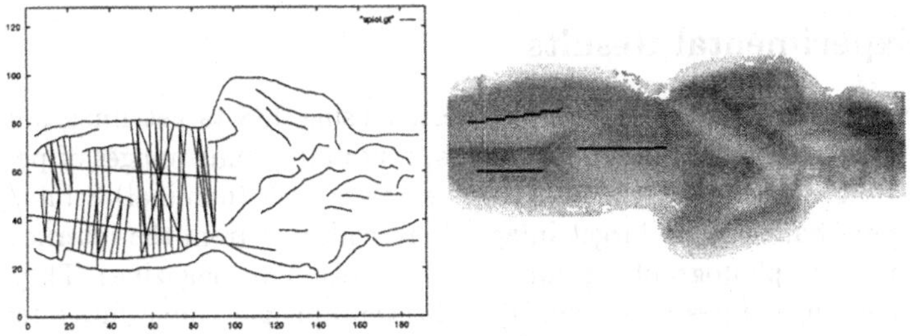

Fig. 2. Grouping a spine and two thighs: *Top left* the segment axes that will be grouped into a spine-thigh group, overlaid on the edges, showing the upper bounds on segment length and the their associated symmetries; *Top right* the spine and thigh group assembled from these segments, overlaid on the image.

Specifically, two ribbons can only form part of the same segment if they have similar widths and axes. Two segments may form a limb if: their search intervals intersect; there is skin in the interior of both ribbons; their average widths are similar; and in joining their axes, not too many edges must be crossed. There is no angular constraint on axes in grouping limbs. The output of this stage contains many groups that do not form parts of human-like shapes: they are unlikely to survive as grouping proceeds to higher levels.

The limbs and segments are then assembled into putative girdles. There are grouping procedures for two classes of girdle; one formed by two limbs, and one formed by one limb and a segment. The latter case is important when one limb segment is hidden by occlusion or by cropping. The constraints associated with these girdles are derived from the case of the hip girdle, and use the same form of interval-based reasoning as used for assembling limbs.

Limb-limb girdles must pass three tests. The two limbs must have similar widths. There must be a line segment (the pelvis) between their ends, whose position is bounded at one end by the upper bound on aspect ratio, and at the other by the symmetries forming the limb and whose length is similar to twice the average width of the limbs. Finally, occlusion constraints rule out certain types of configurations : limbs in a girdle may not cross each other, they may not cross other segments or limbs, and there is a forbidden configuration of kneecaps (see figure 1). A limb-segment girdle is formed using similar constraints, but using a limb and a segment.

Spine-thigh groups are formed from two segments serving as upper thighs, and a third, which serves as a trunk. The thigh segments must have similar average widths, and it must be possible to construct a line segment between their ends to represent a pelvis in the manner described above. The trunk segment must have an average width similar to twice the average widths of the thigh segments. Finally, the whole configuration of trunk and thighs must satisfy geometric constraints depicted in figure 1. The grouper asserts that human figures are present if it can assemble either a spine-thigh group or a girdle group. Figure 2 illustrates the process of assembling a spine-thigh group.

4 Experimental Results

The performance of the system was tested using 138 target images of naked people and 1401 assorted control images, containing some images of people but none of naked people. Most images were taken with (nominal) 8 bits/pixel in each color channel. The target images were collected from the internet and by scanning or re-photographing images from books and magazines. They show a very wide range of postures. Some depict several people, sometimes intertwined. Some depict only small parts of the bodies of one or more people. Most of the people in the images are Caucasians; a small number are Blacks or Asians.

Five types of control image were used

- 1241 images sampled[4] from an image database produced by California Department of Water Resources (DWR), including landscapes, pictures of animals, and pictures of industrial sites,
- 58 images of clothed people, a mixture of Caucasians, Blacks, Asians, and Indians, largely showing their faces, 3 re-photographed from a book and the rest photographed from live models at the University of Iowa,
- 44 assorted images from a CD included with an issue of MacFormat [11],
- 11 assorted personal photos, re-photographed with our CCD camera, and
- 47 pictures of objects and textures taken in our laboratory for other purposes.

The DWR images are 128 by 192 pixels. The images from other sources were reduced to approximately the same size. Table 1 summarizes the performance of each stage of the system.

Mistakes by the skin filter occur for several reasons. In some test images, the naked people are very small. In others, most or all of the skin area is desaturated, so that it fails the first-stage skin filter. Some control images pass the skin filter because they contain people, particularly several close-up portrait shots. Other control images contain material whose color closely resembles that of human skin: particularly wood and the skin or fur of certain animals. All but 8 of our 58 control images of faces and clothed people failed the skin filter primarily because many of the faces occupy only a small percentage of the image area. In 18 of these images, the face was accurately marked as skin. In 12 more, a substantial portion of the face was marked, suggesting that the approach provides a useful pre-filter for programs that mark faces. Failure on the remaining images is largely due to the small size of the faces, desaturation of skin color, and fragmentation of the face when eye and mouth areas are rejected by the skin filter.

[4] The sample consists of every tenth image; in the full database, images with similar numbers tend to have similar content.

Figures 3-4 illustrate its performance on the test images. Configurations marked by the spine-thigh detector are typically spines. The girdle detector often marks structures which are parts of the human body, but not hip or shoulder girdles. This presents no major problem, as the program is trying to detect the presence of humans, rather than analyze their pose in detail.

False negatives occur for several reasons. Some close-up or poorly cropped images do not contain arms and legs, vital to the current geometrical analysis algorithm. Regions may have been poorly extracted by the skin filter, due to desaturation. The edge finder may fail due to poor contrast between limbs and their surroundings. Structural complexity in the image, often caused by strongly colored items of clothing, confuses the grouper. Finally, since the grouper uses only segments that come from bottom up mechanisms and does not predict the presence of segments which might have been missed by occlusion, performance is notably poor for side views of figures with arms hanging down. Figures 5-6 show typical performance on control images. The current implementation is frequently confused by groups of parallel edges, as in industrial scenes, and sometimes accepts ribbons lying largely outside the skin regions. We believe the latter problem can easily be corrected.

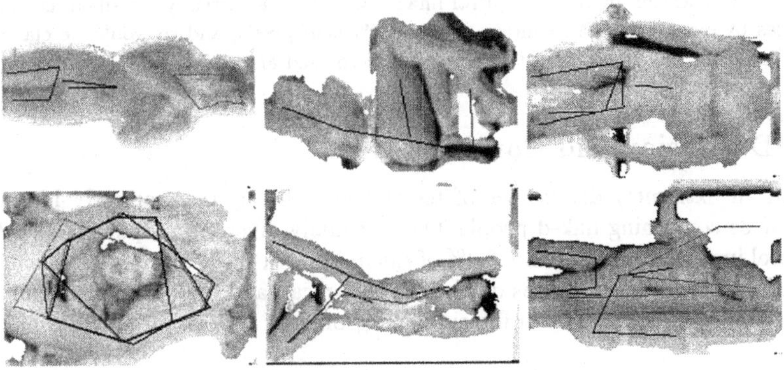

Fig. 3. Typical images correctly classified as containing naked people. The output of the skin filter is shown, with spines overlaid in red, limb-limb girdles overlaid in blue, and limb-segment girdles overlaid in blue.

	eliminated by skin filter	eliminated by geometrical analysis	marked as containing naked people
test images	13.8% (19)	34.1% (47)	52.2% (72)
control images	92.6% (1297)	4.0% (56)	3.4% (48)

Table 1. Overall classification performance of the system.

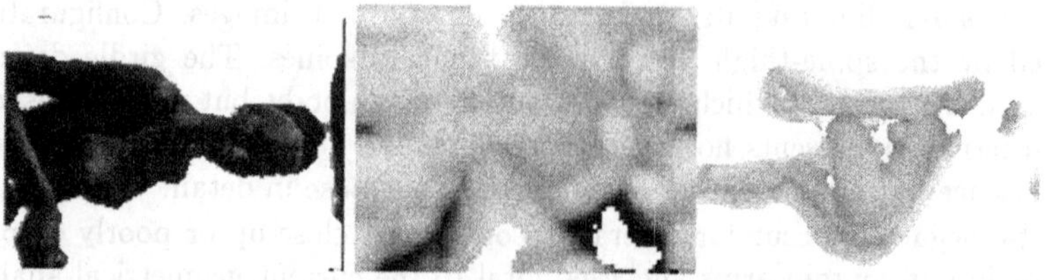

Fig. 4. Typical false negatives: the skin filter marked significant areas of skin, but the geometrical analysis could not find a girdle or a spine. Failure is often caused by absence of limbs, low contrast, or configurations not included in the geometrical model (notably side views).

Fig. 5. A collection of typical control images which were correctly classified as control images by our system. All contain at least 30% skin pixels, and so would be classified as containing naked people if the skin filter were used alone.

5 Discussion and Conclusions

From an extremely diverse set of test images, this system correctly identifies 52.2% as containing naked people. On an equally diverse and quite large set of control images, it returns only 3.4% of the images. In the terminology of content-based retrieval, the system is displaying 52% recall and 60% precision against a large control set[5]. Both skin filtering and geometric processing are required for

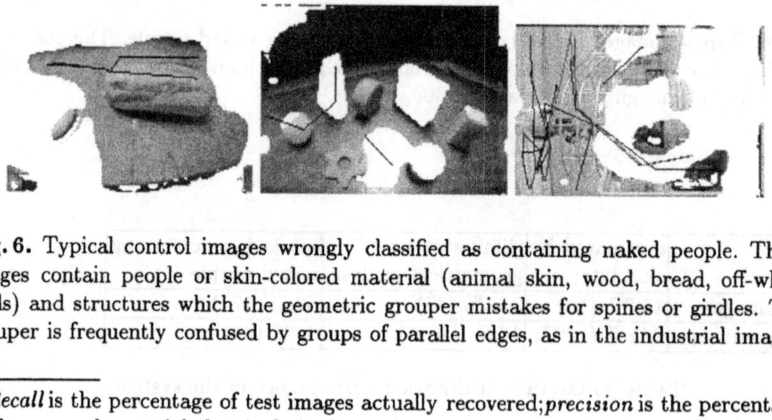

Fig. 6. Typical control images wrongly classified as containing naked people. These images contain people or skin-colored material (animal skin, wood, bread, off-white walls) and structures which the geometric grouper mistakes for spines or girdles. The grouper is frequently confused by groups of parallel edges, as in the industrial image.

[5] *Recall* is the percentage of test images actually recovered; *precision* is the percentage of recovered material that is desired.

this level of performance: the skin filter by itself has better recall (86.2%), but returns twice as many false positives. This is an extremely impressive result for a high-level query ("find naked people") on a very large (1539 image) database with no manual intervention and almost no control over the content of the test and control images.

This system demonstrates detection of jointed objects of highly variable shape, in a diverse range of poses, seen from many different camera positions. It also demonstrates that color cues can be very effective in recognizing objects that whose color is not heavily saturated and whose surfaces display significant specular effects, under diverse lighting conditions, without relying on preprocessing to remove specularities. While the current implementation uses only very simple geometrical grouping rules, covering only poses with visible limbs, the performance of this stage could easily be improved. In particular: the ribbon detector should be made more robust; detectors should be added for non-ribbon features (for example, faces); grouping rules for structures other than spines and girdles should be added; grouping rules should be added for close-up views of the human body.

The reason we have achieved such good performance, and expect even better performance in the future, is that we use object models quite different from those commonly used in computer vision (though similar to proposals in [2, 14]). In the new system, an object is modelled as a loosely coordinated collection of detection and grouping rules. The object is recognized if a suitable group can be built. These grouping rules incorporate both surface properties (color and texture) and simple shape information. In the present system, the integration of different cues is simple (though effective), but a more sophisticated recognizer would integrate them more closely. This type of model gracefully handles objects whose precise geometry is extremely variable, where the identification of the object depends heavily on non-geometrical cues (e.g. color) and on the interrelationships between parts. While our present model is hand-crafted and is by no means complete, there is good reason to believe that an algorithm could construct a model of this form, automatically or semi-automatically, from a 3D object model.

Finally, as this paper goes to press, a second experimental run using a substantially improved version of the grouper has displayed 44 % recall and an extraordinary 74 % precision on a set of 355 test images and 2782 control images from extremely diverse sources.

Acknowledgements: We thank Joe Mundy for suggesting that the response of a grouper may indicate the presence of an object and Jitendra Malik for helpful suggestions. This research was supported by the National Science Foundation under grants IRI-9209728, IRI-9420716, IRI-9501493, under a National Science Foundation Young Investigator award, an NSF Digital Library award IRI-9411334, and under CDA-9121985, an instrumentation award.

References

1. Brady, J. Michael and Haruo Asada (1984) "Smoothed Local Symmetries and Their Implementation," *Int. J. Robotics Res.* 3/3, 36–61.

2. Brooks, Rodney A. (1981) "Symbolic Reasoning among 3-D Models and 2-D Images," *Artificial Intelligence* 17, pp. 285–348.
3. Canny, John F. (1986) "A Computational Approach to Edge Detection," *IEEE Patt. Anal. Mach. Int.* 8/6, pp. 679–698.
4. Cedras C., Shah M., (1994) "A Survey of Motion Analysis from Moving Light Displays" *Computer Vision and Pattern Recognition* pp 214-221.
5. Fleck,Margaret M. (1994) "Practical edge finding with a robust estimator," *Proc. of the IEEE Conf. on Computer Vision and Pattern Recognition*, pp. 649–653.
6. Gershon, Ron, Allan D. Jepson, and John K. Tsotsos (1986) "Ambient Illumination and the Determination of Material Changes," *J. Opt. Soc. America* A 3/10, pp. 1700–1707.
7. Gong, Yihong and Masao Sakauchi (1995) "Detection of Regions Matching Specified Chromatic Features," *Comp. Vis. Im. Underst.* 61/2, pp. 263–269.
8. Jacobs, C.E., Finkelstein, A., and Salesin, D.H., "Fast Multiresolution Image Querying," *Proc SIGGRAPH-95*, 277-285, 1995.
9. Lee, H.-J. and Chen, Z. "Determination of 3D human body postures from a single view," CVGIP, **30**, 148-168, 1985
10. Lowe, David G. (1987) "The Viewpoint Consistency Constraint," *Intern. J. of Comp. Vis*, 1/1, pp. 57–72.
11. MacFormat, issue no. 28 with CD-Rom, September, 1995.
12. Mundy, J.L. and Zisserman, A. *Geometric Invariance in Computer Vision*, MIT press, 1992
13. Murase, H. and Nayar, S.K., "Visual learning and recognition of 3D objects from appearance," to appear, *Int. J. Computer Vision*, 1995.
14. Nevatia, R. and Binford, T.O., "Description and recognition of curved objects," *Artificial Intelligence*, **8**, 77-98, 1977
15. Niblack, W., Barber, R, Equitz, W., Flickner, M., Glasman, E., Petkovic, D., and Yanker, P. (1993) "The QBIC project: querying images by content using colour, texture and shape," *IS and T/SPIE 1993 Intern. Symp. Electr. Imaging: Science and Technology, Conference 1908, Storage and Retrieval for Image and Video Databases.*
16. Ogle, Virginia E. and Michael Stonebraker (1995) "Chabot: Retrieval from a Relational Database of Images," *Computer* 28/9, pp. 40–48.
17. Pentland, A., Picard, R.W., and Sclaroff, S. "Photobook: content-based manipulation of image databases," MIT Media Lab Perceptual Computing TR No. 255, Nov. 1993.
18. Rossotti, Hazel (1983) *Colour: Why the World isn't Grey*, Princeton University Press, Princeton, NJ.
19. Treisman, Anne (1985) "Preattentive Processing in Vision," *Com. Vis. Grap. Im. Proc.* 31/2, pp. 156–177.
20. Ullman, S. and Basri, R. (1991). Recognition by linear combination of models, *IEEE PAMI*, **13**, 10, 992-1007.
21. Zisserman, A., Mundy, J.L., Forsyth, D.A., Liu, J.S., Pillow, N., Rothwell, C.A. and Utcke, S. (1995) "Class-based grouping in perspective images",*Intern. Conf. on Comp. Vis.*
22. Zerroug, M. and Nevatia, R. "From an intensity image to 3D segmented descriptions," ICPR, 1994.

Acquiring Visual-Motor Models for Precision Manipulation with Robot Hands*

Martin Jägersand, Olac Fuentes, Randal Nelson

Department of Computer Science
University of Rochester
Rochester, N.Y. 14627
U.S.A.
{jag,fuentes,nelson}@cs.rochester.edu
http://www.cs.rochester.edu/u/{jag,fuentes,nelson}/

Abstract. Dextrous high degree of freedom (DOF) robotic hands provide versatile motions for fine manipulation of potentially very different objects. However, fine manipulation of an object grasped by a multifinger hand is much more complex than if the object is rigidly attached to a robot arm. Creating an accurate model is difficult if not impossible. We instead propose a combination of two techniques: the use of an approximate estimated motor model, based on the grasp tetrahedron acquired when grasping an object, and the use of visual feedback to achieve accurate fine manipulation. We present a novel active vision based algorithm for visual servoing, capable of learning the manipulator kinematics and camera calibration online while executing a manipulation task. The approach differs from previous work in that a full, coupled image Jacobian is estimated online without prior models, and that a trust region control method is used, improving stability and convergence. We present an extensive experimental evaluation of visual model acquisition and visual servoing in 3, 4 and 6 DOF.

1 Introduction

In an active or behavioral vision system the acquisition of visual information is not an independent open loop process, but instead depends on the active agent's interaction with the world. Also the information need not represent an abstract, high level model of the world, but instead is highly task specific, and represented in a form which facilitates particular operations. Visual servoing, when supplemented with online visual model estimation, fits into the active vision paradigm. Results with visual servoing and varying degrees of model adaption have been presented for robot arms [23, 4, 21, 15, 16, 13, 2, 11][2]. Visual models suitable for specifying visual alignments have also been studied [9, 1, 10], but it remains to be proven that the approach works on more complex manipulators than the serial link robot arm. In this paper we present an active vision technique, having interacting action (control), visual sensing and model building modules, which allows the simultaneous visual-motor model estimation and control (visual servoing) of a dextrous multi finger robot hand. We investigate experimentally how well our algorithms work on a 16 DOF Utah/MIT hand.

A combined model acquisition and control approach has many advantages. In addition to being able to do uncalibrated visual servo control, the online estimated models are useful for instance for: (1) Prediction and constraining search in visual tracking [16, 4] (2) Doing local coordinate transformations between manipulator (joint), world and visual frames[15, 16]. (3) Synthesizing views from other agent poses [14]. In robot

* Support was provided by the Fulbright Commission and ONR grant N00014-93-I-0221.

[2] For a review of this work we direct the reader to [15] or [2].

arm manipulation we have found such a fully adaptive approach to be particularly helpful when carrying out difficult tasks, such as manipulation of flexible material [15, 16], or performing large rotations for visually exploring object shape [17]. For the Utah/MIT dextrous robot hand the fully adaptive approach is appealing because precisely modeling manipulation of a grasped object is much harder than for a typical robot arm, where the object is rigidly attached to the end effector. The four fingers of the Utah/MIT hand have a total of 16 actuated joints. When grasping a rigid object the fingers form a complex parallel kinematic chain. Although a few experimental systems that demonstrate the capabilities of dextrous robot hands have been implemented (e.g. [18, 20, 22]), their potential for performing high-precision tasks in unmodelled environments has not been fully realized. We have proposed an approach to dextrous manipulation [7], which does not require a-priori object models; all the required information can be read directly from the hand's sensors. This *alignment virtual tool* [19] allows versatile manipulation as shown in fig. 1 by providing an approximate transformation between the 6 DOF pose space of the grasped object and the 16 DOF joint control space. But the transformation is only approximate, and there is no way to account for slips in the grasp and other inaccuracies, so some external, object pose based feedback is needed to achieve high precision object fine manipulation.

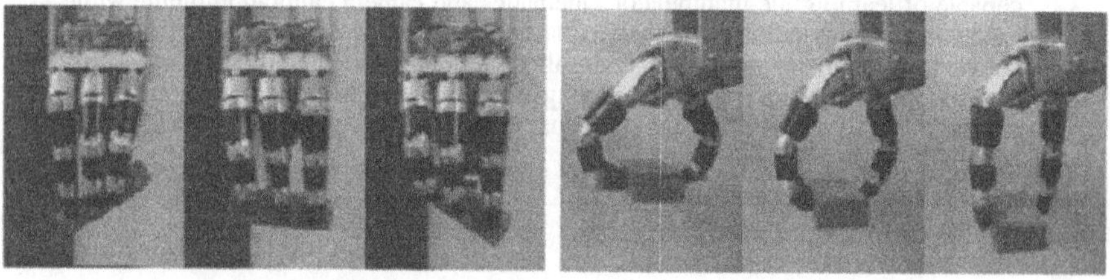

Fig. 1. Examples of manipulation with Utah/MIT hand.

In this paper we show how to learn online, and adaptively refine a local estimate of a visual-motor model, and to use that estimate for stable, convergent control even in difficult non-linear cases. Our objectives are to completely do away with prior calibration, either experimental "test movements" or model based, and to improve the convergence to allow solving manipulation tasks in larger workspaces, and/or more difficult cases than previously possible. The difficulty in the problem we solve lies much in the - for manipulator control purposes - extremely low sampling frequency imposed by the video rate of cameras. With very few samples, we want to control over large portions of the visual-motor work space and deal with large initial deviations between desired and actual states. This is different from what we feel mainstream control theory is meant to solve. In the paper we instead cast the estimation and control problem in an optimization theory framework, where we can draw on previous experience on estimating with very sparse data, and obtaining convergence despite large initial errors and large step sizes.

The overall structure of our vision based manipulation system is shown in fig. 2. In the two following sections we describe the online model acquisition. In sections 4 and 5 we describe the visual and motor control of the hand. Visual goal assignment, trajectory generation and what kind of visual features to use are task dependent. In [16] we describe the high (task) level parts of the system and how to solve real world manipulation problems with this system, such as solving a shape sorting puzzle, handling flexible materials and exchanging a light bulb. In section 6 we study low level aspects of visual control, and use simple random trajectories to quantitatively evaluate positioning and model estimation.

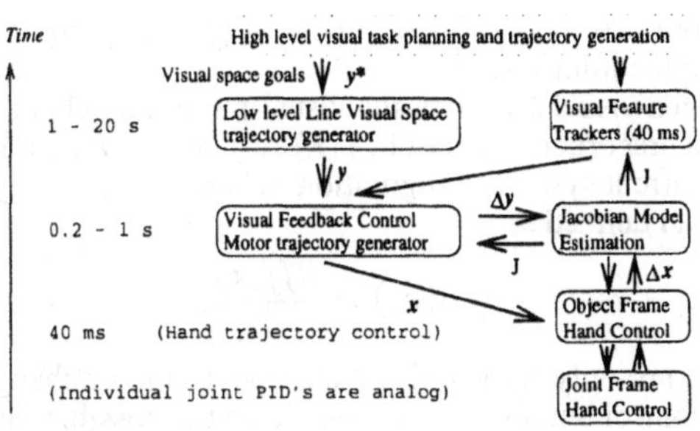

Fig. 2. Structure of the system, and the time frames each part runs in. Arrows indicate information exchange.

2 Viewing Model

We consider the robot hand[3] as an active agent in an unstructured environment. The hand manipulates a grasped object in approximate motor pose space[4] $\mathbf{x} \in \Re^n$ through the transformations derived in section 5, and observes the results of its actions as changes in a visual perception or "feature" vector $\mathbf{y} \in \Re^m$. Visual features can be drawn from a large class of visual measurements[23, 15], but we have found that the ones which can be represented as point positions or point vectors in camera space are suitable[16]. We track features such as boundary discontinuities (lines,corners) and surface markings. Redundant visual perceptions ($m \gg n$) are desirable as they are used to constrain the raw visual sensory information.

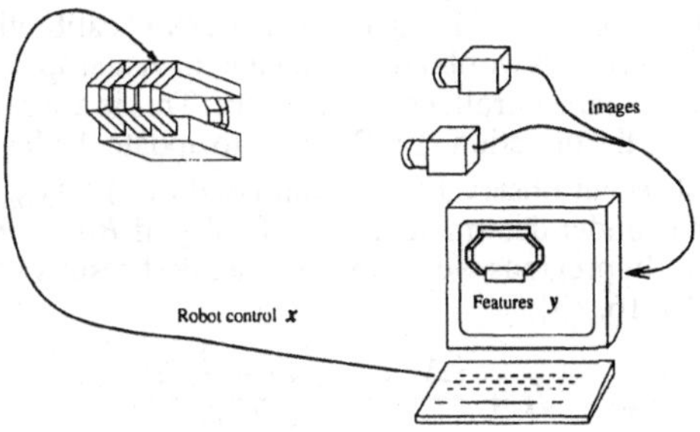

Fig. 3. Visual control setup using two cameras.

The visual features and the agent's actions are related by some visual-motor model f, satisfying $\mathbf{y} = f(\mathbf{x})$. The goal for the control problem is, given current state $\mathbf{x}_0$ and $\mathbf{y}_0$, the desired final state in visual space $\mathbf{y}^*$, to find a motor command, or sequence thereof, $\Delta\mathbf{x}_k$ s.t. $f(\mathbf{x}_0 + \sum_k \Delta\mathbf{x}_k) = \mathbf{y}^*$. Alternatively, one can view this problem as the minimization of the functional $\phi = \frac{1}{2}(f - \mathbf{y}^*)^T(f - \mathbf{y}^*)$. In a traditional visual control setup we have to know two functions, camera calibration h and robot kinematics g, either a priori, or through some calibration process. The final robot positioning accuracy

[3] The visual feedback methods derived here also apply to other holonomic manipulators.

[4] Vectors are written bold, scalars plain and matrices capitalized.

depends on the accuracy of both of these functions ($f^{-1} = g(h(\cdot))$), since feedback is only performed over the joint values.

In our uncalibrated visual servoing, the visual-motor model is unknown and at any time k we estimate a first order model of it, $f(\mathbf{x}) \approx f(\mathbf{x}_k) + J(\mathbf{x}_k)(\mathbf{x} - \mathbf{x}_k)$. The model is valid around the current system configuration $\mathbf{x}_k$, and described by the "image"[2] or visual-motor Jacobian defined as

$$(J_{j,i})(\mathbf{x}_k) = \frac{\partial f_j(\mathbf{x}_k)}{\partial x_i} \tag{1}$$

The image Jacobian not only relates visual changes to motor changes, as is exploited in visual feedback control but also highly constrains the possible visual changes to the n–D subspace $\mathbf{y}_{k+1} = J\Delta\mathbf{x} + \mathbf{y}_k$ of $\mathbf{y}_{k+1} \subset \Re^m$ (remember $m \gg n$). Thus the Jacobian J is also a visual model, parameterized in exactly the degrees of freedom our system can change in. A collection of such Jacobians represents a piecewise linear visual-motor model of the part of the workspace the agent has explored so far. In our case f is linearized on an adaptive size mesh, as explained in section 4.

3 Visual-Motor Model Estimation

In previous visual servoing work a Jacobian has been either (1) derived analytically, (2) derived partially analytically and partially estimated (eg. [4, 21]), or (3) determined experimentally by physically executing a set of orthogonal calibration movements $\mathbf{e}_i$ (eg. [1, 9]) and approximating the Jacobian with finite differences:

$$\hat{J}(\mathbf{x}, \mathbf{d}) = (f(\mathbf{x} + d_1\mathbf{e}_1) - f(\mathbf{x}), \ldots, f(\mathbf{x} + d_n\mathbf{e}_n) - f(\mathbf{x}))D^{-1} \tag{2}$$

where $D = \text{diag}(\mathbf{d}), \mathbf{d} \in \Re^n$.

We instead seek an online method, which estimates the Jacobian by just observing the process, without a-priori models or introducing any extra "calibration" movements. In observing the object movements we obtain the changes in visual appearance $\Delta\mathbf{y}_{measured}$ corresponding to a particular controller command $\Delta\mathbf{x}$. This is a secant approximation of the derivative of f along the direction $\Delta\mathbf{x}$. We want to update the Jacobian in such a way as to satisfy the most recent observation (secant condition): $\Delta\mathbf{y}_{measured} = \hat{J}_{k+1}\Delta\mathbf{x}$ The above condition is under determined, and a family of *Broyden* updating formulas can be defined [5, 13]. In previous work we have had best results with this asymmetric correction formula [15, 16, 13]:

$$\hat{J}_{k+1} = \hat{J}_k + \frac{(\Delta\mathbf{y}_{measured} - \hat{J}_k\Delta\mathbf{x})\Delta\mathbf{x}^T}{\Delta\mathbf{x}^T\Delta\mathbf{x}} \tag{3}$$

This is a rank 1 updating formula in the *Broyden* hierarchy, and it converges to the Jacobian after n orthogonal moves $\{\Delta\mathbf{x}_k, k = 1\ldots n\}$. It trivially satisfies the secant condition, but to see how it works let's consider a change of basis transformation P to a coordinate system O' aligned with the last movement so $\Delta\mathbf{x}' = (\Delta x_1, 0, \ldots, 0)$. In this basis the correction term in eq. 3 is zero except for the first column. Thus our updating schema does the minimum change necessary to fulfill the secant condition (minimum change criterion). Let $\Delta\mathbf{x}_1 \ldots \Delta\mathbf{x}_n$ be n orthogonal moves, and $P_1 \ldots P_n$ the transformation matrices as above. Then $P_2 \ldots P_n$ are just circular permutations of P_1, and the updating schema is identical to the finite differences in eq. 2, in the O'_1 frame.

Note that the estimation in eq. 3 accepts movements along arbitrary directions $\Delta\mathbf{x}$ and thus needs no additional data other than what is available as a part of the manipulation task we want to solve. In the more general case of a set of non-orthogonal moves $\{\Delta\mathbf{x}_k\}$ the Jacobian gets updated along the dimensions spanned by $\{\Delta\mathbf{x}_k\}$.

4 Trust Region Control

While previous work in visual servoing have mostly used simple proportional control, typically with slight modifications to account for some dynamics, we have found that for hand manipulation a more sophisticated approach is needed. We use two ideas from optimization: (1) A trust region method [3] estimates the current model validity online, and controller response is restricted to be inside this "region of trust". (2) A homotopy or path following method [8] is used to divide a potentially non-convex problem into several smaller convex problems by creating subgoals along trajectories planned in visual space. For details see [16], in which we describe a method for high level visual space task specification, planning, and trajectory generation. For details on the low level control properties see [13]. In addition to allowing convergent control, these two methods serve to synchronize the model acquisition with the control, and cause the model to be estimated on an adaptive size mesh; dense when the visual-motor mapping is difficult, sparse when it is near linear.

The trust region method adjusts a parameter α so that the controller never moves out of the validity region of the current Jacobian estimate. To do that we solve a constrained problem for $||\delta_k|| < \alpha$ instead of taking the whole Newton step $\Delta \mathbf{x}$ of a proportional controller.

$$\min_{||\delta_k||<\alpha_k} ||\mathbf{y}_k - \mathbf{y}^* + \hat{J}_k \delta_k||^2 \tag{4}$$

Define a model agreement as $d_k = \frac{||\hat{J}\delta||}{||\Delta \mathbf{y}_{measured}||}$ and adjust α according to (d_{lower} and d_{upper} are predefined bounds):

$$\alpha_{k+1} = \begin{cases} \frac{1}{2}\alpha_k & \text{if } d_k \leq d_{lower} \\ \alpha_k & \text{if } d_{lower} < d_k \leq d_{upper} \\ \max(2||\delta||, \alpha) & \text{if } d_k > d_{upper} \end{cases} \tag{5}$$

5 Non-Model-Based Dextrous Manipulation

We propose a non-model-based approach to manipulation for the Utah/MIT hand, a 16 DOF four-fingered dextrous manipulator [12]. The method does not require any prior information about the object; all the required information can be read directly from the hand's sensors. The method allows arbitrary (within the robot's physical limits) translations and rotations of the object being grasped. We rely on the compliance of the Utah/MIT hand to maintain a stable grasp during manipulation, instead of attempting to compute analytically the required forces.

We assume that prior to manipulation the object has been stably grasped. The basic idea of our technique is that the commanded fingertip positions (setpoints) define a rigid object in three-dimensional space, the *grasp tetrahedron*. The contact points of the fingers on the object define another object that we refer to as the *contact tetrahedron*. Because of the compliant control system, we can model the statics of the situation by regarding each vertex of the contact tetrahedron as being attached to the corresponding vertex of the grasp tetrahedron by a virtual spring. In the case of an object free to move, net wrench is constrained to be zero. Moreover, for a well conditioned grasp, the wrench zero coincides with a deep local minimum of the spring energy function on a manifold defined by the rigid displacements of the grasp tetrahedron with respect to the contact tetrahedron. This means that if we treat the fingertip contacts as fixed points on the object, then the object can be rotated and translated by executing the desired rigid transformation on the grasp tetrahedron. Although, due to non-zero finger size and other effects, the assumption of fixed contact points is not strictly correct, our experimental results show that the compliance of the Utah/MIT hand compensates for the errors introduced by the use of this assumption.

Let $\mathbf{c} = [c_x, c_y, c_z]^T$ be the coordinates of the centroid of the grasp tetrahedron. We define an object-centered frame of reference C, with its axes parallel to the fixed hand-centered reference frame and with origin at c. Let $\mathbf{q}_0, \ldots, \mathbf{q}_3$ be the coordinates of the vertices of the grasp tetrahedron; given the manipulation command $\mathbf{x} = \langle x, y, z, \alpha, \beta, \gamma \rangle$, where x, y and z represent displacements of the object with respect to its initial position and α, β, γ are interpreted as sequential rotations about the x, y and z axes of C, the resulting setpoints $\mathbf{q'}_0, \ldots, \mathbf{q'}_3$ are given by:[5]

$$\begin{bmatrix} q'_{ix} \\ q'_{iy} \\ q'_{iz} \\ 1 \end{bmatrix} = \begin{bmatrix} c\alpha c\beta & c\alpha s\beta s\gamma - s\alpha c\gamma & c\alpha s\beta c\gamma + s\alpha s\gamma & x + c_x \\ s\alpha c\beta & s\alpha s\beta s\gamma + c\alpha c\gamma & s\alpha s\beta c\gamma - c\alpha s\gamma & y + c_y \\ -s\beta & c\beta s\gamma & c\beta c\gamma & z + c_z \\ 0 & 0 & 0 & 1 \end{bmatrix} \begin{bmatrix} q_{ix} - c_x \\ q_{iy} - c_y \\ q_{iz} - c_z \\ 1 \end{bmatrix}$$

Addition of a Cartesian Controller At the lowest level, each joint in the Utah/MIT hand is controlled using a standard PD controller. In principle, the errors in object position and orientation can be reduced by increasing the gains in these controllers. However, this would also result in a decrease in compliance. As a way to reduce position errors without sacrificing compliance, we implemented a higher-level PID Cartesian controller to correct errors directly in the 6-dimensional position-orientation space, using an estimate of the errors in $\langle x, y, z, \alpha, \beta, \gamma \rangle$ as input. For details see [6].

6 Experimental Results

The Utah/MIT hand is a relatively imprecise and difficult to control manipulator. Each joint is actuated by a pair of pneumatically-driven antagonist tendons. This results in a compliant response, but it also makes reliable and consistent positioning difficult to achieve due to hysteresis and friction effects. Furthermore, especially during long manipulation sequences, slipping in the grasp points add to these errors.

We tested repeatability under closed loop visual control and compared the results to traditional joint control. The test object was a piece of standard 2 by 4 wood construction stud, approximately 2 inches long. Positions were measured through a 0.001" accuracy dial meter. Accurate visual measurements were provided by tracking special surface markings in the first experiments, see fig. 4. The markers could be placed arbitrarily on the object, since the servoing algorithm is self calibrating. Placement near the corners gives better conditioned visual measurements with respect to rotations. In the second experiment, for improved visual accuracy, we track LED's mounted on wood screws. Two cameras were positioned approximately 90 degrees apart. 1 pixel corresponds approximately to a 0.3mm object movement. Repeatability was measured by moving the object on a random trajectory towards the dial meter until it was in contact, reading the dial value and visual or joint values, then retreating on a second random trajectory, and then trying to reachieve the visual or joint goal. In fig. 4 let x be the optic ray, y the image row and z the column. Around the x, y and z axis respectively, a grasped object can be translated about 40mm, 60mm and 30mm, and rotate 70, 45 and 90 degrees (see fig. 1). We tested but did not find any correlation between trajectory length and positioning error.

6.1 Repeatability in 3 and 4 DOF

In table 1 we compare repeatability for four different positioning methods, based on 50 trials with each method. The 3 DOF visual feedback controls only translations of

---

[5] we use $c\theta$ and $s\theta$ as shorthand for $cos(\theta)$ and $sin(\theta)$, respectively.

Fig. 4. Setup for hand fine manipulation repeatability experiments

the object. The 4 DOF visual feedback controls the translations and rotation around the optical axis of fig 4. Visual feedback in both cases consist of $m = 16$ feature values from 8 markers, and 2 cameras. The two joint feedback methods differ in that one only uses joint feedback, while in the Cartesian one, the error vector is projected onto the 6DOF space defined by the grasp tetrahedron. As seen in table 1 visual servoing performs on the average a little better than joint feedback methods. When studying maximum errors visual servoing is significantly better. This is because the visual feedback can compensate for slip in the grasp contact points.

Error	Visual feedback		Joint feedback	
	3 DOF	4 DOF	Joint	Cartesian
Mean	0.3mm	0.4mm	0.9mm	0.5mm
Max	1.0mm	1.3mm	2.5mm	2.9mm

	Controlled DOF's			
	3 DOF	4 DOF	5 DOF	6 DOF
κ	1.6	2.6	11	30

Table 1. Left: Measured repeatability of the Utah/MIT hand fine manipulating a rigid object under visual and joint feedback control. **Right:** Condition numbers κ of different degree of freedom (DOF) Jacobians.

Visual feedback gets more difficult in high DOF systems. One reason is that the number of parameters estimated in the Jacobian increases. Another is that high DOF tasks are often more ill conditioned. Particularly, there is a difference between pure translations (3 DOF) and combinations of translations and rotations (4-6 DOF). The condition numbers of the 3 to 6 DOF estimated Jacobian are shown in table 1. We tried 5 and 6 DOF visual servoing using the setup described above, but could not get reliable performance. For 5 DOF 37 % of the trials failed (diverged), and for 6 DOF nearly all failed. This is not surprising given the mechanical difficulties in controlling the hand. When trying to make the small movements needed for convergence the actual response of the manipulator has significant random component. The signal to error ratio in visual control space is then further decreased by the bad condition number of the transformation between motor and visual space. The high condition numbers in the 5 and 6 DOF cases stem from two distinct problems. With the two camera setup we use, the rotation around the z-axis is (visually) ill conditioned (differentially similar to the translation along the optic axis.) This problem occurs with a two camera setup, when the points tracked in each camera are planar. The other problem is that two of the manipulations are differentially very similar in Cartesian motor space. By taking out the two problematic DOF's we get the low condition numbers of the 3 and 4 DOF manipulations.

6.2 6 DOF visual feedback control of the Utah/MIT hand

In order to evaluate visual servoing in the full 6 DOF pose space we improved the visual measurements by tracking LED's mounted on adjustable screws attached to the test objects. The LED's facililiate more precise tracking than the surface markings, and using the screws we adjusted them into a non-coplanar configuration, avoiding the singularity we got from tracking the planar surfaces of the object. With this setup we were able to do both 5 and 6 DOF visual servoing, achieving accurate end positions in over 90% of the trials, and only diverging completely in two out of 67 trials.

We performed the same positioning experiment as described in section 6.1, but here in the full 6 DOF of the rigid object. Positioning errors were measured along the 3 major axes. For the Y and Z axes the measuring axis coincides with the center of rotation, so the errors measured are purely translational. For the X axis, the fingers are in the way, and we used a measuring point below the fingers, and thus measure the sum of rotational and translational error components.

The results are shown in table 2. The positioning error along the X-axis is much worse than the two others. We propose two reasons for this: (1) The difference in the measuring location discussed above and (2) along the Y and Z axes the translation is effected by all fingers moving in parallel, which tend to average random disturbances in an individual finger, while along the X axis only the thumb opposes the force of the three other fingers, so positioning is no more precise than for one finger.

Visual feedback improves the positioning in all cases. The improvement is significant along the Y and Z axes, but only marginal along the X axis. This result is consistent with our earlier observations that visual feedback improves an already precise manipulator more than an imprecise one. We don't have a good analysis of the exact reasons for this. In part it can probably be attributed to the visual feedback needing a monotonic response to very small movements, while in an imprecise manipulator the response is random. For a more precise manipulator the response may be monotonic, but non-linear, which affects only open loop accuracy.

	Visual feedback			Cartesian joint feedback		
Axis:	X	Y	Z	X	Y	Z
Mean error:	1.0mm	0.2mm	0.2mm	1.1mm	0.7mm	0.6mm
Max error:	2.4mm	0.5mm	0.5mm	2.9mm	4.9mm	2.3mm

Table 2. Measured repeatability of the Utah/MIT hand fine manipulating a rigid object under 6 DOF visual and Cartesian joint feedback control.

Error distributions for the 6 DOF positioning experiment are shown in fig. 5. For open loop joint feedback, the object positioning errors are of two distinct types. The first is due to the inaccurate response of the fingers themselves, the second is due to slippage in the grasp points. The mode around 0.5 mm is due to manipulator inaccuracies. The errors caused by slippage are represented in the mode around 2 mm.

In the visual servoing behavior we noted that trials fell into three categories. About half the trials converged very fast. Most of the rest converged, initially fast, but much slower towards the end. Last a small percentage did not converge at all (2 out of 67 visual feedback trials). We have not observed this difference in convergence speed when using the same visual feedback controller on robot arms. We think a partial explanation may be that mechanical phenomena in the hand and remotizer linkage cause some positions to be harder to reach than others. We have also observed that when the visual servo controller is turned off, the hand often does not remain in the same position, but drifts significantly, despite the joint setpoints remaining the same.

Not all visual servoing trials converged satisfactory. Specifically along the x axis errors were sometimes as large as 2mm (see table 2). The controller convergence criterion is partly causing this. When convergence is unsatisfactory over several sample intervals, the controller adjusts the expected positioning accuracy down, and also the derived gain. This is to stop long oscillatory behaviors near the goal. In some cases the adjustment is too large, impairing accuracy.

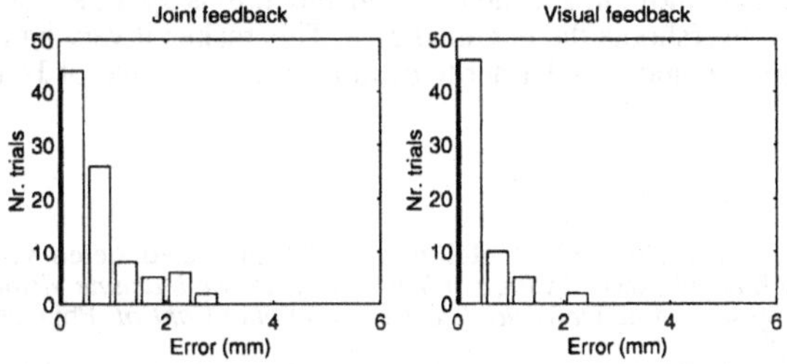

Fig. 5. Distribution of translational positioning errors for joint and visual feedback positioning of the Utah/MIT hand.

7 Discussion

We have been doing visual servo control on robot arms in high DOF spaces for several years[15, 16, 13]. Getting it to work well on a multi finger gripper like our Utah/MIT dextrous hand was much harder than we had expected. For simple robot arm control (eg. move along a line in 3 DOF) we have found that almost any controller based on the Newton method will work[15, 16]. There are also many accounts of success in practical application for low DOF manipulation in the literature [4, 21, 2, 11]. Robot arm control is also fairly insensitive to somewhat ill conditioned visual measurements, such as caused by bad camera placement, and the particular choice of visual features to track.

To successfully apply visual servo control on the hand we found that we had to be more careful in choosing the camera placement, and which features to track. Since it is the 6 DOF pose of the object we are controlling, rather than the individual fingers, we have to find viewpoints where the object is relatively unobstructed by the fingers so we can reliably see object features, and track them possibly through large rotations. The relative inaccuracy of the hand responses to motion commands, together with a, for 6 DOF manipulations, often somewhat ill-conditioned visual-motor Jacobian demand some special measures in the controller. The restricted step in our trust region controller proved to be crucial. Intuitively, both the trust region controller and the visual space trajectory planning serve to synchronize model acquisition with the actions, so that the actions never run too far ahead into a region where the model estimate is no longer accurate. Without the step restriction, and when the system is in a near singular state (which is not uncommon, see section 6.1), the controller may make a move of uncontrolled length and speed. This often causes the grasp to shift unfavorably, the object to be dropped, or the visual trackers to loose track. With the step restriction, the movement is still erroneous, but small. In most cases, this small movement is enough to get out of the singularity, so that normal servoing can be resumed.

The estimation and control algorithms we developed have strong theoretical properties (see [13]). It is still very important to experimentally evaluate how they perform in real environments, with real process disturbances/noise and manipulators with real mechanically caused errors and finite precision. Our results show that visual feedback control yields significant repeatability improvement (5 times) in an imprecise PUMA robot arm[15], while only a smaller improvement in the Utah/MIT hand. This is explained by the different characteristics of the manipulators. The inaccuracies in positioning the PUMA arm stem mainly from backlash, which the visual feedback can correct, and the model adaption is made robust to. In the hand a combination of sticktion and flexibility makes it much harder to control accurately. When trying to make small movements, the hand will initially not move at all, and the controller ramps up the signal. When the hand finally unsticks it overshoots the intended goal. This makes it very hard to control with a feedback controller, and even harder to estimate the visual-motor Jacobian.

References

1. R. Cipolla, P. A. Hadfield and N. J. Hollinghurst "Uncalibrated Stereo Vision with Pointing for a Man-Machine Interface" In *Proc of IAPR workshop on Machine Vision*, Tokyo, 1994.
2. P. I. Corke. *High-Performance Visual Closed-Loop Robot Control.* PhD thesis, University of Melbourne, 1994.
3. G. Dahlquist and A. Björck. *Numerical Methods*. Prentice Hall, 199x, Preprint.
4. J. T. Feddema and C. S. G. Lee. Adaptive image feature prediction and control for visual tracking with a hand-eye coordinated camera. *IEEE Tr. Sys, man and Cyber*, 20(5), 1990.
5. R. Fletcher. *Practical Methods of Optimization*. Chichester, 1987.
6. O. Fuentes and R. C. Nelson. Experiments on dextrous manipulation without prior object models. TR 606, Computer Science, U of Rochester, 1996.
7. O. Fuentes and R. C. Nelson. Morphing hands and virtual tools (or what good is an extra degree of freedom?). Technical Report 551, Computer Science U of Rochester, 1994.
8. Garcia and Zangwill. *Pathways to solutions, fixed points, and equilibria*. Prentice-Hall, 1981.
9. G. Hager. Calibration-free visual control using projective invariance. In *ICCV* 1995.
10. M. Harris. Vision guided part alignment with degraded data. TR-AI 615, U Edinburgh 93.
11. K. Hosoda, M. Asada. Versatile visual servoing without knowledge of true jacobian. In *Proc of IROS*, 1994.
12. S. Jacobsen, E. Iversen, D. Knutti, R. Johnson, and K. Bigger. Design of the Utah/MIT Dextrous Hand. In *Proc ICRA* 1986.
13. M. Jägersand. Visual servoing using trust region methods and estimation of the full coupled visual-motor jacobian. In *Proc of IASTED Conf Applications of Robotics and Control*, 1996.
14. M. Jägersand. *Model Free View Synthesis of an Articulated Agent*, Technical Report 595, Computer Science Department, University of Rochester, Rochester, New York, 1995.
15. M. Jägersand and R. C. Nelson. Adaptive differential visual feedback for uncalibrated hand-eye coordination and motor control. TR 579, Computer Science U of Rochester, 1994.
16. M. Jägersand and R. C. Nelson. Visual space task specification, planning and control. In *Proceedings of the 1995 IEEE Symposium on Computer Vision*, 1995.
17. K. Kutulakos and M. Jägersand. Exploring objects by purposive viewpoint control and invariant-based hand-eye coordination. In *Workshop on Vision for Robots*, IROS, 1995.
18. P. Michelman and P. Allen. Complaint manipulation with a dexterous robot hand. In *Proc 1993 IEEE Int. Conference on Robotics and Automation*, pages 711–716, 1993.
19. R. Nelson, M. Jagersand, O. Fuentes *Virtual Tools: A Framework for Simplifying Sensory-Motor Control in Complex Robotic Systems* TR 576 University of Rochester, 1995.
20. W. Paetsch and G. von Wichert. Solving insertion tasks with a multifingered gripper by fumbling. In *Proc IEEE Int. Conference on Robotics and Automation*, pages 173–179, 1993.
21. N. Papanikolopoulos and P. Khosla. Adaptive robotic visual tracking: Theory and experiments. *IEEE Transactions on Automatic Control*, 98(3), 1993.
22. T. H. Speeter. Primitive based control of the Utah/MIT dextrous hand. In *Proc IEEE International Conference on Robotics and Automation*, pages 866–877, Sacramento, 1991.
23. L. Weiss, A. Sanderson, and C. P. Neuman. Dynamic Sensor-Based Control of Robots with Visual Feedback. *IEEE Journal of Robotics and Automation*, RA-3(5), October 1987.

A System for Reconstruction of Missing Data in Image Sequences Using Sampled 3D AR Models and MRF Motion Priors

Anil C. Kokaram and Simon J. Godsill

Signal Processing and Communications Group,
Cambridge University Engineering Dept.,
Cambridge CB2 1PZ, England

Abstract. This paper presents a new technique for interpolating missing data in image sequences. A 3D autoregressive (AR) model is employed and a sampling based interpolator is developed in which reconstructed data is generated as a typical realization from the underlying AR process. rather than e.g. least squares (LS). In this way a perceptually improved result is achieved. A hierarchical gradient-based motion estimator, robust in regions of corrupted data, employing a Markov random field (MRF) motion prior is also presented for the estimation of motion before interpolation.

1 Introduction

The problem of missing data reconstruction in image sequences has traditionally not been fully addressed by the computer vision and video processing communities in the past. Various order statistic operations have been proposed for the suppression of impulsive noise in image sequences [1] but in general the problem of reconstructing missing data has been seen as a subset of the impulsive noise problem. An important example of missing data degradation is found in the motion picture industry. Particles caught in the film transport mechanism can damage the image information. The missing data regions manifest themselves as 'blotches' of random intensity in the sequence, known as 'Dirt and Sparkle'. The problem also occurs in film from high speed cameras used to record the evolution of short duration events such as explosions or impacts. In all these cases, the image information in the corrupted area is largely destroyed.

This work considers the development of spatio-temporal processes for detection/interpolation. The detection/estimation approach is adopted here in order to treat only suspected areas of distortion. This is an alternative philosophy to the usual global application of median filters employed to solve this problem. In addition, for good detail preservation of texture, it is necessary to look beyond the use of median filters to a model based approach for texture generation.

The paper is organized as follows: section 2 describes the new robust motion estimation/correction methods and also briefly describes the blotch detection scheme (See [2]); section 3 describes the 3D AR model and robust parameter estimation for that model; section 4 describes the sampling based interpolation scheme; section 5 presents results obtained from processing degraded film image sequences; and finally section 6 concludes the paper and discusses future directions for the work.

2 Motion estimation and the detection of blotches

Corrupted pixels, which are part of 'blotches' in the image sequence, generally do not occur in the same spatial location and with the same brightness in consecutive frames. They are therefore well defined as temporal discontinuities in the image sequence. Blotches can be distinguished from sites of occlusion and uncovering because they are at sites which are 'occluded' in both the next and previous frames (i.e. this image information does not exist in either of the two surrounding frames). Sites of occlusion and uncovering, however, represent discontinuities in either the backward or forward temporal direction, but never both. A simple but effective detector (*SDIa*) for corrupted pixels was presented in [2]. It flags pixels as corrupted, when *both* the squared motion compensated pixel differences (forward and backward in time) are larger than some user defined threshold. The reader is encouraged to refer to [2] for details.

Robust motion estimation is important for the correct operation of the detector. The next section presents a new technique for gradient based motion estimation, using a combination of low-level video processing algorithms.

2.1 Multiresolution Wiener Based Motion Estimation

There exist many formulations for pel-recursive, translational, motion estimators which successively refine an estimate for the displacement between frames n and $n-1$ at location $\vec{x}$, $\vec{d}_{n,n-1}(\vec{x})$. This is achieved via updates calculated through a Taylor series expansion of the image function around the current estimated displacement. These appear to have been somewhat overlooked by the computer vision community. Biemond [3] presented a Wiener solution for an update displacement, $\hat{\vec{u}}_i$, which is more robust to noise. Errors in this motion estimator can be linked directly to the extent of ill–conditioning in a gradient matrix, $\mathbf{M}_g = [\mathbf{G}^T\mathbf{G}]$ (see below). At an edge in the image, it is clear that the motion estimate is most confident in the perpendicular direction. Yet it is at just such locations that $\mathbf{M}_g$ can be ill-conditioned. This fact was recognized independently by Martinez, [4], who employed the SVD of $\mathbf{M}_g$ in an earlier work on a non-iterative gradient based motion estimation scheme. In the case where the estimator was at an edge, it was possible to salvage some useful information by aligning the extracted motion vector with the direction of maximum image gradient, using the SVD.

This paper therefore proposes a combined strategy for an adaptive Wiener based (AWB) motion estimation scheme in which the update displacement, $\vec{u}_i$, is generated as follows,

$$\vec{u}_i = \begin{cases} \alpha_{max}\vec{k}_{max} & \text{if } \frac{\lambda_{max}}{\lambda_{min}} > \alpha \\ \left[\mathbf{G}^T\mathbf{G} + \mu\mathbf{I}\right]^{-1}\mathbf{G}^T\mathbf{z} & \textit{otherwise} \end{cases} \tag{1}$$

$$\mu = |\mathbf{z}|\frac{\lambda_{max}}{\lambda_{min}} \text{ if } \frac{\lambda_{max}}{\lambda_{min}} \leq \alpha$$

$$\alpha_{max} = \frac{\vec{k}_{max}^T}{\lambda_{max}}\mathbf{G}^T\mathbf{z}$$

where λ, $\vec{k}$ refer to the eigenvalues and eigenvectors of $\mathbf{G}^T\mathbf{G}$, and α_{max} is a scalar variable introduced to simplify the final expression. In this combined strategy, the condition of $\mathbf{M}_g$ is monitored through α. When the condition number ($\frac{\lambda_{max}}{\lambda_{min}}$) is larger than this value, the SVD of $\mathbf{M}_g$ is used to generate the 'valid' motion component. Otherwise, $\mathbf{M}_g$ is assumed to be well conditioned and the regularized wiener solution for the update is used in which μ is proportional to the product of the magnitude of the current DFD and the condition of the matrix[5].

2.1.1 Implementation

The AWB estimator is incorporated into a block based scheme where each block in the image is assigned one motion vector. In order to reduce computation and the occurrence of spurious vectors, the AWB estimator is only employed in blocks where motion is *detected.* This consists of thresholding the mean absolute error (MAE) between the current block and the block at the same location in the previous and next frame. An MAE larger than the threshold is assumed to indicate motion and only in that case is motion estimation engaged.

Gradient based motion estimation schemes only work when the Taylor series expansion of the image function is valid i.e. when estimating small displacements. In most interesting image sequences, especially those in movies, the assumption of small motion is not valid. This can be overcome through a multiresolution strategy for motion estimation in a similar way to [6, 7]. For the implementation in this paper, an L level image pyramid (2:1 subsampling) is generated using an FIR gaussian kernel with variance 1.5 and window size 9×9. After n iterations of the AWB motion estimator, the vectors are propagated down to the next level and used as initial estimates for motion estimation at that level. At the original resolution level, it is typical that some areas which are not moving are assigned motion vectors only because of their proximity to moving regions in the upper levels of the pyramid. This motion halo effect is reduced in the manner of [8] by double checking for motion at level 0 in the pyramid.

2.2 Vector field correction

After motion estimation is complete, the *SDIa* detector can be used to flag pixels which are detected as corrupted. These pixels can be grouped together as necessary to measure the spatial extent of each blotch. The problem now is to fill the indicated region with some realistic estimate for the missing image data. This requires using information from both the next and previous frames. But the motion estimates are detrimentally affected by the presence of a Blotch. Therefore, an interpolated vector is required at this site. Also important is to ensure that the interpolation process is robust enough to ignore or to *de-emphasize* data collected using an incorrect motion vector. This is discussed in the next section.

It is assumed that motion vectors in frame n constitute a Markov Random Field (MRF)[1]. The conditional probability of the vector $\vec{d}_{n,n-1}(\vec{x})$ given the frames I_n and I_{n-1} and some neighborhood subset of motion vectors, $\mathbf{S}_n(\vec{x})$, can be written using Bayes theorem as

$$p(\vec{d}_{n,n-1}(\vec{x})|I_n, I_{n-1}, \mathbf{S}_n(\vec{x})) = \frac{p(I_n, I_{n-1}|\vec{d}_{n,n-1}(\vec{x}), \mathbf{S}_n(\vec{x}))p(\vec{d}_{n,n-1}(\vec{x})|\mathbf{S}_n(\vec{x}))}{p(I_n, I_{n-1}|\mathbf{S}_n(\vec{x}))} \quad (2)$$

The situation between frames $n, n+1$ can be written similarly. The likelihood is taken as a zero mean gaussian distribution of DFD's, with variance σ_e^2. Note that the vector field considered is defined on the block lattice and not the pixel lattice. Therefore the likelihood should contain a contribution from every pixel in the block centered on location $\vec{x}$ as follows (dropping $\mathbf{S}_n(\vec{x})$ for brevity),

$$p(\mathbf{I}|\vec{d}_{n,n-1}(\vec{x})) \propto \exp - \left(\frac{1}{2\sigma_e^2} \sum_{\vec{x} \in B} w^2(\vec{x})[I_n(\vec{x}) - I_{n-1}(\vec{x} + \vec{d}_{n,n-1}(\vec{x}))]^2 \right) \quad (3)$$

where B is the set of all locations in the block and $\mathbf{I}$ represents I_n, I_{n-1}. A weight $w(\vec{x})$ is associated with each DFD measurement. This weight is set to 0 wherever the pixel site is flagged as corrupted by the detector, and set to 1 otherwise. N_w is the sum of these weights over the block.

In the manner of e.g. [11, 12] a Gibbs Energy prior is used for $\vec{d}_{n,n-1}(\vec{x})$ as follows

$$p(\vec{d}_{n,n-1}(\vec{x})|\mathbf{S}_n(\vec{x})) \propto \exp - \left(\sum_{\vec{v} \in \mathbf{S}_n(\vec{x})} \lambda(\vec{v})[\vec{d}_{n,n-1}(\vec{x}) - \vec{v}]^2 \right) \quad (4)$$

where $\vec{v}$ is each vector in the neighborhood represented by $\mathbf{S}_n(\vec{x})$, and $\lambda(\vec{v})$ is the weight associated with each clique. The situation is illustrated in the left hand portion of figure 1. Note that the cliques employed here assume first order interactions even though the eight connected neighborhood can involve some second order cliques [9].

In order to discourage 'smoothness' over too large a range, $\lambda(\vec{v})$ is defined as $\lambda(\vec{v}) = N_w/|\vec{X}(\vec{v}) - \vec{x}|$ where $\vec{X}(\vec{v})$ is the location of the block providing the

[1]It is assumed that the reader is familiar with the concept of Markov Random Fields. See [9, 10, 11].

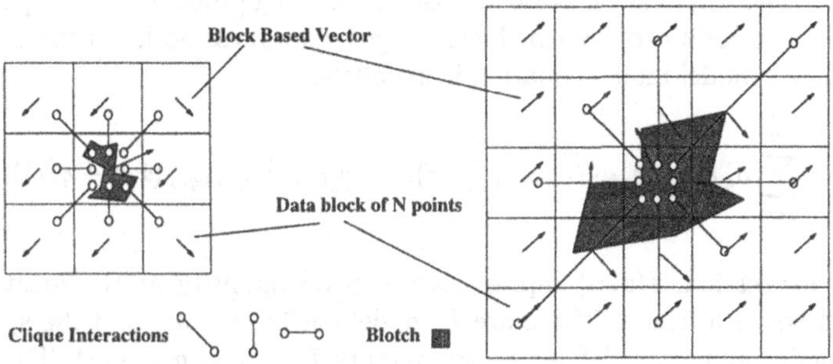

Figure 1: Left : Neighborhood and cliques used for $p(\mathbf{d}_{n,n-1}(\vec{x}))$. Right : Altered Neighborhood used with a large blotch.

neighborhood vector $\vec{v}$. This location is measured in terms of block lengths. As before, N_w is the number of uncorrupted pixels in the block. Large $\lambda(\vec{v})$ encourages motion vector smoothness, and small σ_e^2 encourages vectors which minimize the DFD.

It is true that equations 2, 3, 4 are sufficient to estimate the motion field itself [9, 11], and the prior can be altered to account for motion discontinuities. However, a direct solution for the MAP estimate (2) with respect to the vector field (via some Monte Carlo technique) is computationally demanding [9]. In practice, after the use of the AWM estimator and the blotch detector, it is already possible to make a very confident assessment of the locations of corruption. Therefore, there is no longer interest in the uncorrupted regions. Rather than relax the vector field around the corrupted location using, e.g. the Gibbs Sampler [9], it is found sufficient to reduce the solution search space[2] to the vectors in the neighborhood of the blotch. Each vector in turn is tested as a candidate solution to the corrected vector by substitution in equation 2. The candidate which maximizes 2 is selected as a working approximation to the MAP estimate. Note that the denominator of equation 2 is constant and can be ignored. An estimate for σ_e^2 is made from the measured DFD for each vector candidate. In the case where the DFD does not vary much with different candidate vectors, the operation reduces to a type of weighted vector median. When the blotch engulfs several blocks, the candidate vector set is chosen from an altered neighborhood where the blocks concerned contain less than 10% corruption. This strategy is illustrated in figure 1.

3 The 3D AR Model

The structure of the AR model allows efficient computational algorithms to be developed, and it is this, together with its spatiotemporal nature which is of

[2]A similar simplification was made by Stiller [12] for motion field smoothing.

interest. The physical basis for its use as an image model for interpolation is limited to its ability to describe local image smoothness both in time and space. The 3D AR model model equation is as follows.

$$I(\vec{x},n) = \sum_{k=1}^{\mathcal{P}} a_k I(x + q_{xk} + \mathrm{d}x_{n,n+q_{nk}}(\vec{x}), y + q_{yk} + \mathrm{d}y_{n,n+q_{nk}}(\vec{x}, n + q_{nk})) + \epsilon(\vec{x},n) \tag{5}$$

In this expression, $I(\vec{x},n)$ represents the pixel intensity at the location $\vec{x} = (x,y)$ in the nth frame. There are $\mathcal{P}$ model coefficients a_k, and the spatiotemporal model support is defined by the vectors $\vec{q}_k = [q_{xk},\ q_{yk},\ q_{nk}]$. The support locations are offset by the relative displacement between the predicted pixel location and the support location. The displacement between frame n and frame m is $\vec{d}_{n,m}(x,y) = [\mathrm{d}x_{n,m}(x,y),\ \mathrm{d}y_{n,m}(x,y)]$. Finally, $\epsilon(x,y,n)$ is the prediction error at location $(x,\ y,\ n)$. Figure 2 shows a temporally causal 3D AR model with 5 pixels support.

In this paper the parameters of the model are estimated using weighted least squares [2]. The weights assigned to each prediction equation are 0 where the blotch detector has flagged a corrupted pixel site and 1 otherwise. $\epsilon(x,y,n)$ is assumed to be drawn from a white Gaussian noise process with variance σ_ϵ^2.

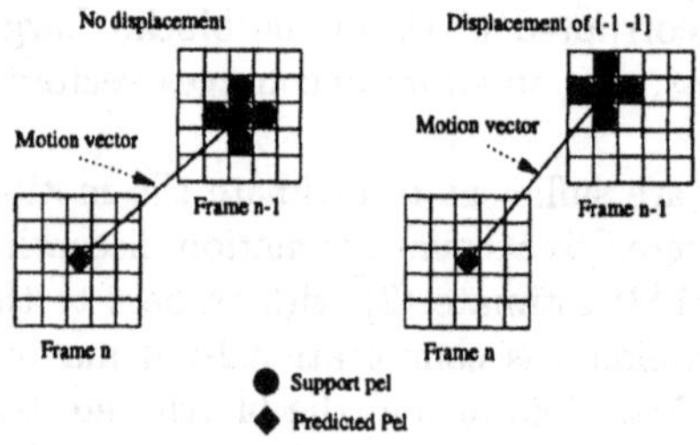

Figure 2: Handling motion with the 3D AR model.

4 Interpolation

Given the position of missing pixels, motion estimates and AR parameters for a sub-block of the image, the missing information is now interpolated. The methods currently proposed [2] reconstruct the missing data with an interpolation which minimizes the excitation energy in a least-squares sense. However, this solution, which is equivalent to the maximum *a posteriori* (MAP) estimate under Gaussian assumptions [13], tends to be *oversmooth* compared with surrounding pixels, especially when the missing area is large. The problem is well illustrated in figures 3, 6. Oversmoothing in the reconstructed image occurs because estimation techniques which use these or other familiar objective functions do not make allowance for the random component in image sequences, which cannot be predicted exactly from surrounding pixel values.

We propose an interpolator which draws the missing pixel values as a *random sample* from their posterior probability distribution conditional upon the known pixel values which surround the missing region. In this way the interpolation will be typical of the AR process under consideration and should not exhibit the oversmooth nature of other interpolators. Similar principles have been successfully applied to the interpolation of missing samples from audio signals which can be modelled as a 1-d AR process [14, 15, 16].

4.1 Sampled interpolations

The vector of excitation values $\mathbf{e}$ corresponding to a block of data $\mathbf{i}$ is written in matrix-vector notation as $\mathbf{e} = \mathbf{A}\mathbf{i}$, where $\mathbf{A}$ is constructed from the AR parameter vector $\mathbf{a}$ in such a way as to generate $\epsilon(\vec{r})$ (see (5)) for N distinct values of $\vec{r}$. This expression can be partitioned into a part corresponding to known data pixels $\mathbf{i}_k$ and unknown data $\mathbf{i}_u$, leading to $\mathbf{A}\mathbf{i} = \mathbf{A}_k\mathbf{i}_k + \mathbf{A}_u\mathbf{i}_u$, where $\mathbf{A}_k$ and $\mathbf{A}_u$ are the corresponding columnwise partitions of $\mathbf{A}$.

Under Gaussian and independence assumptions for the excitation process the posterior distribution for the missing pixels is given by (see appendix A)

$$p(\mathbf{i}_u|\mathbf{i}_k, \sigma_e^2, \mathbf{a}) = \frac{|\mathbf{A}_u^T\mathbf{A}_u|^{1/2}}{(2\pi\sigma_e^2)^{l/2}} \exp\left(-\frac{1}{2\sigma_e^2}(\mathbf{i}_u - \mathbf{i}_u^{\mathrm{MAP}})^T(\mathbf{A}_u^T\mathbf{A}_u)(\mathbf{i}_u - \mathbf{i}_u^{\mathrm{MAP}})\right) \tag{6}$$

which we note is in the form of a multivariate normal distribution. $\mathbf{i}_u^{\mathrm{MAP}}$ is the standard MAP/least squares (see [13, 2]) interpolator, given by:

$$\mathbf{i}_u^{\mathrm{MAP}} = -(\mathbf{A}_u^T\mathbf{A}_u)^{-1}\mathbf{A}_u^T\mathbf{A}_k\mathbf{i}_k \tag{7}$$

An estimate for σ_e^2 in equation (6) can be made from observations of the excitation in the uncorrupted region around the missing pixels following AR parameter estimation.

Drawing a random sample from the multivariate normal distribution of equation (6) can be achieved using well known procedures and may be summarized as:

$$u_i \sim \mathrm{N}(0,1), \ (i = 1 \ldots l) \tag{8}$$

$$\mathbf{i}_u^{\mathrm{samp}} = \mathbf{i}_u^{\mathrm{MAP}} + \mathbf{S}^{-1}\mathbf{u} \tag{9}$$

where '$\sim$' denotes a random draw from the distribution to the right, $\mathrm{N}(0,1)$ is the standard normal distribution and $\mathbf{S}$ is any convenient $(l \times l)$ matrix square-root factorization which satisfies $\mathbf{A}_u^T\mathbf{A}_u/\sigma_e^2 = \mathbf{S}^T\mathbf{S}$. $\mathbf{u}$ is the column vector formed from the elements u_i. The sampled interpolation $\mathbf{i}_u^{\mathrm{samp}}$ can then be substituted for the missing pixels in the restored image.

Drawing a sample from the conditional distribution can be seen to involve calculation of the MAP estimate $\mathbf{i}_u^{\mathrm{MAP}}$ and adding an appropriately coloured noise term $\mathbf{S}^{-1}\mathbf{u}$. Calculation of $\mathbf{S}^{-1}$ need not involve any significant overhead over MAP interpolation if a matrix square-root factorization method such as Cholesky Decomposition is used in the inversion of $\mathbf{A}_u^T\mathbf{A}_u$ (equation (7)).

5 Results

Because the sampling based interpolator draws a typical sample for the interpolated data, it is not possible to measure the performance of the interpolator

Figure 3: Degraded Frame 2 of FRANK with large blotches boxed.

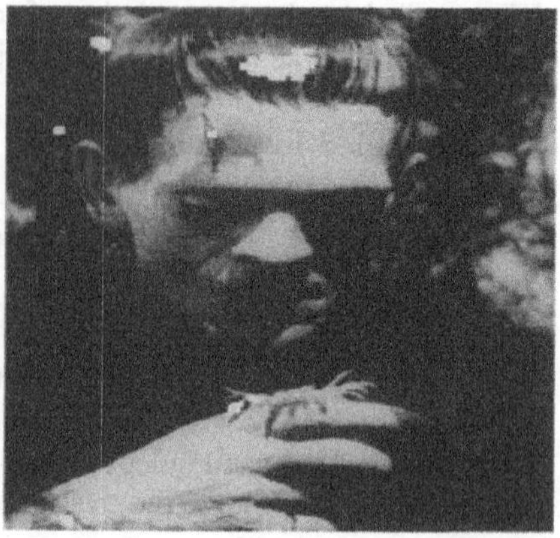

Figure 4: Detected Blotches (bright white) in Frame 2, after dilation of detection field.

by using a standard distortion measure such as MSE with artificially degraded sequences, since the MAP estimate is likely to give the give the lowest MSE. Indeed that is its shortcoming. Therefore it is best to illustrate performance by using visual comparisons.

Figure 3 shows a frame of a real degraded sequence. The overall motion in the sequence is a rapid vertical pan, with some fast motion in the petals of the flower and some complicated motion in the background trees. The blotches to be considered are highlighted in figure 3. Each frame is of resolution 256×256. A 2 level pyramid was employed for motion estimation. The block size used was 9×9, the motion threshold was 10.0 grey levels, 10 iterations of the AWB estimator were used at each level, and $\alpha = 100.0$. The blotch detection threshold, e_t was set at 25.0. The areas which were then flagged as corrupted (using *SDIa*, see [2]) are shown as bright white pixels in figure 4, superimposed on a darkened version of frame 2 (figure 3). Note the false alarms in the region of the petals of the flower - the petals themselves are quite impulsive features which do not move smoothly from frame to frame. The block based motion estimation technique does not function well here.

Figures 6 and 5 show a zoomed version of the MAP reconstruction with and without vector field correction respectively. The interpolated regions are boxed in white. The improvement is illustrated clearly in the large blotch (shown on frame 2). The interpolated data is much better in keeping with the rest of the hairstyle when the motion vector used at the blotch site has been 'corrected'. The 3D AR model used here was causal with support only in the previous frame in a 3×3 block of 9 support points. The improvement in quality with motion field correction is the same whatever the interpolation employed and so no more examples are given.

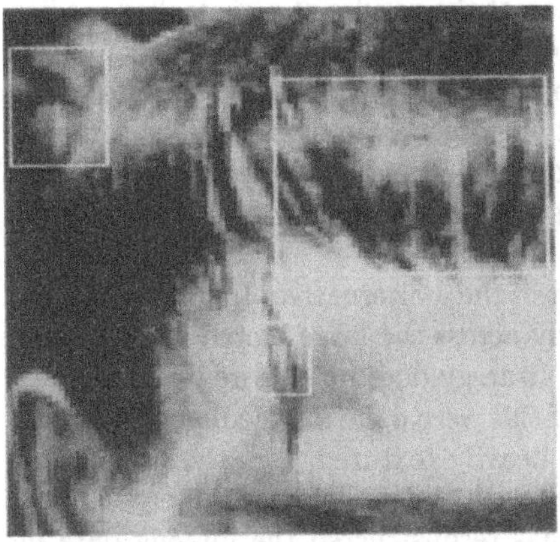

Figure 5: Interpolation using MAP estimate without vector correction. (Interpolated regions boxed in white.)

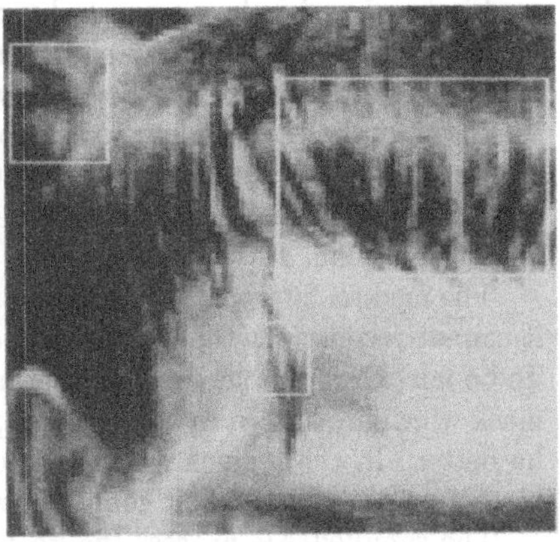

Figure 6: Interpolation using MAP estimate *with* vector correction. (Interpolated regions boxed in white.)

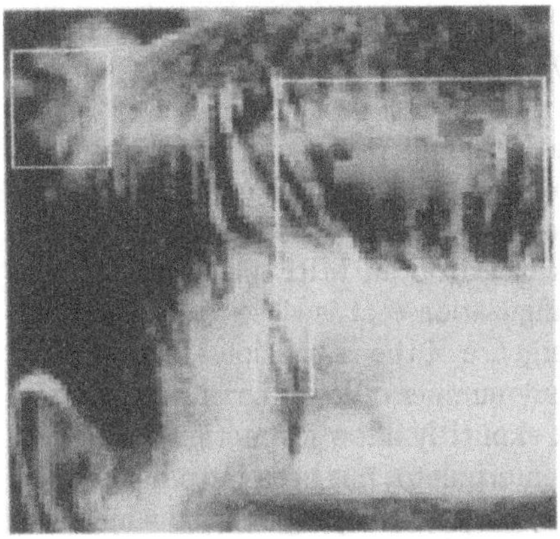

Figure 7: Interpolation using spatio-temporal median filter. (Interpolated regions boxed in white.)

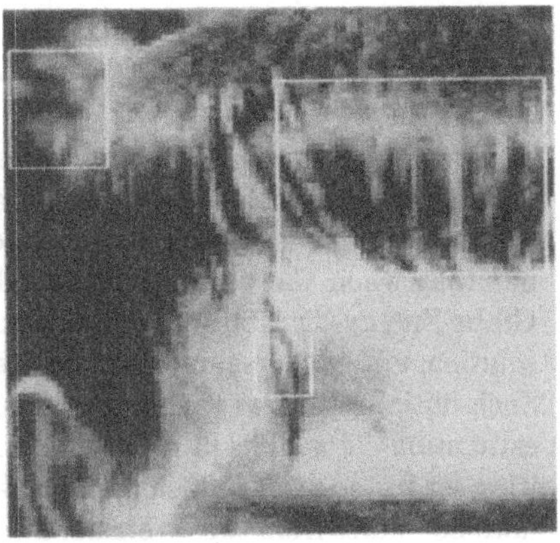

Figure 8: Sampled Interpolation. (Interpolated regions boxed in white.)

Figures 7 and 8 show a zoomed version of the results of a controlled median filtered operation and a sampled AR interpolation (section 4) respectively. The interpolated regions are also boxed in white. The ML3Dex filter as defined in [2] was used at the sites of detected distortion to generate figure 7. The same 3D AR model as for the MAP interpolation of figure 6 was used. Both results employed the corrected vectorfield in assembling the motion compensated data for interpolation.

The median filtered result is the worst of the 3 alternatives (figures 6, 8, 7) as it cannot reconstruct the texture properly across the large blotch in particular. To be fair, however, the median filtering strategy does not ignore pixels from its mask which are known to be corrupted; if this were done the median result could be better. In the regions which are not heavily textured, e.g. the background blotches, the median result compares well with the model based interpolations.

The sampled AR process (figure 8) has reconstructed the missing data including the detail extremely well. Where the MAP interpolation has introduced a slight smoothing effect, the sampled interpolator has recreated the random 'graininess' which is typical of the surrounding area in the image. This is seen best if the interpolated regions in the large blotch are compared in figures 6 and 8. Furthermore, note that even though it is clear that the 'corrected' motion vector used at the blotched location is not necessarily 100% accurate[3], this has not detrimentally affected the model based interpolation schemes. In fact, because of the spatial extent of the model support in the previous frame, the model can cope, in this case, with inaccuracies of up to ± 1 pixel in the motion vector used.

6 Conclusions and Further Work

This work has presented a new scheme for detail preserving interpolation of missing data in image sequences. In achieving this goal, it has also introduced a new technique for gradient based motion estimation. It has also been pointed out that when the vector field model is based on an MRF prior employing a Gibbs Energy distribution, an initial configuration that is close to the final true solution will of course improve the convergence of the relaxation algorithms [9]. Such initial estimates can be had using any number of lower complexity motion estimation algorithms [3, 7] which do not explicitly allow for motion discontinuities, for instance. A new sampling based interpolator has been introduced which does not suffer from the 'oversmoothing' of large missing regions and which does not substantially increase the computation required.

Finally, it must be noted that in achieving the goal of missing data interpolation, this paper has employed 3 different models of the image sequence in order to address conveniently each sub-problem as it arises. Translational motion is a convenient model for generating a fast initial configuration for the motion field. Equation 2 is a probabilistic formulation which then allows the correction of the vector field through an image sequence model which imposes an implicit

[3]Essentially, it has been estimated by replacing it with one of the surrounding vectors.

constraint on the smoothness of the motion field. The AR model equation 5 then imposes some constraint on the image values to allow the interpolation of the missing region. All of these formulations emphasize a different aspect of the image sequence and it is possible to combine them all into a single Bayesian framework for the estimation of the various parameters, including the missing data itself. This is the current focus of our research.

A Posterior probability for interpolated data

We derive here the conditional posterior probability expression for the missing pixels, given in equation (6).

Assuming a Gaussian independent excitation with variance σ_e^2 we can write down the probability for $\mathbf{e}$ as

$$p(\mathbf{e}) = (2\pi\sigma_e^2)^{-N/2} \exp\left(-\frac{\mathbf{e}^T\mathbf{e}}{2\sigma_e^2}\right)$$

The distribution for the corresponding block of image pixels $\mathbf{i}$ is then obtained by the change of variables $\mathbf{e} = \mathbf{A}\mathbf{i}$ (see section 4.1), giving:

$$p(\mathbf{i}) = p(\mathbf{e} = \mathbf{A}\mathbf{i}) = (2\pi\sigma_e^2)^{-N/2} \exp\left(-\frac{\mathbf{i}^T\mathbf{A}^T\mathbf{A}\mathbf{i}}{2\sigma_e^2}\right) \tag{10}$$

Note that this distribution is strictly conditional upon a minimal region of AR support pixels at the edges of the block (see e.g. [17] for the 1-d case), but we omit this dependence here for clarity of exposition. The form of the final result for missing pixels is unchanged by this simplification provided the region of support contains only known pixels. The above expression also assumes a causal AR model (in which case the Jacobian for the variable change is always unity).

The conditional distribution for missing pixels $\mathbf{i}_u$ is then obtained from the probability chain rule as:

$$p(\mathbf{i}_u|\mathbf{i}_k) = \frac{p(\mathbf{i})}{p(\mathbf{i}_k)}$$

Note that the denominator term in this expression is constant for any given image and AR model. Hence the final result can be determined by rearrangement of (10) in terms of $\mathbf{i}_u$ and noting that the resulting distribution must be normalized w.r.t. $\mathbf{i}_u$. Substituting $\mathbf{A}\mathbf{i} = \mathbf{A}_k\mathbf{i}_k + \mathbf{A}_u\mathbf{i}_u$ (see section 4.1) into equation (10) and rearranging to give a normalized distribution leads to the final result of equation (6).

References

[1] G.R. Arce. Multistage order statistic filters for image sequence processing. *IEEE Transactions on Signal Processing*, 39:1146–1161, May 1991.

[2] A. Kokaram, R. Morris, W. Fitzgerald, and P. Rayner. Detection/interpolation of missing data in image sequences.. *IEEE Image Processing*, pages 1496–1519, Nov. 1995.

[3] J. Biemond, L. Looijenga, D. E. Boekee, and R.H.J.M. Plompen. A pel–recursive wiener based displacement estimation algorithm. *Signal Processing*, 13:399–412, 1987.

[4] D. M. Martinez. *Model–based motion estimation and its application to restoration and interpolation of motion pictures.* PhD thesis, Massachusetts Institute of Technology, 1986.

[5] L. Boroczky J. Driessen and J. Biemond. Pel–recursive motion field estimation from image sequences. *Visual Communication and Image Representation*, 2:259–280, 1991.

[6] W. Enkelmann. Investigations of multigrid algorithms for the estimation of optical flow fields in image sequences. *Computer Vision graphics and Image Processing*, 43:150–177, 1988.

[7] J. Kearney, W.B. Thompson, and D. L. Boley. Optical flow estimation: An error analysis of gradient based methods with local optimisation. *IEEE PAMI*, pages 229–243, March 1987.

[8] M. Bierling. Displacement estimation by heirarchical block matching. In *SPIE VCIP*, pages 942–951, 1988.

[9] R. D. Morris. *Image Sequence Restoration using Gibbs Distributions.* PhD thesis, Cambridge University, England, 1995.

[10] S. Geman and D. Geman. Stochastic relaxation, gibbs distributions and the bayesian restoration of images. *IEEE PAMI*, 6:721–741, 1984.

[11] J. Konrad and E. Dubois. Bayesian estimation of motion vector fields. *IEEE Trans PAMI*, 14(9), September 1992.

[12] C. Stiller. Motion–estimation for coding of moving video at 8kbit/sec with gibbs modeled vectorfield smoothing. In *SPIE VCIP.*, volume 1360, pages 468–476, 1990.

[13] R. Veldhuis. *Restoration of Lost Samples in Digital Signals.* Prentice-Hall., 1990.

[14] P. J. W. Rayner and S. J. Godsill. The detection and correction of artefacts in archived gramophone recordings. In *Proc. IEEE Workshop on Audio and Acoustics.*, 1991.

[15] J. J. K. Ó Ruanaidh and W. J. Fitzgerald. Interpolation of missing samples for audio restoration.. *Electronic Letters.*, 30(8), April 1994.

[16] J. J. Rajan, P. J. W. Rayner, and S. J. Godsill. A Bayesian approach to parameter estimation and interpolation of time-varying autoregressive processes using the Gibbs sampler. *Submitted to IEEE Trans. on Signal Processing.*, June 1995.

[17] M. B. Priestley. *Spectral Analysis and Time Series.* Academic Press, 1981.

Elimination of Specular Surface-Reflectance Using Polarized and Unpolarized Light

Volker Müller
MAZ Mikroelektronik Anwendungszentrum Hamburg GmbH
Harburger Schloßstraße 6-12, D-21079 Hamburg, Germany
e-mail: vm@maz-hh.de, Tel.: +49/40/76629-1421, FAX: +49/40/76629-199

Abstract

Highlights are an unwanted phenomenon in computer vision, they may severely hamper the use of standard image processing algorithms. Highlights are caused by specular reflectance, thus the objective is to eliminate this type of reflectance. In this paper, a new polarization-based method is introduced to separate the diffuse and specular component of reflection.

A polarisation filter can reduce the intensity of higlights but can not completly eliminate this troublesome effect. Two images input with different orientation of a polarisation filter in front of a camera provide the necessary information to calculate the intensity of specular reflectance on plane surfaces. A third image with different orientation is necessary to determine specular reflectance in three-dimensional scenes. The specular reflectance is substracted from the original image, providing an image without highlights.

Both polarized and unpolarized light is used in this paper. Polarized light has the advantage, that it reduces the degrees of freedom of reflected light, thus no knowlegde about surface properties and image aquisition geometry is required, when specular reflectance shall be removed.

The method introduced in this paper can be applied to all kinds of dielectrics including textured surfaces. Highlights can be removed both from greylevel images and colour images, in the latter case yielding colour constancy.

If the inspected objects are placed at a definite position, it is possible to aquire most of the required data for highlight-elimination during a training phase. In this way, highlights can be removed with just one image input, even if the source of illumination is changing. Thus the hardware-requirements to the vision system are considerably reduced, narrowing the gap between research and practical applications.

Keywords: Physics-based vision, elimination of highlights, polarization, image-restauration, Fresnel reflectance model

1. Introduction

Camera-based computer vision almost exclusivly deals with images that are generated by light reflected from surfaces. Analysis of optical reflection on surfaces is an important branch of physics-based vision, a field of research that aims at incorporating the physics of the image-generating process into computer vision. Estimation of optical surface parameters will provide additional information of the inspected scene and hence increase the performance of artificial vision systems.

Optical reflection on surfaces can be separated into two categories: diffuse reflection and specular reflection. The intensity of diffuse reflectance I_d is independend from the viewing direction of the image sensor, whereas specular reflectance I_s is concentrated in a compact lobe around the direction, where the angle of reflection equals the angle of incidence [11]. Light intensity I is a simple superposition of both kinds of reflectance:

$$I = I_d + I_s \tag{1}$$

Specular reflectance is the physical cause of highlights, a rather unwanted phenomenon in computer vision. In the area of highlights, image intensity may exceed the dynamics of standard CCD image-sensors. Highlights may produce artificial edges, that are not caused by geometrical properties of the object but just by local changes of reflectance. In colour images, highlights have the colour of the source of illumination, not that of the object. For these reasons, elimination of specular reflectance is an important step in image preprocessing. Subsequent image processing algorithms will produce more reliable results.

Different approaches have been introduced to separate diffuse and specular reflectance. Several authors [3,4,8,12] use the different colours of diffuse and specular reflectance to identify highlights. However this method does not work on textured surfaces and if the colour of the object is similar to the colour of the source of illumination. In [6] a method is introduced, that analyses spatial distribution of reflectance to calculate a compensation function removing specular reflectance. That method is limited to plane surfaces and it requires the knowledge of image aquisition geometry and of the reflection parameters of the surface.

Use of polarized light and/or polarization filters can partly remove highlights simply by means of optics. Based on experience without mathematical analysis, this effect is widly use in computer vision applications [9].

Systematic research on polarisation is a rather new topic in computer vision. Koshikawa [5] and Wolff [13,14] used analysis of polarization for shape estimation and material classification. Nayar et. al. [10] introduced a combination of colour analysis and polarisation in order to remove highlights and obtain colour constancy. In [7] a polarisation-based approach is introduced, that can be applied to both colour and grey-level images. However, in three-dimensional scenes that method requires knowledge of material properties and it can not be applied to rather weak highlights.

2. Polarization and Reflection of Light

In the following analysis of the polarisation state of light, we distinguish three different sections that light is passing during its way from the source of illumination to the image sensor (fig. 1): between source of illumination and surface (incident light, marked with superscript i), between surface and polarization filter (reflected light, superscript r) and between polarisation filter and camera (super-script f).

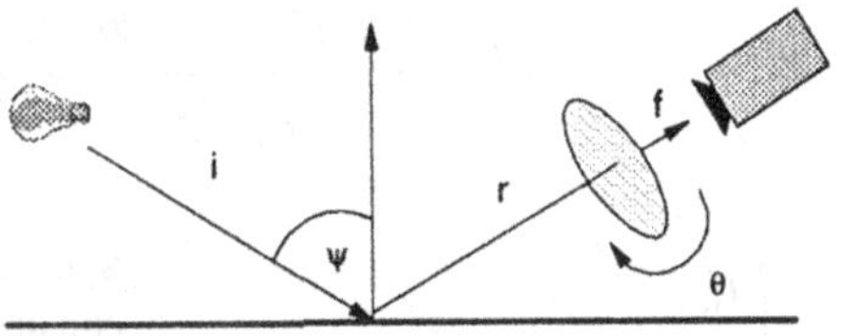

Fig. 1: Polarization by reflection

Light from standard technical and most natural sources[1] of illumination is unpolarized, i.e. the electromagnetic wave has a constant amplitude in each direction perpendicular to the direction of propagation. We define a specular plane determined by the ray of incident light, the surface normal and the ray of ideal[2] specular reflected light. A projection to two axes parallel and perpendicular to the specular plane provides two components of light, that in the case of unpolarized light have the same image intensity.

$$I^i = I^i_{\parallel} + I^i_{\perp} \qquad I^i_{\parallel} = I^i_{\perp} \text{ if unpolarized} \tag{2}$$

After reflection on a surface, the diffuse component of reflected light remains unpolarized, but the specular component of reflectance becomes partly polarized[1], thus analysis of the state of polarisation can be used to distinguish this two kinds of reflectance.

The degree of polarisation of specularly reflected light depends on the Fresnel coefficients $F_{\parallel}(\eta,\psi)$ and $F_{\perp}(\eta,\psi)$, which are determined by the material-depending index of refraction η and the angle of incidence ψ [15]:

$$I^r_{\perp}(\eta,\psi) = \frac{F_{\perp}(\eta,\psi)}{F_{\perp}(\eta,\psi)+F_{\parallel}(\eta,\psi)} * I^r_s \qquad I^r_{\parallel}(\eta,\psi) = \frac{F_{\parallel}(\eta,\psi)}{F_{\perp}(\eta,\psi)+F_{\parallel}(\eta,\psi)} * I^r_s \tag{3}$$

In this paper we only consider linear polarized light, i.e. our analysis of surface reflection only covers dielectrics. Reflection on metals may cause an additional phase difference beween $I^r_{\parallel}$ and $I^r_{\perp}$, producing circular polarized light [1].

Light intensity behind the polarisation filter depends on on the degree of polarisation of reflected light I^r and in addition, it may depend on orientation θ of the polarization filter.

Unpolarized light stemming from diffuse reflection is reduced by one half independent of the orientation of the polarisation filter:

$$I^f_d = \frac{1}{2} I^r_d \tag{4}$$

On the other hand, the intensity of specularly reflected and hence partly polarized light that has passed through a polarisation filter depends on orientation θ of the polarisation filter:

---

[1] However, light from blue sky, that is far away from the sun, is polarized [14]

[2] i.e. angle of incidence equals angle of reflection

$$I_s^f = \frac{F_{\parallel} \cos^2 \theta + F_{\perp} \sin^2 \theta}{F_{\perp} + F_{\parallel}} * I_s^r = I_{s,c} + I_{s,v} \cos 2\theta \qquad (5)$$

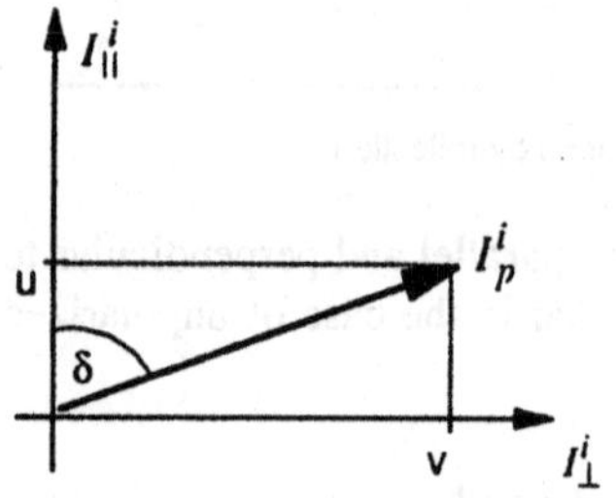

Fig. 2: Projection of incident polarized light to the specular plane

Specularly reflected light can be expressed by a constant part $I_{s,c}$ and a variable part $I_{s,v}$, the intensity of the latter depending on the orientation of the polarisation filter. $I_{s,c}$ stems from the unpolarized part of specular reflctance, $I_{s,v}$ from the polarized part.
These formulae provide the necessary information to separate diffuse and specular reflectance as we will show in section 3.
Polarized light may be used in controlled machine vision applications by mounting a polarisation filter in front of the source of illumination. If polarized light is used, reflected light has less degrees of freedom and subsequently less knowledge about the reflecting surface is required as it will be shown in section 4.
After reflection on a surface, diffuse reflectance will be completly depolarized [16], i.e. eq. 4 is valid for polarized incident light, too.
Polarized incident light will remain completly polarized after specular reflection. It can be decomposed into two components parallel and perpendicular to the specular plane (Fig. 2). The magnitude of the components depends on angle δ between incident polarized light and the specular plane (index p denoting polarized light):

$$I_p^i = u\, I_{\parallel,p}^i + v\, I_{\perp,p}^i \qquad \tan\delta = \frac{v}{u} \qquad (6)$$

The intensity of reflected light is calculated by multiplication of each Fresnel-coefficient with the actual component in its direction. Thus eq. 3 becomes

$$I_{\perp,p}^r = \frac{v\, F_{\perp}}{v\, F_{\perp} + u\, F_{\parallel}} * I_s^r \qquad I_{\parallel,p}^r = \frac{u\, F_{\parallel}}{v\, F_{\perp} + u\, F_{\parallel}} * I_s^r \qquad (7)$$

and eq. 5 describing the light intensity behind the polarisation filter becomes

$$I_{s,p}^f = \frac{u\, F_{\parallel} \cos^2 \theta + v\, F_{\perp} \sin^2 \theta}{v\, F_{\perp} + u\, F_{\parallel}} * I_s^r = I_{s,v} \cos 2\theta \qquad (8)$$

In contrast to eq. 5, specular reflectance of incident polarized light has only a variable part $I_{s,v}$, since the light remains completly polarized after reflection.
Hence we have a comprehensive set of formulae decribing the polarizing effect of optical reflection on dielectric surfaces.

3. Elimination of Highlights Using Unpolarized Incident Light

Unpolarized light is to be applied, if the vision system including illumination, inspected object and illumination can not be sealed up in a black box, for example, if ordinary ceiling lighting is used for illumination.
The Fresnel-coefficients $F_{\|}(\eta,\psi)$ and $F_{\perp}(\eta,\psi)$ depend on the index of refraction η and the angle of incidence ψ. We assume, that the inspected object consists of one material, i.e. η is constant[3]. If the object has a plane surface, ψ and thus $F_{\|}$ and $F_{\perp}$ are constant, too. The Fresnel-coefficients can be calculated for dielectrics according to [7]:

$$F_{\perp}(\eta,\psi)=\frac{a^2-2a\cos\psi+\cos^2\psi}{a^2+2a\cos\psi+\cos^2\psi} \quad (9)$$

$$F_{\|}(\eta,\psi)=\frac{a^2-2a\sin\psi\tan\psi+\sin^2\psi\tan^2\psi}{a^2+2a\sin\psi\tan\psi+\sin^2\psi\tan^2\psi}F_{\perp}(\eta,\psi) \quad (10)$$

where

$$a=\sqrt{n^2-\sin^2\psi}$$

Superposing diffuse (eq. 4) and specular reflectance (eq. 5), we get a total light intensity of

$$I^f(\theta)=\frac{1}{2}I_d^r+\frac{F_{\|}\cos^2\theta+F_{\perp}\sin^2\theta}{F_{\perp}+F_{\|}}*I_s^r \quad (11)$$

There remain two unknown variables: I_d and I_s. By input of two images with a different angle θ of the polarisation filter, the diffuse reflectance I_d can be calculated, thus providing an image, in which the highlight has been mathematically eliminated.
In threedimensional scenes, the angle of incidence ψ is unknown and thus $F_{\|}$ and $F_{\perp}$ can not be determined. In order to eliminate I_s, we assume that in highlights specular reflection is much larger than diffuse reflection $I_s \gg I_d$. This assumption is reasonable for many dielectrics with a smooth surface, if the specular reflectance is concentrated in one or more bright spots.
From eq. 5, we can get the relation $F_{\perp}/F_{\|}$, if we have two images with maximum intensity (angle of polarisation filter $\theta = 90^0$) and minimum intensity ($\theta = 0^0$).

$$\frac{I_{max}(\theta+\alpha=90^\circ)}{I_{min}(\theta+\alpha=0^\circ)}=\frac{F_{\perp}}{F_{\|}} \quad (12)$$

---

[3] for known materials, the value of η can be looked up in tables. Otherwise it can be got by a simple optical experiment measuring the Brewster-angle[1]

The orientation of the polarisation filter, in which I_{max} and I_{min} can be obtained, depends on the direction of the surface normal. This effect can be illustrated by slowly turning the polarisation filter. In one position of the filter, one hightlight may be optically eliminated, in a different position, another highlight is eliminated and the first one appears again (fig. 3). The surface orientation in each pixel is expressed by a phase angle α.

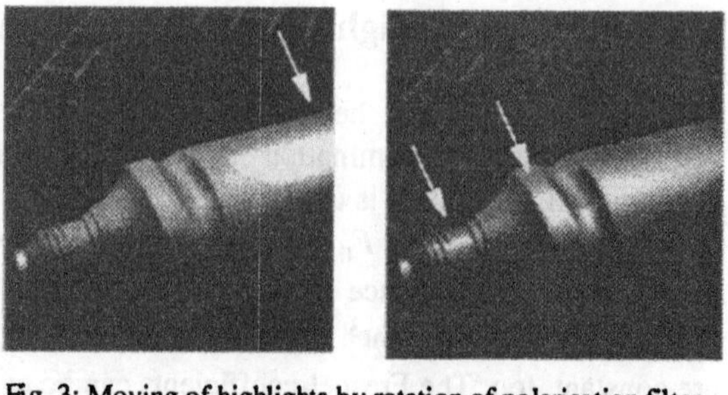

Fig. 3: Moving of highlights by rotation of polarisation filter (contrast enhanced for printing)

Thus we have three unknown variables: I_{max}, I_{min} and α. Input of three images with three different orientations of the polarisation filter provides the necessary number of equations to obtain I_{max} and I_{min} [7]. Using eq. 10, we get

$$\frac{I_{max}}{I_{min}} = \frac{F_{\perp}}{F_{\parallel}} = \frac{a^2 + 2a \sin\psi \tan\psi + \sin^2\psi \tan^2\psi}{a^2 - 2a \sin\psi \tan\psi + \sin^2\psi \tan^2\psi} \tag{13}$$

This equation can be numerically solved to get ψ for each pixel of the image. When ψ is known, by using eq. 9 and eq. 10, it is in turn possible to calculate $F_{\parallel}$ and $F_{\perp}$. Now the Fresnel-coefficients are known for each pixel of the image and thus we can calculate I_d, obtaining an image without specular reflectance.

4. Elimination of Highlights Using Polarized Incident Light

Use of polarized incident light reduces the degrees of freedom of reflected light as described in section 2. As a result, we can eliminate specular reflection with no or little knowledge of the surface properties of the inspected object. Polarized light can be produced by mounting a polarisation sheet between the source of illumination and the inspected object. This only poses a problem, if the source of illumination is very large. Another minor disadvantage is the reduction of light intensity by one half according to eq. 4.

If a plane surface is to be inspected, we have a known geometry of image aquisition. Since the surface normal is constant in the whole image, the specular plane has a constant orientation, too. The specular reflectance remains completly

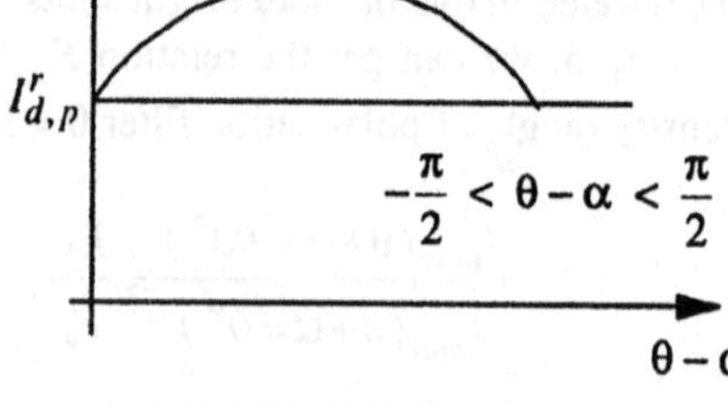

Fig. 4: light intensity as function of $\theta - \alpha$

polarized after reflection. Assume the angle between polarisation of incident light and the specular plane is δ, then it is possible to mount a second polarisation filter in front of the camera with an orientation of $90^0 + \delta$, that can completly remove specular reflectance by means of optics. This is an effect frequently used in computer vision [9].
This method can not be applied to three-dimensional scenes, since the orientation of the specular plane is unknown. This is especially true for rough surfaces, when specular reflectance may occur quite far away from the point, where the angle of reflection equals the angle of incidence.
However, since the reflected specular reflectance is completly polarised, at each pixel there is one orientation of the polarisation filter, in which specular reflectance is completly removed. Total image intensity can be expressed as a constant term I_d plus a cosine term dependend on the orientation of the polarisation filter θ and a phase angle α representing the orientation of the surface normal as introduced in section 3, eq. 8 (fig. 4):

$$I_p^f = \frac{1}{2} I_{d,p}^r + I_{s,p}^r \cos 2(\theta - \alpha) \tag{14}$$

Similar to a method described in [10], using the trigonometric expression

$$\cos(x - y) = \cos x \cos y + \sin x \sin y$$

eq. 14 can be expressed as a dot matrix of two vectors:

$$\begin{aligned} I_{p,i}^f &= f_i \cdot v \\ v &= \left(\frac{1}{2} I_{d,p}^r, I_{s,p}^r \cos 2\alpha, I_{s,p}^r \sin 2\alpha\right) \\ f_i &= (1, \cos 2\theta, \sin 2\alpha) \end{aligned} \tag{15}$$

where index i represents images with different filter position. In this way, we get three equations with three unknown variables I_d, I_s and α. Thus the required diffuse reflectance I_d can directly be calculated.
Unlike the method using unpolarized light in section 3, we do not need to know the index of refraction of the surface and relativly weak highlights can be eliminated, too, since no assumption about the magnitudes of I_d and I_s has been made. In addition, we can avoid the troublesome numerical calculation of eq. 13.

5. Reduction of Number of Images

To eliminate specular reflection in three-dimensional scenes according to the methods described in section 3. and 4., we have to input three images with different orientations of the polarisation filter. Up to now, there is no camera available, that

allows parallel input of images with different angles of a polarisation sheet. Mechanical rotation of a polarisation filter is slow and susceptible to trouble. In moving productions lines, several cameras arranged one behind the other have to be used, creating synchronisation problems.
In this section, we want to create sets of data during installation of a vision system that allow to reduce the number of images. It is assumed, that specular reflection has a constant value, that does not change during operation. The following conditions must be met:
- during inspection, the object is in a fixed position, or there is only a small translation, that is easy to measure and that does not change the geometry of image aquisition.
- the index of refraction η does not change during operation, i.e. there is no change in material.
- the form of the object remains constant.

What we can inspect are greylevel and colour characteristics of the surface, such as texture, that may be difficult to control in presence of specular reflectance. Changes in colour and brightness of the surface effect diffuse reflectance, the specular reflectance remains unchanged.

In the first case we assume, that the source of illumination remains constant. Using eq. 14 and eq. 15, we can calculate a compensation function K, that removes specular reflectance:

$$\frac{1}{2} I_d^r = I_p^f + K$$
$$K = - I_{s,p}^r \cos 2(\theta - \alpha) \qquad (16)$$

Compensation function K is calculated once for each pixel for a definite orientation θ of the polarisation filter during installation of the vision system. It remains unchanged during operation. Thus only one image has to be input to inspect the object.

In the second case, the source of illumination remains constant, but the polarisation filter in front of it is removed, so that we use unpolarized light during inspection, thus eq. 14 becomes

$$I^f = \frac{1}{2} I_d^r + I_s^r \qquad (17)$$

Since the intensity of incident light doubles when removing the polarisation sheet, the intensity of diffuse reflected light doubles, too:

$$\frac{1}{2} I_d^r = I_{d,p}^r \qquad (18)$$

Thus we get a compensation function

$$K = -I_s^r = -(I^f - I_{d,p}^r) \tag{19}$$

Like in the first case, it is sufficient to input one image for inspection of the object.

In the last case, we want to use a different source of illumination during operation. Polarized light is used for training only. This means, that during operation the light intensity of each reflection component of eq. 17 will be multiplied with an unknown factor a:

$$I_a^f = a\frac{1}{2}I_d^r + a\,I_s^r \tag{20}$$

In this equation, we have two unknown variables: a and I_S. I_S can be determined in a separate measurement as described in the case above (eq. 19). Thus we get a compensation function

$$K = -a\,I_s^r = -\frac{I_a^f}{\frac{1}{2}I_d^r + I_s^r}I_s^r \tag{21}$$

Like before, it is possible to inspect an object with input of just one image, if the necessary measurements have been performed before, thus considerably reducing the hardware requirements for practical applications.

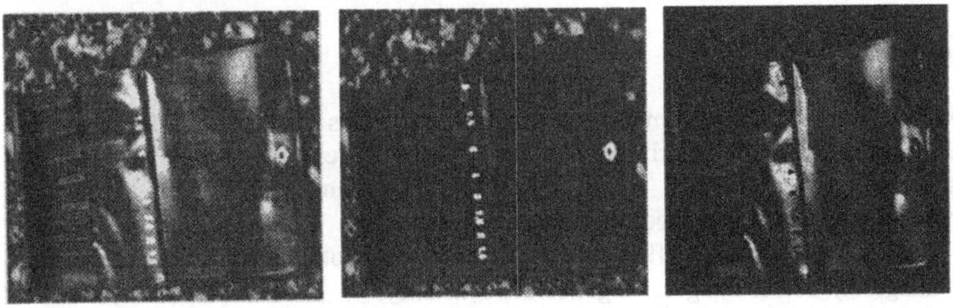

a) image with pol. filter b) diffuse reflectance c) partial specular reflectance

Fig. 5: elimination of specular reflectance on a soft plastic case

6. Experimental Results

In this section we will show some results of experiments, that demonstrate the applicabiltity of the methods described above.

Fig. 5. show an example of a soft plastic case with several bumps and dents. The scene is illuminated with unpolarized light. Fig. 5a. shows an image input with a polarisation filter in front of the camera, i.e. a part of the specular reflectance has been removed by means of optics. In Fig. 5b. the highlights have computationally been removed with the method introduced in section 3. Fig. 5c. shows the difference between 5a and 5b. Image 5c holds the information that is required to calculate a compensation function according to eq. 16.

Fig. 6 shows a plastic bottle with a rather rough surface. Thus specular reflectance occurs quite for away from the point, where the angle of incidence equals the angle of reflectance. The bottle is illuminated with polarized light. Like in fig. 5, fig. 6a shows the image input with a polarisation filter in front of the camera. In fig. 6b, specular reflectance is removed as described in section 4. Fig. 6c shows the difference between fig. 6a and 6b.

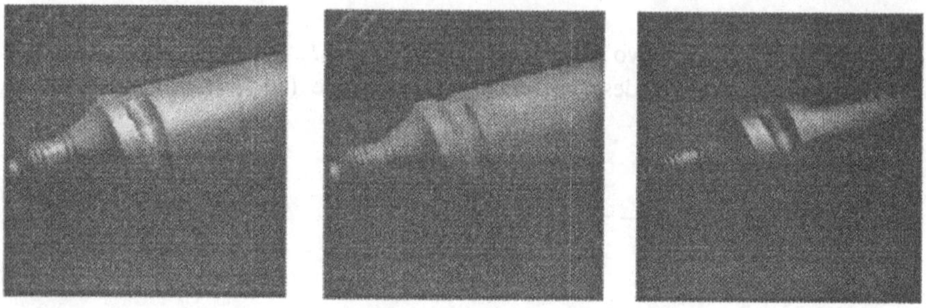

a) image with pol. filter b) diffuse reflectance c) partial specular reflectance

Fig. 6: Removal of higlight on plastic bottle using polarized light

7. Summary

A new polarization-based method to eliminate highlights has been introduced. Either unpolarized light or polarized light can be used to remove highlights, that may severly hamper standard image processing algorithms. The main advantages compared with other methods described in literature are:

- specular and diffuse reflection can be separated both in colour and greylevel images. In colour images, we get colour constancy.
- the algorithms can be applied to single pixels, they do not require information obtained from larger areas of the image, thus they can be applied to textured surfaces
- a polarization filter absorbs a great part of specular reflectance. Therefore the problem of limited dynamics of the camera is reduced.
- If polarized light is used, no information about geometry of image aquisition and surface characteristics is necessary.
- If the required measurements are made during installation of the vision system, only one image of the inspected object is necessary. The remaining computations are simple and can be performed very quickly.

Especially the last point helps to close the gap between research and application, since the hardware requirements to the vision system are greatly reduced.

Further research work should concentrate on combining analysis of polarisation with analysis of spatial distribution of reflection in order to gain more information from a scene.

Up to know, highlights have just been seen as an unwanted phenomenon, that should be eliminated during image-preprocessing. However, the magnitude and distribution of specular reflection is the result of surface characteristics such as gloss and roughness. Through Maxwell's law, we can get direct access to electric properties of the inspected surface. These are considerations, that may open the way for new inspection tasks of computer vision systems.

References

[1] M. Born, E. Wolf: Principles of Optics, Pergamon, London 1965

[2] F.R. Kessler: Physik: Optik II. Deutsches Institut für Fernstudien, Tübingen 1978

[3] G.J. Klinker, S.A. Shafer, T. Kanade: Using a color reflection model to separate highlights from object color. 1. ICCV: pp. 145-150, London 1987

[4] G.J. Klinker: A Physical Approach to Color Image Understanding. Wellesley, USA, 1993, A.K. Peters Ltd.

[5] K. Koshikawa: A polarimetric approach to shape understanding of glossy objects. 3. International Joint Conference on Artificial Intelligence, pp. 493-495, 1979

[6] V. Müller: Analysis of Optical Reflection - A new Approach to Surface Inspection, In: W.G. Kropatsch, H. Bischof (Eds.): Mustererkennung 1994. 16. Symposium of the German Working Group on Pattern Recognition (DAGM) and 18. Workshop of the Austrian Working Group on Pattern Recognition ÖAGM, pp. 74-80, Wien 1994

[7] V. Müller: Polarization-Based Separation of Diffuse and Specular Surface-Reflection, In G.Sagerer, S.Posch, F.Kummert (Eds.): Mustererkennung 1995, pp. 202-209, Springer, Berlin 1995

[8] S.W. Lee: Understanding of Surface Reflection in Computer Vision by Color and Multiple Views, PhD thesis, University of Pennsylvania, 1991

[9] H. Li, H. Burkhardt: Prototypentwicklung von Algorithmen zur Orientierungsschätzung von Pralinen aufbauend auf Grauwertmerkmalen (I) - Prinzip und Implementierung. Interner Bericht 3/91, Technische Informatik I, TU-HH, Dezember 1991

[10] S.K. Nayar, X.S. Fang, T. Boult: Removal of Specularities Using Color and Polarization, IEEE Conference on Computer Vision and Pattern Recognition, pp. 583-590, New York 1993

[11] B.T. Phong: Illumination for computer generated pictures. Commun. ACM 18, (6), pp. 311-317, June 1975

[12] S. Shafer: Using Color to Separate Reflection Compenents. Color Research and Applications, Vol. 10, pp. 210-218, 1985

[13] R.Siegel,J.R.Howell: Thermal radiation heat transfer. Taylor&Francis,Washington 1992

[14] L.B. Wolff: Spectral and polarization stereo methods using a single light source. 1. ICCV, pp. 708-715, London 1987

[15] L.B. Wolff: Polarization Methods in Computer Vision, PhD thesis, Columbia University, 1990

[16] L.B. Wolff, T.E. Boult: Constraining object features using a polarization reflectance model, IEEE Trans. on Pattern Analysis and Machine Intelligence Vol. 13 (6), pp. 635-657, 1991

Separating Real and Virtual Objects from Their Overlapping Images

Noboru Ohnishi[1,2], Kenji Kumaki[1], Tsuyoshi Yamamura[1], and Toshimitsu Tanaka[3]

[1] Department of Information Engineering, Nagoya University, Furo-cho, Chikusa-ku, Nagoya 464-01 Japan
[2] Bio-Mimetic Control Research Center, RIKEN, 8-31, Rokuban 3-chome, Atsuta-ku, Nagoya, 456 Japan
[3] Computation Center, Nagoya University, Furo-cho, Chikusa-ku, Nagoya 464-01 Japan

Abstract. We often see scenes where an object's virtual image is reflected on window glass and overlaps with an image of another object behind the glass. This paper proposes a method for separating real and virtual objects from the overlapping images. Our method is based on the optical property that light reflected on glass is polarized, while light transmitted through glass is less polarized. It is possible to eliminate reflected light with a polarizing filter. The polarization direction, however, changes even for planar glass and is not easily determined without information about the position and orientation of the glass and objects relative to the camera. Our method uses a series of images obtained by rotating a polarizing filter. Real objects are separated by selecting the minimum image intensity among a series of images for each pixel. The virtual image of objects is obtained by subtracting the image of the real objects from the image having the maximum image intensity among a series of images for each pixel. We present experiments with actual scenes to demonstrate the effectiveness of the proposed method.

1 Introduction

In real scenes indoors and outdoors there are many transparent objects with glossy surfaces, such as glass and water. We often see the reflected (virtual) image of an object on these surfaces located on the same side as we are, relative to the surface. The reflected image overlaps with other objects behind the transparent object. For example, we cannot see a person in a car clearly, because the surrounding scene is reflected on the window glass. Thus, reflected images hinder our perception of what is behind a transparent object.

It is important to separate the real from the reflected components in overlapping images. Eliminating reflected images improves the quality of TV camera images. Detecting reflected images helps a mobile robot perceive a transparent object obstructing its advance; a mobile robot can also use reflected images to detect the existence of objects outside the camera's field of view.

This paper proposes a method for separating real and virtual objects from their overlapping images. Our method is based on the polarization of light reflected on a specular surface, using a series of images captured by rotating a polarizing filter; the polarization direction is not required in advance. Using simple comparison and subtraction, we easily separated real objects from overlapping reflected images.

There are a few related studies [1]-[4]. Wolf [1][2] used polarization to separate reflection components, and developed a polarization camera. Yamada et al.[2] developed a vision system for removing the specular reflection component so that mobile robots could correctly locate a white guide line. Polarization was used to detect wet roads [4]. In these studies the polarization direction was known in advance, and only the reflection on the surface of an opaque object, such as a floor or a road, was introduced. In contrast, this paper deals with reflections on the surface of a transparent object behind which other objects exist. We do not require the polarization direction in advance.

In section 2 we describe optics, especially polarization. In section 3 we present a model of the image formation in terms of polarization and propose a method based on the model for separating overlapping objects without prior information about the polarization direction. In section 4 we show the experiment results in real scenes and demonstrate the method's effectiveness. Finally, we offer our conclusions.

2 Optics on Polarization

2.1 Polarization of Reflected and Transmitted Light

Consider the light reflection between two materials, A and B, whose refraction coefficients are n_A and n_B (Fig. 1). In the figure, θ_A is the angle of incidence and θ_B the angle of refraction. There is the following relation between θ_A and θ_B (Snell law):

$$n_A \sin\theta_A = n_B \sin\theta_B \tag{1}$$

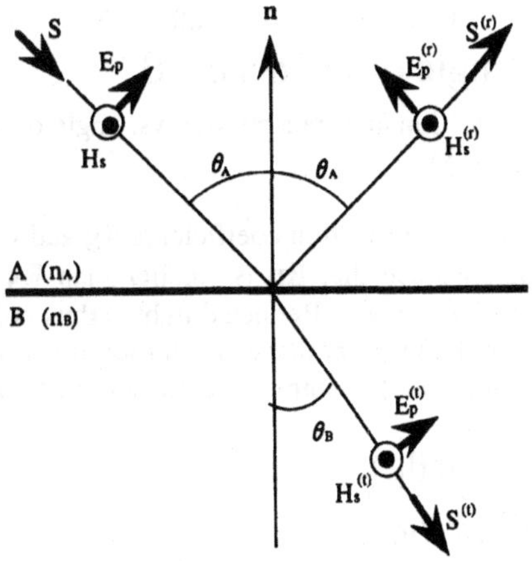

Fig. 1 Reflection and transmission

If the boundary of the two materials is completely specular, the angle of reflection is equal to the angle of incidence. As in optics, the Fresnel reflection coefficients are represented as

$$R_s = \frac{\sin^2(\theta_A - \theta_B)}{\sin^2(\theta_A + \theta_B)},$$
$$R_p = \frac{\tan^2(\theta_A - \theta_B)}{\tan^2(\theta_A + \theta_B)}, \tag{2}$$

where subscripts s and p indicate parallel and orthogonal components to the incident plane consisting of the incident light vector **S** and surface normal **n**.

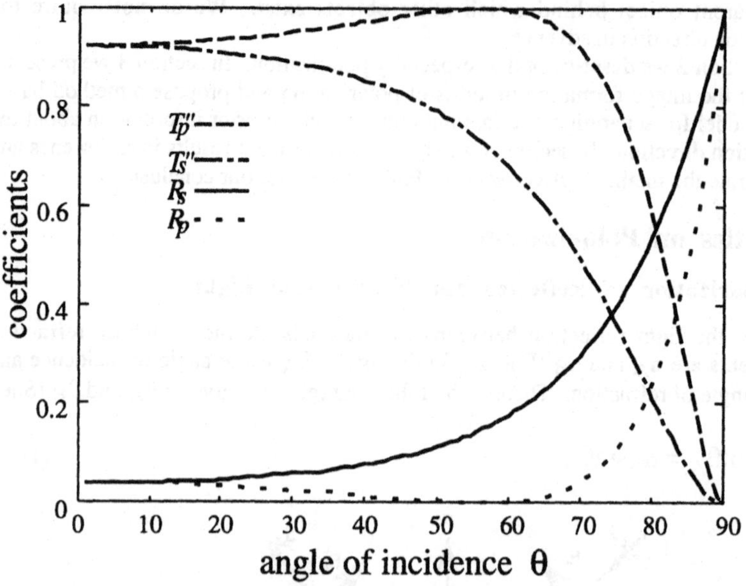

Fig. 2 Coefficients of reflection and transmission vs. angle of incidence (n_A=1.0, n_B=1.5)

Figure 2 shows the changes of reflection coefficients R_S and R_P with the angle of incidence. On the graph we see that R_P is smaller than R_S for an angle of incidence ranging from 30 to 80 degrees. Reflected light is thus polarized and has a smaller component parallel to the incident plane. Therefore, it is possible to remove light reflected on a specular surface. In contrast, the transmission coefficients in Fig. 1 are given as:

$$T_s = \frac{n_B\cos\theta_B}{n_A\cos\theta_A}\left\{\frac{2\cos\theta_A\sin\theta_B}{\sin(\theta_A + \theta_B)}\right\},$$
$$T_p = \frac{n_B\cos\theta_B}{n_A\cos\theta_A}\left\{\frac{2\cos\theta_A\sin\theta_B}{\sin(\theta_A + \theta_B)\cos(\theta_A - \theta_B)}\right\} \tag{3}$$

Now consider light transmitted from material A through thin material B to material A. The transmission coefficient from material B to A is easily obtained using Eq. (3) as follows:

$$T'_s = \frac{n_A \cos\theta_A}{n_B \cos\theta_B}\left\{\frac{2\cos\theta_B \sin\theta_A}{\sin(\theta_A+\theta_B)}\right\}^2,$$
$$T'_p = \frac{n_A \cos\theta_A}{n_B \cos\theta_B}\left\{\frac{2\cos\theta_B \sin\theta_A}{\sin(\theta_A+\theta_B)\cos(\theta_A-\theta_B)}\right\}^2 \quad (4)$$

Then the total transmission coefficient is obtained as:

$$T''_s = T_s T'_s, \quad T''_p = T_p T'_p \quad (5)$$

Figure 2 shows the changes of coefficients T''_s and T''_p. Transmitted light is less polarized than reflected light for an angle of incidence smaller than 80 degrees.

2.2 Direction of Polarization

This subsection describes how the polarization angle of reflected light changes depending on the position of a point on a glass surface.

Suppose that a camera locates in the *z-x* plane and its optical axis directs toward the coordinate origin O. The point C' is the projection of the center of camera lens C (See Fig.3(a)). The incident plane at point P in Fig.3(a) is obtained as

$$\boldsymbol{n}_{if} = \boldsymbol{n} \times \boldsymbol{p}, \quad (6)$$

where $\boldsymbol{n}$ is the normal vector of glass, $\boldsymbol{p}$ the vector from the point C to the point P, and the symbol 'x' denotes cross product of vectors. Because $\boldsymbol{p} = \boldsymbol{p}' + \mathbf{d}$ and $\boldsymbol{n} \times \boldsymbol{p}' = \boldsymbol{0}$, Eq.(6) is rewritten as :

$$\boldsymbol{n}_{if} = \boldsymbol{n} \times (\boldsymbol{p}' + \boldsymbol{d}) = \boldsymbol{n} \times \boldsymbol{d}. \quad (7)$$

Eq.(7) says that the direction of incident plane, orthogonal to the direction of polarization, changes depending on the vector $\boldsymbol{d}$, the position of a point on the surface. Figure 3(b) shows schematically the change of polarization direction.

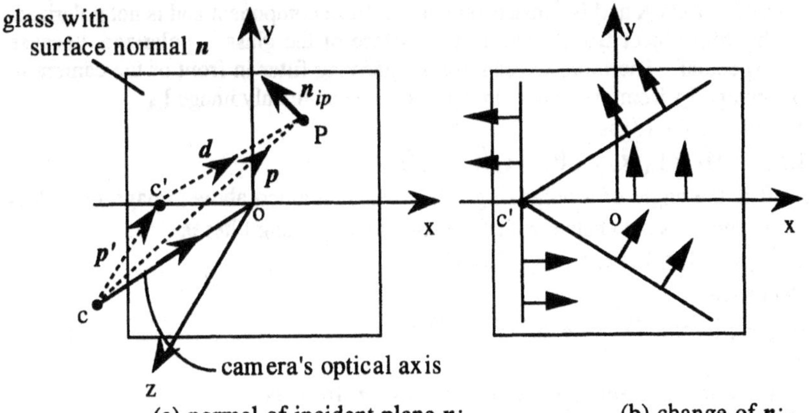

(a) normal of incident plane $\boldsymbol{n}_{ip}$ (b) change of $\boldsymbol{n}_{ip}$

Fig. 3 The normal of incident plane (polarization direction) and its change on a surface

3 Separation Method

In the previous section, we demonstrated that light reflected on a specular surface is polarized, while transmitted light is less polarized. Based on this physical property, we made a model of image formation under polarization and developed a method based on the model for separating overlapping objects.

3.1 Model of Image Formation

In the situation shown in Fig. 4, the camera captured the image of object A transmitted through the glass, and the reflected image of the object, giving us the following:

$$I(x, y) = I_A(x, y) + I_B(x, y) \tag{8}$$

where $I(x, y)$ is the camera image at each pixel (x, y), $I_A(x, y)$ the transmitted image, and $I_B(x, y)$ the reflected image.

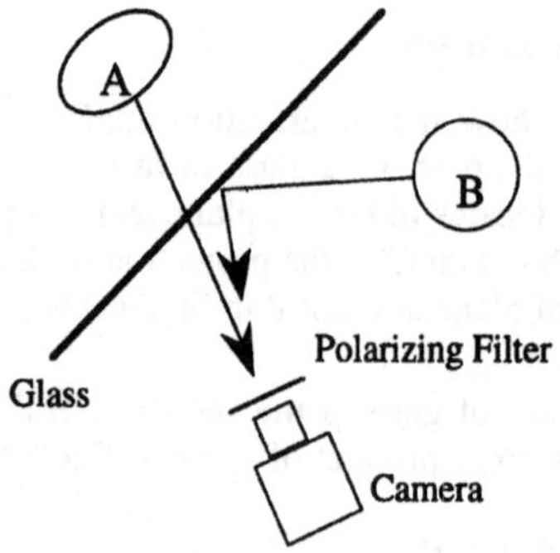

Fig. 4 Situation causing overlap of reflected and transmitted images

General light sources, such as the sun and incandescent light, emit unpolarized light. In this paper, we assume that the specular component is small enough to be neglected in the reflection of any object other than glass. Light reflected on the surfaces of objects A and B consists only of a diffuse component and is not polarized. The light from object B reflected on the surface of the glass is polarized, because glass is specular. Therefore we can use a polarizing filter in front of the camera to remove image I_B from the camera image and to separate only image I_A.

We rewrite Eq. (8) as

$$I(x, y ; \theta) = I_A(x, y ; \theta) + I_B(x, y ; \theta), \tag{9}$$

where θ is the angle of a polarizing filter. As mentioned above, image I_A is less polarized and does not change with the angle of a polarizing filter θ;

$$I_A(x, y ; \theta) \cong I_A(x, y) \qquad \text{for all } \theta.$$

We then have:

$$I(x, y ; \theta) = I_A(x, y) + I_B(x, y ; \theta). \tag{10}$$

3.2 Method for Separating Overlapping Images

As mentioned in **2.2**, the polarizing direction (angle) depends on the relative positions of the object, glass, and camera. It changes depending on the position of the glass, especially curved glass. Thus, measuring the relative positions and the 3D shape of the glass is complicated and unrealistic. By rotating a polarizing filter, we captured a series of images at each polarizing filter angle θ; for example, every 10 degrees.

The real image of object A can be obtained by selecting a minimum image intensity of $I(x, y; \theta)$ for each pixel:

$$\hat{I}_A(x, y) = \min_{\theta \in \Theta} I(x, y; \theta), \tag{11}$$

where Θ is the set of the filter angle. Next, we can easily obtain the maximum intensity image:

$$I_{MAX} = \max_{\theta \in \Theta} I(x, y; \theta).$$

This is ideally represented as

$$I_{MAX} = I_A(x, y) + I_B(x, y). \tag{12}$$

By subtracting Eq. (11) from Eq. (12), we obtain the virtual image of the object B:

$$\hat{I}_B(x, y) = I_{MAX}(x, y) - \hat{I}_A(x, y). \tag{13}$$

4 Experiments

We conducted experiments both in a dark room and outdoors. In a dark laboratory room, we set up a sheet of glass with two balls in front of it and a ball behind it and illuminated them with a fluorescent lamp located in front of the glass. The angle between the camera's optical axis and the normal surface of the glass was 56 degrees. By changing the angle of the polarizing filter attached to the camera, we captured a series of images (18 images, one every 10 degrees) at a fixed iris. The images are shown in Fig. 5.

Using Eqs. (11) and (13), we obtained the real image of the object behind the glass as shown in Fig. 6(a) and the virtual image of the two objects in front of the glass (Fig. 6(b)) which were not seen directly in the camera's field of view. In Fig. 6(b), the right-side contour of the ball behind the glass was detected because of the slight sway of the ball, which was hanging from a string.

We conducted another experiment outdoors, at the entrance to our faculty building. The entrance is partitioned by wide glass, hindering us from seeing inside the entrance. We set a camera outside the entrance, with the optical axis creating an angle of 54 degrees. As with the previous experiment, we captured a series of images (18 images, one every 10 degrees) at a fixed iris. The images are shown in Fig. 7.

In Fig. 7, we know that the bicycles and trees are reflected on the glass and they corrupt the view inside the entrance. We separated the inside entrance scene and the outside scene (reflected on the glass) using Eqs. (11) and (13). The results are shown in Fig. 8. Figure 8(a) shows real objects and we can see inside the entrance clearly: three boards and light switches on the tile wall. In contrast, Figure 8(b) shows the reflected images of a bicycle, a motorcycle, trees and a wall with the same tile pattern as inside the entrance. But we can see the black-and-white reversed image of the boards on the wall. This results from over-subtracting the board image from the maximum image I_{MAX}. Although we neglected the polarization in Eq. (13), transmitted light is somewhat polarized.

The experiment results demonstrate the effectiveness of the proposed method.

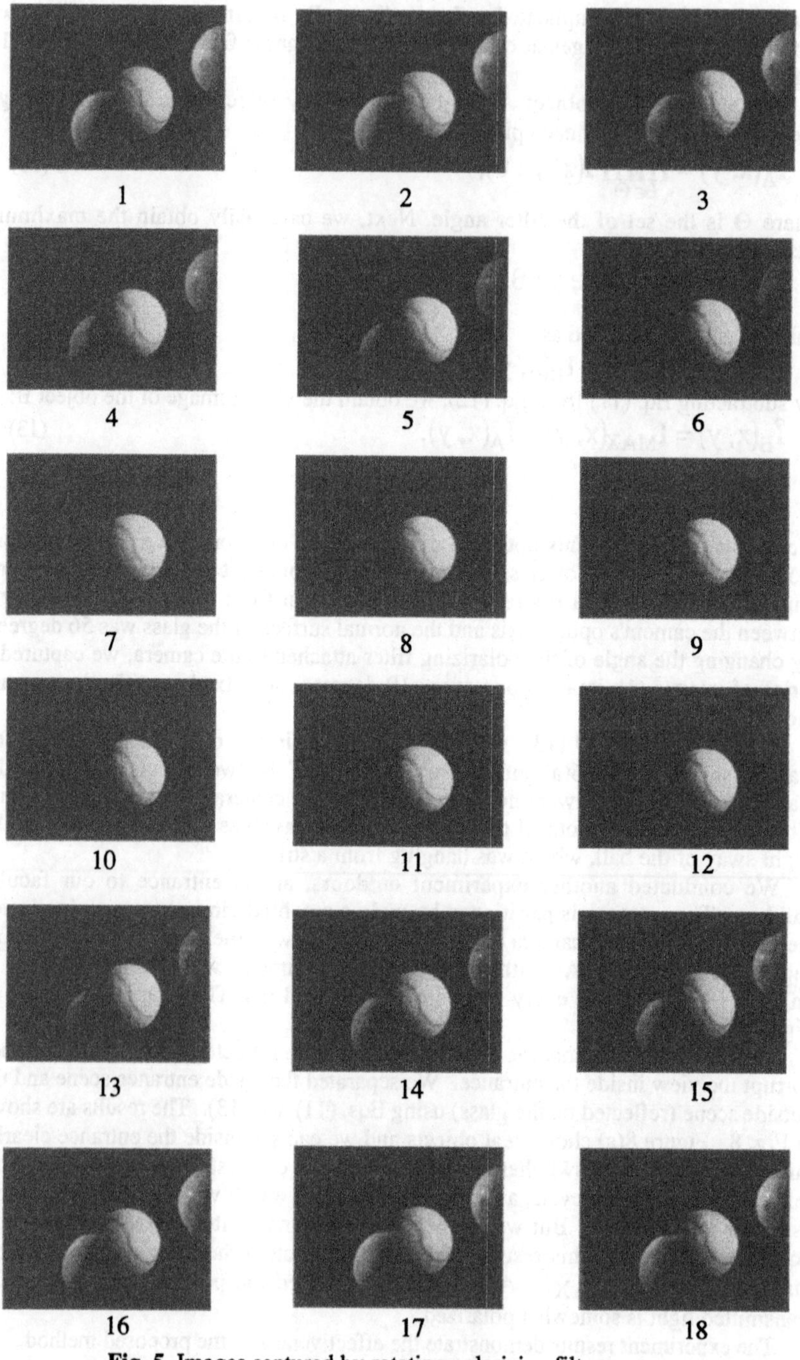

Fig. 5 **Images captured by rotating polarizing filter (every 10 degrees ; left to right)**

(a) real object

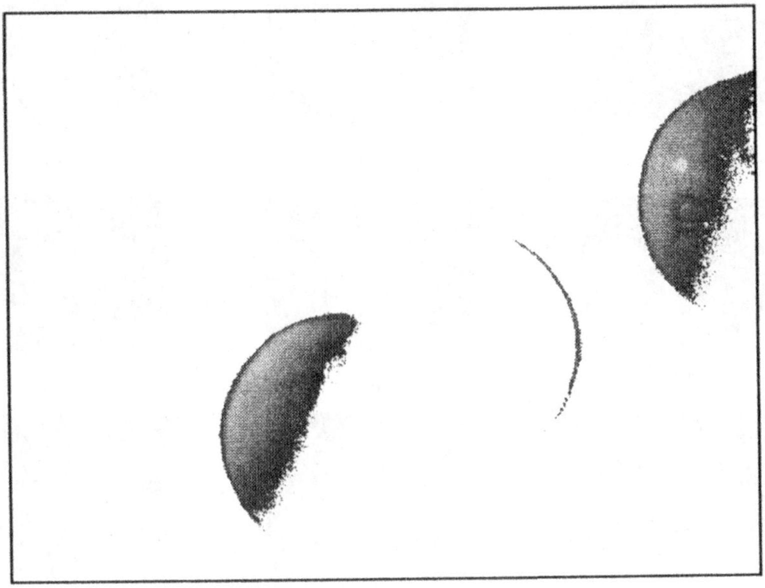

(b) virtual objects (reflected image)

Fig. 6 Separated images

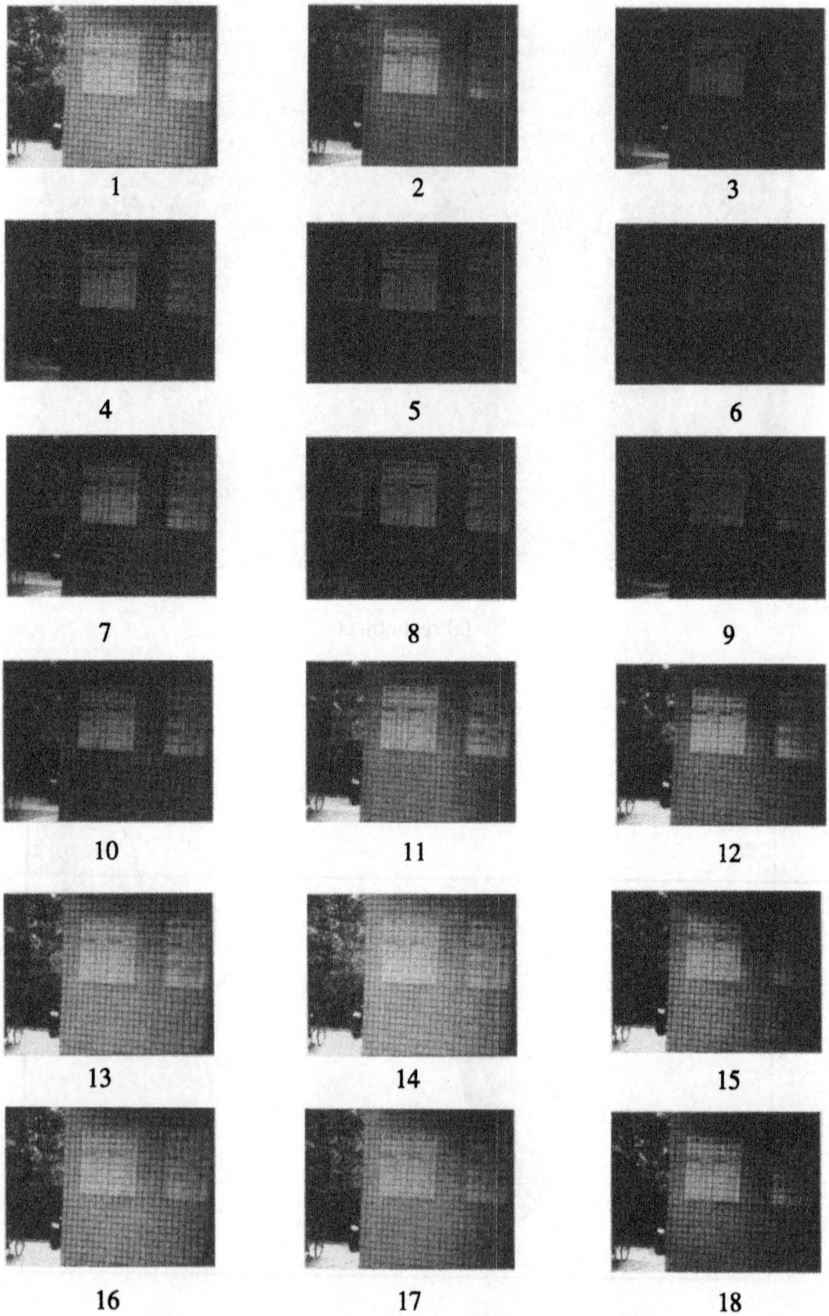

Fig. 7 Images captured by rotating polarizing filter (every 10 degrees ; left to right)

(a) real objects

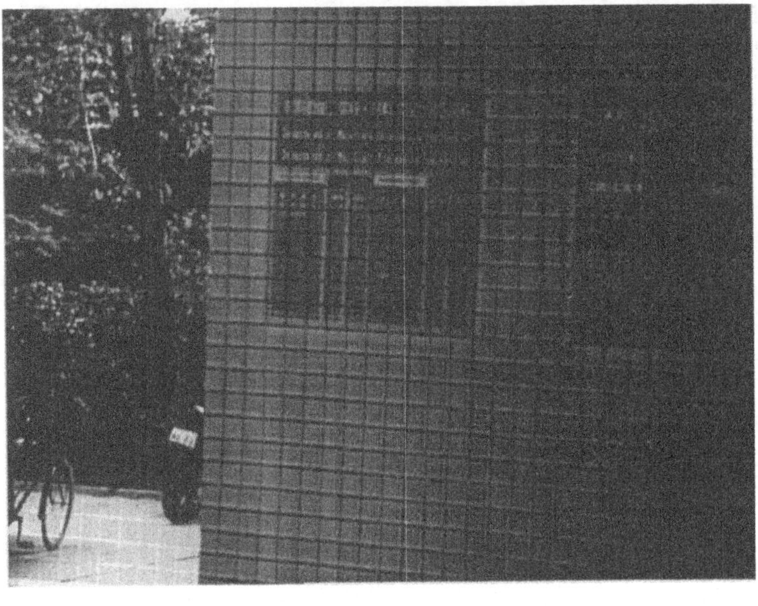

(b) virtual objects (reflected image)

Fig. 8 Separated images

5 Conclusion

We have presented a method for separating the real image transmitted through glass from the virtual image reflected on it. Our method uses the optical property of polarization and does not require any information about the object's position, orientation, or shape. Furthermore, the method requires only simple calculation, comparison, and subtraction for each pixel. It is possible, therefore, to design hardware to perform the separation in real time. Refinement of the separation model (including improving the separated image quality) and the design of a real-time separation hardware are future projects.

Acknowledgment

We would like to thank Mr. Masaki Iwase, Nagoya University graduate student, for assistance in preparing the figure and the members of our laboratory for their valuable input.

References

[1]L. B. Wolf; Using Polarization to Separate Reflection Components, Proc. IEEE Computer Society Conf. CVPR'89, pp.363-369 (1989).

[2]L. B. Wolf and T. A. Mancini; Liquid Crystal Polarization Camera, Proc. of IEEE Workshop on Application of Computer Vision, pp.120-126 (1992.12).

[3]K. Yamada, T. Nakano and S. Yamamoto; A Vision System for Removal of Specular Reflection Component Using a Liquid Crystal, Trans. of IEE Japan, Vol. 113-C, No. 12, pp. 1087-1093 (1993.12).

[4]K. Ueda, I. Horiba, K. Ikegaya and F. Ooi; A Detecting Method of Wet Condition on Road Using Image Processing, Trans. of Information Processing Society of Japan, Vol. 35, No. 6, pp.1072-1080 (1994.02).

Goal-directed Video Metrology

Ian Reid and Andrew Zisserman

Dept of Engineering Science, University of Oxford, Oxford, OX1 3PJ

Abstract. We investigate the general problem of accurate metrology from uncalibrated video sequences where only partial information is available. We show, via a specific example – plotting the position of a goal-bound soccer ball – that accurate measurements can be obtained, and that both qualitative and quantitative questions about the data can be answered.
From two video sequences of an incident captured from different viewpoints, we compute a novel (overhead) view using pairs of corresponding images. Using projective constructs we determine the point at which the vertical line through the ball pierces the ground plane in each frame. Throughout we take care to consider possible sources of error and show how these may be eliminated, neglected, or we derive appropriate uncertainty measures which are propagated via a first-order analysis.

1 Introduction

The 1966 World Cup Final at Wembley Stadium, between England and West Germany, produced what is arguably the best known and most controversial goal in football history. In extra time, with the score at 2-2, Geoff Hurst the England number 10, received the ball from the right, turned, and struck a shot towards the German goal. With the goal-keeper beaten, the ball cannoned down from the crossbar, hit the ground and bounced back out into play (whence it was cleared by the German defence). English players claimed a goal – that the ball had passed completely over the line – and after consultation with his linesman, the referee concurred. England went on to win 4-2, but the controversy has never been satisfactorily resolved.

Here we resolve this controversy using video sequences of the goal. Two monocular sequences acquired from substantially different viewpoints are used for the analysis. Figure 1 shows a series of frames from each of the sequences. Using these, the question we wish to answer is *Did the ball cross the goal line?* And, if not, *How close did it come to crossing the goal line?*

The question is challenging because of the lack of available calibration:

1. The internal calibration of the cameras is unknown (and free — i.e. may well change during the sequence).
2. The motion of the cameras is unknown, and relative orientation (between stereo pairs) changes.
3. For many frames of the sequence there are few features available off the ground plane, other than moving objects — the players and the ball.

From an uncalibrated monocular sequence projective 3D measurements can be made [2, 4, 7] and upgraded to Euclidean (angles, lengths) [6, 12] for unchanging

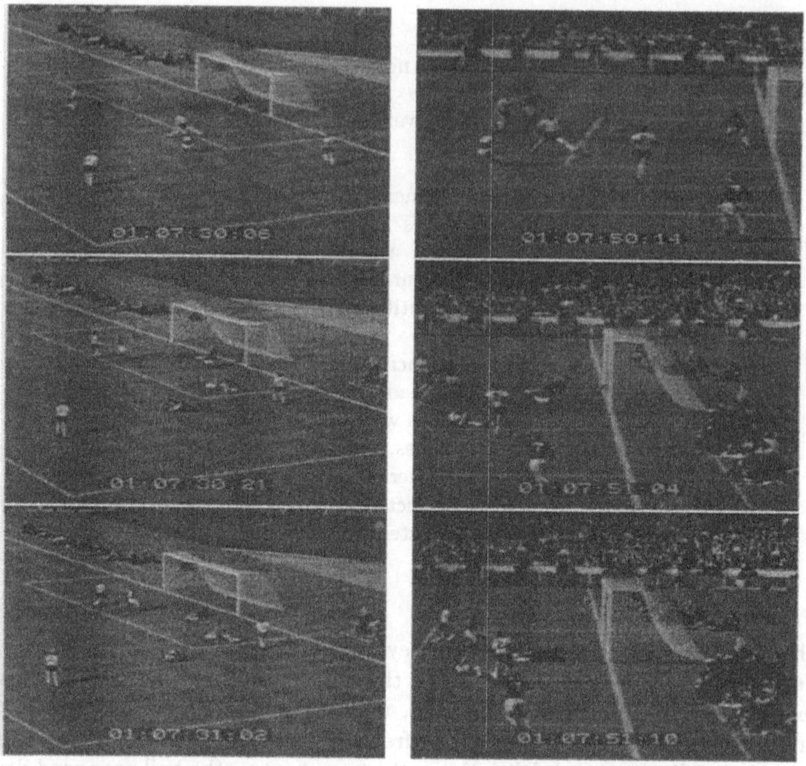

Fig. 1. Images from two available sequences of the incident.

internal parameters. These methods are not applicable for two reasons: first, because of the free internal parameters, and second because the object of interest (the ball) is moving relative to other objects in the image (in particular the ground). A binocular view (images acquired simultaneously) avoids the second problem, and, if the internal parameters and relative orientation of the cameras were also unchanging for a few frames, then 3D Euclidean structure could be recovered [1, 18, 19]. Again, the free parameters and changing relative motion prevent this. In the light of this paucity of information how can the question be answered?

It is answered by projecting the ball vertically onto the ground plane and charting the projected position and uncertainty in this position relative to the goal line. Vertical is defined by the goal posts, and metric information is provided by the dimensions of the ground plane markings. We employ the ground plane homography between views, together with vertical vanishing points, which can be computed even though point features off the ground plane are often not available. The technique is related to Quan and Mohr's [13] "shadow" algorithm for computing, from two images acquired from different viewpoints, the imaged intersection of a line with a plane. This algorithm was subsequently used to compute invariants of 3D objects [3], and for specifying points for robotic grasping [8].

The procedure here is a development on these papers in two ways: first, the intersection is not between an actual line and a plane, but between a virtual line constructed using the vertical direction vanishing point; and, second (and more importantly), a full error analysis is given for the projective transfer. The error analysis takes into account three sources of error: first, the localisation error of the points used to define the projective transformation; second, the localisation error of the computed vanishing point; and, third, the localisation error of the ball. This case study exemplifies the measurement of relative positions and their uncertainty, from uncalibrated image sequences, when there is insufficient information for a full 3D reconstruction.

We begin by describing details of the construction in section 2. The sources of error are outlined in section 3 and the implementational details, including treatment of errors, are given in section 4. Finally, results are presented in section 5.

2 Outline of method

We determine the vertical projection of the ball onto the ground plane from two images acquired simultaneously from different viewpoints. To visualise this, imagine dropping a (vertical) "plumb-line" from the ball to the ground [10]. We show that,

Given
1. *two images acquired simultaneously from different viewpoints,*
2. *the vertical vanishing point in each image,*
3. *the homography (see below) between the images induced by the ground plane,*
4. *the images of a 3D point* **B**.

then the intersection, **P**, *with the ground plane of a vertical line through the point* **B** *can be computed uniquely.*

In the following we denote world entities by upper case, and their corresponding images by lower case 3-vectors, e.g. $\mathbf{x}$ and $\mathbf{x}'$ for points, and $\mathbf{l}$ and $\mathbf{l}'$ for lines. Matrices are denoted by teletype capital letters. For homogeneous quantities, $=$ indicates equality up to a non-zero scale factor.

The geometry of the construction is illustrated in figure 2, where the line is defined by two points, **B** and **V**, in 3D. Actually, **V** is a point on the plane at infinity (an ideal point), and its images $\mathbf{v}, \mathbf{v}'$ are vanishing points, but this does not affect the projective construction. The plane projective transformation (homography) $\mathtt{T}$ between the two images via the ground plane provides a means of transferring lines between the two images. If $\mathbf{l}$ and $\mathbf{l}'$ are images of a line on the ground plane, then $\mathbf{l}' = \mathtt{T}^{-\top}\mathbf{l}$, where $\mathtt{T}$ is a 3×3 point transformation matrix: $\mathbf{x}' = \mathtt{T}\mathbf{x}$ for images of points on the ground plane.

There are four steps in the algorithm for computing $\mathbf{p}$, the image of **P**:

1. Compute the plane projective transformation, $\mathtt{T}$ between the two images from the correspondence of four lines (no three concurrent) i.e. $\mathbf{l}'_i = \mathtt{T}^{-\top}\mathbf{l}_i, i \in$

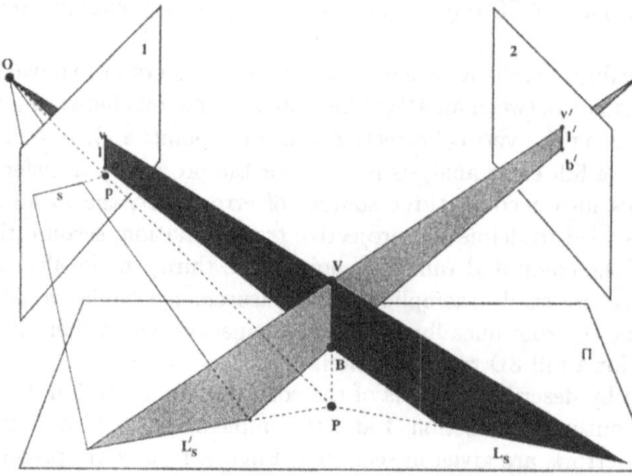

Fig. 2. **L** is a line passing through the points **V** (the vertical ideal point) and **B** (the ball), which intersects the plane π (the ground plane) in the point **P**. **l** and **l′** are images of **L**, which does not lie on π. Equivalently, however, they are the images of lines $\mathbf{L}_s$ and $\mathbf{L'}_s$, respectively, which are on the ground plane — the line **L** casts a "shadow" $\mathbf{L}_s$ from view 1, where the plane π intersect the backprojection of the line **l** from the first image. A similar shadow, $\mathbf{L'}_s$, is generated from view 2. Since $\mathbf{L'}_s$ is on the ground plane, its image is $\mathtt{T}^\top \mathbf{l}'$ in image 1, where T is the point projective transformation between the images induced by the plane π. Since the lines $\mathbf{L}_s$ and $\mathbf{L'}_s$ intersect at **P**, their images **l** and $\mathbf{s} = \mathtt{T}^\top \mathbf{l}'$ respectively, intersect at the image **p** of **P**.

{1, .., 4}. Details of this computation are given in section 4.

2. Compute the lines through the images of **V** and **B**. These lines are given by $\mathbf{l} = \mathbf{v} \times \mathbf{b}$, $\mathbf{l}' = \mathbf{v}' \times \mathbf{b}'$ in the first and second images respectively.
3. Transfer the line $\mathbf{l}'$ from the second onto the first image as $\mathbf{s} = \mathtt{T}^\top \mathbf{l}'$.
4. Then the image of the intersection point is $\mathbf{p} = \mathbf{s} \times \mathbf{l}$ in the first image. A similar construction determines $\mathbf{p}'$, the imaged intersection in the second image, as $\mathbf{p}' = (\mathtt{T}^{-\top} \mathbf{l}) \times \mathbf{l}'$.

This computation can also be transferred to a plan (rectified) view of the ground plane using the projective transformation between the points/lines on the ground plane and their images. In this case the six-yard goal markings are known (up to a plane Euclidean transformation), and these relative measurements provide metric calibration, and a 2D frame in which to evaluate uncertainty.

3 Sources of error

Potential sources of error are discussed in the following sub-sections. In each case it is demonstrated that these potential errors did not arise, or could be accounted for.

Synchronisation

The synchronisation of the two sequences used is an essential assumption of the method. The time difference between frames is assessed by computing the ground plane homography between the two images using features corresponding to *fixtures* (such as marked lines on the ground, and goal posts), and measuring the error when this transformation is applied to features corresponding to moving objects (player's shadows). The error is the pixel distance between the actual and transferred point. The synchronisation error could be up to 20ms between video frames (1/48 s between film frames).

Eight points are obtained to sub-pixel accuracy by intersecting straight lines fitted to the ground plane markings. The homography is then computed using a combination of linear and non-linear minimisation where the cost function is the transfer error.

The accuracy of the transformations is first assessed by measuring the error for fixtures not used in the computation of the transformation. Errors are typically less than two pixels (e.g. for the ground plane computation 8 matches are available, 6 are used to compute the homography, and the error measured on the remaining two). For moving objects the error between *corresponding* frames of sequences are similar to the fixture error, whilst for a *near corresponding* frame, the errors exceed 10 pixels. Figure 3 illustrates these cases. In summary, the ground plane homography is used to establish that the two sequences are "perfectly" synchronised.

Radial distortion

In order to take advantage of projective geometry we require that the image formation process be described accurately by a central projection model. This model is invalid if there is any significant lens distortion, the most common type of which is radial distortion. One manifestation of radial distortion is bending of straight lines near the periphery. We have therefore tested its effect by fitting lines to known straight features in the periphery of images in each sequence. Figure 4 shows two typical images and corresponding residuals after an orthogonal regression fit to putative straight edge data. The lines fitted are superimposed on the images. The side view shows no distortion (residuals are distributed evenly either side of the fitted line), while a small, but for our purposes insignificant, amount of distortion is apparent in the three quarter view (obtained with a wider angle lens).

Straightness of lines/planarity of ground "plane"

If an imaged line remains straight through a range of viewpoints then this is compelling evidence that the world line is straight. Similarly, the straightness of a number of transverse lines on a surface is evidence that the surface is planar. The image measured straightness of all lines of the six-yard markings throughout both sequences indicates the planarity of the ground.

Motion blur

One further potential source of error is motion blur. As the ball moves (while the camera is stationary) it is significantly blurred in the direction of motion.

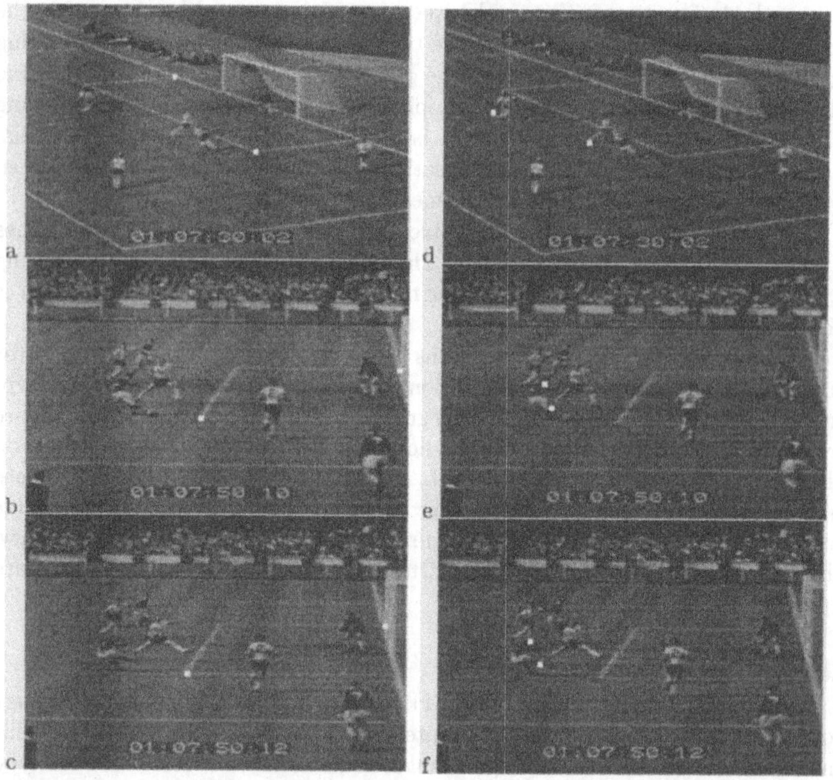

Fig. 3. Assessing the synchronisation of the sequences. (a) is a frame from sequence 1, and (b) (c) are near corresponding frames from sequence 2. The ground plane homographies between the frames (a & b, and a & c) are computed from imaged **fixtures**. Two points (fixtures) are marked on the ground plane of frame (a) and transferred to frames (b) and (c) using the appropriate homography. The disparity between transferred and actual position is negligible (i.e. less that a pixel), indicating the accuracy of the computed homography. (d) (e) (f) are the same frames with points corresponding to **moving** objects marked in (d). The points chosen are the left most point of the shadow of each player. The transferred points are superimposed on (e) and (f). In (e) the correspondence between transferred and actual position is again negligible, indicating that the frames are synchronised. However, in (f) there is a significant discrepancy (10 pixels) indicating that the frames are not synchronised.

Similarly, as the ball is tracked by the cameraman, the stationary features in the environment are observed to blur.

Fortunately during the crucial frames in which the ball is close to crossing the line, there is little blur due to camera motion. That which there is, is accounted for in the line uncertainty by inflating the line covariance appropriately. The significant blur is due almost entirely to the motion of the ball. In this case we take the blur into account by a greater uncertainty in the ball location in the direction of motion.

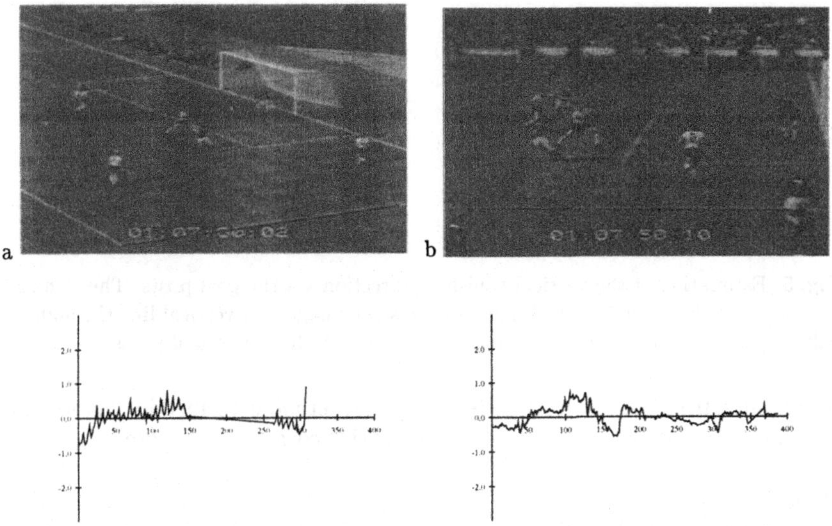

Fig. 4. Testing for radial distortion. (a) an image from the three-quarter view sequence with a detected line (shown black) near the periphery. The residuals of an orthogonal regression fit are shown in the graph below. The axes are in pixels. (b) is for a line from the side sequence.

4 Implementation and error analysis

In this section we discuss the representation, computation, and uncertainty of the geometric primitives and transformations required for the analysis. We follow the approach of [5, 11] computing uncertainty propagation via first-order approximations. We have verified the validity of the first-order model using Monte Carlo techniques. Full details are given in [15]. The covariance of a vector is denoted $\Lambda_{\mathbf{x}}$. The dimension of a matrix is indicated where necessary, in parentheses: e.g. $\Lambda_{\mathbf{x}}(2 \times 2)$.

Lines

Line segments are computed using orthogonal regression on a set of canny edge strings. Each (manually selected) set of edge strings is processed using the RANSAC algorithm [17] to enforce collinearity, adding greatly to the robustness of the line fitting by providing rigorous outlier rejection and by enabling multiple strings to contribute to one line segment.

Lines are represented both as homogeneous three vectors and by (inhomogeneous) two-parameter representations:

$$\mathbf{a} = [a, c]^\top \text{ such that } \begin{array}{ll} ax + y + c = 0, & \text{if the line is closer to horizontal} \\ x + ay + c = 0, & \text{if the line is closer to vertical} \end{array}$$

The uncertainty of a line is represented by a 2×2 covariance matrix, $\Lambda_{\mathbf{a}}$, which

Fig. 5. Estimation of the vertical vanishing direction via the goal-posts. The × marks the ball, and the extended black line indicates the image of a vertical line through the ball (it joins the ball to the vanishing point computed from the goal-posts).

is computed from edgel uncertainties using the method described in Chapter 5 of [5], and where necessary, by a 3 × 3 form for this covariance, denoted $\Lambda_{\mathbf{a}}(3 \times 3)$.

Points

Points are generally localised by intersecting lines, and their covariance computed from the line uncertainty. Where this is not possible, e.g. the ball centre, the point is picked with a mouse, in which case the uncertainty is estimated as the mouse precision (about ±1 pixel in each direction). The uncertainty in a point's position is represented by the 2 × 2 covariance matrix $\Lambda_{\mathbf{x}}$.

The vertical vanishing point is obtained by intersecting lines computed from the obvious vertical cues in each image – the goal-posts. Figure 5 shows an example. Care must be exercised when intersecting lines to find vanishing points, since the final component of the homogeneous representation may be close to zero, rendering the computations of the inhomogeneous coordinates and covariance unstable. For the case of vertical vanishing points which are of special interest here, we derive inhomogeneous coordinates from $\mathbf{v} = [v_x, v_y, v_z]^\top$ as $[v_x/v_y, v_z/v_y]^\top$ and the covariance calculation is modified appropriately.

Intersections

The intersection of two lines is given by the cross-product of homogeneous lines, $\mathbf{v} = \mathbf{l}_1 \times \mathbf{l}_2$. Letting $\mathbf{x} = [v_x/v_z, v_y/v_z]^\top$, we obtain the uncertainty in the location of the intersection using a first-order error analysis:

$$\Lambda_{\mathbf{x}} = \mathtt{D} \begin{bmatrix} \Lambda_{a_1}(3 \times 3) & 0 \\ 0 & \Lambda_{a_2}(3 \times 3) \end{bmatrix} \mathtt{D}^\top, \quad \text{where} \quad \mathtt{D}(2 \times 6) = \left[\frac{\partial \mathbf{x}}{\partial \mathbf{l}_1} \mid \frac{\partial \mathbf{x}}{\partial \mathbf{l}_2} \right]$$

Ground plane homography

The homography between a camera view and the plan view, T, is obtained from the corners of the six-yard area[1]. It is computed from image positions $\mathbf{p}$, which are estimated accurately by the intersection of extended lines, and

[1] We also compute the line transformation using four line correspondences, but have omitted the discussion here since it is analogous to that for points, although complicated by the need for two different parameterisations of lines.

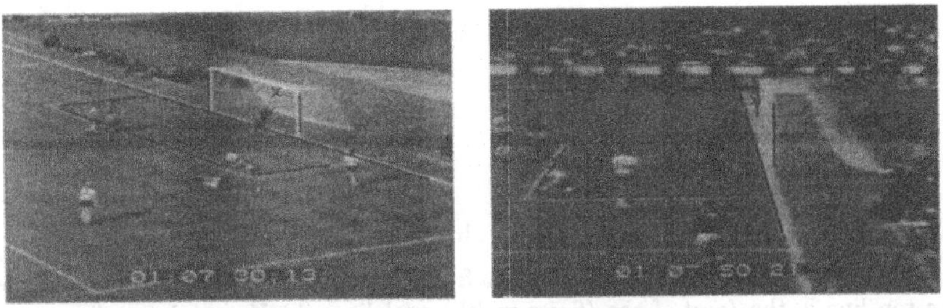

Fig. 6. The three-quarter and side views of the action just before the ball strikes the crossbar, showing the positions of lines used for registration.

corresponding plan view positions $\mathbf{P}$, which have known (exact) values. Each correspondence gives rise to two independent linear equations in the eight unknowns of $\mathtt{T}$, so from four such correspondences we can construct and solve the 8×8 matrix equation $\mathtt{A}\mathbf{t} = \mathbf{b}$, where $\mathbf{t}$ is an 8-vector of the free parameters of $\mathtt{T}$.

Defining the vector $\mathbf{z}$ to be the 8-vector containing the inhomogeneous co-ordinates of the image points $\mathbf{p}_{1\ldots4}$ (hence $\Lambda_{\mathbf{z}}$ is a block diagonal matrix consisting of the 2×2 submatrices $\Lambda_{\mathbf{x}_i}$ where $\mathbf{x} = [p_x/p_z, p_y/p_z]^\top$), it is straightforward to show that:

$$\Lambda_{\mathbf{t}} = \mathtt{D}\,\Lambda_{\mathbf{z}}\,\mathtt{D}^\top \quad \text{where} \quad \mathtt{D}(8 \times 8) = \frac{\partial \mathbf{t}}{\partial \mathbf{z}} = -\mathtt{A}^{-1}\frac{\partial \mathtt{A}}{\partial \mathbf{z}}\mathbf{t}$$

Transforming primitives

The final aspect of uncertainty which must be considered, is how to compute the uncertainty of a transformed primitive, when both the primitive and the transformation are uncertain. In the case of points (lines are analogous but once again, complicated slightly by the necessity for two different representations) the transformation is given by $\mathbf{P} = \mathtt{T}\mathbf{p}$. Thus the uncertainty in $\mathbf{X} = [P_x/P_z, P_y/P_z]^\top$ depends on $\Lambda_{\mathbf{x}}$ and $\Lambda_{\mathbf{t}}$ the covariances of the image position $\mathbf{x} = [p_x/p_z, p_y/p_z]^\top$ and homography $\mathtt{T}$ respectively, and is given by

$$\Lambda_{\mathbf{X}} = \mathtt{D}\begin{bmatrix}\Lambda_{\mathbf{x}} & 0 \\ 0 & \Lambda_{\mathbf{t}}\end{bmatrix}\mathtt{D}^\top \quad \text{where} \quad \mathtt{D}(2 \times 10) = \frac{\partial \mathbf{X}}{\partial \{\mathbf{x}, \mathbf{t}\}}$$

5 Results

As indicated previously, the four lines of the six-yard area are used to register each image with the plan view, and the goal-posts are used to determine the vertical vanishing direction. The centre of the ball is picked manually with a mouse (and its uncertainty set to reflect the error introduced by this process). Figure 6 shows one pair from the sequence with the lines and ball position superimposed.

The rectified six-yard area is shown in figure 7a, with the *centre* of the ball, and its covariance, for the frame before the ball strikes the crossbar. The ellipse represents the 3σ limit, meaning there is, practically speaking, no chance that the centre of the ball is outside this ellipse. The transferred vertical lines used

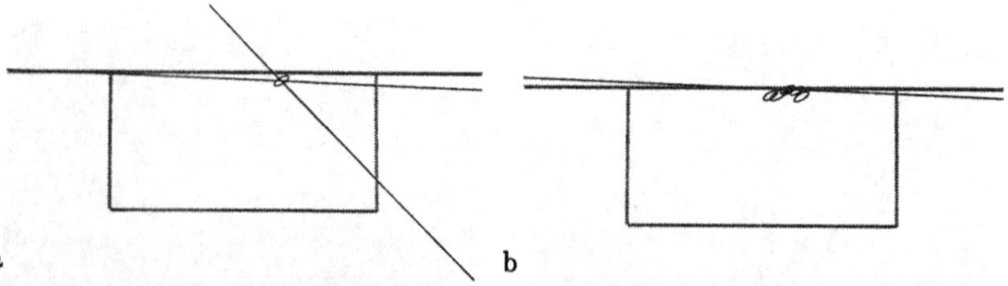

a b

Fig. 7. (a) One ball position on the rectified frame, together with its uncertainty ellipse. The top line is the *front* of the (finite width) goal line; (b) Uncertainty ellipses and constraint lines for the crucial frames of the sequence.

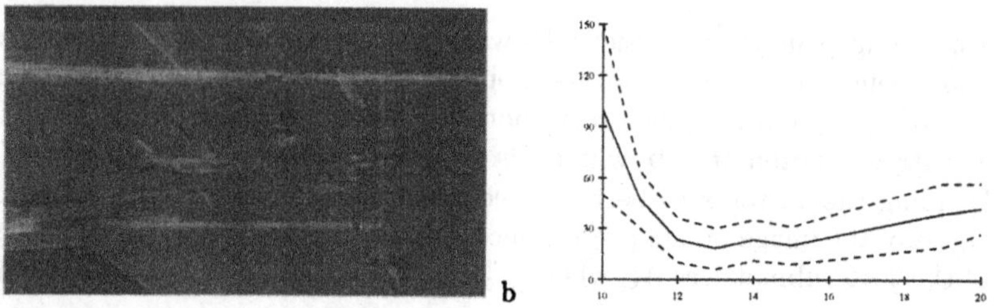

a b

Fig. 8. (a) The computed position of the ball throughout the sequence; (b) distance away from being a goal (in cm) versus frame number for the crucial frames of the sequence.

to compute the position are shown, with the ball position being their point of intersection. The uncertainty ellipses for all of the computed ball positions from just before the ball struck the crossbar to the point where it reenters play can be seen in figure 7b. One point of note here is that there are four frames in the middle of this sequence when a ball position cannot be computed because the goal-keeper obscures the ball in the three quarter view. This does not affect our ability to decide the question of whether it crossed the line or not, since we still have one constraint on the ball position: The transferred vertical line which represents this constraint has been drawn in the figure for the "missing" frames in which the ball strikes the ground, clearly showing that wherever the ball is placed along this line, we can still say with certainty that it was not across the line.

The complete set of computed ball positions from the moment Hurst shot for goal, to the point where it hit the crossbar, then the ground, and finally back out into play is shown in figure 8(a), rendered with the rectified texture (from the last image pair). The answer to the question, "did the ball cross the line?" must also take the ball radius into account, and a more quantitative analysis is given in the graph of figure 8(b) which shows the distance of the ball from being a goal (taking its radius into account) plotted against frame number. The dotted lines indicate three standard deviations from the estimate, thus a conservative estimate has the ball still 6cm from being a goal.

6 Conclusions

While it has been known for some time that 3D structure can be computed from uncalibrated views of a scene given sufficient correspondences in general position, this has rarely been used to answer specific, metric questions about the data. The approach taken here has been to make use of plane projective homographies to compute an overhead view of the action from a sequence of disparate image pairs. An alternative approach might have used virtual parallax, as described in [16], but this could only give a qualitative answer and would have required that the whole of the goal-mouth be visible in all frames. Another, and more convenient method, would have been to use an affine approximation to the imaging geometry — since fewer features would have been required. However, this approximation was found to be insufficiently accurate for these sequences as the images exhibit non-negligible perspective effects. Thus, although in the past the use of affine structure has proved fruitful for various active vision tasks [8, 14] such as tracking or visual servoing, it is not well suited to tackling quantitative measurements tasks unless the projection model truly is affine.

The application we have presented is one of a wider class of problems (such as traffic monitoring) in which the ground plane trajectory of a target is desired, but in which the camera calibration is difficult or impossible to obtain with sufficient accuracy. Sporting domains are often ground-plane orientated, with well known regular marking which can be used for registration, and so are particularly suited to this analysis [9].

While at the same time providing a compelling example of the power of uncalibrated techniques, this work has made a tangible contribution in settling, once and for all, the argument over whether or not the ball crossed the line in the most famous goal of all. In describing the closing seconds of the match, commentator Kenneth Wolstenholme said of celebrating English fans: *They think it's all over. It is now!* Nearly thirty years on, his words are once again appropriate.

Acknowledgements

We are very grateful to Adam Jones and John Davison of the Sunday Times for supplying the video sequences, and for their suggestions together with those of Mike Brooks, Roberto Cipolla and Peter Sander. Financial support was provided by Esprit Project VIVA, ACTS Project VANGUARD and a Glasstone Fellowship.

References

1. P. A. Beardsley, I. D. Reid, A. Zisserman, and D. W. Murray. Active visual navigation using non-metric structure. In *Proc. 5th Int'l Conf. on Computer Vision, Boston*, pages 58–65. IEEE Computer Society Press, 1995.
2. P. A. Beardsley, A. Zisserman, and D. W. Murray. Navigation using affine structure from motion. In *Proc. 3rd European Conf. on Computer Vision, Stockholm*, volume 2, pages 85–96, 1994.
3. S. Demey, A. Zisserman, and P. Beardsley. Affine and projective structure from motion. In D. Hogg and R. Boyle, editors, *Proc. 3rd British Machine Vision Conf., Leeds*, pages 49–58. Springer-Verlag, September 1992.

4. O. D. Faugeras. What can be seen in three dimensions with an uncalibrated stereo rig? In G. Sandini, editor, *Proc. 2nd European Conf. on Computer Vision, Santa Margharita Ligure, Italy*, pages 563–578. Springer-Verlag, 1992.
5. O.D. Faugeras. *Three-Dimensional Computer Vision.* MIT Press, 1993.
6. R. I. Hartley. Self-calibration from multiple views with a rotating camera. In *Proc. 3rd European Conf. on Computer Vision, Stockholm*, volume 1, pages 471–478, 1994.
7. R.I. Hartley, R. Gupta, and T. Chang. Stereo from uncalibrated cameras. In *Proc. of the IEEE Conf. on Computer Vision and Pattern Recognition*, pages 761–764, 1992.
8. N. Hollinghurst and R. Cipolla. Uncalibrated stereo hand/eye coordination. In J. Illingworth, editor, *Proc. 4th British Machine Vision Conf., Guildford*, pages 389–398. BMVA Press, 1993.
9. S. S. Intille and A. F. Bobick. Closed-world tracking. In *Proc. 5th Int'l Conf. on Computer Vision, Boston*, pages 672–678, 1995.
10. A. Jones and J. Davison. Sport: how science can end disputes. *Sunday Times*, 23 July, 1995.
11. K. Kanatani. *Statistical optimization for geometric computation: theory and practice.* AI Lab, Dept of Computer Science, Gunma University, Japan, 1995.
12. S.J. Maybank and O. Faugeras. A theory of self-calibration of a moving camera. *International Journal of Computer Vision*, 8(2):123–151, 1992.
13. L. Quan and R. Mohr. Towards structure from motion for linear features through reference points. In *Proc. IEEE Workshop on Visual Motion*, 1991.
14. I. D. Reid and D. W. Murray. Tracking foveated corner clusters using affine structure. In *Proc. 4th Int'l Conf. on Computer Vision, Berlin*, pages 76–83, Los Alamitos, CA, 1993. IEEE Computer Society Press.
15. I. D. Reid and A. Zisserman. Accurate metrology in uncalibrated video sequences. Technical report, Oxford University, Dept. of Engineering Science, 1996.
16. L. Robert and O. Faugeras. Relative 3d positioning and 3d convex hull computation from a weakly calibrated stereo pair. In *Proc. 4th Int'l Conf. on Computer Vision, Berlin*, pages 540–544, 1993.
17. P.H.S. Torr and D.W. Murray. Outlier detection and motion segmentation. In *Proc SPIE Sensor Fusion VI*, pages 432–443, Boston, September 1993.
18. Z. Zhang, Q.-T. Luong, and O. Faugeras. Motion of an uncalibrated stereo rig: self-calibration and metric reconstruction. Technical Report 2079, INRIA Sophia-Antipolis, October 1993.
19. A. Zisserman, P. A. Beardsley, and I. D. Reid. Metric calibration of a stereo rig. In *Proc. IEEE Workshop on Representations of Visual Scenes, Boston*, pages 93–100. IEEE Computer Society Press, 1995.

Reconstructing Polyhedral Models of Architectural Scenes from Photographs

Camillo J. Taylor and Paul E. Debevec and Jitendra Malik

EECS Department, U.C. Berkeley
Berkeley, CA 94720-1776
Fax: (510) 642 5775
email: {*camillo,debevec,malik*}*@cs.berkeley.edu*

Abstract. This paper presents a new image-based modeling method that facilitates the recovery of accurate polyhedral models of architectural scenes. The method is particularly effective because it exploits many of the constraints that are characteristic of architectural scenes. This work is placed in the context of the Façade project, whose goal is to use images to produce photo-realistic novel views of architectural scenes.

1 Introduction

The goal of our research is to develop a system that starts with multiple photographs of an architectural scene, such as King's College at Cambridge University, and produces photo-realistic images of the scene as it would appear from arbitrary virtual camera positions. Such a system has numerous potential applications: architectural walk-throughs, virtual tourism, virtual museums, and video games. Current methods for producing models of existing architectural sites are extremely labor intensive, error-prone, and do not produce photorealistic results. Our aim is to apply modern computer vision techniques to make the reconstruction problem tractable and photorealism achievable.

In the computer vision community, this problem statement evokes the use of techniques for scene reconstruction from multiple views. While the genesis of these techniques goes back to the photogrammetric literature, there has been a considerable development in the computer vision community under the topics of stereopsis and structure from motion. A good survey of the knowledge in this field up to 1992 is provided in Faugeras's book [Fau93]. Later work has focused on the use of uncalibrated cameras [FLR$^+$95].

Most of the work that has been done on recovering the geometry of a scene from multiple images tackles the problem in its most general form. Their goal is typically to recover the 3D positions of a set of features, points or lines, from multiple image measurements. The mathematics of reconstruction from multiple views of a set of points or lines does not depend on whether these points/lines constitute a disembodied cloud of features, or lie on a smooth surface, or lie on a well constrained volumetric primitive. This generality comes at a price–it is well known that the recovery of scene structure from multiple views is sensitive to noise. Long image sequences help [WHA89, WHA93, TK92], but we believe that at least part of the difficulty comes from the fact that the traditional formulation does not exploit all of the available constraints.

The main insight of this work is that architectural scenes can be well approximated as a collection of volumetric primitives linked together by specific geometrical relationships. Any remaining structure that is not captured in the coarse model can be expressed in terms of depth deviations from the coarse model. This suggests a two stage process for modeling the scene:

1. Recover a coarse model of the building – We have formulated this problem as one of recovering the parameters of a model supplied by the user rather than recovering the positions of individual points or lines. The advantage of modeling the scene with blocks instead of individual points and lines is that the scene can be represented with far fewer parameters (45 for the clock tower model in Fig. 1) which makes the reconstruction problem simpler. These models capture the constraints that are characteristic of architectural scenes such as rectangularity and parallelism.
2. Recover residual detail – Once we have obtained this coarse description of the scene, we apply a *model-based* stereo algorithm to pairs of images in our data set to recover the remaining geometric detail. Unlike previous approaches, we exploit the availability of a coarse model to simplify the problem of computing stereo-correspondences: the model is used to *pre-warp* the images in such a way that factors out foreshortening differences between the images and eliminates all image disparity except those resulting from deviations from the model.

Our approach splits the task of modeling from images into sub-tasks which are easy for a computer algorithm (but not a person), and sub-tasks which are easy for a computer algorithm (but not a person). The user chooses a parameterized model for the scene, which is precisely the type of information that would be difficult to recover automatically. Conversely, the computer recovers the parameters of the model from the image measurements which would be a very difficult task for a person. The geometric detail recovery is done by an automated stereo correspondence algorithm, which has been made feasible and robust by a pre-warping step based on the coarse geometric model. In this case, corresponding points must be computed for a dense sampling of image pixels, a job far too tedious to assign to a human, but feasible for a computer to perform using model-based stereo. Novel synthetic views of the scene are produced by combining the results of this two stage modeling process with the photometric information contained in the original images.

This paper describes the techniques we have developed to solve the first stage of our modeling process: to interactively obtain polyhedral models of architectural scenes from images. A complete description of the entire system can be found in [DTM96]. Section 2 of this paper describes the interactive polyhedral modeling program, Façade, along with the algorithms that have been developed to recover the parameters of a polyhedral model from image measurements. Some of the results obtained with the Façade system are presented in this section. In Section 3 we discuss our conclusions and future work.

2 The Interactive Modeling System

In this section we present *Façade*, a simple interactive modeling system that allows a user to construct a geometric model of a scene from a set of digitized photographs. In Façade, the user constructs a parameterized model of the scene and the program computes the parameters that best make the model conform to the photographs. We first describe the user interface, and then describe the model representation and algorithms involved.

2.1 User Interface

Constructing a geometric model of an architectural scene using Façade is an incremental process. Typically, the user selects a small number of photographs to begin with, and models the scene one piece at a time. The user may refine the model and include more images in the project until the model meets the desired level of detail.

Figure 1 shows the two types of windows used in the Façade program: image viewers and model viewers. The user supplies input to the program by instancing the components of the model, marking features of interest in the images, and indicating which features in the images correspond to which features in the model. Façade then computes the sizes and relative positions of the components of the model that best fit the features marked in the photographs.

Components of the model, called *blocks*, are parameterized geometric primitives such as boxes, prisms, and surfaces of revolution. A box, for example, is parameterized by its length, width, and height. The user models each part of the scene as such a primitive; the user may also create new classes of blocks if desired. What the user does not need to specify are the numerical values of the blocks' parameters; these parameters are recovered automatically from the digitized photographs.

The user may choose to constrain the sizes and positions of any of the blocks. In Figure 1, most of the blocks have been constrained to have equal length and width. Additionally, the four pinnacles have been constrained to have the same proportions. Blocks may also be placed in constrained relations to one other. For example, many of the blocks in Fig. 1 have been constrained to sit centered and on top of the block below. Such constraints are easily specified using a graphical 3D interface. When such constraints are given, they are automatically used by Façade to simplify the reconstruction problem.

The user marks edge features in the images using a simple point-and-click interface; features may be marked with sub-pixel accuracy by zooming into the images. Façade uses edge rather than point features since they are easier to localize and less likely to be completely obscurred. Only a section of any particular edge needs to be marked, so it is possible to make use of partially visible edges. Façade is able to compute accurate reconstructions with only a portion of the visible edges marked in any particular image, particularly when the user has provided constraints on the model.

Fig. 1. An image viewer and a model viewer from the Façade modeling system. The left window shows a single photograph of a clock tower, with features the user has marked shown in green. The right window shows the recovered model, the parameters of which have been computed by Façade's reconstruction algorithm. Note that only the left pinnacle has been marked - the remaining three (including one not visible) have been recovered by encoding symmetry into the model. Although Façade allows any number of images to be used, in this case constraints in the model have made it possible to recover 3D structure from a single photograph.

With the edges marked, the user needs to specify which features in the model correspond to which features in the images. This is accomplished by clicking on an edge in the model and then clicking the corresponding edge in one of the images. The user can rotate his view of the model into a perferred orientation to make the process of forming correspondences easier.

At any time, the user may instruct the computer to reconstruct the model. The computer then solves for the parameters of the model that cause it to align with the observed features in the images. During the reconstruction, the computer also determines and displays the locations from which the original photographs were taken.

2.2 Model Representation

In Façade, an architectural scene is represented as a set of polyhedral blocks. Each block has a set of parameters which define its size and shape. The coordinates of the vertices of these polyhedra with respect to the blocks internal frame of reference can be expressed as a linear function of the block's parameters. For example, in the wedge primitive shown in figure 2, the coordinates of the vertex P_o can be computed using the expression $P_o = (-width, -height, length)^T$. Each block has an associated bounding box, the maximum and minimum extent of this bounding box along each dimension can also be expressed as linear functions of the blocks parameters. For example, the minimum extent of the block shown in Figure 2 along the x dimension, $wedge_x^{MIN}$, can be computed from the expression $wedge_x^{MIN} = -width$;

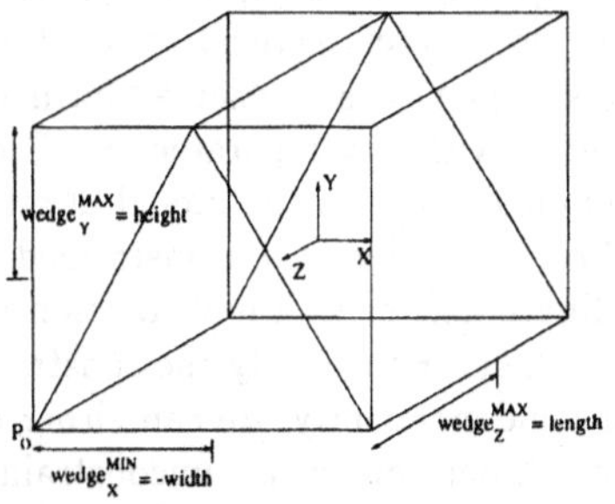

Fig. 2. A wedge primitive and its associated internal parameters.

Each class of block used in our modeling program is defined in a simple block template file. This file specifies the parameters of the blocks, provides the linear equations used to compute the vertices of the block and the limits of the bounding box, and defines how the vertices are connected by edges and faces. The user can add custom blocks to her project simply by creating appropriate template files.

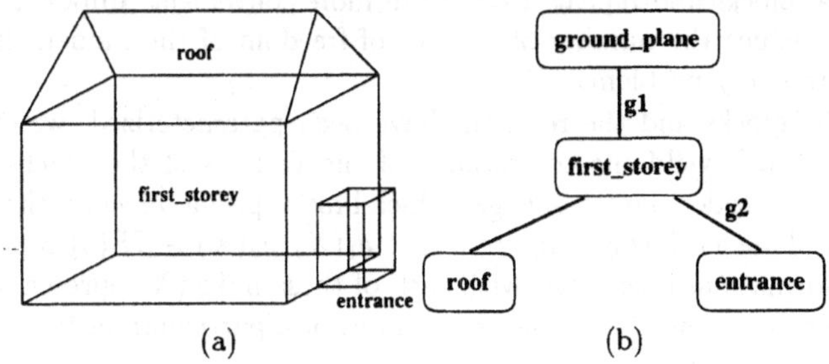

Fig. 3. (a) A simple geometric model of a building (b) The tree representation used in Façade. The nodes in this tree represent geometric primitives while the edges encode spatial relationships between the blocks.

The blocks in Façade are organized in a hierarchical tree structure as shown in figure 3(b). Each node of the tree represents an individual polyhedral block. The root of the tree represents the ground plane, which defines the world coordinate frame of reference. The links in the tree encode spatial relationships amongst the blocks - each edge defines how a particular block is situated with respect to its parent. Similar tree representations are used in a number of commercial computer graphics packages, such as SGI's Inventor, to model complicated objects in terms of simple geometrical primitives.

The relationships between blocks in this model can be represented in terms of a rotation matrix R and a translation vector t. This type of representation would usually involve 6 degrees of freedom, 3 to specify the rotation, and three for the translation. However, in architectural scenes the relationships between the blocks are often constrained to a particular form which means that they can typically be represented with fewer parameters. Facade allows the user to capture these constraints on R and t in the model. First, the rotation R can be represented in one of three ways: by an unconstrained rotation, as a rotation about a particular coordinate axis or as a null rotation i.e. $R = I$.

Second, Facade allows the user to specify the constraints on each component of the translation vector t independently. He can either choose to represent the translation along a given dimension by an unconstrained variable, or he can choose to specify how the bounding boxes of the two blocks are aligned along that dimension. For example, in order to ensure that the roof block in Figure 3 lies on top of the first storey block the user would specify that the maximum y extent of the first storey block should be aligned with the minimum y extent of the roof block. This would imply that the translation term along the y axis would be computed from the following expression $t_y = (first\_storey_y^{MAX} - roof_y^{MIN})$.

Each parameter of each instantiated block is actually a reference to a named symbolic variable, as illustrated in Figure 4. As a result, two parameters of different blocks (or of the same block) can be equated by having each parameter reference the same symbolic variable. This facility allows the user to specify that two or more of the dimensions in a model are the same, which makes modeling symmetrical blocks and repeated structure more convenient. Importantly, these constraints reduce the number of degrees of freedom of the model, simplifying the model recovery problem.

Once the blocks and the relations have been parameterized, we can derive expressions which yield the coordinates of the vertices of the blocks in world coordinates. Consider the set of edges which link a specific block in the model to the ground plane as shown in figure 3. Let $g_1(X), g_2(X) \in SE(3)$ represent the rigid transformations associated with each of these links (X represents a vector of the unknown parameters). The coordinates of a particular vertex $P_w(X)$ are given by the expression:

$$P_w(X) = g_1(X) g_2(X) P(X) \tag{1}$$

Similarly, the orientation of a particular line segment with respect to the ground plane $v_w(X)$ could be computed with:

$$v_w(X) = g_1(X) g_2(X) v(X) \tag{2}$$

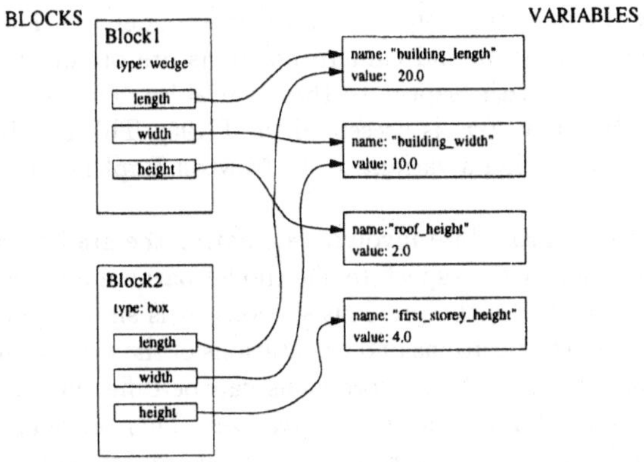

Fig. 4. Implementation of the dimensions of a block primitive in terms of symbol references. A single variable can be referenced in multiple places in the model, allowing constraints of symmetry.

In these equations, the point vectors, $P(X)$ and $P_w(X)$, and the orientation vectors, $v(X)$ and $v_w(X)$, are represented in homogeneous coordinates.

Modeling the scene with these polyhedral blocks, as opposed to points, line segments, surface patches, or polygons, offers a number of advantages which are listed below:

- Most buildings can be readily decomposed into a set of blocks.
- Blocks implicitly model common architectural constraints such as parallel lines and right angles.
- The user can conveniently manipulate the block primitives since they are at a suitably high level of abstraction, whereas individual features such as points and lines can be cumbersome.
- A surface model of the scene is readily obtained from the blocks, so there is no need to infer surfaces from discrete features.
- Modeling in terms blocks and relationships can greatly reduce the number of parameters that the reconstruction algorithm needs to recover. For example, the model in Fig. 1 is parameterized by just 45 variables. If each block in the scene were unconstrained (in its dimensions and position), it would require 240 parameters; if each line segment in the scene were treated independently, it would require 2896 parameters. This reduction in the number of parameters greatly enhances the robustness of the system relative to traditional structure from motion algorithms.

2.3 Reconstruction Algorithm

The reconstruction algorithm used in Façade is posed in terms of an objective function $\mathcal{O}$ which measures the disparity between the projected edges of the recovered model and the edges observed in the images, that is, $\mathcal{O} = \sum Err_i$ where

Err_i represents the disparity computed for correspondence i. Estimates for the unknown model parameters and camera positions are obtained by minimizing the objective function with respect to these variables. The error function used to measure the disparity Err_i is presented in [TK95]. This non-linear objective function is minimized using a variant of the Newton-Raphson method described in [TK94].

The minimization procedure involves calculating the gradient and hessian of the objective function with respect to the model parameters and camera positions. As we have shown in the previous sections, it is simple to construct symbolic expressions for the positions and orientations of the model edges in terms of the unknown parameters - these expressions can be differentiated symbolically to obtain expressions for the desired derivatives. This procedure is computationally inexpensive since the expressions given in equations 2 and 1 exhibit a particularly simple form.

The objective function described above is non-linear and can exhibit multiple local minima. If no initial estimate were available, the algorithm could converge to a local minimum instead of the global minimum. In order to overcome this problem we have developed effective methods for computing acceptable initial estimates for the model parameters and camera positions.

Before the non-linear minimization procedure described above is applied, two separate procedures are invoked to obtain initial estimates for the the unknown parameters. The first procedure recovers initial estimates for the camera rotations while the second provides initial estimates for the camera translations and the parameters of the model. Both of these procedures are based on techniques described in [TK95].

Once initial estimates have been obtained, the non-linear minimization over the entire parameter space is applied to produce a more accurate estimate for all of the unknown parameters. This stage typically reduces the error in the reconstruction by a factor of 2 to 4.

2.4 Results

Figure 1 showed the results of using Façade to reconstruct a clock tower from a single image. Figures 5 and 6 show the results of using Façade to reconstruct a high school building from twelve photographs. The photographs were taken with a calibrated Canon AE-1 35mm still camera with a standard 50mm lens and digitized with the PhotoCD process. Images at the 1536×1024 pixel resolution were processed to correct for lens distortion, then filtered down to 768×512 pixels for use in the modeling system. Fig. 5 shows that the recovered model conforms to the photographs to within a pixel, indicating an accurate reconstruction.

3 Conclusion

We have developed a two stage procedure, embodied in a system called Façade, for recovering geometric models of architectural scenes from photographs [DTM96].

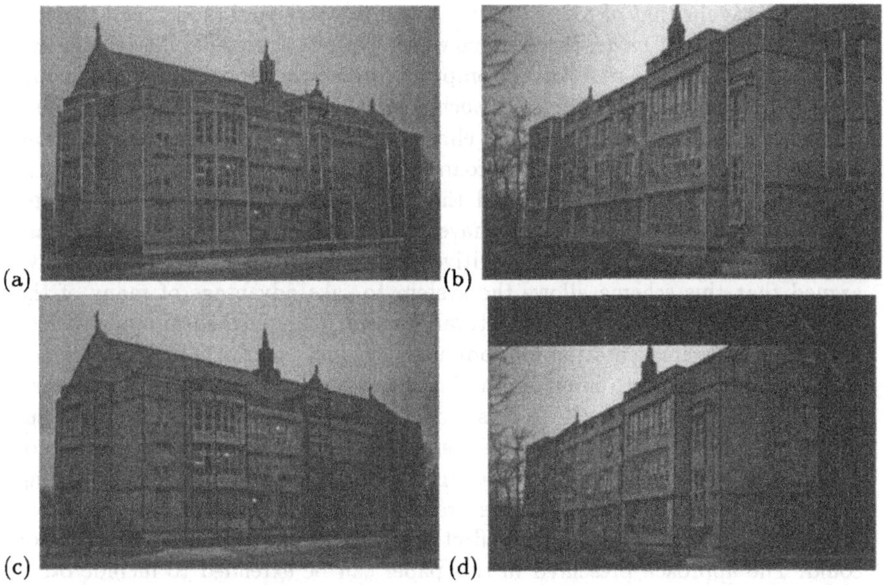

Fig. 5. Two of the twelve photographs used to reconstruct a high school building are shown in (a) and (b). The overlaid lines indicate the edges the user has marked. The edges of the reconstructed model, projected through the recovered camera positions and overlaid on the corresponding images are shown in (c) and (d). The recovered model conforms to the photographs to within a pixel in all twelve images, indicating that the building has been accurately reconstructed.

Fig. 6. Aerial view showing the recovered model and camera positions.

The first stage is that of reconstructing a coarse polyhedral model, and the second stage is that of *model-based stereo* which exploits the availability of a coarse model to simplify the problem of computing stereo-correspondences: the model is used to *pre-warp* the images in such a way that factors out foreshortening differences between the images and eliminates all image disparity except those resulting from deviations from the coarse model.

In this paper we have presented the first stage of the system: the polyhedral reconstruction algorithm. We have described how architectural scenes are modeled in terms of geometric primitives and spatial relationships and we have argued that this scheme allows the system to take advantage of many of the constraints inherent in the architectural domain. This polyhedral model is also a very natural representation for computer graphics and CAD programs which makes it much easier to interface the Façade system to these types of systems.

Allowing the user to specify this model to the system is an appropriate approach to this reconstruction problem since the human operator can easily provide a qualitative description based on her understanding of the scene being viewed. The Façade system can then recover the dimensions of the scene from image measurements much more effectively and accurately than the operator could. The approach presented in this paper can be extended to include other parameterized primitives, such as surfaces of revolution.

References

[DTM96] Paul E. Debevec, Camillo J. Taylor, and Jitendra Malik. Modeling and rendering architecture from photographs: A hybrid geometry- and image-based approach. Technical Report UCB//CSD-96-893, U.C. Berkeley, CS Division, January 1996.

[Fau93] Olivier Faugeras. *Three-Dimensional Computer Vision.* MIT Press, 1993.

[FLR+95] Olivier Faugeras, Stéphane Laveau, Luc Robert, Gabriella Csurka, and Cyril Zeller. 3-d reconstruction of urban scenes from sequences of images. Technical Report Rapport de recherche 2572, INRIA Sophia-Antipolis, June 1995.

[TK92] Carlo Tomasi and Takeo Kanade. Shape and motion from image streams under orthography: a factorization method. *International Journal of Computer Vision*, 9(2):137–154, November 1992.

[TK94] Camillo J. Taylor and David J. Kriegman. Minimization on the lie group so(3) and related manifolds. Technical Report 9405, Center for Systems Science, Dept. of Electrical Engineering, Yale University, New Haven, CT, April 1994.

[TK95] Camillo J. Taylor and David J. Kriegman. Structure and motion from line segments in multiple images. *IEEE Trans. Pattern Anal. Machine Intell.*, 17(11), November 1995.

[WHA93] Juyang Weng, Thomas S. Huang, and Narendra Ahuja. *Motion and Structure from Image Sequences.* Springer Series on Information Sciences. Springer-Verlag, Berlin, 1993.

[WHA89] J. Weng, T.S. Huang, and N. Ahuja. Motion and structure from two perspective views: Algorithms, error analysis, and error estimation. *IEEE Trans. Pattern Anal. Machine Intell.*, 11(5):451–476, May 89.

Structure from Motion (3)

Algebraic Varieties in Multiple View Geometry*

Anders Heyden and Kalle Åström

Dept of Mathematics, Lund University
Box 118, S-221 00 Lund, Sweden
email: heyden@maths.lth.se, kalle@maths.lth.se

Abstract. In this paper we will investigate the different algebraic varieties and ideals that can be generated from multiple view geometry with uncalibrated cameras. The natural descriptor, $\mathcal{V}_n$, is the image of $\mathcal{P}^3$ in $\mathcal{P}^2 \times \mathcal{P}^2 \times \cdots \times \mathcal{P}^2$ under n different projections. However, we will show that $\mathcal{V}_n$ is not a variety.
Another descriptor, the variety $\mathcal{V}_b$, is generated by all bilinear forms between pairs of views and consists of all points in $\mathcal{P}^2 \times \mathcal{P}^2 \times \cdots \times \mathcal{P}^2$ where all bilinear forms vanish. Yet another descriptor, the variety, $\mathcal{V}_t$, is the variety generated by all trilinear forms between triplets of views. We will show that when $n = 3$, $\mathcal{V}_t$ is a reducible variety with one component corresponding to $\mathcal{V}_b$ and another corresponding to the trifocal plane. In ideal theoretic terms this is called a primary decomposition. This settles the discussion on the connection between the bilinearities and the trilinearities.
Furthermore, we will show that when $n = 3$, $\mathcal{V}_t$ is generated by the three bilinearities and one trilinearity and when $n \geq 4$, $\mathcal{V}_t$ is generated by the $\binom{n}{2}$ bilinearities. This shows that four images is the generic case in the algebraic setting, because $\mathcal{V}_t$ can be generated by just bilinearities.

1 Introduction

When estimating structure and motion from an uncalibrated sequence of images, the bilinear and the trilinear constraints play an important role, see [4], [5], [6], [7], [12], [13] and [14]. One difficulty encountered when using these multilinear constraints is that they are not independent, some of them may be calculated from the others, see [2], [3] and [6]. Thus there is a need to investigate the relations between them and to determine the minimal number of multilinear constraints that are needed to generate all multilinear constraints. These multilinear functions have not before been studied using an ideal theoretic approach.

The simplest multilinear constraint is the bilinear constraint, described by the fundamental matrix between two views. The next step is to consider three images at the same instant. At this stage the so called trilinear functions appear, see [12], [13], [5] [4] and [6]. The coefficients of the trilinearities are elements of the so called trifocal tensor.

* This work has been supported by the ESPRIT project BRA EP 6448, VIVA, and the Swedish Research Council for Engineering Sciences (TFR), project 95-64-222.

The obvious extension of the trilinear constraints is to consider four or more images at the same instant. It turns out that there exist quadrilinear constraints between four different views, see [2] and [14]. However, these constraints follows from the trilinear ones, cf. [6], [2] and [14]. It also became apparent that multilinear constraints between more than four views contain no new information.

One strange thing encountered with the bilinearities and trilinearities is that given the three bilinearities, corresponding to three different views, it is in general possible to calculate the camera matrices and the trilinearities from the components of the fundamental matrices, as described in [6] and [9]. But algebraically the trilinear constraints do not follow from the bilinear ones in the following sense. Consider points on the trifocal plane. The bilinear constraints impose only the condition that the three image points are on the trifocal lines, but the trilinear constraints impose one further condition. The question is now how the fact that it is possible to calculate the trilinear constraints from the bilinear ones, via the camera matrices, correspond to the fact that the trilinear constraints do not follow algebraically from the bilinear ones, that is the trilinearities do not belong to the ideal generated by the bilinearities. This is the key question we will try to answer in this paper. We will try to clarify the meaning of the statement 'the trilinear constraints follows from the bilinear ones, when the camera does not move on a line', where the statement is right or wrong depending of what kind of operations we are allowed to do on the bilinearities. The statement is true if we are allowed to pick out coefficients from the bilinearities and use them to calculate camera matrices and then the trilinearities, but the statement is wrong if we are just allowed to make algebraic manipulations of the bilinearities, where the image coordinates are considered as variables.

In order to understand the relations between the bilinearities and the trilinearities we have to use some algebraic geometry and commutative algebra. A general reference for the former is [10] and for the latter [11].

2 Problem Formulation

Consider the following problem: Given n images taken by uncalibrated cameras of a rigid object, describe the possible locations of corresponding points in the different images. Throughout this paper it is assumed that the views are generic, i.e. the focal points are in general position. Mathematically, this can be formulated as follows. Let $\mathcal{P}^2$ and $\mathcal{P}^3$ denote the projective spaces of dimension 2 and 3 respectively. Denote points in $\mathcal{P}^3$ by $\mathbf{X} = (X, Y, Z, W)$ and points in the i:th $\mathcal{P}^2$ by $\mathbf{x}_i = (x_i, y_i, z_i)$. Let A_i, $i = 1, \ldots, n$ be projective transformations, that is linear transformations in projective coordinates, from $\mathcal{P}^3$ to $\mathcal{P}^2$,

$$A_i : \mathcal{P}^3 \ni \mathbf{X} \mapsto A_i\mathbf{X} \in \mathcal{P}^2, \quad i = 1, \ldots, n \ . \tag{1}$$

In (1) each A_i is described by a 3×4 matrix or rank 3. The mapping is undefined on the nullspace of this matrix, corresponding to the **focal point**, f_i, of camera i, that is $A_i f_i = 0$. This can be regarded as one transformation,

$\Phi_n = (A_1, A_2, \ldots, A_n)$, from $\dot{\mathcal{P}}^3$ to $\mathcal{P}^2 \times \mathcal{P}^2 \times \cdots \times \mathcal{P}^2 = (\mathcal{P}^2)^n$,

$$\Phi_n : \dot{\mathcal{P}}^3 \ni \mathbf{X} \mapsto (A_1\mathbf{X}, A_2\mathbf{X}, \ldots, A_n\mathbf{X}) \in (\mathcal{P}^2)^n \ , \tag{2}$$

where $\dot{\mathcal{P}}^3 = \mathcal{P}^3 \setminus \{f_1, f_2, \ldots, f_n\}$, that is $\mathcal{P}^3$ with the camera centres omitted. This removal of a finite set of points from $\mathcal{P}^3$ gives a quasiprojective variety, i.e. an open subset of a projective variety. We want to describe the range of Φ_n as a subset of $(\mathcal{P}^2)^n$.

It can be shown, see [10], that $(\mathcal{P}^2)^n$ is indeed a projective variety. It can be embedded in $\mathcal{P}^{3^n-1}$ as a projective subvariety, using the Segre embedding. We will call this projective subvariety $\mathcal{S}_n$, and think of it as n copies of $\mathcal{P}^2$, and do not bother about the actual embedding. However, it is essential to know that $(\mathcal{P}^2)^n$ is indeed a projective variety.

Moreover, this fact has a very important implication on the functions and ideals generating varieties in $(\mathcal{P}^2)^n$. These functions must be homogeneous of the same degree in every triplet of variables corresponding to a factor $(\mathcal{P}^2)$, see [10], pp. 56. For example, there is no meaning in asking the question if the bilinear constraint between two images is contained in some ideal generating a variety in $(\mathcal{P}^2)^3$ for three images. The reason for this is that variables from the third image are not present in the bilinearity between the first two images. Thus this bilinear constraint is not homogeneous of the same degree in every triplet of variables. This difficulty will be overcome in the sequel by considering every multilinear constraint in its homogenised forms, and when we speak of generators of an ideal, describing a variety in $(\mathcal{P}^2)^n$, we implicitly assume that the generators are replaced by their homogenised equivalents.

2.1 Choice of Coordinates

Since we are only interested in algebraic relations between different ideals, we have the freedom to choose coordinates in $\mathcal{P}^3$ and in every $\mathcal{P}^2$ as we like. Consider the n projective transformations A_i in (1) and the n focal points $f_i \in \mathcal{P}^3$. We choose coordinates in $\mathcal{P}^3$ such that the first five points f_i constitute a projective basis with coordinates $f_1 = (0,0,0,-1)$, $f_2 = (1,0,0,-1)$, $f_3 = (0,1,0,-1)$, $f_4 = (0,0,1,-1)$ and $f_5 = (1,1,1,-1)$, where the minus sign in the fourth component will be convenient later. Furthermore we choose coordinates in each $\mathcal{P}^2$ such that the first three columns of each A_i are the columns of the identity matrix. This means that the projection matrices can be written

$$\begin{aligned} A_1 &= \begin{bmatrix} 1&0&0&0\\ 0&1&0&0\\ 0&0&1&0 \end{bmatrix}, \quad A_2 = \begin{bmatrix} 1&0&0&1\\ 0&1&0&0\\ 0&0&1&0 \end{bmatrix}, \quad A_3 = \begin{bmatrix} 1&0&0&0\\ 0&1&0&1\\ 0&0&1&0 \end{bmatrix}, \\ A_4 &= \begin{bmatrix} 1&0&0&0\\ 0&1&0&0\\ 0&0&1&1 \end{bmatrix}, \quad A_5 = \begin{bmatrix} 1&0&0&1\\ 0&1&0&1\\ 0&0&1&1 \end{bmatrix}, \quad A_n = \begin{bmatrix} 1&0&0&a_n\\ 0&1&0&b_n\\ 0&0&1&c_n \end{bmatrix}, \quad n \geq 6 \ , \end{aligned} \tag{3}$$

where $A_n f_n = 0$ with $f_n = (-a_n, -b_n, -c_n, 1)$, $n \geq 6$. This coordinate system chosen in (3) will be called a **normalised coordinate system** for the multiple view geometry. This choice of coordinates can be done if the matrices A_i are assumed to be in general position.

The **epipole**, $e_{i,j}$, from camera j in image i is defined by $e_{i,j} = A_i f_j$. For example, with our choice of coordinates, $e_{1,2} = (1,0,0)$, $e_{1,3} = (0,1,0)$, $e_{2,1} = (1,0,0)$, $e_{2,3} = (1,-1,0)$, $e_{3,1} = (0,1,0)$ and $e_{3,2} = (1,-1,0)$. The **trifocal plane**, $TP_{i,j,k}$, for images i, j and k is the plane containing f_i, f_j and f_k. For example, with our choice of coordinates, $TP_{1,2,3}$ is described by $Z = 0$. The **epipolar line**, $EL_{i,j}$, is the line in $\mathcal{P}^3$ containing f_i and f_j. The **trifocal line**, $tl_{i,j,k}$, in image i from the triplet of images i, j and k is the intersection of the trifocal plane, $TP_{i,j,k}$, and image plane i. With our choice of coordinates, $tl_{1,2,3}$ is described by $z_1 = 0$, $tl_{2,1,3}$ by $z_2 = 0$ and $tl_{3,1,2}$ by $z_3 = 0$.

2.2 Multilinear Forms

Consider the equations, obtained from (1),

$$A_i \mathbf{X} = \lambda_i \mathbf{x}_i, \quad i = 1, \dots n \ , \tag{4}$$

where the λ_i:s are needed because of the homogeneity of the coordinates. These equations can be written

$$Mu = \begin{bmatrix} A_1 & \mathbf{x}_1 & 0 & 0 & \dots & 0 \\ A_2 & 0 & \mathbf{x}_2 & 0 & \dots & 0 \\ A_3 & 0 & 0 & \mathbf{x}_3 & \dots & 0 \\ \vdots & \vdots & \vdots & \vdots & \ddots & \vdots \\ A_n & 0 & 0 & 0 & \dots & \mathbf{x}_n \end{bmatrix} \begin{bmatrix} \mathbf{X} \\ -\lambda_1 \\ -\lambda_2 \\ -\lambda_3 \\ \vdots \\ -\lambda_n \end{bmatrix} = \begin{bmatrix} 0 \\ 0 \\ 0 \\ \vdots \\ 0 \end{bmatrix} \ . \tag{5}$$

Since M has a nontrivial nullspace, it follows that

$$\operatorname{rank}(M) \leq 3 + n \ . \tag{6}$$

The matrix M contains one block with three rows for each image. The **bilinear constraints** for two images are obtained by taking a subdeterminant containing all three rows for the two images and the corresponding nonzero columns. The **trilinear constraints** for three images are obtained from subdeterminants containing three rows from one of the three images and two rows from each of the other two images and the corresponding nonzero columns. The **quadrilinear constraints** are obtained from subdeterminants containing two rows from each of the four images and the corresponding nonzero columns. Observe that all determinants of $(n+4) \times (n+4)$ submatrices are multihomogeneous of degree $(1,1,1,\dots,1)$, that is of the same degree in every triplet of image coordinates. For example the bilinearity, $b_{1,2}$ between image 1 and 2 can be obtained as $x_3 x_4 \dots x_n b_{1,2}$. This formulation is the same as the one used by Triggs, in [14] and is equivalent to the one used in [6].

2.3 The Varieties

In the sequel we are going to investigate the following subsets of $\mathcal{S}_n$:

Definition 1. The **natural descriptor**, $\mathcal{V}_{\mathrm{n}}$, is the range of Φ_n in (2), i.e. $\mathcal{V}_{\mathrm{n}} = \Phi_n(\dot{\mathcal{P}}^3) \subseteq \mathcal{S}_n$. □

Definition 2. The **bilinear descriptor** or **bilinear variety** $\mathcal{V}_{\mathrm{b}}$, is defined as the projective subvariety in $\mathcal{S}_n$, generated by all bilinear constraints. □

Definition 3. The **trilinear descriptor** or **trilinear variety**, $\mathcal{V}_{\mathrm{t}}$, is defined as the projective subvariety in $\mathcal{S}_n$, generated by all trilinear constraints. □

These definitions raise several questions. It is obvious that $\mathcal{V}_{\mathrm{b}}$ and $\mathcal{V}_{\mathrm{t}}$ are projective subvarieties, since they are defined by homogeneous polynomials. $\mathcal{V}_{\mathrm{n}}$ is a constructible set, see [10], i.e. a rational image of a quasiprojective variety, but is it a variety?. How can these varieties be described as the set of zeros to an ideal of polynomials? Are they irreducible? If not, what are the irreducible components? What are the connections between them? We will answer these questions later.

It is also possible to generate a variety by combining the bilinear and trilinear forms. The projective subvariety, $\mathcal{V}_{\mathrm{bt}}$, in $\mathcal{S}_n$, generated by all bilinear and trilinear constraints, is called the **bitrilinear variety**. It follows that $\mathcal{V}_{\mathrm{bt}} = \mathcal{V}_{\mathrm{b}} \cap \mathcal{V}_{\mathrm{t}}$. When more than three images are available it is possible to generate a variety from the quadrilinear constraints. The projective subvariety, $\mathcal{V}_{\mathrm{q}}$, in $\mathcal{S}_n$, generated by all quadrilinear constraints, is called the **quadrilinear variety**. Again, it is obvious that $\mathcal{V}_{\mathrm{bt}}$ and $\mathcal{V}_{\mathrm{q}}$ are projective subvarieties and we can of course ask the same questions about connections and of irreducibility.

3 Two Images

Things start to be complicated already in the case of two images. Consider Φ_2 in (2) for $n = 2$. If we make a suitable restriction of Φ_2, we get the following well known theorem, see [1].

Theorem 4 (Fundamental theorem of epipolar geometry). *The mapping*

$$\tilde{\Phi}_2 : \mathcal{P}^3 \setminus \{EL_{1,2}\} \ni \mathbf{X} \mapsto (A_1\mathbf{X}, A_2\mathbf{X}) \in (\mathcal{P}^2 \setminus \{e_{1,2}\}) \times (\mathcal{P}^2 \setminus \{e_{2,1}\}) \qquad (7)$$

is a birational map between the quasiprojective varieties $\mathcal{P}^3 \setminus \{EL_{1,2}\}$ *and* $((\mathcal{P}^2 \setminus \{e_{1,2}\}) \times (\mathcal{P}^2 \setminus \{e_{2,1}\})) \cap \mathcal{V}_{\mathrm{b}}$.

Proof. $((\mathcal{P}^2 \setminus \{e_{1,2}\}) \times (\mathcal{P}^2 \setminus \{e_{2,1}\})) \cap \mathcal{V}_{\mathrm{b}}$ is a quasiprojective variety since it is the intersection of two quasiprojective varieties. Obviously, $\tilde{\Phi}_2$ is rational and surjective. It remains to prove that its inverse exists and is rational. This can be seen from the fact that given a point $(\mathbf{x}_1, \mathbf{x}_2)$ in the range of $\tilde{\Phi}_2$, it is possible to reconstruct the point in $\mathcal{P}^3 \setminus \{EL_{1,2}\}$ by intersecting the rays $A_1^{-1}(\mathbf{x}_1)$ and $A_2^{-1}(\mathbf{x}_2)$. The reconstructed point can be written as the intersection of these lines. This is clearly a rational map, which concludes the proof. □

The mapping, $\tilde{\Phi}_2$, in (7) is called the **birational restriction** of Φ_2. Note that the inverse image of Φ_2 is 1-dimensional at the point $(e_{1,2}, e_{2,1})$, because every point on the epipolar line projects to $(e_{1,2}, e_{2,1})$. This means that Φ_2 is not bijective between $\mathcal{V}_b$ and $\dot{\mathcal{P}}^3$.

We now turn to the natural descriptor, $\mathcal{V}_n \subseteq \mathcal{P}^2 \times \mathcal{P}^2$. It consists of the following pairs of points:

- one arbitrary point, (x_1, y_1, z_1), in the first $(\mathcal{P}^2 \setminus \{e_{1,2}\})$ and one point, (x_2, y_2, z_2) in the second $(\mathcal{P}^2 \setminus \{e_{2,1}\})$ on the line $b_{1,2} = 0$ (or vice versa),
- the epipole $e_{1,2}$ in the first $\mathcal{P}^2$ and the epipole $e_{2,1}$ in the second $\mathcal{P}^2$,

corresponding to images of points not on the epipolar line, and points on the epipolar line except the focal points.

Next, we turn to the bilinear variety, $\mathcal{V}_b$, generated by the bilinear forms. For two images there is just one bilinear form,

$$b_{1,2}(x_1, y_1, z_1, x_2, y_2, z_2) = y_1 z_2 - z_1 y_2 \ . \tag{8}$$

The projective subvariety $\mathcal{V}_b \subseteq \mathcal{P}^2 \times \mathcal{P}^2$ generated by $b_{1,2}$ consists of the following pairs of points:

- one arbitrary point, (x_1, y_1, z_1), in the first $(\mathcal{P}^2 \setminus \{e_{1,2}\})$ and one point, (x_2, y_2, z_2) in the second $(\mathcal{P}^2 \setminus \{e_{2,1}\})$ on the line $b_{1,2} = 0$ (or vice versa),
- the epipole $e_{1,2}$ in the first $\mathcal{P}^2$ and an arbitrary point in the second $\mathcal{P}^2$,
- the epipole $e_{2,1}$ in the second $\mathcal{P}^2$ and an arbitrary point in the first $\mathcal{P}^2$.

This includes also the point $(e_{1,2}, e_{2,1})$, consisting of the two epipoles. This shows the following theorem

Theorem 5. *For two images we have, with strict inclusion,* $\mathcal{V}_n \subset \mathcal{V}_b$.

We now return to the natural descriptor $\mathcal{V}_n \subseteq \mathcal{P}^2 \times \mathcal{P}^2$. Consider the ideal $\mathcal{I}_n = \mathcal{I}(\mathcal{V}_n) \subseteq \mathbb{R}[x_1, y_1, z_1, x_2, y_2, z_2]$, where $\mathcal{I}(\mathcal{V})$ denotes the ideal generated by all polynomial functions that vanish at all points in $\mathcal{V}$. Thus $\mathcal{I}_n$ is the ideal generated by the polynomial functions in $(x_1, y_1, z_1, x_2, y_2, z_2)$ that vanish at all points in $\mathcal{V}_n$. Since $\mathcal{V}_n$ is a subset of $\mathcal{P}^2 \times \mathcal{P}^2$, these functions must be bihomogeneous of the same degree in (x_1, y_1, z_1) and (x_2, y_2, z_2), see [10], pp. 56. The bilinearity in (8), of degree $(1, 1)$, generates $\mathcal{I}_n$ according to the following lemma.

Lemma 6. *The ideal $\mathcal{I}_n$, generated by the natural descriptor, can be described as*

$$\mathcal{I}_n = \mathcal{I}_b = (b_{1,2}) \ .$$

Proof. Use (4) for $n = 2$ and eliminate the scale factors, λ_i, and object coordinates, $\mathbf{X}$. For details, see [8]. □

Remark. This means that the closure of $\mathcal{V}_n$ in the Zariski topology equals $\mathcal{V}_b$ since $\mathcal{V}(\mathcal{I}(\mathcal{V}_n))$ is the closure of $\mathcal{V}_n$. □

Theorem 7. *The natural descriptor $\mathcal{V}_n \in \mathcal{P}^2 \times \mathcal{P}^2$ is not a variety, in the sense that it can not be described as the common zeroes to a system of polynomial equations.*

Proof. Every variety, $\mathcal{V}$, fulfils $\mathcal{V} = \mathcal{V}(\mathcal{I}(\mathcal{V}))$. However, Theorem 5 and Lemma 6 show that $\mathcal{V}_n \subset \mathcal{V}_b = \mathcal{V}(\mathcal{I}(\mathcal{V}_n))$, with strict inclusion. □

Remark. We can also describe $\mathcal{V}_b$ as the range of an extension of the map Φ_2 to a multivalued map $\hat{\Phi}_2$ from the whole $\mathcal{P}^3$ defined as

$$\hat{\Phi}_2(\mathbf{X}) = \begin{cases} (A_1\mathbf{X}, A_2\mathbf{X}) & ; A_1\mathbf{X} \neq 0,\ A_2\mathbf{X} \neq 0 \\ \{(A_1\mathbf{X}, \mathbf{x}_2) \,|\, \mathbf{x}_2 \in \mathcal{P}^2\} & ; A_2\mathbf{X} = 0 \\ \{(\mathbf{x}_1, A_2\mathbf{X}) \,|\, \mathbf{x}_1 \in \mathcal{P}^2\} & ; A_1\mathbf{X} = 0\ . \end{cases} \tag{9}$$

Then the range of $\hat{\Phi}_2$ equals exactly the variety, $\mathcal{V}_b$, generated by the bilinear constraint. □

It is obvious that $\mathcal{V}_b$ is irreducible because it is generated by a single irreducible polynomial. Thus we have answered all questions raised above for the two image case.

4 Three Images

4.1 Varieties

Consider Φ_3 in (2) for $n = 3$. Making a suitable restriction of Φ_3, we get the following well known theorem.

Theorem 8 (Fundamental theorem of trifocal geometry). *The mapping*

$$\tilde{\Phi}_3 : \begin{cases} \mathcal{P}^3 \setminus \{TP_{1,2,3}\} \to (\mathcal{P}^2 \setminus \{tl_{1,2,3}\}) \times (\mathcal{P}^2 \setminus \{tl_{2,1,3}\}) \times (\mathcal{P}^2 \setminus \{tl_{3,1,2}\}) \\ \mathbf{X} \mapsto (A_1\mathbf{X}, A_2\mathbf{X}, A_3\mathbf{X}) \end{cases} \tag{10}$$

is a birational map between the quasiprojective varieties $\mathcal{P}^3 \setminus \{TP_{1,2,3}\}$ and $((\mathcal{P}^2 \setminus \{tl_{1,2,3}\}) \times (\mathcal{P}^2 \setminus \{tl_{2,1,3}\}) \times (\mathcal{P}^2 \setminus \{tl_{3,1,2}\})) \cap \mathcal{V}_b$.

Proof. In the same way as Theorem 7, see [8]. □

The mapping, $\tilde{\Phi}_3$, in (10) is called the **birational restriction** of Φ_3. It is always possible to reconstruct a point in $((\mathcal{P}^2 \setminus \{e_{1,2}, e_{1,3}\}) \times (\mathcal{P}^2 \setminus \{e_{2,1}, e_{2,3}\}) \times (\mathcal{P}^2 \setminus \{e_{3,1}, e_{3,2}\})) \cap \mathcal{V}_b$, which is in the range of Φ_3, using a rational map by intersecting two lines. However, we can not in advance choose a function for this. For example, if the point $\mathbf{X}$ is on the epipolar line between image 1 and 2 we can not use just A_1 and A_2 to make a reconstruction. We have to use A_3 also. This indicates why it is impossible to find an inverse rational map which would give a birational equivalence between $\dot{\mathcal{P}}^3$ and $\mathcal{V}_n$.

The natural descriptor $\mathcal{V}_n$ consists of the following points:

- one arbitrary point in the first $(\mathcal{P}^2 \setminus \{tl_{1,2,3}\})$, one point in the second $(\mathcal{P}^2 \setminus \{tl_{2,1,3}\})$ on the line $b_{1,2} = 0$ and one point in the third $(\mathcal{P}^2 \setminus \{tl_{3,1,2}\})$ on the intersection between the lines $b_{1,3} = 0$ and $b_{2,3} = 0$ (and any permutation of the three images),
- one arbitrary point on $(tl_{1,2,3} \setminus \{e_{1,3}, e_{1,2}\})$ in the first $\mathcal{P}^2$, one arbitrary point on $(tl_{2,1,3} \setminus \{e_{2,3}, e_{2,1}\})$ in the second $\mathcal{P}^2$ and the unique point on $(tl_{3,1,2} \setminus \{e_{3,1}, e_{3,2}\})$ in the third $\mathcal{P}^2$ given by the trilinear constraints or as a projection of the reconstructed point from image 1 and image 2 onto the third image (and any permutation of the three images),
- the epipole $e_{1,3}$ in the first $\mathcal{P}^2$, the epipole $e_{2,3}$ in the second $\mathcal{P}^2$ and an arbitrary point on the trifocal line $tl_{3,1,2}$ in the third $(\mathcal{P}^2 \setminus \{e_{3,1}, e_{3,2}\})$ (and any permutation of the three images),

corresponding to images of points not on the trifocal plane, points in the trifocal plane, not on an epipolar line and points on the epipolar lines.

Consider the variety $\mathcal{V}_t \subseteq \mathcal{P}^2 \times \mathcal{P}^2 \times \mathcal{P}^2$ generated by the trilinear forms. The projective subvariety in $\mathcal{P}^2 \times \mathcal{P}^2 \times \mathcal{P}^2$ corresponding to these forms is given by the same points as $\mathcal{V}_n$ plus the following triplets of points:

- the epipole $e_{1,3}$ in the first $\mathcal{P}^2$, the epipole $e_{2,3}$ in the second $\mathcal{P}^2$ and an arbitrary point, (x_3, y_3, z_3), in the third $\mathcal{P}^2$ (and any permutation of the three images).

Remark. Just as in the case of two images we can also describe this variety, $\mathcal{V}_t$ as the range of an extension of the map Φ_3 to a multivalued map $\hat{\Phi}_3$ from the whole $\mathcal{P}^3$ defined as the obvious extension of (9). Then the range of $\hat{\Phi}_3$ equals exactly the variety, $\mathcal{V}_t$, generated by the trilinear constraints. □

We now turn to the variety, $\mathcal{V}_b$, generated by the bilinear forms. In this case each image pair contributes with a bilinear form,

$$b_{1,2} = y_1 z_2 - z_1 y_2, \; b_{1,3} = x_1 z_3 - z_1 x_3, \; b_{2,3} = (x_2 + y_2) z_3 - z_2 (x_3 + y_3) \; . \quad (11)$$

The projective subvariety $\mathcal{V}_b \subseteq \mathcal{P}^2 \times \mathcal{P}^2 \times \mathcal{P}^2$ generated by these forms consists of the same triplets of points as $\mathcal{V}_t$ plus the following triplets of points:

- one arbitrary point on $tl_{1,2,3}$ in the first $\mathcal{P}^2$, one arbitrary point on $tl_{2,1,3}$ in the second $\mathcal{P}^2$ and one arbitrary point on $tl_{3,1,2}$ in the third $\mathcal{P}^2$.

Theorem 9. *For three images we have, with strict inclusions,*

$$\mathcal{V}_n \subset \mathcal{V}_t \subset \mathcal{V}_b \; . \quad (12)$$

Furthermore, the variety $\mathcal{V}_b$ is reducible and can be written as a union of two irreducible varieties as

$$\mathcal{V}_b = \mathcal{V}_t \cup \mathcal{V}_{tp} \; , \quad (13)$$

where $\mathcal{V}_{tp}$ is the variety containing one point on each trifocal line.

One consequence of this theorem is that the bitrilinear variety, $\mathcal{V}_{\rm bt}$, and the trilinear variety, $\mathcal{V}_t$, coincide, i.e. $\mathcal{V}_{\rm bt} = \mathcal{V}_t$. In fact, it follows from Theorem 9 that $\mathcal{V}_t \subset \mathcal{V}_b$. Since $\mathcal{V}_{\rm bt} = \mathcal{V}_t \cap \mathcal{V}_b$, from the definition of $\mathcal{V}_{\rm bt}$, it follows that $\mathcal{V}_{\rm bt} = \mathcal{V}_t$.

We now return to the natural descriptor $\mathcal{V}_{\rm n} \subsetneq \mathcal{P}^2 \times \mathcal{P}^2 \times \mathcal{P}^2$. Consider the ideal $\mathcal{I}_{\rm n} = \mathcal{I}(\mathcal{V}_{\rm n}) \subsetneq \mathbb{R}[x_1, y_1, z_1, x_2, y_2, z_2, x_3, y_3, z_3]$. Since $\mathcal{V}_{\rm n}$ is a subset of $\mathcal{P}^2 \times \mathcal{P}^2 \times \mathcal{P}^2$, the functions in $\mathcal{I}_{\rm n}$ must be trihomogeneous in (x_1, y_1, z_1), (x_2, y_2, z_2) and (x_3, y_3, z_3), see [10], pp. 56. The bilinearities in (11) are functions of degree $(1,1,0)$, $(1,0,1)$ and $(0,1,1)$, which can be extended to functions of degree $(1,1,1)$ as described above. There are also trilinear functions of degree $(1,1,1)$. We have the following lemma, which states that the closure of $\mathcal{V}_{\rm n}$ is $\mathcal{V}_{\rm t}$, and theorem (for the proofs, see [8]).

Lemma 10. *The ideal defined by the natural descriptor, can be described as*

$$\mathcal{I}_{\rm n} = \mathcal{I}(\mathcal{V}_{\rm t}) = \mathcal{I}_{\rm t} \ .$$

Theorem 11. *The natural descriptor $\mathcal{V}_{\rm n} \subsetneq \mathcal{P}^2 \times \mathcal{P}^2 \times \mathcal{P}^2$ is not a variety, in the sense that it can not be described as the common zeroes to a system of polynomial equations.*

4.2 Ideals

All multilinear constraints are obtained from (5) with $n = 3$. The trilinearities are obtained as subdeterminants involving at least two rows from each image, for example

$$\begin{aligned} t_{5,7} &= y_1 z_2 z_3 + z_1 x_2 z_3 - x_1 z_2 z_3 - z_1 z_2 y_3 \\ t_{5,9} &= x_1 z_2 x_3 + x_1 z_2 y_3 - y_1 z_2 x_3 - z_1 x_2 x_3 \\ t_{6,7} &= y_1 x_2 z_3 + y_1 y_2 z_3 - x_1 y_2 z_3 - z_1 y_2 y_3 \\ t_{6,9} &= x_1 y_2 x_3 + x_1 y_2 y_3 - y_1 x_2 x_3 - y_1 y_2 x_3 \ , \end{aligned} \tag{14}$$

where $t_{i,j}$ denotes the subdeterminant obtained after removing rows i and j.

We will now describe the relations between the ideals generated by the trilinearities and the bilinearities. The ideal, $\mathcal{I}_{\rm b}$, in $\mathbb{R}[x_1, y_1, z_1, x_2, y_2, z_2, x_3, y_3, z_3]$, generated by the bilinearities in (11), is called the **bilinear ideal**. In the same way, the ideal, $\mathcal{I}_{\rm t}$, generated by the trilinearities, is called the **trilinear ideal**. Finally, the ideal, $\mathcal{I}_{\rm bt}$, generated by the bilinearities and the trilinearities, is called the **bitrilinear ideal**.

First we are going to study different ways of generating $\mathcal{I}_{\rm b}$ and $\mathcal{I}_{\rm t}$. It is obvious that $\mathcal{I}_{\rm b}$ is generated by the 3 bilinearities, but that no 2 of them are sufficient to generate $\mathcal{I}_{\rm b}$. Things are more complicated for $\mathcal{I}_{\rm t}$. Although the trilinear constraint, locally, can be written as the vanishing of 3 trilinear forms among all trilinearities we need 4 forms to generate $\mathcal{I}_{\rm t}$. Consider first the following simple example, which reveals the difference between a minimal generating set and the codimension of the corresponding variety.

Example 1. The condition that two vectors, $u = (a, b, c)$ and $v = (d, e, f)$, in $\mathbb{R}^3$ are parallel can be written rank $\left[\begin{smallmatrix} a & b & c \\ d & e & f \end{smallmatrix}\right] < 2$, which is equivalent to $p_1 = p_2 = p_3 = 0$, where $p_1 = ae - bd$, $p_2 = bf - ce$ and $p_3 = cd - af$. Introduce the ideal $\mathcal{I}_{\text{ex}} = (p_1, p_2, p_3) \subseteq \mathbb{R}[a, b, c, d, e, f]$. The codimension of the variety $\mathcal{V}(\mathcal{I}_{\text{ex}})$ is 2, since the rank condition above can locally be obtained from 2 polynomial equations. This can be seen from $fp_1 + dp_2 + ep_3 = 0$. However, it is not possible to generate the ideal (p_1, p_2, p_3) by any two of the polynomials, p_1 and p_2, for example, because $u = (1, 0, 1)$ and $v = (1, 0, 2)$ obeys both $p_1 = 0$ and $p_2 = 0$ but $p_3 = -1$. This means that the codimension of the variety $\mathcal{V}(\mathcal{I}_{\text{ex}})$ is 2 and $\{p_1, p_2, p_3\}$ is a minimal generating set for $\mathcal{I}_{\text{ex}}$. □

Theorem 12. *The ideal $\mathcal{I}_{\text{t}}$ can be generated by the bilinearities and one trilinearity, $t_{6,9}$. $\mathcal{I}_{\text{t}}$ can not be generated by three multilinear functions.*

Proof. Using Gröbner basis calculations, see [8]. □

Remark. Using our choice of coordinates, the trilinearity needed apart from the bilinearities can be any of $t_{3,6}$, $t_{3,9}$ or $t_{6,9}$ (they are in fact the same polynomial). Using an arbitrary coordinate system, it is in general possible to choose an arbitrary non-degenerate trilinearity. The condition that must be fulfilled is that the trilinearity does not vanish on the trifocal lines. □

We are now ready to prove our key result describing the relations between $\mathcal{I}_{\text{b}}$ and $\mathcal{I}_{\text{t}}$. First observe that if $z_1 = z_2 = z_3 = 0$, then all bilinearities in (11) vanish but the trilinearity $t_{6,9}$ does not vanish. This means that the trilinear constraint $t_{6,9} = 0$ imposes one further condition on the other image coordinates. The conditions $z_1 = z_2 = z_3 = 0$ describe the intersection of the trifocal plane with the three images, which indicates that it could correspond to an associate prime ideal of $\mathcal{I}_{\text{b}}$. For the proof of the following theorem, see [8].

Theorem 13 (Primary decomposition of the bilinear ideal). *The ideal $\mathcal{I}_{\text{b}}$ is reducible and can be decomposed as*

$$\mathcal{I}_{\text{b}} = \mathcal{I}_{\text{t}} \cap \mathcal{I}_{\text{tp}} \ , \tag{15}$$

where $\mathcal{I}_{\text{tp}} = (z_1, z_2, z_3)$ is the ideal corresponding to the trifocal plane. In (15), *$\mathcal{I}_{\text{t}}$ and $\mathcal{I}_{\text{tp}}$ are prime ideals and thus irreducible.*

This theorem shows that the trilinear ideal can be obtained from the bilinear ideal in the following way. First make a primary decomposition of the bilinear ideal. This gives two unique primary ideals. Then throw away the ideal that can be generated by linear functions. The remaining one is the trilinear ideal. It follows that the ideal $\mathcal{I}_{\text{bt}}$ generated by the bilinearities and the trilinearities is the same as the ideal $\mathcal{I}_t$ generated by the trilinearities, i.e. $\mathcal{I}_{\text{bt}} = \mathcal{I}_{\text{t}}$.

We conclude this section with the observation that the dimension of the varieties $\mathcal{V}_{\text{t}}$ and $\mathcal{V}_{\text{b}}$ is $3 = 9 - 3 - 3$. The number of variables is 9, they are divided into 3 groups of projective vectors and the constraints can locally be written as the vanishing of 3 polynomial equations. This means that the codimension is 3. Thus we would like to have 3 polynomials to generate the variety $\mathcal{V}_{\text{t}}$, unfortunately this is not possible according to Theorem 12.

5 More than Three Images

Because of lack of space we only give the results here, for proofs see [8].

Theorem 14. *For n images, $n \geq 4$, we have with strict inclusion,*

$$\mathcal{V}_{\mathrm{n}} \subset \mathcal{V}_{\mathrm{b}} = \mathcal{V}_{\mathrm{t}} = \mathcal{V}_{\mathrm{q}} \ . \tag{16}$$

Geometrically this can be seen as follows. When we have three images the bilinear constraints fail to distinguish between correct and incorrect point correspondences on the trifocal lines, but when we have another image outside the trifocal plane it is possible to resolve this failure by using the three new bilinear constraints involving the fourth image. Again the closure of $\mathcal{V}_{\mathrm{n}}$ is $\mathcal{V}_{\mathrm{t}}$. Observe that the bilinearities are sufficient to generate the closure of $\mathcal{V}_{\mathrm{n}}$, that is no trilinearities are needed.

Theorem 15. *For n images, $n \geq 4$, we have*

$$\mathcal{I}_{\mathrm{b}} = \mathcal{I}_{\mathrm{t}} = \mathcal{I}_{\mathrm{q}} = \mathcal{I}(\mathcal{V}_{\mathrm{n}}) \ . \tag{17}$$

In the case of 4 images we have 6 bilinearities and

Theorem 16. *The ideal $\mathcal{I}_{\mathrm{b}}$, for 4 images, can be generated by all 6 bilinearities but not by any 5 multilinear functions.*

In the case of 5 images we have 10 bilinearities and since the codimension of $\mathcal{V}_{\mathrm{b}}$ is 3, it would be nice to have $2 * 5 - 3 = 7$ bilinear forms generating $\mathcal{I}_{\mathrm{b}}$. However, this is not sufficient and neither is 8 bilinear forms. In fact, only 1 bilinear form can be removed.

Theorem 17. *The bilinear ideal $\mathcal{I}_{\mathrm{b}}$ for 5 images can be generated by 9 bilinear forms, but not by any 8 multilinear functions.*

Conjecture 18. *It is not possible to generate $\mathcal{I}_{\mathrm{t}}$ for $n \geq 3$ images by $2n - 3$ bilinearities or by $2n - 3$ other multilinear functions.*

6 Conclusions

In this paper we have shown that the image of $\mathcal{P}^3$ in $\mathcal{P}^2 \times \mathcal{P}^2 \times \cdots \times \mathcal{P}^2$ under an n-tuple of projections $\Phi_n = (A_1, A_2, \ldots, A_n)$ is not an algebraic variety, i.e. it can not be described as the set of common zeros to a system of polynomial equations. We have described two different approaches to obtain an algebraic variety. The first one is to extend Φ_n to a multivalued map, defining the image of the focal point f_i of camera i to be the set of points corresponding to the actual epipoles in the other images and an arbitrary point in image i. The second one is to restrict Φ_n by removing an epipolar line or a trifocal plane and the corresponding image points. Moreover, the closure of this image is equal to the trilinear variety.

We have shown that for three images the variety defined by the bilinearities is reducible and can be written as a union of two irreducible varieties; the variety defined by the trilinearities and a variety corresponding to the trifocal lines. For the ideals the situation can be described by saying that the ideal generated by the bilinearities can be written in a primary decomposition as an intersection of two prime ideals; the ideal generated by the trilinearities and an ideal corresponding to the trifocal lines.

Finally, if four or more images are available the ideal generated by the bilinearities is the same as the ideal generated by the trilinearities. This means that it is possible to use only bilinearities to generate the algebraic variety defined by all multilinear forms. We have also shown that the ideal generated by the quadrilinearities is the same as the ideal generated by the trilinearities.

Acknowledgements

The authors would like to thank Gunnar Sparr and Sven Spanne for guidance and support and the reviewers for valuable suggestions.

References

1. Faugeras, O., D., What can be seen in three dimensions with an uncalibrated stereo rig?, *ECCV'92, Lecture notes in Computer Science, Vol 588. Ed. G. Sandini, Springer-Verlag,* 1992, pp. 563-578.
2. Faugeras, O., D., Mourrain, B., On the geometry and algebra on the point and line correspondences between N images, *Proc. ICCV'95, IEEE Computer Society Press,* 1995, pp. 951-956.
3. Faugeras, O., D., Mourrain, B., About the correspondences of points between N images, *Proc. IEEE Workshop on Representation of Visual Scenes,* 1995.
4. Hartley, R., I., Projective Reconstruction and Invariants from Multiple Images, *IEEE Trans. Pattern Anal. Machine Intell.,* vol. 16, no. 10, pp. 1036-1041, 1994.
5. Hartley, A linear method for reconstruction from lines and points, *Proc. ICCV'95, IEEE Computer Society Press,* 1995, pp. 882-887.
6. Heyden, A., Reconstruction from Image Sequences by means of Relative Depths, *Proc. ICCV'95, IEEE Computer Society Press,* 1995, pp. 1058-1063, Also to appear in *IJCV, International Journal of Computer Vision.*
7. Heyden, A., Åström, K., A Canonical Framework for Sequences of Images, *Proc. IEEE Workshop on Representation of Visual Scenes,* 1995.
8. Heyden, A., Åström, K., Algebraic Properties of Multilinear Constraints, *Technical Report, CODEN: LUFTD2/TFMA--96/7001--SE, Lund, Sweden,* 1996.
9. Luong, Q.-T., Vieville, T., Canonic Representations for the Geometries of Multiple Projective Views, *ECCV'94, Lecture notes in Computer Science, Vol 800. Ed. Jan-Olof Eklund, Springer-Verlag,* 1994, pp. 589-599.
10. Schafarevich, I., R., *Basic Algebraic Geometry I - Varieties in Projective Space,* Springer Verlag, 1988.
11. Sharp, R., Y., *Steps in Commutative Algebra,* London Mathematical Society Texts, 1990.
12. Shashua, A., Trilinearity in Visual Recognition by Alignment, *ECCV'94, Lecture notes in Computer Science, Vol 800. Ed. Jan-Olof Eklund, Springer-Verlag,* 1994, pp. 479-484.
13. Shahsua, A., Werman, M., Trilinearity of Three Perspective Views and its Associated Tensor, *Proc. ICCV'95, IEEE Computer Society Press,* 1995, pp. 920-925.
14. Triggs, B., Matching Constraints and the Joint Image, *Proc. ICCV'95, IEEE Computer Society Press,* 1995, pp. 338-343.

3D Model Acquisition from Extended Image Sequences

Paul Beardsley, Phil Torr and Andrew Zisserman

Dept of Engineering Science, University of Oxford, Oxford OX1 3PJ

Abstract. A method for matching image primitives through a sequence is described, for the purpose of acquiring 3D geometric models. The method includes a novel robust estimator of the trifocal tensor, based on a minimum number of token correspondences across an image triplet; and a novel tracking algorithm in which corners and line segments are matched over image triplets in an integrated framework. The matching techniques are both robust (detecting and discarding mismatches) and fully automatic.
The matched tokens are used to compute 3D structure, which is initialised as it appears and then recursively updated over time. The approach is uncalibrated - camera internal parameters and camera motion are not known or required.
Experimental results are provided for a variety of scenes, including outdoor scenes taken with a hand-held camcorder. Quantitative statistics are included to assess the matching performance, and renderings of the 3D structure enable a qualitative assessment of the results.

1 Introduction

The aim of this work is to recover 3D models from long uncalibrated monocular image sequences. These models will be used for graphics and virtual reality applications. The sequences are generated by circumnavigating the object of interest (e.g. a house) acquiring images with a camcorder. Neither the internal calibration of the camera nor the motion of the camera are known. In particular the motion is unlikely to be smooth. The focus of this paper is on the matching and tracking of image primitives which underpins the structure recovery.

We build on the work of a number of previous successful systems which have recovered structure and motion from tracked image primitives (tokens). Coarsely these systems can be divided into those that use sequential [1, 3, 4, 10, 19, 28], and those that use batch updates [13, 15, 23]. Here the matching and structure update are sequential. However, the basic unit is an image triplet, rather than the more usual image pair. It is in this area of tracking technology [26, 27] that we have made the most significant innovations. A finite displacement (several cms) between views prohibits the utilisation of simple nearest neighbour token matching strategy between consecutive images. In this work:

1. Corner and line segments are matched simultaneously over image triplets in an integrated framework, by employing the trifocal tensor [14, 21, 22].
2. A robust estimator for the trifocal tensor is developed, with the tensor instantiated over three views using a minimal set (six) of point matches and a RANSAC scheme.

The use of a robust scheme provides protection against mismatches [6, 24] and independently moving objects [25].

Two important advantages of the method described here are that the camera model covers a full perspective projection, not its affine approximation (weak or para-perspective) as in [23], and no knowledge of camera internal parameters or relative motion is required. However, a consequence is that the 3D structure recovered is up a projective transformation, rather than Euclidean [7, 12].

Results are presented for a variety of real scenes, with an assessment of matching performance (lifetime of tracked tokens, total number of matches), and examples of the recovered structure. All of the processing (matching, structure recovery etc), is automatic, involving no hand picked points.

Notation and multiple view geometry

The representations of multiple view geometry are based on [3, 7, 8, 12, 14, 16].

For a triplet of images the image of a 3D point $\mathbf{X}$ is $\mathbf{x}$, $\mathbf{x}'$ and $\mathbf{x}''$ in the first, second and third images respectively, and similarly the image of a line is $\mathbf{l}$, $\mathbf{l}'$ and $\mathbf{l}''$, where $\mathbf{x} = (x_1, x_2, x_3)^\top$ and $\mathbf{l} = (l_1, l_2, l_3)^\top$ are homogeneous three vectors.

Image pairs - bilinear relations Corresponding points in two images satisfy the epipolar constraint

$$\mathbf{x}'^\top \mathbf{F} \mathbf{x} = 0 \tag{1}$$

where $\mathbf{F}$ is the 3×3 fundamental matrix, with maximum rank 2. This is the bilinear relation in the homogeneous coordinates of the corresponding points in two images.

Image triplets - trilinear relations Corresponding points in three images, and corresponding lines in three images, satisfy trilinear relations which are encapsulated in the trifocal tensor, T, a $3 \times 3 \times 3$ homogeneous tensor. Using the tensor a point can be transferred to a third image from correspondences in the first and second:

$$x_l'' = x_i' \sum_{k=1}^{k=3} x_k \mathrm{T}_{kjl} - x_j' \sum_{k=1}^{k=3} x_k \mathrm{T}_{kil},$$

for all $i, j = 1 \ldots 3$. Similarly, a line can be transferred as

$$l_i = \sum_{j=1}^{j=3} \sum_{k=1}^{k=3} l_j' l_k'' \mathrm{T}_{ijk}$$

i.e. the same tensor can be used to transfer both points and lines.

2 Matching strategies

Simple matching based only on similarity of image primitive attributes will inevitably produce mismatches. For image *pairs*, the fundamental matrix provides a constraint for identifying mismatches between image corners: corresponding corners are constrained to lie on (epipolar) lines. For *triplets* of images, the trifocal tensor provides a more powerful constraint for identifying mismatches for both points and lines: a primitive matched in two images defines the position of the corresponding primitive in the third image. It is a more powerful constraint because the position of a match is completely constrained, rather than just restricted to a line, and also because it applies to both corners and lines, rather than just corners. There is a natural symbiosis between a 2-image matching scheme and the robust computation of the fundamental matrix, and also between a 3-image matching scheme and the robust computation of the trifocal tensor.

In the following subsections we describe three robust matching schemes applicable to a camera moving through a scene that is largely static. No *a priori* information on camera internal parameters or motion is assumed, other than a threshold on the maximum disparity between images. The methodology for matching is essentially the same for all three schemes and involves three distinct stages. The stages are motivated in the description of the first matching scheme.

2.1 Matching corners between image pairs

1. **Seed correspondences by unguided matching**
   The aim is to obtain a small number of reliable seed correspondences. Given a corner at position (x, y) in the first image, the search for a match considers all corners within a region centred on (x, y) in the second image with a threshold on maximum disparity. The strength of candidate matches is measured by cross-correlation. The threshold for match acceptance is deliberately conservative at this stage to minimise incorrect matches.
2. **Robust computation of a geometric constraint**
   There is potentially a significant presence of mismatches amongst the seed matches. Correct matches will obey a geometric constraint, in this case the epipolar geometry. The aim then is to obtain a set of "inliers" consistent with the geometric constraint using a robust technique — RANSAC has proved the most successful [6, 9, 17]: A putative fundamental matrix (up to three solutions) is computed from a random set of seven corner correspondences (the minimum number required to compute a fundamental matrix). The support for this fundamental matrix is determined by the number of correspondences in the seed set within a threshold distance of their epipolar lines. This is repeated for many random sets, and the fundamental matrix with the largest support is accepted. The outcome is a set of corner correspondences consistent with the fundamental matrix, and a set of mismatches (outliers). The fundamental matrix is then re-estimated using all of its associated inliers to improve its accuracy.

3. **Guided matching**
   The aim here is to obtain additional matches consistent with the geometric constraint. The constraint provides a far more restrictive search region than that used for unguided matching. Consequently, a less severe threshold can be used on the matching attributes. In this case, matches are sought for unmatched corners searching only epipolar lines. This generates a larger set of consistent matches.

The final two steps are repeated until the number of matches stabilises.

Typically the number of corners in a 512×512 image of an indoor scene is about 300, the number of seed matches is about 100, and the final number of matches is about 200. Using corners computed to sub-pixel accuracy, the typical distance of a point from its epipolar line is ~0.2-0.4 pixels.

2.2 Matching points and lines between image triplets

The same three steps are used over image triplets, with the geometric constraint provided by the trifocal tensor.

1. **Seed correspondences by unguided matching**
   For lines, seed correspondences over the three images are obtained by matching on a number of attributes (see [2]). For corners, seed correspondences between images 1 & 2, and 2 & 3 are obtained using the fundamental matrix as described above.
2. **Robust computation of a geometric constraint**
   A full description of this method is given in section 3. Briefly, a putative trifocal tensor (up to three solutions) is computed from a random set of six seed point correspondences. The putative tensor is evaluated by measuring its support in the seed set, utilising both corners and lines. The tensor with the greatest support is chosen, and re-estimated using its consistent point and line correspondences. Inconsistent matches are assumed to be mismatches and are marked as unmatched.
3. **Guided matching**
   Corner and line matching is resumed, but now with a far more restrictive search area — for a putative match between a pair of tokens, only a region about the predicted position in the third image need be searched. This generates a larger set of consistent matches.

Here, both points and lines contribute to the estimate of the geometric constraint, and in turn the one constraint is used to search for both corner and line correspondences. In this manner matched corners provide support for line segment matching and vice-versa.

Typically the number of seed matches over a triplet is about 100 corners, and 10-20 lines. The final number of matches is about 150 and 10-50 respectively. Using corners computed to sub-pixel accuracy, the typical distance of a corner/line from its transferred position is ~1.0 pixel.

2.3 Matching between image primitives and 3D structure

The previous two matching schemes were for image to image matching. Once an estimate of 3D structure is available however (at any stage in the image sequence after the initialisation phase is completed) then it is possible to use the 3D structure to aid the matching. This augmented scheme is carried out whenever a new image arrives, to obtain matches between the last image of the sequence and the new image. The result provides both token matches in the images, and also a correspondence between existing 3D structure and tokens in the new image.

1. **Seed correspondences by unguided matching**
   As in the matching of corners between image pairs, section 2.1.
2. **Robust computation of a geometric constraint**

   a. Robust computation of the fundamental matrix
   As in the matching of corners between image pairs, section 2.1.
   b. Robust computation of the camera matrix
   The set of matches obtained above provide a correspondence between the existing 3D point structure and the new image corners. RANSAC is used to compute the 3×4 camera matrix $\mathtt{P}$, which projects the 3D points onto their correspondences in the new image. A putative projection matrix is computed from a random sample of six correspondences. Support for this matrix is given by those correspondences whose projection lies within a threshold distance of the associated image primitive. The projection matrix with greatest support is re-estimated using all consistent point and line matches. Inconsistent matches are assumed to be mismatches and are marked as unmatched.
3. **Guided matching**
   Corner matching is resumed. $\mathtt{P}$ is used to project any unmatched 3D points and lines onto the new image, and a match is searched for around the projected position. This generates a larger set of consistent matches.

Typically, we find that the majority of matches are obtained in the initial matching stage when the fundamental matrix is used. However, the use of the camera matrix computation can add 5-10 matches in a total of 200. The r.m.s. error between projected 3D structure and actual image tokens in the new image is ~0.2-0.5 pixels.

2.4 Implementation details

Two types of image primitives are used - corners and line segments - extracted independently in each image. Corners are detected to sub-pixel accuracy using the Harris corner detector [11]. Lines are detected by the standard procedure of: Canny edge detection [5]; edge linking; segmentation of the chain at high curvature points; and finally, straight line fitting to the resulting chain segments. The straight line fitting is by orthogonal regression, with a tight tolerance to ensure that only actual line segments are extracted, i.e. that curves are not piecewise linear approximated. Further implementation details are given in [2].

2.5 Comparison of pairwise and triplet based matching

We compare two schemes for matching between images:

1. **Method 1: Pairwise based** Corners are matched between image pairs 1 & 2 and 2 & 3 as in section 2.1; 3D point structure is instantiated from the matches between 1 & 2; based on the 2 & 3 matches, the matrix P which projects the 3D structure to image 3 is computed as in section 2.3.
2. **Method 2: Triplet based** This is the matching scheme described in section 2.2. In this scheme both corners and lines are matched.

Experiment I: Number of matches/mismatches for image triplets We assess and compare the two matching schemes by two measures — the number of matches, and the number of mismatches.

Figure 1 shows three consecutive images of a model house, processed by the 2-image and 3-image schemes, with matched corners superimposed on the images. For the 2-image scheme, only those points which survived over all three images are shown, to enable a proper comparison with the 3-image approach. There is little difference in the distribution of the matches found. Furthermore, there are no mismatches under either scheme. Figure 2 shows the same information for an outdoor scene. In this case, there are a few mismatches under both schemes.

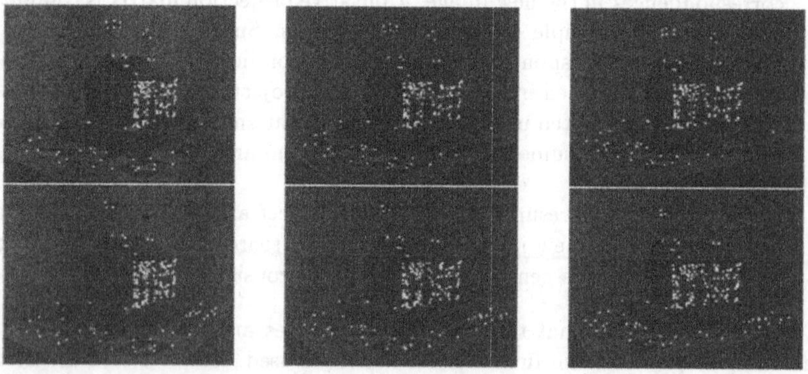

Fig. 1. Three images from a sequence of a model house. Camera motion is a lateral displacement of about 3-4cm between frames. The image size is 760x550 pixels, and about 400 corners are detected in each image. **Upper 3 images:** In the 2-image scheme, about 200 matches are obtained between each pair. The r.m.s. perpendicular distance of points from epipolar lines is about 0.4 pixels. About 160 matches survive across all three frames. R.m.s. error between projected 3D points and corresponding image corners is about 0.5 pixels. **Lower 3 images:** In the 3-image scheme, about 180 matches are obtained across all three images. The r.m.s. error between transferred points (using the trifocal tensor) and actual points in the third image is about 1.0 pixel. R.m.s. projected error is again about 0.5 pixels.

The 2-image matching scheme gives rise to some image matches which exist only between image 1-2, or image 2-3. This can be because a number of the proposed matches (actually mismatches) accidentally agree with the epipolar

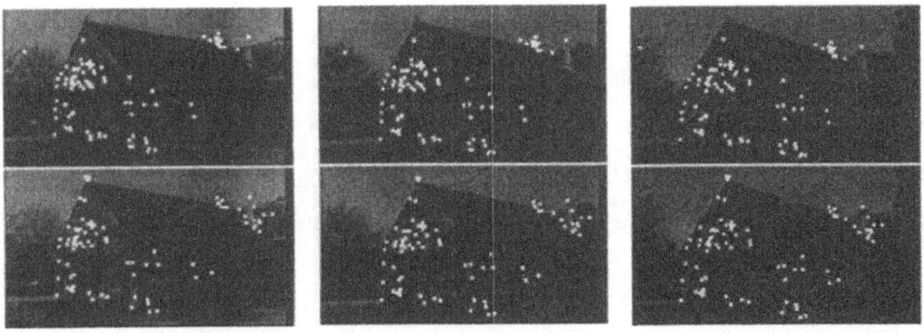

Fig. 2. Three images from a sequence of a chapel, acquired by a hand-held camcorder. Camera motion is lateral and a few centimetres between frames. Image size and corner count as in Figure 1. **Upper 3 images:** In the 2-image scheme, about 150-200 matches are obtained between each pair, with r.m.s. distance of points to epipolar lines about 0.4 pixels. About 80 matches survive across all three frames. **Lower 3 images:** In the 3-image scheme, the number of tokens matched across all three images is again about 80. The r.m.s. error between transferred points (using the trifocal tensor) and actual points in the third image is about 1.5 pixels. R.m.s. projected error is about 0.5 pixels.

geometry. (Other reasons for matches existing only between image 1-2 or image 2-3 are that the corner detector does not continue to find the point, or the physical point moves out of view.) A mismatch which accidentally agrees with the epipolar geometry generates a meaningless 3D point, which cannot project to a potential match in the third image, so the corner is not matched across the triplet. Figure 3 shows the full set of matches for images 1-2 of the outdoor scene (a superset of the matches in Figure 2, all consistent with the estimated fundamental matrix), and the mismatches present in this set. Epipolar mismatches of this type occur particularly in an area of texture which gives many corners of similar appearance along the epipolar line (such as trees and bushes in outdoor scenes).

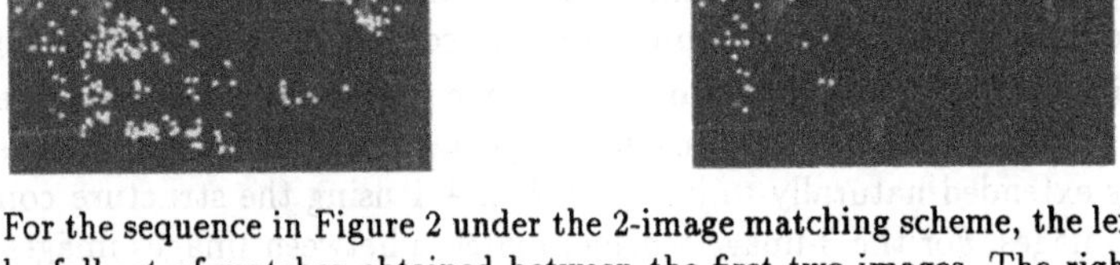

Fig. 3. For the sequence in Figure 2 under the 2-image matching scheme, the left image shows the full set of matches obtained between the first two images. The right image shows those matches which are in actuality outliers. They are congregated mainly on the ambiguous texture area created by the trees in the background, and are accepted by the 2-image matching because they accidentally agree with the epipolar geometry. Only one of these outliers survives when a third image is processed, however - see Figure 2.

Figure 4 indicates how the robust trifocal tensor computation in the 3-image matching scheme enables direct identification of mismatches in the set of seed point and line matches.

Fig. 4. Results for matching a triplet of images using the trifocal tensor, superimposed onto the last image. Top **left:** point correspondences over the three images, with small squares indicating the previous positions. **centre:** matches consistent with the computed trifocal tensor, **right:** outlying matches (mismatched points). At bottom, the same information for line matches. In all, there are 101 seed point correspondences over the three images, 76 are indicated as inliers, and 25 as outliers. There are 15 seed line correspondences, 11 are indicated as inliers, and 4 as outliers.

In summary, the experimental results do not suggest that the 3-image matching scheme produces a marked improvement over the 2-image scheme. There are still good reasons for favouring the 3-image approach, however. Firstly, it is computationally much more elegant and efficient to use the trifocal tensor to match over three images and eliminate mismatches, rather than creating 3D structure from the first two images and then projecting it to the third. Secondly, mismatches in the 2-image scheme which accidentally agree with the epipolar geometry are only detected after processing has moved to the third image. By this stage, it is cumbersome to return to the first two images, remove the mismatch and attempt to rematch the points. Furthermore, the mismatches may have adversely affected the computation of the fundamental matrix, leading to missed matches. In contrast, the 3-image scheme offers the possibility of detecting suspect pairwise matches immediately by using the trifocal tensor. Finally, the 3-image approach and the trifocal tensor allow the integrated use of points and lines, unlike the 2-image approach where corners drive the processing.

Experiment II: Track Lifetime The image matches are extended over a sequence as follows. For the 2-image scheme, the method for images 2 & 3 described above is extended naturally to images n & $n+1$ using the structure computed from n images. For the 3-image scheme, matches between images images $n-1$, n & $n+1$ are generated using the robust estimation of the trifocal tensor for this triplet.

Figure 5 shows two comparisons of the 2-image and 3-image schemes, in terms of overall matching statistics along a sequence. As expected, the total number of matches is generally higher for the 2-image scheme because it includes transient corners which only appear for 2 images, as well as a small number of mismatches which accidentally agree with the epipolar geometry. The 2-image scheme also has more long surviving matches; this is because the 3-image scheme applies a stronger geometric constraint, and so is less tolerant of localisation errors in the features.

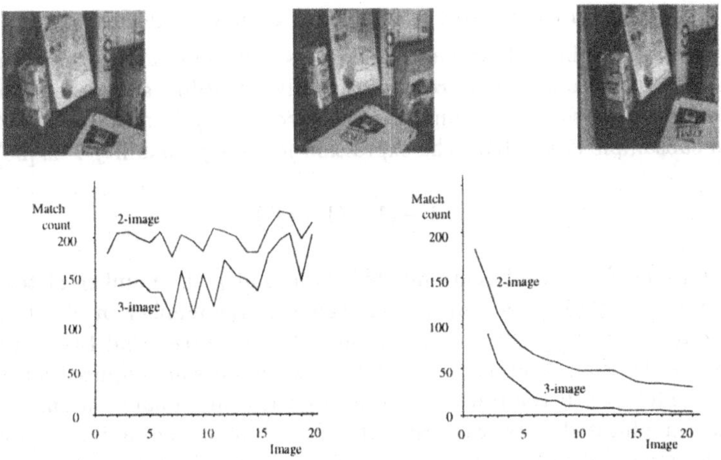

Fig. 5. Three images from a sequence of twenty, taken as the camera moves while fixating a set of objects. The left graph shows the total number of matches at each stage, as an indicator of overall performance. The right graph shows the number of matches which have been tracked continuously from the start of the sequence at each point.

3 Robust computation of the trifocal tensor

In this section we describe the robust computation of the trifocal tensor from a set of putative corner and line correspondences over three images, obtained as described in the previous section. The trifocal tensor has 27 elements, but only their ratio is significant, leaving 26 that must be specified. Each triplet of point correspondences provides four independent linear equations for the elements of the tensor, and each triplet of line correspondences provides two linear equations. Therefore provided that $2n_l + 4n_p \geq 26$ (where n_l is the number of lines, and n_p is the number of points), the tensor can be determined uniquely (up to scale) using a linear algorithm. Consequently, the tensor can be computed linearly from a minimum of 7 points or 13 lines or a combination of the two. However, the tensor has only 18 independent degrees of freedom, which means that it can be computed from 6 point correspondences, though not uniquely — there are 1 or 3 solutions, according to the number of real roots of an associated cubic.

For random sampling methods, such as Least Median Squares (LMS) [20] or RANSAC it is extremely important that the minimum number of correspondences are used, so as to reduce the probability of a mismatch being included in the random sample of correspondences, furthermore a six point solution will be exact, having only 18 degrees of freedom in the coefficients. This is why the novel six point solution used here is important. It is not a problem that there are three solutions, since the correct solution of the three is identified when measuring support for the solution from the full set of putative matches. The method for finding the trifocal tensor from six points uses the theory of Quan [18] for computing an invariant of six points from 3 views, and is described in [2].

3.1 Comparison of six and seven point robust schemes

Ideally every possible subsample of the full set of putative matches would be considered, but this is usually computationally infeasible, so m the number of samples, is chosen sufficiently high to give a probability Υ in excess of 95% that a good subsample is selected. The expression for this probability Υ is [20]

$$\Upsilon = 1 - (1 - (1 - \epsilon)^p)^m, \tag{2}$$

where ϵ is the fraction of contaminated data, and p the number of tokens in each sample. Table 1 gives some sample values of the number m of subsamples required to ensure $\Upsilon \geq 0.95$ for given p and ϵ. It can be seen that the smaller the data set needed to instantiate a model, the less samples are required for a given level of confidence. If the fraction of data that is contaminated is unknown, as is usual, an educated worst case estimate of the level of contamination must be made in order to determine the number of samples to be taken. It can be seen that as the proportion of outliers increases many more samples need to be taken for the seven point algorithm then for the six point method.

Features	Fraction of Contaminated Data, ϵ						
p	5%	10 %	20 %	25 %	30 %	40 %	50 %
6	3	4	10	16	24	63	191
7	3	5	13	21	35	106	382

Table 1. *The number m of subsamples required to ensure $\Upsilon \geq 0.95$ for given p and ϵ, where Υ is the probability that all the data points selected in one subsample are non-outliers.*

The six point algorithm gives better results than the seven point algorithm described in [25], when tested on both real data and synthetic data (where the ground truth is known). This is for two reasons, the first being that that the six point algorithm requires fewer correspondences, and so has less chance of including an outlier as evinced by Table 1; the second, and perhaps more important, is that the six point algorithm exactly encodes the constraints on the parameters of the trifocal tensor. The seven point algorithm on the other hand has too many degrees of freedoms, 27, when there should only be 18. This means that the tensor is over parameterised and a least squares solution will usually give a result that violates the constraints on its coefficients leading to a solution that is implicitly incorrect.

The six point algorithm is also considerably faster. In the case of the seven point algorithm the eigenvector of a 27×27 matrix must be found, which is slower than the solution of a cubic. Furthermore far fewer six point samples need to be taken to get a given degree of confidence in the result.

4 Structure Recovery

Camera matrices are generated at the start of a sequence, using either the fundamental matrix in a 2-image scheme, or the trifocal tensor in a 3-image scheme, and matched corners and line segments are then used to instantiate estimates of 3D point and line structure. An update process is subsequently employed for each new image added to the sequence. Matching between the last image and the new image provides a correspondence between existing 3D structure and the new image primitives, enabling computation of the camera matrix P for the new image. Once P is known, existing estimates of 3D structure are updated from the new observations using an Extended Kalman Filter. Then, P is recomputed using a non-linear computation which minimises the squared distance between the projection of the updated 3D structure and the actual image observations on the image plane for the new image. New points and line segments in 3D are initialised whenever new structure becomes visible in the course of the sequence.

4.1 Results

This section contains experimental results for the estimated 3D structure. The structure is "Quasi-Euclidean" (a form which is close to being true Euclidean) which can be obtained given approximate knowledge of the camera internal parameters as described in [3].

Figures 6 and 7 show results for the sequence of a model house, and the outdoor scene of a chapel. Point and line structure is shown. The recovered structure is best illustrated by using Delaunay triangulation to obtain image connectivity of the structure, and then mapping image intensity onto the triangular facets in 3D. Lines significantly improve the rendering since they often mark boundaries of object planes.

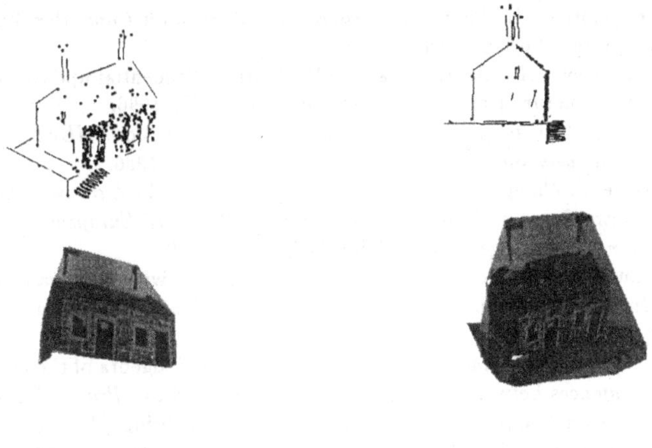

Fig. 6. At top is the point and line structure recovered for the model house of figure 1. Top-right shows the front wall and roof viewed edge on. The bottom images are obtained by rendering image intensity onto the 3D structure and viewing it from novel viewpoints (viewpoints which were never seen in the original sequence).

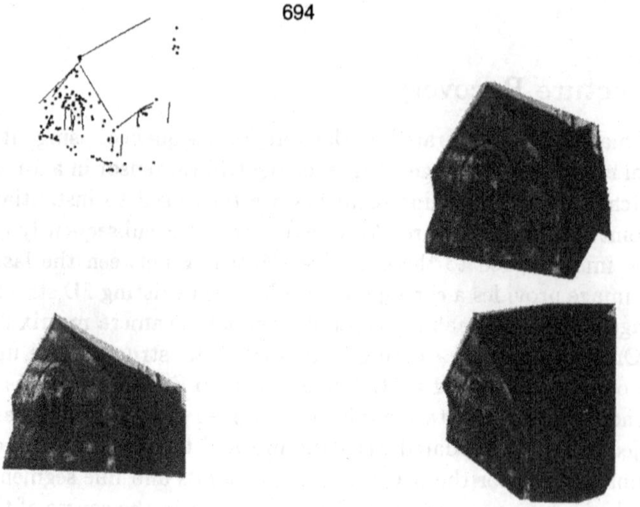

Fig. 7. Results for the outdoor scene of a chapel of Figure 2. Details are as for the previous figure. The dappling effect on the front of the chapel is sunshine through trees.

Acknowledgements
Thanks to Phil McLauchlan for the images in Figure 5. Financial support for this work was provided by EU ACTS Project VANGUARD.

References

1. N. Ayache and F. Lustman. Fast and reliable passive trinocular stereovision. *Proc. International Conference on Computer Vision*, 1987.
2. P.A. Beardsley, P. Torr, and A. Zisserman. 3D model acquisition from extended image sequences. Technical report, Dept of Eng Science, University of Oxford, 1996.
3. P.A. Beardsley, A.P. Zisserman, and D.W. Murray. Navigation using affine structure and motion. In *Proc. 3rd European Conference on Computer Vision*, pages 85–96. Springer-Verlag, 1994.
4. P.A. Beardsley, A.P. Zisserman, and D.W. Murray. Sequential update of projective and affine structure from motion. To appear in IJCV, 1996.
5. J.F. Canny. A computational approach to edge detection. *IEEE Transactions Pattern Analysis and Machine Intelligence*, 8:769–798, 1986.
6. R. Deriche, Z. Zhang, Q.T. Luong, and O. Faugeras. Robust recovery of the epipolar geometry for an uncalibrated stereo rig. In *Proc. 3rd European Conference on Computer Vision*, pages 567–576. Springer-Verlag, 1994.
7. O.D. Faugeras. What can be seen in three dimensions with an uncalibrated stereo rig? In *Proc. 2nd European Conference on Computer Vision*, pages 563–578. Springer-Verlag, 1992.
8. O.D. Faugeras and B. Mourrain. On the geometry and algebra of the point and line correspondences between N images. In E. Grimson, editor, *Proc. 5th International Conference on Computer Vision*, pages 951–956, Cambridge, MA, June 1995.
9. M. A. Fischler and R. C. Bolles. Random sample consensus: a paradigm for model fitting with application to image analysis and automated cartography. *Commun. Assoc. Comp. Mach.*, vol. 24:381–95, 1981.

10. C.G. Harris and J.M. Pike. 3D positional integration from image sequences. In *Third Alvey Vision Conference*, pages 233–236, 1987.
11. C.G. Harris and M. Stephens. A combined corner and edge detector. In *Fourth Alvey Vision Conference*, pages 147–151, 1988.
12. R. Hartley. Invariants of points seen in multiple images. GE internal report, to appear in PAMI, GE CRD, Schenectady, NY 12301, USA, 1992.
13. R.I. Hartley. Euclidean reconstruction from uncalibrated views. In J.L. Mundy, A. Zisserman, and D. Forsyth, editors, *Applications of invariance in computer vision*, pages 237–256. Springer-Verlag, 1994.
14. R.I. Hartley. A linear method for reconstruction from points and lines. In E. Grimson, editor, *Proc. 5th International Conference on Computer Vision*, pages 882–887, Cambridge, MA, June 1995.
15. R. Mohr, F. Veillon, and L. Quan. Relative 3D reconstruction using multiple uncalibrated images. *Proc. Conference Computer Vision and Pattern Recognition*, pages 543–548, 1993.
16. J.L. Mundy and A.P. Zisserman. *Geometric invariance in computer vision*. MIT Press, 1992.
17. L. Quan. Invariants of 6 points from 3 uncalibrated images. In J. O. Eckland, editor, *Proc. 3rd European Conference on Computer Vision*, pages 459–469. Springer-Verlag, 1994.
18. L. Robert, M.Buffa, and M.Hebert. Weakly-calibrated stereo perception for robot navigation. In E. Grimson, editor, *Proc. 5th International Conference on Computer Vision*, pages 46–51, Cambridge, MA, June 1995.
19. P.J. Rousseeuw. *Robust Regression and Outlier Detection*. Wiley, New York, 1987.
20. A. Shashua. Trilinearity in visual recognition by alignment. In *Proc. 3rd European Conference on Computer Vision*, pages 479–484. Springer-Verlag, 1994.
21. M.E. Spetsakis and J. Aloimonos. Structure from motion using line correspondences. *International Journal of Computer Vision*, pages 171–183, 1990.
22. C. Tomasi and T. Kanade. Shape and motion from image streams under orthography: a factorisation method. *International Journal of Computer Vision*, pages 137–154, 1992.
23. P.H.S. Torr. *Motion segmentation and outlier detection*. PhD thesis, Dept. of Engineering Science, University of Oxford, 1995.
24. P.H.S. Torr, A. Zisserman, and D.W. Murray. Motion clustering using the trilinear constraint over three views. In R. Mohr and C. Wu, editors, *Europe-China Workshop on Geometrical Modelling and Invariants for Computer Vision*, pages 118–125. Springer–Verlag, 1995.
25. P.H.S.Torr and D.W.Murray. Outlier detection and motion segmentation. In *SPIE 93*, 1993.
26. X.Hu and N.Ahuja. Matching point features with ordered geometric, rigidity and disparity constraints. *IEEE Transactions Pattern Analysis and Machine Intelligence*, 16(10):1041–1048, 1994.
27. C. Zeller and O. Faugeras. Projective, affine and metric measurements from video sequences. In *Proceedings of the International Symposium on Optical Science, Engineering and Instrumentation*, 1995.
28. Z. Zhang and O. Faugeras. *3D Dynamic Scene Analysis*. Springer-Verlag, 1992.

Computing Structure and Motion of General 3D Curves from Monocular Sequences of Perspective Images*

Théo Papadopoulo and Olivier Faugeras

INRIA Sophia Antipolis, 2004 Route des Lucioles, BP 93, 06902 SOPHIA-ANTIPOLIS Cedex, France

Abstract. *This paper[2] discusses the problem of recovering 3D motion and structure for rigid curves. For general 3D rigid curves, there is exactly one constraint for each image point that relates the so-called spatio-temporal derivatives to the instantaneous motion parameters. However, this constraint directly derives from the perspective equations and thus ignores the fact that the viewed object must be entirely in front of the focal plane. This paper provides a description of some constraints insuring the visibility of the curve and examines their geometrical consequences for 3D reconstructions. Ignoring them leads to erroneous solutions with a high probability even when using perfect data. An algorithm which takes advantage of these constraints is proposed. The accuracy of the obtained results is demonstrated on both synthetic and real image sequences.*

1 Introduction

Among the methods which attempt to recover three-dimensional information about an observed scene from images, those involving only monocular sequences are some of the most difficult to bring to life. Many of the methods proposed up to now are either point or line based (e.g. [ZHAH88, WHA92]). Most of the general methods for edges are using the point equations and it is necessary to compute the full motion field along the edges; in general this is known to be theoretically impossible. To solve this problem, researchers have developed a number of techniques such as imposing some additional properties on the geometry of the 3D curve [WWB91, PF94b], imposing some knowledge about the motion or "inventing" a full motion field compatible with the normal one and that obeys some smoothness constraint [Hil83, WW88].

None of these techniques are really satisfying from a theoretical point of view, and some have never been tested on real images. The work described here attempts to solve the same kind of problem but does not suppose that point matches are available, and does not compute an estimate of the full motion field prior to the 3D motion computation (it corresponds essentially to the implementation of the theory described in [Fau93]). From this point of view, we show how crucial it is to take into account the visibility constraints described in [Car94] for the motion computation. Moreover, it is demonstrated on examples that these are still not sufficient if an accurate 3D structure has to be computed from the motion since small errors in the motion parameters induce discontinuities in the reconstructions. More recently, [CÅG95] have presented a generalization of the work of [Car94] to compute the motion of dynamic silhouettes.

* This work was partially funded under Esprit Basic Research Project 6019, Insight 2

[2] The results in this paper are a summary of those given in [PF95] available as a technical report from the URL: <URL ftp://ftp.inria.fr/INRIA/publication/publi-ps-gz/RR/RR-2765.ps.gz>.

This paper is organised into four parts: the first section introduces some notation and terms, and gives a brief reminder of the underlying theory. The second section focusses on some singular situations resulting from this theory whereas the third and fourth sections describe respectively an algorithm to solve the considered problem and the results which have been obtained with it.

2 The Theory of Motion Fields for Curves

The goal of this section is to recall briefly some results about motion fields of curves and to introduce notation which will be used in the remainder of this paper. Full descriptions of the used quantities can be found in [Fau93].

2.1 The Camera Model

We assume that the camera obeys the pinhole model with unit focal length. We denote O as the focal center point and $\mathcal{R}$ as the retinal plane. A frame (O, X, Y, Z) is attached naturally to the camera by supposing that the plane (O, X, Y) is parallel to $\mathcal{R}$. As such, all equations involving 3D parameters will be written in this frame.

Given a 3D point $\mathrm{M} = (X, Y, Z)$ and its 2D perspective projection $\mathbf{m} = (x, y, 1)$ on $\mathcal{R}$, we express their relationship by the perspective equation $\mathrm{M} = Z\mathbf{m}$. This equation is fundamental as all constraints we use are derived directly from it. The concept of temporal changes can be incorporated with the introduction of a time parameter τ.

2.2 Definitions

We now assume that we observe in a sequence of images a family (c_τ) of curves, where τ denotes time. For each time instant τ, c_τ is supposed to be the perspective projection in the retina of a 3D curve (C) that moves in space. Furthermore, for each time instant τ, c_τ is assumed to be a regular curve. This family of curves sweeps out a surface (Σ) (the spatio-temporal surface) in the three-dimensional space (x, y, τ). Fig. 1 illustrates an example of such a spatio-temporal surface.

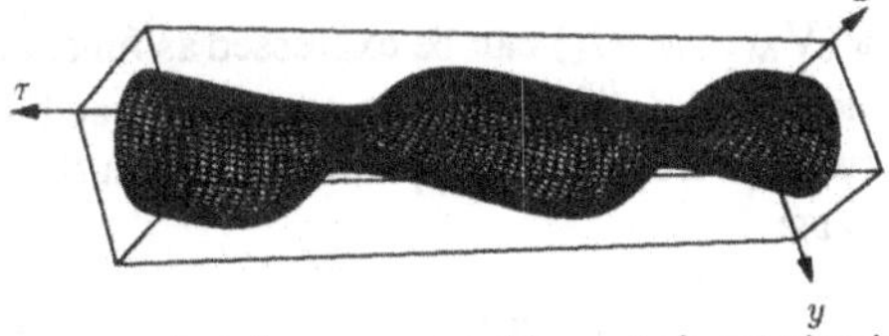

Fig. 1. *The spatio-temporal surface generated by a circle rotating in front of the camera.*

At a given time instant τ, let s be the Euclidean arclength of (c_τ) and S the Euclidean arclength of (C). We further suppose that S is not a function of time (i.e. the motion is *isometric*). To define the speed of a point of the image curve, we have to specify a way to define the path followed by that point over time. Two natural strategies can be adopted toward this goal: firstly, such a path can be defined by considering the points for which S is constant over time, or alternatively we can keep s constant. But only the first alternative has a physical meaning as it means that a point is fixed on the 3D curve. This leads to the definition of the *real motion field* $\mathbf{v}_\mathbf{m}^r$ which is the partial derivative of $\mathbf{m}(s,\tau)$ with respect to time when S is kept constant. This field is the projection of the 3D velocity field in the retina.

Introducing the Frenet frame $(\mathbf{t},\mathbf{n})$, where $\mathbf{t}$ and $\mathbf{n}$ are respectively the unit tangent and normal vectors to (c_τ) at $\mathbf{m}$, we can write $\mathbf{v}_\mathbf{m}^r = w\mathbf{t}+\beta\mathbf{n}$. Under the weak assumption of *isometric* motion, it can be proven [Fau93] that only the normal motion field β can be recovered from the spatio-temporal surface (Σ). The tangential real motion field w *cannot* be recovered from (Σ).

Therefore, the full real motion field is not computable from the observation of the image of a moving curve under the isometric assumption. This can be considered as a new statement of the so-called *aperture* problem [Hil83]. In order to solve it, we *must* add more constraints, for example that the 3D motion is rigid.

2.3 The Case of a Rigid 3D Curve

Assuming now that the curve (C) is moving rigidly, let $(\Omega,\mathbf{V})$ be its kinematic screw at the optical center O of the camera. Additionally, recall that the velocity $\mathbf{V_M} = \dot{\mathbf{M}}$ of any point $\mathbf{M}$ attached to the rigid curve is given by $\mathbf{V_M} = \mathbf{V} + \Omega \times \mathbf{M}$. Taking the total derivative of the equation $\mathbf{M} = Z\mathbf{m}$ with respect to time, and projecting the resulting equation onto $\mathbf{t}$ and $\mathbf{n}$ yields two scalar equations:

$$Z(w + \Omega\cdot\mathbf{b}) = -\mathbf{U_n}\cdot\mathbf{V}\,, \tag{1}$$

$$Z(\beta - \Omega\cdot\mathbf{a}) = \mathbf{U_t}\cdot\mathbf{V}\,, \tag{2}$$

where $\mathbf{U_t} = \mathbf{m}\times\mathbf{t}$, $\mathbf{U_n} = \mathbf{m}\times\mathbf{n}$, $\mathbf{a} = \mathbf{m}\times\mathbf{U_t}$ and $\mathbf{b} = \mathbf{m}\times\mathbf{U_n}$.

Notice that these equations are not new, they are just a reformulation (in the basis $\mathbf{t},\mathbf{n}$) of the standard *image field* equations derived by Longuet-Higgins and Prazdny [LHP80]. Eq. (1) can be considered as the *definition* of w given Z, Ω and $\mathbf{V}$, whereas Eq. (2) can be considered as a constraint equation should the unknown Z not be present. The solution is thus to find a new estimate for Z as a function of measurable values and of the unknown quantities which are constant along the curve. One way to obtain such an estimate without adding new constraints is to differentiate Eq. (2) with respect to time. This operation introduces the new unknowns $\dot\Omega$ and $\dot{\mathbf{V}}$. All the other new quantities appearing in this formula ($\mathbf{V}_{\mathbf{M}_Z}$, $\dot\beta$, $\dot{\mathbf{U}}_\mathbf{t}$) can be expressed as functions of the unknowns $\Omega,\mathbf{V},\dot\Omega,\dot{\mathbf{V}}$, of the values $\mathbf{m},\mathbf{n},\beta,\frac{\partial\beta}{\partial s}$, κ (the curvature at $\mathbf{m}$), and $\partial_{\mathbf{n}_\beta}\beta$ (the normal acceleration) which can be recovered from the spatio-temporal surface (Σ). The resulting equation can be written as:

$$(\mathbf{V}_Z+Z(\Omega\times\mathbf{m})\cdot\mathbf{k})(\beta-\Omega\cdot\mathbf{a})+Z\left(w\frac{\partial\beta}{\partial s}+\partial_{\mathbf{n}_\beta}\beta-\dot\Omega\cdot\mathbf{a}-\Omega\cdot\dot{\mathbf{a}}\right)=\dot{\mathbf{U}}_\mathbf{t}\cdot\mathbf{V}+\mathbf{U_t}\cdot\dot{\mathbf{V}}\,. \tag{3}$$

Replacing w and $\dot{\mathbf{U}}_\mathbf{t}$ by their expressions in terms of the other quantities yields a linear equation in Z; it is thus easy to eliminate this quantity between this new equation and Eq. (2). This can be done *iff.* the coefficient of Z in these equations is non-zero. Points of the image curve for which this condition is true are called *non-degenerate*. We can thus state the following theorem.

Theorem 1 *At each non-degenerate point of an observed curve* (c_τ) *evolving during time, it is possible to write one polynomial equation in the coordinates* $\Omega, \mathbf{V}, \dot{\Omega}$ *and* $\dot{\mathbf{V}}$. *The coefficients of this equation are polynomials in the quantities* $\mathbf{m} \,.\, \mathbf{n} \,.\, \kappa \,.\, \beta \,.\, \frac{\partial \beta}{\partial s} \,.\, \partial_{\mathbf{n}_\beta} \beta$, *which can be measured from the spatio-temporal surface* (Σ).

This equation is denoted by L_1 hereafter (it is too big an equation to be given here). It can be proved that it is the simplest equation relating the motion to some spatio-temporal parameters for general rigid curves. Thus, only L_1 is considered in the remainder of this paper. Once the motion has been computed, Eq. (2) provides a way to compute the depth Z along the curve. The perspective equation then yields the structure.

3 Visibility Conditions

Recovering the 3D motion by minimizing the set of equations L_1 expressed at many different points of a curve is very difficult even with perfect data: the optimizing process almost always becomes lost in a local minimum. These solutions correspond to 3D reconstructions for which differents parts of the curves lie on the two different sides of the focal plane. This is not so strange since the pinhole camera model makes no distinction between the part of the space which is in front of the camera and the one that is behind! Taking account of this physical constraint needs a closer look to the degenerate points introduced in the previous section: these may appear at places on the image curve for which the coefficients of the depth Z in either Eq. (2) or Eq. (3) is zero. The next paragraphs study respectively the constraints arising in these two cases. The main conclusion of this section is that at such points (when they exist), there is a local loss of depth information, however they yield an important constraint on the 3D motion.

3.1 The First Order Visibility Constraints

The first case arises from the situation where the coefficient in Z in Eq. (2) is zero: $\beta - \Omega \cdot \mathbf{a} = 0$. From Eq. (2), this implies that the quantity $\mathbf{U_t} \cdot \mathbf{V}$ must also vanish at the same point. Physically (but not mathematically), the converse is also true since if $\mathbf{U_t} \cdot \mathbf{V} = 0$, then either $Z = 0$ or $\beta - \Omega \cdot \mathbf{a} = 0$. The first case is impossible (all the viewed point must have $Z > 0$), this implies that $\beta - \Omega \cdot \mathbf{a} = 0$. It is easy to prove that, ensuring that at each point for which one of the two condition $\beta - \Omega \cdot \mathbf{a} = 0$ or $\mathbf{U_t} \cdot \mathbf{V} = 0$ holds then the other one holds too, is equivalent to ensuring that the the depth function $Z(s)$ is of constant sign along the curve: thus the reconstructed 3D curve lies totally on the same side of the focal plane. Consequently, we can state the following proposition:

Proposition 1 *Ensuring that the conditions* $\beta - \Omega \cdot \mathbf{a} = 0$ *and* $\mathbf{U_t} \cdot \mathbf{V} = 0$ *are true at the same point of the curve is equivalent to ensuring that the reconstructed curve will be entirely in front or behind the focal plane. For this reason, and because they come from the study of Eq. (2) which involves only first order derivatives, we call these degeneracy conditions the* ***first order visibility conditions****. The points at which these conditions hold are called* ***first order degenerate points****.*

Remark 1 *Trying to compute the depth at a degenerate point using Eq. (2) yields an underdeterminacy of type 0/0. Of course, in practice, this situation never occurs exactly due to the noise on the data. However, at these points the depth information is unstable.*

The importance of these constraints for the motion analysis of curves has already been recognized in [Car94]. It is quite easy to show that degenerate points are generically isolated points on the curve, the only possible configurations of curve and motion for which non isolated points appear being the case for which $\mathbf{V} = \mathbf{0}$ and the case for which the curve is either reduced to a point or is locally part of a straight line passing through the projective point $\mathbf{V}$ (the focus of expansion). All these cases are non-generic, and so are not taken into consideration hereafter.

Following [Car94], it is important to notice that the condition $\mathbf{U_t} \cdot \mathbf{V} = 0$ has a very nice geometrical interpretation corresponding to the fact that $\mathbf{V}$ is the continuous analog to the epipole for a camera in a stereo rig. Actually, $\mathbf{U_t}$ is the projective representative of the tangent line to the curve (c_τ) at point $\mathbf{m}$: thus the condition $\mathbf{U_t} \cdot \mathbf{V} = 0$ just states that the focus of expansion $\mathbf{V}$ is on the tangent $\mathbf{U_t}$ at the considered point $\mathbf{m}$! The left part of Fig. 2 shows two points satisfying this condition for a closed curve. Unfortunately, there is no such easy geometric interpretation for the condition $\beta - \Omega \cdot \mathbf{a} = 0$.

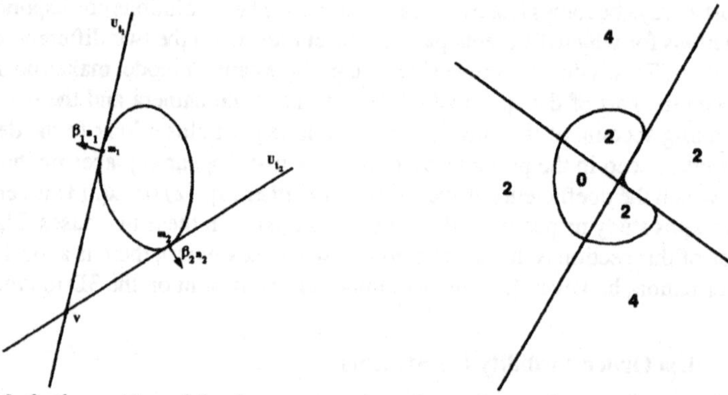

Fig. 2. *Left: the position of the degenerate points given a focus of expansion* $\mathbf{V}$ *and the curve* (c_τ). *Right: The number of degenerate points as a function of the relative position of* $\mathbf{V}$ *to the closed curve* (c_τ). *The different regions are separated by the tangent lines at the inflexion points of the curve and by the curve itself.*

Number of First Order Visibility Conditions The previous interpretation leads also to a way of counting the number of degenerate points on the curve as a function of the position of the focus of expansion and of the geometry of the image curve. Since these points correspond to the boundary points[3] of the curve seen by an observer placed at $\mathbf{V}$, it is easy to understand that the actual number of points depends on the concavities of the curve and on the position of $\mathbf{V}$. The right part of Fig. 2 shows the number of degenerate points as a function of the position of $\mathbf{V}$ relatively to the curve for a fairly complex closed curve. A consequence of this is that the number of visibility conditions depends not only on the actual motion but also on the geometry of the projected curve. This means that the search space for the motion parameters varies with these two parameters.

[3] The boundary points are the curve points where the viewing direction is tangent to the curve.

3.2 The Second Order Visibility Constraints

Eq. (3) can also exhibit a degenerate situation for which the coefficient of Z is zero. Obviously, since Eq. (3) stems directly from (2), it becomes degenerate at first order degenerate points; but it also introduces some new degenerate points which we call *second order degenerate points*. Thus, the conditions for which a second order degenerate point appears are:

$$\left\{\begin{array}{lr} \mathbf{U_t}\cdot\mathbf{V} & \neq 0\,, \\ \beta - \Omega\cdot\mathbf{a} & \neq 0\,, \\ \dot{\mathbf{U}}_\mathbf{t}\cdot\mathbf{V} + \mathbf{U_t}\cdot\dot{\mathbf{V}} - \mathbf{V}_Z(\beta - \Omega\cdot\mathbf{a}) & = 0 \\ \Updownarrow & \\ w\frac{\partial\beta}{\partial s} + \partial_{\mathbf{n}_\beta}\beta - \dot{\Omega}\cdot\mathbf{a} - \Omega\cdot\dot{\mathbf{a}} + ((\Omega\times\mathbf{m})\cdot\mathbf{k})(\beta - \Omega\cdot\mathbf{a}) = 0\,. & \end{array}\right. \tag{4}$$

If we had replaced w, $\dot{\mathbf{U}}_\mathbf{t}$ and $\mathbf{V}_Z$ by their values (not done here because of the size of the result), we would have seen that the equivalence between the third and fourth equations of (4) is always true for all first order degenerate points. The purpose of the first two equations is thus just to select the points that are specific to the loss of depth information in Eq. (3). These conditions can be considered as two constraints between $\Omega, \mathbf{V}, \dot{\Omega}$ and $\dot{\mathbf{V}}$. Since they are linear in $\dot{\Omega}$ and $\dot{\mathbf{V}}$, it is convenient to consider that they force $\dot{\Omega}$ and $\dot{\mathbf{V}}$ to lie each in a plane defined by the spatio-temporal surface, Ω and $\mathbf{V}$.

Unlike the first order visibility constraints, the second order visibility constraints are not easy to cope with. The fundamental reason for this is numerical. These equations actually involve the second order derivatives $\frac{\partial\beta}{\partial s}$, $\partial_{\mathbf{n}_\beta}\beta$ and κ (arising in $\dot{\mathbf{U}}_\mathbf{t}$). The accuracy of the values computed for these parameters is much lower than the accuracy obtained for the values x, y, $\mathbf{n}$ and β. Moreover since the visibility conditions are somewhat exceptional events along the image curve, it is not possible to count on statistical techniques such as least-squares to stabilize the information involved in these constraints. For these reasons, the second order visibility constraints were not used in the algorithm which follows. It is however important to keep them in mind during the analysis of the obtained results especially for the 3D reconstructions.

4 Algorithm

We now assume that the spatio-temporal parameters $\mathbf{m}, \mathbf{n}, \kappa, \beta, \frac{\partial\beta}{\partial s}, \partial_{\mathbf{n}_\beta}\beta$ have been computed from the long monocular sequence at a given time instant[4] τ_0. This section shows how to recover the instantaneous motion (given by Ω, $\mathbf{V}$, $\dot{\Omega}$ and $\dot{\mathbf{V}}$) at time τ_0.

From our experiments, implementing a minimization scheme for the recovery of motion and structure based only on the equation L_1 leads to a process that almost never converges toward the right solution. Similarly, Carlsson [Car94] found that the method was impractical (because it involved the localization of the set of all possible FOE's), and non-robust (because only a small amount of the image information is used).

In this section, we show how it is possible to combine the visibility conditions and the system based on the L_1 equation to recover the 3D motion of the curve. This approach overcomes all the previous objections.

[4] Due to lack of space, this problem is not adressed here. The reader is referred to [PF94a].

- The L_1 based system can help the localization of the set of the possible FOE's, since the potential solutions are zeros of L_1. As the search is embedded in a least-squares approach, the solutions are local minima of some objective function depending on all the equations obtained by writing L_1 for all the pixels of a given image curve.
- Equation L_1 also allows us to take into account all the available information along the image curve of interest. Of course, there are many cases for which there is yet more information to use (special points on the contours or that belong to the object surface delimited by the contour), but we believe that the contour is a good compromise between the reliability and the quantity of information.
- Finally, adding the visibility constraint to the minimization scheme might avoid most of the local minima corresponding to wrong solutions.

Basically, this section presents a modified gradient-descent method which simultaneously takes account of the first order visibility constraints.

4.1 Minimizing Solving for the System of Equations L_1

The system of equations L_1 is strongly over determined (hundreds of equations for only twelve unknowns). Therefore a minimisation scheme must be implemented. The adopted method is a gradient descent where the descent direction is determined by a least square approach using an SVD decomposition of the Jacobian of the system (this is very similar to a Levenberg-Marquardt step). The convergence toward a minimum is estimated with the merit criterion:

$$C_1 = \sum_{\text{all the points}} \frac{|L_1|}{|L_1| + \sum_{\text{for each parameter } p} |\frac{\partial L_1}{\partial p}|} .$$

However, some care needs to be taken because the system is homogeneous in $(\mathbf{V}, \dot{\mathbf{V}})$, to avoid the solution $\mathbf{V} = \dot{\mathbf{V}} = \mathbf{0}$. The adopted solution is to impose that $||\mathbf{V}|| = 1$ and to remove the component of the gradient with respect to $\mathbf{V}$ that is orthogonal to the sphere defined by this constraint. In addition, we use a relaxation technique: the system is first minimized while keeping $\dot{\Omega} = \dot{\mathbf{V}} = \mathbf{0}$[5]. Then, once this minimization has converged, the values of $\dot{\Omega}$ and $\dot{\mathbf{V}}$ that minimize the criterion C_1 are computed. This trick favors the motions with small accelerations among all the motions parameterized by Ω, $\mathbf{V}$, $\dot{\Omega}$ and $\dot{\mathbf{V}}$. Consequently, as the dimensionality of the search space is divided by two, the gradient descent is much less prone to be caught in a wrong local minima due to the noise on the estimated parameters.

4.2 Measuring the Violation of the First Order Visibility Constraints

As presented above, the constraints are either satisfied (in which case the 3D curve is totally visible) or they are not (and part of it is behind the camera). Having such a binary criterion is not very useful in practice, so we need to define a quantity giving some "continuous" measure of the violation of these constraints. This is not always possible

[5] This is a reasonable thing to do as we are dealing in practice with motions for which accelerations must always be very small.

so the measure consists of two parts. The first of these measures is binary and is true *iff.* $\beta - \Omega \cdot \mathbf{a}$ and $\mathbf{U_t} \cdot \mathbf{V}$ have the same number of zeros. This measure is called the *compatibility* of the solution. The second one is continuous:

$$C_2 = \frac{Length(\textit{Part of the curve for which } Z < 0)}{Length(curve)} .$$

The smaller this measure is, the better the situation. Notice that this has two advantages:

- It permits to fix the sign of $\mathbf{V}$ (since C_2 enforce that $Z > 0$) which was not the case with the constraint $||\mathbf{V}|| = 1$;
- If the solution is not compatible, then the criterion C_2 cannot be zero.

Enforcing the visibility constraints to be true by modifying Ω and $\mathbf{V}$ can be done in many different ways, but always requires the localization of the zeros of $\beta - \Omega \cdot \mathbf{a}$ and of $\mathbf{U_t} \cdot \mathbf{V}$. Once that has been done, the easiest method to ensure that the visibility constraints become true is simply to optimize for Ω (resp. $\mathbf{V}$) with $\mathbf{V}$ being fixed (resp. Ω). Between the many different potential candidates the solution chosen is the one that perturbs the current values of Ω and $\mathbf{V}$ as little as possible. At present, this is the best method we have found to cope with the fact that the actual number of visibility constraints may change during the minimization procedure.

4.3 Integrating the Visibility Constraints with the Minimization Scheme

The main difficulty in achieving such an integration is the different nature of the information brought by the two types of information sources.

- There is a small number of visibility but they need to be verified as exactly as possible. This number depends on the 3D motion and the geometry of the image curve.
- The equations L_1 are defined at almost all the points of the curve. However, they involve second order derivatives that are prone to errors so that noise affects them and some might exhibit values not exactly equal to zero.

It is therefore difficult to treat all the different constraints in an homogeneous way. Thus, the adopted solution is to perturb periodically the evolution of the minimization process, by enforcing the visibility constraint to be true. The periodicity of these perturbations is governed by the criterion C_2 (C_2 exceeds 20%). From the experiments, this gives a good compromise between a smooth evolution based on the dense information (along the edge) and a catastrophic one (i.e. non-smooth changes) based on the visibility conditions. The allowance for having visibility constraints only partially verified (C_2 is authorized to be significantly different from 0) during the minimization is justified by the fact that imposing them too strongly is quite expensive (because the minimization evolves very slowly).

As presented, the scheme gives a convergence rate toward the right solution of 70% with perfect data and of 50% with noisy data (obtained from noisy synthetic images or from real ones) given random initialisations. The remaining 50% fall into three classes:

- The solution is not compatible. In such a case, it is not considered at all.

- The solution is compatible but not stable with respect to the visibility conditions. This means that if we let the system evolve without imposing the visibility of the 3D curve, the evolution of the system leads to solutions for which the criterion C_2 becomes larger and larger. These spurious solutions can be detected by letting the L_1 minimization process evolve freely (without imposing the visibility constraints) after the convergence of the process.

- In some cases, there are some other solutions that appear and these look like good solutions. These are often connected to situations that are difficult to distinguish from the real solution.

In order to obtain a convergence rate close to 100%, many different initial random points are given to the previous scheme. In most cases, 5 trials are enough to warrant that the correct solution has been found. Among the different solutions found, the one that is compatible and that minimizes the criterion $C_1 + C_2$ is selected.

5 Results

5.1 Methodology

In order to assess the quality of the obtained results, three approaches were used:

- First, a noisy synthetic sequence of a planar curve was used. This allows full control over the whole process. Theoretical values of all the estimated data are known and the two sets can be easily compared. Moreover, the camera calibration data are perfectly known which avoids the problem of their accuracy.

- Second, real images of planar curves can be used. Then, another method can be used to estimate Ω and V. This method is more robust since it uses only the first order spatio-temporal derivatives [PF94b].

- Finally, real images of 3D curves with controlled motion can be acquired. However, it is always quite difficult to control the relative position of the camera accurately with respect to the scene and thus it is difficult to compute the theoretical values of Ω, V and of their derivatives. Thus, since we have also the previous tests, we verify the accuracy of the results by looking at the characteristics of the motion that are invariant to the position of the camera. In our case, these measures will essentially be the rotational velocity and the angle between Ω and V.

Fig. 3 shows images extracted from these sequences. The motions used are a combination of a rotation and of the translation. The speeds were chosen so that the image curve does not move more than 3 or 4 pixels between two frames.

For the results shown here, from 13 up to 41 successive images were used to compute the spatio-temporal parameters. Successfull results — that is the convergence has occured for at least 90% of the tries — were obtained. Of course, in general, the greater number of images considered, the better the results (better convergence rate, better accuracy of the solution). The presented results consider only successive images, but the same tests have been made starting with subsamples of these sequences, and the method seems to be fairly robust to a subsampling ratio of 1:3.

Fig. 3. *This figure shows some images of the sequences used in this paper. The leftmost image comes from a synthetic sequence of a planar curve with an added Gaussian noise of signal-to-noise ratio of 20%, the second image is a real image of a planar curve. The two right real images are taken from a sequence of images of a non planar curve. Each of the image curves contains about one thousand points.*

5.2 Results

Table 1 shows a typical result obtained from the synthetic sequence. It is easy to see that the obtained result are *very* accurate for Ω and $\mathbf{V}$ and much worse for $\dot{\Omega}$ and $\dot{\mathbf{V}}$. Table 2 shows a comparison of the results obtained by the planar and the general method.

	Ω	$\mathbf{V}$
theoretical	$[0, 0.02, 0]$	$[-0.9911, 0.0690, 0.1138]$
measured	$[3.04e^{-4}, 0.0204, -1.46e^{-4}]$	$[-0.9916, 0.0526, 0.1179]$
angular error	0.95°	0.97°
norm error	1.92%	—
	$\dot{\Omega}$	$\dot{\mathbf{V}}$
theoretical	$[0, 0, 0]$	$[-8.3e^{-4}, 0, 1.4e^{-3}]$
measured	$[-4.3e^{-6}, -9.2e^{-5}, -4.0e^{-6}]$	$[4.6e^{-3}, -3.1e^{-4}, 1.1e^{-3}]$
angular error	—	107.6°
norm error	$9.2e^{-5}$ (absolute)	191.76%

Table 1. *A typical motion computation with the synthetic noisy sequence. One can observe that the accelerations (especially $\dot{\mathbf{V}}$) are* ***extremely*** *noisy, whereas the kinematic screw is very accurate. This shows that the errors on the acceleration do not affect much the accuracy of the motion and justifies a posteriori the choice of minimizing only with respect to the motion parameters (keeping the accelerations null) since this reduces dramatically the number of solutions. The dash denotes errors that are not relevant.*

When dealing with real images of non-planar curves, a simple measure of the accuracy of the results is to compute some invariants of the 3D motion with respect to the viewing geometry. There are two such invariants (since $\mathbf{V}$ is known only up to a scale factor) that are the signed norm of Ω (the angular speed around the rotation axis) and the angle between Ω and $\mathbf{V}$ (which is related to the translation speed along the instantaneous rotation axis). Fig. 4 shows the results obtained all along a sequence of 100

	Ω	$\mathbf{V}$
planar method	$[2.67e^{-4}, 5.46e^{-3}, 1.25e^{-3}]$	$[-0.9947, 0.0955, 0.0393]$
general method	$[2.92e^{-4}, 5.03e^{-3}, 9.78e^{-4}]$	$[-0.9938, 0.1112, 0.0047]$
angular error	1.99°	2.18°
norm error	8.56%	—

Table 2. *A comparison between the results obtained from the planar method with those obtained with the general one. Since the former does not compute $\dot{\Omega}$ and $\dot{\mathbf{V}}$, no comparison can be made for these values but in our experiment $\dot{\Omega} = \mathbf{0}$ and we find $\dot{\Omega} = [-6.08e^{-8}, 1.50e^{-9}, 3.56e^{-8}]$. These comparison show that the general method seems to behave well on real data at least on the parameters Ω and $\mathbf{V}$. The dash denotes the errors that are not relevant since $\mathbf{V}$ is always normalized to one.*

images. Among the 60 independant results[6] that were obtained only 3 are wrong (basically they roughly correspond to the solution for which $\Omega = -\Omega_r$ and $\mathbf{V} = -\mathbf{V}_r$ where Ω_r and $\mathbf{V}_r$ denote the true solution). This means that the overall approach fail to converge to the right solution only 5% of the time.

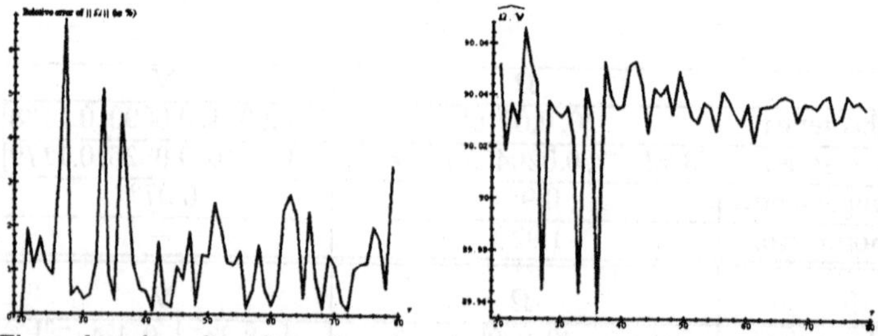

Fig. 4. *These plots show the values of two invariant measures of the 3D motion all along a sequence of 100 real images of a non-planar curve. The motion was computed independently for all the time instant from 20 to 80. The horizontal axis represents time. Notice that the 3 results at time 28, 34 and 37 are wrong (see text) and indeed correspond to the peaks in the error measures. The left plot show the relative error of the norm of Ω (the angular speed): not taking into account for the 3 wrong solutions the maximum relative error is of about 3.3%. The right plot shows the angle between Ω and $\mathbf{V}$: because of the scale factor ambiguity the real value of this invariant is not known but it should be constant which seems to be quite well verified.*

Finally, Fig. 5 shows some 3D reconstructions obtained from equation L_1 once Ω and $\mathbf{V}$ have been computed. Notice that these reconstructions are discontinuous around the degenerate points. This is not so surprising since the reconstruction is pointwise and since nothing in our approach imposes that the 3D curve be continuous: the only constraint is that all the points must be in front of the camera. Actually very small errors on the motion parameters result in discontinuities in the 3D reconstruction. Thus, the

[6] For this experiment, 41 successive images were used to compute the spatio-temporal parameters for each time instant: this is why the horizontal axis of Fig. 4 displays a variation of τ that goes from 20 to 80.

obvious way for obtaining better reconstructions would be to take into account the continuous nature of the 3D curve. It is important to notice that these discontinuities are not specific to the problem of motion and structure: such discontinuities also appear with stereo reconstructions at the points at which the epipolar lines are tangent to the curves.

Fig. 5. *This figure shows some reconstructions obtained once the motion has been computed. The first and third plots represent more or less front views whereas the viewpoint of the other plots has been chosen to show that the global structure of the curve is recovered. The left plots correspond to the real sequence of planar curve (and indeed the top view shows that the reconstruction is almost planar). The right plots correspond to the sequence in which the sheet of paper representing the curve is pasted on the calibration pattern (the top view clearly shows that there are two part almost planar and that the angle between those is about* 90°.

6 Conclusion

It has been shown how to derive an equation that relates the 3D motion of a curve and its accelerations with the spatio-temporal parameters arising from the deformation of the corresponding image curve. An analysis of all the situations in which this equation becomes degenerate (under the hypothesis of a regular image curve) has been given. Some of these situations give rise to conditions known as the visibility constraints. From this study, an algorithm to compute the 3D motion of a rigid 3D general curve that combines both the motion equations and the visibility constraints has been proposed.

This approach is based on equations which are somewhat more complex than the standard equations, but that have the advantages of being mathematically exact and of involving only some quantities which can be computed from image sequences (i.e. there is no need for "inventing" the full motion field from the normal one along the curve as done in most of the papers dealing with motion based on edges). This makes possible the computation of the motion of a single curve in the image. The limitation is that the curve has to be quite large in the image for the estimation of the second order spatio-temporal derivatives to be accurate enough. Practically with curves which are 900–1200 pixels long, the algorithm gives quite accurate results and fails only about 5% of the time. This seems reasonable since this corresponds to a bootstrapping mode and does not take into account the fact that the motion is continuous. Once initialized, the solution can be tracked. The size of the curves used might seem un-practical with the current capture devices, however the resolution of cameras is currently increasing so fast that this might not be a problem anymore in two or three years time. Besides this fact, this algorithm has readily demonstrated that it is important to state that the observed objects are visible: we have seen this for the problem of motion of curves but we suspect that this is more generally true for other types of computations and other primitives.

Finally, we have shown that away from the degenerate situations the 3D structure of the curve can be obtained pointwise. However, the approach does not enforce the continuity of the 3D curve and small errors on the motion parameters usually generate large discontinuities in the 3D reconstruction. Thus, to obtain better 3D reconstructions this continuity must be taken into account. We intend to do so in the near future.

Acknowledgements

We are very grateful to Charlie Rothwell for all his improvements to the text wording.

References

[AM92] Emmanuel Arbogast and Roger Mohr. An egomotion algorithm based on the tracking of arbitrary curves. In *Proceedings 2nd European Conference on Computer Vision*, volume 588 of *Lecture Notes on Computer Science*. Springer-Verlag, 1992.

[CÅG95] Roberto Cipolla, Kalle E. Åstrom, and Peter J. Giblin. Motion from the frontier of curved surfaces. In *Proceedings of the 5th Proc. International Conference on Computer Vision*.

[Car94] Stefan Carlsson. Sufficient image structure for 3-d motion and shape estimation. In Jan-Olof Eklundh, editor, *Proceedings of the 3rd European Conference on Computer Vision*, volume 800 of *Lecture Notes in Computer Science*. Springer-Verlag, 1994.

[Fau93] Olivier D. Faugeras. *Three-Dimensional Computer Vision: a Geometric Viewpoint*. The MIT Press, 1993.

[Hil83] Ellen Catherine Hildreth. *The Measurement of Visual Motion*. The MIT Press, 1983.

[LHP80] H. C. Longuet-Higgins and K. Prazdny. The interpretation of moving retinal images. *Proceedings of the Royal Society of London*, B 208:385–387, 1980.

[PF94a] Théo Papadopoulo and Olivier Faugeras. Estimation of the second order spatio-temporal derivatives of deforming image curves. In *Proc. International Conference on Pattern Recognition*, volume 1, October 1994. IEEE Computer Society Press.

[PF94b] Théo Papadopoulo and Olivier Faugeras. Motion field of curves: Applications. In Jan-Olof Eklundh, editor, *Proceedings of the 3rd European Conference on Computer Vision*, volume 801 of *Lecture Notes on Computer Science*, 1994. Springer-Verlag.

[PF95] Théo Papadopoulo and Olivier Faugeras. Computing structure and motion of general 3d rigid curves from monocular sequences of perspective images. Technical Report 2765, INRIA, December 1995.

[WHA92] Juyang Weng, Thomas S. Huang, and Narendra Ahuja. Motion and structure from line correspondences: Closed-form solution, uniqueness, and optimization. *IEEE Transactions on Pattern Analysis and Machine Intelligence*, 14(3), March 1992.

[WW88] Allen M. Waxman and Kwangyoen Wohn. Image flow theory: a framework for 3-d inference from time-varying imagery. In C. Brown, editor, *Advances in Computer Vision – vol. 1*, volume 1, chapter 3. Laurence Erlbaum Associates, Hillsdale NJ, 1988.

[WWB91] Kwang Yoen Wohn, Jian Wu, and Roger W. Brockett. A contour-based recovery of image flow: Iterative transformation method. *IEEE Transactions on Pattern Analysis and Machine Intelligence*, 13(8), August 1991.

[ZHAH88] Xinhua Zhuang, Thomas S. Huang, Narendra Ahuja, and Robert M. Haralick. A simplified linear optic flow-motion algorithm. *Computer Vision, Graphics, and Image Processing*, 42(3):334–344, June 1988.

A Factorization Based Algorithm for Multi-Image Projective Structure and Motion

Peter Sturm and Bill Triggs

GRAVIR-IMAG & INRIA Rhône-Alpes*
46, Avenue Félix Viallet, 38031 Grenoble, France
email: Peter.Sturm@imag.fr, Bill.Triggs@imag.fr

Abstract. We propose a method for the recovery of projective shape and motion from multiple images of a scene by the factorization of a matrix containing the images of all points in all views. This factorization is only possible when the image points are correctly scaled. The major technical contribution of this paper is a practical method for the recovery of these scalings, using only fundamental matrices and epipoles estimated from the image data. The resulting projective reconstruction algorithm runs quickly and provides accurate reconstructions. Results are presented for simulated and real images.

1 Introduction

In the last few years, the geometric and algebraic relations between uncalibrated views have found lively interest in the computer vision community. A first key result states that, from two uncalibrated views, one can recover the 3D structure of a scene up to an unknown projective transformation [Fau92, HGC92]. The information one needs to do so is entirely contained in the fundamental matrix, which represents the epipolar geometry of the 2 views.

Up to now, projective reconstruction has been investigated mainly for the case of 2 views. Faugeras [Fau92] studied projective reconstruction using 5 reference points. Hartley [HGC92] derives from the fundamental matrix 2 projection matrices, equal to the true ones up to an unknown projective transformation. These are then used to perform reconstruction by triangulation[HS94]. As for multiple images, most of the current methods [MVQ93, Har93, MM95] initially privilege a few views or points and thus do not treat all data uniformly.

Recently, multi-linear matching constraints have been discovered that extend the epipolar geometry of 2 views to 3 and 4 views. Shashua [Sha95] described the trilinear relationships between 3 views. Faugeras and Mourrain [FM95], and independently Triggs [Tri95a] have systematically studied the relationships between N images. Triggs introduced a new way of thinking about projective reconstruction. The image coordinates of the projections of a 3D point are combined into a single "joint image vector". Then, projective reconstruction consists essentially of rescaling the image coordinates in order to place the joint image vector in a certain 4-dimensional subspace of the joint image space called the *joint image*. This subspace is characterized by the multi-linear matching constraints between the views.

* This work was performed within a joint research programme between CNRS, INPG, INRIA, UJF.

The projective reconstruction method we propose in this paper is based on the joint image formalism, but it is not necessary to understand this formalism to read the paper. We show that by rescaling the image coordinates we can obtain a *measurement matrix* (the combined image coordinates of all the points in all the images), which is of rank 4. Projective structure and motion can then be obtained by a singular value factorization of this matrix. So, in a sense this work can be considered as an extension of Tomasi-Kanade's and Poelman-Kanade's factorization methods [TK92, PK94] from affine to perspective projections.
The paper is organized as follows. (1) We motivate the idea of reconstruction through the rescaling of image coordinates. Throughout this paper we will restrict attention to the case of bilinear matching constraints (fundamental matrix), although the full theory [Tri95b] also allows tri- and quadrilinear matching constraints to be used. (2) We discuss some numerical considerations and describe the proposed projective reconstruction algorithm. (3) We show results that we have obtained with real and simulated data. (4) We conclude and discuss several open issues, which will be part of our future work.

2 Projective Reconstruction from Multiple Views

2.1 The Projective Reconstruction Problem

Suppose we have a set of n 3D points visible in m perspective images. Our goal is to recover 3D structure (point locations) and motion (camera locations) from the image measurements. We will assume no camera calibration or additional 3D information, so we will only be able to reconstruct the scene up to an overall projective transformation of the 3D space [Fau92, HGC92].
We will work in homogeneous coordinates with respect to arbitrary projective coordinate frames. Let $\boldsymbol{Q}_p$ be the unknown homogeneous coordinate vectors of the 3D points, $\boldsymbol{P}_i$ the unknown 3×4 image projection matrices, and $\boldsymbol{q}_{ip}$ the measured homogeneous coordinate vectors of the image points, where $p = 1, \ldots, n$ labels points and $i = 1, \ldots, m$ labels images. Each object is defined only up to an arbitrary nonzero rescaling, *e.g.* $\boldsymbol{Q}_p \sim \mu_p \boldsymbol{Q}_p$. The basic image projection equations say that — up to a set of unknown scale factors — the $\boldsymbol{q}_{ip}$ are the projections of the $\boldsymbol{Q}_p$:

$$\lambda_{ip} \boldsymbol{q}_{ip} = \boldsymbol{P}_i \boldsymbol{Q}_p$$

We will call the unknown scale factors λ_{ip} **projective depths**[2]. If the $\boldsymbol{Q}_p$ and the $\boldsymbol{q}_{ip}$ are chosen to have affine normalization ('weight' components equal to 1) and the $\boldsymbol{P}_i$ are normalized so that the vectorial part of the 'weight' component row has norm 1, the projective depths become true optical depths, *i.e.* true orthogonal distances from the focal plane of the camera.
The complete set of image projections can be gathered into a single $3m \times n$ matrix equation:

$$\boldsymbol{W} \equiv \begin{pmatrix} \lambda_{11}\boldsymbol{q}_{11} & \lambda_{12}\boldsymbol{q}_{12} & \cdots & \lambda_{1n}\boldsymbol{q}_{1n} \\ \lambda_{21}\boldsymbol{q}_{21} & \lambda_{22}\boldsymbol{q}_{22} & \cdots & \lambda_{2n}\boldsymbol{q}_{2n} \\ \vdots & \vdots & \ddots & \vdots \\ \lambda_{m1}\boldsymbol{q}_{m1} & \lambda_{m2}\boldsymbol{q}_{m2} & \cdots & \lambda_{mn}\boldsymbol{q}_{mn} \end{pmatrix} = \begin{pmatrix} \boldsymbol{P}_1 \\ \boldsymbol{P}_2 \\ \vdots \\ \boldsymbol{P}_m \end{pmatrix} (\boldsymbol{Q}_1 \; \boldsymbol{Q}_2 \cdots \boldsymbol{Q}_n)$$

[2] This is not the same notion as the "projective depth" of Shashua, which is a cross ratio of distances along epipolar lines [Sha94]

Notice that *with the correct projective depths* λ_{ip}, the $3m \times n$ **rescaled measurement matrix** $\boldsymbol{W}$ has rank at most 4. If we could recover the depths, we could apply an SVD based factorization technique similar to that used by Tomasi and Kanade [TK92] to $\boldsymbol{W}$, and thereby recover both 3D structure and camera motion for the scene. The main technical advance of this paper is a practical method for the recovery of the unknown projective depths, using fundamental matrices and epipoles estimated from the image data.

Taken individually, the projective depths are arbitrary because they depend on the arbitrary scale factors chosen for the $\boldsymbol{P}_i$, the $\boldsymbol{Q}_p$ and the $\boldsymbol{q}_{ip}$. However taken as a whole the rescaled measurements $\boldsymbol{W}$ have a strong internal coherence. The overall scale of each triple of rows and each column of $\boldsymbol{W}$ can be chosen arbitrarily (*c.f.* the arbitrary scales of the projections $\boldsymbol{P}_i$ and the 3D points $\boldsymbol{Q}_p$), but once these $m+n$ overall scales have been fixed there is no further freedom of choice for the remaining $mn - m - n$ scale factors in λ_{ip}. Hence, the projective depths really do contain useful information.

2.2 Recovery of Projective Depths

Now we will show how the projective depths can be recovered from fundamental matrices and epipoles, modulo overall row and column rescalings. The point projection equation $\lambda_{ip}\boldsymbol{q}_{ip} = \boldsymbol{P}_i\boldsymbol{Q}_p$ implies that the 6×5 matrix

$$\begin{pmatrix} \boldsymbol{P}_i & \lambda_{ip}\boldsymbol{q}_{ip} \\ \boldsymbol{P}_j & \lambda_{jp}\boldsymbol{q}_{jp} \end{pmatrix} = \begin{pmatrix} \boldsymbol{P}_i & \boldsymbol{P}_i\boldsymbol{Q}_p \\ \boldsymbol{P}_j & \boldsymbol{P}_j\boldsymbol{Q}_p \end{pmatrix} = \begin{pmatrix} \boldsymbol{P}_i \\ \boldsymbol{P}_j \end{pmatrix} \left(\boldsymbol{I}_{4\times 4} \middle| \boldsymbol{Q}_p \right)$$

has rank at most 4. Hence, all of its 5×5 minors vanish. We can expand these by cofactors in the last column to get homogeneous linear equations in the components of $\lambda_{ip}\boldsymbol{q}_{ip}$ and $\lambda_{jp}\boldsymbol{q}_{jp}$. The coefficients are 4×4 determinants of projection matrix rows. These turn out to be just fundamental matrix and epipole components [Tri95a, FM95]. In particular, if abc and $a'b'c'$ are even permutations of 123 and $\boldsymbol{P}_i^a$ denotes row a of $\boldsymbol{P}_i$, we have:

$$[\boldsymbol{F}_{ij}]_{aa'} = \begin{vmatrix} \boldsymbol{P}_i^b \\ \boldsymbol{P}_i^c \\ \boldsymbol{P}_j^{b'} \\ \boldsymbol{P}_j^{c'} \end{vmatrix} \qquad [\boldsymbol{e}_{ij}]^a = \begin{vmatrix} \boldsymbol{P}_i^a \\ \boldsymbol{P}_j^1 \\ \boldsymbol{P}_j^2 \\ \boldsymbol{P}_j^3 \end{vmatrix} \tag{1}$$

Applying these relations to the three 5×5 determinants built from two rows of image i and three rows of image j gives the following fundamental relation between epipolar lines:

$$(\boldsymbol{F}_{ij}\boldsymbol{q}_{jp})\,\lambda_{jp} = (\boldsymbol{e}_{ij} \wedge \boldsymbol{q}_{ip})\,\lambda_{ip} \tag{2}$$

This relation says two things:

- **Equality up to scale:** The epipolar line of $\boldsymbol{q}_{jp}$ in image i is the line through the corresponding point $\boldsymbol{q}_{ip}$ and the epipole $\boldsymbol{e}_{ij}$. This is just a direct re-statement of the standard epipolar constraint.
- **Equality of scale factors:** If the correct projective depths are used in (2), the two terms have *exactly the same size* — the equality is exact, not just up to scale. This is the new result that allows us to recover projective depths using fundamental matrices and epipoles. Analogous results based on higher order matching tensors can be found in [Tri95b], but in this paper we will use only equation (2).

Our strategy for the recovery of projective depths is quite straightforward. Equation (2) relates the projective depths of a single 3D point in two images. By estimating a sufficient number of fundamental matrices and epipoles, we can amass a system of homogeneous linear equations that allows the complete set of projective depths of a given point to be found, up to an arbitrary overall scale factor. At a minimum, this can be done with any set of $m-1$ fundamental matrices that link the m images into a single connected graph. If additional fundamental matrices are available, the equations become redundant and (hopefully) more robust. In the limit, all $m(m-1)/2$ fundamental matrices and all $m(m-1)$ equations could be used to find the m unknown depths for each point, but this would be computationally very expensive. We are currently investigating policies for choosing economical but robust sets of equations, but in this paper we will restrict ourselves to the simplest possible choice: the images are taken pairwise in sequence, $\boldsymbol{F}_{12}, \boldsymbol{F}_{23}, \ldots, \boldsymbol{F}_{m-1\,m}$.

This is almost certainly not the most robust choice, but it (or any other minimal selection) has the advantage that it makes the depth recovery equations trivial to solve. Solving the vector equation (2) in least squares for λ_{ip} in terms of λ_{jp} gives:

$$\lambda_{ip} = \frac{(\boldsymbol{e}_{ij} \wedge \boldsymbol{q}_{ip}) \cdot (\boldsymbol{F}_{ij}\boldsymbol{q}_{jp})}{\|\boldsymbol{e}_{ij} \wedge \boldsymbol{q}_{ip}\|^2}\,\lambda_{jp} \tag{3}$$

Such equations can be recursively chained together to give estimates for the complete set of depths for point p, starting from some arbitrary initial value such as $\lambda_{1p} = 1$.

However there is a flaw in the above argument: fundamental matrices and epipoles can only be recovered up to an unknown scale factor, so we do not actually know the scale factors in equations (1) or (2) after all! In fact this does not turn out to be a major problem. It is a non-issue if a minimal set of depth-recovery equations is used, because the arbitrary overall scale factor for each image can absorb the arbitrary relative scale of the $\boldsymbol{F}$ and $\boldsymbol{e}$ used to recover the projective depths for that image. However if redundant depth-recovery equations are used it is essential to choose a self-consistent scaling for the estimated fundamental matrices and epipoles. We will not describe this process here, except to mention that it is based on the quadratic identities between matching tensors described in [Tri95b].

Note that with unbalanced choices of scale for the fundamental matrices and epipoles, the average scale of the recovered depths might tend to increase or decrease exponentially during the recursive chaining process. Theoretically this is not a problem because the overall scales are arbitrary, but it could well make the factorization phase of the reconstruction algorithm numerically ill-conditioned. To counter this we re-balance the recovered matrix of projective depths after it has been built, by judicious overall row and column scalings.

2.3 Projective Shape and Motion by Factorization

Once we have obtained the projective depths, we can extract projective shape and motion from the rescaled measurement matrix $\boldsymbol{W}$.

Let

$$\boldsymbol{W} = \boldsymbol{U}\ \mathrm{diag}(\sigma_1, \sigma_2, \ldots, \sigma_s)\ \boldsymbol{V}$$

be a Singular Value Decomposition (SVD) of $\boldsymbol{W}$, with $s = \min\{3m, n\}$ and singular values $\sigma_1 \geq \sigma_2 \geq \ldots \geq \sigma_s \geq 0$. Since $\boldsymbol{W}$ is of rank 4, the σ_i for $i > 4$ vanish. Thus, only the first 4 columns (rows) of U (V) contribute to this matrix product. Let U' (V') the matrix of the first 4 columns (rows) of U (V). Then,

$$W = U'_{3m\times4} \underbrace{\mathrm{diag}(\sigma_1, \sigma_2, \sigma_3, \sigma_4)}_{\Sigma} V'_{4\times n} = U' \, \Sigma \, V' \ .$$

Any factorization of Σ into two 4×4 matrices Σ' and Σ'', $\Sigma = \Sigma' \Sigma''$, leads to

$$W = \underbrace{U'\Sigma'}_{\hat{U}} \underbrace{\Sigma''V'}_{\hat{V}} = \hat{U}_{3m\times4}\hat{V}_{4\times n} \ .$$

We can interpret the matrix $\hat{U}$ as a collection of m (3×4) projection matrices $\hat{P}_i$ and $\hat{V}$ as collection of n 4-vectors $\hat{Q}_p$, representing 3D shape:

$$W = \hat{U}\hat{V} = \begin{pmatrix} \hat{P}_1 \\ \hat{P}_2 \\ \vdots \\ \hat{P}_m \end{pmatrix}_{3m\times4} \left(\hat{Q}_1 \; \hat{Q}_2 \cdots \hat{Q}_n\right)_{4\times n} \tag{4}$$

Equation (4) shows that the $\hat{P}_i$ and $\hat{Q}_p$ represent at least projective motion and shape, since

$$\hat{P}_i\hat{Q}_p = \lambda_{ip} q_{ip} \sim q_{ip} \ .$$

Unlike the case of orthographic projections [TK92], there are no further constraints on the $\hat{P}_i$ or $\hat{Q}_p$: we can *only* recover projective shape and motion. For any non singular projective transformation $T_{4\times4}$, $\hat{P}_i T$ and $T^{-1}\hat{Q}_p$ is an equally valid factorization of the data into projective motion and shape:

$$(\hat{P}_i T)(T^{-1}\hat{Q}_p) = \hat{P}_i\hat{Q}_p \sim q_{ip} \ .$$

A consequence of this is that the factorization of Σ is arbitrary. For the implementation, we chose $\Sigma' = \Sigma'' = \Sigma^{1/2} = \mathrm{diag}(\sigma_1^{1/2}, \sigma_2^{1/2}, \sigma_3^{1/2}, \sigma_4^{1/2})$.

3 The Algorithm

Based on the observations made above, we have developed a practical algorithm for projective reconstruction from multiple views. Besides the major two steps, determination of the scale factors λ_{ip} and factorization of the rescaled measurement matrix, the outline of our algorithm is based on some numerical considerations.

3.1 Normalization of Image Coordinates

To ensure good numerical conditioning of the method, we work with normalized image coordinates, as described in [Har95]. This normalization consists of applying a similarity transformation (translation and uniform scaling) T_i to each image, so that the transformed points are centered at the origin and the mean distance from the origin is $\sqrt{2}$.

All of the remaining steps of the algorithm are done in normalized coordinates. Since we actually compute projective motion and shape for the transformed image points $\boldsymbol{T}_i \boldsymbol{q}_{ip}$, $\hat{\boldsymbol{P}}_i \hat{\boldsymbol{Q}}_p = \lambda_{ip} \boldsymbol{T}_i \boldsymbol{q}_{ip} \sim \boldsymbol{T}_i \boldsymbol{q}_{ip}$, the resulting projection estimates $\hat{\boldsymbol{P}}_i$ must be corrected : $\hat{\boldsymbol{P}}_i' = \boldsymbol{T}_i^{-1} \hat{\boldsymbol{P}}_i$. The $\hat{\boldsymbol{P}}_i'$ and $\hat{\boldsymbol{Q}}_p$ then represent projective motion and shape corresponding to the measured image points $\boldsymbol{q}_{ip}$.

Our results show that this simple normalization drastically improves the results of the projective reconstruction.

3.2 Balancing the Rescaled Measurement Matrix

Consider the factorization of the rescaled measurement matrix W in projective motion and shape :

$$\boldsymbol{W} = \begin{pmatrix} \lambda_{11}\boldsymbol{q}_{11} & \lambda_{12}\boldsymbol{q}_{12} & \cdots & \lambda_{1n}\boldsymbol{q}_{1n} \\ \lambda_{21}\boldsymbol{q}_{21} & \lambda_{22}\boldsymbol{q}_{22} & \cdots & \lambda_{2n}\boldsymbol{q}_{2n} \\ \vdots & \vdots & \ddots & \vdots \\ \lambda_{m1}\boldsymbol{q}_{m1} & \lambda_{m2}\boldsymbol{q}_{m2} & \cdots & \lambda_{mn}\boldsymbol{q}_{mn} \end{pmatrix} = \begin{pmatrix} \hat{\boldsymbol{P}}_1 \\ \hat{\boldsymbol{P}}_2 \\ \vdots \\ \hat{\boldsymbol{P}}_m \end{pmatrix} \begin{pmatrix} \hat{\boldsymbol{Q}}_1 \; \hat{\boldsymbol{Q}}_2 \cdots \hat{\boldsymbol{Q}}_n \end{pmatrix}$$

Multiplying column l of $\boldsymbol{W}$ by a non zero scalar ν_l corresponds to multiplying $\hat{\boldsymbol{Q}}_l$ by ν_l. Analogously, multiplying the image k rows $(3k-2, 3k-1, 3k)$ by a non zero scalar μ_k corresponds to multiplying the projection matrix $\hat{\boldsymbol{P}}_k$ by μ_k. Hence, point-wise and image-wise rescalings of $\boldsymbol{W}$ do not affect the recovered projective motion and shape. However, these considerations are only valid in the absence of noise. In presence of noise, $\boldsymbol{W}$ will only be approximately of rank 4, and scalar multiplications of $\boldsymbol{W}$ as described above *will* affect the results. We therefore aim to improve the results of the factorization by applying appropriate point- and image-wise rescalings to $\boldsymbol{W}$. The goal is to ensure good numerical conditioning by rescaling so that all rows and columns of $\boldsymbol{W}$ have on average the same order of magnitude. To do this we use the following iterative scheme :

1. Rescale each column l so that $\sum_{r=1}^{3m} (w_{rl})^2 = 1$.
2. Rescale each triplet of rows $(3k-2, 3k-1, 3k)$ so that $\sum_{l=1}^{n} \sum_{i=3k-2}^{3k} w_{il}^2 = 1$.
3. If the entries of $\boldsymbol{W}$ changed significantly, repeat 1 and 2.

Note that, since we work with normalized image coordinates $\boldsymbol{q}_{ip}$, it would be sufficient to balance only the $m \times n$ matrix (λ_{ip}) instead of $\boldsymbol{W}$.

3.3 Outline of the Algorithm

The complete algorithm is composed of the following steps.

1. Normalize the image coordinates, by applying transformations $\boldsymbol{T}_i$.
2. Estimate the fundamental matrices and epipoles with the method of [Har95].
3. Determine the scale factors λ_{ip} using equation (3).
4. Build the rescaled measurement matrix $\boldsymbol{W}$.
5. Balance $\boldsymbol{W}$ by column-wise and "triplet-of-rows"-wise scalar mutliplications.
6. Compute the SVD of the balanced matrix $\boldsymbol{W}$.
7. From the SVD, recover projective motion and shape.
8. Adapt projective motion, to account for the normalization transformations $\boldsymbol{T}_i$ of step 1.

4 Experimental Evaluation of the Algorithm

4.1 Experiments with Simulated Images

We conducted a large number of experiments with simulated images to quantify the performance of the algorithm. The simulations used three different configurations: lateral movement of a camera, movement towards the scene, and a circular movement around the scene (see figure 1). In configuration 2, the depths of points lying on the line joining the projection centers can not be recovered. Reconstruction of points lying close to this line is extremely difficult, as was confirmed by the experiments, which resulted in quite inaccurate reconstructions for this configuration.
For the circular movement, the overall trajectory of the camera formed a quarter circle, centered on the scene. For each specific experiment, the trajectory length was the same for all three configurations. The m different viewing positions were equidistantly distributed along the trajectory.

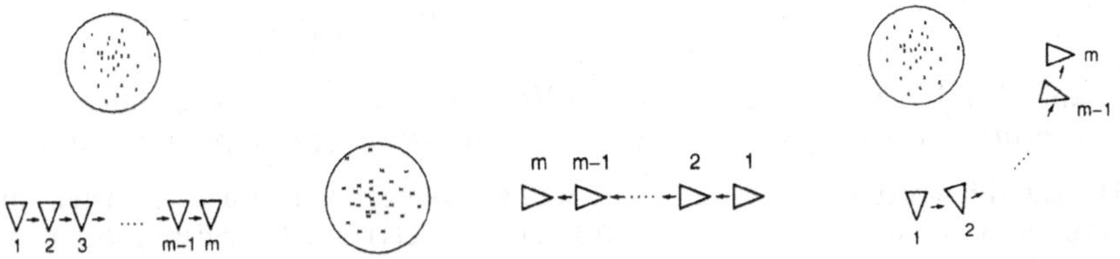

Fig. 1. The 3 configurations for simulation. *(1) Lateral movement. (2) Translation towards the scene. (3) Circular movement.*

In order to simulate realistic situations, we adopted the following parameters: the camera's calibration matrix was diag(1000, 1000, 1). The scene was composed of points distributed uniformly in a sphere of radius 100. The distance between the camera and the center of the sphere was 200 (for configuration 2 this was the distance with respect to the view m).
For each configuration, the following experiment was conducted 50 times:

1. Determine at random 50 points in the sphere.
2. Project the points into the m views.
3. Add Gaussian noise of levels $0.0, 0.5, \ldots, 2.0$ to the image coordinates.
4. Carry out projective reconstruction with our algorithm.
5. Compute the image distance error of the backprojected points (2D error): $\frac{1}{mn}\sum_{i=1}^{m}\sum_{p=1}^{n}||\boldsymbol{P}_i\hat{\boldsymbol{Q}}_p - \boldsymbol{q}_{ip}||$, where $||.||$ means the Euclidean vector norm.
6. Align the projective reconstruction with the Euclidean model and compute the distance error in the Euclidean frame (3D error).

The results of these experiments were analyzed with respect to several variables, as reported in the following subsections. All values represented in the graphs are the mean result over 50 trials. To monitor the effect of outliers on the results, we also computed the median values. These gave graphs similar to those for the means, which we will not show here.

2D errors are given in pixels, whereas 3D errors are given relative to the scene's size, in percent.

Sensitivity to Noise Graphs 1 and 2 show the behavior of the algorithm with respect to different noise levels for the three configurations. For this experiment, reconstruction was done from 10 views.

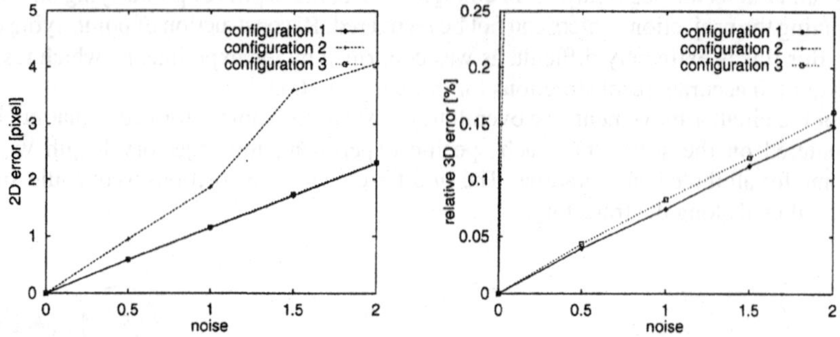

Graphs 1 and 2 : Sensitivity to noise. *The 2D error curves for the configurations 1 and 3 are nearly undistinguishable. 3D error for configuration 2 goes rapidly off scale.*

The algorithm performed almost equally well for configurations 1 and 3, whereas the 3D error for configuration 2 exceeds 100 % for 2.0 pixels noise. Considering the graphs of configuration 2, we also see that 2D and 3D error are not always well correlated. For configurations 1 and 3, the 2D error is of the same order as pixel noise. Note also the linear shape of the graphs.

Number of Views The image noise for this experiment was 1.0 pixel.

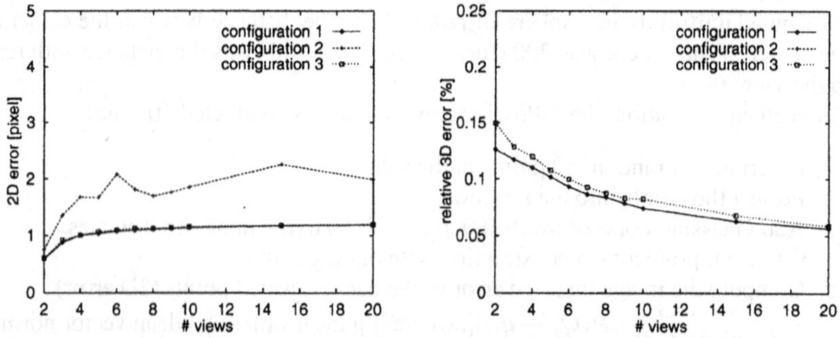

Graphs 3 and 4 : Behavior with respect to number of views. *The 2D error curves for the configurations 1 and 3 are nearly undistinguishable. The 3D error for configuration 2 lies above 5 %. The curve is thus not visible in the graph.*

The graphs show the expected behavior : when more views are used for reconstruction, the structure is recovered more accurately. Secondly, 2D error augments with increasing number of views, but shows a clearly asymptotic behavior. 1. Note that the use of 20 views reduces the 3D error to 50 % of that for 2 views.

Importance of Normalization and Balancing The error values in the previous graphs were obtained with the algorithm as described in subsection 3.3. To underline the importance of using normalized image coordinates, we also ran the algorithm using unnormalized ones. The effects of not balancing the rescaled measurement matrix before factorization were also examined.

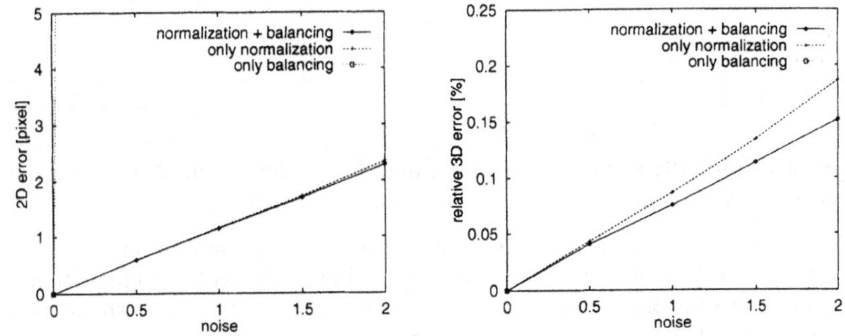

Graphs 5 and 6 : Influence of normalization and balancing. *The results presented here were obtained for configuration 1. The 2D error curve for "only balancing" goes off scale even for 0.5 pixels noise and the 3D curve is so steep that it is not even visible.*

When the image coordinates are not normalized, the error is already off scale for 0.5 pixel noise. An explanation for this is the bad conditioning of the rescaled measurement matrix (see also next paragraph). As for balancing, we see that this improves 3D errors up to 20 %, and hence should always be part of the algorithm.

Robustness of the Factorization The applicability of our factorization method is based on the rank 4-ness of the rescaled measurement matrix $\boldsymbol{W}$ (in the noiseless case). To test the robustness of this property, we evaluated how close $\boldsymbol{W}$ is to rank 4 in practice. To be close to rank 4, the ratio of the 4th and 5th largest singular values , $\sigma_4 : \sigma_5$, should be large with respect to the ratio of the 1st and 4th largest, $\sigma_1 : \sigma_4$. In the following graphs, these two ratios are represented, for configurations 1 and 2 and for 2 and 20 views. Note that the y-axes are scaled logarithmically.

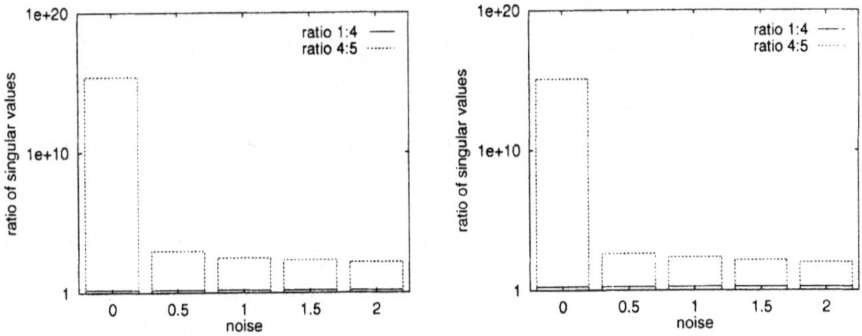

Graphs 7 and 8 : Ratios of singular values for configuration 1. *The graph on the left shows the situation for 2 views, on the right for 20 views.*

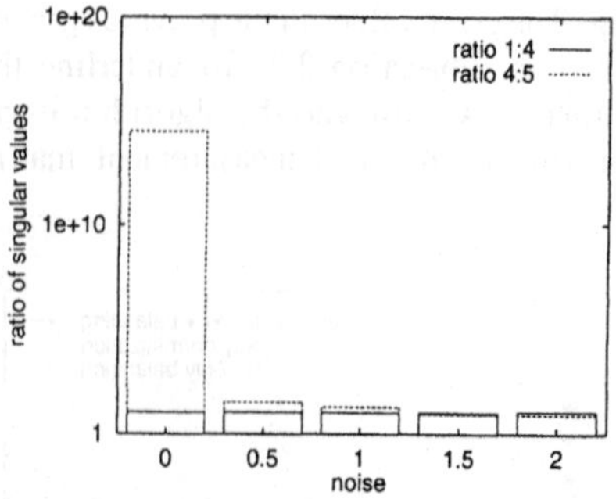

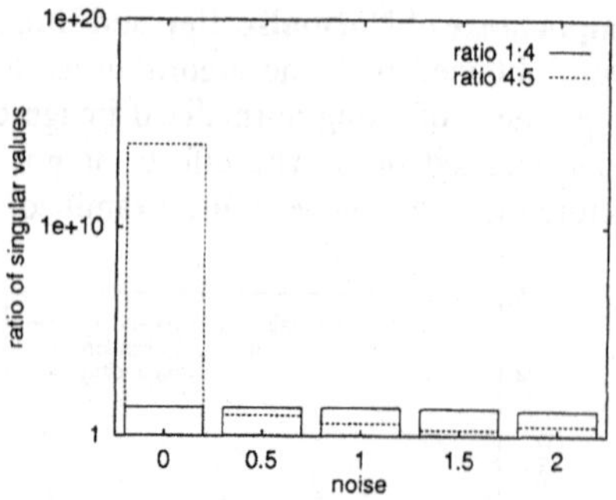

Graphs 9 and 10: Ratios of singular values for configuration 2. *The graph on the left shows the situation for 2 views, on the right for 20 views.*

We see that for configuration 1, the matrix is always very close to rank 4: $(\sigma_1 : \sigma_4)$ is lower than 2, whereas $(\sigma_4 : \sigma_5)$ lies clearly above 100. As for configuration 2, the graphs reflect the bad performance in 3D reconstruction. $(\sigma_1 : \sigma_4)$ is about 10, while for high noise levels or many views $(\sigma_4 : \sigma_5)$ is close to 1.

4.2 Evaluation with Real Images

The algorithm has also been tested on several sequences of real images. For 2 of them we show results.

The House Sequence Figure 2 shows the first and last image of a sequence of 6 images of a scene with a wooden house. 38 points were tracked over the whole sequence, but only extracted with ± 1 pixel accuracy.

Fig. 2. First and last image of the house sequence and one image of the castle sequence.

To estimate the quality of the projective reconstruction, we aligned it with an approximate Euclidean model of the scene obtained from calibrated views (see figure 3). Lines have been drawn between some of the points to aid visualization.
In the side and front views we see that right angles are approximately conserved, and that the windows are coplanar with the wall. The bumpiness on the left side of the roof is due to the fact that the roof stands out slightly from the house's front wall (see figure 2), thus causing occlusion in the last view of the edge point between roof and wall.

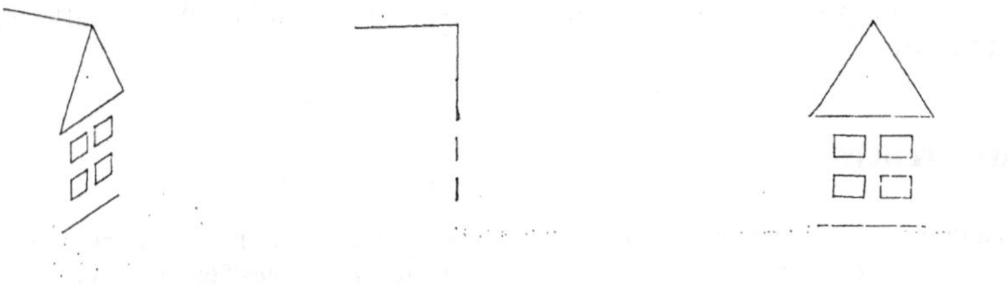

Fig. 3. Three views of the reconstructed house. *(1) "General view". (2) Side view. (3) Front view.*

The Castle Sequence 28 points have been tracked through the 11 images of the scene shown in the right part of figure 2. 3D ground truth is available, and the reconstruction errors have been evaluated quantitatively. The projective reconstruction was aligned with the Euclidean model and the resulting RMS error was 4.45 mm for an object size of about 220mm × 210mm × 280mm. The RMS error of the reprojected structure with respect to the measured image points was less than 0.02 pixels.
We also applied a Levenberg-Marquardt nonlinear least-squares estimation algorithm, with the results of our method as initialization. This slightly improved the 2D reprojection error, however the 3D reconstruction error was not significantly changed.

5 Discussion and Further Work

In this paper, we have proposed a method of projective reconstruction from multiple uncalibrated images. The method is very elegant, recovering shape and motion by factorization of one matrix, containing all image points of all views. This factorization is only possible when the image points are correctly scaled. We have proposed a very simple way to obtain the individual scale factors, using only fundamental matrices and epipoles estimated from the image data.
The algorithm proves to work well with real images. Quantitative evaluation by numerical simulations shows the robustness of the factorization and the good performance with respect to noise. The results also show that it is essential to work with normalized image coordinates.
Some aspects of the method remain to be examined. In the current implementation, we recover projective depths by chaining equation (2) for pairs of views $(12), (23), \ldots, (m-1, m)$. However, it would be worth investigating whether other kinds of chaining are not more stable. Furthermore, uncertainty estimates on the fundamental matrices should be considered when choosing which of the equations (2) to use. To run the algorithm in practice, it should also be able to treat points which are not visible in all images. Finally the method could be extended to use trilinear and perhaps even quadrilinear matching tensors.
Acknowledgements. This work was partially supported by INRIA France and E.C. projects HCM and SECOND. Data for this research were partially provided by the Calibrated Imaging Laboratory at Carnegie Mellon University, supported by ARPA,

NSF, and NASA (the castle sequence can be found at http://www.cs.cmu.edu/ cil/cil-ster.html).

References

[Fau92] O. Faugeras. What can be seen in three dimensions with an uncalibrated stereo rig? In *Proc. 2nd ECCV, Santa Margherita Ligure, Italy*, pages 563–578, May 1992.

[FM95] O. Faugeras and B. Mourrain. On the geometry and algebra of the point and line correspondences between N images. In *Proc. 5th ICCV, Cambridge, Massachusetts*, pages 951–956, June 1995.

[Har93] R.I. Hartley. Euclidean reconstruction from uncalibrated views. In *Proc.* DARPA–ESPRIT *Workshop on Applications of Invariants in Computer Vision, Azores, Portugal*, pages 187–202, October 1993.

[Har95] R. Hartley. In Defence of the 8-point Algorithm. In *Proc. 5th ICCV, Cambridge, Massachusetts*, pages 1064–1070, June 1995.

[HGC92] R. Hartley, R. Gupta, and T. Chang. Stereo from uncalibrated cameras. In *Proc. CVPR, Urbana-Champaign, Illinois*, pages 761–764, 1992.

[HS94] R. Hartley and P. Sturm. Triangulation. In *Proc.* ARPA *IUW, Monterey, California*, pages 957–966, November 1994.

[MM95] P.F. McLauchlan and D.W. Murray. A unifying framework for structure and motion recovery from image sequences. In *Proc. 5th ICCV, Cambridge, Massachusetts*, pages 314–320, June 1995.

[MVQ93] R. Mohr, F. Veillon, and L. Quan. Relative 3D reconstruction using multiple uncalibrated images. In *Proc. CVPR, New York*, pages 543–548, June 1993.

[PK94] C. J. Poelman and T. Kanade. A paraperspective factorization method for shape and motion recovery. In *Proc. 3rd ECCV, Stockholm, Sweden*, pages 97–108, May 1994.

[Sha94] A. Shashua. Projective structure from uncalibrated images: Structure from motion and recognition. IEEE *Trans. on PAMI*, 16(8):778–790, August 1994.

[Sha95] A. Shashua. Algebraic functions for recognition. IEEE *Trans. on PAMI*, 17(8):779–789, August 1995.

[TK92] C. Tomasi and T. Kanade. Shape and motion from image streams under orthography: A factorization method. *IJCV*, 9(2):137–154, 1992.

[Tri95a] B. Triggs. Matching constraints and the joint image. In *Proc. 5th ICCV, Cambridge, Massachusetts*, pages 338–343, June 1995.

[Tri95b] B. Triggs. The geometry of projective reconstruction I: matching constraints and the joint image. *IJCV*, 1995. submitted.

Author Index

Springer-Verlag and the Environment

We at Springer-Verlag firmly believe that an international science publisher has a special obligation to the environment, and our corporate policies consistently reflect this conviction.

We also expect our business partners – paper mills, printers, packaging manufacturers, etc. – to commit themselves to using environmentally friendly materials and production processes.

The paper in this book is made from low- or no-chlorine pulp and is acid free, in conformance with international standards for paper permanency.

Lecture Notes in Computer Science

For information about Vols. 1–992

please contact your bookseller or Springer-Verlag

Vol. 993: T.C. Fogarty (Ed.), Evolutionary Computing. Proceedings, 1995. VIII, 264 pages. 1995.

Vol. 994: M. Hebert, J. Ponce, T. Boult, A. Gross (Eds.), Object Representation in Computer Vision. Proceedings, 1994. VIII, 359 pages. 1995.

Vol. 995: S.M. Müller, W.J. Paul, The Complexity of Simple Computer Architectures. XII, 270 pages. 1995.

Vol. 996: P. Dybjer, B. Nordström, J. Smith (Eds.), Types for Proofs and Programs. Proceedings, 1994. X, 202 pages. 1995.

Vol. 997: K.P. Jantke, T. Shinohara, T. Zeugmann (Eds.), Algorithmic Learning Theory. Proceedings, 1995. XV, 319 pages. 1995.

Vol. 998: A. Clarke, M. Campolargo, N. Karatzas (Eds.), Bringing Telecommunication Services to the People – IS&N '95. Proceedings, 1995. XII, 510 pages. 1995.

Vol. 999: P. Antsaklis, W. Kohn, A. Nerode, S. Sastry (Eds.), Hybrid Systems II. VIII, 569 pages. 1995.

Vol. 1000: J. van Leeuwen (Ed.), Computer Science Today. XIV, 643 pages. 1995.

Vol. 1001: M. Sudan, Efficient Checking of Polynomials and Proofs and the Hardness of Approximation Problems. XIV, 87 pages. 1995.

Vol. 1002: J.J. Kistler, Disconnected Operation in a Distributed File System. XIX, 249 pages. 1995.

VOL. 1003: P. Pandurang Nayak, Automated Modeling of Physical Systems. XXI, 232 pages. 1995. (Subseries LNAI).

Vol. 1004: J. Staples, P. Eades, N. Katoh, A. Moffat (Eds.), Algorithms and Computation. Proceedings, 1995. XV, 440 pages. 1995.

Vol. 1005: J. Estublier (Ed.), Software Configuration Management. Proceedings, 1995. IX, 311 pages. 1995.

Vol. 1006: S. Bhalla (Ed.), Information Systems and Data Management. Proceedings, 1995. IX, 321 pages. 1995.

Vol. 1007: A. Bosselaers, B. Preneel (Eds.), Integrity Primitives for Secure Information Systems. VII, 239 pages. 1995.

Vol. 1008: B. Preneel (Ed.), Fast Software Encryption. Proceedings, 1994. VIII, 367 pages. 1995.

Vol. 1009: M. Broy, S. Jähnichen (Eds.), KORSO: Methods, Languages, and Tools for the Construction of Correct Software. X, 449 pages. 1995. Vol.

Vol. 1010: M. Veloso, A. Aamodt (Eds.), Case-Based Reasoning Research and Development. Proceedings, 1995. X, 576 pages. 1995. (Subseries LNAI).

Vol. 1011: T. Furuhashi (Ed.), Advances in Fuzzy Logic, Neural Networks and Genetic Algorithms. Proceedings, 1994. (Subseries LNAI).

Vol. 1012: M. Bartošek, J. Staudek, J. Wiedermann (Eds.), SOFSEM '95: Theory and Practice of Informatics. Proceedings, 1995. XI, 499 pages. 1995.

Vol. 1013: T.W. Ling, A.O. Mendelzon, L. Vieille (Eds.), Deductive and Object-Oriented Databases. Proceedings, 1995. XIV, 557 pages. 1995.

Vol. 1014: A.P. del Pobil, M.A. Serna, Spatial Representation and Motion Planning. XII, 242 pages. 1995.

Vol. 1015: B. Blumenthal, J. Gornostaev, C. Unger (Eds.), Human-Computer Interaction. Proceedings, 1995. VIII, 203 pages. 1995.

VOL. 1016: R. Cipolla, Active Visual Inference of Surface Shape. XII, 194 pages. 1995.

Vol. 1017: M. Nagl (Ed.), Graph-Theoretic Concepts in Computer Science. Proceedings, 1995. XI, 406 pages. 1995.

Vol. 1018: T.D.C. Little, R. Gusella (Eds.), Network and Operating Systems Support for Digital Audio and Video. Proceedings, 1995. XI, 357 pages. 1995.

Vol. 1019: E. Brinksma, W.R. Cleaveland, K.G. Larsen, T. Margaria, B. Steffen (Eds.), Tools and Algorithms for the Construction and Analysis of Systems. Selected Papers, 1995. VII, 291 pages. 1995.

Vol. 1020: I.D. Watson (Ed.), Progress in Case-Based Reasoning. Proceedings, 1995. VIII, 209 pages. 1995. (Subseries LNAI).

Vol. 1021: M.P. Papazoglou (Ed.), OOER '95: Object-Oriented and Entity-Relationship Modeling. Proceedings, 1995. XVII, 451 pages. 1995.

Vol. 1022: P.H. Hartel, R. Plasmeijer (Eds.), Functional Programming Languages in Education. Proceedings, 1995. X, 309 pages. 1995.

Vol. 1023: K. Kanchanasut, J.-J. Lévy (Eds.), Algorithms, Concurrency and Knowlwdge. Proceedings, 1995. X, 410 pages. 1995.

Vol. 1024: R.T. Chin, H.H.S. Ip, A.C. Naiman, T.-C. Pong (Eds.), Image Analysis Applications and Computer Graphics. Proceedings, 1995. XVI, 533 pages. 1995.

Vol. 1025: C. Boyd (Ed.), Cryptography and Coding. Proceedings, 1995. IX, 291 pages. 1995.

Vol. 1026: P.S. Thiagarajan (Ed.), Foundations of Software Technology and Theoretical Computer Science. Proceedings, 1995. XII, 515 pages. 1995.

Vol. 1027: F.J. Brandenburg (Ed.), Graph Drawing. Proceedings, 1995. XII, 526 pages. 1996.

Vol. 1028: N.R. Adam, Y. Yesha (Eds.), Electronic Commerce. X, 155 pages. 1996.

Vol. 1029: E. Dawson, J. Golić (Eds.), Cryptography: Policy and Algorithms. Proceedings, 1995. XI, 327 pages. 1996.

Vol. 1030: F. Pichler, R. Moreno-Díaz, R. Albrecht (Eds.), Computer Aided Systems Theory - EUROCAST '95. Proceedings, 1995. XII, 539 pages. 1996.

Vol.1031: M. Toussaint (Ed.), Ada in Europe. Proceedings, 1995. XI, 455 pages. 1996.

Vol. 1032: P. Godefroid, Partial-Order Methods for the Verification of Concurrent Systems. IV, 143 pages. 1996.

Vol. 1033: C.-H. Huang, P. Sadayappan, U. Banerjee, D. Gelernter, A. Nicolau, D. Padua (Eds.), Languages and Compilers for Parallel Computing. Proceedings, 1995. XIII, 597 pages. 1996.

Vol. 1034: G. Kuper, M. Wallace (Eds.), Constraint Databases and Applications. Proceedings, 1995. VII, 185 pages. 1996.

Vol. 1035: S.Z. Li, D.P. Mital, E.K. Teoh, H. Wang (Eds.), Recent Developments in Computer Vision. Proceedings, 1995. XI, 604 pages. 1996.

Vol. 1036: G. Adorni, M. Zock (Eds.), Trends in Natural Language Generation - An Artificial Intelligence Perspective. Proceedings, 1993. IX, 382 pages. 1996. (Subseries LNAI).

Vol. 1037: M. Wooldridge, J.P. Müller, M. Tambe (Eds.), Intelligent Agents II. Proceedings, 1995. XVI, 437 pages. 1996. (Subseries LNAI).

Vol. 1038: W: Van de Velde, J.W. Perram (Eds.), Agents Breaking Away. Proceedings, 1996. XIV, 232 pages. 1996. (Subseries LNAI).

Vol. 1039: D. Gollmann (Ed.), Fast Software Encryption. Proceedings, 1996. X, 219 pages. 1996.

Vol. 1040: S. Wermter, E. Riloff, G. Scheler (Eds.), Connectionist, Statistical, and Symbolic Approaches to Learning for Natural Language Processing. IX, 468 pages. 1996. (Subseries LNAI).

Vol. 1041: J. Dongarra, K. Madsen, J. Waśniewski (Eds.), Applied Parallel Computing. Proceedings, 1995. XII, 562 pages. 1996.

Vol. 1042: G. Weiß, S. Sen (Eds.), Adaption and Learning in Multi-Agent Systems. Proceedings, 1995. X, 238 pages. 1996. (Subseries LNAI).

Vol. 1043: F. Moller, G. Birtwistle (Eds.), Logics for Concurrency. XI, 266 pages. 1996.

Vol. 1044: B. Plattner (Ed.), Broadband Communications. Proceedings, 1996. XIV, 359 pages. 1996.

Vol. 1045: B. Butscher, E. Moeller, H. Pusch (Eds.), Interactive Distributed Multimedia Systems and Services. Proceedings, 1996. XI, 333 pages. 1996.

Vol. 1046: C. Puech, R. Reischuk (Eds.), STACS 96. Proceedings, 1996. XII, 690 pages. 1996.

Vol. 1047: E. Hajnicz, Time Structures. IX, 244 pages. 1996. (Subseries LNAI).

Vol. 1048: M. Proietti (Ed.), Logic Program Syynthesis and Transformation. Proceedings, 1995. X, 267 pages. 1996.

Vol. 1049: K. Futatsugi, S. Matsuoka (Eds.), Object Technologies for Advanced Software. Proceedings, 1996. X, 309 pages. 1996.

Vol. 1050: R. Dyckhoff, H. Herre, P. Schroeder-Heister (Eds.), Extensions of Logic Programming. Proceedings, 1996. VII, 318 pages. 1996. (Subseries LNAI).

Vol. 1051: M.-C. Gaudel, J. Woodcock (Eds.), FME'96: Industrial Benefit and Advances in Formal Methods. Proceedings, 1996. XII, 704 pages. 1996.

Vol. 1052: D. Hutchison, H. Christiansen, G. Coulson, A. Danthine (Eds.), Teleservices and Multimedia Communications. Proceedings, 1995. XII, 277 pages. 1996.

Vol. 1053: P. Graf, Term Indexing. XVI, 284 pages. 1996. (Subseries LNAI).

Vol. 1054: A. Ferreira, P. Pardalos (Eds.), Solving Combinatorial Optimization Problems in Parallel. VII, 274 pages. 1996.

Vol. 1055: T. Margaria, B. Steffen (Eds.), Tools and Algorithms for the Construction and Analysis of Systems. Proceedings, 1996. XI, 435 pages. 1996.

Vol. 1056: A. Haddadi, Communication and Cooperation in Agent Systems. XIII, 148 pages. 1996. (Subseries LNAI).

Vol. 1057: P. Apers, M. Bouzeghoub, G. Gardarin (Eds.), Advances in Database Technology — EDBT '96. Proceedings, 1996. XII, 636 pages. 1996.

Vol. 1058: H. R. Nielson (Ed.), Programming Languages and Systems – ESOP '96. Proceedings, 1996. X, 405 pages. 1996.

Vol. 1059: H. Kirchner (Ed.), Trees in Algebra and Programming – CAAP '96. Proceedings, 1996. VIII, 331 pages. 1996.

Vol. 1060: T. Gyimóthy (Ed.), Compiler Construction. Proceedings, 1996. X, 355 pages. 1996.

Vol. 1061: P. Ciancarini, C. Hankin (Eds.), Coordination Languages and Models. Proceedings, 1996. XI, 443 pages. 1996.

Vol. 1062: E. Sanchez, M. Tomassini (Eds.), Towards Evolvable Hardware. IX, 265 pages. 1996.

Vol. 1063: J.-M. Alliot, E. Lutton, E. Ronald, M. Schoenauer, D. Snyers (Eds.), Artificial Evolution. Proceedings, 1995. XIII, 396 pages. 1996.

Vol. 1064: B. Buxton, R. Cipolla (Eds.), Computer Vision – ECCV '96. Volume I. Proceedings, 1996. XXI, 725 pages. 1996.

Vol. 1065: B. Buxton, R. Cipolla (Eds.), Computer Vision – ECCV '96. Volume II. Proceedings, 1996. XXI, 723 pages. 1996.

Vol. 1066: R. Alur, T.A. Henzinger, E.D. Sontag (Eds.), Hybrid Systems III. IX, 618 pages. 1996.

Vol. 1067: H. Liddell, A. Colbrook, B. Hertzberger, P. Sloot (Eds.), High-Performance Computing and Networking. Proceedings, 1996. XXV, 1040 pages. 1996.

Vol. 1068: T. Ito, R.H. Halstead, Jr., C. Queinnec (Eds.), Parallel Symbolic Languages and Systems. Proceedings, 1995. X, 363 pages. 1996.